Birkhäuser

Advances in Mathematical Fluid Mechanics

The *Advances in Mathematical Fluid Mechanics* series is a forum for the publication of high-quality, peer-reviewed research monographs and edited collections on the mathematical theory of fluid mechanics, with special regards to the Navier-Stokes equations and other significant viscous and inviscid fluid models. Titles in this series consider theoretical, numerical, and computational methods, as well as applications to science and engineering. Works in related areas of mathematics that have a direct bearing on fluid mechanics are also welcome. All manuscripts are peer-reviewed to meet the highest standards of scientific literature.

Sunčica Čanić • Jeffrey Kuan • Boris Muha •
Krutika Tawri

Deterministic and Stochastic Fluid-Structure Interaction

Birkhäuser

Sunčica Čanić
Mathematics
University of California, Berkeley
Berkeley, CA, USA

Jeffrey Kuan
Mathematics
University of Maryland, College Park
College Park, MD, USA

Boris Muha
Department of Mathematics
Faculty of Science, University of Zagreb
Zagreb, Croatia

Krutika Tawri
Mathematics
University of California, Berkeley
Berkeley, CA, USA

ISSN 2297-0320 ISSN 2297-0339 (electronic)
Advances in Mathematical Fluid Mechanics
ISBN 978-3-032-00897-8 ISBN 978-3-032-00898-5 (eBook)
https://doi.org/10.1007/978-3-032-00898-5

Mathematics Subject Classification: 76M35, 35Q30, 76D05, 35A01, 35D30, 76-02, 76-10

This book is published under the imprint Birkhäuser, www.birkhauser-science.com by the registered company Springer Nature Switzerland AG
The registered company address is: Gewerbestrasse 11, 6330 Cham, Switzerland

Preface

This book provides an in-depth review of the latest mathematical advances in the field of deterministic and stochastic fluid-structure interaction (FSI). FSI problems describe the dynamic interaction between fluids and structures. In this book, fluids will be considered incompressible, and structures will include elastic, viscoelastic, and poroelastic solids. These types of problems arise in many real-life applications including modeling the flow of blood in the human cardiovascular system, modeling bioartifical organs, and modeling large-scale structures such as wings, bridges, and dams. FSI is prevalent in science, engineering, and technology, making the rigorous analysis of such coupled fluid-structure systems important for continued technological development and progress in engineering.

The main focus of this book is on the analysis of the well-posedness of FSI problems. Benchmark models of FSI are presented and analyzed in this book. Most results presented here address existence, uniqueness, and stability of solutions to benchmark FSI problems in the context of finite energy solutions (weak solutions). For this class of solutions, a constructive well-posedness theory based on an operator splitting strategy has been recently developed, which has proven to be a robust way to study FSI problems both in the deterministic and probabilistic settings. An appealing feature of this operator-splitting proof strategy is that it can serve as a backbone for a design of numerical schemes to calculate solutions to these exceedingly complicated multiphysics problems.

Due to their mathematical complexity, stochastic FSI has only recently become the focus of mathematical studies. Although random noise is ubiquitous in nature, the study of mathematical well-posedness for stochastic FSI systems influenced by random noise is wide open. Understanding these systems has the potential to provide insight into fundamental questions such as turbulence and stochastic mixing, as well as various real-world phenomena. There is currently only a handful of results that discuss stochastic FSI in terms of the mathematical well-posedness theory. In this book we review those recent results. The chapters on stochastic FSI are written as an introduction to the use of stochastic techniques to study stochastic FSI, and are aimed at audiences that may have experience with deterministic FSI, but may be new to stochastic analysis.

This book is written for advanced undergraduate students, PhD students, postdocs, and researchers with a background in partial differential equations and fluid dynamics. The content is structured into four main parts. Part I, titled Preliminaries, covers the essential deterministic and stochastic concepts from analysis that will be foundational for studying coupled FSI problems throughout the book. Part II introduces prototypes of *deterministic* FSI problems and the basic concepts of operator splitting strategy used in the proofs of existence of finite energy solutions throughout this book. Part III is dedicated to *stochastic* FSI. This part is written as an introduction to the use of stochastic techniques to study stochastic FSI, and is aimed at audiences that may have experience with deterministic FSI (from Part II), but may be new to stochastic analysis. Part IV focuses on *deterministic* FSI with *poroelastic* structures. Each chapter typically ends with Bibliographical Notes, which provide an in-depth review of mathematical literature pertinent to the chapter's topic, offering readers a comprehensive overview of foundational works and recent developments in the field.

Berkeley, CA, USA Sunčica Čanić
College Park, MD, USA Jeffrey Kuan
Zagreb, Croatia Boris Muha
Berkeley, CA, USA Krutika Tawri

Declarations

Competing Interests The authors have no competing interests to declare that are relevant to the content of this manuscript.

Contents

List of Figures

List of Tables

Chapter 1
Introduction and Outline

1.1 Introduction

Fluid-structure interaction (FSI) describes systems in which fluids and structures (solids) interact dynamically. These are multiphysics systems, governed by multiple partial differential equations (PDEs), each representing a distinct physical subsystem. FSI problems arise across a range of fields, including aerospace engineering [134, 154, 155, 204, 211, 238, 315, 316, 327, 349] biomedical engineering [29, 30, 43, 70, 187, 252, 341], civil and structural engineering [3, 8, 10, 86, 115, 156, 229, 261, 277, 347], marine and ocean engineering [136, 241, 328], the energy industry [189, 228, 280], and automotive engineering [61, 111, 253]. The prevalence of FSI in practical applications makes rigorous analysis of these coupled fluid-structure systems essential for advancing technology and engineering.

This monograph focuses on the mathematical theory of FSI problems involving viscous, incompressible fluids interacting with deformable structures. The fluid flow will be modeled by the time-dependent Stokes equations or the Navier-Stokes equations for an incompressible, viscous fluid. The deformable structure will be elastic, viscoelastic, or poroelastic modeled by linear or nonlinear elasticity—including reduced models of shells and plates, and by the Biot equations of poroelasticity. The fluid and structure are coupled through two types of coupling conditions: the kinematic and dynamic conditions, evaluated at the interface where the fluid and structure meet. Depending on the specific problem, these conditions may be evaluated at a fixed interface (linear coupling) or a moving interface (nonlinear coupling). Nonlinear coupling introduces additional complexity, as the fluid domain becomes time-dependent and not known in advance, creating geometric nonlinearity that poses significant mathematical challenges.

This monograph addresses both **linearly and nonlinearly coupled FSI problems** in both the **deterministic and stochastic settings**.

S. Čanić et al., *Deterministic and Stochastic Fluid-Structure Interaction*, Advances in Mathematical Fluid Mechanics, https://doi.org/10.1007/978-3-032-00898-5_1

Despite the fact that stochasticity plays a critical role in modeling real-life problems, the analysis of stochastic FSI problems has only recently become the focus of mathematical studies [213–216, 323]. Stochasticity accounts for real-world variability and uncertainty, offering more flexible, realistic, and robust solutions across diverse applications. There are many applications that suggest and provide evidence to treating moving surfaces as random boundaries in order to deal with geometric uncertainties due to insufficient data or measurement errors; see, e.g., [73] in the context of wind-engineering and [284] in the context of blood flow. In general, studying stochastic FSI is important because well-posedness of stochastic FSI models provides confidence that the deterministic FSI models are, indeed, robust to stochastic noise that occurs naturally in real-life problems. Additionally, understanding stochastic FSI systems has the potential to provide insight into fundamental questions such as turbulence and stochastic mixing.

We will examine stochastic benchmark FSI problems, where both the fluid and structure equations are subject to stochastic forcing. In stochastic FSI, the unknowns (such as fluid velocity and structure displacement) are random variables, which further complicate the analysis of these problems. In particular, in nonlinearly coupled stochastic FSI, the fluid domain itself is a random variable not known a priori and is a part of the solution itself. This introduces additional challenges in the analysis of such systems.

This monograph focuses on analyzing the well-posedness of both deterministic and stochastic FSI problems, with a particular emphasis on finite energy (weak) solutions. The analysis centers on proving existence, uniqueness, and stability within this class of solutions. For these finite energy solutions, a constructive well-posedness theory based on time discretization via operator splitting strategy has been recently developed, which has proven to be a robust way to study FSI problems in both deterministic and probabilistic contexts. An appealing feature of this operator splitting approach is its potential to inform the design of numerical schemes, enabling practical computation of solutions for these highly complex multiphysics problems. A central theme of this book is the application of constructive time discretization via operator splitting techniques to establish the existence of finite energy solutions across a hierarchy of benchmark FSI problems.

To the best of our knowledge, this book is the first comprehensive monograph to systematically explore both deterministic and stochastic FSI, guiding readers from foundational FSI concepts to advanced topics in stochastic FSI. It includes a chapter summarizing essential analytical background to support understanding of the material covered.

This text serves as an introductory resource on the mathematical analysis of FSI problems, suitable for advanced undergraduates, PhD students, postdoctoral researchers, and professionals with a background in partial differential equations and fluid dynamics.

1.2 Outline

There are four main parts of this monograph: Part I, II, III, and IV. Each part is divided into chapters, which are further divided onto sections and subsections. Chapters are numbered sequentially to ensure clarity when referencing specific chapters.

Part I, called Preliminaries, covers the deterministic and stochastic preliminaries. It contains Chaps. 2 and 3. Chapter 2 discusses basic deterministic preliminaries, and Chap. 3 discusses stochastic preliminaries. More specifically, Chap. 2 focusses on concepts from real analysis and functional analysis that will be important in the analysis of coupled fluid-structure dynamics in the remainder of the book. Chapter 3 focuses on covering basic probabilistic background, with an emphasis on random variables that take values in Banach spaces. This chapter also discusses the three modes of probabilistic convergence of random variables, which will play a key role in the analysis of stochastic FSI: convergence in probability, weak convergence (or convergence in law or distribution), and convergence almost surely. Furthermore, in this chapter, we discuss the relationships between these modes of convergence, and review classical theorems that will be important in our analysis later.

Part II explores deterministic fluid-structure interaction through two benchmark problems: one with linear coupling (Chap. 4) and another with the more complex nonlinear coupling (Chap. 5). In both benchmark problems, the fluid flows through a two-dimensional cylinder with deformable walls, driven by prescribed inlet and outlet pressure data. In Chap. 4, the fluid is modeled by the time-dependent Stokes equations, and the structure is modeled by a thin shell equation. The fluid and structure are linearly coupled. In Chap. 5, the fluid is modeled by the Navier-Stokes equations, and the structure is modeled by the same thin shell equation as in the case of linear coupling. However, in this chapter, the fluid and structure are nonlinearly coupled. In both chapters, fundamental concepts for establishing the existence of weak solutions using time discretization via operator splitting are introduced, a method that allows for systematic analysis of these coupled FSI systems. In Chap. 6, we discuss how the operator splitting scheme can be extended to prove existence of finite energy (weak) solutions across a broader range of FSI problems, including nonlinearly coupled FSI problems with Navier-slip conditions, FSI with nonlinear Koiter shells, interactions with multilayered structures, mesh-supported shells, and 3D fluid-2D structure interactions.

Part III introduces stochastic FSI problems. The material progresses gradually from simpler to more complex stochastic FSI problems. Chapter 7 begins with a discussion of foundational progress in the analysis of stochastic fluid-structure interaction, focusing on a simplified, reduced linearly coupled model. This reduced model captures the essential dynamics of fluid-structure interaction through a single self-contained stochastic PDE governing the structure's behavior, commonly referred to as the stochastic viscous wave equation. Although the analysis of this stochastic viscous wave equation does not need the use of the splitting scheme and relies more on standard tools from stochastic PDEs, a description of the analysis of this equation is included in this book since this was the first equation studied in the context stochastic FSI, and since the results shed some interesting new light on the solution behavior. A fully coupled stochastic benchmark FSI problem is discussed in Chap. 8. Similarly as in Part II of this monograph, we start with the problem with linear coupling. In this context, we introduce a new general framework for analyzing general stochastic fluid-structure systems based on an operator splitting approach. The focus in this chapter is on a benchmark problem in which a stochastically forced linearly elastic shell interacts with the flow of a viscous incompressible Newtonian fluid in two spatial dimensions, modeled by the time-dependent Stokes equations. The problem is forced by a "rough" stochastic forcing given by multiplicative white noise. We prove the existence of a weak solution in the probabilistically weak sense to this stochastic fluid-structure interaction problem, and in the case of additive noise (in contrast to multiplicative noise), we show the stronger result of existence of a unique weak solution in a probabilistically strong sense. The notions of these solutions are defined in the Preliminaries chapter. In Chap. 9, we discuss the well-posedness for a *nonlinearly coupled* stochastic FSI problem driven by a multiplicative white-in-time noise, and by the inlet and outlet data. This benchmark problem incorporates the main mathematical difficulties associated with nonlinearly coupled stochastic FSI problems. The noise depends on structure displacement, structure velocity, and fluid velocity, and the noise is chosen in a way that it affects the fluid-structure system as a whole via the stochastic forcing on the structure and on the bulk fluid flow. The existence of a weak martingale solution is obtained using an operator splitting approach. Several new challenges are addressed to deal with the geometric nonlinearities of the moving domain, which is a *random variable not known a priori*. This result is the state-of-the-art in stochastic FSI and is the culmination of the stochastic FSI chapters.

Part IV of this book, containing Chaps. 10 and 11, is dedicated to the study of deterministic FSI involving *poroelastic* media. Poroelastic structures are porous media that are elastic, in which the fluid flow through the pores is coupled to elastodynamics of the poroelastic matrix (skeleton). The model equations are represented by a set of parabolic-hyperbolic-type PDEs known as the Biot equations. Chapters 10 and 11 present a review of the well-posedness theory for fluid-poroelastic structure interaction problems involving poroelastic structures interacting with "free" flows of viscous incompressible fluids, i.e., with the fluid flows not constrained to the poroelastic medium. These free fluid flows are located outside the poroelastic medium and interacting with the poroelastic medium across a fluid-structure interface. Examples of such flows include blood flow through poroelastic arteries, design of bioartificial organs, groundwater flow in aquifers, coastal and riverbank erosion, and reservoir engineering. Chapter 10 addresses a benchmark problem in this area by considering a fluid-poroelastic structure interaction problem with *linear coupling*, which is assumed across a fixed interface, where the interface itself is a poroelastic medium. In particular, the interface in this problem is a poroelastic plate modeled by the Biot poroelastic plate equations. The poroelastic plate separates the flow of free fluid modeled by the time-dependent Stokes equations, from a poroelastic medium modeled by the Biot equations. The existence and uniqueness of a weak solution is presented for this linear problem. The last chapter, Chap. 11, focuses on a *nonlinearly coupled* fluid-poroelastic structure interaction problem. The problem describes the interaction between the free fluid flow of an incompressible, viscous fluid modeled by the Navier-Stokes equations, coupled across a moving interface to a bulk poroelastic medium modeled by the Biot equations. The moving interface has inertia and elastic energy, and it is modeled by a "reticular" plate equation, "transparent" to the fluid flow. This problem is exceedingly difficult to study mathematically. This chapter presents details of the well-posedness theory developed for this class of nonlinearly coupled FSI problems, for which the proof of the existence of finite energy solutions is based on an operator splitting strategy. This example pushes the boundaries of applicability of weak solutions theory to studying FSI involving bulk poroelasticity where the poroelastic medium region is of the same dimension as the fluid domain.

We conclude this outline by noting that each part of this monograph contains Bibliographical Notes, typically located at the end of each chapter. These notes provide an in-depth review of mathematical literature pertinent to the chapter's topic, offering readers a comprehensive overview of foundational works and recent developments in the field.

Acknowledgments The authors extend their gratitude to Paolo Galdi for his meticulous reading of the monograph and his insightful suggestions. They also acknowledge Felix Brandt for his valuable contributions in identifying numerous typographical errors and providing key mathematical remarks that have enhanced the quality of this work. Additionally, the authors wish to recognize the following research grants that made this work possible: US National Science Foundation grants DMS-2408928, DMS-2247000, DMS-2011319 (Canic), Mathematical Sciences Postdoctoral Research Fellowship grant DMS-2303177 (Kuan), DMS-2407197 (Tawri), and Croatia-USA bilateral grant IP-2022-10-2962 (Muha).

Part I
Preliminaries

Chapter 2
Deterministic Preliminaries

In this chapter, we review fundamental concepts from real and functional analysis that are essential for our study of coupled fluid-structure dynamics throughout the book. We begin by introducing the essential abstract function spaces, namely, Banach and Hilbert spaces. Next, we examine specific examples of these spaces that play a key role in the analysis of FSI, focusing on L^p spaces and Sobolev spaces. As an application of L^p spaces, which will be particularly relevant for studying reduced fluid-structure interaction models in Chap. 7, we explore the properties of the Fourier transform on $L^2(\mathbb{R}^d)$.

Next, we introduce function spaces that account for both spatial and temporal dependence, specifically Bochner spaces of the form $L^p(0, T; B)$, where B is a Banach space. These spaces are essential for analyzing evolutionary partial differential equations (PDEs), namely, the PDEs that are time-dependent. Since this book also explores problems posed on moving domains, we then discuss function spaces adapted to such settings. This includes extensions of Bochner spaces to accommodate time-dependent geometries, key mappings for functions defined on moving domains, and generalizations of classical inequalities to this dynamic framework.

Next, we discuss compact embeddings, presenting several classical results that will play a crucial role throughout our analysis. We then provide an overview of the fundamental equations governing fluid dynamics and elasticity, discussing the concepts of a priori estimates and finite energy weak solutions to these equations. To give some intuition for the analytical techniques used in this book to construct solutions to fluid-structure interaction, we conclude the chapter by introducing two key constructive methods—Rothe's method and Lie operator splitting—at an abstract level. We discuss their role in constructing approximate solutions to partial differential equations and their application to fluid-structure interaction problems.

S. Čanić et al., *Deterministic and Stochastic Fluid-Structure Interaction*, Advances in Mathematical Fluid Mechanics, https://doi.org/10.1007/978-3-032-00898-5_2

2.1 Banach Spaces

In this section, we provide a brief overview of some of the essential spaces in functional analysis, namely, Banach spaces and Hilbert spaces. Here, we present these spaces in their most abstract form, with explicit examples introduced later in the discussions on L^p spaces and Sobolev spaces. We outline key results related to Banach spaces, directing the reader to Appendix A of [324] and Section 1.3 of [314] for a more in-depth treatment. We begin by defining the concept of a Banach space.

Definition 2.1.1 (Banach Space) A **Banach space** B with norm $\|\cdot\|_B$ is a complete normed vector space, where the completeness property refers to the requirement that every Cauchy sequence in B with respect to the norm $\|\cdot\|_B$ is convergent and converges to some element of B. This norm defines a topology of convergence on B, and we say that a sequence of elements $\{x_n\}_{n=1}^{\infty}$ in B **converges strongly** to $x \in B$ if $\lim_{n\to\infty} \|x - x_n\|_B = 0$.

Example 2.1.1 ($\mathbb{R}^d$ as a Banach Space) Any finite-dimensional normed vector space is a Banach space. For example, $\mathbb{R}^d$ with the standard Euclidean norm

$$\|(x_1, x_2, \ldots, x_d)\| = \left(\sum_{i=1}^{d} x_i^2\right)^{1/2}$$

is a Banach space.

We denote by $C^k(\mathcal{O})$ the space of functions $f : \mathcal{O} \to \mathbb{R}$ with continuous derivatives up to order k. Moreover, we define α-Hölder continuous as follows.

Definition 2.1.2 A real-valued function $f : \mathcal{O} \to \mathbb{R}$ is α-**Hölder continuous** for $\alpha \in (0, 1)$ if

$$\sup_{x,y\in\mathcal{O},x\neq y} \frac{|f(x) - f(y)|}{|x - y|^{\alpha}} < \infty.$$

A function is locally α-Hölder continuous if $\sup_{x,y\in K,x\neq y} \frac{|f(x)-f(y)|}{|x-y|^{\alpha}} < \infty$ for every compact set $K \subset \mathcal{O}$.

Example 2.1.2 ($C(\overline{\mathcal{O}})$ as a Banach Space) The more interesting case is the case of Banach spaces that are infinite-dimensional, as these can include infinite-dimensional vector spaces of functions. As an example, given a bounded open set $\mathcal{O} \subset \mathbb{R}^d$, we can define the space $C(\overline{\mathcal{O}})$ of continuous (and thus bounded) functions $f : \overline{\mathcal{O}} \to \mathbb{R}$, endowed with the norm

$$\|f\|_{C(\overline{\mathcal{O}})} = \max\{|f(x)| : x \in \overline{\mathcal{O}}\}.$$

Here, $\overline{\mathcal{O}}$ refers to the topological closure of $\mathcal{O}$, which is a compact set, which allows the norm to be well-defined. Note that unlike the case of finite-dimensional Euclidean space with the standard Euclidean norm, the Banach space $C(\overline{\mathcal{O}})$ is an infinite-dimensional vector space.

Since Banach spaces can potentially be infinite-dimensional, one important property of Banach spaces is whether they are separable, namely, whether they can be approximated by countably many elements in the Banach space.

Definition 2.1.3 (Separability of Banach Spaces) A Banach space B with norm $\|\cdot\|_B$ is **separable** if it contains a countable dense subset, namely, a countable collection of elements $\{x_n\}_{n=1}^{\infty}$ in B such that for any arbitrary $x \in B$ and $\delta > 0$, there exists some x_n in the countable collection such that $\|x - x_n\|_B < \delta$.

Example 2.1.3 (Examples of Separable Banach Spaces) One can easily check that finite-dimensional normed vector spaces are separable, and one can also show that $C(\overline{\mathcal{O}})$ is separable when $\mathcal{O}$ is a bounded open subset of $\mathbb{R}^d$. However, not all Banach spaces are separable, and we will see such an example when we later consider L^p spaces in Sect. 2.4.

Next, we discuss the concept of continuous linear maps between Banach spaces. We first recall the following example from linear algebra. Suppose that V and W are two finite-dimensional vector spaces. Then, a linear map (a linear transformation) between V and W is a map that satisfies $T(c_1v_1 + c_2v_2) = c_1T(v_1) + c_2T(v_2)$ for any vectors $v_1, v_2 \in V$ and scalars $c_1, c_2 \in \mathbb{R}$. In the case of linear maps between finite-dimensional vector spaces, a map that is linear is automatically continuous. However, because Banach spaces can be infinite-dimensional, this is no longer true, and we must define the concept of a **bounded linear map**.

Definition 2.1.4 (Bounded Linear Maps) Let B_1 and B_2 be two Banach spaces with norms $\|\cdot\|_{B_1}$ and $\|\cdot\|_{B_2}$, respectively. Then, a linear map $T : B_1 \to B_2$ is a **bounded linear map** if there exists a positive constant C such that

$$\|T(x)\|_{B_2} \leq C\|x\|_{B_1} \qquad \text{for all } x \in B_1. \tag{2.1}$$

We denote the space of bounded linear maps between B_1 and B_2 by $\mathcal{L}(B_1, B_2)$, and we note that this is also a Banach space, with a norm defined for $T \in \mathcal{L}(B_1, B_2)$ by

$$\|T\|_{\mathcal{L}(B_1,B_2)} := \inf\left\{\|T(x)\|_{B_2} : \|x\|_{B_1} = 1\right\},$$

so that $\|T\|_{\mathcal{L}(B_1,B_2)}$ represents the *smallest* constant C for which the inequality (2.1) holds.

Remark 2.1.1 (Bounded Linear Maps as Continuous Linear Maps) The inequality (2.1) refers to the word *bounded* in the definition of a bounded linear functional. This condition is important because while a general linear map between two Banach spaces is not necessarily continuous due to infinite-dimensionality, a *bounded* linear map between two Banach spaces *must be a continuous linear map.*

One essential linear map between Banach spaces is given by a linear functional, which is a real-valued bounded linear map on a Banach space B.

Definition 2.1.5 (Bounded Linear Functionals and Dual Space) Let B be a Banach space with norm $\|\cdot\|_B$. A **bounded linear functional** on B is a bounded linear map from B to $\mathbb{R}$. We denote the Banach space of bounded linear functionals on B by B^*, and we refer to B^* as the **dual space of** B. Thus, a linear map $\ell : B \to \mathbb{R}$ is an element of the dual space B^* of a Banach space B if there exists a constant $C > 0$ such that

$$|\ell(x)| \leq C\|x\|_B \quad \text{for all } x \in B,$$

where the norm on B^* is given by $\|\ell\|_{B^*} = \inf\{|\ell(x)| : \|x\|_B = 1\}$.

We can use the dual space to define another type of convergence on Banach spaces, besides strong convergence, which involves testing elements against linear functionals in the dual space.

Definition 2.1.6 (Weak Convergence) A sequence $\{x_n\}_{n=1}^{\infty}$ in a Banach space **converges weakly** to $x \in B$ if $\ell(x_n) \to \ell(x)$ as $n \to \infty$, for all bounded linear functionals $\ell \in B^*$.

As a more delicate type of convergence, sometimes, it is not easy to identify the dual B^* of a Banach space, but perhaps it is easier to consider B itself to be the dual of some other Banach space B'. In this case, we can define the following notion of weak-star convergence.

Definition 2.1.7 (Weak-Star Convergence) Suppose that a Banach space B can be identified as the dual space of another Banach space B'. Then, a sequence $\{x_n\}_{n=1}^{\infty}$ in B **converges weakly star (weakly*)** to $x \in B$ if $x_n(y) \to x(y)$ as $n \to \infty$, for all $y \in B'$, where we are viewing x_n and x as bounded linear functionals on B' sending $y \in B'$ to a real number.

Note that for any Banach space B, we can identify every element of B with a bounded linear functional on B^* (as an element of B^{**}) as follows. Given $\ell \in B^*$, which is a bounded linear functional on B, we can define for each $x \in B$ a bounded linear functional $x^{**} \in B^{**}$ by

$$x^{**}(\ell) = \ell(x) \quad \text{for } \ell \in B^*, \tag{2.2}$$

so that B can be identified with a subset of B^{**}. This motivates the following definition of a reflexive Banach space.

Definition 2.1.8 (Reflexivity of Banach Spaces) A Banach space B is **reflexive** if B can be identified with B^{**}, via the identification (2.2).

Remark 2.1.2 The theory of finite-dimensional vector spaces shows that every finite-dimensional vector space is reflexive. However, this is not necessarily true for infinite-dimensional vector spaces.

The reason why weak* convergence is useful in Banach spaces is because it has nice sequential compactness properties, which are not true for strong convergence. In particular, while the bounded closed ball centered at the origin in a finite-dimensional normed vector space such as $\mathbb{R}^d$ is compact, this is no longer true for a general Banach space, because Banach spaces can be potentially infinite-dimensional. Namely, given a Banach space B, and a bounded sequence $\{x_n\}_{n=1}^{\infty}$ in B, there does not necessarily exist a strongly convergent subsequence. However, the following *Banach-Alaoglu theorem* shows that there does exist a weakly* convergent subsequence, which is useful for obtaining a notion of (weak*) sequential compactness for sequences of functions that are uniformly bounded. We refer the reader to pg. 567 of Appendix A in [324] and Theorem 3.16 in [60] for further discussion of the Banach-Alaoglu theorem.

Theorem 2.1.1 (Banach-Alaoglu Theorem) *The closed unit ball in a Banach space B that can be identified as the dual of some Banach space B' is weakly* compact. If B is reflexive, then the closed unit ball in B is weakly sequentially compact.*

Remark 2.1.3 The corresponding result for weak convergence in the case where B is reflexive follows from the fact that for a reflexive Banach space B, one can easily verify that the notions of weak and weak* convergence coincide with each other. The proof of sequential compactness requires some additional analysis of separable spaces; see [60], Section 3.6.

We will see several examples of weak/weak* convergence and reflexive Banach spaces in Sect. 2.4, which discusses an important class of Banach spaces known as L^p spaces.

2.2 Hilbert Spaces

Next, we discuss a special type of Banach space known as a Hilbert space, which is essentially a Banach space with the additional structure given by an inner product. This inner product allows us to discuss concepts such as orthogonality in the context of function spaces and gives Hilbert spaces additional structure. We give a broad overview of Hilbert space theory in this subsection and refer the reader to Appendix A of [324], Chapter 4 of [313], and Chapter 5 of [60] for more details on Hilbert spaces.

Definition 2.2.1 (Hilbert Space) A Hilbert space H is a Banach space with an inner product $(\cdot, \cdot)_H$, which is a bilinear, symmetric, and positive definite map $H \times H \to \mathbb{R}$, which defines the norm of the Hilbert space $\|\cdot\|_H$ via $\|h\|_H := (h, h)_H^{1/2}$.

With the additional structure of an inner product, we have the essential notion of orthogonality, which is not present in the context of Banach spaces. In particular, we will say that two elements $h_1, h_2 \in H$ in a Hilbert space are **orthogonal** if $(h_1, h_2)_H = 0$.

Example 2.2.1 ($\mathbb{R}^d$ as a Hilbert Space) All finite-dimensional inner product spaces are Hilbert spaces. The typical example is Euclidean space $\mathbb{R}^d$ with the inner product given by the usual dot product, where for $\boldsymbol{x} = (x_1, x_2, \dots, x_d)$ and $\boldsymbol{y} = (y_1, y_2, \dots, y_d)$, we define

$$(\boldsymbol{x}, \boldsymbol{y})_{\mathbb{R}^d} := \boldsymbol{x} \cdot \boldsymbol{y} = \sum_{i=1}^{d} x_i y_i.$$

We note that the standard norm on Euclidean space is compatible with this inner (dot) product, in the sense that $\|\boldsymbol{x}\| = (\boldsymbol{x} \cdot \boldsymbol{x})^{1/2}$.

As with Banach spaces, we can consider the dual of a Hilbert space, since Hilbert spaces are just a special class of Banach spaces. The set of linear functionals on a Hilbert space H is the set of all bounded linear functionals $\ell : H \to \mathbb{R}$, with respect to the norm $\|\cdot\|_H$ defined via the inner product on the Hilbert space H. However, it is well-known that Hilbert spaces are *self-dual* as a result of their special inner product structure, with an explicit identification given by the inner product. This is a well-known and fundamental result known as the **Riesz representation theorem**. We refer the reader to Proposition 2.2 on pg. 559 in [324], Theorem 5.3 in [313], and Theorem 5.5 in [60] for the proof of the Riesz representation theorem and further discussion.

Theorem 2.2.1 (Riesz Representation Theorem) *For a Hilbert space H with an inner product $(\cdot, \cdot)_H$, given any bounded linear functional $\ell : H \to \mathbb{R}$, there exists a unique element $h_0 \in H$ such that $\ell(h) = (h, h_0)_H$ for all $h \in H$. Hence, the dual of H can be identified with the Hilbert space H itself via the following identification:*

$$h_0 \in H \quad \Longleftrightarrow \quad \ell_{h_0}(h) = (h, h_0) \in H^*.$$

In the well-posedness theory of partial differential equations, the Riesz representation theorem is used in a fundamental way to establish existence for elliptic partial differential equations, using a result known as the **Lax-Milgram theorem**. We refer the interested reader to Theorem 1 on pg. 317–319 in Section 6.2.1 in [132] and Corollary 5.8 in [60] for a proof of the Lax-Milgram theorem.

Lemma 2.2.1 (Lax-Milgram) *Let H be a Hilbert space. Assume that $a : H \times H \to \mathbb{R}$ is a bilinear map that is continuous and coercive, that is, there exist constants $\alpha, \beta > 0$ such that*

$$|a(u, v)| \leq \beta \|u\|_H \|v\|_H, \qquad u, v \in H$$

and

$$a(u,u) \geq \alpha \|u\|_H^2, \qquad u \in H.$$

Assume that $f : H \to \mathbb{R}$ is a bounded linear functional on H. Then there exists a unique element $u \in H$ such that

$$a(u,v) = (f,v)_H, \qquad \forall v \in H.$$

2.3 Measure Spaces and Measurable Functions

Since in this book we will often consider measurable functions on $\mathbb{R}^d$, and we will also consider random variables, which are measurable functions on probability spaces, we recall here the definition of a **measurable space**.

We first consider a set of objects, which we denote abstractly by Ω, and we define the notion of a σ-algebra. A collection $\mathcal{F}$ of subsets of Ω is a σ**-algebra** if it is closed under complements and countable unions, namely, if $E \in \mathcal{F}$, then $E^c = \Omega \setminus E \in \mathcal{F}$ also, and if $\{E_i\}_{i=1}^{\infty}$ is a countable sequence of sets in $\mathcal{F}$, then $\bigcup_{i=1}^{\infty} E_i \in \mathcal{F}$ also. We can define a **measurable space** as an ordered pair $(\Omega, \mathcal{F})$ consisting of a space Ω and its σ-algebra $\mathcal{F}$ of measurable sets. We can then define a **measure space** to be an ordered triple $(\Omega, \mathcal{F}, \mu)$ consisting of:

- A set Ω
- A σ-algebra $\mathcal{F}$ of measurable subsets of Ω
- A measure $\mu : \mathcal{F} \to [0, \infty)$ that assigns each set in the collection $\mathcal{F}$ to a nonnegative real number, which we refer to as the measure of the set.

Therefore, a measure space adds information about the measure of measurable sets, to the definition of a measurable space. We note that the measure μ must obey certain properties, such as countable sub-additivity, which we will not explicitly state here. In addition, while one can consider measures (such as signed measures) that take values that are not just nonnegative numbers, we will exclusively consider nonnegative measures throughout this book. We refer the reader to Chapters 1, 2, and 6 of [313] for further discussion of measure spaces in analysis and to Chapter 1 of [128] for further discussion of measure spaces in probability theory.

Example 2.3.1 (Probability Spaces) An abstract **probability space** $(\Omega, \mathcal{F}, \mathbb{P})$ is an example of a measure space, where Ω is an abstract set representing all possible outcomes, $\mathcal{F}$ is a σ-algebra of subsets of Ω describing possible events, and $\mathbb{P}$ is a probability measure, which assigns a probability between 0 and 1 inclusive to each possible event. A simple example of such a measure space is the outcome of a single dice roll, where $\Omega = \{1, 2, 3, 4, 5, 6\}$ describes all possible outcomes, the

σ-algebra of all possible events $\mathcal{F}$ is the power set of Ω, and $\mathbb{P}$ is equal to $1/6$ times the counting measure. We refer the reader to Chap. 3 on probability preliminaries, where we focus on measure spaces, which are specifically probability spaces.

Example 2.3.2 (Euclidean Space and Lebesgue Measure) Another canonical example of a measure space is the case of Euclidean space. To define Euclidean space as a measure space, let $\Omega = \mathbb{R}^d$, and let $\text{Borel}(\mathbb{R}^d)$ denote the σ-algebra of **Borel measurable sets** in $\mathbb{R}^d$, which is the smallest σ-algebra containing all open and closed sets of $\mathbb{R}^d$. Then, letting dx denote Lebesgue measure, we have that $(\Omega, \text{Borel}(\mathbb{R}^d), dx)$ is a measure space, where the Lebesgue measure is a natural extension of the standard measure on rectangles:

$$dx([a_1, b_1] \times [a_2, b_2] \times \ldots \times [a_d, b_d]) = \prod_{i=1}^{d} (b_i - a_i)$$

to the Borel measurable sets. In addition, there is a broader σ-algebra of sets $\mathcal{L}(\mathbb{R}^d)$, known as the **Lebesgue measurable sets**, on which the Lebesgue measure is defined. It is well-known that the Lebesgue measurable sets $\mathcal{L}(\mathbb{R}^d)$ form a σ-algebra containing more sets than the Borel measurable sets $\text{Borel}(\mathbb{R}^d)$, so that $\text{Borel}(\mathbb{R}^d) \subset \mathcal{L}(\mathbb{R}^d)$ and that in fact, $\mathcal{L}(\mathbb{R}^d)$ is the essentially the completion of the Borel measurable sets $\text{Borel}(\mathbb{R}^d)$ under all subsets of Borel measurable sets with zero measure; see [313]. In this case, one can consider the measure space of $(\mathbb{R}^d, \text{Borel}(\mathbb{R}^d), dx)$ with the Borel measurable sets in $\mathbb{R}^d$, or $(\mathbb{R}^d, \mathcal{L}(\mathbb{R}^d), dx)$ with the Lebesgue measurable sets in $\mathbb{R}^d$.

Example 2.3.3 (Banach Spaces and Borel Measurable Subsets) Recall that a **Banach space** is a normed vector space, which is complete with respect to its norm. We discussed Banach spaces previously in this chapter, in Sect. 2.1. A Banach space B has a notion of a norm $\| \cdot \|_B : B \to [0, \infty)$, which defines a metric on B, used to measure distances between elements of B. The norm, treated as the metric in a metric space defined by B and its norm, generates a topology of open and closed sets on B. Then, we can consider the set of Borel measurable subsets of B, denoted by $\text{Borel}(B)$, which is the smallest σ-algebra of subsets of B containing all open and closed subsets of B. We denote this measure space represented by a Banach space and its set of Borel sets by the ordered pair $(B, \text{Borel}(B))$.

Now that we have discussed Lebesgue measure on $\mathbb{R}^d$, we introduce the following terminology.

Definition 2.3.1 (Almost Everywhere) We say that a property holds **almost everywhere (a.e.)** on a measurable subset $\mathcal{O} \subset \mathbb{R}^d$ if the property holds for all $x \in E$ for some measurable subset $E \subset \mathcal{O}$ such that the Lebesgue measure of the complement $\mathcal{O} \setminus E$, denoted by $|\mathcal{O} \setminus E|$, equals zero.

We now recall the concept of a measurable function. Given two measurable spaces with σ-algebras $(\Omega_1, \mathcal{F}_1)$ and $(\Omega_2, \mathcal{F}_2)$, a function $f : \Omega_1 \to \Omega_2$ is a

measurable function if $f^{-1}(E) \in \mathcal{F}_1$ for all $E \in \mathcal{F}_2$, namely, the preimage of every measurable subset in the range is a measurable subset of the domain. The most commonly used examples of measurable functions will be random variables from a probability space to some function space, which will be discussed in the chapter on probabilistic preliminaries in Sect. 3, and real-valued functions $f : \mathcal{O} \to \mathbb{R}$ for a simply connected open set $\mathcal{O} \subset \mathbb{R}^d$ with Lipschitz boundary $\partial\mathcal{O}$.

2.4 L^p Spaces

We now discuss a special class of real-valued measurable functions, known as an L^p space, that has a central role in the analysis of partial differential equations, as these L^p spaces provide a rigorous functional framework for measuring integrability of functions. We refer the reader to Chapter 1 of [314], Chapter 4 of [60], and Appendix A of [324] for further discussion of L^p spaces.

For the discussion of L^p spaces in this subsection, let $\mathcal{O}$ be a simply connected open set in $\mathbb{R}^d$, which we will assume is either $\mathbb{R}^d$ itself or a bounded open subset $\mathcal{O} \subset \mathbb{R}^d$ with a Lipschitz boundary $\partial\mathcal{O}$.

Definition 2.4.1 (Definition of $L^p(\mathcal{O})$) For $1 \leq p < \infty$, we define $L^p(\mathcal{O})$ to be the set of (equivalence classes of) measurable functions $f : \mathcal{O} \to \mathbb{R}$ such that $\displaystyle\int_{\mathcal{O}} |f|^p dx < \infty$, where two measurable functions $f, g : \mathcal{O} \to \mathbb{R}$ are in the same equivalence class if $f = g$ almost everywhere on $\mathcal{O}$. We define the norm on $L^p(\mathcal{O})$ by

$$\|f\|_{L^p(\mathcal{O})} = \left(\int_{\mathcal{O}} |f|^p dx\right)^{1/p}.$$

For $p = \infty$, we define $L^\infty(\mathcal{O})$ to be the set of (equivalence classes of) measurable functions $f : \mathcal{O} \to \mathbb{R}$ such that there exists some constant $M > 0$ such that $|f(x)| \leq M$ almost everywhere. The infimum of all possible constants M such that $|f(x)| \leq M$ almost everywhere is defined to be the **essential supremum** (ess sup) of $|f|$, and we let

$$\|f\|_{L^\infty(\mathcal{O})} = \text{ess sup}|f|.$$

In the definition of an $L^p(\mathcal{O})$ space, we identify real-valued measurable functions $f : \mathcal{O} \to \mathbb{R}$ that are equal to each other almost everywhere to be in the same equivalence class, but we will not explicitly notate elements of $L^p(\mathcal{O})$ as equivalence classes, for convenience of notation. Hence, when we discuss a specific element $f \in L^p(\mathcal{O})$, we are implicitly considering a measurable function $f : \mathcal{O} \to \mathbb{R}$ that is only defined up to a set of measure zero in $\mathcal{O}$.

For any $1 \leq p < \infty$, we denote by $L^p_{\text{loc}}(\mathcal{O})$ the set of measurable functions for which the integral of the p-th power of the absolute value is finite over all compact subsets of $\mathcal{O}$. Similarly, we denote by $L^\infty_{\text{loc}}(\mathcal{O})$ the set of measurable functions that are essentially bounded over all compact subsets of $\mathcal{O}$.

We can use the norm on $L^p(\mathcal{O})$ to define a topology of convergence on $L^p(\mathcal{O})$, where we say that a sequence $\{f_n\}_{n=1}^\infty$ of functions in $L^p(\mathcal{O})$ **converges strongly in** $L^p(\mathcal{O})$ to f if

$$\lim_{n\to\infty} \|f - f_n\|_{L^p(\mathcal{O})} \to 0.$$

We now collect several important results about $L^p(\mathcal{O})$, the first of which is Hölder's inequality. For a given exponent of $1 \leq p \leq \infty$, we define the conjugate exponent p' by

$$\frac{1}{p} + \frac{1}{p'} = 1, \quad \Longrightarrow \quad p' = \frac{p}{p-1},$$

where we formally set $1/\infty = 0$ and $1/0 = \infty$. Using conjugate exponents, we have the following statement of Hölder's inequality for L^p spaces.

Proposition 2.4.1 (Hölder's Inequality) *Suppose that $f \in L^p(\mathcal{O})$ and $g \in L^{p'}(\mathcal{O})$. Then,*

$$\int_\mathcal{O} |fg| \leq \left(\int_\mathcal{O} |f|^p dx\right)^{1/p} \left(\int_\mathcal{O} |g|^{\frac{p}{p-1}} dx\right)^{\frac{p-1}{p}} = \|f\|_{L^p(\mathcal{O})}\|g\|_{L^{p'}(\mathcal{O})}.$$

More generally, for $\frac{1}{p} + \frac{1}{q} = \frac{1}{r}$, $1 \leq p, q, r \leq \infty$, and $f \in L^p(\mathcal{O})$, $g \in L^q(\mathcal{O})$, we have that $fg \in L^r(\mathcal{O})$ and furthermore,

$$\|fg\|_{L^r(\mathcal{O})} \leq \|f\|_{L^p(\mathcal{O})}\|g\|_{L^q(\mathcal{O})}.$$

As a corollary of Hölder's inequality, we obtain the following important result relating L^p spaces on bounded domains.

Corollary 2.4.1 *Let $\mathcal{O}$ be a bounded domain in $\mathbb{R}^d$. Then for $1 \leq p_1 < p_2 \leq \infty$,*

$$L^{p_2}(\mathcal{O}) \subset L^{p_1}(\mathcal{O}).$$

Proof By using Hölder's inequality with the conjugate exponents $\dfrac{p_2}{p_1}$ and $\dfrac{p_2}{p_2 - p_1}$, we obtain

$$\int_\mathcal{O} |f|^{p_1} \leq \left(\int_\mathcal{O} |f|^{p_2}\right)^{\frac{p_1}{p_2}} |\mathcal{O}|^{1-\frac{p_2}{p_1}} < \infty,$$

where $|\mathcal{O}|$ denotes the Lebesgue measure of the set $\mathcal{O}$, which is finite since $\mathcal{O}$ is a bounded domain. □

However, we remark that in the statement of this corollary, it is essential that $\mathcal{O}$ is a *bounded domain*, as it is well-known that the corresponding result is not true for L^p spaces on $\mathbb{R}^d$. For example, for any given spatial dimension d, one can easily check that $(1+|x|)^{-1/2} \in L^p(\mathbb{R}^d)$ for all $p > 2$ but $(1+|x|)^{-1/2} \notin L^p(\mathbb{R}^d)$ for all $1 \leq p \leq 2$.

Another corollary of Hölder's inequality is interpolation for L^p spaces. Note that in the interpolation lemma, the "relative distance" between L^p spaces is measured by the reciprocal of the exponent, $1/p$.

Corollary 2.4.2 (Interpolation Inequality for $L^p(\mathcal{O})$ Spaces) *Suppose that* $1 \leq p \leq r \leq q \leq \infty$ *and* $f \in L^p(\mathcal{O}) \cap L^q(\mathcal{O})$. *Then,* $f \in L^r(\mathcal{O})$ *with the estimate*

$$\|f\|_{L^r(\mathcal{O})} \leq \|f\|^{\theta}_{L^p(\mathcal{O})} \|f\|^{1-\theta}_{L^q(\mathcal{O})},$$

where $\theta \in [0, 1]$ *is determined by* $\dfrac{1}{r} = \dfrac{\theta}{p} + \dfrac{1-\theta}{q}$.

Proof Since $\dfrac{1}{r} = \dfrac{\theta}{p} + \dfrac{1-\theta}{q}$, we obtain that $\dfrac{\theta r}{p} + \dfrac{(1-\theta)r}{q} = 1$ and hence, we can use the conjugate exponents $\dfrac{p}{\theta r}$ and $\dfrac{q}{r(1-\theta)}$ in Hölder's inequality to obtain

$$\begin{aligned}\|f\|_{L^r(\mathcal{O})} &= \left(\int_{\mathcal{O}} |f|^r dx\right)^{1/r} = \left(\int_{\mathcal{O}} |f|^{r\theta} |f|^{r(1-\theta)} dx\right)^{1/r} \\ &\leq \left(\int_{\mathcal{O}} |f|^p dx\right)^{\theta/p} \left(\int_{\mathcal{O}} |f|^q dx\right)^{(1-\theta)/q} = \|f\|^{\theta}_{L^p(\mathcal{O})} \|f\|^{1-\theta}_{L^q(\mathcal{O})}.\end{aligned}$$

□

Next, we discuss duality in $L^p(\mathcal{O})$ spaces. We recall that given $1 \leq p < \infty$, the **dual space of** $L^p(\mathcal{O})$, denoted by $(L^p(\mathcal{O}))^*$, is the set of all bounded linear functionals $\ell : L^p(\mathcal{O}) \to \mathbb{R}$, which means that ℓ is a linear map such that $|\ell(f)| \leq C\|f\|_{L^p(\mathcal{O})}$ for all $f \in L^p(\mathcal{O})$ for some constant C. Note that Hölder's inequality implies that given $L^p(\mathcal{O})$, for the conjugate exponent $p' = \dfrac{p}{p-1}$ and any fixed $g \in L^{p'}(\mathcal{O})$, we in fact have that the map

$$f \in L^p(\mathcal{O}) \to \int_{\mathcal{O}} fg dx \in \mathbb{R}$$

is a bounded linear functional on $L^p(\mathcal{O})$. In fact, the following result shows that every linear functional on $L^p(\mathcal{O})$ has this form, when $1 \leq p < \infty$. See Proposition 4.1 (pg. 565, Appendix A) in [324] and Theorem 4.1 (pg. 13, Chapter 1) in [314] for further discussion.

Proposition 2.4.2 (Dual of $L^p(\mathcal{O})$) *For $1 \leq p < \infty$, the dual of $L^p(\mathcal{O})$ is isomorphic to $L^{p'}(\mathcal{O})$ for $p' = \dfrac{p}{p-1}$, where the isomorphism is given by*

$$\ell_g(f) \in (L^p(\mathcal{O}))^* \quad \Longleftrightarrow \quad \left(f \to \int_{\mathcal{O}} fg dx\right),$$

for a fixed but arbitrary $g \in L^{p'}(\mathcal{O})$.

Remark 2.4.1 It is important to remark that the dual space of $L^\infty(\mathcal{O})$ cannot be identified with $L^1(\mathcal{O})$, as this is the only exception to the previous result. In particular, we can identify the dual space of $L^\infty(\mathcal{O})$ with the space of bounded measures $\mathcal{M}(\mathcal{O})$, which is a larger space than $L^1(\mathcal{O})$.

We recall that we can use duality to define a weaker type of convergence than strong convergence of L^p norms. For $1 \leq p < \infty$, we can use the fact that the dual of $L^p(\mathcal{O})$ is $L^{p'}(\mathcal{O})$ for the conjugate exponent $p' = \frac{p}{p-1}$ to define the following notion of weak convergence. We say that a sequence $\{f_n\}_{n=1}^\infty$ of functions in $L^p(\mathcal{O})$ for $1 \leq p < \infty$ **converges weakly in** $L^p(\mathcal{O})$ to f, denoted by the notation $f_n \rightharpoonup f$, if

$$\lim_{n\to\infty} \int_{\mathcal{O}} f_n g dx = \int_{\mathcal{O}} fg dx, \qquad \text{for all } g \in L^{p'}(\mathcal{O}),\ p' = \frac{p}{p-1}.$$

While the dual of $L^\infty(\mathcal{O})$ is not $L^1(\mathcal{O})$, we can define a notion of weak-star convergence for $L^\infty(\mathcal{O})$, by viewing $L^\infty(\mathcal{O})$ as the dual of $L^1(\mathcal{O})$. A sequence of functions $\{f_n\}_{n=1}^\infty$ in $L^\infty(\mathcal{O})$ **converges weakly star (weakly*) in** $L^\infty(\mathcal{O})$, also denoted by $f_n \rightharpoonup f$, if

$$\lim_{n\to\infty} \int_{\mathcal{O}} f_n g dx = \int_{\mathcal{O}} fg dx, \qquad \text{for all } g \in L^1(\mathcal{O}).$$

Note here that the distinction between weak and weak-star convergence is subtle. While weak convergence of f_n to f in some general Banach space B involves checking that $\ell(f_n) \to \ell(f)$ for all linear functionals ℓ in the dual space B^*, in weak-star convergence, we instead view the functions f_n converging weakly star to f as elements of a dual space to some other Banach space (rather than considering the dual of the space in which f_n reside).

The following proposition is a consequence of the Banach-Alaoglu theorem, stated in Theorem 2.1.1. In particular, the following result states that having uniform bounds on a sequence in $L^p(\mathcal{O})$ is sufficient for obtaining weak (or weak-star for $p = \infty$) convergence of some subsequence in $L^p(\mathcal{O})$.

Proposition 2.4.3 (Weak Subsequential Compactness in $L^p(\mathcal{O})$) *Suppose that $\{f_n\}_{n=1}^\infty$ is a bounded sequence in $L^p(\mathcal{O})$, in the sense that there exists a positive constant C such that*

$$\|f_n\|_{L^p(\mathcal{O})} \leq C, \qquad \text{for all } n.$$

Then, there exists a subsequence $\{f_{n_k}\}_{k=1}^{\infty}$ *and* $f \in L^p(\mathcal{O})$ *such that* $f_{n_k} \rightharpoonup f$ *weakly in* $L^p(\mathcal{O})$ *if* $1 \leq p < \infty$, *or such that* $f_{n_k} \rightharpoonup f$ *weakly star in* $L^\infty(\mathcal{O})$ *if* $p = \infty$.

Remark 2.4.2 We emphasize that the same result is not true for strong convergence, as the closed unit ball in $L^p(\mathcal{O})$ is *not compact with respect to the strong topology.* This is what makes weak convergence particularly useful, as it gives us a weaker mode of convergence that will still be a result of uniform bounds on approximate solutions to partial differential equations, for example. A simple example that shows that the closed unit ball is not strongly compact is the following example in $L^2(0, 1)$, where we can consider the sequence of functions

$$f_n = \sin(2\pi n x).$$

One can check that $\|f_n\|_{L^2(0,1)} = \frac{1}{\sqrt{2}}$ for all n, so that the sequence $\{f_n\}_{n=1}^{\infty}$ is uniformly bounded in $L^2(0, 1)$. However, it does not have any *strongly* convergent subsequence in $L^2(0, 1)$, as one can verify that $\|f_m - f_n\|_{L^2(0,1)} = 1$ for all $m \neq n$, so no subsequence could ever be a Cauchy sequence in $L^2(0, 1)$. By the previous proposition, which is a consequence of the Banach-Alaoglu theorem (Theorem 2.1.1), it does have a weakly convergent subsequence. In fact, one can show that the original sequence itself, $\{f_n\}_{n=1}^{\infty}$, actually converges weakly in $L^2(0, 1)$ to the zero function: $f_n \rightharpoonup 0$ in $L^2(0, 1)$.

2.5 Sobolev Spaces

As in the exposition for $L^p(\mathcal{O})$ spaces, let $\mathcal{O}$ be either $\mathbb{R}^d$ or a bounded domain $\mathcal{O} \subset \mathbb{R}^d$ with Lipschitz boundary. The analysis of partial differential equations involves looking at equations involving partial derivatives, and hence, we must use a functional analytic framework for analyzing derivatives of functions, which is given by the classical theory of Sobolev spaces. We refer the reader to numerous well-written accounts of Sobolev space theory in the references [1, 60, 132, 161] and aim to give a brief but self-sufficient exposition of the main results from the theory of Sobolev spaces in this current section.

The purpose of Sobolev spaces is to give a functional framework for understanding derivatives of functions, but we have to clarify the notion of derivative that we are considering. Given any function $f : \mathcal{O} \to \mathbb{R}$, there is of course the standard notion of a derivative, given by $f'(x)$, defined in the classical way in a pointwise fashion when f is regular enough, $f \in C^1(\overline{\mathcal{O}})$, for example. We refer to such a notion of a derivative as a **strong derivative**. While we may be interested often in solutions to partial differential equations where the derivatives can be computed in a pointwise fashion and where the partial differential equation also holds in a pointwise fashion, the space $C^1(\overline{\mathcal{O}})$ is not a Banach space (rather it is what is known as a Fréchet space), and hence, it is not the best functional framework for standard procedures such as taking limits of sequences and extracting convergent subsequences from uniform bounds.

This motivates weakening the notion of a derivative to what is known as a **weak derivative**, as these weak derivatives will allow us to invoke the elegant functional frameworks of Banach and Hilbert spaces, for example. The notion of a weak derivative is based upon noting that by integration by parts, for a function $f \in C^1(\overline{\mathcal{O}})$, we have that for any compactly supported function $\varphi \in C_c^\infty(\mathcal{O})$:

$$\int_{\mathcal{O}} \frac{\partial f}{\partial x_i} \varphi dx = - \int_{\mathcal{O}} f \frac{\partial \varphi}{\partial x_i} dx,$$

where we make the crucial observation that the right-hand side quantity makes sense, even if $f \in L^p(\mathcal{O})$ and even if f does not possess a strong (classical) derivative. This motivates the following definition of a *weak derivative.*

Definition 2.5.1 (Weak Derivative) A function $g \in L^1_{\text{loc}}(\mathcal{O})$ is the **weak partial derivative with respect to** x_i of $f \in L^1_{\text{loc}}(\mathcal{O})$, if

$$\int_{\mathcal{O}} g\varphi dx = - \int_{\mathcal{O}} f \frac{\partial \varphi}{\partial x_i} dx \quad \text{for all } \varphi \in C_c^\infty(\mathcal{O}).$$

We refer to a weak derivative of f as *the* weak derivative, because it is easy to verify that if a weak derivative exists, it must be unique (up to a set of measure zero). We use the notation $\partial_{x_i} f$ to refer to this weak derivative and the notation of the **gradient** $\nabla f : \mathcal{O} \to \mathbb{R}^d$ to refer to the collection of all weak first-order partial derivatives $(\partial_{x_1} f, \partial_{x_2} f, \dots, \partial_{x_d} f)$, when it is clear from context that these weak partial derivatives will not be confused with the corresponding strong partial derivative. By iterating weak derivatives, we can define the **α-weak derivative** for a multi-index $\alpha = (\alpha_1, \alpha_2, \dots, \alpha_d)$ by

$$\partial^\alpha f = \partial_{x_1}^{\alpha_1} \partial_{x_2}^{\alpha_2} \cdots \partial_{x_d}^{\alpha_d} f,$$

where the differentiation operators on the right-hand side are interpreted as weak partial derivatives. We define the **order of the multi-index** $\alpha = (\alpha_1, \alpha_2, \dots, \alpha_d)$ as $|\alpha| = \alpha_1 + \alpha_2 + \ldots + \alpha_d$.

Example 2.5.1 (Examples of Weak Derivatives) We give some brief examples. If $f \in C_1(\overline{\mathcal{O}})$, it is easy to check that $\frac{\partial f}{\partial x_i}$ defined classically is the weak derivative of f. However, weak derivatives also allow us to define derivatives of classically non-differentiable functions. For example, the weak derivative of $f(x) = |x|$ defined on $(-1, 1)$ is the function

$$g(x) = -1 \text{ on } -1 < x < 0, \quad g(x) = 1 \text{ on } 0 < x < 1,$$

where we refer to g as f', and where we say that we interpret this derivative in a *weak (or distributional) sense*. Although g is ambiguously defined at $x = 0$, we recall that this is not an issue, as functions in $L^p(\mathcal{O})$ are defined almost everywhere up to sets of measure zero anyway.

By combining the notion of L^p spaces with the notion of a weak derivative, we obtain the following definition of a Sobolev space, first for positive integer order.

Definition 2.5.2 (Sobolev Space, $W^{k,p}(\mathcal{O})$) Let k be a positive integer, and let $1 \leq p \leq \infty$. Then, we define the **Sobolev space** $W^{k,p}(\mathcal{O})$ to be the set of functions $f \in L^p(\mathcal{O})$ for which all weak partial derivatives of order less than or equal to k are in $L^p(\mathcal{O})$:

$$W^{k,p}(\mathcal{O}) := \{f \in L^p(\mathcal{O}) : \partial^\alpha f \in L^p(\mathcal{O}) \text{ for all } 1 \leq |\alpha| \leq k\}.$$

The Sobolev space $W^{k,p}(\mathcal{O})$ is a Banach space with norm given by

$$\|f\|_{W^{k,p}(\mathcal{O})} := \begin{cases} \left(\|f\|_{L^p(\mathcal{O})} + \sum_{1 \leq |\alpha| \leq k} \|\partial^\alpha f\|_{L^p(\mathcal{O})} \right)^{1/p} & \text{for } 1 \leq p < \infty, \\ \|f\|_{L^\infty(\mathcal{O})} + \sum_{1 \leq |\alpha| \leq k} \|\partial^\alpha f\|_{L^\infty(\mathcal{O})}, & \text{for } p = \infty. \end{cases}$$

In the special case where $p = 2$, we use the notation that $H^k(\mathcal{O}) := W^{k,p}(\mathcal{O})$, and we note that in this case, $H^k(\mathcal{O})$ is more specifically a Hilbert space, with an inner product given by

$$(f, g)_{H^k(\mathcal{O})} = \int_{\mathcal{O}} fg dx + \sum_{1 \leq |\alpha| \leq k} \int_{\mathcal{O}} (\partial^\alpha f)(\partial^\alpha g) dx.$$

We make a comment about the special case of $p = \infty$. Functions in $W^{1,\infty}(\mathcal{O})$ are functions that are bounded almost everywhere and whose (weak) derivatives of first order are also bounded almost everywhere. It is classically known that, when $\mathcal{O}$ is an open and bounded domain with Lipschitz boundary, we can identify $W^{1,\infty}(\mathcal{O})$ with the space of **Lipschitz continuous functions**; see the proof of Theorem 4 on pg. 296 in Section 5.8.2b of [132].

Proposition 2.5.1 ($W^{1,\infty}(\mathcal{O})$ and Lipschitz Functions) *Let $\mathcal{O}$ be a Lipschitz domain in $\mathbb{R}^d$. Then, every element $f \in W^{1,\infty}(\mathcal{O})$ is equal to a Lipschitz function on $\mathcal{O}$ up to a set of measure zero (almost everywhere).*

While Sobolev spaces can be initially defined using weak derivatives as above, this only allows us to define Sobolev spaces $W^{k,p}(\mathcal{O})$ of positive integer order k. Next, we define **fractional Sobolev spaces**.

Definition 2.5.3 (Fractional Sobolev Space) For an open set $\mathcal{O}$ in $\mathbb{R}^d$, $0 < s < 1$, and $1 \le p < \infty$, we define the fractional Sobolev space $W^{s,p}(\mathcal{O})$ as follows:

$$W^{s,p}(\mathcal{O}) = \left\{ u \in L^p(\mathcal{O}) : \int_{\mathcal{O}} \int_{\mathcal{O}} \frac{|u(x) - u(y)|^p}{|x - y|^{sp+d}} dxdy < \infty \right\},$$

endowed with the norm

$$\|u\|_{W^{s,p}(\mathcal{O})} = \left(\|u\|^p_{L^p(\mathcal{O})} + \int_{\mathcal{O}} \int_{\mathcal{O}} \frac{|u(x) - u(y)|^p}{|x - y|^{sp+d}} dxdy \right)^{\frac{1}{p}}.$$

In the case where $p = 2$, we simply use the notation $H^s(\mathcal{O}) = W^{s,2}(\mathcal{O})$ for all $0 < s < 1$. This definition cannot simply be extended to $s \ge 1$.

Instead, for any positive integer k, any $0 < s < 1$ and $1 \le p < \infty$, we define $W^{k+s,p}(\mathcal{O})$ as

$$W^{k+s,p}(\mathcal{O}) := \{ f \in W^{k,p}(\mathcal{O}) : \partial^\alpha f \in W^{s,p}(\mathcal{O}) \text{ for all } |\alpha| = k \}$$

endowed with the norm

$$\|f\|_{W^{k+s,p}(\mathcal{O})} := \left(\|f\|^p_{W^{k,p}(\mathcal{O})} + \sum_{|\alpha|=k} \|\partial^\alpha f\|^p_{W^{s,p}(\mathcal{O})} \right)^{1/p}.$$

The following result gives a relation between Definitions 2.5.2 and 2.5.3: If $\mathcal{O}$ is a bounded domain with Lipschitz boundary, then for any $0 < s < 1$ and $1 \le p < \infty$,

$$W^{1,p}(\mathcal{O}) \subset W^{s,p}(\mathcal{O}).$$

For a proof of this embedding, see, e.g., Proposition 2.2 in [119].

Next, we recall the trace theorem, which we will make frequent use of in the study of fluid-structure interaction in the subsequent chapters. The trace theorem is motivated by taking restrictions along boundaries of domains. Given a bounded domain $\mathcal{O} \subset \mathbb{R}^d$ with a piecewise Lipschitz boundary $\partial\mathcal{O}$, there is a classically defined restriction map:

$$R : \varphi \in C^\infty(\overline{\mathcal{O}}) \to \varphi|_{\partial\mathcal{O}} \in C^\infty(\partial\mathcal{O}),$$

for smooth functions on $\overline{\mathcal{O}}$. The trace operator essentially is an extension of the restriction map to Sobolev spaces, where such a trace map is well-defined for Sobolev spaces of sufficiently large order. We state the trace theorem as follows. See Theorem 1.5.2.1 in [174].

Theorem 2.5.1 *Let $\mathcal{O} \subset \mathbb{R}^d$ be a bounded domain with a piecewise Lipschitz boundary. Then, for $s > 1/p$, there exists a trace map $T : W^{s,p}(\mathcal{O}) \to$*

$W^{s-\frac{1}{p},p}(\partial\mathcal{O})$, *which is a bounded linear operator such that* $T(\varphi) = \varphi|_{\partial\mathcal{O}}$ *for all* $\varphi \in C^\infty(\overline{\mathcal{O}})$.

Given the importance of the trace map and of the zero (Dirichlet) boundary condition for many partial differential equations, it is often useful to consider the trace-free functions in a Sobolev space, namely, those functions in $W^{k,p}(\mathcal{O})$, which have zero trace along $\partial\mathcal{O}$, when $\mathcal{O}$ is a bounded domain in $\mathbb{R}^d$.

Definition 2.5.4 (Definition of $W_0^{k,p}(\mathcal{O})$) For a bounded domain $\mathcal{O} \subset \mathbb{R}^d$, we define $W_0^{k,p}(\mathcal{O})$ to be the closure of the compactly supported functions $C_c^\infty(\mathcal{O})$ in $W^{k,p}(\mathcal{O})$ with respect to the norm $\|\cdot\|_{W^{k,p}(\mathcal{O})}$.

Finally, we can use duality to define the notion of a **Sobolev space of negative order**.

Definition 2.5.5 (Definition of Sobolev Space of Negative Order) For $s > 0$ and $1 < p < \infty$, we define the Sobolev space $W^{-s,p}(\mathcal{O})$ to be the dual space of $W_0^{s,p'}(\mathcal{O})$ for the conjugate exponent $p' = \frac{p}{p-1}$. Namely, $W^{-s,p}(\mathcal{O})$ consists of all bounded linear functionals on $W_0^{s,p'}(\mathcal{O})$. The $W^{-s,p}(\mathcal{O})$-norm is given by the usual norm for dual spaces of Banach spaces:

$$\|\ell\|_{W^{-s,p}(\mathcal{O})} = \inf\{|\ell(f)| : f \in W_0^{s,p'}(\mathcal{O}), \|f\|_{W^{s,p'}(\mathcal{O})} = 1\}.$$

We remark that even though $W^{-s,p}(\mathcal{O})$ is a space of bounded linear functionals, we can often identify certain elements of $W^{-s,p}(\mathcal{O})$ with functions g by considering a bounded linear functional whose action on some $f \in W_0^{s,p'}(\mathcal{O})$ is via integration over $\mathcal{O}$, for example, $f \in W_0^{s,p'}(\mathcal{O}) \to \int_{\mathcal{O}} fg dx$. For example, in this way, one can easily verify that any element $g \in L^2(\mathcal{O})$ is an element of $H^{-s}(\mathcal{O})$ for any $s > 0$, since the map

$$f \in H_0^s(\mathcal{O}) \to \int_{\mathcal{O}} fg dx$$

is a bounded linear functional on $H_0^s(\mathcal{O})$ whenever $g \in L^2(\mathcal{O})$ by Hölder's inequality.

Next, we recall some essential properties of Sobolev spaces, which we will make frequent use of in the subsequent chapters. An underlying heuristic principle in the analysis of partial differential equations is the principle that having integrability of higher derivatives implies stronger integrability of lower-order derivatives. In particular, one can gain additional stronger information about the regularity of a function by showing regularity of higher-order weak derivatives, or essentially showing that a given function belongs to some Sobolev space of positive order. This is a general principle in the analysis of partial differential equations that can be made precise through the following Sobolev embedding theorem (see, e.g., Corollary 4.53 in [114]), which we state in full generality for fractional Sobolev spaces.

Theorem 2.5.2 (Continuous Embedding) *Let $0 < s < 1$ and $1 < p < \infty$ and let $\mathcal{O}$ be an open set in $\mathbb{R}^d$ with Lipschitz boundary. Then the following embeddings are continuous:*

1. *If $sp < d$, then $W^{s,p}(\mathcal{O}) \hookrightarrow L^q(\mathcal{O})$ for any $q \leq \frac{dp}{d-sp}$.*
2. *If $sp = d$, then $W^{s,p}(\mathcal{O}) \hookrightarrow L^q(\mathcal{O})$ for any $q < \infty$.*
3. *If $sp > d$, then $W^{s,p}(\mathcal{O}) \hookrightarrow L^\infty(\mathcal{O})$ and in fact $W^{s,p}(\mathcal{O}) \hookrightarrow C^{0,s-\frac{d}{p}}(\mathcal{O})$.*

As a corollary, we state an explicit case of the Sobolev embeddings that is frequently used in the case where $p = 2$.

Corollary 2.5.1 ($H^s(\mathcal{O}) \subset C(\mathcal{O})$ for $s > \frac{d}{2}$) *Let $\mathcal{O} \subset \mathbb{R}^d$ be an open set with Lipschitz boundary. Then, for $s > \frac{d}{2}$, the Sobolev space $H^s(\mathcal{O})$ embeds continuously into $C(\mathcal{O})$.*

Finally, we give two important results on interpolation in Sobolev spaces. See Theorem 5.2 in [2].

Theorem 2.5.3 (Gagliardo Nirenberg Inequality) *Let $\mathcal{O} \subset \mathbb{R}^d$ be a bounded domain with Lipschitz boundary. Let $p \geq 1$ and let j, m be two nonnegative integers. There exists a constant C depending only on d, m, and p, such that if $0 < j < m$, and $u \in W^{m,p}(\mathcal{O})$, then*

$$\|u\|_{W^{j,p}(\mathcal{O})} \leq C\|u\|_{W^{m,p}(\mathcal{O})}^{\frac{j}{m}} \|u\|_{L^p(\mathcal{O})}^{\frac{(m-j)}{m}}.$$

Theorem 2.5.4 (Ladyzenskaya Inequality) *Let $\mathcal{O} \subset \mathbb{R}^d$ be a bounded domain with Lipschitz boundary. Then there exists a constant $C > 0$ depending only on $\mathcal{O}$ such that*

$$\|u\|_{L^4(\mathcal{O})} \leq C\|u\|_{L^2(\mathcal{O})}^{\frac{1}{2}} \|u\|_{H^1(\mathcal{O})}^{\frac{1}{2}}, \qquad \text{if } d = 2,$$

$$\|u\|_{L^4(\mathcal{O})} \leq C\|u\|_{L^2(\mathcal{O})}^{\frac{1}{4}} \|u\|_{H^1(\mathcal{O})}^{\frac{3}{4}}, \qquad \text{if } d = 3,$$

2.6 The Fourier Transform on $\mathbb{R}^d$

As a specific application of the theory of L^p spaces, we will consider the Fourier transform, which is a fundamental transformation that associates a function $f(x)$ in spatial variables $x \in \mathbb{R}^d$ to a function $\hat{f}(\xi)$ in frequency space $\xi \in \mathbb{R}^d$. The Fourier transform plays an essential role in the analysis of partial differential equations, especially in the context of linear partial differential operators, due to its useful properties, such as exchanging differentiation for multiplication by polynomials (see Proposition 2.6.1). We refer the reader to Chapter 3, Section 3 in [324], and Chapters 5 and 6 in [312] for further discussion.

Given a function $f : \mathbb{R}^d \to \mathbb{R}$, we define the **Fourier transform** of f, denoted by $\hat{f} : \mathbb{R}^d \to \mathbb{R}$, by

$$\mathcal{F}(f)(\xi) := \hat{f}(\xi) = \frac{1}{(2\pi)^{d/2}} \int_{\mathbb{R}^d} f(x) e^{-ix\cdot\xi} dx, \tag{2.3}$$

where we use the variable x as in $f(x)$ to denote the physical variables, while we use the variable ξ as in $\hat{f}(\xi)$ to denote the transformed (frequency) variables. In a physical interpretation, f represents a function in physical space, while $\hat{f}$ represents a (continuous) frequency decomposition representing the intensity of different frequencies in f. One can heuristically think of the Fourier transform as a continuous analogue of Fourier series, which gives a convergent frequency decomposition of complex periodic functions into simpler periodic functions, which is discrete in frequency space.

When considering the Fourier transform as defined in (2.3), it is not a priori apparent which classes of functions f have a well-defined Fourier transform, as one would require that the integral defining the Fourier transform must be convergent. One can define a useful class of functions known as Schwartz functions, for which the Fourier transform is a linear operator.

Definition 2.6.1 (The Schwartz Space $\mathcal{S}(\mathbb{R}^d)$) The **Schwartz space** $\mathcal{S}(\mathbb{R}^d)$ consists of all smooth functions $f : \mathbb{R}^d \to \mathbb{R}$ that are rapidly decreasing in the sense that

$$\sup_{x\in\mathbb{R}^d} |x|^k |\partial_x^\alpha f(x)| \leq C_{k,\alpha},$$

for all positive integers k and all multi-indices α, for a constant $C_{k,\alpha}$ that depends on k and the multi-index α.

Example 2.6.1 (The Gaussian) The standard example of a Schwartz function is the Gaussian function $f(x) = e^{-\frac{1}{2}|x|^2}$.

Remark 2.6.1 (The Schwartz Class as a Fréchet Space) The Schwartz class is not a Banach space, but it is an example of a more general function space known as a Frechet space, with a countable family of associated seminorms indexed by nonnegative integers k and multi-indices α, which are given by

$$[f]_{k,\alpha} = \sup_{x\in\mathbb{R}^d} |x|^k |\partial_x^\alpha f(x)|,$$

where a function $f \in \mathcal{S}(\mathbb{R}^d)$ if and only if every seminorm $[f]_{k,\alpha}$ is finite for each k, α. These seminorms define a topology of convergence on this Fréchet space, where we say that $f_n \to f$ in $\mathcal{S}(\mathbb{R}^d)$ if $[f - f_n]_{k,\alpha} \to 0$ for each k, α.

It is a well-known and important property that the Fourier transform *exchanges differentiation and multiplication by polynomials*, and this property is a rigorous demonstration of the heuristic idea that the Fourier transform exchanges regularity for decay at infinity and vice versa.

Proposition 2.6.1 (Interchange of Differentiation and Multiplication) *For a Schwartz function $f \in \mathcal{S}(\mathbb{R}^d)$,*

$$\mathcal{F}(x_k f)(\xi) = i\partial_{\xi_k}\hat{f}(\xi), \qquad \mathcal{F}(\partial_{x_k} f)(\xi) = i\xi_k \hat{f}(\xi).$$

In addition, the Fourier transform interchanges two additional operations of multiplication of functions and convolution.

Definition 2.6.2 Given two functions $f, g : \mathbb{R}^d \to \mathbb{R}$, the **convolution**, denoted by $f * g : \mathbb{R}^d \to \mathbb{R}$, is defined by

$$(f * g)(x) := \int_{\mathbb{R}^d} f(y)g(x-y)dy.$$

A simple change of variables shows that the convolution operation is symmetric, namely, $(f * g)(x) = (g * f)(x)$. A crucial property of the Fourier transform is that the Fourier transform interchanges convolution with multiplication.

Proposition 2.6.2 (Interchange of Multiplication and Convolution of Functions) *For Schwartz functions $f, g \in \mathcal{S}(\mathbb{R}^d)$,*

$$\widehat{f * g}(\xi) = (2\pi)^{d/2}\hat{f}(\xi)\hat{g}(\xi) \qquad \textit{and} \qquad \widehat{fg}(\xi) = \frac{1}{(2\pi)^{d/2}}(\hat{f} * \hat{g})(\xi).$$

Using the interchange of differentiation and multiplication in Proposition 2.6.1, one can easily show the following result, which explains why the Schwartz class is a natural space to consider for the Fourier transform.

Proposition 2.6.3 (Fourier Transform as a Continuous Operator on $\mathcal{S}(\mathbb{R}^d)$) *The Fourier transform $\mathcal{F}$ is a continuous linear map $\mathcal{F} : \mathcal{S}(\mathbb{R}^d) \to \mathcal{S}(\mathbb{R}^d)$.*

Even more importantly, not only is the Fourier transform a continuous linear operator on the Schwartz class $\mathcal{S}(\mathbb{R}^d)$, but it is in fact bijective, with a continuous inverse, given by the **Fourier inversion formula**.

Theorem 2.6.1 (Fourier Inversion) *The Fourier transform $\mathcal{F}(f) = \frac{1}{(2\pi)^{d/2}} \int_{\mathbb{R}^d} f(x)e^{-ix\cdot\xi}dx$ is a continuous linear isomorphism $\mathcal{F} : \mathcal{S}(\mathbb{R}^d) \to \mathcal{S}(\mathbb{R}^d)$, with continuous inverse $\mathcal{F}^{-1} : \mathcal{S}(\mathbb{R}^d) \to \mathcal{S}(\mathbb{R}^d)$ given by the following formula:*

$$\mathcal{F}^{-1}(f) = \check{f}(x) = \frac{1}{(2\pi)^{d/2}} \int_{\mathbb{R}^d} f(\xi)e^{ix\cdot\xi}d\xi.$$

Remark 2.6.2 This gives rise to the Fourier inversion formula:

$$f(x) = \frac{1}{(2\pi)^{d/2}} \int_{\mathbb{R}^d} \hat{f}(\xi) e^{ix\cdot\xi} d\xi \qquad \text{for } f \in \mathcal{S}(\mathbb{R}^d).$$

Next, we use the theory of L^p spaces to reveal additional structure to the Fourier transform. We first make the preliminary observation that given the theory of L^p functions, the integral defining the Fourier transform in (2.3) actually converges for a broader class of functions than the Schwartz class $\mathcal{S}(\mathbb{R}^d)$. Namely, we note that

$$|\hat{f}(\xi)| \leq \frac{1}{(2\pi)^{d/2}} \left| \int_{\mathbb{R}^d} f(x) e^{-ix\cdot\xi} dx \right| \leq \frac{1}{(2\pi)^{d/2}} \int_{\mathbb{R}^d} |f(x)| dx,$$

so that

$$\|\hat{f}\|_{L^\infty(\mathbb{R}^d_\xi)} \leq C \|f\|_{L^1(\mathbb{R}^d_x)}.$$

More generally, we have the following result for the Fourier transform acting on L^p spaces.

Proposition 2.6.4 (Hausdorff-Young Inequality) *The Fourier transform $\mathcal{F}$ is a bounded linear map $\mathcal{F} : L^p(\mathbb{R}^d_x) \to L^{p'}(\mathbb{R}^d_\xi)$ for $1 \leq p \leq 2$ and $p' = \dfrac{p}{p-1}$.*

An important result known as **Plancherel's theorem** says that in the case of $L^2(\mathbb{R}^d)$, the Fourier transform is actually an isometry from $L^2(\mathbb{R}^d)$ to itself, with respect to the $L^2(\mathbb{R}^d)$ inner product.

Theorem 2.6.2 (Plancherel's Theorem) *For any $f \in L^2(\mathbb{R}^d_x)$, we have that*

$$\int_{\mathbb{R}^d} |f(x)|^2 dx = \int_{\mathbb{R}^d} |\hat{f}(\xi)|^2 d\xi,$$

and hence the Fourier transform $\mathcal{F} : L^2(\mathbb{R}^d_x) \to L^2(\mathbb{R}^d_\xi)$ is an isometric bounded linear map, in the sense that $(f, g)_{L^2(\mathbb{R}^d_x)} = (\hat{f}, \hat{g})_{L^2(\mathbb{R}^d_\xi)}$ for all $f, g \in L^2(\mathbb{R}^d_x)$.

We end this section by giving some simple applications of Fourier transforms to the theory of partial differential equations, by showing how using a frequency-space interpretation can help to easily solve partial differential equations such as the heat equation and the wave equation.

An Application of the Fourier Transform to the Linear Heat Equation As a first application of the Fourier transform, we consider the heat equation for $(t, x) \in \mathbb{R}^+ \times \mathbb{R}^d$, which is a linear parabolic equation of the form

$$\partial_t f - \Delta f = F \tag{2.4}$$

for an inhomogeneous source term F, with initial data given by

$$f(0,\cdot) = f_0(x).$$

We formally take the Fourier transform of (2.4) using the fact that the Fourier transform exchanges multiplication by polynomials and differentiation, as discussed in Proposition 2.6.1. We emphasize that we will take the Fourier transform only in the spatial variable x, while keeping the time variable in the physical domain, so that we are essentially taking a spatial Fourier transform along each time slice of constant time.

For simplicity, let us first consider $F = 0$ so that we are considering the homogeneous heat equation:

$$\partial_t f - \Delta f = 0, \qquad \text{on } \mathbb{R}^+ \times \mathbb{R}^d.$$

After taking a spatial Fourier transform along each fixed time slice, we obtain

$$\partial_t \hat{f}(t,\xi) + |\xi|^2 \hat{f}(t,\xi) = 0,$$

with initial condition given by

$$\hat{f}(0,\cdot) = \hat{f}_0(\xi).$$

Notice that by taking the Fourier transform, the spatial derivatives become multiplication by a polynomial in frequency space, so that the resulting equation in frequency space is just an ordinary differential equation in the time variable t for each fixed but arbitrary $\xi \in \mathbb{R}^d_\xi$, which can be solved explicitly as

$$\hat{f}(t,\xi) = e^{-t|\xi|^2} \hat{f}_0(\xi). \tag{2.5}$$

Assuming that f has sufficient decay so that $\hat{f}$ and all quantities are well-defined (e.g., as elements of the Schwartz class), we can then use the Fourier inversion formula to conclude that the solution to (2.4) is

$$f(t,x) = \frac{1}{(2\pi)^{d/2}} \int_{\mathbb{R}^d} e^{-t|\xi|^2} \hat{f}_0(\xi) e^{ix\cdot\xi}\, d\xi.$$

We note that the representation formula (2.5) reveals that in frequency space, the solution to the homogeneous heat equation is really given by a Fourier multiplier, namely, multiplication by a function of ξ composed with a Fourier inversion. This motivates the formal notation for the solution to the homogeneous heat equation as a Fourier multiplier with $e^{t\Delta}$:

$$e^{t\Delta}(f_0) := \frac{1}{(2\pi)^{d/2}} \int_{\mathbb{R}^d} e^{-t|\xi|^2} \hat{f}_0(\xi) e^{ix\cdot\xi}\, d\xi,$$

where this operator $e^{t\Delta}$, which forms a C^0-*linear semigroup*, is a linear operator on the Schwartz class $\mathcal{S}(\mathbb{R}^d)$, or more generally on $L^2(\mathbb{R}^d)$.

To handle the case of the general inhomogeneous wave equation with a source term F as in (2.4), we can use a general principle for linear partial differential equations known as **Duhamel's principle**. The general idea behind Duhamel's principle is that one can regard a source term $F(t,x)$ as an infinite collection of external impulses up to time t. Hence, for a fixed time s, $0 \leq s \leq t$, and a final time t, the contribution of the source term $F(s,x)$ at the previous time s to the final time t can be thought of as the solution to the heat equation with initial data $F(s,x)$, propagated by time $t-s$ to the final time t. Then, one can sum up the contribution from all of the individual impulses of the source term $F(s,x)$ from $s=0$ to $s=t$, which are all propagated by a time frame of $t-s$ from the initial time of impulse at time s to the final time t. This gives rise to the following formula for the solution to (2.4), where the last term is the contribution from the source term given by Duhamel's principle:

$$\begin{aligned} f(t,x) &= \frac{1}{(2\pi)^{d/2}} \int_{\mathbb{R}^d} e^{-t|\xi|^2} \hat{f}_0(\xi) e^{ix\cdot\xi}\, d\xi \\ &\quad + \frac{1}{(2\pi)^{d/2}} \int_0^t \int_{\mathbb{R}^d} e^{-(t-s)|\xi|^2} \hat{F}(s,\xi) e^{ix\cdot\xi}\, d\xi ds. \end{aligned}$$

Although we have obtained this formula using the heuristic provided by Duhamel's principle, this principle can be rigorously verified by substituting this formula and explicitly checking that it satisfies the original partial differential equation with the inhomogeneous source term F. Using formal operator notation, we can express the solution to the inhomogeneous heat equation via Fourier multipliers as

$$f(t,x) = e^{t\Delta} f_0 + \int_0^t e^{(t-s)\Delta} F(s,\cdot) ds.$$

An Application of the Fourier Transform to the Linear Wave Equation
Next, we use Fourier transform methods to study the linear inhomogeneous wave equation:

$$\partial_{tt} f - \Delta f = F, \tag{2.6}$$

with initial conditions for the initial displacement $f(0,\cdot)$ and the initial velocity $\partial_t f(0,\cdot)$ given by

$$f(0,\cdot) = f_0(x), \qquad \partial_t f(0,\cdot) = f_1(x),$$

which is a linear hyperbolic partial differential equation.

We first consider the homogeneous wave equation with no external source term:

$$\partial_{tt} f - \Delta f = 0,$$

and as for the heat equation, we take the Fourier transform in only the spatial variables keeping the time variable in the physical domain to obtain

$$\partial_{tt} \hat{f} + |\xi|^2 \hat{f} = 0$$

with initial data

$$\hat{f}(0, \cdot) = \hat{f}_0(\xi), \qquad \partial_t \hat{f}(0, \cdot) = \hat{f}_1(\xi).$$

In frequency space, this can be solved explicitly as

$$\hat{f}(t, \xi) = \hat{f}_0(\xi) \cos(|\xi| t) + \hat{f}_1(\xi) \frac{\sin(|\xi| t)}{|\xi|}.$$

Therefore, assuming sufficient decay and regularity of the initial data f_0, f_1, we can obtain an explicit solution for the linear homogeneous wave equation by the Fourier inversion formula:

$$f(t, x) = \frac{1}{(2\pi)^{d/2}} \int_{\mathbb{R}^d} \hat{f}_0(\xi) \cos(|\xi| t) e^{ix\cdot\xi} d\xi + \frac{1}{(2\pi)^{d/2}} \int_{\mathbb{R}^d} \hat{f}_1(\xi) \frac{\sin(|\xi| t)}{|\xi|} e^{ix\cdot\xi} d\xi,$$

and since we can interpret the solution to the homogeneous wave equation in time as a Fourier multiplier, this motivates the following notation for the solution, where we recall that the symbol of $\sqrt{-\Delta}$ in Fourier space is $|\xi|$:

$$f(t, x) = \cos(t\sqrt{-\Delta}) f_0 + \frac{\sin(t\sqrt{-\Delta})}{\sqrt{-\Delta}} f_1.$$

We can then obtain the solution for the inhomogeneous linear wave equation with a general source term F via Duhamel's principle, to obtain that the solution to (2.6) is given by

$$\begin{aligned} f(t, x) &= \cos(t\sqrt{-\Delta}) f_0 + \frac{\sin(t\sqrt{-\Delta})}{\sqrt{-\Delta}} f_1 + \int_0^t \frac{\sin((t-s)\sqrt{-\Delta})}{\sqrt{-\Delta}} F(s, \cdot) ds \\ &= \frac{1}{(2\pi)^{d/2}} \int_{\mathbb{R}^d} \hat{f}_0(\xi) \cos(|\xi| t) e^{ix\cdot\xi} d\xi + \frac{1}{(2\pi)^{d/2}} \int_{\mathbb{R}^d} \hat{f}_1(\xi) \frac{\sin(|\xi| t)}{|\xi|} e^{ix\cdot\xi} d\xi \\ &+ \frac{1}{(2\pi)^{d/2}} \int_0^t \int_{\mathbb{R}^d} \frac{\sin(|\xi|(t-s)}{|\xi|} \hat{F}(s, \xi) e^{ix\cdot\xi} d\xi. \end{aligned}$$

2.7 Bochner Spaces

We have so far discussed functional spaces such as Banach spaces and Hilbert spaces, which are useful for defining functions existing on spatial domains. However, many of the fluid dynamics and fluid-structure interaction problems that we will consider throughout this book will be time-dependent dynamic problems, where quantities, such as the fluid velocity, will be functions of *both time and space*, where the regularity and integrability of these functions in time and space may be different. Thus, we need a framework for understanding functions in spaces of the form $L^p(0, T; B)$ for Banach spaces B (representing a spatial functional space), and this rigorous framework is provided by **Bochner spaces**. In this section, we give a brief overview of the main results concerning Bochner spaces and refer the interested reader to the exposition in Chapter 1 of [191] and in [120] for further details.

The motivation behind Bochner spaces is to consider functions of time and space, where we can consider a function $f(t, x) : [0, T] \times \mathcal{O} \to \mathbb{R}$ for $\mathcal{O} \subset \mathbb{R}^d$ alternatively as functions of time, taking values in some function space B of spatial functions defined on the spatial domain $\mathcal{O}$, $f : [0, T] \to B$. We would like to be able to perform standard operations of analysis on such functions of time and space, such as taking limits and taking integrals. Recalling the standard Lebesgue integration theory of real-valued functions, the definition of integrals is based fundamentally on the integration of simple functions and the use of simple functions to approximate arbitrary Lebesgue measurable functions $f : \mathbb{R} \to \mathbb{R}$. However, because Bochner spaces will have functions $f : [0, T] \to B$ taking values in (potentially) infinite-dimensional spaces B, the measurability properties of these functions are more subtle. Since we will eventually want to take integrals of these spacetime functions, this motivates using *simple functions* analogously as the basis for measurability and integrability in Bochner spaces, where intuitively, the functions that are measurable as functions $f : [0, T] \to B$ with respect to the Lebesgue measurable subsets of $[0, T]$ and the Borel sigma algebra of B might be strictly larger than the set of functions that are almost everywhere limits on $[0, T]$ of simple functions (while these two collections of functions coincide when B is separable, the set of measurable functions is strictly larger than the functions that can be approximated by simple functions when B is not separable). This highlights the importance of simple functions, and of the functions that can be approximated by simple functions that are called *strongly measurable functions*, which we define below; see also Section 1.1 in [191].

Definition 2.7.1 (Simple Function) For a Banach space B, a **simple function** $f : [0, T] \to B$ is a function of the form

$$f(t) = \sum_{i=1}^{N} 1_{A_i}(t) b_i,$$

where $A_1, A_2, \ldots, A_N \subset [0, T]$ are disjoint Lebesgue measurable sets and $b_i \in B$ for all i. A function $f : [0, T] \to B$ is **strongly measurable** if there exists a sequence of simple functions $\{f_k\}_{k=1}^{\infty}$ such that $f_k(t) \to f(t)$ strongly in B, namely, $\lim_{k\to\infty} \|f(t) - f_k(t)\|_B = 0$, for almost every $t \in [0, T]$.

In the same way that integration of Lebesgue measurable functions $f : \mathbb{R} \to \mathbb{R}$ is based on simple functions, the integration theory for Bochner spaces can be carried out for strongly measurable functions $f : [0, T] \to B$ via approximation by simple functions. Because the strongly measurable functions $f : [0, T] \to B$ are those that can be approximated pointwise by simple functions, the strongly measurable functions are the natural setting for defining integration in Bochner spaces. Since Banach spaces are vector spaces and hence are closed under scalar multiplication and addition, we can define the integral of a simple function $f : [0, T] \to B$ of the form $f(t) = \sum_{i=1}^{N} 1_{A_i}(t) b_i$ for a Banach space B by

$$\int_0^T f(t)dt = \sum_{i=1}^{N} \mu(A_i) b_i,$$

where $\mu(A_i)$ denotes the Lebesgue measure of the set $A_i \subset [0, T]$. Then, since strongly measurable functions $f : [0, T] \to B$ can be approximated pointwise almost everywhere on $[0, T]$ by an approximating sequence of simple functions $\{f_k\}_{k=1}^{\infty}$ with $\|f(t) - f_k(t)\|_B \to 0$ as $k \to \infty$ for almost every $t \in [0, T]$, we say that a strongly measurable function $f : [0, T] \to B$ is **Bochner integrable** if $\lim_{k\to\infty} \int_0^T f_k(t)dt$ exists in B, and if so, we define

$$\int_0^T f(t)dt = \lim_{k\to\infty} \int_0^T f_k(t)dt,$$

where this definition is well-defined in the sense that if the limit exists for one approximating sequence of simple functions, then the value of the integral is *independent of the specific choice of approximating sequence of simple functions*.

The following generalization of the triangle inequality gives a useful criterion for Bochner integrability of strongly measurable functions, which can be found in Proposition 1.2.2 in [191].

Theorem 2.7.1 (Triangle Inequality) *Consider a strongly measurable function $f : [0, T] \to B$, where B is a Banach space. Then, f is Bochner integrable if and only if*

$$\int_0^T \|f(t)\|_B dt < \infty$$

as a usual integral of a real-valued measurable function on $[0, T]$*, and we furthermore have the following triangle inequality:*

$$\left\| \int_0^T f(t)dt \right\|_B \leq \int_0^T \|f(t)\|_B dt.$$

Now that we have a well-defined notion of integration of strongly measurable functions, we can define the following **Bochner spaces**, which will play an essential role in our analysis of time-dependent dynamic partial differential equations; see also Section 1.2.b in [191].

Definition 2.7.2 (Definition of Bochner Space $L^p(0, T; B)$) Let B be a Banach space (not necessarily separable), and let $1 \leq p \leq \infty$. Then, the **Bochner space** $L^p(0, T; B)$ consists of all strongly measurable functions $f : [0, T] \to B$ such that $\|f\|_{L^p(0,T;B)} < \infty$, where

$$\|f\|_{L^p(0,T;B)} = \left(\int_0^T \|f(t)\|_B^p dt \right)^{1/p} \qquad \text{for } 1 \leq p < \infty,$$

and

$$\|f\|_{L^\infty(0,T;B)} = \text{ess sup}\{\|f(t)\|_B : t \in [0, T]\}.$$

Furthermore, the Bochner space $L^p(0, T; B)$ for $1 \leq p \leq \infty$ is a Banach space with respect to the norm $\|\cdot\|_{L^p(0,T;B)}$.

Finally, we discuss the notions of duality for Bochner spaces, reflexivity, and separability. We begin with the concept of duality, by stating the following result, which can be deduced from Theorem 1.3.10 combined with Corollary 1.3.22 in [191].

Theorem 2.7.2 (Dual Space of $L^p(0, T; B)$) *Let* $1 \leq p < \infty$ *and let* B *be a Banach space with a dual space* B^**. Then,* $(L^p(0, T; B))^*$ *is isomorphic to* $L^{p'}(0, T; B^*)$ *for* $p' = \dfrac{p}{p-1}$ *with the natural identification of* $\varphi \in L^{p'}(0, T; B^*)$ *with a linear functional on* $L^p(0, T; B)$ *via*

$$f \in L^p(0, T; B) \to \int_0^T \langle f(t), \varphi(t) \rangle dt,$$

where the notation $\langle \cdot, \cdot \rangle : B \times B^* \to \mathbb{R}$ *denotes the natural pairing of an element of the Banach space* B *with a linear functional from the dual space* B^**. In particular, if* $1 < p < \infty$ *and* B *is reflexive, then* $L^p(0, T; B)$ *is a reflexive Banach space also.*

Note that the preceding theorem gives us a notion of weak and weak-star convergence in Bochner spaces. For example, $f_n \rightharpoonup f$ weakly (weakly star) in

$L^p(0, T; L^q(\mathcal{O}))$ for a spatial domain $\mathcal{O} \subset \mathbb{R}^d$ if

$$\int_0^T \int_{\mathcal{O}} f_n g dx dt \to \int_0^T \int_{\mathcal{O}} f g dx dt \quad \text{for all } g \in L^{p'}(0, T; L^{q'}(\mathcal{O})),$$

weakly if $1 \le p, q < \infty$, and instead weakly star if either p or q is equal to ∞.

Finally, we conclude by stating a final result on separability of Bochner spaces, which is the content of Proposition 1.2.29 in [191], specialized to the specific measure space consisting of $[0, T]$ with Lebesgue measure.

Theorem 2.7.3 (Separability of $L^p(0, T; B)$) *Suppose that B is a separable Banach space and that $1 \le p < \infty$. Then, $L^p(0, T; B)$ is also a separable Banach space.*

Many classical results about real-valued functions $f : \mathbb{R}^d \to \mathbb{R}$ extend to Banach space-valued functions $f : \mathbb{R}^d \to B$, where B is a Banach space. One example of such a classical result that we will use in the more general Banach space-valued functional setting is the *Lebesgue differentiation theorem*, which is stated as follows (see Theorem 3.21 in [140] for the statement for real-valued functions and Section 3.4 in [179] for the statement for Banach space-valued functions):

Theorem 2.7.4 (Lebesgue Differentiation Theorem) *Let B be a Banach space, and let $f : \mathbb{R}^d \to B$ be a locally integrable function, namely, $f \in L^1_{loc}(\mathbb{R}^d; B)$ (which means that $\int_K \|f(x)\|_B dx < \infty$ for any compact subset $K \subset \mathbb{R}^d$). Then, almost every $x \in \mathbb{R}^d$ is a **Lebesgue point** of f in the sense that*

$$\lim_{r \to 0} \frac{1}{|B(x, r)|} \int_{|x-y|<r} \|f(x) - f(y)\|_B = 0,$$

where $B(x, r) := \{y \in \mathbb{R}^d : |x - y| < r\}$.

2.8 Varying Geometry

So far, we have defined Bochner spaces of the form $L^p(0, T; B)$ for some Banach space B, where the Banach space B is independent of time, and the function $f \in L^p(0, T; B)$ can be interpreted as a strongly measurable map from $[0, T]$ to the fixed Banach space. Since B encapsulates the spatial features of the solution to the partial differential equation in any problem we are studying, having B independent of time is the usual and useful functional framework for problems posed on fixed spatial domains. However, in fluid-structure interaction, there are typically moving domains that are time-dependent. In the moving domain framework, the fluid interacts dynamically with a structure, and the displacement of the structure determines the fluid domain, which is unknown a priori and evolves dynamically. Consequently, the associated functional framework must account for this time dependence, since the spatial function spaces are defined on a moving domain. In this section, we explore several approaches to handling these challenges, which will be essential for the remainder of the book in studying *nonlinearly coupled moving boundary* models of fluid-structure interaction. These include:

- **Definition of maps to a fixed domain.** We will use maps such as the *Arbitrary Lagrangian-Eulerian (ALE) map*, in some cases along with the *Piola transform*, to move scalar functions, vector fields, and tensors between the *Eulerian (moving domain) formulation* involving time-dependent domains and the *Lagrangian (fixed domain) formulation*. These maps allow us to move between a fixed spatial domain $\mathcal{O}$ and time-dependent spatial domains $\mathcal{O}(t)$, where when we pull back functions to the fixed domain from the time-dependent domain, we then get functions all defined on the same spatial domain, at the expense of introducing new geometric nonlinearities into the equation, which can be useful for the analysis.
- **Time-dependent Bochner spaces.** If we consider functions on the time-dependent (moving) spatial domain, instead of Bochner spaces of the form $L^p(0, T; B)$, we will have to consider functions existing in Bochner spaces of the form $L^p(0, T; B(t))$, where the evolution of the time-dependent functional space $B(t)$ is appropriately defined.
- **Uniformity of classical inequalities on moving domains.** Many classical inequalities, such as Poincaré's inequality and Korn's inequality, are useful in the study of elasticity and fluid dynamics on fixed domains, but in order to use these for moving boundary problems, we have to establish versions of these inequalities for moving domains that are uniform in parameters related to the motion of the fluid domain, such as the Lipschitz constant describing the regularity of the Lipschitz boundary of $\partial\mathcal{O}(t)$, if $\mathcal{O}(t)$ is a time-dependent spatial domain.

We discuss those next.

2.8.1 Maps Between Fixed and Moving Domains

Reynolds Transport Formula We begin by stating the Reynolds transport formula, which will be necessary in analyzing PDEs on non-cylindrical domains.

Theorem 2.8.1 (Reynolds Transport Formula) *Consider a time-dependent spatial domain $\mathcal{O}(t) = A(\mathcal{O}, t)$, where A is a smooth map. Let f be a continuously differentiable field. Then,*

$$\frac{d}{dt}\int_{\mathcal{O}(t)} f\,dx = \int_{\mathcal{O}(t)} \left(\frac{\partial f}{\partial t} + \text{div}(fv)\right) dx = \int_{\mathcal{O}(t)} \frac{\partial f}{\partial t} dx + \int_{\partial\mathcal{O}(t)} fv \cdot n dS,$$

where $v = \frac{\partial A}{\partial t}$ *and* $n(t)$ *is the unit outward normal to* $\partial\mathcal{O}(t)$.

Now we will describe maps that allow us to transfer functions, vector fields, and tensors between fixed domains and moving domains. We will primarily use the Arbitrary Lagrangian-Eulerian (ALE) map and the Piola transform for these purposes, which we describe below.

Arbitrary Lagrangian-Eulerian (ALE) Maps To prove the existence of solutions on a moving domain, in certain situations, it may be more convenient to map the moving-domain formulation to a fixed domain. We will do so by introducing the so-called Arbitrary Lagrangian-Eulerian (ALE) maps that are ubiquitous in numerical analysis of fluid-structure interaction. The ALE map is a C^1-diffeomorphism between the moving spatial domain and a fixed domain. Depending on the geometries of the moving domain, these maps are not always easy to construct. However, we will often consider the case where the top boundary of the time-dependent spatial domain deforms only in one fixed direction, typically normal to some reference configuration that we will denote by Γ. In such cases, the following setup will be useful.

Assume that Γ is flat and can be identified with $(0, L)^{d-1}$ and that the time-dependent spatial domain, denoted by $\mathcal{O}_\eta(t)$, at any time $t \geq 0$, is a subgraph of some function $R + \eta(t) : \Gamma \to \mathbb{R}$. That is,

$$\mathcal{O}_\eta(t) := \{(x', x_d) \in \Gamma \times \mathbb{R} : 0 < x_d < R + \eta(t, x')\}. \tag{2.7}$$

Then we will consider $\mathcal{O} = \Gamma \times (0, 1)$, and the most natural definition for ALE map $A_\eta(t) : \mathcal{O} \to \mathcal{O}_\eta(t)$ is as follows:

$$A_\eta(t)(x', x_d) = (x',\ (R + \eta(t, x'))x_d) \quad x' \in \Gamma. \tag{2.8}$$

The Jacobian of this ALE mapping is

$$J_\eta(t) = \det \nabla A_\eta(t) = R + \eta(t, x'). \tag{2.9}$$

Observe that if $\eta(t)$ is differentiable, then for any differentiable function f defined on $\mathcal{O}_\eta(t)$, its transformation $f^\eta(t) := f \circ A_\eta(t)$ is differentiable, and we have

$$\nabla f \circ A_\eta(t) = \nabla f^\eta (\nabla A_\eta)^{-1}(t).$$

Remark 2.8.1 In the setup where we allow only the transverse displacement to be different from zero, the ALE mapping was easy to construct explicitly. If we assume that the longitudinal displacement of the structure is non-negligible, i.e., at any time $t \geq 0$, we have $\eta(t) : \Gamma \subset \mathbb{R}^{d-1} \to \mathbb{R}^d$, then the ALE maps may not be as readily available. In such situations, various mappings based on elliptic extensions of the boundary data to the fluid domain may be designed to deal with this issue. For example, the ALE maps can be obtained by considering the harmonic extension of the structure configuration $\mathrm{id} + \eta$ in the fixed domain $\mathcal{O} = \Gamma \times (0, 1)$, i.e., by solving for each $t \geq 0$ the following equations:

$$\begin{aligned} \Delta A_\eta(t) = 0, &\quad \text{in } \mathcal{O}, \\ A_\eta(t) = \mathrm{id} + \eta(t) \text{ on } \Gamma, \quad \text{and} &\quad A_\eta(t) = \mathrm{id} \text{ on } \partial\mathcal{O} \setminus \Gamma, \end{aligned} \tag{2.10}$$

where $\mathrm{id} : \Gamma \to \Gamma$ is the identity map, if η has Dirichlet boundary conditions.

The existence and uniqueness of the solution to (2.10) is classical. However, more work is required to prove that these maps are well-defined, i.e., $A_\eta : \mathcal{O} \to \mathcal{O}_\eta$ is indeed a C^1-diffeomorphism for every $t \geq 0$. This can be established if, for example, the Lipschitz norm of η is small enough to guarantee that $A_\eta \in C^1(\bar{\mathcal{O}})$ is injective (see Theorem 5.5-1 in [91] for more details).

Piola Transform Throughout this monograph, we will consider incompressible fluids, that is, fluids with divergence-free velocity. Note that working with the ALE maps (2.8) can be challenging as such maps may not preserve the divergence-free property of a vector field. In other words, given a structure displacement η and the associated fluid domain $\mathcal{O}_\eta$, for any vector field $\boldsymbol{u}$ defined on $\mathcal{O}_\eta$ such that $\mathrm{div}_{\mathcal{O}_\eta}\boldsymbol{u} = 0$, it may *not* be true that $\mathrm{div}_{\mathcal{O}}(\boldsymbol{u} \circ A_\eta) = 0$. For that purpose, we define the **Piola transform** between tensor fields defined over the deformed and the reference domains as follows:

$$\mathcal{P}(\boldsymbol{u}) = (\det\nabla A_\eta)\nabla A_\eta^{-1}(\boldsymbol{u} \circ A_\eta). \tag{2.11}$$

The following lemma states that the Piola transform (2.11) preserves the divergence-free property of a vector field defined on the deformed domain. The proof of this lemma can be found in [91].

Lemma 2.8.1 (Properties of the Piola Transform) *Let* $\boldsymbol{u} : \mathcal{O}_\eta \to \mathbb{R}^d$. *Then* $\mathcal{P}(\boldsymbol{u}) : \mathcal{O} \to \mathbb{R}^d$ *defined in* (2.11) *satisfies*

$$div_{\mathcal{O}}\mathcal{P}(\boldsymbol{u}) = (det\nabla A_\eta)\left((div_{\mathcal{O}_\eta}\boldsymbol{u}) \circ A_\eta\right).$$

2.8.2 *Time-Dependent Bochner Spaces*

In the previous section, we have discussed Bochner spaces of functions taking values in a fixed Banach space B, and in contrast, we now consider the case of Bochner spaces of functions from $t \in [0, T]$ to time-dependent Banach spaces $B(t)$. These spaces will play an important role in the subsequent analysis of fluid-structure interaction, because in many of these fluid-structure models, the fluid domain is a moving time-dependent domain, and hence the fluid velocity takes values in a time-evolving Banach space. Because for these Bochner spaces of the form $L^p(0, T; B(t))$, the Banach space in which the function $t \to B(t)$ takes values depends on the time $t \in [0, T]$, the explicit definition of such functions is significantly more technical than for Bochner spaces of the form $L^p(0, T; B)$ for a fixed Banach space B. However, since the definition itself is intuitive and since the focus of this book will be on the application of these Bochner spaces in time-dependent spaces to problems of practical significance, we content ourselves with the case when $B(t)$ are function spaces (such as L^p spaces and Sobolev spaces) where the definitions do not involve abstract theory of Bochner spaces. For an abstract approach to develop a functional framework involving time-dependent Banach spaces $B(t)$, we refer the interested reader to [5].

For any time-dependent spatial domain $\mathcal{O}(t)$, $t \in [0, T]$ (e.g., see, (2.7)), we will define functional spaces for the non-cylindrical domains $\bigcup_{t\in[0,T]}\{t\} \times \mathcal{O}(t)$.

Definition 2.8.1 Let $Q \subset \mathbb{R}^{d+1}$ be given by $\bigcup_{t\in[0,T]}\{t\}\times\mathcal{O}(t)$. For $1 \leq r, p \leq \infty$, we set

$$
\begin{aligned}
&L^r(0, T; L^p(\mathcal{O}(t))) := \\
&\left\{u \in L^1(Q) \,\middle|\, u(t) \in L^p(\mathcal{O}(t)) \text{ for a.e. } t \in [0, T],\right. \\
&\left. t \mapsto \|u(t)\|_{L^p(\mathcal{O}(t))} \in L^r(0, T)\right\},
\end{aligned}
$$

and

$$
\|u\|_{L^r(0,T;L^p(\mathcal{O}(t)))} := \left(\int_0^T \|u(t)\|^r_{L^p(\mathcal{O}(t))}\, dt\right)^{1/r}, \quad 1 \leq r < \infty,
$$

$$
\|u\|_{L^\infty(0,T;L^p(\mathcal{O}(t)))} := \operatorname*{ess\ sup}_{t\in[0,T]} \|u(t)\|_{L^p(\mathcal{O}(t))}.
$$

In the study of fluid dynamics on moving time-dependent domains, we will often use the following time-dependent Bochner spaces as functional spaces in our analysis:

$$
\begin{aligned}
L^r(0, T; W^{1,p}(\mathcal{O}(t))) &:= \left\{v \in L^r(0, T; L^p(\mathcal{O}(t))) \,\middle|\, \nabla v \in L^r(I, L^p(\mathcal{O}(t)))\right\}, \\
L^r(0, T; W^{1,p}_0(\mathcal{O}(t))) &:= \left\{v \in L^r(0, T; W^{1,p}(\mathcal{O}(t)))\right. \\
&\qquad \left.\middle|\, v(t) \in W^{1,p}_0(\mathcal{O}(t)) \text{ a.e. } t \in [0, T]\right\}, \\
L^r(0, T; W^{1,p}_{0,\sigma}(\mathcal{O}(t))) &:= \left\{v \in L^r(0, T; W^{1,p}_0(\mathcal{O}(t)))\right. \\
&\qquad \left.\middle|\, \operatorname{div} v(t) = 0 \text{ a.e. } t \in [0, T]\right\}.
\end{aligned}
$$

We state the following result, proved in Proposition 3.17 and Corollary 3.25 of [276], which justifies all functional operations we will use for these generalized Bochner spaces.

Lemma 2.8.2 *The space $L^r(0, T; L^p(\mathcal{O}(t))^d)$ is a Banach space. Moreover, let $X(t)$ denote either of the spaces $W^{1,p}_{0,\sigma}(\mathcal{O}(t))^d$, $W^{1,p}_0(\mathcal{O}(t))^d$, or $W^{1,p}(\mathcal{O}(t))^d$. Then, the generalized Bochner spaces $L^r(0, T; X(t))$, $1 < r < \infty$, are reflexive Banach spaces.*

2.8.3 Classical Inequalities on Moving Domains

Finally, we conclude this section about functions on moving domains by discussing the analogue of classical inequalities such as the Poincaré and Korn inequalities, which are essential for the analysis of fluid dynamics, to the context of moving domains. This will involve keeping track of the (uniform) dependence of constants on the geometric properties of the domain, for example, the Lipschitz constant describing the regularity of the boundary of a domain.

Trace on Moving Interface We will now define the trace for functions of both time and space, which are spatially in Sobolev spaces defined on time-varying domains, and we give an equivalent analogue of the trace theorem (compare this to the fixed domain trace theorem in Theorem 2.5.1). Let us consider the linear mapping

$$v|_{\Gamma_{\eta(t)}} := \gamma_{\eta(t)} : v \mapsto v(x, y, R + \eta(t, x, y)), \tag{2.12}$$

defined for $v \in C(\bar{\mathcal{O}}_{\eta(t)})$, where $\mathcal{O}_\eta$ is defined in (2.7) for any function $\eta : \Gamma \to \mathbb{R}$, Γ being identified with $(0, L)^{d-1}$. Then Lemma 2 in [167] gives the following trace theorem. For a trace theorem on more general moving geometries, see [263].

Lemma 2.8.3 (Trace Theorem, Time-Dependent Domains) *Assume that $\eta(\eta \in C([0, T] \times \Gamma) \cap L^\infty(0, T; H_0^2(\Gamma)))$ such that for some $\alpha > 0$, we have $R + \eta(t) \geq \alpha$ for all $t \in [0, T]$. Then, for any $v \in L^2(0, T; H^1(\mathcal{O}_{\eta(t)}))$, the time-dependent trace $\gamma_\eta(v)$, defined in (2.12), satisfies*

$$\gamma_\eta(v) \in L^2(0, T; W^{1-1/p,p}(\Gamma)), \quad \textit{for any} \begin{cases} 1 \leq p < 2 \text{ if } d = 3, \\ 1 \leq p \leq 2 \text{ if } d = 2. \end{cases} \tag{2.13}$$

Proof We will outline the proof here, which starts by defining the following auxiliary function:

$$w(t, x, y, s) = v(t, x, y, R + \eta(t, x, y) - s), \quad 0 \leq s \leq \alpha.$$

Note that when $d = 2$, Sobolev embedding Theorem 2.5.2 implies that $\eta \in L^\infty(0, T; W^{1,\infty}(\Gamma))$ and in the case of $d = 3$, $\eta \in L^\infty(0, T; W^{1,q}(\Gamma))$ for any

$q < \infty$. Hence, calculating the gradient of w shows us that there exists a universal constant $C > 0$ that

$$\|w\|_{L^2(0,T;W^{1,p}(\Gamma\times[0,\alpha]))} \le C(1 + \|\eta\|_{L^\infty(0,T;H^2(\Gamma))})\|v\|_{L^2(0,T;H^1(\mathcal{O}_\eta(t)))},$$

for any $1 \le p < 2$ if $d = 3$ and for any $1 \le p \le 2$ if $d = 2$. Hence, the classical trace theorem tells us that the trace $\gamma_\eta(v) = w|_{s=0}$ is well-defined and belongs to the space $L^2(0, T; W^{1-\frac{1}{p},p}(\Gamma))$. Hence, we conclude that there exists a constant that depends only on α (recall $R + \eta(t) \ge \alpha$), such that

$$\|v|_{\Gamma_\eta}\|_{L^2(0,T;W^{1-\frac{1}{p},p}(\Gamma))} \le C(1 + \|\eta\|_{L^\infty(0,T;H^2(\Gamma))})\|v\|_{L^2(0,T;H^1(\mathcal{O}_\eta(t)))}, \tag{2.14}$$

for any $1 \le p < 2$ if $d = 3$ and for any $1 \le p \le 2$ if $d = 2$.

□

Korn Inequality on Moving Domains We will obtain energy estimates for the fluid-structure system. In doing so, we will need to understand the relationship between the L^2-bounds for the symmetrized gradient of the fluid velocity and its H^1-bounds on moving domains, where the **symmetrized gradient** of $u : \mathcal{O} \subset \mathbb{R}^d \to \mathbb{R}^d$ is defined by

$$\boldsymbol{D}(u) = \frac{1}{2}\left(\nabla u + (\nabla u)^t\right),$$

where the superscript t denotes the transpose of a $d \times d$ matrix.

The following version of the Korn inequality in two dimensions, suitable for our moving domain setting, proved in Lemma 1 in [342], will be used.

Lemma 2.8.4 (Korn's Inequality) *Let $\mathcal{O} \subset \mathbb{R}^2$ be an open bounded set with Lipschitz boundary. Then there exists a constant $K(\mathcal{O})$, depending only on the domain $\mathcal{O}$, such that*

$$\|u\|_{H^1(\mathcal{O})} \le K(\mathcal{O})\left(\left|\int_{\mathcal{O}} u\,dx_1\,dx_2\right| + \left|\int_{\mathcal{O}} (x_1u_2 - x_2u_1)\,dx_1\,dx_2\right| + \|\boldsymbol{D}(u)\|_{L^2(\mathcal{O})}\right), \tag{2.15}$$

for any $u \in (H^1(\mathcal{O}))^d$. Let us suppose that the domains $\mathcal{O}_\eta(t)$ are changing in the sense that they are equal to $\mathcal{O}_\eta(t) = A_\eta(t)(\mathcal{O})$, where A_η is a family of bi-Lipschitz mappings such that the bi-Lipschitz constants of $A_\eta(t)$ and $A_\eta(t)^{-1}$ are uniform in t, and such that the family $\{A_\eta(t)\}$ is strongly compact in $W^{1,\infty}(\mathcal{O})$. Then the constant in the inequality (2.15) *can be chosen independently of t.*

Extension of Functions by Zero We will also use the following theorem to extend a function by 0 outside of the moving domain with Lipschitz boundary, which will be useful for extending fluid quantities, defined on time-dependent varying domains, to a fixed larger maximal domain that contains all of the moving domains. This will allow us to consider functions defined on the same fixed maximal domain, rather than on varying moving domains. For a more general theorem in this context for subgraphs of functions that are Hölder continuous, see Appendix B in [215] (see also Theorem 3.33 in [254] and Theorem 2.7 in [257]).

Theorem 2.8.2 *Assume that $\eta : \Gamma \to \mathbb{R}$ is a Lipschitz continuous function with Lipschitz constant L such that $\sup_\Gamma |\eta| \le M$. Consider $u \in H^s(\mathcal{O}_\eta)$ for some $s < \frac{1}{2}$, where $\mathcal{O}_\eta$ is defined in (2.7). We define the extension of u by 0 as*

$$\tilde{u} = u \qquad \text{in } \mathcal{O}_\eta,$$
$$\tilde{u} = 0 \qquad \text{in } \mathbb{R}^d \setminus \mathcal{O}_\eta.$$

Then, there exists a constant $C_{s,L,M}$ depending only on s, M and L such that

$$\|\tilde{u}\|_{H^s(\mathbb{R}^d)} \le C_{s,L,M} \|u\|_{H^s(\mathcal{O}_\eta)}, \qquad s < \frac{1}{2}. \tag{2.16}$$

2.9 Compactness Results for Function Spaces

Now that we have defined all of the function spaces that we will consider, we consider mappings between function spaces, in particular compact operators. We first define the concept of a **compact operator**, which gives a natural way of extracting convergent subsequences in function spaces. See Chapter 6 in [60], Section 6 in Chapter 4 of [313], and Section 5.7 in [132] for further details.

To define the concept of a compact operator, suppose that B_1 and B_2 are two Banach spaces with norms $\|\cdot\|_{B_1}$ and $\|\cdot\|_{B_2}$. Then, we recall from the discussion of Banach spaces that a bounded linear map $T : B_1 \to B_2$ is a linear function such that

$$\|T(x)\|_{B_2} \le C\|x\|_{B_1} \qquad \text{for all } x \in B_1,$$

for some constant C, where we define $\|T\|_{\mathcal{L}(B_1,B_2)} = \inf\{\|T(x)\|_{B_2} : \|x\|_{B_1} = 1\}$. Bounded linear maps between two Banach spaces are important because they are equivalent to the set of *continuous linear maps* between two Banach spaces. So if $\{x_n\}_{n=1}^\infty$ is a sequence in B_1 converging strongly to $x \in B_1$, then $\{T(x_n)\}_{n=1}^\infty$ is a sequence in B_2 converging strongly to $T(x)$ in B_2, since

$$\|T(x) - T(x_n)\|_{B_2} = \|T(x - x_n)\|_{B_2} \le \|T\|_{\mathcal{L}(B_1,B_2)}\|x - x_n\|_{B_1} \to 0.$$

However, in the analysis of partial differential equations, we will often require stronger properties for maps between Banach spaces in order to show well-posedness of partial differential equations. In particular, we will often have a situation where to analyze a partial differential equation, we consider a sequence $\{f_n\}_{n=1}^{\infty}$ to an approximate problem that we want to show converge along a subsequence as $n \to \infty$. However, we will have only uniform boundedness of $\{f_n\}_{n=1}^{\infty}$ in some Banach space B, while we want to obtain strong convergence of this sequence in some space. Recall that the closed unit ball in a Banach space is not strongly compact, so we cannot deduce that there is a subsequence of $\{f_n\}_{n=1}^{\infty}$ that converges strongly in the space B for which we have uniform bounds on B. This is why we need to consider *compact linear maps* between Banach spaces, as having a compact embedding of B into another Banach space B_0 will allow us to extract a strongly convergent subsequence from the sequence of approximate solutions, which converges strongly with respect to the norm of B_0 (rather than with respect to the norm of the original space B).

We have the following definition of a compact linear map between two function spaces.

Definition 2.9.1 Let B_1 and B_2 be two Banach spaces with norms $\|\cdot\|_{B_1}$ and $\|\cdot\|_{B_2}$. Then, a bounded linear map $T : B_1 \to B_2$ is **compact** if for every bounded sequence $\{x_n\}_{n=1}^{\infty}$ in B_1, there exists a subsequence $\{x_{n_k}\}_{k=1}^{\infty}$ and some $y \in B_2$ such that $T(x_{n_k}) \to y$ strongly in B_2.

Of particular importance are compact embeddings, where we can embed a Banach space compactly into a larger Banach space.

Definition 2.9.2 Suppose that B_0 is a Banach space with norm $\|\cdot\|_{B_0}$ that can be identified with a closed subspace of another larger Banach space B with norm $\|\cdot\|_B$, in such a way that the inclusion map of B_0 to B is a bounded linear map. Then, we say that B_0 is **continuously embedded** in B, denoted as $B_0 \hookrightarrow B$, where we denote the inclusion map by $\iota : B_0 \to B$. We say that B_0 is **compactly embedded** in B, denoted as $B_0 \subset\subset B$, if the inclusion map $\iota : B_0 \to B$ is a compact linear map from B_0 to B, so that any sequence $\{x_n\}_{n=1}^{\infty}$ that is bounded with respect to the norm of B_0 has a convergent subsequence that converges strongly, when considered as a sequence in B, with respect to the norm in B.

Next, we discuss some important examples of compact embeddings of function spaces that are classically known and will be used in subsequent chapters. We first discuss the Rellich-Kondrachov theorem, which states a classical compact embedding for Sobolev spaces; see, for example, Section 5.7 in [132].

Theorem 2.9.1 (Rellich-Kondrachov Theorem) *For any $1 \le p < d$ and any bounded open set $\mathcal{O}$ with Lipschitz boundary in $\mathbb{R}^d$,*

$$W^{1,p}(\mathcal{O}) \subset\subset L^q(\mathcal{O}) \quad \text{for all } 1 \le q < \frac{dp}{p-d}.$$

In particular, $W^{1,p}(\mathcal{O}) \subset\subset L^p(\mathcal{O})$ for $1 \le p < d$.

Remark 2.9.1 It is important in the Rellich-Kondrachov theorem that the open set $\mathcal{O}$ is bounded, as the compact embedding in the Rellich-Kondrachov theorem does not hold for general unbounded domains.

Next, we state a result about embeddings of Sobolev spaces into Sobolev spaces of lower order, for $p = 2$.

Theorem 2.9.2 *Let* $-\infty < s_2 < s_1 < \infty$ *and let* $\mathcal{O}$ *be a bounded open set in* $\mathbb{R}^d$ *with Lipschitz boundary. Then,* $H^{s_1}(\mathcal{O}) \subset\subset H^{s_2}(\mathcal{O})$.

We now discuss general compactness results for functions of both space and time. We first discuss two classical compactness results, the Arzela-Ascoli theorem and the Aubin-Lions compactness theorem.

Theorem 2.9.3 (Arzela-Ascoli Theorem) *Let* $T > 0$ *and let* $B_0 \subset\subset B$ *be Banach spaces such that* B_0 *embeds compactly into* B*. Then, if* $\{f_n\}_{n=1}^{\infty}$ *is a sequence of continuous functions* $f_n : [0, T] \to B_0$ *such that*

- $\|f_n\|_{L^{\infty}(0,T;B_0)} \leq C$ *for a constant* C *that is independent of* n,
- *The sequence* $\{f_n\}_{n=1}^{\infty}$ *is* ***equicontinuous*** *as a family of functions taking values in* B*, in the sense that for each* $\epsilon > 0$*, there exists* δ *(independent of* n*) such that for all* $t_1, t_2 \in [0, T]$ *such that* $|t_1 - t_2| < \delta$*, we have that* $\|f_n(t_1) - f_n(t_2)\|_B < \epsilon$ *for all* n.

Then, there exists a subsequence $\{f_{n_k}\}_{k=1}^{\infty}$ *such that* $f_{n_k} \to f$ *strongly in* $C(0, T; B)$.

A related result is the Aubin-Lions compactness theorem, which states that for functions that are uniformly bounded in the Bochner space $L^p(0, T; B)$ where B is a Banach space, one can obtain some notion of strong convergence if we also have in addition a uniform bound on the weak time derivatives in a potentially weaker space. The Aubin-Lions compactness theorem is one of the fundamental compactness results in the theory of nonlinear evolution equations.

We begin our exposition of compactness theorems by starting with the following lemma that forms the basis of Aubin-Lions theorem that follows.

Lemma 2.9.1 (Ehrling's Lemma) *Let* $B_0 \subset B_1 \subset B_2$ *be Banach spaces such that the embedding of* B_0 *in* B_1 *is compact and the embedding of* B_1 *in* B_2 *is continuous. Then, for any* $\delta > 0$*, there exists* $C(\delta)$ *such that for all* $u \in B_0$*, we have*

$$\|u\|_{B_1} \leq \delta \|u\|_{B_0} + C(\delta)\|u\|_{B_2}.$$

Theorem 2.9.4 (Aubin-Lions Theorem) *Assume that* $B_0 \subset B_1 \subset B_2$ *are Banach spaces such that the embedding of* B_0 *in* B_1 *is compact and the embedding of* B_1 *in* B_2 *is continuous. For any* $1 \leq p, r \leq \infty$*, consider the set*

$$E_{p,r} := \left\{ v \in L^p(0, T; B_0); \frac{dv}{dt} \in L^r(0, T; B_2) \right\}.$$

Then,

1. *If $p < \infty$, the embedding of $E_{p,r}$ in $L^p(0, T; B_1)$ is compact.*
2. *If $p = \infty$ and $r > 1$, the embedding of $E_{p,r}$ in $C(0, T; B_1)$ is compact.*

Since it will be used often when passing to the limit of approximate solutions, we also discuss a formulation of the Aubin-Lions compactness lemma for piecewise constant functions. Namely, consider a sequence $\{u_N\}_{N=1}^{\infty}$ such that each u_N takes values $u_N : [0, T] \to H$ in some Hilbert space H. Furthermore, let $\Delta t = T/N$ be a time discretization parameter with $t_k = k\Delta t$, and suppose that the function u_N for each positive integer N is piecewise constant in time with respect to Δt, in the sense that u_N is constant-valued on each subinterval (t_k, t_{k+1}) for $0 \leq k \leq N-1$. In practice, these u_N will often be approximate solutions, defined via an approximation scheme such as Rothe's method or Lie operator splitting, and one would want to pass to the limit in these approximate solutions as $N \to \infty$, or equivalently as $\Delta t \to 0$. We can therefore use the following Aubin-Lions-type compactness lemma for piecewise constant functions; see Theorem 1 in [126].

Theorem 2.9.5 (Dreher-Jüngel Compactness Criterion) *Let $V \subset H \subset W$ be Hilbert spaces, where $V \subset H$ is a compact embedding and $H \subset W$ is a continuous embedding. Consider piecewise constant functions u_N on $[0, T]$, defined to be piecewise constant on each subinterval (t_k, t_{k+1}) for $t_k = k\Delta t$ and $0 \leq k \leq N - 1$, which take values in H for each time t. Then, if there exists a constant C (independent of N) such that*

$$\left\| \frac{\tau_{\Delta t} u_N - u_N}{\Delta t} \right\|_{L^1(\Delta t, T; W)} + \|u_N\|_{L^2(0,T;V)} \leq C,$$

where $\tau_{\Delta t} u_N(t, \cdot) := u_N(t - \Delta t, \cdot)$ for $\Delta t \leq t \leq T$, then u_N is precompact in $L^2(0, T; H)$.

We now discuss some classical compact embeddings between function spaces, starting first with compact embeddings involving Sobolev spaces of fractional order in the time variable. Recalling the definition of the fractional Sobolev space in Definition 2.5.3, we consider compact embeddings involving the spaces $W^{s,p}(0, T; B)$ for a Banach space B, with norm defined by

$$\|f\|_{W^{s,p}(0,T;B)} = \left(\|f\|^p_{L^p(0,T;B)} + \int_0^T \int_0^T \frac{\|f(t) - f(\tau)\|_B^p}{|t - \tau|^{1+sp}} d\tau dt \right)^{1/p}.$$

We state the following compact embedding result, which is Theorem 2.1 in [139].

Theorem 2.9.6 *Suppose that we have the following chain of embeddings of Banach spaces: $B_0 \subset\subset B \subset B_1$, where B_0 and B_1 are reflexive Banach spaces, with B_0 embedding compactly into B. Then, for all $1 < p < \infty$ and all $0 < s < 1$, we have the following compact embedding:*

$$L^p(0, T; B_0) \cap W^{s,p}(0, T; B_1) \hookrightarrow L^p(0, T; B).$$

We also have the following compact embedding of certain fractional Sobolev spaces into spaces of continuous functions in time taking values in a Banach space; see, e.g., Theorem 2.2 [139].

Theorem 2.9.7 *Assume that $B_0 \subset B_1$ are Banach spaces such that the embedding of B_0 in B_1 is compact. Then for any $p > 1$ and $0 < s < 1$ such that $sp > 1$, the embedding*

$$W^{s,p}(0, T; B_0) \hookrightarrow C(0, T; B_1),$$

is compact.

By using the Arzela-Ascoli theorem, we also have the following compact embedding into the space of continuous functions in time taking values in a Banach space.

Theorem 2.9.8 *Assume that $B \subset\subset B_0$ is a compact embedding of Banach spaces. Then for any $p > 1$ and $0 < s < 1$ such that $sp > 1$, we have the following compact embedding:*

$$W^{s,p}(0, T; B_0) \cap C(0, T; B) \hookrightarrow C(0, T; B_0).$$

Proof As discussed in the proof of Theorem 2.2 in [139], when $sp > 1$, we have that $W^{s,p}(0, T; B_0) \subset C^\beta(0, T; B_0)$ for $0 < \beta < sp - 1$, where $C^\beta(0, T; B_0)$ is the space of Hölder continuous functions. Therefore, boundedness in $C^\beta(0, T; B_0)$ gives equicontinuity in B_0, while boundedness in $C(0, T; B)$ gives boundedness in a Banach space B that compactly embeds into B_0, so the result follows immediately by the Arzela-Ascoli theorem; see Theorem 2.9.3. □

In certain situations, especially when we consider stochastic problems, we do not expect the solutions to be differentiable in time. For that purpose, we introduce the following spaces.

Definition 2.9.3 (Nikolskii Space) For a Banach space B, consider $u \in L^1(0, T; B)$. Let the translation of u be denoted by $\tau_h u(t) = u(t - h)$. For any $1 \leq p < \infty$ and $0 < s < 1$, we define the Nikolskii space $N^{s,p}(0, T; B)$ by

$$N^{s,p}(0, T; B) = \left\{ u \in L^p(0, T; B) : \sup_{0<h<T} \frac{\|\tau_h u - u\|_{L^p(h,T;B)}}{h^s} < \infty \right\},$$

endowed by the norm

$$\|u\|_{N^{s,p}(0,T;B)} = \left(\|u\|^p_{L^p(0,T;B)} + \left(\sup_{0<h<T} \frac{\|\tau_h u - u\|_{L^p(h,T;B)}}{h^s} \right)^p \right)^{\frac{1}{p}}.$$

The fractional Sobolev space defined in Definition 2.5.3 is related to the Nikolskii space in the following way: For any $s \in (0, 1)$ and $p \in [1, \infty)$, we have

$$W^{s,p}(0, T; B) \subset N^{s,p}(0, T; B).$$

We will find bounds for fractional time derivatives of the solution in the sense of Definition 2.9.3 and use the following theorem to establish compactness, the proof of which can be found in [307, 325].

Theorem 2.9.9 *Assume that $B_0 \subset B_1 \subset B_2$ are Banach spaces such that the embedding of B_0 in B_1 is compact and the embedding of B_1 in B_2 is continuous. Then for any $1 \le p \le \infty$ and $0 < s < 1$, the embedding*

$$L^p(0, T; B_0) \cap N^{s,p}(0, T; B_2) \hookrightarrow L^p(0, T; B_1),$$

is compact.

Finally, we conclude by stating two more results about continuous and compact embeddings. The first result is a result on interpolation of Sobolev spaces for functions of both space and time and their compact embeddings; see [6].

Theorem 2.9.10 *Suppose that $1 < p_0, p_1, q_0, q_1 < \infty$ and $s_0, s_1, m_0, m_1 \ge 0$. Define for any $\theta \in (0, 1)$*

$$s_\theta = (1 - \theta)s_0 + \theta s_1, \qquad \frac{1}{p_\theta} = \frac{1 - \theta}{p_0} + \frac{\theta}{p_1},$$

and similarly

$$m_\theta = (1 - \theta)m_0 + \theta m_1, \qquad \frac{1}{q_\theta} = \frac{1 - \theta}{q_0} + \frac{\theta}{q_1}.$$

Assume that $\mathcal{O} \subset \mathbb{R}^d$ is an open domain with Lipschitz boundary. Then,

1. *The embedding*

$$W^{s_0,p_0}(0, T, W^{m_0,q_0}(\mathcal{O})) \cap W^{s_1,p_1}(0, T, W^{m_1,q_1}(\mathcal{O})) \hookrightarrow W^{s,p}(0, T, W^{m,q}(\mathcal{O})),$$

is compact, provided

$$s < s_\theta, \qquad s - \frac{1}{p} < s_\theta - \frac{1}{p_\theta} \quad \text{and} \quad m < m_\theta, \qquad m - \frac{d}{q} < m_\theta - \frac{d}{q_\theta}.$$

2. *The embedding*

$$W^{s_0,p_0}(0, T, W^{m_0,q_0}(\mathcal{O})) \cap W^{s_1,p_1}(0, T, W^{m_1,q_1}(\mathcal{O})) \hookrightarrow C^{0,s}([0, T], W^{m,q}(\mathcal{O})),$$

is compact, provided

$$0 \leq s < s_\theta - \frac{1}{p_\theta} \quad \text{and} \quad m - \frac{d}{q} < m_\theta - \frac{d}{q_\theta}.$$

The last embedding result we discuss in this section is a result that shows how obtaining additional regularity for both weak time and spatial derivatives allows us to obtain Hölder continuity of functions in time and space, and it can be thought of intuitively also as an interpolation result. See Lemma 4 in [95] for a proof (see also [167]).

Theorem 2.9.11 *Assume that* $\mathcal{O} \subset \mathbb{R}^d$ *is an open domain with Lipschitz boundary. The following embedding is continuous for any* $\alpha \in (0, 1)$*:*

$$L^\infty(0, T; H^2(\mathcal{O})) \cap W^{1,\infty}(0, T; L^2(\mathcal{O})) \hookrightarrow C^{0,\alpha}([0, T]; H^{2-2\alpha}(\mathcal{O})).$$

Moreover, for any $p > d + 2$*, there exists* $\alpha \in (0, 1)$ *such that the embedding*

$$L^p(0, T; W^{2,p}(\mathcal{O})) \cap W^{1,p}(0, T; L^p(\mathcal{O})) \hookrightarrow C^{0,\alpha}([0, T]; C^{1,\alpha}(\mathcal{O}))$$

is compact.

2.10 Generalized Aubin-Lions-Simon Compactness Results for Time-Dependent Banach Spaces

In moving boundary problems, such as those discussed in Chaps. 5 and 11 of this book, compactness arguments are crucial to show the existence of a weak solution to the underlying nonlinear fluid-structure interaction problem. In those cases, the fluid velocities are defined on moving domains, not known a priori, and classical compactness arguments cannot be directly applied. To handle compactness for problems on moving domains, in 2019, Muha and Čanić introduced a generalization of the Aubin-Lions-Simon compactness result [270], which we state here.

The generalized Aubin-Lions-Simon compactness theorem by Muha and Čanić holds for the following general scenario. Suppose that we are interested in "solving" the following evolution problem:

$$\begin{aligned} \frac{d\boldsymbol{u}}{dt} &= A^t \boldsymbol{u}, \ t \in (0, T), \\ \boldsymbol{u}(0) &= \boldsymbol{u}_0, \end{aligned} \tag{2.17}$$

where $A^t : V(t) \to W(t)$ is a family of (nonlinear) spatial differential operators that depend on t. A way to solve this problem is to discretize the problem in time by sub-dividing the time interval into N subintervals, each of width $\Delta t = T/N$, and introducing the piecewise constant approximate functions:

$$\boldsymbol{u}_N = \boldsymbol{u}_N^n \quad \text{for } t \in ((n-1)\Delta t, n\Delta t],\ n = 1, \ldots, N, \tag{2.18}$$

which satisfy, e.g., a backward Euler approximation of the problem on (t_n, t_{n+1}):

$$\frac{\boldsymbol{u}_N^{n+1} - \boldsymbol{u}_N^n}{\Delta t} = A^{t_{n+1}} \boldsymbol{u}_N^{n+1} \quad \text{or} \quad \frac{\boldsymbol{u}_N^{n+1} - \boldsymbol{u}_N^n}{\Delta t} = A^{t_n} \boldsymbol{u}_N^{n+1}, \tag{2.19}$$

where the choice of $A^{t_{n+1}}$ or A^{t_n} depends on the problem at hand.

This approach to solving the evolution problem (2.17) is sometimes called Rothe's method.

Let us consider the following weak formulation for problem (2.19):

$$\left\langle \frac{\boldsymbol{u}_N^{n+1} - \boldsymbol{u}_N^n}{\Delta t}, \boldsymbol{q} \right\rangle = \left\langle A^{t^n} \boldsymbol{u}_N^{n+1}, \boldsymbol{q} \right\rangle \quad \forall \boldsymbol{q} \in Q_N^n, \tag{2.20}$$

where Q_N^n is a test space and V_N^n is the solution space. We will be assuming that the test space and the solution space are continuously embedded in a Hilbert space V:

$$V_N^n \hookrightarrow V,\ Q_N^n \hookrightarrow V,$$

where V is a Hilbert space compactly embedded into a Hilbert space H. Then, the following generalization of the Aubin-Lions-Simon compactness lemma holds true for sequences of approximate solutions defined by (2.20).

Theorem 2.10.1 (Aubin-Lions-Simon Compactness Theorem for Moving Domains [270]) *Let V and H be Hilbert spaces such that $V \subset\subset H$. Furthermore, let V_N^n and Q_N^n be Hilbert spaces such that $(V_N^n, Q_N^n) \hookrightarrow V \times V$, where the embeddings are uniformly continuous w.r.t. N and n, and*

$$V_N^n \subset\subset \overline{Q_N^n}^H \hookrightarrow (Q_N^n)'.$$

Let $\boldsymbol{u}_N^n \in V_N^n$, $n = 1, \ldots, N$, and define $(\boldsymbol{u}_N)_{N \in \mathbb{N}}$ via $\boldsymbol{u}_N^n$ so that (2.18) *holds. Suppose that $(\boldsymbol{u}_N)_{N \in \mathbb{N}} \subset L^2(0, T; H)$ and that properties (A)–(C) hold true:*

(A) *There exists a universal constant $C > 0$ such that for every $\Delta t > 0$*

A1. $\|\boldsymbol{u}_N\|_{L^2(0,T;V)} \le C$,
A2. $\|\boldsymbol{u}_N\|_{L^\infty(0,T;H)} \le C$,
A3. $\|\tau_{\Delta t}\boldsymbol{u}_N - \boldsymbol{u}_N\|^2_{L^2(\Delta t,T;H)} \le C\Delta t$.

(B) *There exists a universal constant $C > 0$ such that*

$$\left\| P_N^n \frac{\boldsymbol{u}_N^{n+1} - \boldsymbol{u}_N^n}{\Delta t} \right\|_{(Q_N^n)'} \leq C(\|\boldsymbol{u}_N^{n+1}\|_{V_N^{n+1}} + 1), \; n = 0, \dots, N-1,$$

where P_N^n *is the orthogonal projector onto* $\overline{Q_N^n}^H$.

(C) *The function spaces* Q_N^n *and* V_N^n *depend smoothly on time in the following sense:*

C1. *For every* $\Delta t > 0$, *and for every* $l \in \{1, \dots, N\}$ *and* $n \in \{1, \dots, N-l\}$, *there exists a space* $Q_N^{n,l} \subset V$ *and the operators* $J_{N,l,n}^i : Q_N^{n,l} \to Q_N^{n+i}$, $i = 0, 1, \dots, l$, *such that* $\|J_{N,l,n}^i \boldsymbol{q}\|_{Q_N^{n+i}} \leq C\|\boldsymbol{q}\|_{Q_N^{n,l}}$, $\forall \boldsymbol{q} \in Q_N^{n,l}$, *and*

$$\left((J_{N,l,n}^{j+1}\boldsymbol{q} - J_{N,l,n}^{j}\boldsymbol{q}), \boldsymbol{u}_N^{n+j+1}\right)_H \leq C\Delta t \|\boldsymbol{q}\|_{Q_N^{n,l}} \|\boldsymbol{u}_N^{n+j+1}\|_{V_N^{n+j+1}}, \quad j \in \{0, \dots, l-1\}, \tag{2.21}$$

$$\|J_{N,l,n}^i \boldsymbol{q} - \boldsymbol{q}\|_H \leq C\sqrt{l\Delta t}\|\boldsymbol{q}\|_{Q_N^{n,l}}, \quad i \in \{0, \dots, l\}, \tag{2.22}$$

where $C > 0$ *is independent of* $\Delta t(N)$, n *and* l.

C2. *Let* $V_N^{n,l} = \overline{Q_N^{n,l}}^V$. *There exist the functions* $I_{N,l,n}^i : V_N^{n+i} \to V_N^{n,l}$, $i = 0, 1, \dots, l$, *and a universal constant* $C > 0$, *such that for every* $\boldsymbol{v} \in V_N^{n+i}$

$$\|I_{N,l,n}^i \boldsymbol{v}\|_{V_N^{n,l}} \leq C\|\boldsymbol{v}\|_{V_N^{n+i}}, \quad i \in \{0, \dots, l\}, \tag{2.23}$$

$$\|I_{N,l,n}^i \boldsymbol{v} - \boldsymbol{v}\|_H \leq g(l\Delta t)\|\boldsymbol{v}\|_{V_N^{n+i}}, \quad i \in \{0, \dots, l\}, \tag{2.24}$$

where $g : \mathbb{R}_+ \to \mathbb{R}_+$ *is a universal, monotonically increasing function such that* $g(h) \to 0$ *as* $h \to 0$.

C3. Uniform Ehrling property: *For every* $\delta > 0$, *there exists a constant* $C(\delta)$ *independent of* n, l *and* $\Delta t(N)$, *such that*

$$\|\boldsymbol{v}\|_H \leq \delta\|\boldsymbol{v}\|_{V_N^{n,l}} + C(\delta)\|\boldsymbol{v}\|_{(Q_N^{n,l})'}, \quad \boldsymbol{v} \in V_N^{n,l}. \tag{2.25}$$

Then $(\boldsymbol{u}_N)_{N\in\mathbb{N}}$ *is relatively compact in* $L^2(0, T; H)$.

Remark 2.10.1 Conditions (A) and (B) correspond to the "classical" conditions in the Aubin-Lions lemma for problems on *fixed domains* [126], with $V_N^n = V$, $Q_N^{n,l} = Q_N^n = Q$. Namely, in that case, condition (C) is trivially satisfied. Condition (A3) is not necessary; however, it is usually satisfied as a by-product of the time discretization approach (Rothe's method); see [270]. Conditions (C) are crucial for compactness arguments on moving domains. They are directly related to making sense of taking the time derivative on different domains. Conditions (C1) and (C2) are "local" in the sense that they require smooth dependence of the test spaces and solution spaces on time, locally in time, namely, for the time shifts from $n\Delta t$ to time $(n+l)\Delta t$. Condition (C3), i.e., the uniform Ehrling property, provides a "global" estimate. More precisely, while the classical Ehrling lemma (see Lemma 2.9.1 in Chap. 2) holds for any set of function spaces parameterized by l, n and Δt, i.e., $\{V_N^{n,l}, H_N^{n,l}, (Q_N^{n,l})'\} = \{V_N^{n,l}, \overline{Q_N^{n,l}}^H, (Q_N^{n,l})'\}$, $V_N^{n,l} \hookrightarrow H_N^{n,l} \simeq (H_N^{n,l})' \hookrightarrow (Q_N^{n,l})'$ where the first embedding is compact and the last is injective, with the constants $C(\delta)$ depending on l, n and Δt, the uniform Ehrling property, i.e., (C3) requires that those constants *do not depend* on the parameters n, l and N, which is a condition on the entire family of function spaces. Thus, the uniform Ehrling property requires that the classical Ehrling lemma holds *uniformly* for the entire family of function spaces $\{V_N^{n,l}, H_N^{n,l}, (Q_N^{n,l})'\}$ parameterized by l, n and N.

Condition (C1) states that there exists a "common" test space $Q_N^{n,l}$, which is continuously embedded in all the test spaces $Q_N^{n,i}, i = 1, \ldots, l$ for the time shifts $(n+i)\Delta t$, $i = 1, \ldots, l$, and such that properties (2.21) and (2.22) hold on $Q_N^{n,l}$, which is related to making sense of approximate time derivatives.

Condition (C2) is similar in the sense that it states certain properties of a "common" solution space $V_N^{n,l}$, common for the time shifts $(n+i)\Delta t, i = 1, \ldots, l$, satisfying properties (2.23) and (2.24), which state continuous dependence of the solutions spaces on time, locally near time $n\Delta t$, for an arbitrary n.

2.11 Mollifiers

We will define a regularizing operator called a mollifier. Given any radial function

$$\rho(|x|) \in C_0^\infty(\mathbb{R}^d), \quad \rho \geq 0, \quad \int_{\mathbb{R}^d} \rho\, dx = 1, \tag{2.26}$$

and $\lambda > 0$, we define the mollification $\mathcal{J}_\lambda v$ of functions $v \in L^p(\mathbb{R}^d)$, $1 \leq p \leq \infty$, by

$$(\mathcal{J}_\lambda v)(x) = \lambda^{-d} \int_{\mathbb{R}^d} \rho\left(\frac{x-y}{\lambda}\right) v(y)\, dy. \tag{2.27}$$

We will list some properties of the mollification operator defined above in (2.27) (see, e.g., [132], Lemma 3.5 in [249])

Lemma 2.11.1 (Properties of Mollifiers) *Let $\mathcal{J}_\lambda$ be the mollification operator defined in* (2.27). *Then $\mathcal{J}_\lambda v \in C^\infty(\mathbb{R}^d)$ and:*

1. *For all $v \in C(\mathbb{R}^d)$, $\mathcal{J}_\lambda v \to v$ uniformly on any compact set $K \subset \mathbb{R}^d$ and*

$$\|\mathcal{J}_\lambda v\|_{L^\infty(\mathbb{R}^d)} \leq \|v\|_{L^\infty(\mathbb{R}^d)}.$$

2. *Mollifiers commute with weak distributional derivatives,*

$$D^\alpha \mathcal{J}_\lambda v = \mathcal{J}_\lambda D^\alpha v, \quad \forall |\alpha| \le m,\ v \in H^m(\mathbb{R}^d).$$

3. *For all* $u \in L^p(\mathbb{R}^d)$, $v \in L^q(\mathbb{R}^d)$, *with* $\frac{1}{p} + \frac{1}{q} = 1$,

$$\int_{\mathbb{R}^d} (\mathcal{J}_\lambda u) v \, dx = \int_{\mathbb{R}^d} u (\mathcal{J}_\lambda v) \, dx.$$

4. *For all* $v \in H^s(\mathbb{R}^d)$,

$$\lim_{\lambda \to 0} \|\mathcal{J}_\lambda v - v\|_{H^s(\mathbb{R}^d)} = 0,$$

and,

$$\|\mathcal{J}_\lambda v - v\|_{H^{s-1}(\mathbb{R}^d)} \le C\lambda \|v\|_{H^s(\mathbb{R}^d)}.$$

5. *For all* $v \in H^m(\mathbb{R}^d)$, $k \in \mathbb{Z}^+ \cup \{0\}$,

$$\|\mathcal{J}_\lambda v\|_{H^{m+k}(\mathbb{R}^d)} \le c_{m,k} \lambda^{-k} \|v\|_{H^m(\mathbb{R}^d)}.$$

2.12 Gronwall's Lemma

Frequently, in the analysis of partial differential equations, we will obtain estimates for various quantities by rearranging the equation or by integrating the quantities in the given equation against a given strategically chosen test function. The aim of such computations is to obtain a *differential inequality*, namely, an inequality describing the net change in time of some quantity of interest, for example, the total energy of the system, in terms of given parameters and given data (such as initial data, boundary data such as inlet/outlet pressure) and the quantity itself integrated in time. We will hence encounter formal inequalities of the form

$$E(t) \le C + \int_0^t E(s) ds, \quad \text{for all } t \ge 0,$$

for example, where C can be a constant (or even generalized to a time-dependent nonnegative function), and $E(t)$ is some nonnegative quantity of interest, such as the total energy of some system at time t. It is useful to then be able to deduce from this differential inequality that the quantity of interest is finite for all time. The basic idea behind this is given by the Gronwall's inequality, which gives an exponential bound for $E(t)$, where the Gronwall's inequality is an extension of the basic fact that a nonnegative function satisfying $E'(t) \leq CE(t)$ must satisfy $0 \leq E(t) \leq E(0)e^{CT}$, where $E(0)$ is the initial value of $E(t)$. The statement of Gronwall's inequality, which can be found in Lemma 1.4.2 in [52], is as follows:

Theorem 2.12.1 (Gronwall's Inequality) *Let $a(t) \geq 0$ and $\beta(t) \geq 0$ be integrable real-valued functions on $[0, T]$. Then, if $r(t) \in L^\infty(0, T)$ is a nonnegative function satisfying the following differential inequality:*

$$r(t) \leq b_0 + \int_0^t \beta(s)ds + \int_0^t a(s)r(s)ds \qquad \text{for all } t \in [0, T],$$

where b_0 is a nonnegative constant that is independent of $t \in [0, T]$, then $r(t) \leq C$ for almost every $t \in [0, T]$, with the explicit bound:

$$r(t) \leq b_0 exp\left(\int_0^t a(s)ds\right) + \int_0^t \beta(s) exp\left(\int_s^t a(\tau)d\tau\right) ds.$$

Often, in the analysis of partial differential equations, we will discretize in time in order to construct approximate solutions to a partial differential equation, which will involve studying estimates on rates of change that are discrete in time. For this purpose, we will use a discrete form of Gronwall's inequality, which we refer to as the *discrete Gronwall inequality*. We state the discrete Gronwall inequality, which can be found in Lemma 5.1 in [180] and Lemma 4.1 in [279], as follows.

Theorem 2.12.2 *Let $\{a_n\}_{n=1}^\infty$, $\{b_n\}_{n=1}^\infty$, $\{c_n\}_{n=1}^\infty$, and $\{\gamma_n\}_{n=1}^\infty$, be sequences of nonnegative numbers, for which the inequality*

$$a_n + (\Delta t)\sum_{i=1}^n b_i \leq B + (\Delta t)\sum_{i=1}^n c_i + (\Delta t)\sum_{i=1}^n \gamma_i a_i,$$

holds for all positive integers n, and for some constants $B \geq 0$ and $\Delta t > 0$, which satisfy

$$(\Delta t)\gamma_n < 1 \qquad \text{for all positive integers } n.$$

Then, for all positive integers n, a_n is locally bounded in the sense that $0 \leq a_n \leq C_N$ for $n = 1, 2, ..., N$ for a constant C_N depending on N, with the explicit bound:

$$a_n + (\Delta t)\sum_{i=1}^n b_i \leq exp\left((\Delta t)\sum_{i=1}^n \frac{\gamma_i}{1-(\Delta t)\gamma_i}\right)\left(B + (\Delta t)\sum_{i=1}^n c_i\right),$$

which holds for all positive integers n.

2.13 The Equations of Fluid Dynamics

Next, we give a brief overview of the partial differential equations in fluid dynamics that will be used in the fluid-structure interaction models we will consider throughout the book. We will discuss the Navier-Stokes equations and illustrate the concept of a priori energy estimates and weak formulations to partial differential equations for the Navier-Stokes equations. This will lead naturally to the classical notion of a finite energy solution to the Navier-Stokes equations, and we will give a brief overview of well-posedness results for the Navier-Stokes equations. We will also discuss variants of the Navier-Stokes equations, including various linearized and stationary models of fluid flow, such as the linear Stokes equations and the stationary Stokes equations. Throughout this chapter and throughout the book, we will exclusively consider **incompressible fluids**, which are fluids where the volume of fluid regions is preserved by the flow of the fluid, though we remark that there is a very rich mathematical literature on compressible fluid flows and the corresponding compressible Navier-Stokes equations.

2.13.1 The Incompressible Navier-Stokes Equations

The incompressible Navier-Stokes equations model the macroscopic dynamics of an incompressible fluid, in terms of the evolution of the fluid velocity $\boldsymbol{u} : \mathcal{O} \to \mathbb{R}^d$ and the fluid pressure $p : \mathcal{O} \to \mathbb{R}$ where $\mathcal{O}$ is a sufficiently smooth, bounded fluid domain on which the equations are posed. We will consider the case in which the fluid has constant density, which we normalize to be equal to one, and in which the fluid occupies a domain $\mathcal{O} \subset \mathbb{R}^d$ for $d = 2$ or $d = 3$, where $\mathcal{O}$ is a bounded simply connected open set in $\mathbb{R}^d$ with Lipschitz boundary. In this case, the **incompressible Navier-Stokes equations** are given by

$$\begin{cases} \rho_f(\partial_t \boldsymbol{u} + (\boldsymbol{u} \cdot \nabla)\boldsymbol{u}) - \nabla \cdot \boldsymbol{\sigma}(\boldsymbol{u}, p) = 0 \\ \nabla \cdot \boldsymbol{u} = 0 \end{cases}, \qquad \text{on } \mathcal{O},$$

where $\rho_f > 0$ is a constant fluid density. The $\mathbb{R}^{d\times d}$-valued (matrix-valued) **Cauchy stress tensor** $\boldsymbol{\sigma}$, which describes the force exerted by a test volume of the fluid on its surroundings, is given by

$$\boldsymbol{\sigma}(\boldsymbol{u}, \pi) = 2\mu \boldsymbol{D}(\boldsymbol{u}) - p\boldsymbol{I}, \tag{2.28}$$

where μ is a physical constant representing the **fluid viscosity** and $\boldsymbol{D}(\boldsymbol{u})$ denotes the symmetrized gradient:

$$\boldsymbol{D}(\boldsymbol{u}) = \frac{1}{2}(\nabla \boldsymbol{u} + (\nabla \boldsymbol{u})^t).$$

By using the divergence-free condition, one can rewrite the momentum equation so that the Navier-Stokes equations can be expressed alternatively as

$$\begin{cases} \rho_f(\partial_t \boldsymbol{u} + (\boldsymbol{u} \cdot \nabla)\boldsymbol{u}) - \mu\Delta\boldsymbol{u} + \nabla p = 0 \\ \nabla \cdot \boldsymbol{u} = 0 \end{cases}, \quad \text{on } \mathcal{O}. \tag{2.29}$$

Next, we need to supply boundary conditions for the Navier-Stokes equations. The classical boundary condition is the **no-slip condition**, which states that at the boundary of the fluid domain, the particles of the fluid stick to the boundary and hence, have zero fluid velocity, so that

$$\boldsymbol{u}|_{\partial\mathcal{O}} = \boldsymbol{0}.$$

We also supply the Navier-Stokes equations with an initial condition:

$$\boldsymbol{u}(0) = \boldsymbol{u}_0 \in L^2(\mathcal{O}),$$

where the presence of the initial data in $L^2(\mathcal{O})$ makes sense physically, as representing finite initial kinetic energy. We make several remarks about the Navier-Stokes equations:

1. *The physical interpretation of the equations.* The first equation in the Navier-Stokes equations is the *momentum equation*, representing balance of momentum via Newton's second law of motion. This can be seen by interpreting the term $\partial_t \boldsymbol{u} + (\boldsymbol{u} \cdot \nabla)\boldsymbol{u}$ as a *material derivative* $\frac{D\boldsymbol{u}}{dt}$, which is the derivative of the fluid velocity along the integral curve representing the trajectory of a fixed particle, and by interpreting the Cauchy stress tensor term as the force acting on a test volume of the fluid. The second equation represents *conservation of mass* for an incompressible, viscous fluid, and it is the infinitesimal differential form of the statement that any (sufficiently regular) test volume in the fluid domain has its volume conserved by the action of the flow map in time generated by the time-dependent fluid velocity. The second equation is also known as *the divergence-free condition.* The divergence-free condition $\nabla \cdot \boldsymbol{u} = 0$, which is also referred to as the *incompressibility condition*, is the essential feature of incompressible fluids. This requires the fluid velocity to be divergence-free, which is a constraint in the system. The divergence-free condition complicates the analysis of the equations, especially when analyzing the Navier-Stokes equations in broader settings such as on time-dependent moving domains.
2. *Pressure as a Lagrange multiplier.* First, from the statement of the Navier-Stokes equations, we can immediately note that the pressure $p : \mathcal{O} \to \mathbb{R}$ is determined only up to an additive constant, which reflects the physical property that *on fixed domains* it is *relative differences in the pressure* rather than *absolute pressure* that drives the dynamics of the fluid. Next, while both the fluid velocity $\boldsymbol{u} : \mathcal{O} \to \mathbb{R}^3$ and the fluid pressure $p : \mathcal{O} \to \mathbb{R}$ are unknown functions in the Navier-Stokes equations, when considering such an equation *weakly* (with a particular weak formulation which we will define shortly), one really only solves for the fluid velocity, as the fluid pressure will vanish from the weak formulation by the choice

of divergence-free test functions. In particular, if one takes the *Leray projection*, which is a projection onto the space of divergence-free functions, the pressure term p will disappear from the fluid momentum equation, leaving an equation of just $\boldsymbol{u}$ itself. It is usually said that the pressure p is a *Lagrange multiplier*, as it is just a function that enforces the divergence-free constraint in the momentum equation. While $\boldsymbol{u}$ is divergence-free, the nonlinear *advection term* $(\boldsymbol{u} \cdot \nabla)\boldsymbol{u}$ is not necessarily divergence-free, and hence the pressure is a Lagrange multiplier that compensates for this.

3. *The nonlinear advection term and the fluid viscosity.* We remark that one of the primary difficulties in the analysis of the Navier-Stokes equations is understanding the impact of the nonlinear advection term $(\boldsymbol{u} \cdot \nabla)\boldsymbol{u}$. This is a quadratic nonlinearity in the equation, and it has been shown heuristically to transfer energy from low frequencies to high frequencies in the fluid velocity. The destabilizing effect of the advection term is balanced out by the regularizing effects of the fluid viscosity, which gives rise to the parabolic regularizing term $-\mu\Delta\boldsymbol{u}$, where the strength of viscous effects is related to the fluid viscosity constant μ. The balance between the nonlinear advection effects and the viscous effects is reflected by the dimensionless **Reynolds number**, which is defined as $\text{Re} = \dfrac{LU\rho_f}{\mu}$, where L is the **characteristic length** and U is the **characteristic velocity** of the system (where the characteristic length/velocity is a representation of the typical length and velocity scales found in a given physical system). The Reynolds number can be derived as a natural dimensionless quantity of inherent interest that naturally arises when one expresses the Navier-Stokes equations in their dimensionless form by rewriting the equations in terms of intrinsic dimensionless variables. When the Reynolds number is small, the viscous effects dominate in the fluid, giving rise to smooth **laminar** or sheet-like flows, while at high Reynolds numbers, inertial effects due to the advection term are the dominant contribution, which gives rise to **turbulent** flows. The mathematical analysis of turbulent flows at high Reynolds numbers, and the study of how incompressible fluid flows evolve as the Reynolds number increases, is a highly challenging problem of substantial mathematical interest.

Remark 2.13.1 Throughout most of the book, we will set the fluid density $\rho_f = 1$ in the momentum equation (2.28) and (2.29), though we remark that the physical value of the fluid density ρ_f is important in the consideration of physically relevant quantities, such as the Reynolds number.

A Priori Energy Estimate for the Navier-Stokes Equations For any partial differential equation, it is useful to first establish an a priori estimate for its solutions. This involves assuming, a priori, the existence of a smooth solution and deriving bounds on its behavior. These estimates typically rely on integration by parts to establish boundedness properties in specific functional spaces. Such analysis provides insight into the appropriate function spaces where solutions may exist. Solutions that belong to these finite energy spaces are referred to as *finite energy solutions*.

We hence carry out an a priori energy estimate for the Navier-Stokes equations, assuming that we have the a priori existence of some (divergence-free) fluid velocity function $\boldsymbol{u} : \mathcal{O} \to \mathbb{R}^d$ and pressure function $p : \mathcal{O} \to \mathbb{R}$ that are sufficiently smooth, where $\mathcal{O} \subset \mathbb{R}^d$ is a bounded domain with a Lipschitz boundary. Given such a solution, we multiply the momentum equation in the Navier-Stokes equations by $\boldsymbol{u}$ and integrate in space and time to obtain

$$\int_0^T \int_{\mathcal{O}} (\partial_t \boldsymbol{u}) \cdot \boldsymbol{u} + \int_0^T \int_{\mathcal{O}} [(\boldsymbol{u} \cdot \nabla)\boldsymbol{u}] \cdot \boldsymbol{u} - \int_0^T \int_{\mathcal{O}} (\nabla \cdot \boldsymbol{\sigma}(\boldsymbol{u}, p)) \cdot \boldsymbol{u} = 0,$$

where we integrate up to an arbitrary (but fixed) time $T > 0$. Since we have a priori assumed that the fluid velocity and pressure are smooth, we can integrate by parts to obtain

$$\frac{1}{2}\int_{\mathcal{O}} |\boldsymbol{u}(T)|^2 + \int_0^T \int_{\mathcal{O}} \boldsymbol{\sigma}(\boldsymbol{u}, p) : \nabla \boldsymbol{u} = \frac{1}{2}\int_{\mathcal{O}} |\boldsymbol{u}_0|^2,$$

where we used the no-slip condition $\boldsymbol{u}|_{\partial\mathcal{O}} = \boldsymbol{0}$ to eliminate any boundary terms on $\partial\mathcal{O}$, and where we computed that

$$\begin{aligned}\int_0^T \int_{\mathcal{O}} [(\boldsymbol{u} \cdot \nabla)\boldsymbol{u}] \cdot \boldsymbol{u} &= \sum_{i,j=1}^d \int_0^T \int_{\mathcal{O}} u_i u_j (\partial_i u_j) = \sum_{i,j=1}^d \int_0^T \int_{\mathcal{O}} u_i \partial_i \left(\frac{1}{2} u_j^2\right) \\ &= -\frac{1}{2}\sum_{i,j=1}^d \int_0^T \int_{\mathcal{O}} \partial_i u_i u_j^2 = -\frac{1}{2}\sum_{i,j=1}^d \int_0^T \int_{\mathcal{O}} (\nabla \cdot \boldsymbol{u})|\boldsymbol{u}|^2 = 0\end{aligned}$$

We note that in this computation, we used the divergence-free condition $\nabla \cdot \boldsymbol{u} = 0$ and the no-slip condition to show that the boundary term from integration by parts along $\partial\mathcal{O}$ is equal to zero. We can also use the definition of the Cauchy stress tensor $\boldsymbol{\sigma}(\boldsymbol{u}, p) = 2\mu \boldsymbol{D}(\boldsymbol{u}) - p\boldsymbol{I}$, to compute that

$$\begin{aligned}\int_0^T \int_{\mathcal{O}} \boldsymbol{\sigma}(\boldsymbol{u}, p) : \nabla \boldsymbol{u} &= 2\mu \int_0^T \int_{\mathcal{O}} \boldsymbol{D}(\boldsymbol{u}) : \nabla \boldsymbol{u} - \int_0^T \int_{\mathcal{O}} p(\nabla \cdot \boldsymbol{u}) \\ &= 2\mu \int_0^T \int_{\mathcal{O}} |\boldsymbol{D}(\boldsymbol{u})|^2,\end{aligned}$$

by the divergence-free condition and the definition of the symmetric gradient $\boldsymbol{D}(\boldsymbol{u})$. We hence obtain the final energy estimate:

$$\frac{1}{2}\int_{\mathcal{O}} \boldsymbol{u}(T) + 2\mu \int_0^T \int_{\mathcal{O}} |\boldsymbol{D}(\boldsymbol{u})|^2 = \frac{1}{2}\int_{\mathcal{O}} |\boldsymbol{u}_0|^2.$$

As a heuristic, the uniform bounds for the a priori estimate (assuming a priori existence of a solution with enough temporal and spatial derivatives with sufficient regularity) define a **finite energy space**, and this finite energy space is the natural space for which we can look for (weak) solutions of the partial differential equation. Thus, motivated by the previous a priori energy estimate, we define the following finite energy space for the Navier-Stokes equations.

Definition 2.13.1 Define the **finite energy space** of solutions for the Navier-Stokes equations by

$$V = L^\infty(0, T; L^2(\mathcal{O})) \cap L^2(0, T; \mathcal{V}_f),$$

where $\mathcal{V}_f$ is the closed subspace of (weakly) divergence-free functions in $H_0^1(\mathcal{O})$, given by

$$\mathcal{V}_f = \{\boldsymbol{u} \in H_0^1(\mathcal{O}) : \nabla \cdot \boldsymbol{u} = 0 \text{ on } \mathcal{O}\}.$$

The Weak Formulation of the Navier-Stokes Equations We will see throughout the analysis of fluid-structure models in this book that finite energy methods are closely related to methods for showing existence of weak solutions, which are solutions that satisfy the partial differential equation in the sense of weak derivatives. Namely, the partial differential equation is satisfied in a weak sense of testing against a sufficiently smooth test function in time and space, in order to transfer all the derivatives onto the test function (as in the sense of a weak derivative). To make this rigorous, we say that a weak solution to a partial differential equation satisfies a **weak formulation** for all admissible test functions, which is obtained by formally multiplying the partial differential equation with a sufficiently smooth test function and integrating by parts to obtain a resulting expression called the weak formulation. For the Navier-Stokes equations, the fluid velocity is required to be divergence-free, so for the purposes of deriving the weak formulation, it is advantageous to impose the divergence-free condition on the test function space also:

$$\mathcal{Q} = C_c^\infty([0, T); \mathcal{V}_f).$$

Hence, the test space consists of all test functions $\boldsymbol{q}$ on $[0, T] \times \mathcal{O}$ that are smooth and compactly supported in time on $[0, T)$, such that the divergence-free condition $\nabla \cdot \boldsymbol{q}(t) = 0$ is satisfied on all of $\mathcal{O}$ for all $t \in [0, T]$.

To derive the weak formulation, we hence consider a test function $\boldsymbol{q} \in \mathcal{Q}$, multiply the momentum balance equation of the Navier-Stokes equations by $\boldsymbol{q}$, and integrate in space and time to obtain

$$\int_0^T \int_{\mathcal{O}} \partial_t \boldsymbol{u} \cdot \boldsymbol{q} + \int_0^T \int_{\mathcal{O}} [(\boldsymbol{u} \cdot \nabla)\boldsymbol{u}] \cdot \boldsymbol{q} - \int_0^T \int_{\mathcal{O}} (\nabla \cdot \boldsymbol{\sigma}(\boldsymbol{u}, p)) \cdot \boldsymbol{q} = 0.$$

We then integrate by parts, using analogous calculations as in the derivation of the a priori energy estimate to obtain

$$-\int_0^T \int_{\mathcal{O}} \boldsymbol{u} \cdot \partial_t \boldsymbol{q} - \int_0^T \int_{\mathcal{O}} \boldsymbol{u} \cdot [(\boldsymbol{u} \cdot \nabla)\boldsymbol{q}] + 2\mu \int_0^T \int_{\mathcal{O}} \boldsymbol{D}(\boldsymbol{u}) : \boldsymbol{D}(\boldsymbol{q}) = \int_{\mathcal{O}} \boldsymbol{u}_0 \cdot \boldsymbol{q}(0).$$

This is the weak formulation, and we can use this weak formulation to define a weak solution as follows.

Definition 2.13.2 (Weak Solution to Navier-Stokes Equations) We say that a function $\boldsymbol{u}$ belonging to the finite energy space V is a **weak solution** to the Navier-Stokes equations on $[0, T]$ if for all $\boldsymbol{q} \in \mathcal{Q}$,

$$-\int_0^T \int_{\mathcal{O}} \boldsymbol{u} \cdot \partial_t \boldsymbol{q} - \int_0^T \int_{\mathcal{O}} \boldsymbol{u} \cdot [(\boldsymbol{u} \cdot \nabla)\boldsymbol{q}] + 2\mu \int_0^T \int_{\mathcal{O}} \boldsymbol{D}(\boldsymbol{u}) : \boldsymbol{D}(\boldsymbol{q}) = \int_{\mathcal{O}} \boldsymbol{u}_0 \cdot \boldsymbol{q}(0),$$

and in addition, the following energy inequality holds for all $t \in [0, T]$:

$$\frac{1}{2} \int_{\mathcal{O}} |\boldsymbol{u}(t)|^2 + 2\mu \int_0^t \int_{\mathcal{O}} |\boldsymbol{D}(\boldsymbol{u})|^2 \leq \frac{1}{2} \int_{\mathcal{O}} |\boldsymbol{u}_0|^2.$$

We emphasize that because we are considering *divergence-free test functions* in the definition of the test space as a result of the solution space having the divergence-free constraint, the pressure term $\int_0^T \int_{\mathcal{O}} p(\nabla \cdot \boldsymbol{q})$ in the weak formulation vanishes, so that the fluid pressure does not appear at all in the weak formulation, and rather the fluid velocity is the only unknown function in the weak formulation. This is consistent with our interpretation of the fluid pressure as a Lagrange multiplier that enforces the divergence-free condition on the fluid velocity, and once the weak solution for the fluid velocity is obtained, one can actually recover the pressure via a nonlocal operator.

The consideration of *weak solutions* for the Navier-Stokes equations was a significant mathematical insight in the analysis of these equations, accomplished by Leray in [236]. While it is clear that any pointwise solution to the Navier-Stokes equations is a weak solution, weak solutions are genuinely weaker types of solutions than strong or (pointwise) classical solutions to the Navier-Stokes equations, where the equation is considered in a distributional sense. While analysis prior to Leray only obtained results for existence of strong solutions locally in time to the Navier-Stokes equations, Leray was able to in fact establish the existence of solutions of the type described in Definition 2.13.2 globally in time in three spatial dimensions for the Navier-Stokes equations, and Hopf in [185] later obtained the corresponding result on global existence of solutions in the sense of Definition 2.13.2 for bounded domains in $\mathbb{R}^3$. Leray and Hopf's results together form the basis of the following fundamental theorem in the analysis of incompressible fluid flows:

Theorem 2.13.1 (Existence of Weak Solutions [185, 236]) *Let $\mathcal{O}$ be $\mathbb{R}^d$ or a bounded open domain in $\mathbb{R}^d$ for $d = 2, 3$. Then, there exists a weak solution to the Navier-Stokes equations on $[0, T]$ in the sense of Definition 2.13.2, for any arbitrary $T > 0$, for arbitrary initial data $\boldsymbol{u}_0 \in L^2(\mathcal{O})$.*

Finally, we remark that in the case of two spatial dimensions, there are significantly stronger results on the existence of solutions to the Navier-Stokes equations, due to the fact that the equations have stronger properties in two spatial dimensions, such as the fact that vorticity, given by the curl $\nabla \times \boldsymbol{u}$, is transported by the flow in two spatial dimensions (while this is not true in three dimensions due to a phenomenon known as **vortex stretching**). Due to the special nature of the Navier-Stokes equations in two spatial dimensions, one obtains global existence and uniqueness of strong solutions to the Navier-Stokes equations in two spatial dimensions, due to Ladyzhenskaya [230].

For clarity of presentation and to avoid problems related to fluid domains that are not Lipschitz, most of the fluid models that we consider for the fluid-structure interaction problems in this book will be in two spatial dimensions. However, we emphasize that our approach to such fluid-structure interaction models will be in the spirit of Leray-Hopf solutions, as this approach will carry over to analysis of more complex higher-dimensional fluid-structure interaction models, consisting of three-dimensional fluid models interacting with two-dimensional structures. In particular, while there are many approaches to fluid-structure interaction (e.g., in terms of existence of weak or strong solutions), our goal in obtaining existence of solutions to a fluid-structure interaction problem in this book will be to obtain an analogue of Leray-Hopf solutions to fluid-structure systems, where such solutions are finite energy weak solutions that exist globally up until a time of potential domain degeneracy (such as self-collision of the structure with itself), or globally on arbitrary time intervals $[0, T]$ for $T > 0$ if there is no such possibility of domain degeneracy.

2.13.2 Linear and Stationary Stokes Equations

In several of the linear fluid-structure interaction models that we will consider later in the book, we will consider linearized models of fluid flow. Taking the Navier-Stokes equations and linearizing around the zero solution $\boldsymbol{u} = 0$ and $p = 0$, we obtain the following linearized equation, known as the **linear Stokes equations**, which are the following set of parabolic partial differential equations:

$$\begin{cases} \partial_t \boldsymbol{u} - \nabla \cdot \boldsymbol{\sigma}(\boldsymbol{u}, p) = 0 \\ \nabla \cdot \boldsymbol{u} = 0 \end{cases}, \quad \text{on } \mathcal{O}, \tag{2.30}$$

where as before, $\boldsymbol{\sigma}(\boldsymbol{u}, p) = 2\mu \boldsymbol{D}(\boldsymbol{u}) - p\boldsymbol{I}$ is the Cauchy stress tensor with symmetrized gradient $\boldsymbol{D}(\boldsymbol{u})$ and $\mathcal{O}$ is a bounded domain in $\mathbb{R}^d$ with Lipschitz boundary. This system can be equivalently written in terms of the fluid pressure, using the definition of the Cauchy stress tensor, as

$$\begin{cases} \partial_t \boldsymbol{u} - \mu\Delta\boldsymbol{u} + \nabla p = 0 \\ \nabla \cdot \boldsymbol{u} = 0 \end{cases}, \quad \text{on } \mathcal{O}.$$

We emphasize that in contrast with the Navier-Stokes equations, the linear Stokes equations are linear partial differential equations, as they do not have the quadratic nonlinearity in the advection term present in the (nonlinear) Navier-Stokes equations. In this sense, linear Stokes flows are good approximations of laminar flows at low Reynolds numbers, in which viscous effects dominate over inertial (advection) effects. Because the Stokes equations are linear, they are well-studied and their well-posedness properties are well-understood by now. One can verify through a standard energy estimate analysis that is analogous to that for the Navier-Stokes equations that the linear Stokes equations have the same *finite energy space* as the Navier-Stokes equations.

The analysis of long-time behavior of fluid flows is also of considerable interest, and hence, it is interesting to consider the long-time behavior of linear Stokes flows, in which the fluid flow attains a stationary state or an equilibrium. These time-independent stationary states, which represent fluid flow configurations over large time scales, can be modeled by the **stationary Stokes equations**, which are the following set of elliptic linear partial differential equations:

$$\begin{cases} \nabla \cdot \boldsymbol{\sigma}(\boldsymbol{u}, p) = 0 \\ \nabla \cdot \boldsymbol{u} = 0 \end{cases}, \quad \text{on } \mathcal{O}.$$

Using the definition of the Cauchy stress tensor, one can rewrite the stationary Stokes equations as

$$\begin{cases} \mu\Delta\boldsymbol{u} = \nabla p, \\ \nabla \cdot \boldsymbol{u} = 0 \end{cases}, \quad \text{on } \mathcal{O}.$$

We can see that the stationary Stokes equations are similar in spirit to Laplace's equation in each scalar component of the fluid velocity, except with the additional contribution of the fluid pressure, which enforces a divergence-free constraint on the vector-valued fluid velocity. As a set of elliptic equations, the stationary Stokes equations have been well-studied in various contexts, including on bounded domains, on unbounded domains, and on $\mathbb{R}^d$; see [145, 326]. The stationary Stokes equations are useful for modeling viscosity-dominated flows at low Reynolds numbers over large enough time scales over which the system has enough time to reach a stationary equilibrium state of fluid flow.

2.14 The Equations of Linear Elasticity

Next, we discuss the hyperbolic partial differential equations of linear elasticity that will appear often in our analysis of fluid-structure interaction. These describe the displacement of an elastic body from a given reference configuration, in terms of a vector-valued structure displacement $\boldsymbol{\eta}$. We discuss two specific models: models of thin elastic and viscoelastic membranes and plates (such as Koiter shells) and models of bulk elasticity of solids (described by the Lamé equations), which will both play a role in our subsequent analysis. For these models, we will carry out a priori energy estimates and define finite energy spaces and weak solutions, as in our corresponding analysis of fluid dynamical models, to illustrate some fundamental aspects of the analysis of these partial differential equations. For a comprehensive treatment of the mathematical theory of elasticity, we refer the interested reader to [91–93].

2.14.1 The Koiter Shell Equation

The Koiter shell equation describes the elastodynamics of a thin shell, via a fourth-order hyperbolic differential equation.

While there is a well-studied formulation of the Koiter shell equations for shells with general vector-valued structure displacements, we study a simplified model of Koiter shell dynamics, as this is the model that will appear in our later formulations of fluid-structure interaction problems; see, e.g., Chaps. 4 and 5. The derivation of the linearly elastic and linearly viscoelastic cylindrical Koiter shell equations accounting for the full vector-valued structure displacements $\boldsymbol{\eta}$ can be found in, e.g., [66].

To this end, we consider a one-dimensional model of Koiter shell dynamics, where the reference configuration of the Koiter shell is given by the closed interval $\Gamma = [0, L] \times \{1\}$ and the Koiter shell exists in two-dimensional space with coordinates $(z, r) \in \mathbb{R}^2$. Thus, the reference configuration of the Koiter shell, which is the physical location of the Koiter shell with zero structural displacement, is given by $0 \leq z \leq L$ and $r = 1$, where the choice of $r = 1$ is an arbitrary choice that will be useful for the specific geometry of our subsequent fluid-structure interaction models.

For simplicity, we assume that the Koiter shell has clamped ends and that the Koiter shell displaces in only the transverse direction (in the r direction), so that we can represent the vector-valued structure displacement $\boldsymbol{\eta}$ instead as a scalar (time-dependent) transverse r displacement $\eta : \Gamma \times [0, T] \to \mathbb{R}$. Then, the equation describing the elastic Koiter shell is given by the following equation:

$$\partial_t^2 \eta + C_0 \eta - C_1 \partial_z^2 \eta + C_2 \partial_z^4 \eta = F \text{ on } \Gamma, \tag{2.31}$$

supplied with clamped shell boundary conditions:

$$\eta|_{z=0} = \eta|_{z=L} = \partial_z \eta|_{z=0} = \partial_z \eta|_{z=L} = 0 \quad \text{for all } t,$$

and with initial data for the initial displacement and the initial velocity:

$$\eta|_{t=0} = \eta_0, \quad \partial_t \eta|_{t=0} = v_0.$$

We emphasize that the equation is posed as an equation on the *fixed reference (the so-called Lagrangian) configuration* Γ rather than on the moving physical (the so-called Eulerian) configuration $\Gamma(t)$. By interpreting $\eta \boldsymbol{e}_r$ as the transverse displacement of the Koiter shell from its reference configuration, we can define the time-dependent physical configuration of the Koiter shell by

$$\Gamma(t) = \{(z, r) : 0 \leq z \leq L, r = 1 + \eta(t, z)\}.$$

The assumption that the Koiter shell displaces in only the vertical/transverse direction guarantees that the physical time-dependent configuration of the moving Koiter shell always remains a subgraph. More specifically, by the regularity properties of η for appropriate finite energy initial data, it will be a Lipschitz subgraph when considering a Koiter shell in one spatial dimension. We remark that the consideration of general vector-valued displacements of Koiter shells (not constrained to displace in just the radial direction) is more challenging in terms of defining the physical time-dependent location of the structure, due to potential self-intersection of the shell with itself (though since the equations are posed on the reference configuration rather than the moving time-dependent configuration $\Gamma(t)$, the actual formulation of the equations of elasticity themselves would not be significantly more complex).

As a final comment, the constants C_0, C_1, C_2 are physical constants, depending on the physical elastic properties of the shell. For the exact form of these constants, we refer the reader to [274, Appendix] and [66].

Energy Estimate To obtain an a priori energy estimate and hence the definition of the finite energy space, we multiply the Koiter shell equation by the transverse structure velocity $\partial_t \eta$ and integrate over Γ and over $[0, T]$ to obtain

$$\int_0^T \int_\Gamma \partial_t^2 \eta \partial_t \eta + C_0 \int_0^T \int_\Gamma \eta \partial_t \eta - C_1 \int_0^T \int_\Gamma \partial_z^2 \eta \partial_t \eta$$
$$+ C_2 \int_0^T \int_\Gamma \partial_z^4 \eta \partial_t \eta = \int_0^T \int_\Gamma F \partial_t \eta.$$

We integrate by parts to obtain

$$\frac{1}{2}\int_\Gamma |\partial_t \eta(T)|^2 + \frac{1}{2}C_0 \int_\Gamma |\eta(T)|^2 + \frac{1}{2}C_1 \int_\Gamma |\partial_z \eta(T)|^2 + \frac{1}{2}C_2 \int_\Gamma |\partial_z^2 \eta(T)|^2$$
$$= \frac{1}{2}\int_\Gamma |v_0|^2 + \frac{1}{2}C_0 \int_\Gamma |\eta_0|^2 + \frac{1}{2}C_1 \int_\Gamma |\partial_z \eta_0|^2 + \frac{1}{2}C_2 \int_\Gamma |\partial_z^2 \eta_0|^2 + \int_0^T \int_\Gamma F \partial_t \eta.$$

Define the total energy:

$$E(t) := \frac{1}{2}\int_\Gamma |\partial_t \eta(t)|^2 + \frac{1}{2}C_0 \int_\Gamma |\eta(t)|^2 + \frac{1}{2}C_1 \int_\Gamma |\partial_z \eta(t)|^2 + \frac{1}{2}C_2 \int_\Gamma |\partial_z^2 \eta(t)|^2,$$

which is the sum of the kinetic energy, which is the first term, and the elastic potential energy, which are the remaining terms. Then, we obtain the energy estimate:

$$E(T) = E(0) + \int_0^T \int_\Gamma F \partial_t \eta.$$

In the case where there is no external forcing so that $F = 0$, we obtain an energy equality, which gives *conservation of energy*:

$$E(T) = E(0).$$

In the case where there is nonzero external forcing, which we will assume has regularity $F \in L^1(0, T; L^2(\Gamma))$, we note that the energy inequality also holds formally for all $t \in [0, T]$, so that

$$\max_{0 \le s \le t} E(t) \le E(0) + \|F\|_{L^1(0,t;L^2(\Gamma))} \max_{0 \le s \le t} \|\partial_t \eta\|_{L^2(\Gamma)}.$$

Using Cauchy's inequality with ϵ and using the definition of the energy $E(t)$, we obtain that for some positive $\epsilon < 1/2$,

$$\max_{0 \le s \le t} E(s) \le E(0) + \epsilon \max_{0 \le s \le t} E(t) + C(\epsilon)\|F\|^2_{L^1(0,T;L^2(\Gamma))}.$$

Therefore, moving the term with $\epsilon > 0$ to the other side, we obtain that

$$\max_{0 \le s \le t} E(s) \le C\left(E(0) + \|F\|^2_{L^1(0,T;L^2(\Gamma))}\right),$$

for some constant C. Hence, we obtain that for some constant $C(E(0), \|F\|_{L^1(0,T;L^2(\Gamma))})$ that depends only on the initial total energy $E(0)$, and the norm $\|F\|_{L^1(0,T;L^2(\Gamma))}$ of the inhomogeneous source term F, the following estimate of $E(t)$ holds:

$$\max_{0 \le t \le T} E(t) \le C \left(E(0), \|F\|_{L^1(0,T;L^2(\Gamma))} \right).$$

From this a priori estimate, we can define the space of finite energy solutions, which is the space of functions for which the final a priori estimate (energy inequality) $E(t) \le C$ holds for almost all $t \in [0, T]$.

Definition 2.14.1 The finite energy space for the elastic Koiter shell equation is given by

$$\mathcal{V}_s = L^\infty(0, T; H_0^2(\Gamma)) \cap W^{1,\infty}(0, T; L^2(\Gamma)).$$

The Weak Formulation Next, we define the weak formulation, and we define the test space for the elastic Koiter shell equation by

$$\mathcal{Q}_s = C_c^\infty([0, T); H_0^2(\Gamma)).$$

We multiply the Koiter shell equation by a test function $\varphi \in \mathcal{Q}_s$ to obtain the following weak formulation.

Definition 2.14.2 A function $\eta \in \mathcal{V}_s$ is a **weak solution** to the Koiter shell equation if for all test functions $\varphi \in \mathcal{Q}_s$,

$$-\int_0^T \int_\Gamma \partial_t \eta \partial_t \varphi + C_0 \int_0^T \int_\Gamma \eta \varphi + C_1 \int_0^T \int_\Gamma \partial_z \eta \partial_z \varphi$$
$$+ C_2 \int_0^T \int_\Gamma \partial_z^2 \eta \partial_z^2 \varphi = \int_\Gamma v_0 \varphi(0) + \int_0^T \int_\Gamma F \varphi.$$

It is easy to show the following existence result for the linear elastic Koiter shell equation in one spatial dimension.

Theorem 2.14.1 *Let the initial data satisfy $\eta_0 \in H_0^2(\Gamma)$ and let $v_0 \in L^2(\Gamma)$. Then, there exists a unique weak solution $\eta \in \mathcal{V}_s$ to the Koiter shell equation.*

2.14.2 The Lamé Equations of Bulk Linear Elasticity

Next, we discuss the Lamé equations, which are linear equations describing bulk elasticity. We will pose these equations in two spatial dimensions, as this is the specific context we will need for the future discussion of fluid-poroviscoelastic structure interaction, though the extension of these equations to three spatial dimensions is analogous. We let $\mathcal{O}_s$ be an open domain in $\mathbb{R}^2$ with Lipschitz boundary, representing the reference configuration of some two-dimensional elastic solid, and let $\boldsymbol{\eta} : \mathcal{O}_s \to \mathbb{R}^2$ denote the vector-valued displacement of the elastic solid from its reference configuration $\mathcal{O}_s$.

The elastodynamics of the linearly elastic solid is given by the following linear hyperbolic partial differential equation for the structure displacement $\boldsymbol{\eta}$, also known as the Lamé system of equations:

$$\partial_t^2 \boldsymbol{\eta} - \nabla \cdot S_b(\boldsymbol{\eta}, \boldsymbol{\eta}_t) = 0, \quad \text{in } \mathcal{O}_s, \tag{2.32}$$

where the **Piola-Kirchhoff stress tensor** is given by

$$S_b(\boldsymbol{\eta}, \boldsymbol{\eta}_t) = 2\mu_e \boldsymbol{D}(\boldsymbol{\eta}) + \lambda_e (\nabla \cdot \boldsymbol{\eta}) \boldsymbol{I} + 2\mu_v \boldsymbol{D}(\boldsymbol{\eta}_t) + \lambda_v (\nabla \cdot \boldsymbol{\eta}_t) \boldsymbol{I}. \tag{2.33}$$

The positive constants μ_e and λ_e are elasticity parameters, known as the Lamé parameters, and the nonnegative constants μ_v and λ_v are the viscoelasticity parameters. We emphasize that the elasticity equation is posed on the reference domain $\mathcal{O}_s$ (in the Eulerian formulation). We will consider two cases: (1) the purely elastic case (where $\mu_v = \lambda_v = 0$) and (2) the viscoelastic case (where $\mu_v > 0$ and $\lambda_v > 0$). We will consider the Lamé equations of elasticity with zero forcing on the right-hand side, as this will parallel the elasticity equations that we will see in the study of fluid-poroviscoelastic structure interaction.

We supplement the Lamé equations (2.32) with zero Dirichlet boundary conditions:

$$\boldsymbol{\eta}|_{\partial \mathcal{O}_s} = \boldsymbol{0},$$

and with initial data for the initial displacement and initial velocity:

$$\boldsymbol{\eta}(0) = \boldsymbol{\eta}_0, \qquad \partial_t \boldsymbol{\eta}(0) = \boldsymbol{v}_0.$$

Energy Estimate We derive an a priori energy estimate for the Lamé system as follows. We multiply (2.32) by the structure velocity $\partial_t \boldsymbol{\eta}$ and integrate in space over $\mathcal{O}_s$, and in time over $[0, T]$. After integrating by parts and using the Dirichlet boundary condition $\boldsymbol{\eta}|_{\partial \mathcal{O}_s} = \boldsymbol{0}$ for the structure displacement, we obtain

$$\frac{1}{2} \int_{\mathcal{O}_s} |\partial_t \boldsymbol{\eta}(T)|^2 + \int_0^T \int_{\mathcal{O}_s} S_b(\boldsymbol{\eta}, \boldsymbol{\eta}_t) : (\nabla \partial_t \boldsymbol{\eta}) = \frac{1}{2} \int_{\mathcal{O}_s} |\boldsymbol{v}_0|^2.$$

Using the definition of the Piola-Kirchhoff stress tensor (2.33), we compute the corresponding integrals:

$$
\begin{aligned}
2\mu_e \int_0^T \int_{\mathcal{O}_s} \boldsymbol{D}(\boldsymbol{\eta}) : \nabla \boldsymbol{\eta}_t &= 2\mu_e \int_0^T \int_{\mathcal{O}_s} \boldsymbol{D}(\boldsymbol{\eta}) : \partial_t \boldsymbol{D}(\boldsymbol{\eta}) \\
&= \mu_e \int_{\mathcal{O}_s} |\boldsymbol{D}(\boldsymbol{\eta})(T)|^2 - \mu_e \int_{\mathcal{O}_s} |\boldsymbol{D}(\boldsymbol{\eta}_0)|^2, \\
\lambda_e \int_0^T \int_{\mathcal{O}_s} (\nabla \cdot \boldsymbol{\eta}) \boldsymbol{I} : \nabla \boldsymbol{\eta}_t &= \frac{1}{2}\lambda_e \int_{\mathcal{O}_s} |(\nabla \cdot \boldsymbol{\eta})(T)|^2 - \frac{1}{2}\lambda_e \int_{\mathcal{O}_s} |(\nabla \cdot \boldsymbol{\eta}_0)|^2, \\
2\mu_v \int_0^T \int_{\mathcal{O}_s} \boldsymbol{D}(\boldsymbol{\eta}_t) : \nabla \boldsymbol{\eta}_t &= 2\mu_v \int_0^T \int_{\mathcal{O}_s} |\boldsymbol{D}(\boldsymbol{\eta}_t)|^2, \\
\lambda_v \int_0^T \int_{\mathcal{O}_s} (\nabla \cdot \boldsymbol{\eta}_t) \boldsymbol{I} : \nabla \boldsymbol{\eta}_t &= \lambda_v \int_0^T \int_{\mathcal{O}_s} |\nabla \cdot \boldsymbol{\eta}_t|^2.
\end{aligned}
$$

We combine the integrals above to obtain the final energy estimate:

$$
E(T) + 2\mu_v \int_0^T \int_{\mathcal{O}_s} |\boldsymbol{D}(\boldsymbol{\eta}_t)|^2 + \lambda_v \int_0^T \int_{\mathcal{O}_s} |\nabla \cdot \boldsymbol{\eta}_t|^2 = E(0),
$$

where $E(t)$ is the total energy defined by

$$
E(t) = \frac{1}{2} \int_{\mathcal{O}_s} |\partial_t \boldsymbol{\eta}(t)|^2 + \mu_e \int_{\mathcal{O}_s} |\boldsymbol{D}(\boldsymbol{\eta})(t)|^2 + \frac{1}{2}\lambda_e \int_{\mathcal{O}_s} |(\nabla \cdot \boldsymbol{\eta})(t)|^2.
$$

We can see the impact of viscoelasticity on the energy inequality, as the viscoelasticity terms in the case when $\mu_v > 0$ and $\lambda_v > 0$ give rise to a dissipation in the elastic energy of the system over time, while the case of pure elasticity gives conservation of energy (without dissipation of energy). The energy estimate motivates the following definition of the finite energy spaces.

Definition 2.14.3 The finite energy space for the purely elastic case of the linear Lamé equations is

$$
V_s = L^\infty(0, T; H_0^1(\mathcal{O}_s)) \cap W^{1,\infty}(0, T; L^2(\mathcal{O}_s)),
$$

and the finite energy space for the viscoelastic case of the Lamé equations is

$$
V_s = L^\infty(0, T; H_0^1(\mathcal{O}_s)) \cap W^{1,\infty}(0, T; L^2(\mathcal{O}_s)) \cap H^1(0, T; H^1(\mathcal{O}_s)).
$$

The Weak Formulation We define the test space to be

$$
\mathcal{Q}_s = C_c^\infty([0, T); H_0^1(\mathcal{O}_s)),
$$

and by testing the Lamé equation with an arbitrary test function $\boldsymbol{\varphi} \in \mathcal{Q}_s$ and integrating by parts, we obtain the following weak formulation and the associated notion of a weak solution.

Definition 2.14.4 A function $\boldsymbol{\eta} \in V_s$ (in the corresponding purely elastic or viscoelastic finite energy space) is a **weak solution** to the linear Lamé equation if for all test functions $\boldsymbol{\varphi} \in \mathcal{Q}_s$, the following weak formulation is satisfied:

$$-\int_0^T \int_{\mathcal{O}_s} \partial_t \boldsymbol{\eta} \cdot \partial_t \boldsymbol{\varphi} + 2\mu_e \int_0^T \int_{\mathcal{O}_s} \boldsymbol{D}(\boldsymbol{\eta}) : \boldsymbol{D}(\boldsymbol{\varphi}) + \lambda_e \int_0^T \int_{\mathcal{O}_s} (\nabla \cdot \boldsymbol{\eta})(\nabla \cdot \boldsymbol{\varphi})$$

$$+ 2\mu_v \int_0^T \int_{\mathcal{O}_s} \boldsymbol{D}(\boldsymbol{\eta}_t) : \boldsymbol{D}(\boldsymbol{\varphi}) + \lambda_v \int_0^T \int_{\mathcal{O}_s} (\nabla \cdot \boldsymbol{\eta}_t)(\nabla \cdot \boldsymbol{\varphi}) = \int_{\mathcal{O}_s} \boldsymbol{v}_0 \cdot \boldsymbol{\varphi}(0).$$

We have the following existence and uniqueness results for weak solutions to the Lamé equations.

Theorem 2.14.2 *For initial data $\boldsymbol{\eta}_0 \in H_0^1(\mathcal{O}_s)$ and $\boldsymbol{v}_0 \in L^2(\mathcal{O}_s)$, there exists a unique weak solution $\boldsymbol{\eta} \in V_s$ to the Lamé equations in both the purely elastic and the viscoelastic case.*

2.14.3 *Injectivity of Lagrangian Maps*

In the previous subsection, we discussed the equations of mathematical elasticity, which describe the elastodynamics of elastic or deformable structures by specifying a vector-valued displacement $\boldsymbol{\eta} : \mathcal{O}_s \to \mathbb{R}^d$ from a reference configuration $\mathcal{O}_s \subset \mathbb{R}^d$ of an elastic solid in $\mathbb{R}^d$ for $d = 2$ or $d = 3$. This means that given a point $\boldsymbol{x} \in \mathcal{O}_s \subset \mathbb{R}^d$, the physical location of $\boldsymbol{x}$ after the elastic deformation described by the vector-valued displacement function $\boldsymbol{\eta}$ is $\boldsymbol{x} + \boldsymbol{\eta}(x) = (\boldsymbol{I} + \boldsymbol{\eta})(\boldsymbol{x})$, where $\boldsymbol{I}(\boldsymbol{x}) = \boldsymbol{x}$ is the identity map. Therefore, the physical configuration of the elastic solid, specified by a vector-valued structure displacement function, would formally be given by

$$(\boldsymbol{I} + \boldsymbol{\eta})(\mathcal{O}_s) = \{\boldsymbol{x} + \boldsymbol{\eta}(\boldsymbol{x}) : \boldsymbol{x} \in \mathcal{O}_s\}.$$

We thus observe that the map $\boldsymbol{\Phi}^\eta$, which we refer to as the **Lagrangian map**, and which has the explicit form

$$\boldsymbol{\Phi}^\eta : \mathcal{O}_s \to \mathbb{R}^d, \qquad \boldsymbol{\Phi}^\eta(\boldsymbol{x}) = (\boldsymbol{I} + \boldsymbol{\eta})(\boldsymbol{x}), \quad \boldsymbol{x} \in \mathcal{O}_s,$$

describes how points in the reference configuration of the elastic material, are mapped to their physical locations after the elastic deformation $\boldsymbol{\eta}$.

While the correspondence of points in the reference configuration and in physical space via the Lagrangian map $\boldsymbol{\Phi}^\eta$ makes sense intuitively by considering $\boldsymbol{\eta}$ as a vector-valued displacement of the elastic material, we note that this correspondence on a mathematical level is formal, and not inherently rigorous in the following sense. In particular, it is unclear a priori, for a reference configuration $\mathcal{O}_s$ and a given vector-valued displacement $\boldsymbol{\eta}$, whether the image $\boldsymbol{\Phi}^{\eta}(\mathcal{O}_s)$ representing the physical configuration of the elastic solid makes sense. For example, the Lagrangian map $\boldsymbol{\Phi}^{\eta}$ could fail to be injective, which describes potential self-intersection or geometric degeneracy of the elastic structure, and even for smooth vector-valued displacement functions $\boldsymbol{\eta}$, the Lagrangian map could fail to be injective. Hence, the issue of injectivity of the Lagrangian map extends beyond just having enough regularity for $\boldsymbol{\eta}$, and it will thus be useful to have explicit conditions on $\boldsymbol{\eta}$ that will guarantee that the Lagrangian map is injective, so that we can make proper sense of the physical configuration $\boldsymbol{\Phi}^{\eta}(\mathcal{O}_s)$ of the elastic solid. We will state the following criterion for injectivity of the Lagrangian map, which is Theorem 5.5–2 on pg. 225 of [91], which gives a criterion for injectivity of a given map if one can find another map (known a priori already to be injective) that matches boundary values with the given map in question.

Theorem 2.14.3 (Injectivity of Lagrangian Maps) *Let $\mathcal{O}_s$ be a bounded Lipschitz domain. Consider a map $\boldsymbol{\Phi} : \overline{\mathcal{O}_s} \to \mathbb{R}^d$, which is C^1 as a map from $\mathcal{O}_s$ to $\mathbb{R}^d$, which is continuous as a map from $\overline{\mathcal{O}_s} \to \mathbb{R}^d$, and which satisfies*

$$det(\nabla \boldsymbol{\Phi}) > 0 \qquad \textit{pointwise on } \mathcal{O}_s.$$

If there exists an injective continuous map $\boldsymbol{\Phi}_0 : \overline{\mathcal{O}_s} \to \mathbb{R}^d$ such that

$$\boldsymbol{\Phi}|_{\partial \mathcal{O}_s} = \boldsymbol{\Phi}_0|_{\partial \mathcal{O}_s},$$

then $\boldsymbol{\Phi} : \overline{\mathcal{O}_s} \to \mathbb{R}^d$ is an injective homeomorphism with $\boldsymbol{\Phi}(\overline{\mathcal{O}_s}) = \boldsymbol{\Phi}_0(\overline{\mathcal{O}_s})$, and $\boldsymbol{\Phi} : \mathcal{O}_s \to \mathbb{R}^d$ is an injective diffeomorphism with $\boldsymbol{\Phi}(\mathcal{O}_s) = \boldsymbol{\Phi}_0(\mathcal{O}_s)$.

2.15 Approximation Schemes for Partial Differential Equations

In this section, we discuss several methods for constructing approximate solutions to partial differential equations. In general, to show existence of solutions to partial differential equations, there is a standard and well-established framework consisting of the following essential steps:

1. **Construction of approximate solutions.** The goal here is to approximate the original partial differential equation in a way that makes it possible to show the existence of solutions to some approximate form of the problem. Standard approximation methods for partial differential equations include finite-dimensional Galerkin approximations, Picard iteration, or numerical schemes.

2. **Uniform boundedness of approximate solutions.** By carefully constructing the approximate problems, one can obtain uniform boundedness of the approximate solutions that is uniform in the approximation parameter. These uniform bounds in finite energy spaces allow us to obtain weak and weak-star convergence of the approximate solutions along a subsequence. Consequently, this allows us to pass to the limit in any linear terms in the weak formulation.
3. **Compactness arguments.** However, if the partial differential equation is nonlinear, we will need strong convergences of the approximate solutions, rather than weak or weak-star convergences, in order to rigorously pass to the limit in the approximate weak formulations satisfied by the approximate solutions. To this end, we can use compact embeddings and results such as the Aubin-Lions lemma and the Arzela-Ascoli theorem to deduce strong convergence of approximate solutions along subsequences via compact embeddings.
4. **Pass to the limit in the approximate weak formulations.** The approximate solutions satisfy approximate weak formulations, which are approximations to the limiting weak formulation. Given all of the weak, weak-star, and strong convergences of the approximate solutions from the previous two steps (establishing uniform boundedness of approximate solutions and compactness arguments), we can then pass to the limit in the approximate weak formulations to obtain a limiting weak solution that satisfies a limiting weak formulation.

The main challenge in applying this framework to the analysis of various fluid-structure interaction problems lies in overcoming the inherent mathematical complexities of FSI systems. These include handling highly coupled, nonlinear systems of partial differential equations characteristic of such multiphysics problems, addressing geometric nonlinearities, and dealing with a priori unknown, time-dependent moving domains.

The first step of constructing approximate solutions for these highly coupled fluid-structure interaction problems can be quite challenging and delicate, and in this section, we give a broad overview of the two main techniques that one can use to approximate these highly coupled systems: (1) Rothe's method (time discretization) and (2) Lie operator splitting.

As the general setting, we consider a first-order partial differential equation in the abstract form with initial data:

$$\frac{d\phi}{dt} + A(\phi, t) = 0, \qquad \phi(0) = \phi_0, \tag{2.34}$$

where $\phi : [0, T] \to H$ is an unknown function taking values in some Hilbert space H, and $A(\phi, t)$ is a (potentially nonlinear and time-dependent) operator on the Hilbert space H. While this equation is written in the first-order form, we note that one can write higher-order equations (with higher-order derivatives in time) as a first-order system of equations by including time derivatives of the unknown function into a vector of unknown quantities, so the current abstract formulation is quite robust. In this section, we introduce the conceptual ideas behind the approximation methods used in the analysis of fluid-structure interaction in the context of the abstract equation (2.34).

2.15.1 Rothe's Method

Rothe's method is a semidiscretization scheme for constructing approximate solutions to (2.34), where the term *semidiscretization* refers to the fact that we will discretize in the time variable, but solve the spatial problem continuously (e.g., in terms of a weak formulation). To approximate solutions on a given time interval $[0, T]$ up until a predetermined final time T, we introduce a time discretization parameter $\Delta t = T/N$, where N is the number of subintervals, which will be notated as $[t_n, t_{n+1}]$ for $t_n = n\Delta t$ and $0 \leq n \leq N - 1$. For each time discretization parameter N, we will construct approximate solutions to the abstract equation ϕ_N^n for $1 \leq n \leq N$, where ϕ_N^n will be the approximate solution generated on the time interval $[t_{n-1}, t_n]$, and where the output ϕ_N^n of the approximate problem solved on one subinterval $[t_{n-1}, t_n]$ is used as the input to the approximate problem solved on the next subinterval $[t_n, t_{n+1}]$. More explicitly, Rothe's method proceeds as follows:

1. **Initialization of scheme.** Given a time discretization parameter N, initialize the scheme with the initial data so that $\phi_N^0 = \phi_0$.
2. **Generating approximate solution on a subinterval.** Inductively, for the time subinterval $[t_n, t_{n+1}]$, we are given ϕ_N^n from either the initial data (if $n = 0$) or from the result of the iteration of the method on the previous subinterval $[t_{n-1}, t_n]$. We then consider the time-discretized problem (e.g., using a backward Euler finite difference approximation):

$$\frac{\phi_N^{n+1} - \phi_N^n}{\Delta t} + A(\phi_N^{n+1}, t_{n+1}) = 0, \tag{2.35}$$

or equivalently

$$\phi_N^{n+1} = \phi_N^n - (\Delta t) A(\phi_N^{n+1}, t_{n+1}).$$

We see therefore that after time discretization, ϕ_N^{n+1} is retrieved as the solution to an approximate form of the problem, where the time variable is discretized, and we have to solve an abstract spatial problem, as ϕ_N^{n+1} appears in the operator A in this approximation and hence the method is still *implicit* (as opposed to *explicit*). Therefore, to retrieve ϕ_N^{n+1} as a solution of the approximate equation (2.35), one often will need to study the weak formulation of (2.35) (which is now only a spatially dependent problem instead of a problem in both time and space), and use abstract tools such as the Lax-Milgram method to extract a solution from this implicit problem.

3. **Construction of time-dependent approximate solution.** After the scheme is performed on all time subintervals for each N, one can construct a time-dependent approximate solution ϕ_N on the full interval $[0, T]$, by gluing together the static solutions ϕ_N^n for each of the time subintervals $[t_n, t_{n+1}]$ by using a piecewise constant function, such as

$$\phi_N(t) = \phi_N^n \text{ or } \phi_N^{n+1}, \qquad \text{for } n\Delta t \le t < (n+1)\Delta t,$$

for example, or a linear interpolation

$$\overline{\phi}_N(t) = \frac{(n+1)\Delta t - t}{\Delta t}\phi_N^n + \frac{t - n\Delta t}{\Delta t}\phi_N^{n+1}, \qquad \text{for } n\Delta t \le t \le (n+1)\Delta t.$$

Depending on the specific problem, it can sometimes be advantageous to use the piecewise linear approximate solution since it is continuous in time, whereas the piecewise constant approximation has potential jump discontinuities at every time $t = n\Delta t$. Gluing together the static approximate solutions ϕ_N^n on the subintervals $[t_n, t_{n+1}]$ into an approximate solution, which depends on both space and time, such as ϕ_N or $\overline{\phi}_N$, then gives us an approximate solution to the abstract problem (2.34) on $[0, T]$, for each time discretization parameter N.

Upon obtaining an approximate solution ϕ_N or $\overline{\phi}_N$ for each time discretization parameter N to (2.34), we can then obtain uniform bounds on ϕ_N that are independent of the parameter N, and then use these uniform bounds to pass to the limit in (2.35) to obtain a limiting solution ϕ to the original abstract equation (2.34).

We remark that one can see Rothe's method as a time discretization in the following sense. In the special finite-dimensional case where $\phi : [0, T] \to \mathbb{R}^d$ and $A : \mathbb{R}^d \to \mathbb{R}^d$ is given by a diagonalizable matrix, one can see that the solution to (2.34) is given by

$$\phi(T) = e^{AT}\phi_0.$$

Then, Rothe's method is essentially discretizing this as

$$e^{AT}\phi_0 = (e^{A\Delta t})^N \phi_0,$$

and approximating on each time subinterval the action of $e^{A\Delta t}$ through time discretization using a backward Euler approximation.

2.15.2 *The Lie Operator Splitting Method*

Next, we discuss another method for constructing approximate solutions, which works well for highly coupled problems of a multiphysics nature, such as fluid-structure interaction or stochastic partial differential equations. The Lie operator splitting method, also known as the Marchuk-Yanenko splitting scheme, takes a highly coupled problem and splits it into its different physical components via subproblems on discretized subintervals, where the subproblems are successively run on each subinterval to approximate the full coupled problem on that subinterval. Because the original coupled problem is being split into its different physical components, one has to carefully handle coupling conditions to develop a stable scheme that respects the energy inequality and gives uniformly bounded approximate solutions. This will be the main content of much of the analysis of fluid-structure interaction problems in this book.

In this subsection, we will give a general overview of the splitting method for the abstract equation (2.34), and we will consider specific examples for fluid-structure interaction systems in later chapters of the book. The Lie operator splitting is also a time discretization scheme, so we consider a time discretization parameter N, which is a positive integer defining a time step via $\Delta t = T/N$. We use this time step to discretize the entire time interval $[0, T]$ into subintervals of the form $[t_n, t_{n+1}]$ for $t_n = n\Delta t$ and nonnegative integers $n = 0, 1, ..., N-1$. The intuition behind the Lie operator splitting is to approximate the Eq. (2.34) on each time subinterval $[t_n, t_{n+1}]$ by the action of successive subproblems solved on that subinterval $[t_n, t_{n+1}]$, with subproblem operators A_i that altogether approximate the operator A appearing in the original abstract problem (2.34).

More explicitly, assume that we can write the operator A in (2.34) as $A = A_1 + A_2$, and the goal is to use an operator splitting scheme to construct approximate solutions ϕ_N^n on each time interval $[t_n, t_{n+1}]$ for a fixed time discretization parameter N. To do this, we consider a fixed (but arbitrary) time discretization parameter N and then initialize the scheme by setting $\phi_N^0 = \phi_0$. Then, for each $n = 0, \dots, N-1$, we inductively generate the static approximate functions $\phi_N^{n+\frac{1}{2}}$ and ϕ_N^{n+1} on the time subinterval $[t_n, t_{n+1}]$, given the result ϕ_N^n from the previous subinterval $[t_{n-1}, t_n]$ (or from the initial data if $n = 0$).

More specifically, to generate $\phi_N^{n+\frac{1}{2}}$ from ϕ_N^n, we solve

$$\textbf{Subproblem1}: \begin{cases} \dfrac{d\phi_1}{dt} + A_1(\phi_1, t) = 0 & \text{on } [t_n, t_{n+1}] \\ \phi_1(t_n) = \phi_N^n \end{cases}.$$

Then set $\phi_N^{n+\frac{1}{2}} = \phi_1(t_{n+1})$. To generate ϕ_N^{n+1} from $\phi_N^{n+\frac{1}{2}}$, we solve

$$\textbf{Subproblem2}: \begin{cases} \dfrac{d\phi_2}{dt} + A_2(\phi_2, t) = 0 & \text{on } [t_n, t_{n+1}] \\ \phi_2(t_n) = \phi_N^{n+\frac{1}{2}} \end{cases}.$$

Then set $\phi_N^{n+1} = \phi_2(t_{n+1})$.

Notice that the subproblems communicate only via the initial data, i.e., the solution of Subproblem 1 is the initial data for the solution of Subproblem 2, while the initial data for Subproblem 1 is the solution of Subproblem 2 from the previous time step.

Formally, if one adds the action of these two subproblems together, we recover the action of the full abstract problem (2.34) approximately, since $A = A_1 + A_2$. We emphasize that this formulation of the Lie operator splitting method that we are considering is continuous in time without time discretization, but as we will discuss in the next subsection, it is possible to combine the time discretization of Rothe's method with the ideas of the Lie operator splitting to obtain a semidiscrete operator splitting method.

Even though $A = A_1 + A_2$ exactly, there is still error that is accrued by the splitting of the operator A. For example, consider even the simple finite-dimensional context of $\phi : [0, T] \to \mathbb{R}^d$ and a diagonalizable matrix defining a linear transformation $A : \mathbb{R}^d \to \mathbb{R}^d$, where $A = A_1 + A_2$ for two linear transformations $A_1, A_2 : \mathbb{R}^d \to \mathbb{R}^d$ that are both also diagonalizable. Then, the solution at the final time T is given by

$$\phi(T) = e^{AT}\phi_0 = (e^{A\Delta t})^N \phi_0,$$

whereas the approximate solution $\tilde{\phi}$ obtained via the Lie operator splitting would be

$$\tilde{\phi}(T) = (e^{A_2\Delta t}e^{A_1\Delta t})^N \phi_0.$$

Unless A_1 and A_2 commute (e.g., if they are simultaneously diagonalizable), we can have that $\phi(T)$ and $\tilde{\phi}(T)$ can be different. Even this simple finite-dimensional example shows that splitting the problem necessarily gives rise to the possibility of introducing additional, splitting error depending on how the splitting is accomplished.

2.15.3 *Semidiscretized Operator Splitting for FSI*

In the study of fluid-structure interaction in later chapters of this book, it will be useful, especially for highly coupled fluid-structure systems, to use a combination of time discretization by Rothe's method and an operator splitting to separate the contributions of the fluid and structure dynamics to the fully coupled dynamics. This gives rise to a semidiscretization scheme, where for each discretization parameter N and associated time step $\Delta t = T/N$, we discretize the interval $[0, T]$ into N subintervals $[t_n, t_{n+1}]$ of length Δt for $t_n = n\Delta t$, and then run the following scheme, which we will state for the abstract equation (2.34):

1. **Initialize the scheme.** Fix a discretization parameter N, and define $\phi_N^0 = \phi_0$.

2. **Run through the subproblems on each subinterval.** Given ϕ_N^n (either from the initial data or from the result of the previous subinterval), we approximate (2.34) on $[t_n, t_{n+1}]$ by solving a structure subproblem and a fluid subproblem, which split the contributions of the structure and the fluid in such a way that respects the balance of energy in the system. We also discretize in time by using a backward Euler approximation of the time derivative to obtain a semidiscrete scheme. We decompose the full fluid-structure operator $A = A_1 + A_2$, and first solve one of the subproblems, for example, a *structure subproblem*, for $\phi_N^{n+\frac{1}{2}}$:
$$\frac{\phi_N^{n+\frac{1}{2}} - \phi_N^n}{\Delta t} + A_1(\phi_N^{n+\frac{1}{2}}, t_{n+1}) = 0,$$
where A_1 is an operator on the Hilbert space H accounting for the contributions of the structure dynamics on the fully coupled problem. Then we solve a fluid subproblem for ϕ_N^{n+1}:
$$\frac{\phi_N^{n+1} - \phi_N^{n+\frac{1}{2}}}{\Delta t} + A_2(\phi_N^{n+1}, t_{n+1}) = 0,$$
where A_2 is an operator on the Hilbert space H accounting for the contributions of the fluid dynamics. We solve each of the subproblems once on each subinterval $[t_n, t_{n+1}]$, and move on to the next subinterval. We iterate the structure and fluid subproblems on each of the N subintervals, to obtain a sequence of approximate functions $\phi_N^{n+\frac{i}{2}}$ for positive integers n and $i = 0, 1$ such that $0 \le n + \frac{i}{2} \le N$, which depend only on the spatial variable and are defined at discrete points in time.
3. **Construction of approximate solutions.** Finally, we can use a piecewise constant function or a piecewise linear interpolation to define approximate solutions ϕ_N for each time discretization parameter N, which are defined on the whole time interval $[0, T]$. For example, we can define piecewise constant approximate solutions ϕ_N (which are potentially discontinuous at each $t_n = n\Delta t$) of the form
$$\phi_N(t) = \phi_N^n, \phi_N^{n+\frac{1}{2}}, \text{ or } \phi_N^{n+1}, \qquad \text{for } n\Delta t \le t < (n+1)\Delta t,$$
or we can define an approximate solution $\overline{\phi}_N$ that is a continuous linear interpolation:
$$\overline{\phi}_N(t) = \frac{(n+1)\Delta t - t}{\Delta t}\phi_N^n + \frac{t - n\Delta t}{\Delta t}\phi_N^{n+1}, \qquad \text{for } n\Delta t \le t \le (n+1)\Delta t.$$

Upon obtaining approximate solutions ϕ_N or $\overline{\phi}_N$, we would like to pass to the limit as $N \to \infty$ in these approximate solutions provided that we can obtain uniform

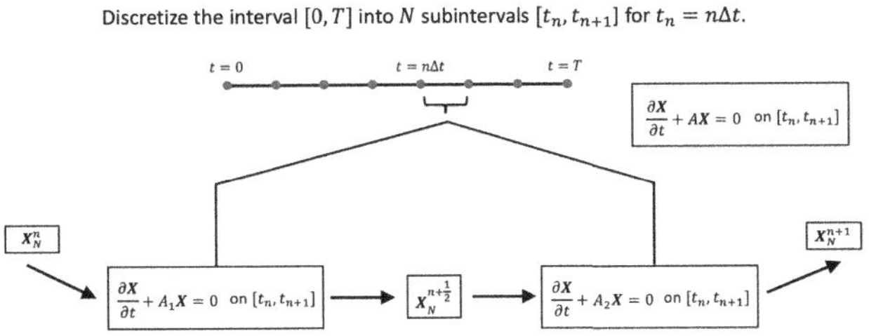

Fig. 2.1 A visual representation of the Lie-Trotter operator splitting method applied to an abstract partial differential equation

bounds on the approximate solutions that are *independent of the time discretization parameter* N. Constructing the splitting scheme so as to have uniform bounds on the resulting approximate solutions independently of N is highly nontrivial, especially for highly coupled systems, as it requires splitting apart the coupling conditions between the various multiphysical problems (e.g., the fluid and structure problems) in a way that properly maintains the total energy balance of the system. In our study of fluid-structure interaction in later chapters, we will see many examples of how it is essential to perform the splitting in a specific way so as to maintain stability of the resulting scheme after time discretization and after splitting the coupled multiphysical dynamics into separate subproblems (Fig. 2.1).

Chapter 3
Probabilistic Preliminaries

In this chapter, we will focus on probabilistic background, with an emphasis on random variables that take values in Banach spaces. This will be important since stochastic fluid-structure systems, which are discussed in Chaps. 8 and 9, involve random quantities, such as the fluid/structure velocity and the structure displacement, which will now be random functions due to the stochasticity in the system. We will summarize results about random variables taking values in Banach spaces and discuss the three modes of probabilistic convergence of such random variables, which will play a key role in the analysis of stochastic FSI: *convergence in probability, weak convergence (or convergence in law or distribution), and convergence almost surely.* We will discuss the relationships between these modes of convergence and review classical theorems that will be important in our analysis later. In Chap. 9, we will need further results that are generalizations of the results and definitions introduced in this chapter. We will introduce these concepts progressively as we proceed in Chap. 9. In the current chapter, we restrict ourselves to developing a vocabulary for discussing stochasticity in random systems and provide a review of the concept of one-dimensional white noise and spacetime white noise, which are prototypical examples of random noise considered in stochastic PDEs. We will conclude by discussing the notion of stochastic integration against these types of random noise, which will be important for quantifying the effects of randomness on PDE dynamics and for rigorously defining the notion of a solution to a system of stochastic PDEs.

Though we will provide a brief review of many of the foundational results from probability theory that will be needed later in this book, we assume that the reader is acquainted with abstract probability spaces and is comfortable with real-valued random variables. Though we will review the basics of probability theory more generally for random variables taking values in Banach spaces, we note that there are many parallels with the case of real-valued random variables. We refer the reader to texts such as [128, 201] for a detailed account of measure theoretic probability theory, and we assume basic familiarity with the general theory of

S. Čanić et al., *Deterministic and Stochastic Fluid-Structure Interaction*, Advances in Mathematical Fluid Mechanics, https://doi.org/10.1007/978-3-032-00898-5_3

abstract measure theoretic probability theory and real-valued random variables in the remainder of this chapter. For more information about Brownian motion and stochastic integration, we refer the reader to [104, 109, 201, 290], and for more information about stochastic PDEs, we refer the reader to the introductory book [109], the following classic book on stochastic analysis in infinite-dimensional spaces [104], and the references [178, 346]. For more details about random variables taking values in general infinite-dimensional Banach spaces, we refer the reader to the reference [104], and for random variables taking values in more general topological spaces, we refer the reader to [52] and the references therein.

3.1 Probability Spaces and Random Variables

Recall that a probability space is an ordered triple $(\Omega, \mathcal{F}, \mathbb{P})$ consisting of a set of outcomes Ω, a sigma algebra of measurable sets $\mathcal{F}$ (which can be thought of as the collection of events), and a probability measure $\mathbb{P}$ on the measurable sets (events) in $\mathcal{F}$. Given a probability space, we can talk about the probability of an event $A \in \mathcal{F}$ happening, and we denote this probability by $\mathbb{P}(A)$.

We will now state the $\pi - \lambda$ theorem, where we define a π-system and a λ-system as follows:

- A π-system is a collection of subsets of Ω that is closed under finite intersections.
- A λ-system (also called a Dynkin system) is a collection $\mathcal{L}$ of subsets of Ω such that:

  1. $\Omega \in \mathcal{L}$,
  2. If $A \in \mathcal{L}$, then $A^c \in \mathcal{L}$,
  3. If $A_1, A_2, \ldots \in \mathcal{L}$ are increasing nested sets, then $\bigcup_{n=1}^{\infty} A_n \in \mathcal{L}$.

We now state the $\pi - \lambda$ theorem; see, e.g., Theorem 3.2 in [38].

Theorem 3.1.1 **($\pi - \lambda$ Theorem)** *Let $\mathcal{P}$ be a π-system of subsets of Ω, and let $\mathcal{L}$ be a λ-system of subsets of Ω such that $\mathcal{P} \subseteq \mathcal{L}$. Then*

$$\sigma(\mathcal{P}) \subseteq \mathcal{L},$$

where $\sigma(\mathcal{P})$ denotes the smallest σ-algebra containing $\mathcal{P}$.

Given a probability space $(\Omega, \mathcal{F}, \mathbb{P})$, there can be random quantities or random functions that exist on this probability space, which we may be interested in observing or measuring. This leads naturally to the concept of a random variable on a topological space S equipped with σ-algebra $\mathcal{A}$.

Definition 3.1.1 (Random Variable) Consider two measurable spaces $(\Omega, \mathcal{F})$ and $(S, \mathcal{A})$. An S-**valued random variable** $X(\omega)$ is a measurable map $X(\omega)$: $(\Omega, \mathcal{F}) \to (S, \mathcal{A})$ i.e., $X^{-1}(A) \in \mathcal{F}$ for any $A \in \mathcal{A}$.

For simplicity of notation, we usually omit the dependence of a random variable on the outcome $\omega \in \Omega$, and hence, we will denote a random variable by X instead of $X(\omega)$, whenever it is clear from context that X is random. In the case where $S = \mathbb{R}$, this coincides with the usual definition of a real-valued random variable. Given a random variable $X(\omega) : \Omega \to S$, we define the **law (or distribution) of** X to be the probability measure μ_X on $(S, \mathcal{A})$ defined by

$$\mu_X(A) = \mathbb{P}(X(\omega) \in A), \qquad \text{for } A \in \mathcal{A}.$$

In the analysis of stochastic partial differential equations, for example, in Chaps. 8 and 9, we will often consider Banach space-valued random variables, and it will thus be important for us to quantify the boundedness properties of random variables taking values in Banach spaces. A Banach space B-valued random variable $X(\omega) : \Omega \to B$ **belongs to the space** $L^p(\Omega; B)$ if one of the following is true:

$$\begin{cases} \mathbb{E}\Big(\|X(\omega)\|_B^p\Big) < \infty, & \text{if } \ 1 \le p < \infty, \\ \|X(\omega)\|_B < C \text{ almost surely for some constant } C \ge 0, & \text{if } \ p = \infty. \end{cases} \tag{3.1}$$

Here, the operation $\mathbb{E}$ of taking an expectation involves integrating over the probability space Ω with respect to the probability measure $\mathbb{P}$.

Knowing that a random variable $X(\omega)$ is in $L^p(\Omega; B)$ gives important information about how likely it is for X to attain large values. This is quantified through **Chebyshev's inequality**, which states that if $X \in L^p(\Omega; B)$ for $1 \le p < \infty$, then

$$\mathbb{P}(\|X\|_B \ge \lambda) \le \frac{\mathbb{E}(\|X\|_B^p)}{\lambda^p}, \qquad \text{for all } \lambda > 0. \tag{3.2}$$

See also Theorem 1.6.4 in [128] for a more general form of Chebyshev's inequality.

A σ-algebra of subsets of Ω is a collection $\mathcal{F}$ of subsets of Ω that contains the empty set $\varnothing$ and is closed under complements and countable unions. Sets in $\mathcal{F}$ are sometimes called $\mathcal{F}$-measurable.

Finally, we mention the concept of **independence of random variables**, in the context of random variables taking values in Banach spaces.

Definition 3.1.2 (Independence of Random Variables) Let X and Y be two B-valued random variables, where B is a Banach space. Then, X and Y are **independent** if for all $A_1, A_2 \in \text{Borel}(B)$,

$$\mathbb{P}(X(\omega) \in A_1, Y(\omega) \in A_2) = \mathbb{P}(X(\omega) \in A_1)\mathbb{P}(Y(\omega) \in A_2).$$

3.2 Probabilistic Convergence

Next, we discuss convergence of random variables and review standard results about probabilistic convergence in the context of random variables taking values in Banach spaces. We discuss convergence in probability, weak convergence, and convergence almost surely and state several fundamental results that relate these different types of convergence with each other and that will be useful in the future chapters about stochastic FSI. To describe these modes of convergence, we consider a general context in which the random variables we consider take values in a topological space S, equipped with Borel σ-algebra $\mathcal{A}$.

3.2.1 *Convergence Almost Surely*

Convergence almost surely is the probabilistic analogue of pointwise convergence. Suppose that $\{X_n\}_{n=1}^{\infty}$ and X are S-valued random variables that are all defined on the same probability space $(\Omega, \mathcal{F}, \mathbb{P})$. Then, X_n **converges to** X **almost surely** if

$$\mathbb{P}\left(\{\omega : \lim_{n\to\infty} X_n(\omega) = X(\omega)\}\right) = 1,$$

or, **equivalently**, if

> $X_n \to X$ as $n \to \infty$ for all outcomes ω in some measurable set $\Omega_0 \in \mathcal{F}$ with $\Omega_0 \subset \Omega$, where $\mathbb{P}(\Omega_0) = 1$.

3.2.2 *Convergence in Probability*

Next, we describe convergence in probability of S-valued random variables, which states informally that random variables are arbitrarily close in value to a limiting random variable with arbitrarily high probabilities as $n \to \infty$.

Given a sequence $\{X_n\}_{n=1}^{\infty}$ of S-valued random variables and an S-valued random variable X, all defined on the same probability space, we say that X_n **converges in probability** to X if, for every neighborhood B_0 of 0, we have

$$\lim_{n\to\infty} \mathbb{P}(X_n \notin X + B_0) = 0.$$

We state the following well-known connection between convergence almost surely and convergence in probability; see Theorem 2.3.2 in [128]. We note that the proof uses a well-known measure theoretic result known as the Borel-Cantelli lemma stated below; see Theorem 2.3.1 in [128].

Lemma 3.2.1 (Borel-Cantelli) *Letting* $(A_n)_{n=1}^{\infty} \subseteq \mathcal{F}$ *s.t.* $\sum_{n\in\mathbb{N}} \mathbb{P}(A_n) < \infty$, *then* $\mathbb{P}(\{\omega \in \Omega : \omega \in A_n \text{ for infinitely many } n\}) = 0$.

Proposition 3.2.1 (Convergence in Probability) *Let* $\{X_n\}_{n=1}^{\infty}$ *be a sequence of* S*-valued random variables such that* X_n ***converges in probability*** *to a limiting* S*-valued random variable* X. *Then, there exists a* ***subsequence*** *such that*

$$X_{n_k} \to X \quad \textbf{almost surely} \text{ as } k \to \infty.$$

We also state the following result on convergence of Banach space-valued random variables in probability, which shows that independence of random variables is "preserved" under convergence of random variables in probability. This result is a simple generalization of a classical result; see, e.g., Lemma 8.3 in [62].

Lemma 3.2.2 *Assume that* $(\Omega, \mathcal{F}, \mathbb{P})$ *is a probability space and* B_1 *and* B_2 *are Banach spaces, and that* $(X_n)_{n\in\mathbb{N}}$ *and* $(Y_n)_{n\in\mathbb{N}}$ *are two sequences of random variables defined on* $(\Omega, \mathcal{F}, \mathbb{P})$ *taking values in* B_1 *and* B_2 *respectively. Assume that* $X_n \to X$ *in* B_1 *and* $Y_n \to Y$ *and* B_2 *in probability. If* X_n *and* Y_n *are independent random variables for every* $n \geq 1$, *then* X *and* Y *are also independent.*

3.2.3 Weak Convergence

Next, we will discuss weak convergence of random variables, which describes convergence of the expected values of observable quantities. For further discussion of weak convergence, we refer the reader to Section 3.2 in [128] for example.

A sequence of **random variables** $\{X_n(\omega)\}_{n=1}^{\infty}$ taking values in a topological space S equipped with Borel σ-algebra $\mathcal{A}$ **converges weakly (or equivalently, converges in law or distribution)** to a limiting random variable $X(\omega)$ if for all continuous bounded functions $f : S \to \mathbb{R}$,

$$\lim_{n\to\infty} \mathbb{E}[f(X_n)] = \mathbb{E}[f(X)].$$

Here, we can intuitively think of the bounded continuous function $f : S \to \mathbb{R}$ as some observable that depends on the random value of X_n or X.

We can also define the notion of **weak convergence for probability measures**, and we will state this specifically for probability measures on Banach spaces. Suppose that $\{\mu_n\}_{n=1}^{\infty}$ and μ are probability measures on a common topological space $(S, \mathcal{A})$. Then, the **measures** μ_n **converge weakly** to μ if for all continuous bounded functions $f : S \to \mathbb{R}$,

$$\lim_{n\to\infty} \int_S f(x) d\mu_n(x) = \int_S f(x) d\mu(x).$$

Note that if S-valued random variables $\{X_n\}_{n=1}^{\infty}$ converge weakly to X, then their laws $\{\mu_{X_n}\}_{n=1}^{\infty}$ also converge weakly (as probability measures on S) to μ_X.

We can also formulate weak convergence equivalently in terms of the topology of the space in which the random variables take values, which is the content of the following **Portmanteau theorem**.

Theorem 3.2.1 (Portmanteau Theorem) *Let S be a metric space and $\mathcal{A}$ be its Borel σ-algebra. Let $(\mu_n)_{n=1}^{\infty}$ and μ be probability measures on $(S, \mathcal{A})$. Then the following are equivalent:*

1. *μ_n converges weakly to μ.*
2. *$\lim_n \mu_n(A) = \mu(A)$ for all sets A for which $\mu(\partial A) = 0$.*
3. *$\limsup \mu_n(C) \le \mu(C)$ for all closed sets C.*
4. *$\liminf \mu_n(U) \ge \mu(U)$ for all open sets U.*
5. *$\int f d\mu_n \to \int f d\mu$ for all bounded, Lipschitz continuous f on S.*

Next, we note a useful criterion for determining whether two sequences of random variables on the same probability space have the same weak limit, given by Theorem 3.1 in [39].

Proposition 3.2.2 *Let B be a Banach space and let $\{X_n\}_{n=1}^{\infty}$ and $\{Y_n\}_{n=1}^{\infty}$ be sequences of B-valued random variables, satisfying*

$$\|X_n - Y_n\|_B \to 0, \qquad \text{almost surely as } n \to \infty.$$

Then, if X_n converges weakly to some B-valued random variable X as $n \to \infty$, then Y_n also converges weakly and has the same limit X as $n \to \infty$.

Weak convergence is a particularly nice form of probabilistic convergence because there is a well-known criterion, known as **tightness**, which will allow us to extract a weakly convergent subsequence from a tight sequence of random variables; see Theorem 3.2.2.

Definition 3.2.1 (Tightness of Probability Measures) Suppose that S is a topological space and that $\mathcal{A}$ is a σ-algebra containing the topology. A sequence of probability measures μ_n on $(S, \mathcal{A})$ is tight if for any $\epsilon > 0$, there exists a compact set $K_\epsilon \in \mathcal{A}$ depending on ϵ such that

$$\mu_n(K_\epsilon) > 1 - \epsilon, \qquad \text{for all } n.$$

A sequence of $(S, \mathcal{A})$-valued random variables $\{X_n\}_{n=1}^{\infty}$ is **tight** if their laws $\mathbb{P}(X_n \in \cdot)$ are tight.

Classical Results About Convergence

We will next discuss a few classical results about different modes of convergence. We start with Prokhorov's theorem that relates the concept of tightness of laws to weak convergence of laws in separable and metrizable spaces. Then we will give a classical almost sure representation theorem, which relates weak convergence and convergence almost surely (see its variant in Sect. 3.2.4). Finally, we will discuss the Gyöngy-Krylov lemma, which relates weak convergence of appropriate joint laws of random variables to convergence in probability, and hence to almost sure convergence along a subsequence.

Note that tightness (see Definition 3.2.1) is a condition that essentially keeps the probabilistic mass of random variables from escaping out to infinity, by requiring the probabilistic mass of a collection of probability measures to be uniformly contained in compact sets. If we specialize to the case of *separable Banach spaces* or even more generally to *separable, complete metric spaces*, then tightness allows us to extract weakly convergent subsequences from tight sequences of random variables. This is the content of the following classical probability result, which is part of a more general result known as Prokhorov's theorem, stated, for example, in Theorem 3.2.13 in [128] (see also Proposition 6.1 in [239]).

Theorem 3.2.2 (Prokhorov's Theorem) *The sequence $\{\mu_n\}_{n=1}^{\infty}$ is a tight sequence of probability measures on a separable Banach space B, if and only if there exists a subsequence μ_{n_k} and a probability measure μ on B such that μ_{n_k} converges weakly to μ as $k \to \infty$.*

We make some additional remarks about tightness, in terms of the connection between uniform boundedness of random variables and tightness, which will be essential in the study of stochastic PDEs. We first establish the following corollary to the previous theorem, which states that uniform boundedness in expectation of *real-valued random variables* is sufficient to establish tightness.

Corollary 3.2.1 *Suppose that $\{X_n\}_{n=1}^{\infty}$ is a sequence of real-valued random variables such that*

$$\mathbb{E}(|X_n|^p) < C, \qquad \textit{for all } n, \textit{ for some } 1 \le p < \infty.$$

Then, the sequence $\{X_n\}_{n=1}^{\infty}$ is tight.

Proof Choose $\epsilon > 0$. Choose λ sufficiently large so that $\dfrac{C}{\lambda^p} < \epsilon$. Then, by Chebyshev's inequality, if we set $K_\epsilon = [-\lambda, \lambda]$, we have that

$$\mathbb{P}(\{\omega : X_n(\omega) \in K_\epsilon\}) = \mathbb{P}(X_n \in K_\epsilon) > 1 - \epsilon, \qquad \text{for all } n.$$

Since K_ϵ is a closed and bounded subset of $\mathbb{R}$, it is a compact set in $\mathbb{R}$, which establishes that $\{X_n\}_{n=1}^{\infty}$ is a tight sequence of real-valued random variables, which establishes the result. □

Remark 3.2.1 However, we note that a similar argument does not work for random variables that take values in general (infinite-dimensional) Banach spaces equipped with strong topology. In particular, if $\{X_n\}_{n=1}^{\infty}$ is a sequence of B-valued random variables where B is a separable Banach space, then having a uniform bound of the form

$$\mathbb{E}(\|X_n\|_B^p) < C, \qquad \text{for all } n, \tag{3.3}$$

does *not* imply that the random variables $\{X_n\}_{n=1}^{\infty}$ are tight. The reason the argument above involving Chebyshev's inequality works for the case of real-valued random variables is because closed and bounded subsets of $\mathbb{R}$ are compact in $\mathbb{R}$. However, because general Banach spaces can be infinite dimensional, closed and bounded subsets of Banach spaces are no longer necessarily compact subsets of the Banach space. In particular, given the uniform bound (3.3) above, we can find a closed ball $\overline{B(R_\epsilon)}$ of radius R_ϵ in B for which

$$\mathbb{P}(X_n \in \overline{B(R_\epsilon)}) > 1 - \epsilon, \qquad \text{for all } n.$$

However, this closed ball $\overline{B(R_\epsilon)}$ is not necessarily compact in B equipped with strong topology, which prevents us from obtaining that the random variables $\{X_n\}_{n=1}^{\infty}$ are tight. Hence, in order to deduce tightness from the uniform bound (3.3), one would need to embed the Banach space B compactly into another Banach space B_0. Then, the closure in B_0 of the image of $\overline{B(R_\epsilon)}$ under this compact embedding from B to B_0 would be a compact subset of B_0, in which case we can show that the random variables $\{X_n\}_{n=1}^{\infty}$ considered instead as B_0-valued random variables are tight.

Next, we discuss the classical Skorokhod representation theorem, which will give a way of "upgrading" weak convergence of random variables to almost sure convergence of random variables, at the expense of moving to a potentially different probability space. However, moving to a different probability space can be done while retaining the laws of the random variables, so that even on the new probability space, we can make statements about the distributions of the random variables on the initial probability space. The statement of the Skorokhod representation theorem (see, e.g., [308] and textbooks such as [39, 332]) is as follows.

Theorem 3.2.3 (Skorokhod Representation Theorem) *Suppose that $\{\mu_n\}_{n=1}^{\infty}$ is a sequence of probability measures on a complete and separable metric space S (e.g., a separable Banach space) that converges weakly to a limiting probability measure μ on S. Then, there exist a probability space $(\tilde{\Omega}, \tilde{\mathcal{F}}, \tilde{\mathbb{P}})$ and S-valued random variables $\{\tilde{X}_n\}_{n=1}^{\infty}$ and $\tilde{X}$ on $(\tilde{\Omega}, \tilde{\mathcal{F}}, \tilde{\mathbb{P}})$ such that*

$$\tilde{X}_n \to \tilde{X}, \qquad \tilde{\mathbb{P}}\text{-almost surely},$$

and the laws of $\tilde{X}_n$ and $\tilde{X}$ for all positive integers n are μ_n and μ, respectively.

However, in applications, we will require a variant of this representation theorem stated later in Theorem 3.2.5.

Example 3.2.1 Since the Skorokhod representation theorem may seem abstract initially, we give a concrete example to demonstrate how this theorem works. Consider the probability space $\Omega = [0, 1]$, where $\mathcal{F}$ is the set of Borel measurable sets and $\mathbb{P}$ is the usual uniform measure on $[0, 1]$ (Lebesgue measure dx). For each k and n, define the random variables

$$X_{n,k}(\omega) = 1_{[(k-1)/n, k/n]}(\omega), \qquad \text{for } \omega \in [0, 1],$$

where $n \geq 1$ and $1 \leq k \leq n$. Then, consider the sequence of random variables $\{X_n\}_{n=1}^{\infty}$, where if N is the largest integer such that $1 + 2 + 3 + \ldots + N < n$ and if $k = n - (1 + 2 + 3 + \ldots + N)$, then

$$X_n = X_{N+1,k}.$$

Explicitly, the sequence we are considering is

$$X_{1,1}, X_{2,1}, X_{2,2}, X_{3,1}, X_{3,2}, X_{3,3}, X_{4,1}, X_{4,2}, X_{4,3}, X_{4,4}, \ldots.$$

One can verify that the sequence $\{X_n\}_{n=1}^{\infty}$ constructed above converges weakly to the zero random variable X, which is zero for all $\omega \in [0, 1]$, but X_n does not converge almost surely.

However, by the Skorokhod representation theorem, we can essentially rearrange how outcomes are mapped to realizations without changing laws to recover almost sure convergence. In the notation of the Skorokhod representation theorem, we will let the tilde probability space $(\tilde{\Omega}, \tilde{\mathcal{F}}, \tilde{\mathbb{P}})$ be the same as the initial probability space $(\Omega, \mathcal{F}, \mathbb{P})$. (However, we remark that in general, the probability space $(\tilde{\Omega}, \tilde{\mathcal{F}}, \tilde{\mathbb{P}})$ can be different from $(\Omega, \mathcal{F}, \mathbb{P})$, but we do not need to do this for this particular example.) Then, we will define $\tilde{X}_n$ by letting N as before be the largest positive integer such that $1 + 2 + \ldots + N < n$, and then we will define

$$\tilde{X}_n = X_{N+1,1}.$$

Thus, the sequence of random variables $\{\tilde{X}_n\}_{n=1}^{\infty}$ is given by

$$X_{1,1}, X_{2,1}, X_{2,1}, X_{3,1}, X_{3,1}, X_{3,1}, X_{4,1}, X_{4,1}, X_{4,1}, X_{4,1}, \ldots.$$

Note that $\tilde{X}_n$ has the same law as X_n for all n, but now we have that $\tilde{X}_n$ converges almost surely to the zero random variable $\tilde{X}$, whereas the original sequence $\{X_n\}_{n=1}^{\infty}$ converges weakly to the zero random variable X, but does not converge almost surely.

Next, we will discuss a result that uses weak convergence in order to show convergence in probability of random variables. This result is known as the Gyöngy-Krylov lemma, and it will be closely connected to the Skorokhod representation theorem. In stochastic PDE problems, what often happens is that one uses the

Skorokhod representation theorem to upgrade weak convergence to almost sure convergence, but the challenge is that the random functions are now defined on an alternate, "tilde" probability space. One often wants to see if the random solution that is constructed to a stochastic problem can be shown to exist on the *original probability space*, and to do this, one usually invokes a standard Gyöngy-Krylov argument that uses a uniqueness result for the stochastic problem to transfer the random functions on the tilde probability space back to the initial probability space. In Chap. 8, we will discuss the full so-called Gyöngy-Krylov *diagonal argument* in the context of fully coupled stochastic FSI problems, but here, we state the relevant lemma for future reference. This lemma can be thought of as a **criterion for showing that random variables converge in probability**, which relies on the weak convergence of the joint laws of ordered pairs of these random variables. For further discussion about the Gyöngy-Krylov lemma and its applications, see Lemma 1.1 in [177] and Proposition 6.3 in [239].

To state the Gyöngy-Krylov lemma, we introduce the following notation. Let $\{X_n\}_{n=1}^{\infty}$ be a sequence of random variables taking values in a separable Banach space B. We will use $\nu_{m,n}$ to denote the joint law of (X_m, X_n), which is a probability measure on $B \times B$ defined by

$$\nu_{m,n}(U \times V) = \mathbb{P}(X_m \in U, X_n \in V), \qquad \text{for } U, V \in \text{Borel}(B).$$

Then the following theorem holds.

Theorem 3.2.4 (Gyöngy-Krylov Lemma) *Let $\{X_n\}_{n=1}^{\infty}$ be a sequence of random variables taking values in a separable Banach space B. The sequence*

$$X_n \text{ converges in probability to some } B\text{-valued random variable } X$$

if and only if for any two subsequences $\{X_{l_k}\}_{k=1}^{\infty}$, $\{X_{m_k}\}_{k=1}^{\infty}$ of $\{X_n\}_{n=1}^{\infty}$, the sequence of joint probability measures $\{\nu_{l_k,m_k}\}_{k=1}^{\infty}$ converges weakly to a probability measure ν on $B \times B$ that is supported on the diagonal, i.e., ν is such that

$$\nu(\{(x, y) \in B \times B : x = y\}) = 1. \tag{3.4}$$

Condition (3.4) is known as the *diagonal condition*.

3.2.4 More General Almost Sure Representation Theorems

So far, we have worked with random variables taking values in separable Banach spaces, and the version of the Skorokhod representation theorem that we have stated in Theorem 3.2.3 is stated in this context, for random variables taking values in separable Banach spaces. However, the restriction of considering separable Banach spaces can be inconvenient in many stochastic PDE problems, as it prohibits the consideration of random variables taking values in spaces with more general topologies, such as spaces of weakly continuous functions or reflexive Banach spaces with the weak topology instead of the strong topology of convergence.

In this section, we give a brief exposition of more general non-metrizable topological spaces known as *Jakubowski* spaces and discuss the formulation of the Skorokhod representation theorem in this context. We will use this generalized Skorokhod representation theorem in Sect. 8.7.2 to handle random variables taking values in function spaces of weakly continuous function-valued processes.

We begin by discussing the definition of a Jakubowski space, which is a topological space generalizing Banach spaces, introduced in [197]. See also the exposition in Appendix B of both [309] and [310] for a summary of properties of Jakubowski spaces.

Definition 3.2.2 (Jakubowski Space) A topological space $(S, \mathcal{A})$ is a *Jakubowski* space if there exists a countable family of continuous functions $\{f_i\}_{i \in I}$ with $f_i : S \to [-1, 1]$ that *separate* points of S in the sense that for each $x, y \in S$, there exists some f_i such that $f_i(x) \neq f_i(y)$.

Remark 3.2.2 Alternatively, as stated explicitly in [197], one can think of this countable family $\{f_i\}$ of continuous separating functions as defining a map $F : S \to [-1, 1]^I$, which is *injective* (which reflects the separating property) and *continuous*, where the ith component of $F(x)$ for $x \in S$ is given by $f_i(x)$.

Remark 3.2.3 Any Polish space, i.e., a separable completely metrizable topological space, such as separable Banach spaces equipped with the weak topology and spaces of weakly continuous functions with values in a separable Banach space, belongs to this class.

Theorem 3.2.5 (Almost Sure Representation Theorem in Jakubowski Spaces [197]) *Let $(\Omega, \mathcal{F}, \mathbb{P})$ be a probability space and let $(S, \mathcal{A})$ be a* ***Jakubowski space****. Let $\mathcal{G}$ be the σ-algebra generated by the family $\{f_i\}$. Let $\{\mu_n\}$ be a tight sequence of probability measures on $(S, \mathcal{G})$. Then there exist a subsequence $\{\mu_{n_k}\}$, a probability space $(\tilde{\Omega}, \tilde{\mathcal{F}}, \tilde{\mathbb{P}})$, and S-valued random variables $\{X_k\}$ and X defined on $(\tilde{\Omega}, \tilde{\mathcal{F}}, \tilde{\mathbb{P}})$ such that:*

1. *Law of X_k is μ_{n_k} for every $k \in \mathbb{N}$*
2. *$(X_k) \to X$, $\tilde{\mathbb{P}}$-almost surely*

Moreover, the probability space $(\tilde{\Omega}, \tilde{\mathcal{F}}, \tilde{\mathbb{P}})$ is such that $\tilde{\Omega} = [0, 1)$, $\tilde{\mathcal{F}}$ is the Borel algebra on $\tilde{\Omega}$ and $\tilde{\mathbb{P}}$ is the Lebesgue measure on $\tilde{\Omega}$.

3.3 Stochastic Processes and Stochastic Integration

Next, we discuss the types of random noise that will be relevant to our analysis of stochastic FSI. We begin with one-dimensional white noise, which is a type of random noise in time whose intensity at every point in time is independent from its intensity at every other point in time. We will formally denote the intensity of a one-dimensional white noise at a given time t by $\dot{W}(t)$, so that we can express this independence property of white noise formally as

$$\mathbb{E}[\dot{W}(s)\dot{W}(t)] = \delta_0(t - s), \qquad s, t \geq 0,$$

where δ_0 is the Dirac delta function.

As the notation suggests, we can think of white noise as the formal time derivative of a stochastic process $(W(t))_{t\geq 0}$, known as a one-dimensional Brownian motion. In this section, we will review basic properties of stochastic processes and review the definition and properties of a one-dimensional Brownian motion. We will then use these ideas to review the construction of the stochastic integral against one-dimensional white noise, which is the Itô integral.

3.3.1 Stochastic Processes

While we previously discussed random variables, the time-dependent nature of PDE dynamics requires us to consider random quantities that evolve over time. To address this, we introduce the concept of a *stochastic process*, or a random process in time. Stochastic processes can be indexed by discrete time steps (such as a random walk) or can be indexed by continuous time (such as a Brownian motion). We will focus on the case of stochastic processes indexed by continuous time, since this will be the case that is most relevant to our analysis later. Consider a measurable space $(S, \mathcal{A})$. In this case, an S-valued **stochastic process** X on a probability space $(\Omega, \mathcal{F}, \mathbb{P})$ is a collection of random variables $(X_t)_{t\geq 0}$ indexed by time $t \geq 0$ taking values in $(S, \mathcal{A})$, i.e., $X_t : (\Omega, \mathcal{F}) \to (S, \mathcal{A})$ is measurable. These random variables all take values in the same space, which can be assumed to be a Banach space.

When we index the random variables $(X_t)_{t\geq 0}$ by time $t \geq 0$, it will be convenient to think of X_t as the random state that we observe at a time $t \geq 0$. Intuitively, when we have such a stochastic process, we can then think of observing X_t for increasing values of t as observing more information in time about a random system. We will mathematically encode this increase in observed information about a random system in time using the mathematical concept of a filtration.

Definition 3.3.1 (Filtration) Suppose that $(\Omega, \mathcal{F}, \mathbb{P})$ is a probability space. A **filtration** $(\mathcal{F}_t)_{t\geq 0}$ is an increasing family of sigma algebras, in the sense that $\mathcal{F}_t \subset \mathcal{F}$ for all $t \geq 0$, each $\mathcal{F}_t$ is a sigma algebra, and $\mathcal{F}_s \subset \mathcal{F}_t$ for all $s \leq t$. When we have a probability space $(\Omega, \mathcal{F}, \mathbb{P})$ and a particular filtration $(\mathcal{F}_t)_{t\geq 0}$ on this probability space, we can refer to the filtered probability space as $(\Omega, \mathcal{F}, (\mathcal{F}_t)_{t\geq 0}, \mathbb{P})$.

We also define the properties of filtrations being right-continuous and complete, which will be useful for defining measurability of stochastic processes, and for passing to almost sure limits in stochastic processes, respectively.

Definition 3.3.2 (Right-Continuous and Complete Filtration) Let $(\Omega, \mathcal{F}, (\mathcal{F}_t)_{t\geq 0}, \mathbb{P})$ be a filtered probability space. For every $t \geq 0$, we define $\mathcal{F}_{t^+} = \bigcap_{s>t}\mathcal{F}_s$. We say that $(\mathcal{F}_t)_{t\geq 0}$ is **right-continuous** if $\mathcal{F}_t = \mathcal{F}_{t^+}$ for all $t \geq 0$. This filtration is a **complete filtration** if every measurable set $A \subset \mathcal{F}$ with $\mathbb{P}(A) = 0$ is included in $\mathcal{F}_t$ for all $t \geq 0$. We say that a filtered probability space satisfies the *usual conditions* if it is both right-continuous and complete.

Given a filtration, we can then discuss whether a given stochastic process is compatible with the information that is observable at each time $t \geq 0$, specified by the filtration $(\mathcal{F}_t)_{t\geq 0}$ on the underlying probability space.

Definition 3.3.3 (Adapted Process) A stochastic process $(X_t)_{t\geq 0}$ taking values in $(S, \mathcal{A})$ defined on a probability space with filtration $(\Omega, \mathcal{F}, (\mathcal{F}_t)_{t\geq 0}, \mathbb{P})$ is **adapted** to the filtration $(\mathcal{F}_t)_{t\geq 0}$ if $X_t : (\Omega, \mathcal{F}_t) \to (S, \mathcal{A})$ is measurable for all $t \geq 0$.

Adaptability is an important property of a stochastic process with respect to a filtration, because if we consider the filtration $(\mathcal{F}_t)_{t\geq 0}$ to be the full "information" that can be observed from a stochastic system at time t or before, then a stochastic process $(X_t)_{t\geq 0}$ being adapted means that the stochastic process respects the structure of the information that can be seen at any given time from the σ-algebra of measurable events $\mathcal{F}$ in the probability space. Given a stochastic process $(X_t)_{t\geq 0}$ defined on a probability space $(\Omega, \mathcal{F}, \mathbb{P})$, a natural way to construct a filtration to which $(X_t)_{t\geq 0}$ is adapted is to define

$$\mathcal{F}_t^0 = \sigma(X_s, 0 \leq s \leq t). \tag{3.5}$$

Then $(\mathcal{F}_t^0)_{t\geq 0}$ is a filtration called the natural filtration of the process $(X_t)_{t\geq 0}$ and $(X_t)_{t\geq 0}$ is $(\mathcal{F}_t^0)_{t\geq 0}$-adapted..

In the following two propositions, we will approximate any process adapted to some filtration with a sequence of processes adapted to a sequence of filtrations "converging" to the aforementioned filtration. First, we will prove a similar result for random variables. The context of the following results is as follows. Let $(\Omega, \mathcal{F}, \mathbb{P})$ be a probability space, and let $(W_n)_{n=1}^{\infty}$ and W be real-valued random variables on this probability space, such that

$$W_n \to W, \qquad \text{almost surely.}$$

Let $\mathcal{F}_W^n$ be the sigma-algebra generated by W_n and let $\mathcal{F}_W$ be the sigma-algebra generated by W.

Proposition 3.3.1 *The following results hold:*

1. *For any measurable set* $S \in \mathcal{F}_W$*, there exist a subsequence* $\{n_k\}$ *and measurable sets* $S_{n_k} \in \mathcal{F}_W^{n_k}$ *such that* $\mathbb{1}_{S_{n_k}} \to \mathbb{1}_S$ *almost surely.*
2. *For any* $\mathcal{F}_W$*-measurable random variable* X *taking values in a separable Banach space* B*, there exists a subsequence of* $\mathcal{F}_W^{n_k}$*-measurable random variables* X_{n_k} *such that* $X_{n_k} \to X$ *almost surely.*

Proof Let $\mathcal{A}$ be the collection of measurable sets of the form $\{\omega : W(\omega) \in (a, b)\}$ for some real numbers $a < b$. Note that $\mathcal{A}$ is a π-system. Let $\mathcal{B}$ be the collection of sets $E \in \mathcal{F}_W$ that satisfy the following property: there exists a subsequence of measurable sets $E_{n_k} \subset \mathcal{F}_W^{n_k}$ such that $\lim_{k\to\infty}\mathbb{P}(E\Delta E_{n_k}) = 0$. It suffices to show that B is a λ-system, as the result will follow from the π-λ theorem, Theorem 3.1.1. We hence verify the properties of a λ-system as follows. Clearly, $\Omega \in \mathcal{B}$ since $\Omega \in \mathcal{F}_W^n$ for all n. Next, it is easy to verify that if $E \in \mathcal{B}$, then $E^c \in \mathcal{B}$. So it suffices to show that if $E_1, E_2, \ldots$ is an increasing sequence of sets in $\mathcal{B}$, then $E := \bigcup_{n=1}^{\infty} E_n \subset \mathcal{B}$. To prove that, we will construct a sequence of sets S_{n_k} such that $1_{S_{n_k}} \to 1_E$ almost surely as follows. Thanks to the definition of E, for a given k, there exists N such that $\mathbb{P}(E \setminus E_N) < 2^{-k}$. Then, given $E_N \in \mathcal{B}$, there exists some set $S_{n_k} \in \mathcal{F}_W^{n_k}$ such that $\mathbb{P}(S_{n_k}\Delta E_N) < 2^{-k}$. This choice of n_k can also be made so that the subsequence n_k is strictly increasing in k. Hence, by construction, we have that for all positive integers k, $\mathbb{P}(E\Delta S_{n_k}) \leq 2^{-k+1}$ for all k. So by the Borel-Cantelli lemma, Lemma 3.2.1, $1_{S_{n_k}} \to 1_E$ almost surely.

For a proof of the second result on random variables, consider a countable dense sequence $\{y_m\}_{m=1}^{\infty}$ in the separable Banach space B. Given k, note that the balls $\{B(y_m, 2^{-k})\}_{m=1}^{\infty}$ cover B for each fixed k. We then define the sets

$$U_{1,k} = B(y_1, 2^{-k}),$$
$$U_{2,k} = B(y_2, 2^{-k}) \setminus B(y_1, 2^{-k}),$$

and more generally,

$$U_{m,k} = B(y_m, 2^{-k}) \setminus \bigcup_{j=1}^{m-1} B(y_j, 2^{-k}).$$

Then, for a fixed k, the sets $U_{m,k}$ for $m = 1, 2, \ldots$ are disjoint and $\bigcup_{m=1}^{\infty} U_{m,k} = B$.

To establish the result on random variable approximation, consider an $\mathcal{F}_W$-measurable random variable X taking values in B. We will construct an approximating sequence of $\mathcal{F}_{W_{n_k}}$-measurable random variables X_{n_k} such that $X_{n_k} \to X$ almost surely, using the following two levels of approximation:

- **Step 1.** Given k, we first approximate the random variable X by an $\mathcal{F}_W$-measurable simple function $\hat{X} := \sum_{n=1}^{N_k} y_n \mathbb{1}_{E_n}$ for appropriate disjoint $\mathcal{F}_W$-measurable events E_n, such that

$$\mathbb{P}\Big(\|X - \hat{X}\|_B > 2^{-k}\Big) \leq 2^{-k}.$$

- **Step 2.** We then approximate each indicator function $\mathbb{1}_{E_n}$ for $n = 1, 2, \ldots N_k$ by some indicator function $\mathbb{1}_{E_{n_k}}$ that is $\mathcal{F}_{W_{n_k}}$-measurable, for n_k sufficiently large so that

$$\mathbb{P}(E_n \Delta E_{n_k}) < \frac{1}{N_k} 2^{-2k} \cdot \left(\max_{i=1,2,\ldots,N_k} \|y_i\|_B \right).$$

Step 1. Approximation of X by an $\mathcal{F}_W$-measurable simple function. Consider the simple functions defined by

$$\hat{X}_{j,k} = \sum_{i=1}^{j} y_i \mathbb{1}_{\{X \in U_{i,k}\}}.$$

Because the sets $\{X \in U_{i,k}\}_{i=1}^{\infty}$ for fixed k are disjoint sets whose entire union is the full probability space Ω, there exists N_k sufficiently large such that

$$\mathbb{P}\Big(\|X - \hat{X}_{N_k,k}\|_B > 2^{-k}\Big) < 2^{-k}. \tag{3.6}$$

We will set $\hat{X} := \hat{X}_{N_k,k}$, and we emphasize that $\hat{X}$ is an $\mathcal{F}_W$-measurable simple function that approximates X in the sense of (3.6), satisfying

$$\mathbb{P}\Big(\|X - \hat{X}\|_B > 2^{-k}\Big) < 2^{-k}. \tag{3.7}$$

Step 2. Approximation by simple functions adapted to approximating filtrations. We have that

$$\hat{X} = \sum_{i=1}^{N_k} y_i \mathbb{1}_{\{X \in U_{i,k}\}} := \sum_{i=1}^{N_k} y_i \mathbb{1}_{E_{i,k}},$$

and we will approximate each measurable set $\mathbb{1}_{\{X \in U_{i,k}\}}$ by $\mathcal{F}_{W_n}$-measurable sets using the previous result on approximation. In particular, for each $\mathbb{1}_{\{X \in U_{i,k}\}}$ for fixed i and k, there is an associated sequence $\{E_{i,k}^n\}_{n=1}^{\infty}$ of sets in $\mathcal{F}$ such that

$$\lim_{n \to \infty} \mathbb{P}(E_{i,k} \Delta E_{i,k}^n) = 0, \quad \text{and } E_{i,k}^n \text{ is } \mathcal{F}_{W_n}\text{-measurable for each positive integer } n.$$

We can then choose n_k sufficiently large such that for all $i = 1, 2, ..., N_k$ (where N_k was defined in Step 1), we have that

$$\mathbb{P}(E_{i,k}\Delta E_{i,k}^{n_k}) < \frac{1}{N_k}2^{-2k} \cdot \left(\max_{i=1,2,\dots,N_k} \|y_i\|_B \right)^{-1}, \qquad \text{for all } i = 1, 2, \dots, N_k$$

Then, if we define

$$X_{n_k} := \sum_{i=1}^{N_k} y_i \mathbb{1}_{E_{i,k}^{n_k}},$$

we have that X_{n_k} is $\mathcal{F}_{W_{n_k}}$-measurable and by the triangle inequality,

$$\mathbb{E}\Big(\|\hat{X} - X_{n_k}\|_B \Big) \le \sum_{i=1}^{N_k} \|y_i\|_B \cdot \mathbb{P}(E_{i,k}\Delta E_{i,k}^{n_k}) < 2^{-2k}.$$

This implies that

$$\mathbb{P}\Big(\|\hat{X} - X_{n_k}\|_B > 2^{-k} \Big) < 2^{-k},$$

and therefore, combining this with (3.7), we have constructed a sequence X_{n_k} of $\mathcal{F}_{W_{n_k}}$-measurable random variables such that

$$\mathbb{P}\Big(\|X - X_{n_k}\|_B > 2^{-k+1} \Big) < 2^{-k+1}.$$

Then, by the Borel-Cantelli lemma, Lemma 3.2.1, we conclude that $X_{n_k} \to X$ almost surely, where each X_{n_k} is $\mathcal{F}_{W_{n_k}}$-measurable, which completes the proof. □

Proposition 3.3.2 *Let $(W_n(t))_{t\in[0,T]}$ and $(W(t))_{t\in[0,T]}$ be stochastic processes on a probability space $(\Omega, \mathcal{F}, \mathbb{P})$ taking values in $C(0, T; \mathbb{R})$, and define the associated natural filtrations:*

$$\mathcal{F}_t^n := \sigma\{W_n(s) : 0 \le s \le t\}, \qquad \mathcal{F}_t := \sigma\{W(s) : 0 \le s \le t\}.$$

Let $\boldsymbol{q} \in C^1(0, T; B)$ be a continuously differentiable process in time $t \in [0, T]$ taking values in a separable Banach space B, which is adapted to the filtration $(\mathcal{F}_t)_{t\in[0,T]}$. Then, for an associated subsequence $\{n_k\}_{k=1}^{\infty}$, there exists a sequence of $\boldsymbol{q}_{n_k} \in W^{1,\infty}(0, T; B)$ such that:

- *$\boldsymbol{q}_{n_k}$ is adapted to the filtration $(\mathcal{F}_t^{n_k})_{t\in[0,T]}$*
- *$\boldsymbol{q}_{n_k} \to \boldsymbol{q}$ strongly in $W^{1,\infty}(0, T; B)$, almost surely*

Proof We construct the sequence $\boldsymbol{q}_{n_k}$ as follows. The goal will be to show that we can construct a sequence $\{\boldsymbol{q}_{n_k}\}_{k=1}^{\infty}$ in $W^{1,\infty}(0, T; B)$ that is adapted to the filtration $(\mathcal{F}_t^{n_k})_{t\in[0,T]}$ such that

$$\mathbb{P}\Big(\|\boldsymbol{q}_{n_k} - \boldsymbol{q}\|_{W^{1,\infty}(0,T;B)} > 2^{-k}\Big) < 2^{-k}. \tag{3.8}$$

To do this, we first approximate $\boldsymbol{q}$ using a time-shift and a linear interpolation. Given k, there exists N_k (depending on k) sufficiently large so that for $\Delta t_k = T/N_k$,

$$\begin{aligned}
&\mathbb{P}\left(\|\tau_{\Delta t_k}\boldsymbol{q} - \boldsymbol{q}\|_{W^{1,\infty}(0,T;B)} > \frac{1}{3}2^{-k}\right) \le \frac{1}{3}2^{-k},\\
&\mathbb{P}\left(\|\overline{\boldsymbol{q}}_k - \tau_{\Delta t_k}\boldsymbol{q}\|_{W^{1,\infty}(0,T;B)} > \frac{1}{3}2^{-k}\right) \le \frac{1}{3}2^{-k},
\end{aligned} \tag{3.9}$$

where the time-shifted function $\tau_{\Delta t_k}\boldsymbol{q}$ is defined by

$$\tau_{\Delta t_k}\boldsymbol{q}(t,\cdot) = \begin{cases} \boldsymbol{q}(0), & \text{if } t \in [0, \Delta t_k],\\ \boldsymbol{q}(t - \Delta t_k, \cdot), & \text{if } t \in [\Delta t_k, T], \end{cases}$$

and where the linear interpolation $\overline{\boldsymbol{q}}_k$ is defined, for $t \in [i\Delta t_k, (i+1)\Delta t_k]$ by

$$\bar{\boldsymbol{q}}_k(t) = \left(\frac{(i+1)\Delta t_k - t}{\Delta t_k}\right)\boldsymbol{q}((i-1)\Delta t_k) + \left(\frac{t - i\Delta t_k}{\Delta t_k}\right)\boldsymbol{q}(i\Delta t_k).$$

The estimates in (3.9) are valid for sufficiently large N_k since $\|\partial_t \boldsymbol{q}\|_{C(0,T;B)} \le C(\omega)$ almost surely, and furthermore, by construction (in particular due to the time-shift), the linear interpolation $\overline{\boldsymbol{q}}_k$ is adapted to the filtration $(\mathcal{F}_t)_{t\ge 0}$ also.

Now, we will use the previous result for random variables to approximate $\overline{\boldsymbol{q}}_k$ (with fixed mesh size Δt_k and N_k subintervals) by some function $\boldsymbol{q}_k$. Recall that $\overline{\boldsymbol{q}}_k$ is a linear interpolation of points $\boldsymbol{q}(i\Delta t_k)$, where for $i = 1, \dots, N_k$, the random variable $\overline{\boldsymbol{q}}_k(i\Delta t)$ is $\mathcal{F}_{(i-1)\Delta t_k}$-measurable by construction. Then, thanks to Proposition 3.3.1, we can construct random variables $\boldsymbol{q}_k^i$ by taking n_k to be sufficiently large, such that:

- $\boldsymbol{q}_k^0$ is $\mathcal{F}_0^{n_k}$-measurable
- $\boldsymbol{q}_k^i$ is $\mathcal{F}_{(i-1)\Delta t_k}^{n_k}$-measurable for every $i = 1, 2, \dots, N_k$
- We have the following probabilistic bound for all $i = 0, 1, \dots, N_k$:

$$\mathbb{P}\left(\|\boldsymbol{q}_k^i - \overline{\boldsymbol{q}}_k(i\Delta t_k)\|_B > \frac{1}{3}2^{-k}\min(1, \Delta t_k/2)\right) \le \frac{1}{3(N_k+1)}2^{-k} \tag{3.10}$$

Then, we define the approximate process $\boldsymbol{q}_{n_k}$ to be a linear interpolation of the $\boldsymbol{q}_k^i$:

$$\boldsymbol{q}_{n_k}(t) = \left(\frac{(i+1)\Delta t_k - t}{\Delta t_k}\right)\boldsymbol{q}_k^i + \left(\frac{t - i\Delta t_k}{\Delta t_k}\right)\boldsymbol{q}_k^{i+1}, \quad \text{if } i\Delta t_k \le t \le (i+1)\Delta t_k,$$

and since $\boldsymbol{q}_{n_k}$ and $\overline{\boldsymbol{q}}_k$ are linear interpolations along common points $i\Delta t_k$, the estimate (3.10) implies that

$$\mathbb{P}\left(\|\boldsymbol{q}_{n_k} - \overline{\boldsymbol{q}}_k\|_{W^{1,\infty}(0,T;B)} > \frac{1}{3}2^{-k}\right) \le \frac{1}{3}2^{-k}. \tag{3.11}$$

Combining the estimates (3.9) and (3.11) gives the desired estimate (3.8) for the constructed approximations $\boldsymbol{q}_{n_k}$, which by construction are adapted to the filtration $(\mathcal{F}_t^{n_k})_{t\in[0,T]}$. □

Remark 3.3.1 Observe that the proof of Proposition 3.3.2 is also valid for processes $\boldsymbol{q}$ taking values in $C(0,T;B_0) \cap C^1(0,T;B_1)$ where B_0, B_1 are Banach spaces satisfying $B_0 \subset B_1$.

Next, we define the notions of conditional expectation and martingales, where martingales will be especially important in the study of stochastic integration.

Definition 3.3.4 (Conditional Expectation) Let $(\Omega, \mathcal{F}, \mathbb{P})$ be a probability space and let H be a real, separable Hilbert space. Let $X : \Omega \to H$ be an integrable H-valued random variable, that is, $\mathbb{E}\|X\|_H < \infty$. Let $\mathcal{G}$ be a sub-σ-algebra of $\mathcal{F}$. An H-valued random variable $Y : \Omega \to H$ is said to be the conditional expectation of X given $\mathcal{G}$ if Y is $\mathcal{G}$-measurable and $\int_G Y d\mathbb{P} = \int_G X d\mathbb{P}$ for every $G \in \mathcal{G}$.

Such a random variable exists and is unique $\mathbb{P}$-almost surely, and we denote it by $Y = \mathbb{E}[X|\mathcal{G}]$.

Definition 3.3.5 (Martingale) Let $(X_t)_{t\ge 0}$ be an $(\mathcal{F}_t)_{t\ge 0}$-adapted stochastic process taking values in a real, separable Hilbert space H. We say that $(X_t)_{t\ge 0}$ is an $(\mathcal{F}_t)_{t\ge 0}$-**martingale**, provided

$$\mathbb{E}[X_t \mid \mathcal{F}_s] = X_s \quad \mathbb{P}\text{-almost surely}$$

for all $0 \le s \le t < \infty$.

By viewing a stochastic process $(X_t)_{t\ge 0}$ as a collection of random variables on a probability space $(\Omega, \mathcal{F}, \mathbb{P})$ indexed in time, we are viewing the stochastic process as a collection of random variables $X_t : \Omega \to S$. However, instead of viewing time as just an index set, we can also view the entire stochastic process, which we will refer to as X, as a random function $X : [0,\infty)\times\Omega \to S$, where $X(t,\omega)$ is a function of both time $t \in [0,\infty)$ and the outcome $\omega \in \Omega$. We can then define measurability properties of stochastic processes that refer to how the measurability properties of the stochastic process evolve in time and interact with the time parameter. We will consider the following measurability properties in time that will be important in our future analysis: joint measurability, progressive measurability, and predictability. We start with joint measurability.

Definition 3.3.6 (Jointly Measurable Process) Given a stochastic process $(X_t)_{t\geq 0}$ on a probability space $(\Omega, \mathcal{F}, \mathbb{P})$ taking values in a measurable space $(S, \mathcal{A})$, the stochastic process is **jointly measurable** if the map

$$(t, \omega) \mapsto X(t, \omega), \qquad X : ([0, \infty) \times \Omega, \ \text{Borel}([0, \infty)) \otimes \mathcal{F}) \to (S, \mathcal{A}),$$

is measurable. Here $\text{Borel}([0, \infty))$ is the Borel σ-algebra of $[0, \infty)$.

We move onto the definition of a progressively measurable process. Progressive measurability is a stronger condition than joint measurability and is particularly useful in defining stochastic integrals.

Definition 3.3.7 (Progressively Measurable Process) Let $(\Omega, \mathcal{F}, (\mathcal{F}_t)_{t\geq 0}, \mathbb{P})$ be a filtered probability space. A stochastic process $(X_t)_{t\geq 0}$ taking values in $(S, \mathcal{A})$ is $(\mathcal{F}_t)_{t\geq 0}$-**progressively measurable** if for every fixed time $t \in [0, \infty)$ the map

$$(s, \omega) \mapsto X(s, \omega), \qquad X : ([0, t] \times \Omega, \text{Borel}([0, t]) \otimes \mathcal{F}_t) \to (S, \mathcal{A}),$$

is measurable. Here $\text{Borel}([0, t])$ is the Borel σ-algebra of $[0, t]$.

An $(\mathcal{F}_t)_{t\geq 0}$-progressively measurable process is jointly measurable and $(\mathcal{F}_t)_{t\geq 0}$-adapted.

We move onto predictable processes.

Definition 3.3.8 (Predictable Process) Let $(\Omega, \mathcal{F}, (\mathcal{F}_t)_{t\geq 0}, \mathbb{P})$ be a filtered probability space. A stochastic process $(X_t)_{t\geq 0}$ is **predictable** if it is measurable with respect to the **predictable σ-algebra**, which is the sigma algebra on $[0, \infty) \times \Omega$ generated by subsets of the form

$$\{[a, b) \times A : 0 \leq a < b \text{ and } A \in \mathcal{F}_a\}.$$

We can use the following criterion to classify stochastic processes as predictable; see Proposition 5.1 in Chapter IV, Section 5 of [290].

Proposition 3.3.3 *Suppose that $(X_t)_{t\geq 0}$ is a stochastic process on a probability space $(\Omega, \mathcal{F}, (\mathcal{F}_t)_{t\geq 0}, \mathbb{P})$ that is adapted to the filtration $(\mathcal{F}_t)_{t\geq 0}$ taking values in a Banach space B. Suppose that $(X_t)_{t\geq 0}$ also has left continuous paths almost surely so that with probability one, $X(t, \omega) : [0, \infty) \to B$ is left continuous. Then, $(X_t)_{t\geq 0}$ is predictable as a stochastic process taking values in B.*

Thus, a predictable stochastic process is a special type of adapted process whose value at any given time can be "predicted" using the information available just before that time. We remark that, in particular, predictable processes are progressively measurable. We refer the reader to Chapter I, Section 4 and Chapter IV, Section 5 of [290] for more details about predictable processes.

The criterion for predictability in Proposition 3.3.3 is particularly useful when we can show that the paths of a stochastic process, which is adapted to some filtration, are almost surely continuous. Continuity in time ensures that we can define each path for every point $t \in [0, T]$ almost surely, allowing us to define pointwise random variables $(X_t)_{t\in[0,T]}$ for all t.

However, it is not always feasible to demonstrate such continuity in time. In such cases, we often encounter scenarios where the stochastic process (for a fixed outcome) is defined only *almost everywhere* on $[0, T]$. This limitation makes it advantageous to interpret stochastic processes as random distributions, as pioneered in [52]. We will define a random distribution as a generalization of a stochastic process where instead of assigning random variables to each point in time (or space), the process assigns random variables in the distributional sense. This perspective will become particularly important, for instance, in the study of stochastic fluid-structure interaction on (random) moving domains in Chap. 9.

We define random distributions as in Definition 2.2.1 in [52]:

Definition 3.3.9 (Random Distribution, [52]) Let $(\Omega, \mathcal{F}, \mathbb{P})$ be a probability space and let $\mathcal{D}([0, T])$ be the space of smooth functions compactly supported in $[0, T]$. Denote by $\mathcal{D}'([0, T])$ the dual of $\mathcal{D}([0, T])$ endowed with weak-* topology. A mapping $X : \Omega \to \mathcal{D}'([0, T])$ is called a **random distribution** if for every test function $\varphi \in \mathcal{D}([0, T])$, the mapping $\langle X, \varphi\rangle : \Omega \to \mathbb{R}$ is a measurable function.

Definition 3.3.10 (History of a Random Distribution, [52]) Let X be a random distribution in $\mathcal{D}'([0, T])$. Then:

- We say that X is adapted to $(\mathcal{F}_t)_{t\in[0,T]}$ if $\langle X, \varphi\rangle$ is $\mathcal{F}_t$-measurable for any $\varphi \in \mathcal{D}((0, t))$.
- The family of σ-fields $(\sigma_t[X])_{t\geq 0}$, given as

$$\sigma_t[X] := \bigcap_{s>t} \sigma\left(\bigcup_{\varphi\in\mathcal{D}((0,s))} \{\langle X, \varphi\rangle < 1\} \cup \{N \in \mathcal{F}, \mathbb{P}(N) = 0\} \right),$$

is called the **history** of X.

We will summarize some properties of random distributions here. For a full exposition, see [52].

Lemma 3.3.1 *Let X be an $(\mathcal{F}_t)_{t\in[0,T]}$-adapted random distribution taking values in $L^2(0, T) \subset \mathcal{D}'([0, T])$ in the sense of Definition 3.3.9. Then, for any $T > 0$, there exists a stochastic process $\tilde{X} \in L^2(0, T)$ a.s. that is $(\mathcal{F}_t)_{t\in[0,T]}$-progressively measurable such that*

$$X = \tilde{X} \quad \textit{almost everywhere in } \Omega \times [0, T].$$

We note here that statements equivalent to Definitions 3.3.9, 3.3.10, and Lemma 3.3.1 can be made for Hilbert space-valued random distributions.

In Chaps. 8 and 9, we will consider PDEs driven with a *multiplicative* noise, i.e., the driving force in the PDE will be a stochastic integral with an integrand depending on the solution of the PDE. The following result allows one to establish appropriate measurability properties of the noise coefficient. Moreover, in Chap. 9, when the PDEs are defined on time-dependent random domains, we will work with time-dependent random test functions appearing in the weak formulation of the problem. The following result, proved in Lemma 8.2.3 in [12], will also let us establish appropriate measurability properties of these test functions so that the stochastic integral in the weak formulation is well-defined.

Definition 3.3.11 (Carathéodory Function) Let S, Y be topological spaces and $(\Omega, \mathcal{F})$ a measurable space. We say that a function $\mathcal{C} : \Omega \times S \to Y$ is a *Carathéodory function* if:

1. For every $x \in S$, the map from $(\Omega, \mathcal{F})$ to $(Y, \text{Borel}(Y))$ given by $\omega \mapsto \mathcal{C}(\omega, x)$ is measurable
2. $\mathcal{C}(\omega, \cdot)$ is continuous for almost every $\omega \in \Omega$

Lemma 3.3.2 *Consider two separable metric spaces S and Y, a measurable space $(\Omega, \mathcal{F})$, and a Carathéodory map $\mathcal{C} : \Omega \times S \to Y$. Then for any measurable function $f : \Omega \to S$, the map $\omega \mapsto \mathcal{C}(\omega, f(\omega))$ from $(\Omega, \mathcal{F})$ to $(Y, Borel(Y))$ is measurable.*

We conclude this section with the definition of a stopping time along with a lemma regarding its characterization, which will be used in Chap. 9.

Definition 3.3.12 (Stopping Time) Given a filtered probability space $(\Omega, \mathcal{F}, (\mathcal{F}_t)_{t\geq 0}, \mathbb{P})$, a nonnegative function $\tau : \Omega \to [0, \infty]$ is called an $(\mathcal{F}_t)_{t\geq 0}$-stopping time if for every $t \geq 0$, the event $\{\tau \leq t\}$ is measurable with respect to the σ-algebra $\mathcal{F}_t$, i.e.,

$$\{\omega \in \Omega : \tau(\omega) \leq t\} \in \mathcal{F}_t.$$

A common example of a random time in many applications is a *hitting time*, which is the "first" time a given stochastic process enters a given closed subset in the range. This hitting time is, in fact, a stopping time whenever the stochastic process has *continuous paths* almost surely. The following lemma provides a characterization of stopping time for stochastic processes with continuous paths in a Banach space B.

Lemma 3.3.3 *Let $(X_t)_{t\geq 0}$ be an B-valued stochastic process defined on $(\Omega, \mathcal{F}, \mathbb{P})$. Suppose that X has continuous paths, i.e., $\forall\, \omega \in \Omega$ the function $t \mapsto X_t(\omega)$ is continuous. Then for every closed set C in B, the random variable*

$$\tau_C := \inf\{t \geq 0 : X_t \in C\}$$

is a stopping time with respect to the natural filtration of $(X_t)_{t\geq 0}$.

We conclude this section by stating the following result for "stopped processes," which will be used in Chap. 9, after an application of the Skorohod representation theorem in Jakubowski spaces, i.e., Theorem 3.2.5.

Consider a Banach space B and a function f belonging to the phase space $C(0, T; B)$. Let $R > 0$ be a given positive number. Define

$$\tau_R = \inf\{t \in [0, T] : \|f(t)\|_B \geq R\},$$

along with the stopped process

$$f_R^*(t) = f(t \wedge \tau_R), \qquad \text{for } t \in [0, T],$$

where the wedge sign $a \wedge b$ denotes the minimum of a and b. Then we have the following result:

Proposition 3.3.4 ([215]) *For each $R > 0$, the set B_R defined as*

$$B_R := \{(f, g) \in C(0, T; B)^2 : g = f_R^*\} \tag{3.12}$$

is a measurable subset of $C(0, T; B)^2$.

This result is proved in Proposition 8.7 in [215] for $B = \mathbb{R}$ and can be generalized to any Banach space using the same ideas.

3.4 Kolmogorov Continuity Criterion

The Kolmogorov continuity criterion is a result that provides sufficient conditions for a stochastic process to have almost surely continuous paths. For example, in the case of a stochastic process $(X_t)_{t \in [0,T]}$ taking values in a Banach space B with norm $\|\cdot\|_B$, we can ask whether the paths $t \to X_t(\omega)$ considered as functions from $[0, T] \to \mathbb{R}$ for each fixed outcome ω possess a certain regularity, such as Hölder continuity, almost surely. The Kolmogorov continuity criterion provides Hölder continuity of paths (up to a modification), almost surely, if one can show that a certain estimate on the moments of the increments of the stochastic process is satisfied.

To state the Kolmogorov continuity criterion, we must first define the concept of a modification of a stochastic process, where we will state the definition for a stochastic process indexed by a general index set E.

Definition 3.4.1 A stochastic process $(\tilde{X}_i)_{i \in E}$ is a **modification** of a stochastic process $(X_i)_{i \in E}$ if $\mathbb{P}(X_i = \tilde{X}_i) = 1$ for each $i \in E$.

We first state the Kolmogorov continuity criterion for a Banach space-valued process $(X_t)_{t \in [0,T]}$ indexed by time, taking values in a Banach space B with norm $\|\cdot\|_B$. We will think of this stochastic process as a time-evolutionary random process with values in the Banach space B, which can, for example, be a function space. We have the following result; see Theorem 3.3 in Section 3.2 of [104].

Theorem 3.4.1 (Kolmogorov Continuity Criterion) *Let B be a Banach space with norm $\|\cdot\|_B$. Suppose that $(X_t)_{t\in[0,T]}$ is a B-valued stochastic process on a probability space, which satisfies the following bound for all $s, t \in [0, T]$:*

$$\mathbb{E}\left(\|X_t - X_s\|_B^{\gamma}\right) \le C|t-s|^{1+\epsilon},$$

for constants $\gamma > 1$ and $\epsilon > 0$. Then, $(X_t)_{t\in[0,T]}$ has a modification that is α-Hölder continuous, almost surely, for each α such that $0 \le \alpha < \dfrac{\epsilon}{\gamma}$.

In practice, we can consider most random time-evolving stochastic processes in mathematical problems of interest in the context of Theorem 3.4.1, where for each time, there is an associated function-valued random variable. However, especially when the random noise acting on the system exhibits randomness with variance in both the time *and* spatial variables, it is more useful to regard a random function $f(\omega, t, x)$ for $t \in [0, T]$ and $x \in \mathbb{R}^d$ not as a function-valued stochastic process $f : (\omega, t) \in \Omega \times [0, T] \to B$ in time for a Banach space B, but as a real-valued process $f : (\omega, t, x) \in \Omega \times [0, T] \times \mathbb{R}^d \to \mathbb{R}$ indexed by $(t, x) \in [0, T] \times \mathbb{R}^d$. This will be the particular case in the study of a stochastic viscous wave equation describing a reduced model of stochastic FSI with spacetime white noise (which has random variance in both space and time); see Chap. 7. In these cases, by including the spatial variable (in addition to the time variable) in the index set of the stochastic process, we may be able to show that there is a modification of the real-valued stochastic process $f : (\omega, t, x) \in \Omega \times [0, T] \times \mathbb{R}^d \to \mathbb{R}$ that is Hölder continuous almost surely, not just in the time variable t, but also in the space variable $x \in \mathbb{R}^d$.

In this context, it is useful to state the Kolmogorov continuity criterion more generally for stochastic processes indexed not just in time (as in $(X_t)_{t\in[0,T]}$), but on a subset E of Euclidean space, where $E = [0, 1]^d \subset \mathbb{R}^d$ or $E = \mathbb{R}^+ \times \mathbb{R}^d$. In this more general context, we can consider for each outcome ω the path $i \in E \subset \mathbb{R}^d \to X_i \in \mathbb{R}$ and hence consider the Hölder regularity properties of each path almost surely. The Kolmogorov continuity criterion states that if a certain estimate on the moments of the increments (in the general index set E) of the stochastic process $(X_i)_{i\in E}$ is satisfied, then there exists a **version** of the process $(X_i)_{i\in E}$ with continuous paths, i.e., there exists a **modification** $(\tilde{X}_i)_{i\in E}$ of $(X_i)_{i\in E}$ with continuous paths.

Theorem 3.4.2 (A More General Kolmogorov Continuity Criterion) *Let $\{X_i\}_{i\in[0,1]^d}$ be a real-valued stochastic process. If there exist two constants $\gamma > 1$ and $\epsilon > 0$ such that*

$$\mathbb{E}(|X_{i_1} - X_{i_2}|^{\gamma}) \le C|i_1 - i_2|^{d+\epsilon},$$

then for each α *such that* $0 \leq \alpha < \dfrac{\epsilon}{\gamma}$, *the stochastic process* $\{X_i\}_{i\in[0,1]^d}$ *has a modification that is* α*-Hölder continuous.*

For more details see, for example, Theorem 2.1 in Chapter I, Section 2 of [290].

We can extend this to stochastic processes $(X_i)_{i\in E}$ with index sets $E \subset \mathbb{R}^+ \times \mathbb{R}^d$ on unbounded Euclidean domains, as we can regard these as random processes of time and space. To emphasize this connection, we will reframe the Kolmogorov continuity criterion for general index sets in Theorem 3.4.2 to the context of a stochastic process indexed by $(t, x) \in \mathbb{R}^+ \times \mathbb{R}^d$, in which it will be directly applied to the study of time and spatial Hölder continuity of solutions to the stochastic viscous wave equation with spacetime white noise in Chap. 7. This is similar to Theorem 2.5.1 in [209].

Corollary 3.4.1 *Let* $(X(t,x))_{(t,x)\in\mathbb{R}^+\times\mathbb{R}^d}$ *be a real-valued stochastic process. If there exist two positive constants* γ *and* ϵ *such that for each compact set* $K \subset \mathbb{R}^+ \times \mathbb{R}^d$,

$$\mathbb{E}(|X(t,x) - X(s,y)|^\gamma) \leq C_K |(t,x) - (s,y)|^{d+1+\epsilon} \qquad \textit{for all } (t,x), (s,y) \in K,$$

where C_K *can depend on* K, *then for each* α *such that* $0 \leq \alpha < \dfrac{\epsilon}{\gamma}$, *the stochastic process* $(X(t,x))_{(t,x)\in\mathbb{R}^+\times\mathbb{R}^d}$ *has a modification* $(\tilde{X}(t,x))_{(t,x)\in\mathbb{R}^+\times\mathbb{R}^d}$ *that is locally* α*-Hölder continuous on* $\mathbb{R}^+ \times \mathbb{R}^d$.

Since the Kolmogorov continuity theorem appears more often in the form of Theorem 3.4.2, we provide an explicit proof of Corollary 3.4.1 here. The proof is based on an idea called a patching argument, as described on pg. 160 of [209].

Proof The proof of Corollary 3.4.1 will follow from the Kolmogorov continuity criterion in Theorem 3.4.2 by considering compact cubes that increase to all of $\mathbb{R}^+ \times \mathbb{R}^d$, for example,

$$A_k := [0,k] \times [-k/2, k/2]^d \subset \mathbb{R}^+ \times \mathbb{R}^d, \ k \in \mathbb{R}^+$$

We will construct the desired modification $(\tilde{X}(t,x))_{(t,x)\in\mathbb{R}^+\times\mathbb{R}^d}$ as the limit $k \to \infty$ of α-Hölder continuous modifications $(\tilde{X}_k(t,x))_{(t,x)\in\mathbb{R}^+\times\mathbb{R}^d}$ defined for $(t,x) \in A_k$, constructed as follows.

Fix α such that $0 \leq \alpha < \frac{\epsilon}{\gamma}$. By the usual Kolmogorov continuity criterion given in Theorem 3.4.2, we can construct a modification $(\tilde{X}_k(t,x))_{(t,x)\in A_k}$ of $(X(t,x))_{(t,x)\in A_k}$ that is α-Hölder continuous on A_k. In particular, the modifications $(\tilde{X}_k(t,x))_{(t,x)\in A_k}$ are continuous for all $k \in \mathbb{R}^+$.

We claim that any two of these modifications $\tilde{X}_k(t,x)$ and $\tilde{X}_l(t,x)$ must agree with probability one on their overlap because they are continuous modifications. Otherwise there exists a ball with rational radius and center with rational coordinates on which the two modifications have disjoint range with positive probability. More precisely, consider $k \leq l$ so that $A_k \subset A_l$ is the overlap. We claim that

$$\mathbb{P}(\tilde{X}_k(t,x) = \tilde{X}_l(t,x) \text{ for all } (t,x) \in A_k) = 1.$$

We argue by contradiction. Suppose that

$$\mathbb{P}(\tilde{X}_k(t_0,x_0) \neq X_l(t_0,x_0) \text{ for some } (t_0,x_0) \in A_k) > 0.$$

Since $\tilde{X}_k(t,x)$ and $\tilde{X}_l(t,x)$ are continuous on A_k, for every outcome $\omega \in \Omega$ for which $\tilde{X}_k(t_0,x_0) \neq \tilde{X}_l(t_0,x_0)$ for some $(t_0,x_0) \in A_k$, we can find an open ball $B_r(q)$ with rational radius r centered at a point $q = (t,x) \in \mathbb{Q}^+ \times \mathbb{Q}^d \cap A_k \subset \mathbb{R}^+ \times \mathbb{R}^d$, such that $\tilde{X}_k$ and $\tilde{X}_l$ have *"disjoint" range* on the ball $B_r(q) \cap A_k$ in the sense that there exist two closed intervals $K_1 \subset \mathbb{R}$ and $K_2 \subset \mathbb{R}$ such that

$$\tilde{X}_k(B_r(q) \cap A_k) \subset K_1, \qquad \tilde{X}_l(B_r(q) \cap A_k) \subset K_2, \qquad K_1 \cap K_2 = \varnothing.$$

Hence,

$$\mathbb{P}\Big(\bigcup_{r\in\mathbb{Q}, q\in\mathbb{Q}^+\times\mathbb{Q}^d\cap A_k} \{\tilde{X}_k \text{ and } \tilde{X}_l \text{ have "disjoint" range on } B_r(q) \cap A_k\}\Big)$$
$$\geq \mathbb{P}(\tilde{X}_k(t_0,x_0) \neq \tilde{X}_l(t_0,x_0) \text{ for some } (t_0,x_0) \in A_k) > 0.$$

Therefore, by the countability of the index set, there exist $r_0 \in \mathbb{Q}^+$ and $q_0 \in \mathbb{Q}^+ \times \mathbb{Q}^d \cap A_k$ such that

$$\mathbb{P}(\tilde{X}_k \text{ and } \tilde{X}_l \text{ have "disjoint" range on } B_{r_0}(q_0) \cap A_k) > 0.$$

But this implies that $\mathbb{P}(\tilde{X}_k(q_0) \neq \tilde{X}_l(q_0)) > 0$, which contradicts that $(\tilde{X}_k(t,x))_{(t,x)\in A_k}$ and $(\tilde{X}_l(t,x))_{(t,x)\in A_l}$ are modifications of the same stochastic process on A_k and A_l, respectively, since $q_0 \in A_k \subset A_l$. Therefore, $\tilde{X}_k = \tilde{X}_l$ on A_k almost surely.

This implies that, with probability one (up to a null set), any two modifications from this collection of modifications $(\tilde{X}_k(t,x))_{(t,x)\in A_k}$ on increasing cubes must agree on the overlap of their domains.

To construct the desired modification $\tilde{X}(t,x)$, we now focus on the null sets $E_{k,l}$ for $k < l$, where the modifications $(\tilde{X}_k(t,x))_{(t,x)\in A_k}$ and $(\tilde{X}_l(t,x))_{(t,x)\in A_l}$ do not agree on A_k. Define

$$E = \bigcup_{k<l \text{ and } k,l\in\mathbb{Z}^+} E_{k,l}$$

and note that $\mathbb{P}(E) = 0$. Then, the desired modification $(\tilde{X}(t,x))_{(t,x)\in\mathbb{R}^+\times\mathbb{R}^d}$ is

$$\tilde{X}(t, x, \omega) = \lim_{k\to\infty} \tilde{X}_k(t, x, \omega) \qquad \text{for } \omega \in E^c,$$

$$\tilde{X}(t, x, \omega) = 0, \qquad \text{for } \omega \in E.$$

Indeed, the limit above exists since the sequence $\tilde{X}_k(t, x, \omega)$ for $\omega \in E^c$ is eventually constant. This is because the modifications $(\tilde{X}_k(t, x))_{(t,x)\in A_k}$ all agree pairwise on their common domains for $\omega \in E^c$. Furthermore, it is easy to check that $(\tilde{X}(t, x))_{(t,x)\in\mathbb{R}^+\times\mathbb{R}^d}$ is a modification that is α-Hölder continuous, by the fact that each of the $(\tilde{X}_k(t, x))_{(t,x)\in A_k}$ is a modification on A_k that is α-Hölder continuous. This completes the proof of the corollary. □

3.5 One-Dimensional Brownian Motion

One-dimensional **Brownian motion** (also, more generally, known as **Wiener process**) is a fundamental process in probability theory and is widely used in fields such as physics (e.g., modeling the random motion of particles in a fluid), finance (e.g., modeling stock prices in the Black-Scholes model), and mathematics (e.g., as a building block for stochastic calculus).

Definition 3.5.1 (Normal Distribution) Let $\mu \in \mathbb{R}, \sigma > 0$. We say that a real-valued random variable X is normally distributed mean μ and variance σ^2 and write $X \sim N(\mu, \sigma^2)$ if

$$\mathbb{P}[X \le a] = \frac{1}{\sqrt{2\pi}\sigma}\int_{-\infty}^{a} e^{-\frac{(x-\mu)^2}{2\sigma^2}}\, dx, \qquad \text{for every } a \in \mathbb{R}.$$

Definition 3.5.2 Given a probability space $(\Omega, \mathcal{F}, \mathbb{P})$, a stochastic process $W = (W_t)_{t\ge0}$ taking values in $\mathbb{R}$ is a **one-dimensional Brownian motion or a one-dimensional Wiener process** if the following properties are satisfied:

- The path $W(t, \omega) : [0, \infty) \to \mathbb{R}$ is a continuous path with $W(0) = 0$ almost surely (for all ω in a measurable set having probability one).
- The increments $W(t) - W(s)$ are distributed as $N(0, t-s)$, a normal distribution with mean zero and variance $t - s$.
- The independent increments property is satisfied: if $t_1 < t_2 < \ldots < t_k$, then $W(t_k) - W(t_{k-1}), \ldots, W(t_2) - W(t_1)$ are independent random variables.

The independent increments property has a direct implication, which is crucial for applications in stochastic analysis. More precisely, the following corollary holds true.

Proposition 3.5.1 *Let $(\mathcal{F}_t^W)_{t\ge0}$ denote the natural filtration associated with the Wiener process $(W_t)_{t\ge0}$ defined by*

$$\mathcal{F}_t^W = \sigma(\{W_s : 0 \le s \le t\}),$$

where $\sigma(\{W_s : 0 \le s \le t\}) \subset \mathcal{F}$ denotes the σ-algebra generated by the random variables W_s for $0 \le s \le t$. Then the increment $W_t - W_s$ for $0 \le s < t$ is independent of any random variable that is $\mathcal{F}_s^W$-measurable.

We can extend this independence criterion to general filtrations, as in the following definition.

Definition 3.5.3 Suppose that we are given a probability space $(\Omega, \mathcal{F}, \mathbb{P})$ and a filtration $(\mathcal{F}_t)_{t\ge0}$ on this probability space. Then, we say that a one-dimensional Brownian motion (or Wiener process) $(W_t)_{t\ge0}$ is more specifically a **Brownian motion with respect to the filtration** $(\mathcal{F}_t)_{t\ge0}$ or **an** $(\mathcal{F}_t)_{t\ge0}$**-Brownian motion (or** $(\mathcal{F}_t)_{t\ge0}$**-Wiener process)** if:

- W_t is adapted to $\mathcal{F}_t$ for all $t \ge 0$
- $W(t) - W(s)$ is independent of all $\mathcal{F}_s$-measurable random variables whenever $0 \le s < t$

In particular, note that any one-dimensional Brownian motion $(W_t)_{t\ge0}$ is an $(\mathcal{F}_t^W)_{t\ge0}$-Brownian motion where $(\mathcal{F}_t^W)_{t\ge0}$ is its natural filtration defined above.

In many problems involving stochastic partial differential equations, we will have a probability space and a (random) one-dimensional Brownian motion that drives random stochastic perturbations in the system, which is a Brownian motion with respect to some filtration $(\mathcal{F}_t)_{t\ge0}$. Since the Brownian motion (and its associated filtration) is an essential feature of the problem driving the stochastic dynamics, it is useful to augment the underlying probability space $(\Omega, \mathcal{F}, \mathbb{P})$ with the Brownian motion and its associated filtration, to form a *stochastic basis*.

Definition 3.5.4 (Stochastic Basis) A **stochastic basis** is a tuple $(\Omega, \mathcal{F}, (\mathcal{F}_t)_{t\ge0}, \mathbb{P}, W)$ where:

- Ω is a sample space representing all outcomes
- $\mathcal{F}$ is a sigma-algebra on Ω, representing all measurable events
- $(\mathcal{F}_t)_{t\ge0}$ is a right-continuous and complete filtration, representing information available up to time t
- $\mathbb{P}$ is a probability measure
- W is an $(\mathcal{F}_t)_{t\ge0}$-Wiener process

Brownian motion has many interesting properties, and we refer the reader to the exposition in references, such as [290], Chapter 7 of [128], and Chapter 14 of [201], for more detailed information about Brownian motion and its relevant probabilistic properties. We will conclude our brief summary of Brownian motion by describing some of its fundamental path properties, in particular its properties when considered as a random function of time $t \in [0, \infty)$, that will play a role in describing some of the subtleties underlying the theory of stochastic integration. As we shall see below in Proposition 3.5.2, one of the properties of Brownian motion is that its paths are almost surely α-Hölder continuous for $\alpha \in (0, 1/2)$. Before stating the proposition, we recall here the definition of functions of finite variation.

Definition 3.5.5 A real-valued function $f : [a, b] \to \mathbb{R}$ is **of finite variation** if

$$\sup_{\mathcal{T}} \sum_{i=1}^{|\mathcal{T}|} |f(t_i) - f(t_{i-1})| < \infty,$$

where the supremum is over all finite partitions $\mathcal{T}$ of the interval $[a, b]$ consisting of points:

$$a = t_0 < t_1 < \ldots < t_{N-1} < t_N = b,$$

where $N = |\mathcal{T}|$ is the number of subintervals in the partition.

We then have the following results about the path properties of one-dimensional Brownian motion.

Proposition 3.5.2 *Let $(W_t)_{t \geq 0}$ be a one-dimensional Brownian motion on a probability space $(\Omega, \mathcal{F}, \mathbb{P})$:*

- *With probability 1, the paths of Brownian motion are not differentiable for all $t \geq 0$.*
- *More generally, with probability 1, the paths of Brownian motion are not α-Hölder continuous on any closed interval for $\alpha \in [1/2, 1)$.*
- *The paths of Brownian motion are not of finite variation on any closed interval with probability 1.*
- *The paths of Brownian motion with probability 1 are locally α-Hölder continuous for $\alpha \in (0, 1/2)$ (see Definition 2.1.2).*

For more information about these properties, see Theorem 14.5 and Proposition 14.10 in [201] and Chapter I, Section 2 of [290].

3.6 Stochastic Integration and the Itô Integral

Next, we discuss the construction of the Itô integral in a Hilbert space H, which provides a framework for integrating processes with respect to one-dimensional white noise $dW(t)$. For this construction, we follow the presentation in [104].

As we discussed in the previous subsection, Brownian motion is almost surely not differentiable at any time, meaning that with probability one, the path $W(t, \omega) : [0, \infty) \to \mathbb{R}$ is not differentiable at any time $t \geq 0$, and so the formal expression $dW(t)$ for white noise does not make sense in a classical pathwise way. This complicates the construction of the stochastic integral for the following reason. If we wanted to define the stochastic integral

$$\int_0^T X(t, \omega) dW(t),$$

where $X(t, \omega)$ is a Hilbert space H-valued progressively measurable stochastic process, a natural approach is to first follow the procedure used in Riemann integration. Given a partition $\mathcal{T}$ of the interval $[0, T]$ with points

$$\mathcal{T} = \{t_0, \ldots, t_N\}, \qquad 0 = t_0 < t_1 < \ldots < t_{N-1} < t_N = T, \tag{3.13}$$

we consider the discrete sums:

$$\sum_{i=1}^{N} X(t_{i-1}, \omega) \cdot [W(t_i) - W(t_{i-1})],$$

which define an H-valued random variable. We then aim to take the limit of these sums almost surely, allowing us to establish a pathwise limit for each ω in a measurable subset of Ω of probability one. However, these random sums do not necessarily converge pathwise as the width of the partition

$$\Delta\mathcal{T} := \max_{1 \le i \le N} (t_i - t_{i-1}) \to 0.$$

This is due to the fact that Brownian motion paths are nowhere differentiable and, with probability one, have infinite variation over any finite interval. As a result, the convergence of these sums must be understood in probability rather than almost surely. This distinction highlights that while Brownian motion exhibits well-behaved properties in a probabilistic sense, it does not do so on a path-by-path basis; see Proposition 2.13 in Chapter IV, Section 2 of [290].

We will thus use a different approach to define the stochastic integral, where we will define stochastic integration of elementary integrands and then use a density argument to extend the stochastic integral to a more general class of integrands.

Consider an $(\mathcal{F}_t)_{t\in[0,T]}$-Brownian motion $(W_t)_{t\in[0,T]}$ on a filtered probability space $(\Omega, \mathcal{F}, (\mathcal{F}_t)_{t\in[0,T]}, \mathbb{P})$. Then, we define the collection $\mathcal{S}$ of **elementary integrands**, i.e., processes f of the following form: For some $0 \le a_1 < b_1 < a_2 < b_2 < \ldots < a_n < b_n \le T$ and H-valued $\mathcal{F}_{a_i}$-measurable random variables X_i, we can write f as

$$f(t, \omega) = \sum_{i=1}^{n} X_i(\omega) 1_{(a_i, b_i]}(\omega). \tag{3.14}$$

For **elementary integrands** $f \in \mathcal{S}$, we now define the **stochastic Itô integral** to be given by the following:

$$\int_0^T f(t, \omega) dW(t) = \sum_{i=1}^{n} X_i(\omega) \cdot [W(b_i) - W(a_i)]. \tag{3.15}$$

For more on elementary integrands and elementary processes, see Chapter IV, Section 2 of [290].

Next, we extend the definition of the stochastic Itô integral from elementary predictable integrands to more general integrands by establishing the following identity, which is known as the **Itô isometry**.

Proposition 3.6.1 (Itô Isometry) *For elementary predictable integrands $f \in \mathcal{S}$, the stochastic integral defined in* (3.15) *is a continuous H-valued $(\mathcal{F}_t)_{t\in[0,T]}$–martingale and the following holds true:*

$$\mathbb{E}\left[\left\|\int_0^T f(t,\omega)dW(t)\right\|_H^2\right] = \mathbb{E}\left(\int_0^T \|f(t,\omega)\|_H^2 dt\right).$$

We can extend the definition of the Itô integral to a broader class of integrands by utilizing a density argument that uses the Itô isometry.

The next lemma says that the stochastic integral is also well-defined for the class of progressively measurable square-integrable integrands.

Lemma 3.6.1 *Let W be an $(\mathcal{F}_t)_{t\in[0,T]}$- Brownian motion defined on the probability space $(\Omega, \mathcal{F}, \mathbb{P})$. Let f be an $(\mathcal{F}_t)_{t\in[0,T]}$-progressively measurable stochastic process satisfying*

$$\mathbb{E}\int_0^T \|f\|_H^2 < \infty. \tag{3.16}$$

Then there exists a sequence of elementary $(\mathcal{F}_t)_{t\in[0,T]}$-adapted stochastic processes (f_n) such that

$$f_n \to f \text{ in } L^2(\Omega \times (0,T); H).$$

The stochastic integral $I_t(f) := \int_0^t f(s)\,dW(s)$ is obtained as a limit in $L^2(\Omega; C([0,T]; H))$ of the stochastic integrals $\int_0^t f_n(s)\,dW(s)$.

For any progressively measurable integrand f satisfying (3.16), we can see that the Itô isometry, i.e., Proposition 3.6.1, holds as well and the stochastic integral $I_t(f)$ is a square-integrable H-valued $(\mathcal{F}_t)_{t\in[0,T]}$-martingale.

Moreover, for any partition $\mathcal{T}_N = \{t_0, t_1, \dots, t_N\}$ of $[0,T]$ (see (3.13)) for which the width of the partition $\Delta\mathcal{T}_N \to 0$ as $N \to \infty$, we have the following convergence result for the time-discrete Riemann sums; see Proposition 2.13 in Chapter IV, Section 2 of [290]:

$$\int_0^T f(t,\omega)dW(t) = \mathbb{P}\text{-}\lim_{N\to\infty} \sum_{i=1}^N X(t_{i-1},\omega)\,[W(t_i) - W(t_{i-1})],$$

where "$\mathbb{P}$- lim" denotes convergence in probability.

Remark 3.6.1 Notice that, thanks to Lemma 3.3.1, the stochastic integral

$$I_t(f) = \int_0^t f(s, \omega) dW(s), \tag{3.17}$$

is well-defined for any random distribution f such that $f \in L^2(0, T; H)$ almost surely, given that the σ-algebra $\mathcal{F}_t$ at any $t \in [0, T]$ in the filtration generated by the joint history (see Definition 3.3.10) of W and f, is independent of $\sigma[W(s) - W(t)]$ for any $s > t$.

Now, we recall the following useful inequality, known as the Burkholder-Davis-Gundy (BDG) inequality; see [210] and Chapter IV, Section 4 of [290]. The BDG inequality for martingales is a fundamental result in stochastic analysis.

Theorem 3.6.1 (Burkholder-Davis-Gundy (BDG) Inequality) *For any $p \in [1, \infty)$, there exists a constant $C_p > 0$ such that for every continuous in time martingale M with $M(0) = 0$ such that $\mathbb{E}\left(\sup_{t\in[0,T]} \|M(t)\|_H^2\right) < \infty$ and for every stopping time τ, the following holds true:*

$$\frac{1}{C_p}\mathbb{E}\left([M]_\tau^{\frac{p}{2}}\right) \le \mathbb{E}\left(\sup_{t\in[0,\tau]} \|M(t)\|_H^p\right) \le C_p \mathbb{E}\left([M]_\tau^{\frac{p}{2}}\right),$$

where $[M]_t$ is the quadratic covariation of M. The constant C_p does not depend on M or τ.

The BDG inequality provides bounds on the moments of the supremum in time of martingales in terms of the moments of its quadratic variation. The interested reader can find more details about the BDG inequality and the definition of quadratic covariation in standard textbooks such as [9, 104, 290].

In practice, we will only use the following consequence of Theorem 3.6.1, which plays a crucial role in the study of stochastic integrals and Itô calculus.

Theorem 3.6.2 (BDG Inequality for Stochastic Integrals in Hilbert Spaces) *For each $p > 0$, there exists a positive constant c_p depending only on p (and not on T) such that*

$$\mathbb{E} \sup_{t\in[0,T]} \left(\left\| \int_0^t f(t, \omega) dW(t) \right\|_H^p \right) \le c_p \mathbb{E} \left(\int_0^T \|f(t, \omega)\|_H^2 dt \right)^{p/2},$$

for any progressively measurable $f \in L^2(\Omega; L^2(0, T; H))$.

Moreover, we also have the following discrete version of the BDG inequality given in [281] (cf. the continuous case stated in Theorem 3.6.1).

Theorem 3.6.3 (Discrete Burkholder-Davis-Gundy [281]) *Let $(\Omega, \mathcal{F}, \mathbb{P})$ be a probability space with discrete filtration $(\mathcal{F}_n)_{n=0}^N$. Let $(M_n)_{n=0}^N$ be a discrete $(\mathcal{F}_n)_{n=0}^N$-martingale taking values in a Hilbert space H such that $M_0 = 0$ and*

$\mathbb{E}\|M_n\|_H^2 < \infty$ *for every* $n \geq 0$. *Then for any* $p \geq 1$, *there exists a constant* $C_p > 0$ *such that*

$$\frac{1}{C_p}\mathbb{E}\left(\sum_{n=1}^{N}\|M_n - M_{n-1}\|_H^2\right)^{\frac{p}{2}} \leq \mathbb{E}\max_{0\leq n\leq N}\|M_n\|_H^p$$
$$\leq C_p\mathbb{E}\left(\sum_{n=1}^{N}\|M_n - M_{n-1}\|_H^2\right)^{\frac{p}{2}}. \tag{3.18}$$

Remark 3.6.2 One can additionally prove (see Lemma 2.1 in [139]) that for any $0 < \alpha < \frac{1}{2}$, $q \geq 2$ and any progressively measurable process $f \in L^q(\Omega; L^q(0,T;H))$, we have for the process defined in (3.17) that

$$\mathbb{E}\,\|I(f)\|_{W^{\alpha,q}(0,T;H)}^q \leq C\mathbb{E}\int_0^T \|f(t,\omega)\|_H^q dt, \tag{3.19}$$

for some C independent of f. See Lemma 2.1 in [139] for a proof.

Itô's Formula Itô's formula is a fundamental result in stochastic calculus that extends the classical chain rule of calculus to stochastic processes driven by Brownian motion of the form

$$dX(t) = F(t)dt + G(t)dW(t). \tag{3.20}$$

We remark here that this form (3.20) is an informal way of writing the following integral equation (3.21), where the Hilbert space H-valued process X takes the form

$$X(t) = X_0 + \int_0^t F(s)ds + \int_0^t G(s)dW(s), \tag{3.21}$$

where $X_0 \in L^2(\Omega, \mathcal{F}_0, \mathbb{P}; H)$, $F: \Omega \times [0,T] \to H$ satisfies $F, G \in L^2(0,T;H)$ almost surely, and for some $T > 0$. We will now state the Itô formula for processes with values in a Hilbert space H (see [104, 256]).

Theorem 3.6.4 (Itô's Formula in Hilbert Spaces) *Let* X *be as in* (3.21). *Let* $\psi : H \to \mathbb{R}$ *be a* C^2 *function such that* ψ', ψ'' *are uniformly continuous on bounded subsets of* H. *Then for each* $t \geq 0$, *we have* $\mathbb{P}$-*a.s.:*

$$\psi(X(t)) = \psi(X_0) + \int_0^t \langle\psi'(X(s)), F(s)\rangle_H ds + \int_0^t \langle\psi'(X(s)), G(s)dW(s)\rangle_H$$
$$+ \frac{1}{2}\int_0^t \psi''(X(s))\|G(s)\|_H^2 ds.$$

The following result ensures the robustness of the probabilistic structure underlying the stochastic differential equation (SDE) (3.21). Specifically, the following proposition states that if the solution $X(t)$ and the driving Brownian motion $W(t)$ of the SDE are given on one probability space, and another triplet $(\tilde{X}, \tilde{X}_0, \tilde{W})$ has the same distribution, then there exists a filtration on the new probability space that allows $\tilde{X}$ to also satisfy the same SDE. This result guarantees that the dynamics of the solution are preserved even when reformulated on a different probability space. More precisely, the following classical result of Bensoussan [34] holds.

Proposition 3.6.2 *Suppose that* $X \in L^2(\Omega; L^2(0, T; H))$ *is a solution to* (3.21) *with respect to the stochastic basis* $(\Omega, \mathcal{F}, (\mathcal{F}_t)_{t\geq 0}, \mathbb{P}, W)$. *Suppose that* $(\tilde{X}, \tilde{X}_0, \tilde{W}) \overset{d}{=} (X, X_0, W)$ *where* $(\tilde{X}, \tilde{X}_0, \tilde{W})$ *is defined on a (possibly different) probability space* $(\tilde{\Omega}, \tilde{\mathcal{F}}, \tilde{\mathbb{P}})$. *Then there exists a complete, right-continuous filtration* $(\tilde{\mathcal{F}}_t)_{t\geq 0}$ *such that* $\tilde{X}$ *solves* (3.21) *with respect to the stochastic basis* $(\tilde{\Omega}, \tilde{\mathcal{F}}, (\tilde{\mathcal{F}}_t)_{t\geq 0}, \tilde{\mathbb{P}}, \tilde{W})$ *with initial condition* $\tilde{X}_0$.

In Chaps. 8 and 9, we will construct approximations of solutions of stochastically perturbed partial differential equations that, in the limit, will be the desired solutions. Therefore, it is crucial to ensure that we can pass to the limit in stochastic integrals as we refine the approximation parameters used to construct these approximations. The ability to pass to the limit in stochastic integrals is guaranteed by the following result of Bensoussan [34]:

Lemma 3.6.2 *Let* H *be a Hilbert space. Let* $(\Omega, \mathcal{F}, \mathbb{P})$ *be a complete probability space. For* $n \in \mathbb{N}$, *consider the filtration* $(\mathcal{F}_t^n)_{t\geq 0}$ *and let* W_n *be an* $(\mathcal{F}_t^n)_{t\geq 0}$-*Brownian motion and let* G_n *be an* $(\mathcal{F}_t^n)_{t\geq 0}$-*progressively measurable stochastic process such that* $G_n \in L^2(0, T; H)$ *almost surely. Finally consider a filtration* $(\mathcal{F}_t)_{t\geq 0}$, *an* $(\mathcal{F}_t)_{t\geq 0}$-*Wiener process* W, *and* $G \in L^2(0, T; H)$, *which is* $(\mathcal{F}_t)_{t\geq 0}$-*progressively measurable. Suppose that*

$$W_n \to W \text{ in } C([0, T]; \mathbb{R}) \text{ in probability},$$

$$G_n \to G \text{ in } L^2(0, T; H) \text{ in probability}.$$

Then

$$\int_0^{\cdot} G_n \, dW_n \to \int_0^{\cdot} G \, dW \text{ in } L^2(0, T; H) \text{ in probability}.$$

The proof for this direct technique can be found in Lemma 2.1 in [113] (see also Lemma 2.6.6 in [52]).

3.7 White Noise and Stochastic Integration

We conclude this chapter by extending our past discussion of one-dimensional white noise and stochastic integration to the context of spacetime white noise, which is random noise that is formally independent at every point in space and time.

In this section, we review the concept of spacetime white noise on $\mathbb{R}^+ \times \mathbb{R}^d$ and stochastic integration against white noise. Note that we will use $\mathbb{R}^+$ to denote $[0, \infty)$, which represents the time variable. While we will be primarily concerned with dimensions $d = 1, 2$, we will define white noise in full generality, as the extension to higher dimensions is no more difficult.

We follow the exposition that can be found in [210] about martingale measures and refer the reader to the original reference by Walsh [346] for more details. We note that while the forthcoming analysis can be carried out more generally for martingale measures, we will restrict to the case of white noise for simplicity. The full martingale measure theory can be found in [346] and [210].

Recall that a *Gaussian process* is a process $\{G_i\}_{i \in I}$, such that the finite-dimensional random vectors

$$(G_{i_1}, G_{i_2}, \dots, G_{i_k}), \; i_1, i_2, \dots, i_k \in I$$

have distributions that are multivariable Gaussian, for any finite collection of $i_1, i_2, \dots, i_k \in I$.

We will define the *covariance function* to be the symmetric function $C : I \times I \to \mathbb{R}$ that gives the covariance of any two Gaussians G_{i_1} and G_{i_2}:

$$C(i_1, i_2) = \mathbb{E}[(G_{i_1} - \mathbb{E}(G_{i_1}))(G_{i_2} - \mathbb{E}(G_{i_2}))].$$

For a mean zero Gaussian process, which is a Gaussian process $\{G_i\}_{i \in I}$ such that $\mathbb{E}[G_i] = 0$ for all $i \in I$, this reduces to the simpler formula:

$$C(i_1, i_2) = \mathbb{E}[G_{i_1} G_{i_2}].$$

We will now define white noise as a Gaussian process, taking for granted the existence of such a process.

Definition 3.7.1 (White Noise on $\mathbb{R}^+ \times \mathbb{R}^d$**)** Let $\mathrm{Borel}(\mathbb{R}^+ \times \mathbb{R}^d)$ denote the collection of all Borel subsets of $\mathbb{R}^+ \times \mathbb{R}^d$. White noise on $\mathbb{R}^+ \times \mathbb{R}^d$ is a *mean zero* Gaussian process $\{\dot{W}(A)\}_{A \in \mathrm{Borel}(\mathbb{R}^+ \times \mathbb{R}^d)}$ indexed by the Borel subsets of $\mathbb{R}^+ \times \mathbb{R}^d$, with the covariance function:

$$C(A, B) := \mathbb{E}[\dot{W}(A)\dot{W}(B)] = \lambda(A \cap B), \qquad \text{for } A, B \in \mathrm{Borel}(\mathbb{R}^+ \times \mathbb{R}^d), \tag{3.22}$$

where λ is the Lebesgue measure in $\mathbb{R}^+ \times \mathbb{R}^d$.

Some basic facts about white noise that will be useful later are summarized in the following proposition.

Proposition 3.7.1 *Let* $\{\dot{W}(A)\}_{A \in Borel(\mathbb{R}^+ \times \mathbb{R}^d)}$ *denote white noise defined on a probability space* $(\Omega, \mathcal{F}, \mathbb{P})$*. Then, the following holds true:*

- *For each bounded set* $A \in Borel(\mathbb{R}^+ \times \mathbb{R}^d)$*,* $\dot{W}(A)$ *is normally distributed with mean* 0 *and variance* $\lambda(A)$*, namely,* $\dot{W}(A) \sim N(0, \lambda(A))$*. So* $\dot{W}(A) \in L^2(\Omega)$*.*
- *If* $A \cap B = \varnothing$*, then* $\dot{W}(A)$ *and* $\dot{W}(B)$ *are independent.*
- *Given* $A, B \in Borel(\mathbb{R}^+ \times \mathbb{R}^d)$*,* $\dot{W}(A \cup B) = \dot{W}(A) + \dot{W}(B) - \dot{W}(A \cap B)$*, almost surely (a.s.), as random variables.*
- *White noise is a signed measure taking values in* $L^2(\Omega)$*, namely,* $\dot{W} : Borel(\mathbb{R}^+ \times \mathbb{R}^d) \to L^2(\Omega)$*. Furthermore, white noise considered as a measure is* σ*-finite.*

Proof The first point follows from the fact that white noise is a mean zero Gaussian process, and $\mathbb{E}[(\dot{W}(A))^2] = \lambda(A \cap A) = \lambda(A)$ by (3.22). The second and third points are from Exercise 3.15 in [210]. The second point follows from the fact that $\dot{W}(A)$ and $\dot{W}(B)$ are mean zero Gaussians with zero covariance, by applying (3.22). The third fact follows from the computation of the expectation $\mathbb{E}[(\dot{W}(A \cup B) - \dot{W}(A) - \dot{W}(B) + \dot{W}(A \cap B))^2] = 0$. One can verify this by expanding the square and applying (3.22) repeatedly. Note that the third property gives the finite additivity properties of a measure. For the final property, one must check that white noise has the remaining properties of a measure, and we refer the reader to the proof of Proposition 5.1 in [210]. □

Remark 3.7.1 Heuristically, one thinks of white noise as random noise that is "independent" at every point in time and space. One can then interpret $\dot{W}(A)$ heuristically as being the net contribution of the noise in A. With this heuristic interpretation, it is at least intuitively reasonable that white noise has the properties of a measure. The fact that the noise is independent at every point in time and space is in accordance with the second property in Proposition 3.7.1.

Stochastic Integration Against White Noise We will first define integration of simple functions against white noise and then proceed to the most general case by an approximation argument. For this purpose, we introduce the following nomenclature (see Sec. 5 of [210]):

- For any $A \in \text{Borel}(\mathbb{R}^d)$ and $t \in \mathbb{R}^+$, we use $W_t(A)$ to denote $W_t(A) = \dot{W}([0, t] \times A)$, for $A \in \text{Borel}(\mathbb{R}^d)$, so that $[0, t] \times A \in \text{Borel}(\mathbb{R}^+ \times \mathbb{R}^d)$.
- For $t > 0$, we consider the *filtration* $\{\mathcal{F}_t\}_{t \geq 0}$ associated with white noise where $\mathcal{F}_t$ is the σ-algebra generated by the collection of random variables $\{W_s(A) : s \in [0, t], A \in \text{Borel}(\mathbb{R}^d)\}$.
- We use $\mathcal{S}$ to denote the space of simple functions, which are functions of the form

$$f(t, x, \omega) = \sum_{i=1}^{n} X_i(\omega) 1_{(a_i, b_i]}(t) 1_{A_i}(x), \tag{3.23}$$

where X_i is a bounded, $\mathcal{F}_{a_i}$-measurable random variable with $0 \leq a_i < b_i$, and $A_i \in \text{Borel}(\mathbb{R}^d)$ is bounded.

Definition 3.7.2 Let $f \in \mathcal{S}$ be a simple function. We define

$$\int_0^t \int_{\mathbb{R}^d} f(s, x, \omega) W(dx, ds) = \int_0^t \int_{\mathbb{R}^d} \sum_{i=1}^n X_i(\omega) 1_{(a_i, b_i]}(s) 1_{A_i}(x) W(dx, ds)$$

$$:= \sum_{i=1}^n X_i(\omega)[W_{t \wedge b_i}(A_i) - W_{t \wedge a_i}(A_i)], \tag{3.24}$$

where the "wedge" notation corresponds to $\alpha \wedge \beta = \min\{\alpha, \beta\}$.

It is easy to check that the definition of the integral in (3.24) is independent of the representation of the simple function as (3.23).

We have the following crucial isometry property for the stochastic integral of simple functions against spacetime white noise. This is an extension of the Itô isometry to the stochastic integral against spacetime white noise.

Proposition 3.7.2 *For* $f \in \mathcal{S}$,

$$\mathbb{E}\left[\left(\int_0^\infty \int_{\mathbb{R}^d} f(t, x, \omega) W(dx, dt)\right)^2\right] = \mathbb{E}\left(\int_0^\infty \int_{\mathbb{R}^d} |f(t, x, \omega)|^2 dx dt\right). \tag{3.25}$$

Proof In the case where $f(t, x, \omega)$ is a simple function of the form

$$f(t, x, \omega) = X(\omega) 1_{(a,b]}(t) 1_A(x),$$

one easily checks that

$$\mathbb{E}\left[\left(\int_0^\infty \int_{\mathbb{R}^d} f(t, x, \omega) W(dx, dt)\right)^2\right] = \mathbb{E}\left(X^2(\omega)\, [W_b(A) - W_a(A)]^2\right)$$

$$= \mathbb{E}\left(\mathbb{E}\left[X^2(\omega)\, (W_b(A) - W_a(A))^2\, |\mathcal{F}_a\right]\right)$$

$$= \mathbb{E}\left(X^2(\omega) \mathbb{E}\left[(W_b(A) - W_a(A))^2\, |\mathcal{F}_a\right]\right),$$

where we used the fact that $X \in \mathcal{F}_a$ to take it out of the conditional expectation. Using the third property in Proposition 3.7.1,

$$\mathbb{E}\left[\left(\int_0^\infty \int_{\mathbb{R}^d} f(t, x, \omega) W(dx, dt)\right)^2\right] = \mathbb{E}\left(X^2(\omega) \mathbb{E}\left[\dot{W}^2((a, b] \times A) | \mathcal{F}_a\right]\right).$$

Using the second property in Proposition 3.7.1, we deduce that this is equal to

$$= \mathbb{E}\left[X^2(\omega)\right]\mathbb{E}\left[\dot{W}^2((a,b]\times A))\right] = \lambda(A)(b-a)\mathbb{E}\left[X^2(\omega)\right]$$
$$= \mathbb{E}\left(\int_0^\infty \int_{\mathbb{R}^d} |f(t,x,\omega)|^2 dxdt\right).$$

We note that (3.25) holds for general $f \in \mathcal{S}$, by choosing a representation (3.23) of an arbitrary simple function where the sets $(a_i \times b_i] \times A_i$ are disjoint, and then using the independence property in the second property listed in Proposition 3.7.1.

□

Next, we want to extend the definition of the stochastic integral to more general integrands. For this purpose, we recall the following definitions.

Definition 3.7.3 Let $f(t,x,\omega)$ be a real-valued function $f : \mathbb{R}^+ \times \mathbb{R}^d \times \Omega \to \mathbb{R}$.

1. We say that $f(t,x,\omega)$ is *adapted to the filtration* $(\mathcal{F}_t)_{t\geq 0}$ if the map $\omega \to f(t,x,\omega)$ is $\mathcal{F}_t$ measurable for each $x \in \mathbb{R}^d$ and $t \geq 0$.
2. We say that $f(t,x,\omega)$ is *jointly measurable* if it is measurable as a function in time, space, and the probability space, in particular if it is measurable as a map:

$$f : (\mathbb{R}^+ \times \mathbb{R}^d \times \Omega, \text{Borel}(\mathbb{R}^+ \times \mathbb{R}^d) \times \mathcal{F}) \to (\mathbb{R}, \text{Borel}(\mathbb{R})).$$

3. We say that $f(t,x,\omega)$ is a *predictable process* if it is measurable as a map $f : (\mathbb{R}^+ \times \mathbb{R}^d \times \Omega, \mathcal{P}) \to (\mathbb{R}, \text{Borel}(\mathbb{R}))$, where $\mathcal{P}$ is the **predictable σ-algebra** generated by the sets of the form

$$(a \times b] \times A \times E, \qquad a, b \in \mathbb{R} \text{ with } a < b, \quad A \in \text{Borel}(\mathbb{R}^d), \quad E \in \mathcal{F}_a.$$

Having defined the stochastic integral with respect to spacetime white noise for simple functions in Definition 3.7.2, we can extend the definition of the stochastic integral to integrands $f(t,x,\omega)$, which are predicable processes, satisfying the L^2 bound:

$$\mathbb{E}\left(\int_0^\infty \int_{\mathbb{R}^d} |f(t,x,\omega)|^2 dxdt\right) < \infty. \tag{3.26}$$

This can be done by utilizing a density argument that uses the Itô isometry. In particular, we use the fact that predictable processes $f(t,x,\omega)$ satisfying the bound (3.26) can be approximated in the norm of (3.26) by simple functions (see Proposition 2.3 in [346]). Hence, given such a predictable integrand f, there is a sequence $f_k \in \mathcal{S}$ such that $f_k \to f$ in the sense of

$$\lim_{k\to\infty} \mathbb{E}\left(\int_0^\infty \int_{\mathbb{R}^d} |(f - f_k)(t,x,\omega)|^2 dxdt\right) = 0.$$

Using the isometry relation in Proposition 3.7.2, one can show that the sequence

$$\left\{\int_0^\infty \int_{\mathbb{R}^d} f_k(t,x,\omega)W(dx,dt)\right\}_{k=1}^\infty \tag{3.27}$$

is therefore a Cauchy sequence in $L^2(\Omega)$.

Definition 3.7.4 The random variable obtained in the limit of integrals (3.27) is the *stochastic integral* $\displaystyle\int_0^\infty \int_{\mathbb{R}^d} f(t,x,\omega)W(dx,dt)$.

Proposition 3.7.3 *Given a predictable process* $f(t,x,\omega)$ *satisfying* (3.26)*, the stochastic integral against spacetime white noise* $\displaystyle\int_0^\infty \int_{\mathbb{R}^d} f(t,x,\omega)W(dx,dt)$ *is well-defined and the Itô isometry* (3.25) *holds.*

Proof See Theorem 5.26 in [210] and Theorem 2.5 in [346] for a proof of the Itô isometry, which is carried out more generally in the context of worthy martingale measures. □

We can also define the integral on bounded time intervals, by noting that

$$\int_0^T \int_{\mathbb{R}^d} f(t,x,\omega)W(dx,dt) = \int_0^\infty \int_{\mathbb{R}^d} 1_{(0,T]}(t)f(t,x,\omega)W(dx,dt).$$

More generally, we can identify a broad class of admissible integrands for integration against spacetime white noise. We denote by $\mathcal{P}_W$ the collection of admissible integrands, obtained as the closure of the simple functions within the space of jointly measurable processes $f(t,x,\omega)$ satisfying (3.26), with respect to the norm defined by

$$\|f\|_{\mathcal{P}_W}^2 = \mathbb{E}\left(\int_0^\infty \int_{\mathbb{R}^d} |f(t,x,\omega)|^2 dxdt\right).$$

The usual properties, such as the Itô isometry, hold for this broader class of admissible integrals $\mathcal{P}_W$, and we have already observed that indeed, the collection of predictable processes $f(t,x,\omega)$ satisfying the bound (3.26) is contained within the class of admissible integrands $\mathcal{P}_W$. Since the characterization of admissible integrands in $\mathcal{P}_W$ is at the moment abstract, we list a set of criteria that will help us determine whether a given integrand is admissible. This will be accomplished in the following proposition, which follows directly from Proposition 2 in [108] (see also [108] for additional discussion about the class of integrands $\mathcal{P}_W$):

Proposition 3.7.4 *Let* $\{f(t,x)\}_{t\in[0,T],x\in\mathbb{R}^d}$ *be a stochastic process adapted to the filtration* $\{\mathcal{F}_t\}_{t\in[0,T]}$ *such that the following conditions hold:*

1. *Joint measurability:* $(t,x,\omega) \to f(t,x,\omega)$ *is Borel*$([0,T]\times\mathbb{R}^d)\times\mathcal{F}_T$ *measurable.*

2. *Finite second moments:* $\mathbb{E}\left(|f(t,x)|^2\right) < \infty$ *for all* $t \in [0,T]$, $x \in \mathbb{R}^d$.
3. *Continuity in* $L^2(\Omega)$: *The process* f *considered as a map* $(t,x) \in [0,T]\times\mathbb{R}^d \to L^2(\Omega)$ *is continuous in* $L^2(\Omega)$.
4. *Square integrability:* $\mathbb{E}\left(\int_0^T\int_{\mathbb{R}^d}|f(s,y)|^2 dyds\right) < \infty$.

Then, the stochastic integral

$$\int_0^t\int_{\mathbb{R}^d} f(s,y)W(ds,dy)$$

is defined for all $t \in [0,T]$.

Proof This proposition follows from Proposition 2 of [108], and Proposition 2 of [108] is adapted to the current context. Though Proposition 2 of [108] is stated for the more general case of spatially homogeneous Gaussian noise, the statement of Proposition 2 of [108] specialized to the case of white noise reads as follows:

Let $(f(t,x))_{t\in\mathbb{R}^+,x\in\mathbb{R}^d}$ be a stochastic process adapted to the filtration $(\mathcal{F}_t)_{t\geq 0}$ and define $\mathcal{F} = \bigcup_{t\geq 0}\mathcal{F}_t$. Suppose the following conditions hold:

1. Joint measurability: $(t,x,\omega) \to f(t,x,\omega)$ is Borel$(\mathbb{R}_+\times\mathbb{R}^d)\times\mathcal{F}$ measurable.
2. Finite second moments: $\mathbb{E}\left(|f(t,x)|^2\right) < \infty$ for all $t \in \mathbb{R}^+$, $x \in \mathbb{R}^d$.
3. Continuity in $L^2(\Omega)$: The process f considered as a map $(t,x) \in \mathbb{R}^+\times\mathbb{R}^d \to L^2(\Omega)$ is continuous in $L^2(\Omega)$.
4. Square integrability on a compact set and finite time: There exists a compact set $K \subset \mathbb{R}^d$ and $t_0 > 0$ such that

$$\mathbb{E}\left(\int_0^{t_0}\int_K |f(s,y)|^2 dyds\right) < \infty.$$

Then, $1_{[0,t_0]\times K}(t,x)f(t,x) \in \mathcal{P}_W$.

While the result in Proposition 2 of [108] is stated specifically for spatial dimension two, one can verify that it holds for arbitrary dimension.

To see that the statement of Proposition 2 of [108] implies the result in Proposition 3.7.4, let K_i be a sequence of compact sets that increase to $\mathbb{R}^d$, and consider $(f(t,x))_{t\in[0,T],x\in\mathbb{R}^d}$ satisfying the four conditions in Proposition 3.7.4. We extend $(f(t,x))_{t\in[0,T],x\in\mathbb{R}^d}$ to be defined on all of time $t \geq 0$ by defining

$$\tilde{f}(x,t) = f(x,t) \quad \text{if } t \in [0,T], \qquad \tilde{f}(x,t) = f(x,T) \quad \text{if } t \geq T.$$

Then, $(\tilde{f}(t,x))_{t\in\mathbb{R}^+,x\in\mathbb{R}^d}$ along with $t_0 = T$, and each K_i satisfies the conditions in Proposition 2 in [108]. Therefore, $1_{[0,T]\times K_i}(t,x)f(t,x) = 1_{[0,T]\times K_i}(t,x)\tilde{f}(t,x) \in \mathcal{P}_W$.

Since the fourth condition of Proposition 3.7.4 states that

$$\mathbb{E}\left(\int_0^T \int_{\mathbb{R}^d} |f(s, y)|^2 dyds\right) < \infty,$$

we have that $1_{[0,T]\times K_i}(t, x)f(t, x) \to 1_{[0,T]\times\mathbb{R}^d}(t, x)f(t, x)$ in the norm of $\mathcal{P}_W$, since K_i is a sequence of compact sets in $\mathbb{R}^d$ increasing to all of $\mathbb{R}^d$. Hence, $1_{[0,T]\times\mathbb{R}^d}(t, x)f(t, x) \in \mathcal{P}_W$ since $\mathcal{P}_W$ is complete with respect to its norm. □

A couple of remarks are in order. The first one uses the concept of *modification*, which we now recall.

Definition 3.7.5 Let $\{f(t, x)\}_{t\in[0,T],x\in\mathbb{R}^d}$ be a stochastic process. Then $\{\tilde{f}(t, x)\}$ $_{t\in[0,T],x\in\mathbb{R}^d}$ is a *modification* of $(f(t, x))_{t\in[0,T],x\in\mathbb{R}^d}$ if

$$\mathbb{P}(f(t, x) = \tilde{f}(t, x)) = 1, \qquad \text{for all } t \in [0, T], x \in \mathbb{R}^d.$$

Remark 3.7.2 The third condition in Proposition 3.7.4 implies that there is a jointly measurable modification (see the discussion on pg. 201 of [108], and the proof of Theorem 13 in [106]). Thus, in practice, one does not need to check the first condition, as by taking a modification, the third condition implies the first.

Finally, we recall the following useful inequality, which is a direct consequence of a classical result known as the BDG (Burkholder-Davis-Gundy) inequality, which will be used frequently; see [210] and Chapter IV, Section 4 of [290].

Theorem 3.7.1 *For each $p \geq 2$, there exists a positive constant c_p depending only on p (and not on T) such that*

$$\mathbb{E}\left(\left|\int_0^T \int_{\mathbb{R}^d} f(t, x, \omega)W(dx, dt)\right|^p\right) \leq c_p\mathbb{E}\left(\left(\int_0^T \int_{\mathbb{R}^d} |f(t, x, \omega)|^2 dxdt\right)^{p/2}\right),$$

for all $f \in \mathcal{P}_W$.

Part II
Deterministic Fluid-Structure Interaction

Chapter 4
Deterministic FSI with Linear Coupling

The mathematical analysis of fluid-structure interaction (FSI) began with the study of linearly coupled FSI models. These models refer to mathematical frameworks that describe the interaction between a fluid and a structure, where the coupling is assumed across a fixed interface. In these models, the fluid's motion influences the structure, and the structure's deformation, in turn, affects the fluid's flow. However, the interaction is modeled through linear approximations, making the system more mathematically tractable. The fluid equations are posed on a fixed reference domain, Ω_f, representing a linear approximation of the corresponding moving boundary problem.

Linearly coupled FSI models are useful in situations where the fluid's impact on the structure or the structure's deformation is relatively small, such as slight vibrations or low-speed fluid flows. These models are commonly applied in fields such as aerodynamics, bioengineering (e.g., blood flow in arteries), and engineering (e.g., wind interaction with tall buildings). They provide a valuable approach to studying fluid-structure interactions, especially in cases where more complex nonlinear models are not required.

Despite their simplifications, linearly coupled FSI models still present challenges in terms of analyzing their well-posedness due to the two-way coupling between the fluid and structure and the multiphysics nature of the problem. In this section, we will review significant results in the study of linearly coupled FSI models and introduce a prototypical model of linearly coupled FSI. We describe a Lie operator splitting approach to a constructive proof of existence of weak solutions. Although the Lie operator splitting method was later applied to fluid-structure interactions in the context of nonlinearly coupled models with moving boundaries (see, e.g., [265, 333]), we will summarize its application to the prototypical linearly coupled FSI model, as these splitting schemes will be important for analyzing models presented later in the text.

S. Čanić et al., *Deterministic and Stochastic Fluid-Structure Interaction*, Advances in Mathematical Fluid Mechanics, https://doi.org/10.1007/978-3-032-00898-5_4

In the next section, we present a prototypical linearly coupled FSI problem in which the structure is lower-dimensional, and it is a part of the fluid domain boundary. This problem embodies the main difficulties associated with the analysis of linearly coupled, deterministic FSI problems.

4.1 Problem Description

Consider a prototypical linearly coupled model involving the interaction between a two-dimensional incompressible Newtonian fluid and a one-dimensional elastic structure. The fluid flows through a two-dimensional channel, $[0, L] \times [-R, R]$, with elastic top and bottom walls. The flow is driven by pressure data prescribed at the channel's left and right ends.

We are interested in a problem that is symmetric with respect to the central symmetry axis of the channel $[0, L] \times \{0\}$. In this case, it is sufficient to consider only (top) half of the channel, which we will denote by

$$\Omega_f = \{(z, r) | z \in [0, L], r \in [0, R]\},$$

and impose symmetry boundary conditions (see (4.5)) to the bottom boundary of Ω_f:

$$\Gamma_b = [0, L] \times \{0\}.$$

The inlet and outlet boundary of Ω_f will be denoted by

$$\Gamma_{in} = \{0\} \times [0, R] \qquad \text{and} \qquad \Gamma_{out} = \{L\} \times [0, R].$$

The prescribed inlet/outlet data will be symmetric with respect to the symmetry axis Γ_b, and since the fluid domain is symmetric, we expect the solution to be symmetric so that the dynamics on Ω_f can be reflected across the line Γ_b to retrieve the full (symmetric) dynamics in the full elastic channel.

The elastic walls of the channel will be represented by the top boundary, defined by

$$\Gamma = [0, L] \times \{R\},$$

which will correspond to the reference configuration of the deformable wall. Hence, Γ represents the reference configuration of the fluid-structure interface. See Fig. 4.1.

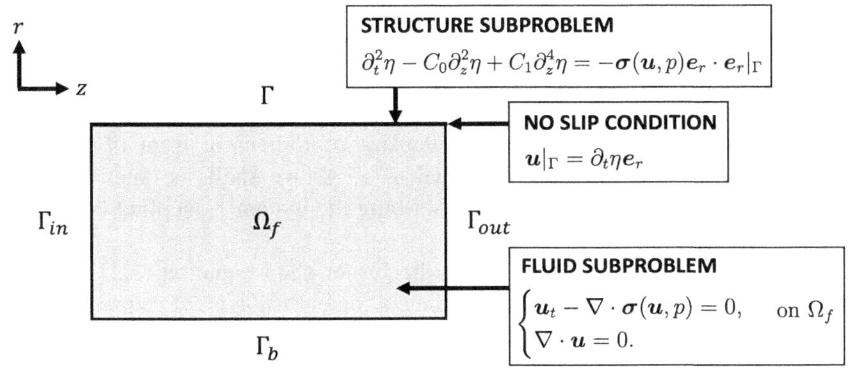

Fig. 4.1 The domain for the linearly coupled prototypical FSI problem in two dimensions, describing the interaction between a fluid modeled by the linear Stokes equations and a thin elastic structure. Though the structure displaces from its reference configuration, by the linear coupling, the interface conditions are evaluated on the reference fluid-structure interface Γ and the linear Stokes equations are defined on the fixed fluid reference domain Ω_f

4.1.1 The Prototypical Model Equations

We present the partial differential equations governing the dynamics of the fluid and structure separately and then discuss the conditions that describe the fluid and structure coupling at the interface.

Structure Subproblem The elastodynamics of the lateral channel walls will be described by the equation(s) written in terms of the displacement $\boldsymbol{\eta} : \Gamma \to \mathbb{R}^2$ of the channel wall (i.e., structure) from its reference configuration Γ. Thus, the equations will be written in Lagrangian formulation as follows:

$$\rho_s h \partial_{tt} \boldsymbol{\eta} + \mathcal{L}_e \boldsymbol{\eta} = \mathbf{f}, \quad \text{on } \Gamma, \ t \in (0, T), \tag{4.1}$$

where ρ_s is the structure density, h is the thin structure thickness, and $\mathbf{f}$ is the loading (force density) driving the motion of the elastic structure, and $\mathcal{L}_e$ is an operator describing the elastic properties of the structure. We will be assuming that operator $\mathcal{L}_e$ is the operator associated with a continuous, coercive (symmetric) bilinear form on the Hilbert space X.

In this text, $\mathcal{L}_e$ will be a fourth-order operator describing a thin elastic plate/shell, as specified in (4.2). Other operators have been considered in the literature, including the operators associated with the linear and nonlinear Koiter shell problems (see [265, 266, 268, 269]) and a generalized string model studied in [285].

In this prototypical example, we will be assuming that the structure **displaces only in the radial/vertical/normal direction**, namely, that $\boldsymbol{\eta} = \eta \boldsymbol{e}_r$ where $\eta : \Gamma \to \mathbb{R}$ is a scalar function and $\boldsymbol{e}_r$ is the unit vector in the r direction.

As a prototypical example, we consider the following elastic plate/shell equation in $1D$:

$$\partial_t^2\eta - C_1\partial_z^2\eta + C_2\partial_z^4\eta = F \qquad \text{on } \Gamma, \tag{4.2}$$

where C_1 and C_2 are positive (constant) elasticity coefficients and F is the source term describing the external load on the structure. The equation describes the second Newton's law of motion. The plate's density coefficient in front of the plate's acceleration $\partial_t^2\eta$ is assumed to be equal to 1. As we shall see later, F will be specified by the coupling conditions describing the dynamic coupling between the fluid and structure.

Notice that Eq. (4.2) corresponds to the Koiter shell equation (2.31), stated in the Preliminaries section, Sect. 2, with the coefficient C_0 in (2.31) equal to zero. In terms of analysis, the term $C_0\eta$ in (2.31) does not play an important role as all the analysis presented below carries over for the Koiter shell model (2.31). See, e.g., [265].

We assume that the elastic structure is clamped at the ends so that the following **boundary conditions** hold:

$$\eta(0) = \eta(L) = \partial_z\eta(0) = \partial_z\eta(L),$$

where $z = 0$ and $z = L$ represent the left and right endpoints of Γ.

In terms of **initial conditions**, we will be assuming that the initial displacement and the initial structure velocity are given and equal to some functions $\eta_0 \in H_0^2(\Gamma)$ and $v_0 \in L^2(\Gamma)$, respectively.

Fluid Subproblem We will be using the linear Stokes equations to describe the dynamics of an incompressible, viscous, Newtonian fluid. The use of the time-dependent Stokes equations, rather than the Navier-Stokes equations, is common in FSI problems with linear coupling, primarily due to energy considerations, which we will explain later. The equations are given in terms of the fluid velocity $\boldsymbol{u} : \Omega_f \to \mathbb{R}^2$ and the fluid pressure $p : \Omega_f \to \mathbb{R}$ as follows:

$$\begin{cases} \boldsymbol{u}_t - \nabla \cdot \boldsymbol{\sigma}(\boldsymbol{u}, p) = 0, \\ \nabla \cdot \boldsymbol{u} = 0. \end{cases} \tag{4.3}$$

The first equation describes the balance of momentum via Newton's second law of motion, where $\boldsymbol{\sigma}(\boldsymbol{u}, p)$ is the Cauchy stress tensor. The second equation in (4.3), known as the incompressibility condition, describes the conservation of mass principle for an incompressible fluid. We will be assuming that the fluid is Newtonian by specifying the Cauchy stress tensor as follows:

$$\boldsymbol{\sigma}(\boldsymbol{u}, p) = 2\mu_f \boldsymbol{D}(\boldsymbol{u}) - p\boldsymbol{I}, \qquad \text{where } \boldsymbol{D}(\boldsymbol{u}) = \frac{1}{2}[(\nabla\boldsymbol{u}) + (\nabla\boldsymbol{u})^t].$$

Here $\boldsymbol{D}(\boldsymbol{u})$ is the symmetrized gradient of fluid velocity, and $\mu_f > 0$ is the (constant) fluid viscosity.

The system (4.3) is supplemented with initial and boundary data. We will be assuming that the **initial data** $\boldsymbol{u}_0$ belongs to $L^2(\Omega_f)$ and that the flow is driven by the following **boundary data**:

$$p|_{\Gamma_{in}} = P_{in}(t) \text{ and } p|_{\Gamma_{out}} = P_{out}(t), \quad u_r = 0 \text{ on } \Gamma_{in} \cup \Gamma_{out}, \tag{4.4}$$

where $P_{in}(t)$ and $P_{out}(t)$ are time-dependent inlet and outlet pressure data, and $u_r = 0$ states that the flow at the inlet and outlet is purely horizontal.

On the bottom boundary Γ_b, we impose the boundary condition, which states that the flow velocity respects the even symmetry at the central axis Γ_b:

$$u_r = 0 \quad \text{and} \quad \partial_r u_z = 0, \qquad \text{on } \Gamma_b. \tag{4.5}$$

The Coupling Conditions The fluid and structure problems described above communicate via a set of coupling conditions. We consider a problem in which the fluid and structure communicate via a *two-way coupling*, which means that the motion of the structure and the dynamics of the fluid both mutually affect each other. Two-way coupling is described by a set of two coupling conditions: the kinematic and dynamic coupling conditions. The kinematic coupling condition couples the kinematic quantities such as velocity. We will be assuming in this prototypical problem that the fluid and structure velocities are continuous at the fluid-structure interface. This is known as the *no-slip condition* since the fluid particles in direct contact with the interface stick to the interface and move with the same velocity as the interface. The dynamic coupling condition describes the balance of forces. In particular, in this prototypical problem, the dynamic coupling condition states that the elastodynamics of the structure is driven by the jump in the normal stress (traction) across the interface. In the case of linear coupling, these conditions are evaluated at the fixed, reference interface Γ. More precisely, we consider the following coupling conditions:

- **Kinematic coupling condition.** Since we assume that the structure displaces in only the radial/vertical direction, the no-slip condition takes the following form:

$$\boldsymbol{u}|_\Gamma = \partial_t \eta \boldsymbol{e}_r. \tag{4.6}$$

- **Dynamic coupling condition.** We will be assuming in this prototypical problem that there is no external stress acting on the structure. Thus, the only force acting on the structure comes from the fluid flow, namely, from its normal Cauchy stress. Since we assume that the structure displaces in only the radial/vertical direction, only the normal component of the fluid Cauchy stress is relevant. Therefore, the force F driving the elastodynamics of (4.2) is given by

$$F = -\boldsymbol{\sigma}(\boldsymbol{u}, p)\boldsymbol{e}_r \cdot \boldsymbol{e}_r|_\Gamma, \tag{4.7}$$

  and so the dynamic coupling condition reads

$$\partial_t^2 \eta - C_1 \partial_z^2 \eta + C_2 \partial_z^4 \eta = -\boldsymbol{\sigma}(\boldsymbol{u}, p)\boldsymbol{e}_r \cdot \boldsymbol{e}_r|_\Gamma. \tag{4.8}$$

Our next step will be to formulate the definition of a weak solution to the linearly coupled prototypical FSI problem. This is done in a classical way by multiplying the structure and fluid equations (4.2) and (4.3), respectively, by sufficiently regular test functions and integrating by parts. The coupling conditions are used to obtain a single weak formulation of the coupled problem.

Before deriving the *weak formulation*, we motivate the weak solution function spaces, i.e., the **finite energy functions spaces** , by deriving an **a priori energy estimate**.

4.2 A Priori Energy Estimate and Weak Formulation

The a priori energy estimate relies on a formal calculation to demonstrate that the system's total energy is governed solely by the work done by the initial and boundary data. Establishing this energy estimate provides preliminary validation that the mathematical model is physically reasonable in terms of bounded energy. Furthermore, the energy estimate offers crucial insight into the *finite energy function spaces* to which we expect our weak solutions to belong.

For the linearly coupled prototypical FSI model, we now formally show that the coupled problem (4.2)–(4.8) satisfies the following energy estimate for all $t \in [0, T]$, up to a fixed but arbitrary final time T:

ENERGY ESTIMATE

$$\begin{aligned}&\frac{1}{2}\|\partial_t \eta(t)\|^2_{L^2(\Gamma)} + \|\eta(t)\|^2_E + \frac{1}{2}\|\boldsymbol{u}(t)\|^2_{L^2(\Omega_f)} + 2\mu_f \int_0^t \|\boldsymbol{D}(\boldsymbol{u})(s)\|^2_{L^2(\Omega_f)} ds \\ &\leq \frac{1}{2}\|v_0\|^2_{L^2(\Gamma)} + \|\eta_0\|^2_E + \frac{1}{2}\|\boldsymbol{u}_0\|^2_{L^2(\Omega_f)} + C\left(\|P_{in}(t)\|^2_{L^2(0,T)} + \|P_{out}\|^2_{L^2(0,T)}\right)\end{aligned} \tag{4.9}$$

where the constant C in the estimate above is independent of the final time T and depends only on the pre-specified geometry of the problem.

We see that on the right-hand side of (4.9), we have the L^2-norms of the initial data and the inlet and outlet boundary data, while on the left-hand side, we have the contribution from the **elastic (potential) energy** defining the "elastic energy norm":

$$\|\eta\|^2_E := \frac{1}{2}C_1\|\partial_z \eta\|^2_{L^2(\Gamma)} + \frac{1}{2}C_2\|\partial_z^2 \eta\|^2_{L^2(\Gamma)}, \tag{4.10}$$

the contribution from the **kinetic energy** of the fluid velocity and the structure motion described by the L^2-norms of the velocities:

$$\frac{1}{2}\|\boldsymbol{u}(t)\|^2_{L^2(\Omega_f)} + \frac{1}{2}\|\partial_t \eta(t)\|^2_{L^2(\Gamma)},$$

plus the contribution of fluid **dissipation** due to fluid viscosity:

$$D(t) := 2\mu_f \|\boldsymbol{D}(\boldsymbol{u})(s)\|^2_{L^2(\Omega_f)}. \tag{4.11}$$

If we denote by $E(t)$ the total energy, i.e., the sum of the kinetic and potential energy,

$$E(t) := \frac{1}{2}\|\boldsymbol{u}(t)\|^2_{L^2(\Omega_f)} + \frac{1}{2}\|\partial_t \eta(t)\|^2_{L^2(\Gamma)} + \|\eta(t)\|^2_E, \tag{4.12}$$

then the energy inequality (4.9) states that the rate of change of the total energy plus dissipation over the time interval $(0, t)$ is bounded only by the work done by the inlet and outlet data:

$$\frac{d}{dt}E(t) + D(t) \leq C(P_{\rm in}(t), P_{\rm out}(t)), \tag{4.13}$$

where $E(t)$ is defined in (4.12), and dissipation in (4.11). Notice that we obtained (4.12) after differentiating the above energy inequality with respect to time.

To show that this energy estimate holds true, we first multiply the momentum equation in (4.3) by $\boldsymbol{u}$ and integrate over Ω_f. Using formal integration by parts, we obtain

$$\int_{\Omega_f} \partial_t \boldsymbol{u} \cdot \boldsymbol{u} + 2\mu_f \int_{\Omega_f} |\mathbf{D}\boldsymbol{u}|^2 - \int_{\partial\Omega_f} (-p\mathbf{I} + 2\mu_f \mathbf{D}(\boldsymbol{u}))\mathbf{n} \cdot \boldsymbol{u} = 0.$$

To deal with the acceleration term, we notice that Ω_f is independent of time, and so we can rewrite this term as

$$\frac{1}{2}\frac{d}{dt}\int_{\Omega_f} |\boldsymbol{u}|^2 + 2\mu_f \int_{\Omega_f} |\mathbf{D}\boldsymbol{u}|^2 - \int_{\partial\Omega_f} (-p\mathbf{I} + 2\mu_f \mathbf{D}(\boldsymbol{u}))\mathbf{n} \cdot \boldsymbol{u} = 0. \tag{4.14}$$

To deal with the boundary integral over $\partial\Omega_f$, we first notice that $\partial\Omega_f$ is composed of three parts: $\partial\Omega_f = \Gamma_{in/out} \cup \Gamma_b \cup \Gamma$. We deal with each segment separately.

On the portion of the fluid domain boundary corresponding to $\Gamma_{in/out}$, the boundary condition (4.4) implies $u_r = 0$. Combined with the divergence-free condition, we obtain $\partial_z u_z = -\partial_r u_r = 0$. Now, using the fact that the normal to $\Gamma_{in/out}$ is $\mathbf{n} = (\mp 1, 0)$, we get

$$\int_{\Gamma_{in/out}} (-p\mathbf{I} + 2\mu_f \mathbf{D}(\boldsymbol{u}))\mathbf{n} \cdot \boldsymbol{u} = \int_{\Gamma_{in}} P_{in} u_z - \int_{\Gamma_{out}} P_{out} u_z. \tag{4.15}$$

In a similar way, using the symmetry boundary conditions (4.5), we get

$$\int_{\Gamma_b} (-p\mathbf{I} + 2\mu_f \mathbf{D}(\boldsymbol{u}))\mathbf{n} \cdot \boldsymbol{u} = 0. \tag{4.16}$$

What is left is to calculate the remaining boundary integral over Γ. For this purpose, we consider the plate equation (4.2), multiplied by $\partial_t \eta$, and integrated by parts to obtain

$$\int_\Gamma F\partial_t\eta = \int_0^L F\partial_t\eta = \frac{1}{2}\frac{d}{dt}\|\partial_t\eta\|^2_{L^2(\Gamma)} + \frac{1}{2}\frac{d}{dt}\left(C_1\|\partial_z\eta\|^2_{L^2(\Gamma)} + C_2\|\partial_z^2\eta\|^2_{L^2(\Gamma)}\right) \tag{4.17}$$

Now we enforce the dynamic coupling condition (4.7) to express F on the left-hand side:

$$\int_0^L F\partial_t\eta = -\int_0^L \boldsymbol{\sigma}\mathbf{n} \cdot \boldsymbol{u},$$

which implies

$$-\int_\Gamma \boldsymbol{\sigma}\mathbf{n} \cdot \boldsymbol{u} = \frac{1}{2}\frac{d}{dt}\|\partial_t\eta\|^2_{L^2(\Gamma)} + \frac{1}{2}\frac{d}{dt}\left(C_1\|\partial_z\eta\|^2_{L^2(\Gamma)} + C_2\|\partial_z^2\eta\|^2_{L^2(\Gamma)}\right). \tag{4.18}$$

By combining (4.15), (4.16), and (4.18), we see that the boundary integral in (4.14) becomes

$$\begin{aligned}
&-\int_{\partial\Omega_f} (-p\mathbf{I} + 2\mu_f\mathbf{D}(\boldsymbol{u}))\mathbf{n} \cdot \boldsymbol{u} \\
&= \frac{1}{2}\frac{d}{dt}\|\partial_t\eta\|^2_{L^2(\Gamma)} + \frac{1}{2}\frac{d}{dt}\left(C_1\|\partial_z\eta\|^2_{L^2(\Gamma)} + C_2\|\partial_z^2\eta\|^2_{L^2(\Gamma)}\right) \\
&+ \int_{\Gamma_{in}} P_{out}u_z - \int_{\Gamma_{out}} P_{in}u_z.
\end{aligned} \tag{4.19}$$

We can now replace the boundary integral in (4.14) with the expression (4.19) to get the **energy equality**:

$$\begin{aligned}
\frac{1}{2}\frac{d}{dt}\left(\|\boldsymbol{u}(t)\|^2_{L^2(\Omega_f)} + \frac{1}{2}\|\partial_t\eta(t)\|^2_{L^2(\Gamma)} + \|\eta(t)\|^2_E\right) + 2\mu_f\|\mathbf{D}(\boldsymbol{u})(s)\|^2_{L^2(\Omega_f)} \\
= \int_{\Gamma_{in}} P_{out}u_z - \int_{\Gamma_{out}} P_{in}u_z.
\end{aligned} \tag{4.20}$$

To estimate the right-hand side and obtain the energy inequality (4.13), we use the trace inequality (see Theorem (2.5.1) in Chap. 2) and Korn's inequality (see (2.15) in Chap. 2) to obtain

$$|P_{in/out}(t)\int_{\Gamma_{in/out}} u_z| \le C|P_{in/out}|\|\boldsymbol{u}\|_{H^1(\Omega_f)} \le \frac{C}{2\epsilon}|P_{in/out}|^2 + \frac{\epsilon C}{2}\|\mathbf{D}(\boldsymbol{u})\|^2_{L^2(\Omega_f)}.$$

By choosing ϵ such that $\frac{\epsilon C}{2} \le \mu_f$, we get the energy inequality (4.13).

From this energy estimate, we can define the **finite energy spaces** for the weak solutions, by noting which norms must be bounded as a result of energy considerations. For example, from the estimate above, we see that formally, we expect the structure displacement η to have the norms $\|\partial_t\eta(t)\|^2_{L^2(\Gamma)}$ and $\|\eta(t)\|^2_E$ be bounded for (almost every) $t \in [0, T]$. Hence, we define the finite energy solution space for the structure displacement η to be

$$V_s = W^{1,\infty}(0, T; L^2(\Gamma)) \cap L^\infty(0, T; \mathcal{V}_s),$$

where $\mathcal{V}_s$ is the following solution space that incorporates the clamped boundary conditions for the structure:

$$\mathcal{V}_s = H_0^2(\Gamma). \tag{4.21}$$

Similarly, we define the solution space for the fluid. For each fixed time, we expect the solution to take values in the following space of divergence-free H^1 functions respecting the boundary conditions of the fluid subproblem:

$$\mathcal{V}_f = \{\boldsymbol{u} \in H^1(\Omega_f) : \nabla\cdot\boldsymbol{u} = 0, u_r = 0 \text{ on } \Gamma_{in} \cup \Gamma_{out} \cup \Gamma_b, u_z = 0 \text{ on } \Gamma\}. \tag{4.22}$$

Then, from the energy estimate, we have the following finite energy space for the fluid velocity $\boldsymbol{u}$:

$$V_f = L^\infty(0, T; L^2(\Omega_f)) \cap L^2(0, T; \mathcal{V}_f).$$

Now that we have derived an energy estimate and the finite energy solution spaces, we can define the weak formulation to the problem. For this purpose, consider a test function $(\boldsymbol{q}, \psi)$ where $\boldsymbol{q}$ is the test function for the fluid velocity and ψ is the test function for the structure velocity, such that $\boldsymbol{q}|_\Gamma = \psi\boldsymbol{e}_r$ (which is the kinematic coupling condition imposed on the test functions). We multiply the fluid momentum equation by $\boldsymbol{q}$ and the structure equation by ψ, and integrate by parts, using the dynamic coupling condition as the boundary condition for the fluid problem. The result is a weak formulation satisfied by the weak solution to the coupled prototypical FSI problem, which is given by the following.

Definition 4.2.1 A fluid velocity $\boldsymbol{u} \in V_f$ and a structure displacement $\eta \in V_s$ define a weak solution to the linearly coupled FSI problem if $\eta(0) = \eta_0$ and furthermore, for all test functions $(\boldsymbol{q}, \psi) \in C_c^\infty([0, T); \mathcal{V}_s \times \mathcal{V}_f)$ such that $\boldsymbol{q}|_\Gamma = \psi\boldsymbol{e}_r$, the following weak formulation holds:

$$-\int_0^T\int_{\Omega_f}\boldsymbol{u}\cdot\partial_t\boldsymbol{q}+2\mu_f\int_0^T\int_{\Omega_f}\boldsymbol{D}(\boldsymbol{u}):\boldsymbol{D}(\boldsymbol{q})-\int_0^T\int_\Gamma\partial_t\eta\cdot\partial_t\psi$$

$$+C_1\int_0^T\int_\Gamma\nabla\eta\cdot\nabla\psi$$

$$+C_2\int_0^T\int_\Gamma\Delta\eta\cdot\Delta\psi=\int_0^T\int_{\Gamma_{in}}P_{in}(t)\cdot q_z-\int_0^T\int_{\Gamma_{out}}P_{out}(t)\cdot q_z$$

$$+\int_{\Omega_f}\boldsymbol{u}_0\cdot\boldsymbol{q}(0)+\int_\Gamma v_0\cdot\psi(0).$$

If we substitute $(\boldsymbol{u}, \partial_t\eta)$ for the test function $(\boldsymbol{q}, \psi)$ in the weak formulation, note that we formally recover the a priori energy estimate above.

The ultimate goal of this chapter is to show that the weak solution defined this way exists and that it is the unique, physical solution to the underlying multiphysics problem. Indeed, the goal of this chapter is to show that this linearly coupled FSI problem has a global weak solution, and we state this result in the following theorem.

Theorem 4.2.1 (Main Existence Result) *Let the initial data $\boldsymbol{u}_0$, η_0 and v_0 be such that $\boldsymbol{u}_0 \in L^2(\Omega_f)$, $\eta_0 \in H_0^2(\Gamma)$, and $v_0 \in L^2(\Gamma)$. Then, there exists a unique weak solution $(\boldsymbol{u}, \eta)$ to the linearly coupled fluid-structure interaction problem specified in Definition 4.2.1.*

The method for showing uniqueness of weak solutions is handled using energy-type arguments, where we have to carefully consider the regularity of the weak solution and the test functions. We will not handle the proof of uniqueness now, as we will demonstrate the mathematical methodology for establishing uniqueness of weak solutions later in Chap. 8 (specifically, see Lemma 8.8.1). We will instead focus on the existence result. For this purpose, we introduce an operator splitting scheme, called the Lie operator splitting scheme, also known as the Marchuk-Yanenko scheme [165], which provides a constructive existence proof and can be used as a backbone for the construction of a numerical scheme to calculate the solutions to this problem. This Lie operator splitting scheme is a robust way to study existence of solutions to FSI problems, and versions of this scheme will be used throughout this text to study a variety of FSI problems, both deterministic and stochastic.

4.3 The Lie Operator Splitting Scheme

The Lie splitting, also known as the Marchuk-Yanenko splitting scheme, is based on solving a fist-order evolution system of the form (4.23) by discretizing the problem in time and splitting the system operator A into N "sub-operators" A_i,

$i = 1, \ldots, N$. Each operator A_i defines a simpler initial value subproblem of the form (4.24). Typically, the splitting is designed to follow some underlying physical principle. In our coupled multiphysics FSI problem, each operator A_i will be determined by the different physics of the subproblem. Thus, we will have two operators A_1 and A_2, one associated with the structure subproblem and the other with a fluid subproblem. The subproblems defined by the operators A_i communicate via the initial data. More precisely, on each time interval $[t_n, t_{n+1}]$, the subproblem (4.24) defined by A_i is solved with the initial data given by the solution of the "previous" subproblem, namely, the subproblem defined by A_{i-1}. This is a generalization of the well-known Lie-Trotter formula [112]. See also Sect. 2.15.2 in Chap. 2 for basics of the Lie splitting strategy.

More precisely, let $N \in \mathbb{N}$, $\Delta t = T/N$, and $t_n = n\Delta t$. We are interested in solving the following initial value problem:

$$\frac{d\phi}{dt} + A\phi = 0 \quad \text{in } (0, T), \quad \phi(0) = \phi_0, \tag{4.23}$$

where A is an operator defined on a Hilbert space, and A can be written as a nontrivial sum $A = A_1 + A_2$. The Lie splitting scheme is defined by first setting $\phi^0 = \phi_0$, and then, for $n = 0, \ldots, N-1$ and $i = 1, 2$, the approximation $\phi^{n+\frac{i}{2}}$ of the solution is obtained by solving

$$\left.\begin{aligned} \frac{d}{dt}\phi_i + A_i\phi_i &= 0 \\ \phi_i(t_n) &= \phi^{n+\frac{i-1}{2}} \end{aligned}\right\} \quad \text{in } (t_n, t_{n+1}), \tag{4.24}$$

and then set $\phi^{n+\frac{i}{2}} = \phi_i(t_{n+1})$, for $i = 1, 2$. It can be shown that this method is at least $\sqrt{\Delta t}$-accurate in time; see e.g., [165].

We apply this splitting to our coupled FSI problem stated in Definition (4.2.1) by first rewriting the problem as a first-order system in time. Then, operators A_1 and A_2 are defined to correspond to a structure subproblem and a fluid subproblem, respectively. The solution of the structure subproblem will be used as the initial data for the fluid subproblem, and the solution of the fluid subproblem will be used as the initial data for the structure subproblem *at the next time step.* No sub-iterations between the fluid and structure subproblems are needed. In this splitting, the fluid and structure communicate only via the initial data. The trick is to define the structure and fluid subproblems in such a way that the resulting scheme is stable and its solution converges to a weak solution of the coupled FSI. This is highly nontrivial and will require some physical intuition to motivate the splitting, as we shall see below.

To apply the Lie splitting strategy, we must first write the coupled FSI problem as a first-order system in time. For this purpose, we introduce a new variable, the structure velocity:

$$\boldsymbol{v} = \partial_t \boldsymbol{\eta} = \partial_t \eta \mathbf{e}_r = v\mathbf{e}_r.$$

Now, the coupled FSI problem in the first-order form reads

$$\left.\begin{aligned} \partial_t \boldsymbol{u} &= \nabla \cdot \boldsymbol{\sigma} \\ \nabla \cdot \boldsymbol{u} &= 0 \end{aligned}\right\} \quad \text{in } \Omega_f,\ t \in (0, T), \tag{4.25}$$

$$\left.\begin{aligned} \partial_t \eta \boldsymbol{e}_r &= \boldsymbol{u}|_\Gamma \\ \partial_t \boldsymbol{v} &= -\mathcal{L}_e \boldsymbol{\eta} - \boldsymbol{\sigma} \boldsymbol{e}_r|_\Gamma \\ \partial_t \boldsymbol{\eta} &= \boldsymbol{v} \end{aligned}\right\} \quad \text{on } \Gamma,\ t \in (0, T), \tag{4.26}$$

$$\left.\begin{aligned} u_r &= 0 \\ \partial_r u_z &= 0 \end{aligned}\right\} \text{on } \Gamma_b,\ t \in (0, T), \tag{4.27}$$

$$\left.\begin{aligned} p + \tfrac{1}{2}|u|^2 &= P_{in/out}(t) \\ u_r &= 0 \end{aligned}\right\} \text{on } \Gamma_{in/out},\ t \in (0, T). \tag{4.28}$$

where $\mathcal{L}_e \boldsymbol{\eta} = \left(-C_1 \partial_z^2 \eta + C_2 \partial_z^4 \eta\right) \boldsymbol{e}_r$.

The problem is solved with the initial data:

$$\boldsymbol{u}(0, .) = \boldsymbol{u}_0,\ \eta(0, .) = \eta_0,\ v(0, .) = v_0. \tag{4.29}$$

The coupled problem (4.25), (4.26) is split into a structure and a fluid subproblems defining the operators A_1 and A_2 in the following way.

The Structure Subproblem (Operator A_1) We consider the structure elastodynamics problem but leave out the forcing term coming from the fluid normal stress. That part will be used in the fluid subproblem as a Robin boundary condition. Thus, the structure subproblem reads

$$\left.\begin{aligned} \partial_t v + \mathcal{L}_e \eta &= 0 \\ \partial_t \eta &= v \end{aligned}\right\} \text{ which implies } \quad \partial_{tt} \eta + \mathcal{L}_e \eta = 0 \text{ on } (0, T) \times \Gamma, \tag{4.30}$$

which will be solved with the initial data for the structure velocity $\partial_t \eta = v$ given by the trace of the fluid velocity $\boldsymbol{u}|_\Gamma \cdot \mathbf{e}_r$ on Γ obtained from the previous time step.

The Fluid Subproblem (Operator A_2) The fluid subproblem consists of solving the time-dependent Stokes equations with the boundary condition on Γ given by the remaining part of the dynamic coupling condition:

$$\left\{\begin{aligned} \partial_t v &= -J\boldsymbol{\sigma}\mathbf{n}|_\Gamma \cdot \mathbf{e}_r \\ \mathbf{u}|_\Gamma &= v\mathbf{e}_r \end{aligned}\right\} \text{ or } \left\{\begin{aligned} \partial_t u_r|_\Gamma &= -J\boldsymbol{\sigma}\mathbf{n}|_\Gamma \cdot \mathbf{e}_r \\ u_z|_\Gamma &= 0 \end{aligned}\right\}, \tag{4.31}$$

where the initial data for the fluid velocity u_r on Γ is taken to be the structure velocity v obtained from the just calculated structure subproblem.

Notice that instead of the "typical" Dirichlet boundary condition for the Stokes equations, we have a Robin-type boundary condition. The boundary condition incorporates the structure inertia term $\partial_t u_r|_\Gamma$ (which is equal to $\partial_t v$), which will be evaluated at time $n + 1$. This will lead to an implicit scheme in the structure subproblem where the fluid loading via the fluid normal stress on Γ will have the information from the time step $n + 1$ and will be implicitly tied to the rest of the structure problem. This approach gives rise to a **stable Lie splitting scheme**.

More precisely, the splitting scheme for the prototypical linearly coupled FSI model is as follows. Fix the target time T and for each positive integer N discretize the time interval $[0, T]$ into N subintervals $[t_n, t_{n+1}]$ of length $\Delta t = T/N$, where the endpoints are defined by $t_n = n\Delta t$. On each time interval $[t_n, t_{n+1}]$, "solve" the two subproblems specified below in weak formulation and update an approximate vector:

$$\begin{pmatrix} \eta_N^{n+\frac{i}{2}} \\ v_N^{n+\frac{i}{2}} \\ \boldsymbol{u}_N^{n+\frac{i}{2}} \end{pmatrix}, \quad i = 1, 2$$

consisting of the approximate state of the structure displacement, the structure velocity, and the fluid velocity at that current time step.

Before we specify the structure and fluid subproblems in weak formulation, we introduce the following notation for the **semidiscrete** versions of the kinetic and elastic **energy and** for **dissipation**:

$$\begin{aligned} E_N^{n+\frac{i}{2}} = \frac{1}{2}\Big(&\|\boldsymbol{u}_N^{n+\frac{i}{2}}\|^2_{L^2(\Omega_f)} + \|v_N^{n+\frac{i}{2}}\|^2_{L^2(\Gamma)} + C_1\|\partial_z \eta_N^{n+\frac{i}{2}}\|^2_{L^2(\Gamma)} \\ &+ C_2\|\partial_z^2 \eta_N^{n+\frac{i}{2}}\|^2_{L^2(\Gamma)}\Big), \end{aligned} \tag{4.32}$$

$$D_N^{n+1} = \Delta t \mu_f \int_{\Omega_f} |D^{\eta_N^n}(\boldsymbol{u}_N^{n+1})|^2, \ n = 0, \dots, N-1, \ i = 0, 1. \tag{4.33}$$

The Structure Subproblem In this subproblem, the fluid velocity stays the same:

$$\boldsymbol{u}_N^{n+\frac{1}{2}} = \boldsymbol{u}_N^n,$$

while the structure displacement $\eta_N^{n+\frac{1}{2}}$ and the structure velocity $v_N^{n+\frac{1}{2}}$ are updated. This entails finding $(\eta_N^{n+\frac{1}{2}}, v_N^{n+\frac{1}{2}}) \in H_0^2(\Gamma) \times H_0^2(\Gamma)$ such that the following system is solved in the weak formulation:

$$\begin{aligned} &\int_\Gamma \frac{\eta_N^{n+\frac{1}{2}} - \eta_N^n}{\Delta t} \varphi dz = \int_\Gamma v_N^{n+\frac{1}{2}} \varphi dz, \qquad \text{for all } \varphi \in L^2(\Gamma), \\ &\int_\Gamma \frac{v_N^{n+\frac{1}{2}} - v_N^n}{\Delta t} \psi + C_1 \int_\Gamma \partial_z \eta_N^{n+\frac{1}{2}} \partial_z \psi \\ &\quad + C_2 \int_\Gamma \partial_z^2 \eta_N^{n+\frac{1}{2}} \partial_z^2 \psi = 0, \ \text{ for all } \psi \in H_0^2(\Gamma). \end{aligned} \tag{4.34}$$

To simplify the notation, we introduce the following bilinear form determined by the elastic energy of the structure:

$$a_S(\eta, \psi) := \langle \mathcal{L}_e \eta, \psi \rangle = \int_\Gamma \big(C_1 \partial_z \eta \partial_z \psi + C_2 \partial_z^2 \eta \partial_z^2 \psi \big). \tag{4.35}$$

Proposition 4.3.1 *For each fixed* $\Delta t > 0$, *problem* (4.34) *has a unique solution* $(v_N^{n+\frac{1}{2}}, \eta_N^{n+\frac{1}{2}}) \in H_0^2(\Gamma) \times H_0^2(\Gamma)$.

Proof The proof is a direct consequence of the Lax-Milgram lemma (see Lemma 2.2.1 in Chap. 2) applied to the weak form:

$$\int_0^L \eta_N^{n+\frac{1}{2}} \psi dz + (\Delta t)^2 a_S(\eta_N^{n+\frac{1}{2}}, \psi) = \int_0^L \big(\Delta t v_N^n + \eta_N^n \big) \psi dz, \;\; \psi \in H_0^2(\Gamma),$$

which is obtained after eliminating $v_N^{n+\frac{1}{2}}$ in the second equation by using the kinematic coupling condition given by the first equation. □

Proposition 4.3.2 *For each fixed* $\Delta t > 0$, *the solution of problem* (4.34) *satisfies the following discrete energy equality:*

$$E_N^{n+\frac{1}{2}} + \frac{1}{2} \big(\|v_N^{n+\frac{1}{2}} - v_N^n\|^2 + C_1 \|\partial_z (\eta_N^{n+\frac{1}{2}} - \eta_N^n)\|^2 + C_2 \|\partial_z^2 (\eta_N^{n+\frac{1}{2}} - \eta^n)\|^2 \big) = E_N^n, \tag{4.36}$$

where the kinetic energy E_N^n *is defined in* (4.32).

Proof From the first equation in (4.34), we immediately get

$$v_N^{n+\frac{1}{2}} = \frac{\eta_N^{n+\frac{1}{2}} - \eta^n}{\Delta t} \in H_0^2(\Gamma).$$

Therefore, we can take $v_N^{n+\frac{1}{2}}$ as a test function in the second equation in (4.34). We replace the test function ψ by $v_N^{n+\frac{1}{2}}$ in the first term on the left-hand side and replace ψ by $(\eta_N^{n+\frac{1}{2}} - \eta^n)/\Delta t$ in the bilinear form a_S. We then use the algebraic identity $(a-b) \cdot a = \frac{1}{2}(|a|^2 + |a-b|^2 - |b|^2)$ to deal with the terms $(v^{n+1/2} - v_N^n) v^{n+1/2}$ and $(\eta^{n+1/2} - \eta_N^n)\eta^{n+1/2}$. After multiplying the entire equation by Δt, the second equation in (4.34) can be written as

$$\begin{aligned} (\|v_N^{n+\frac{1}{2}}\|^2 + \|v_N^{n+\frac{1}{2}} - v_N^n\|^2) + a_S(\eta_N^{n+\frac{1}{2}}, \eta_N^{n+\frac{1}{2}}) + a_S(\eta_N^{n+\frac{1}{2}} - \eta^n, \eta_N^{n+\frac{1}{2}} - \eta^n) \\ = \|v_N^n\|^2 + a_S(\eta_N^n, \eta_N^n). \end{aligned}$$

We then recall that $\boldsymbol{u}^{n+\frac{1}{2}} = \boldsymbol{u}_N^n$ in this subproblem, and so we can add $\|\boldsymbol{u}^{n+1/2}\|^2_{L^2(\Omega_f)}$ on the left-hand side and $\|\boldsymbol{u}^n\|^2_{L^2(\Omega_f)}$ on the right-hand side of the equation, to obtain exactly the energy equality (4.36). □

The Fluid Subproblem In this subproblem, the structure displacement stays the same:

$$\eta_N^{n+1} = \eta_N^{n+\frac{1}{2}},$$

while the structure velocity v_N^{n+1} and the fluid velocity $\boldsymbol{u}_N^{n+1}$ are updated. This entails finding $(\boldsymbol{u}_N^{n+1}, v_N^{n+1}) \in \mathcal{V}_f \times L^2(\Gamma)$ satisfying $\boldsymbol{u}_N^{n+1}|_\Gamma = v_N^{n+1}$ such that

$$\int_{\Omega_f} \frac{\boldsymbol{u}_N^{n+1} - \boldsymbol{u}_N^{n+\frac{1}{2}}}{\Delta t} \cdot \boldsymbol{q} + 2\mu_f \int_{\Omega_f} \boldsymbol{D}(\boldsymbol{u}_N^{n+1}) : \boldsymbol{D}(\boldsymbol{q}) + \int_\Gamma \frac{v_N^{n+1} - v_N^{n+\frac{1}{2}}}{\Delta t} \psi$$
$$= \int_{\Gamma_{in}} P_{N,in}^n q_z - \int_{\Gamma_{out}} P_{N,out}^n q_z, \quad (4.37)$$

where $P_{N,in}^n$ and $P_{N,out}^n$ are the numerical inlet and outlet pressures:

$$P_{N,in/out}^n = \frac{1}{\Delta t} \int_{n\Delta t}^{(n+1)\Delta t} P(t)dt.$$

Proposition 4.3.3 *The fluid subproblem defined by (4.37) has a unique weak solution* $(\boldsymbol{u}_N^{n+1}, v_N^{n+1}) \in \mathcal{V}_f \times L^2(\Gamma)$ *satisfying* $\boldsymbol{u}_N^{n+1}|_\Gamma = v_N^{n+1}$.

Proof The proof is again a consequence of the Lax-Milgram lemma (see Lemma 2.2.1. More precisely, for $(\boldsymbol{u}, v), (\boldsymbol{q}, \psi) \in \mathcal{V}_f \times L^2(\Gamma)$ satisfying $\boldsymbol{u}|_\Gamma = v$ and $\boldsymbol{q}|_\Gamma = \psi$, define the bilinear form associated with problem (4.37):

$$\mathcal{B}((\boldsymbol{u}, v), (\boldsymbol{q}, \psi)) := \int_{\Omega_f} \boldsymbol{u} \cdot \boldsymbol{q} + \Delta t 2\mu_f \int_{\Omega_f} \mathbf{D}(\boldsymbol{u}) : \mathbf{D}(\boldsymbol{q}) + \int_\Gamma v\psi.$$

We need to prove that this bilinear form $\mathcal{B}$ is coercive and continuous. Coercivity is straightforward. Continuity is also straightforward. It follows from the following estimate:

$$\mathcal{B}((\boldsymbol{u}, v), (\boldsymbol{q}, \psi)) \leq C \Big(\|\boldsymbol{u}\|_{L^2(\Omega_f)} \|\boldsymbol{q}\|_{L^2(\Omega_f)} + \|v\|_{L^2(\Gamma)} \|\psi\|_{L^2(\Gamma)}$$
$$+ \Delta t \mu_f \|\boldsymbol{u}\|_{H^1(\Omega)} \|\boldsymbol{q}\|_{H^1(\Omega_f)} \Big).$$

□

Proposition 4.3.4 *For each fixed* $\Delta t > 0$*, the solution of problem (4.37) satisfies the following discrete energy inequality:*

$$E_N^{n+1} + \|\boldsymbol{u}_N^{n+1} - \boldsymbol{u}_N^n\|^2_{L^2(\Omega_f)} + \|v_N^{n+1} - v_N^{n+\frac{1}{2}}\|^2_{L^2(\Gamma)}$$
$$+D_N^{n+1} \le E_N^{n+\frac{1}{2}} + C\left(\|P_{in}(t)\|^2_{L^2(n\Delta t,(n+1)\Delta t)} + \|P_{out}(t)\|^2_{n\Delta t,(n+1)\Delta t)}\right), \tag{4.38}$$

where the kinetic energy E_N^n and dissipation D_N^n are defined in (4.32) and (4.33), and the constant C depends only on the parameters in the problem, and not on Δt (or N).

Proof We replace the test functions $\boldsymbol{q}$ by $\boldsymbol{u}_N^{n+1}$ and ψ by v_N^{n+1} in (4.37) and multiply the resulting equation by Δt. The proof follows after using the algebraic identity $(a-b)\cdot a = \frac{1}{2}(|a|^2 + |a-b|^2 - |b|^2)$ in the same way as in the proof of Proposition 4.3.2. □

Semidiscrete Weak Formulation By combining the semidiscrete versions of the structure and fluid subproblems, we obtain that at each time step, the approximate structure displacement, structure velocity, and fluid velocity satisfy the following **semidiscrete weak formulation** of the coupled problem:

$$\int_{\Omega_f} \frac{\boldsymbol{u}_N^{n+1} - \boldsymbol{u}_N^n}{\Delta t} \cdot \boldsymbol{q} + 2\mu_f \int_{\Omega_f} \boldsymbol{D}(\boldsymbol{u}_N^{n+1}) : \boldsymbol{D}(\boldsymbol{q}) + \int_\Gamma \frac{v_N^{n+1} - v_N^n}{\Delta t}\psi$$
$$+ C_1 \int_\Gamma \partial_z v_N^{n+\frac{1}{2}} \partial_z \psi + C_2 \int_\Gamma \partial_z^2 v_N^{n+\frac{1}{2}} \partial_z^2 \psi = \int_{\Gamma_{in}} P_{N,in}^n q_z - \int_{\Gamma_{out}} P_{N,out}^n q_z. \tag{4.39}$$

4.4 Approximate Solutions and Uniform Boundedness

The approximate functions $\boldsymbol{u}_N^{n+\frac{i}{2}}$, $\eta_N^{n+\frac{i}{2}}$, and $v_N^{n+\frac{i}{2}}$ are functions of spatial variables only and are defined for each time step. To construct approximate solutions as functions of time and space, we use piecewise constant extensions and linear interpolations defined on the entire time interval, as described below.

We define the following piecewise constant approximate solutions η_N, v_N, v_N^*, and $\boldsymbol{u}_N$, which are constant on each time step $[n(\Delta t), (n+1)\Delta t)$:

$$\eta_N(t)=\eta_N^{n+1},\ v_N(t)=v_N^{n+1},\ v_N^*=v_N^{n+\frac{1}{2}},\ \boldsymbol{u}_N = \boldsymbol{u}_N^{n+1},\ \text{ if } t \in [n(\Delta t), (n+1)\Delta t).$$

We also define the approximate solutions $\overline{\eta}_N$, $\overline{v}_N$, and $\overline{\boldsymbol{u}}_N$ by linearly interpolating values at the points $t_n = n\Delta t$, where the values of these linearly interpolated approximate solutions at these discrete time values t_n are

$$\overline{\eta}_N(n\Delta t) = \eta_N^n,\quad \overline{v}_N(n\Delta t) = v_N^n,\quad \overline{\boldsymbol{u}}_N(n\Delta t) = \boldsymbol{u}_N^n,\qquad \text{for } n = 0, 1, \ldots, N.$$

We remark that $\partial_t \overline{\eta}_N = v_N^*$. From the discrete energy estimates, we obtain the following result, which shows that the approximate solutions are uniformly bounded in the finite energy spaces.

Proposition 4.4.1 *The following uniform boundedness results hold for the approximate solutions η_N, $\overline{\eta}_N$, $\boldsymbol{u}_N$, v_N^* and v_N, uniformly in N:*

1. ***Structure displacement*** *η_N is uniformly bounded in $L^\infty(0,T;H_0^2(\Gamma))$ and $\overline{\eta}_N$ is uniformly bounded in $W^{1,\infty}(0,T;L^2(\Gamma)) \cap L^\infty(0,T;H_0^2(\Gamma))$.*
2. ***Fluid velocity*** *$\boldsymbol{u}_N$ is uniformly bounded in $L^\infty(0,T;L^2(\Omega_f)) \cap L^2(0,T;H^1(\Omega_f))$.*
3. ***Structure velocity*** *v_N^* is uniformly bounded in $L^\infty(0,T;L^2(\Gamma))$, and $v_N = \boldsymbol{u}_N|_\Gamma \cdot \boldsymbol{e}_r$ is uniformly bounded in $L^\infty(0,T;L^2(\Gamma)) \cap L^2(0,T;H^{1/2}(\Gamma))$.*

The second part of the last statement follows from the continuity of the trace operator from $H^1(\Omega_f)$ to $H^{1/2}(\Gamma)$ (see the trace theorem stated in Theorem 2.5.1 in Chap. 2).

An important feature of linearly coupled FSI models is that having uniform boundedness of approximate solutions is sufficient for passing to the limit in approximate solutions in order to obtain a weak solution to the continuous problem. This is because uniform boundedness of approximate solutions in appropriate energy-level function spaces allows us to obtain weak and weak-star convergence of approximate solutions along appropriate subsequences. Because the weak formulation to linearly coupled problems involves linear terms, this weak and weak-star convergence is sufficient for passing to the limit.

4.5 Convergence to a Weak Solution

From the uniform boundedness result above, we conclude that there is a limiting structure displacement η, fluid velocity $\boldsymbol{u}$, and structure velocity v, such that the following convergence results hold.

Proposition 4.5.1 *There exist subsequences $(\eta_N)_{N\in\mathbb{N}}$, $(\overline{\eta}_N)_{N\in\mathbb{N}}$, $(v_N^*)_{N\in\mathbb{N}}$, $(v_N)_{N\in\mathbb{N}}$, $(\boldsymbol{u}_N)_{N\in\mathbb{N}}$ and the functions η, $\boldsymbol{u}$ and v, such that as $N \to \infty$:*

$$\begin{aligned}
\eta_N &\rightharpoonup \eta \text{ weakly star in } L^\infty(0,T;H_0^2(\Gamma)),\\
\overline{\eta}_N &\rightharpoonup \eta \text{ weakly star in } W^{1,\infty}(0,T;L^2(\Gamma)) \cap L^\infty(0,T;H_0^2(\Gamma)),\\
v_N^* &\rightharpoonup v \text{ weakly star in } L^\infty(0,T;L^2(\Gamma)),\\
v_N &\rightharpoonup v \text{ weakly in } L^2(0,T;H^{1/2}(\Gamma)) \text{ and weakly star in } L^\infty(0,T;L^2(\Gamma)),\\
\boldsymbol{u}_N &\rightharpoonup \boldsymbol{u} \text{ weakly in } L^2(0,T;H^1(\Omega_f)) \text{ and weakly star in } L^\infty(0,T;L^2(\Omega_f)).
\end{aligned}$$

This is sufficient to pass to the limit in the semidiscrete formulation. We integrate the semidiscrete formulation (4.39) in time from $t = 0$ to $t = T$ in order to obtain

$$\int_0^T \int_{\Omega_f} \partial_t \overline{\boldsymbol{u}}_N \cdot \boldsymbol{q} + 2\mu_f \int_0^T \int_{\Omega_f} \boldsymbol{D}(\boldsymbol{u}_N) : \boldsymbol{D}(\boldsymbol{q}) + \int_0^T \int_\Gamma \partial_t \overline{v}_N \cdot \psi$$
$$+ C_1 \int_0^T \int_\Gamma \partial_z v_N^* \cdot \partial_z \psi$$
$$+ C_2 \int_0^T \int_\Gamma \partial_z^2 v_N^* \cdot \partial_z^2 \psi = \sum_{n=0}^{N-1} \int_{n\Delta t}^{(n+1)\Delta t} \int_{\Gamma_{in}} P_{N,in}^n \cdot q_z$$
$$- \sum_{n=0}^{N-1} \int_{n\Delta t}^{(n+1)\Delta t} \int_{\Gamma_{out}} P_{N,out}^n \cdot q_z,$$

for all test functions $(\boldsymbol{q}, \psi) \in C_c^1([0, T); \mathcal{V}_f \times \mathcal{V}_s)$. From the weak convergence of the approximate structure velocities, we obtain

$$C_1 \int_0^T \int_\Gamma \partial_z v_N^* \cdot \partial_z \psi \to C_1 \int_0^T \int_\Gamma \partial_z v \cdot \partial_z \psi,$$
$$C_2 \int_0^T \int_\Gamma \partial_z^2 v_N^* \cdot \partial_z^2 \psi \to C_2 \int_0^T \int_\Gamma \partial_z^2 v \cdot \partial_z^2 \psi.$$

From the weak convergence of the approximate fluid velocities, we get

$$2\mu_f \int_0^T \int_{\Omega_f} \boldsymbol{D}(\boldsymbol{u}_N) : \boldsymbol{D}(\boldsymbol{q}) \to 2\mu_f \int_0^T \int_{\Omega_f} \boldsymbol{D}(\boldsymbol{u}) : \boldsymbol{D}(\boldsymbol{q}).$$

Finally, we use integration by parts in time, the weak convergence of the fluid velocities, and the fact that $\boldsymbol{q} \in C_c^1([0, T); \mathcal{V}_f)$ in order to deduce that

$$\int_0^T \int_{\Omega_f} \partial_t \overline{\boldsymbol{u}}_N \cdot \boldsymbol{q} = \sum_{n=0}^{N-1} \int_{n\Delta t}^{(n+1)\Delta t} \int_{\Omega_f} \partial_t \overline{\boldsymbol{u}}_N \cdot \boldsymbol{q}$$
$$= \sum_{n=0}^{N-1} \left(- \int_{n\Delta t}^{(n+1)\Delta t} \int_{\Omega_f} \overline{\boldsymbol{u}}_N \cdot \partial_t \boldsymbol{q} + \int_{\Omega_f} \boldsymbol{u}_N^{n+1} \cdot \boldsymbol{q}((n+1)\Delta t) - \int_{\Omega_f} \boldsymbol{u}_N^n \cdot \boldsymbol{q}(n\Delta t) \right)$$
$$= - \int_{\Omega_f} \boldsymbol{u}_0 \cdot \boldsymbol{q}(0) - \int_0^T \int_{\Omega_f} \overline{\boldsymbol{u}}_N \cdot \partial_t \boldsymbol{q} \to - \int_{\Omega_f} \boldsymbol{u}_0 \cdot \boldsymbol{q}(0) - \int_0^T \int_{\Omega_f} \boldsymbol{u} \cdot \partial_t \boldsymbol{q}.$$

An identical argument shows that

$$\int_0^T \int_\Gamma \partial_t \overline{v}_N \cdot \psi \to - \int_\Gamma v_0 \cdot \psi(0) - \int_0^T \int_\Gamma v \cdot \partial_t \psi.$$

Thus, the semidiscrete formulation converges to the weak formulation of the continuous problem, specified in Definition 4.2.1, as $N \to \infty$, proving that the limits of subsequences of approximate solutions constructed above converge to a weak solution of the continuous problem. This establishes the existence of a weak solution to the prototypical linearly coupled FSI problem (4.2)–(4.8) and completes the proof of Theorem 4.2.1.

4.6 Concluding Remarks

In this chapter, we studied a prototypical problem in deterministic FSI with linear coupling. In problems with linear coupling involving incompressible, viscous fluids, the time-dependent Stokes equations, rather than the Navier-Stokes equations, are typically coupled to the elastodynamics of an elastic or viscoelastic structure. The Stokes equations are "more appropriate" for the linearly coupled problems because in this case, it is possible to obtain an energy estimate (see (4.9)), which holds uniformly in time, i.e., the constant C in the energy estimate (4.9) is independent of T. To prove the existence of a weak solution in finite energy spaces, we used a Lie splitting strategy to construct approximate solutions in time for every time step Δt. This method was first considered in this context by Muha and Čanić in [265] and has proven to be so robust that it applies to a large family of FSI problems, as we shall see in this book. Once approximate solutions were constructed, we proved that subsequences of those solutions converge, as the time step goes to zero, to a weak solution of the continuous problem. Due to the linear nature of the entire coupled problem, no compactness arguments were needed to show that the limits of approximate sequences indeed satisfy the weak formulation of the problem. This is, however, not going to be enough when nonlinear FSI problems are considered, as we will see in Chap. 5.

4.7 Bibliographical Notes

The first mathematically rigorous works on linearly coupled fluid-structure interaction involve fluids and structures interacting with each other, where the dimensions of the fluid and structure domains are the same. We will summarize the fundamental developments in linearly coupled models of FSI of this type and then include the results in which the structure is defined on a co-dimension one surface. We note here that the prototypical model that we will consider in this chapter is of that type—the structure is a lower-dimensional membrane or shell that defines the boundary of the fluid domain.

The first result for linearly coupled fluid-structure interaction dates back to [240, Section 9], but the development of modern theory was initiated in the work by Gunzburger et al. [127]. In their study, the linear Stokes equations posed on a fluid domain are coupled with the equations of linear elasticity defined on a solid domain. The fluid and solid domains are in contact with each other along a common interface, and both the fluid and solid are of the same dimension, either 2D or 3D. At the interface, two coupling conditions are prescribed: a kinematic coupling condition prescribing matching of the fluid and structure velocity and a dynamic coupling condition that prescribes continuity of normal stresses. This work proves well-posedness from the perspective of weak solutions, by using a Galerkin method and passing to the limit, and it shows that given initial structure displacement, structure velocity, and fluid velocity in $H^1(\Omega_s)$, $H^1(\Omega_s)$, and $H^1(\Omega_f)$, respectively, there exists a solution where the structure displacement, structure velocity, and fluid velocity are bounded in $H^1(\Omega_s)$, $L^2(\Omega_s)$, and $L^2(\Omega_f)$ for all time, while the fluid velocity is also in $L^2(0, T; H^1(\Omega_f))$. It is shown that with more regular initial data, one can obtain improved regularity for the solution, in addition to being able to recover the fluid pressure, which vanishes in the weak formulation when using divergence-free test functions.

After this initial work [127], there was a growth in interest in linearly coupled models, where the solid and fluid have the same spatial dimension, specifically models in which an elastic solid is immersed in a fluid. For concreteness, in these models, we will refer to the structure displacement defined on the reference structure domain Ω_s by $\boldsymbol{\eta}$, we will refer to the fluid velocity defined on the reference fluid domain Ω_f by $\boldsymbol{u}$, and we will denote the reference configuration of the fluid-structure interface by Γ. In particular, the results of [127] were improved in [18], where a solid was modeled by the linear damped wave equation $\boldsymbol{\eta}_{tt} - \Delta\boldsymbol{\eta} + \boldsymbol{\eta} = 0$ on Ω_s, which was immersed in a fluid modeled by the linear Stokes equations, where the solid and fluid are both 2D or are both 3D. The kinematic coupling condition was a no-slip condition:

$$\boldsymbol{u} = \boldsymbol{\eta}_t, \qquad \text{on } \Gamma,$$

and the dynamic coupling condition was

$$\frac{\partial \boldsymbol{u}}{\partial \nu} - \frac{\partial \boldsymbol{\eta}}{\partial \nu} = p\nu, \qquad \text{on } \Gamma, \tag{4.40}$$

where ν is the normal vector along Γ. As described in [18], this work improves upon the past results in [127], where an initial structure displacement, structure velocity, and fluid velocity in $H^1(\Omega_s)$, $H^1(\Omega_s)$, and $L^2(\Omega_f)$ give rise to a solution that is bounded in time in the weaker spaces $H^1(\Omega_s)$, $L^2(\Omega_s)$, and $L^2(\Omega_f)$. In contrast, this work [18] does not have this decrease in regularity from the initial data to the solutions. The main result of this work is that the linearly coupled problem has an associated evolution that is given by a strongly continuous C_0 semigroup, so that initial data in a finite energy space $\mathcal{H}$ give rise to solutions that are continuous in $\mathcal{H}$. In addition, the semigroup is strongly stable on $\mathcal{H}$ quotiented out by a one-dimensional space of stationary solutions, meaning that on this factor space, the operator norm of the semigroup converges to zero as $t \to \infty$, which implies some long-time decay of solutions. These results were obtained by reformulating the initial fluid-structure interaction problem as an abstract evolution equation, where the fluid pressure is eliminated from the equations by rewriting it using operators applied to the fluid velocity and structure displacement. The study of well-posedness of this system was continued in a later work [16], where it was shown that the solution possesses additional regularity if one assumes that the initial data is in the domain of the generator $\mathcal{A}$ of the associated semigroup and the initial structure displacement is in $H^2(\Omega_s)$, so that the initial data is more regular.

These asymptotic stability results for immersed structures in fluids were later improved in [19], where the same model as in [18] was considered, but instead of a no-slip condition $\boldsymbol{u} = \boldsymbol{\eta}_t$ on Γ, there is a boundary dissipation condition:

$$\boldsymbol{u} = \boldsymbol{\eta}_t + \alpha \frac{\partial \boldsymbol{\eta}}{\partial \nu} \quad \text{on } \Gamma, \tag{4.41}$$

where $\alpha = 1$ for concreteness. For this problem without boundary dissipation, the previous work [18] shows that the associated semigroup is strongly stable, meaning the norm of the solution goes to zero as t goes to infinity. The work in [19] complement this result by showing *uniform stability* of the semigroup, which means that the operator norm of the semigroup decays exponentially as t goes to infinity. This stronger result ensures that the system stabilizes at a uniform rate over time, providing a more robust form of stability compared to the asymptotic stability shown in [18], where the solution norm only goes to zero without any guaranteed rate of decay.

We note that these past results on elastic structures immersed in fluids [16, 18, 19] involve damped wave equations. These results were generalized to elastic structures modeled by elasticity equations, specifically the Lamé equations of linear elasticity, interacting with incompressible Stokes flows in later works [20, 21]. These models involve boundary dissipation in the kinematic coupling condition so that

$$\boldsymbol{\eta}_t - \varphi \boldsymbol{\sigma}(\boldsymbol{\eta}) \cdot \boldsymbol{\nu}|_{\Gamma} = \boldsymbol{u} \quad \text{on } \Gamma,$$

where $\varphi \geq 0$ is a sufficiently regular and nonnegative function, for example, in $C^1(\Gamma)$, and $\boldsymbol{\sigma}(\boldsymbol{\eta})$ is an elasticity stress tensor for the structure. The dynamic coupling condition, instead of being the simpler condition (4.40), is now the full continuity of normal stress between the fluid and structure:

$$2\boldsymbol{D}(\boldsymbol{u}) \cdot \boldsymbol{\nu} = \boldsymbol{\sigma}(\boldsymbol{\eta}) \cdot \boldsymbol{\nu} + p\boldsymbol{\nu}, \qquad \text{on } \Gamma,$$

where $\boldsymbol{D}(\boldsymbol{u}) = \frac{1}{2}(\nabla(\boldsymbol{u}) + (\nabla \boldsymbol{u})^t)$ is the symmetrized gradient for the fluid velocity. We refer the reader to [21] for the full details of the Stokes-Lamé model. In [21], it is shown that this Stokes-Lamé system has an associated strongly continuous C_0 semigroup, which establishes well-posedness, and strong stability properties of the semigroup are established. As was done for the model involving just the damped wave equation in [19], these strong stability results for the Stokes-Lamé system in [21] were strengthened to results on uniform stability of the semigroup in the work [20], in the sense of exponential decay of the operator norms of the semigroup, when the boundary dissipation function φ is strictly positive. The boundary dissipation function $\varphi > 0$ in this case is the analogue of the constant $\alpha = 1$ in (4.41) for the related model involving a linear damped wave equation interacting with Stokes flow.

Nonlinear variants of these models were also considered, in which the Navier-Stokes equations are considered for the fluid interacting with the immersed solid. We emphasize that these models are still *linearly coupled* because the Navier-Stokes equations are posed on a fixed reference fluid domain and the fluid-structure coupling takes place at a fixed reference fluid-structure interface. This model was considered in [25], where the continuity of normal stresses is modified into a *transmission boundary condition* in order to obtain energy balance, due to the effect of the nonlinear advection term in the Navier-Stokes equations and the fact that the nonlinear Navier-Stokes equations interacting with a deformable structure are being posed on a fixed reference fluid domain. In this work, existence of weak solutions that are defined globally in time is established in finite energy spaces by using semigroup methods and microlocal analysis of hyperbolic equations. The proof considers an auxiliary related nonlinear problem where the nonlinearity is of a suitably tractable form, and this auxiliary problem is solved using semigroup methods. The initial problem can be solved by truncating the Navier-Stokes advection nonlinearity to reduce the initial problem to this auxiliary problem and then passing to the limit in the truncation parameter. However, an additional difficulty here is that to show that the solutions that are solutions in the semigroup (mild) formulation are also weak solutions satisfying a variational formulation, we must be able to interpret appropriate terms in the weak formulation, such as the trace of the normal stress of the structure on the boundary. While the finite energy space does not have a structure displacement and velocity with sufficient regularity to classically define a trace of the normal stress of the structure along the fluid-structure interface Γ, the work in [25] shows that one can use microlocal analysis to establish a *hidden regularity* result where if we are given initial structure displacement in $H^1(\Omega_s)$, initial structure velocity in $L^2(\Omega_s)$, and forcing in $L^2(0, T; H^{1/2}(\Omega_s))$, then the corresponding solution to the linear elasticity equation has a normal stress with a well-defined trace on the boundary in $L^2(0, T; H^{-1/2}(\Gamma))$. We remark that this "hidden regularity" result has been fundamental to the development of methods for linearly coupled FSI, as it has resulted in well-posedness results that do not require the addition of (potentially artificial) higher-order terms in the structure equations that regularize the structure dynamics.

The consideration of this linearly coupled model consisting of the nonlinear Navier-Stokes model interacting with an elastic immersed structure is continued in [26], where the existence and uniqueness of strong solutions is considered in 2D and 3D. In terms of strong solutions, there are global unique solutions in 2D, and in 3D, given an initial fluid velocity, structure displacement, and structure velocity in $H^2(\Omega_f)$, $H^2(\Omega_s)$, and $H^1(\Omega_s)$, respectively, one can obtain local existence of strong solutions. It is important to note also that because these solutions are strong, an important component of the analysis in [26] involves obtaining improved regularity estimates that will allow for the recovery of the fluid pressure function. These results on local existence of strong solutions were later improved in subsequent works [223, 224], and [225]. In [223], a variant of the model in [26] is considered with "slab" domains containing flat boundaries, and in this context, the results from [26] are improved. In [223], the regularity of the initial fluid velocity is reduced from H^2 to H^1, and later, in [224], the regularity of the initial structure displacement and velocity is reduced to $H^{\frac{3}{2}+k}$ and $H^{\frac{1}{2}+k}$ for sufficiently small $k > 0$ by using "hidden regularity" and fractional estimates. Finally, in [225], these well-posedness results for strong solutions are extended beyond the model with "slab" domains with a flat fluid-structure interface that was previously found in [223, 224] to a more general geometry involving a potentially non-flat fluid-structure interface.

There is also a large body of literature devoted to the study of linearly coupled FSI problems where the structure is described by a lower-dimensional model. In a series of studies by Chueshov and others, the coupling of linear compressible and incompressible fluids with nonlinear plates, including the von Karman plate, has been rigorously analyzed, with a focus on the well-posedness and long-term behavior of the solutions, such as the existence and properties of global attractors (e.g., [87–90]). These studies focus on both the stability of solutions and the global dynamics, providing critical insights into the interactions between fluids and lower-dimensional elastic structures. A significant area of research in flow-plate interactions is the strong asymptotic stabilization of the flow by the damped structure. A recent paper [24] has made a notable advancement by proving that this stabilization property is indeed achievable. For a more comprehensive review of this topic, we refer the reader to [24].

Since the analysis of qualitative properties of solutions to FSI problems remains highly challenging, even with linear coupling, there has been considerable interest in studying simplified "toy" models that are more tractable but still share key characteristics with more complex FSI systems. A prominent example is the so-called heat-wave system, in which the linear wave equation is coupled with the linear heat equation through kinematic and dynamic coupling conditions. This system, of hyperbolic-parabolic type, can be viewed as a simplified version of FSI systems. In [352], it was demonstrated that this system is strongly stable, though not uniformly stable. Furthermore, under certain geometric conditions, the authors proved that the solution decays polynomially to zero as time t goes to infinity. The decay rate has been improved in subsequent works, e.g., [17, 129, 135]. Additionally, similar results have been proved when the heat equation is replaced by the Stokes equation [13]. Recently, the multilayered heat-wave system, a toy model for multilayered FSI, has garnered interest, and its regularity [264], well-posedness and stability [14], and decay properties [15] have been analyzed. Additionally, the existence of time-periodic solutions was established in [262].

We continue discussing deterministic FSI by focusing on nonlinearly coupled problems.

Chapter 5
Deterministic FSI with Nonlinear Coupling

In this chapter, we discuss a prototypical *nonlinearly coupled* deterministic FSI problem. The problem is characterized by the fact that the fluid domain is moving, and the coupling conditions between the fluid and the structure are evaluated along the current location of the fluid-structure interface. The fluid equations are posed on the time-dependent domain, which is not known *a priori*, and it depends on the solution of the problem, i.e., on the current location of the fluid-structure interface. This geometric nonlinearity is a major complication in the analysis of well-posedness for this class of problems. We study the existence of weak solutions in finite energy spaces. As before, we will use the Lie operator splitting strategy to construct approximate solutions. However, there are many unique challenges in the existence proof that arise when considering the moving boundary, and in this chapter, we will emphasize the main differences between the linearly coupled and nonlinearly coupled FSI problems.

In the next section, we start by describing a prototypical deterministic FSI problem with nonlinear coupling and finish this chapter by proving the existence of a weak solution in the finite energy spaces using a Lie operator splitting approach.

5.1 Model Description

As in the linearly coupled case, we will consider a two-dimensional fluid domain where the reference configuration for the fluid domain is

$$\Omega_f = \{(z, r) | z \in [0, L], r \in [0, R]\}.$$

The top boundary Γ of Ω_f corresponds to the reference configuration for an elastic structure, the left and right boundaries Γ_{in} and Γ_{out} of Ω_f are the inlet and outlet boundaries, and the bottom boundary $\Gamma_b = [0, L] \times \{0\}$ of Ω_f is a line of even

S. Čanić et al., *Deterministic and Stochastic Fluid-Structure Interaction*, Advances in Mathematical Fluid Mechanics, https://doi.org/10.1007/978-3-032-00898-5_5

symmetry of the 2D fluid "channel" $[0, L] \times [-R, R]$ of radius R. As in Sect. 4.1, the fluid flow through this channel will be driven by the time-dependent (dynamic) pressure data prescribed at $\Gamma_{in/out}$.

We denote by $\boldsymbol{u}$ the fluid velocity, and we denote the structure displacement by η. As in the linearly coupled case, we assume that the structure displaces in only the radial direction, so that η is a scalar function that describes the radial displacement of the structure from its reference configuration Γ, $\boldsymbol{\eta} = \eta \boldsymbol{e}_r$.

In the nonlinearly coupled FSI, we will be considering the time-dependent configuration of a thin elastic structure whose location will be denoted by

$$\Gamma(t) = \{(z, r) \in \mathbb{R}^2 : 0 \leq z \leq L, r = R + \eta(t, z)\}.$$

The assumption that the structure displaces in only the radial direction simplifies the analysis by preventing self-intersection of the structure with itself, so that domain degeneracy occurs only when $\eta(t, z) = -R$ at any point $z \in \Gamma$ for any time t. The time-dependent configuration of the structure determines the top boundary of the time-dependent fluid domain $\Omega_f(t)$, which is now defined by

$$\Omega_f(t) = \{(z, r) \in \mathbb{R}^2 : 0 \leq z \leq R, 0 \leq r \leq R + \eta(t, z)\}. \tag{5.1}$$

See Fig. 5.1.

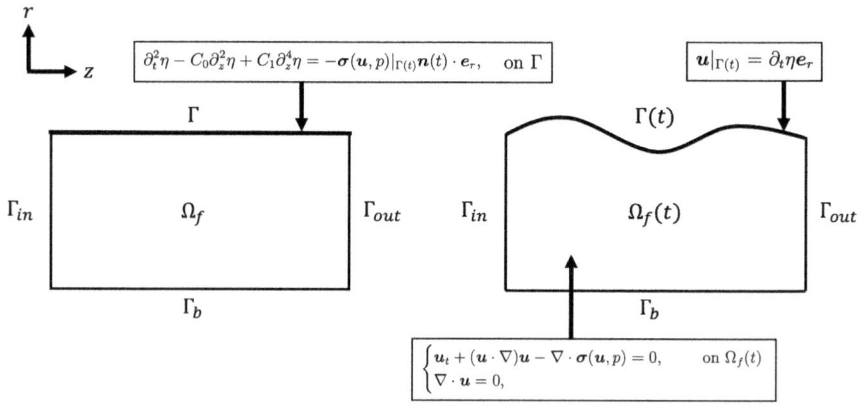

Fig. 5.1 Left: The reference configuration for the nonlinearly coupled prototypical FSI problem in two dimensions. Note that the structure equation is posed on the reference configuration Γ of the structure. Right: The moving fluid domain in physical space for the nonlinearly coupled prototypical FSI problem in two dimensions. Note that the Navier-Stokes equations are posed on the time-dependent a priori unknown fluid domain $\Omega_f(t)$, which is determined by the structure displacement

5.1.1 The Prototypical Model Equations

We begin with describing the fluid flow through $\Omega_f(t)$. In contrast with the linearly coupled prototypical FSI model, the fluid equations in the nonlinearly coupled case are defined on a *moving fluid domain*, whose time-dependent configuration $\Omega_f(t)$ depends on the structure displacement (see (5.1)), which is not known a priori. Additionally, due to the energy considerations, instead of the time-dependent Stokes equations, we consider the full Navier-Stokes equations. The reason for this is that the advection term in the Navier-Stokes equations will cancel out certain terms in the energy estimate that arise from the moving domain, giving rise to a well-defined energy of the problem. More details will be presented below.

The Fluid Subproblem We are looking for the fluid velocity $\boldsymbol{u} : \Omega_f(t) \to \mathbb{R}^2$ and the fluid pressure $p : \Omega_f(t) \to \mathbb{R}$ such that $\boldsymbol{u}$ and p satisfy the Navier-Stokes equations for an incompressible, viscous fluid :

$$\begin{cases} \boldsymbol{u}_t + (\boldsymbol{u} \cdot \nabla)\boldsymbol{u} - \nabla \cdot \boldsymbol{\sigma}(\boldsymbol{u}, p) = 0, \\ \nabla \cdot \boldsymbol{u} = 0, \end{cases} \qquad \text{on } \Omega_f(t), \tag{5.2}$$

where the Cauchy stress tensor is defined by

$$\boldsymbol{\sigma}(\boldsymbol{u}, p) = 2\mu_f \boldsymbol{D}(\boldsymbol{u}) - p\boldsymbol{I}, \qquad \boldsymbol{D}(\boldsymbol{u}) = \frac{1}{2}\left((\nabla \boldsymbol{u}) + (\nabla \boldsymbol{u})^t\right), \tag{5.3}$$

describing the flow of a Newtonian fluid. The Navier-Stokes system (5.2) is supplemented with the following boundary conditions on the inlet and outlet:

$$\begin{aligned} u_r &= 0, \qquad p + \frac{|\boldsymbol{u}|^2}{2} = P_{in}(t), \qquad \text{on } \Gamma_{in}, \\ u_r &= 0, \qquad p + \frac{|\boldsymbol{u}|^2}{2} = P_{out}(t), \qquad \text{on } \Gamma_{out}. \\ u_r &= 0, \qquad \partial_r u_z = 0, \qquad \text{on } \Gamma_b. \end{aligned} \tag{5.4}$$

Notice that the inlet and outlet pressures are now given by the dynamic pressure data:

$$p + \frac{|\boldsymbol{u}|^2}{2} = P_{in/out}(t), \qquad \text{on } \Gamma_{in/out}, \tag{5.5}$$

which have an extra term $\frac{|\boldsymbol{u}|^2}{2}$ due to the advection term that is present in the Navier-Stokes equations but not the linear Stokes equations.

The Structure Subproblem We consider the same linearly elastic plate equation in $1D$, as we did in the linearly coupled problem. The equation is written in terms of the radial/vertical displacement η, which is a scalar function $\eta : \Gamma \to \mathbb{R}$. The equation reads

$$\partial_t^2 \eta - C_1 \partial_z^2 \eta + C_2 \partial_z^4 \eta = F, \qquad \text{on } \Gamma. \tag{5.6}$$

Here, C_1 and C_2 are positive elasticity coefficients associated with the elastic energy of the problem. The load to the structure is given by F. As we shall see later, F in this FSI problem is given by the jump between the external forcing onto the structure and the internal forcing coming from the fluid Cauchy stress. Under the assumption that the external forcing is zero, the structure load F will be given in terms of the normal component of the Cauchy stress coming from the fluid, evaluated at the **current location** of the interface $\Gamma(t)$. This contrasts the expression for F in (4.7), where the normal component of Cauchy stress was evaluated at the fixed, reference configuration Γ. As is typical in elasticity theory, the structure equations are defined in the *Lagrangian formulation* on the reference domain Γ.

We will be considering a clamped plate/shell with the boundary conditions:

$$\eta(0) = \partial_z \eta(0) = \eta(L) = \partial_z \eta(L) = 0. \tag{5.7}$$

The Coupling Conditions Just as in the linearly coupled case, the fluid and the structure are coupled via two coupling conditions, the kinematic and the dynamic coupling conditions. However, in this nonlinearly coupled case, the coupling conditions are evaluated along the current location of the interface $\Gamma(t)$, introducing significant challenges in the analysis of the coupled FSI problem related to this geometric nonlinearity. The two coupling conditions are the following:

- **The kinematic coupling condition.** We consider the no-slip condition as in the linearly coupled case, except that now the trace of the fluid velocity is evaluated at the current location of the interface:

$$\boldsymbol{u}(z, R + \eta(t, z)) = \partial_t \eta(t, z) \boldsymbol{e}_r. \tag{5.8}$$

  We will informally write this condition as $\boldsymbol{u}|_{\Gamma(t)} = \partial_t \eta \boldsymbol{e}_r$.

  Note that this informal way of writing the coupling condition is not entirely precise, as $\partial_t \eta$ is defined on Γ rather than on $\Gamma(t)$, so we are implicitly assuming a composition with the map sending $(z, R) \in \Gamma$ to $(z, R + \eta(t, z)) \in \Gamma(t)$. This will be made more precise when we introduce an arbitrary Lagrangian-Eulerian (ALE) mapping below, which maps the reference fluid domain onto the current fluid domain.
- **The dynamic coupling condition.** As in the linearly coupled case, the dynamic coupling condition states that the elastodynamics of the structure is driven by the jump in the normal stress/force between across the interface (structure). Since we are again assuming that there is no external force acting on the structure, i.e., the external normal stress is zero, the only contribution to the force driving the structure is the fluid normal Cauchy stress. Because we are evaluating the coupling conditions along the *moving interface* $\Gamma(t)$, this condition reads

$$\partial_t^2 \eta - C_1 \partial_z^2 \eta + C_2 \partial_z^4 \eta = -\mathcal{J}^\eta \boldsymbol{\sigma}(\boldsymbol{u}, p)|_{\Gamma(t)} \boldsymbol{n}(t) \cdot \boldsymbol{e}_r, \tag{5.9}$$

where:

- $\boldsymbol{\sigma}(\boldsymbol{u}, p)|_{\Gamma(t)}$ means $\boldsymbol{\sigma}(\boldsymbol{u}(\boldsymbol{z}, \boldsymbol{R} + \boldsymbol{\eta}(\boldsymbol{t}, \boldsymbol{z})), p(z, R + \eta(t, z)))$
- $\boldsymbol{n}(t)$ is the outward unit normal vector to the time-dependent configuration $\Gamma(t)$
- $\mathcal{J}^\eta = \sqrt{1 + (\partial_z \eta)^2}$ is the arc length measure for $\Gamma(t)$, which converts the arc length on the moving interface $\Gamma(t)$ to the arc length on the reference configuration Γ

As we shall see below, the factor $\mathcal{J}^\eta$ is equal to the Jacobian of the transformation mapping the reference fluid domain onto the current fluid domain; see (5.31).

Problem (5.2)–(5.9) is supplemented with the following initial conditions:

$$\boldsymbol{u}(0, .) = \boldsymbol{u}_0, \; \eta(0, .) = \eta_0, \; \partial_t \eta(0, .) = v_0. \tag{5.10}$$

We will be assuming that the initial data satisfy the following compatibility conditions:

$$\begin{aligned} \boldsymbol{u}_0(z, R + \eta_0(z)) &= v_0(z)\mathbf{e}_r, \quad z \in (\Gamma), \\ \eta_0(0) = \eta_0(L) &= v_0(0) = v_0(L) = 0, \\ R + \eta_0(z) &> 0, \quad z \in [0, L]. \end{aligned} \tag{5.11}$$

5.2 A Priori Energy Estimate and Weak Formulation

Just as in the case of the linearly coupled FSI problem, we will derive here a formal energy estimate that will help us motive the definition of weak, finite energy solutions. There are several notable differences in the derivation of an a priori energy estimate in the nonlinearly coupled FSI problem, which are due to the fact that the fluid domain is time-dependent, and the fluid equations we are considering are the Navier-Stokes equations.

As in Sect. 4.2, we will formally show that for the nonlinearly coupled problem (5.2)–(5.11), the following energy estimate holds for all $t \in [0, T]$, up to a fixed but arbitrary final time T:

ENERGY ESTIMATE

$$\begin{aligned} &\frac{1}{2}\|\partial_t \eta(t)\|^2_{L^2(\Gamma)} + \|\eta(t)\|^2_E + \frac{1}{2}\|\boldsymbol{u}(t)\|^2_{L^2(\Omega_f(t))} + 2\mu_f \int_0^t \|\boldsymbol{D}(\boldsymbol{u})(s)\|^2_{L^2(\Omega_f(s))} ds \\ &\leq \frac{1}{2}\|v_0\|^2_{L^2(\Gamma)} + \|\eta_0\|^2_E + \frac{1}{2}\|\boldsymbol{u}_0\|^2_{L^2(\Omega_f(0))} + C\left(\|P_{in}\|^2_{L^2(0,T)} + \|P_{out}\|^2_{L^2(0,T)}\right) \end{aligned} \tag{5.12}$$

where the constant C now depends on the final time T and on the initial data.

A notable difference from the linearly coupled case is that the norms $\|\boldsymbol{u}(t)\|^2_{L^2(\Omega_f(t))}$ and $\|\boldsymbol{D}(\boldsymbol{u})(t)\|^2_{L^2(\Omega_f(t))}$ involve the moving time-dependent fluid domain, so that, for example,

$$\|\boldsymbol{u}(t)\|^2_{L^2(\Omega_f(t))} = \int_{\Omega_f(t)} |\boldsymbol{u}(t)|^2.$$

In the above estimate (5.12), $\|\eta(t)\|_E$ denotes the contribution from the elastic energy of the structure:

$$\|\eta\|_E^2 := \frac{1}{2}C_1\|\partial_z\eta\|^2_{L^2(\Gamma)} + \frac{1}{2}C_2\|\partial_z^2\eta\|^2_{L^2(\Gamma)}, \tag{5.13}$$

and $P_{in/out}$ are the dynamic pressure data given in (5.5).

As in the linearly coupled case (see (4.13)), we can rewrite this energy estimate as

$$\frac{d}{dt}E(t) + D(t) \leq C(P_{in}(t), P_{out}(t)), \tag{5.14}$$

where the total energy $E(t)$ accounts for the kinetic and potential (elastic) energy, defined by

$$E(t) := \frac{1}{2}\|\boldsymbol{u}(t)\|^2_{L^2(\Omega_f(t))} + \frac{1}{2}\|\partial_t\eta(t)\|^2_{L^2(\Gamma)} + \|\eta(t)\|^2_E, \tag{5.15}$$

and dissipation $D(t)$ accounts for dissipation due to fluid viscosity, defined by

$$D(t) := 2\mu_f\|\boldsymbol{D}(\boldsymbol{u})(s)\|^2_{L^2(\Omega_f(t))}. \tag{5.16}$$

As before, this energy estimate states that the rate of change of total energy in the problem, plus dissipation, are bounded by the work done by the inlet and outlet data.

To derive the energy estimate (5.14), we first multiply the momentum equation in (5.2) by $\boldsymbol{u}$, integrate over $\Omega_f(t)$, and formally integrate by parts to obtai

$$\int_{\Omega_f(t)} (\partial_t\boldsymbol{u}\cdot\boldsymbol{u}+(\boldsymbol{u}\cdot\nabla)\boldsymbol{u}\cdot\boldsymbol{u})+2\mu_f\int_{\Omega_f(t)} |\mathbf{D}\boldsymbol{u}|^2-\int_{\partial\Omega_f(t)} (-p\mathbf{I}+2\mu_f\mathbf{D}(\boldsymbol{u}))\mathbf{n}(t)\cdot\boldsymbol{u}=0. \tag{5.17}$$

Notice that the unit normal $\mathbf{n}(t)$ now depends on the current location of the moving boundary $\Gamma(t)$ and is one of the unknowns in the problem, adding to the complexity of the FSI problem due to the geometric nonlinearity of the coupling.

To deal with the inertia term, we first recall that $\Omega_f(t)$ is moving in time and that the velocity of the lateral boundary is given by $\boldsymbol{u}|_{\Gamma(t)}$. The transport theorem (see Theorem 2.8.1 in Chap. 2), applied to the first term on the left-hand side of the above equation, then gives

$$\int_{\Omega_f(t)} \partial_t \boldsymbol{u} \cdot \boldsymbol{u} = \frac{1}{2}\frac{d}{dt}\int_{\Omega_f(t)} |\boldsymbol{u}|^2 - \frac{1}{2}\int_{\Gamma(t)} |\boldsymbol{u}|^2 \boldsymbol{u} \cdot \mathbf{n}(t).$$

The second term on the left-hand side can be rewritten by using integration by parts, and the divergence-free condition, to obtain

$$\int_{\Omega_f(t)} (\boldsymbol{u} \cdot \nabla)\boldsymbol{u} \cdot \boldsymbol{u} = \frac{1}{2}\int_{\partial\Omega_f(t)} |\boldsymbol{u}|^2 \boldsymbol{u} \cdot \mathbf{n}(t) = \frac{1}{2}\Big(\int_{\Gamma(t)} |\boldsymbol{u}|^2 \boldsymbol{u} \cdot \mathbf{n}(t)$$

$$- \int_{\Gamma_{in}} |\boldsymbol{u}|^2 u_z + \int_{\Gamma_{out}} |\boldsymbol{u}|^2 u_z.\Big)$$

These two terms added together give

$$\int_{\Omega_f(t)} \partial_t \boldsymbol{u} \cdot \boldsymbol{u} + \int_{\Omega_f(t)} (\boldsymbol{u} \cdot \nabla)\boldsymbol{u} \cdot \boldsymbol{u} = \frac{1}{2}\frac{d}{dt}\int_{\Omega_f(t)} |\boldsymbol{u}|^2 - \frac{1}{2}\int_{\Gamma_{in}} |\boldsymbol{u}|^2 u_z + \frac{1}{2}\int_{\Gamma_{out}} |\boldsymbol{u}|^2 u_z. \tag{5.18}$$

To deal with the boundary integral over $\partial\Omega_f(t)$ in (5.17), we first notice that $\partial\Omega_f(t) = \Gamma(t) \cup \Gamma_{in/out} \cup \Gamma_b$, and we deal with each part of the boundary separately.

On $\Gamma_{in/out}$, the boundary condition (5.5) implies $u_r = 0$. Combined with the divergence-free condition, we obtain $\partial_z u_z = -\partial_r u_r = 0$. Now, using the fact that the normal to $\Gamma_{in/out}$ is $\mathbf{n} = (\mp 1, 0)$, we get

$$\int_{\Gamma_{in/out}} (-p\mathbf{I} + 2\mu_f \mathbf{D}(\boldsymbol{u}))\mathbf{n} \cdot \boldsymbol{u} = \int_{\Gamma_{in}} P_{in} u_z - \int_{\Gamma_{out}} P_{out} u_z. \tag{5.19}$$

On Γ_b, using the symmetry boundary conditions (5.4), we get

$$\int_{\Gamma_b} (-p\mathbf{I} + 2\mu_f \mathbf{D}(\boldsymbol{u}))\mathbf{n} \cdot \boldsymbol{u} = 0.$$

What is left is to calculate the remaining boundary integral over $\Gamma(t)$, namely, we want to express the integral

$$-\int_{\Gamma(t)} \boldsymbol{\sigma}(\boldsymbol{u}, p)|_{\Gamma(t)} \boldsymbol{n}(t) \cdot \boldsymbol{u} \tag{5.20}$$

in terms of the structure elastodynamics using the dynamic coupling condition (5.9). For this purpose, we consider the plate equation (5.6), multiplied by $\partial_t \eta$, and integrated by parts to obtain

$$\int_\Gamma F \partial_t \eta = \int_0^L F \partial_t \eta = \frac{1}{2}\frac{d}{dt}\|\partial_t \eta\|^2_{L^2(\Gamma)} + \frac{1}{2}\frac{d}{dt}\left(C_1\|\partial_z \eta\|^2_{L^2(\Gamma)} + C_2\|\partial_z^2 \eta\|^2_{L^2(\Gamma)}\right) \tag{5.21}$$

To enforce the dynamic coupling condition (5.9), we notice that this condition is written in Lagrangian formulation, namely, the left- and right-hand sides are defined on the reference configuration Γ. Thus, we transform the integral (5.20) over $\Gamma(t)$ into the corresponding integral over Γ:

$$-\int_{\Gamma(t)} \boldsymbol{\sigma}(\boldsymbol{u}, p)|_{\Gamma(t)}\boldsymbol{n}(t) \cdot \boldsymbol{u} = -\int_{\Gamma} \mathcal{J}^{\eta}\boldsymbol{\sigma}(\boldsymbol{u}, p)|_{\Gamma(t)}\boldsymbol{n}(t) \cdot \boldsymbol{u}$$

where $\mathcal{J}^{\eta} = \sqrt{1 + (\partial_z \eta)^2}$ is the arc length measure for $\Gamma(t)$, as defined earlier just under (5.9).

Now we enforce the dynamic coupling condition (5.9), stated here again for reader's convenience:

$$F = -\mathcal{J}^{\eta}\boldsymbol{\sigma}(\boldsymbol{u}, p)|_{\Gamma(t)}\boldsymbol{n}(t) \cdot \boldsymbol{e}_r \text{ on } \Gamma$$

to express the integral over $\Gamma(t)$. We also recall the kinematic coupling condition that states that the trace of the fluid velocity $\boldsymbol{u}$ at the interface is equal to the structure velocity $\partial_t \eta$: $\boldsymbol{u}(z, R+\eta(t, z)) = \partial_t \eta(t, z)\boldsymbol{e}_r$, and combine this with (5.21) to obtain

$$\begin{aligned} -\int_{\Gamma(t)} \boldsymbol{\sigma}(\boldsymbol{u}, p)|_{\Gamma(t)}\boldsymbol{n}(t) \cdot \boldsymbol{u} &= -\int_{\Gamma} \mathcal{J}^{\eta}\boldsymbol{\sigma}(\boldsymbol{u}, p)|_{\Gamma(t)}\boldsymbol{n}(t) \cdot \boldsymbol{u} = \int_{\Gamma} F\partial_t \eta \\ &= \frac{1}{2}\frac{d}{dt}\|\partial_t \eta\|^2_{L^2(\Gamma)} + \frac{1}{2}\frac{d}{dt}\left(C_1\|\partial_z \eta\|^2_{L^2(\Gamma)} + C_2\|\partial_z^2 \eta\|^2_{L^2(\Gamma)}\right). \end{aligned} \tag{5.22}$$

Finally, by combining (5.22) with (5.19), and by adding the remaining contributions to the energy of the FSI problem, one obtains the following energy equality:

$$\begin{aligned} \frac{1}{2}\frac{d}{dt}\int_{\Omega_f(t)} |\boldsymbol{u}|^2 + \frac{1}{2}\frac{d}{dt}\|\partial_t \eta\|^2_{L^2(\Gamma)} + 2\mu_f \int_{\Omega_f(t)} |\mathbf{D}\boldsymbol{u}|^2 \\ + \frac{1}{2}\frac{d}{dt}\left(C_1\|\partial_z \eta\|^2_{L^2(\Gamma)} + C_2\|\partial_z^2 \eta\|^2_{L^2(\Gamma)}\right) \\ = \int_{\Gamma_{out}} P_{out}u_z - \int_{\Gamma_{in}} P_{in}u_z \end{aligned} \tag{5.23}$$

By using the trace inequality (see Theorem 2.5.1 in Chap. 2) and Korn inequality (see (2.15) in Chap. 2), one can estimate

$$\begin{aligned} |P_{in/out}(t)\int_{\Sigma_{in/out}} u_z| \leq C|P_{in/out}|\|\boldsymbol{u}\|_{H^1(\Omega_f(t))} &\leq \frac{C}{2\epsilon}|P_{in/out}|^2 \\ &+ \frac{\epsilon C}{2}\|\mathbf{D}(\boldsymbol{u})\|^2_{L^2(\Omega_f(t))}. \end{aligned}$$

By choosing ϵ such that $\frac{\epsilon C}{2} \leq \mu_f$, we get the energy inequality (5.14). □

Remark Notice that the constant in the trace inequality depends on the fluid domain, which in our case depends on η. To get an energy estimate in which the constant is independent of η, one can use Gronwall's inequality and obtain the result above in which the constant C depends on time T.

This energy estimate allows us to define the finite energy spaces for the solutions. For the fluid velocity, we define the solution space $\mathcal{V}_f(t)$ (for fixed but arbitrary time), which is time-dependent since the fluid domain is time-dependent:

$$\mathcal{V}_f(t) = \{\boldsymbol{u} \in H^1(\Omega_f(t)) : \nabla \cdot \boldsymbol{u} = 0, u_r = 0 \text{ on } \Gamma_{in} \cup \Gamma_{out} \cup \Gamma_b, u_z = 0 \text{ on } \Gamma\}. \tag{5.24}$$

A precise definition of the time-dependent function spaces is presented in Chap. 2; see Definition 2.8.1.

Since the structure problem is posed on the fixed reference domain Γ, the finite energy space for the structure is the same as in the linearly coupled case:

$$\mathcal{V}_s = H_0^2(\Gamma). \tag{5.25}$$

The solution spaces for the fluid velocity and the structure displacement that involve both space and time are given by

$$V_f = L^\infty(0, T; L^2(\Omega_f(t))) \cap L^2(0, T; \mathcal{V}_f(t)), \tag{5.26}$$

$$V_s = W^{1,\infty}(0, T; L^2(\Gamma)) \cap L^\infty(0, T; \mathcal{V}_s). \tag{5.27}$$

The full coupled solution space is

$$\mathcal{V} = \{(\boldsymbol{u}, \eta) \in V_f \times V_s : \boldsymbol{u}|_{\Gamma(t)} = \partial_t \eta \boldsymbol{e}_r\}, \tag{5.28}$$

and the test space is

$$\mathcal{Q} = \{(\boldsymbol{q}, \psi) \in C_c^\infty([0, T); \mathcal{V}_f(t) \times V_s) : \boldsymbol{q}|_{\Gamma(t)} = \psi \boldsymbol{e}_r\}. \tag{5.29}$$

In both spaces, the kinematic coupling condition is included in the definition, enforcing this condition in a strong sense. This contrasts with the dynamic coupling condition, which is imposed weakly in the problem's weak formulation through the integration-by-parts procedure.

We note that the test space depends on the *a priori unknown structure displacement* because the test space involves the function space $\mathcal{V}_f(t)$, which depends on $\Omega_f(t)$ and hence $\eta(t)$. The fact that the test functions cannot be defined a priori is one of the main challenges that arises more generally in studying nonlinearly coupled FSI problems.

We are now ready to introduce the weak formulation of the problem. To derive the weak formulation, we consider a test function $(\boldsymbol{q}, \psi)$ for the fluid and structure problems, respectively, satisfying $\boldsymbol{q}|_{\Gamma(t)} = \psi \boldsymbol{e}_r$. We multiply the fluid and structure problem by the corresponding test function and use integration by parts to derive a weak formulation. In this case, the weak formulation has an additional symmetrized term arising from the advection term in the Navier-Stokes equation, and furthermore, the integrals over the fluid domain are over the time-dependent fluid domain $\Omega_f(t)$.

Definition 5.2.1 (Moving Domain Weak Formulation) An ordered pair $(\boldsymbol{u}, \eta) \in \mathcal{V}$ is a weak solution to the prototypical nonlinearly coupled FSI problem (5.2)–(5.9) if for all test functions $(\boldsymbol{q}, \psi) \in \mathcal{Q}$, the following weak formulation holds:

$$
\begin{aligned}
&-\int_0^T \int_{\Omega_f(t)} \boldsymbol{u} \cdot \partial_t \boldsymbol{q} + \frac{1}{2} \int_0^T \int_{\Omega_f(t)} \Big([(\boldsymbol{u} \cdot \nabla)\boldsymbol{u}] \cdot \boldsymbol{q} - [(\boldsymbol{u} \cdot \nabla)\boldsymbol{q})\boldsymbol{u}] \Big) \\
&-\frac{1}{2} \int_0^T \int_\Gamma (\partial_t \eta)^2 \cdot \psi + 2\mu_f \int_0^T \int_{\Omega_f(s)} \boldsymbol{D}(\boldsymbol{u}) : \boldsymbol{D}(\boldsymbol{q}) \\
&-\int_0^T \int_\Gamma \partial_t \eta \cdot \partial_t \psi + C_1 \int_0^T \int_\Gamma \partial_z \eta \cdot \partial_z \psi + C_2 \int_0^T \int_\Gamma \partial_z^2 \eta \cdot \partial_z^2 \psi \\
&= \int_0^T \int_{\Gamma_{in}} P_{in}(t) \cdot q_z - \int_0^T \int_{\Gamma_{out}} P_{out}(t) \cdot q_z + \int_{\Omega_f(0)} \boldsymbol{u}_0 \cdot \boldsymbol{q}(0) + \int_\Gamma v_0 \cdot \psi(0).
\end{aligned}
\tag{5.30}
$$

Equation (5.30) is obtained using integration by parts and the following equalities that hold for smooth functions:

$$
\begin{aligned}
\int_{\Omega_f(t)} (\boldsymbol{u} \cdot \nabla)\boldsymbol{u} \cdot \boldsymbol{q} = \frac{1}{2} \int_{\Omega_f(t)} (\boldsymbol{u} \cdot \nabla)\boldsymbol{u} \cdot \boldsymbol{q} - \frac{1}{2} \int_{\Omega_f(t)} (\boldsymbol{u} \cdot \nabla)\boldsymbol{q} \cdot \boldsymbol{u} \\
+ \frac{1}{2} \int_\Gamma (\partial_t \eta)^2 \psi \pm \frac{1}{2} \int_{\Gamma_{out/in}} |u_r|^2 v_r,
\end{aligned}
$$

$$
\int_0^T \int_{\Omega_f(t)} \partial_t \boldsymbol{u} \cdot \boldsymbol{q} = -\int_0^T \int_{\Omega_f(t)} \boldsymbol{u} \cdot \partial_t \boldsymbol{q} - \int_{\Omega_f(0)} \boldsymbol{u}_0 \cdot \boldsymbol{q}(0) - \int_0^T \int_\Gamma (\partial_t \eta)^2 \psi.
$$

Remark Notice how due to the motion of the fluid domain, we get an extra integral over Γ coming from fluid advection, and accounting for the fluid domain boundary motion via the trace of the fluid velocity, which is equal to the velocity of the domain boundary due to the no-slip condition.

This is an important term, which in the derivation of the energy estimate (5.14) cancels out the "bad," cubic term, coming from the application of the transport theorem to the kinetic energy term. Therefore, the presence of the nonlinear fluid advection term in the Navier-Stokes equations is crucial in obtaining the energy estimate (5.14) for fluid-structure interaction problems on moving domains.

To study the well-posedness of moving boundary problems, it is often times more convenient to map the moving boundary problem onto a fixed domain. Although this is not always necessary, in this book, we will follow that approach. See, e.g., [335] for a nonlinearly coupled FSI problem, which is analyzed by using an operator splitting scheme and applying it on current moving domains.

To formulate an equivalent version of the weak formulation that is defined on the fixed reference domain Ω_f, we need to introduce a mapping between the reference fluid domain Ω_f and the time-dependent physical fluid domain $\Omega_f(t)$. For this purpose, we introduce the so-called **arbitrary Lagrangian-Eulerian (ALE) mappings**. See also Sect. 2.8.1 in Chap. 2.

5.2.1 The ALE Mapping

For the given two-dimensional geometry of this particular model, we introduce a family of bijective mappings between Ω_f and $\Omega_f(t)$, known as the arbitrary Lagrangian-Eulerian (ALE) mappings. Assuming sufficient regularity of η, we define a family of bijective mappings:

$$\mathbf{\Phi}^{\eta}_f(t) : \Omega_f \to \Omega_f(t), \quad \left(\mathbf{\Phi}^{\eta}_f\right)(t)(z,r) = \left(z, \frac{R+\eta(t,z)}{R} r\right), \quad (z,r) \in \Omega_f,$$

with an inverse $\left(\mathbf{\Phi}^{\eta}_f\right)^{-1} : \Omega_f(t) \to \Omega_f$ defined by

$$\left(\mathbf{\Phi}^{\eta}_f\right)^{-1}(t)(z,r) = \left(z, \frac{R}{R+\eta(t,z)} r\right), \qquad (z,r) \in \Omega_f(t).$$

We used the superscript η to emphasize the dependency of the definition of this mapping on the structure displacement η. As long as the structure displacement η is sufficiently regular (e.g., if it is a continuous function on Γ) and $\eta > -R$ so that the structure does not come into contact with the bottom boundary, the ALE mapping is a bijective map with a Jacobian given by

$$\mathcal{J}^{\eta}_f = 1 + \frac{\eta}{R}. \tag{5.31}$$

Assuming that the structure displacement η satisfies $\eta > -R$, the Jacobian (5.31) is strictly positive.

Our goal is to map the time-dependent fluid domain onto a fixed reference domain and study the existence of a weak solution defined on the reference domain. For this purpose, we consider how functions and derivatives transform under the ALE mapping.

For a function h defined on the physical domain $\Omega_f(t)$, we define the pullback of the function via composition with the ALE mapping as follows:

$$h^\eta(t,\cdot) = h(t, \mathbf{\Phi}_f^\eta(t,\cdot)).$$

Conversely, we also have that

$$h(t,\cdot) = h^\eta(t, (\mathbf{\Phi}_f^\eta)^{-1}(t,\cdot)).$$

Next, we want to consider how derivatives transform under the ALE mapping. To see why we need to do this, let us consider the fluid velocity $\boldsymbol{u}$, which is defined on $\Omega_f(t)$ and which is required to be divergence-free so that

$$\nabla \cdot \boldsymbol{u} = 0 \qquad \text{on } \Omega_f(t).$$

Consider the pullback via the ALE mapping so that the resulting fluid velocity $\boldsymbol{u}^\eta$ is defined on the fixed reference domain Ω_f. Because the ALE mapping is a nonlinear map, it is no longer true that $\nabla \cdot \boldsymbol{u}^\eta = 0$ on Ω_f, where the divergence is taken with respect to the reference domain variables. Hence, we will need to find an appropriate differential operator that translates the divergence-free condition on the physical domain to an appropriate variant of the divergence-free condition on the fixed reference domain. For this purpose, we use the chain rule to determine how derivatives will transform under the ALE mapping. For clarity, in the upcoming calculations, we will distinguish between the variables $(\tilde{z}, \tilde{r})$ on the reference domain Ω_f and the variables (z, r) on the physical domain $\Omega_f(t)$, though we will drop the tilde notation in the remainder of the section. Given a function h on the physical domain $\Omega_f(t)$, we have that

$$\begin{aligned}\frac{\partial}{\partial z}h(t,z,r) &= \frac{\partial}{\partial z}h^\eta\left(t, z, \frac{R}{R+\eta(t,z)}r\right) = \frac{\partial h^\eta}{\partial \tilde{z}} - \frac{R}{(R+\eta(t,\tilde{z}))^2}(r\partial_z\eta)\frac{\partial h^\eta}{\partial \tilde{r}} \\ &= \frac{\partial h^\eta}{\partial \tilde{z}} - \frac{\tilde{r}\partial_z\eta}{R+\eta(t,\tilde{z})}\frac{\partial h^\eta}{\partial \tilde{r}},\end{aligned}$$

where we used that $r = \left(1 + \frac{\eta}{R}\right)\tilde{r}$. Furthermore,

$$\frac{\partial}{\partial r}h(t,z,r) = \frac{\partial}{\partial r}h^\eta\left(t, z, \frac{R}{R+\eta(t,z)}r\right) = \frac{R}{R+\eta(t,\tilde{z})} \cdot \frac{\partial h^\eta}{\partial \tilde{r}}.$$

So if we define the differential operator

$$\nabla^\eta = \left(\frac{\partial}{\partial \tilde{z}} - \frac{\tilde{r}\partial_z\eta}{R+\eta(t,\tilde{z})}, \frac{R}{R+\eta(t,\tilde{z})}\frac{\partial}{\partial \tilde{r}}\right), \tag{5.32}$$

we have that

$$\nabla^\eta h^\eta = (\nabla h) \circ \boldsymbol{\Phi}_f^\eta \quad \text{and} \quad \nabla^\eta h = \nabla h[\nabla \boldsymbol{\Phi}_f^\eta]^{-1}. \tag{5.33}$$

Similarly,

$$\nabla^\eta \cdot \boldsymbol{u}^\eta = (\nabla \cdot \boldsymbol{u}) \circ \boldsymbol{\Phi}_f^\eta,$$

and

$$\boldsymbol{D}^\eta(\boldsymbol{u}^\eta) = \boldsymbol{D}(\boldsymbol{u}) \circ \boldsymbol{\Phi}_f^\eta, \qquad \text{where } \boldsymbol{D}^\eta(\boldsymbol{u}^\eta) = \frac{1}{2}\Big((\nabla^\eta \boldsymbol{u}^\eta) + (\nabla^\eta \boldsymbol{u}^\eta)^t\Big).$$

Next, we consider the partial derivative with respect to time to obtain

$$\begin{aligned}\frac{\partial}{\partial t} h(t,z,r) = \frac{\partial}{\partial t} h^\eta\left(t, z\frac{R}{R+\eta(t,z)} r\right) &= \frac{\partial h^\eta}{\partial t} - \frac{R}{(R+\eta(t,\tilde{z}))^2} \cdot r \partial_t \eta \frac{\partial h^\eta}{\partial \tilde{r}} \\ &= \frac{\partial h^\eta}{\partial t} - \frac{\tilde{r}\partial_t \eta}{R+\eta(t,\tilde{z})} \frac{\partial h^\eta}{\partial \tilde{r}}.\end{aligned}$$

Let us introduce the ALE velocity:

$$\boldsymbol{w}^\eta = \partial_t \boldsymbol{\Phi}_f^\eta(t)(\tilde{z},\tilde{r}) = \frac{\tilde{r}}{R}\partial_t \eta \boldsymbol{e}_r \tag{5.34}$$

Then, we have that

$$\frac{\partial h}{\partial t} = \frac{\partial h^\eta}{\partial t} - (\boldsymbol{w} \cdot \nabla^\eta) h^\eta. \tag{5.35}$$

5.2.2 Weak Formulation on a Fixed Reference Domain

Using the transformation laws (5.33) and (5.35) for derivatives, we can rewrite the weak formulation for the prototypical nonlinearly coupled FSI problem on the fixed reference domain Ω_f. This introduces additional geometric nonlinearities in the problem, which arise from the Jacobian of the ALE mapping $\boldsymbol{\Phi}_f^\eta(t)$ and from the transformation of derivatives via the ALE mapping.

The solution spaces for the fluid velocity and the structure displacement on the fixed reference domain Ω_f are defined as follows. The fluid velocity space is

$$V_f^\eta = L^\infty(0,T; L^2(\Omega_f)) \cap L^2(0,T; \mathcal{V}_f^\eta), \tag{5.36}$$

where

$$\mathcal{V}_f^\eta = \{\boldsymbol{u} \in H^1(\Omega_f) : \nabla^\eta \cdot \boldsymbol{u} = 0, u_r = 0 \text{ on } \Gamma_{in} \cup \Gamma_{out} \cup \Gamma_b, u_z = 0 \text{ on } \Gamma\}. \tag{5.37}$$

The structure solution space is the same as the structure space for the linearly coupled problem, and is given by V_s, defined in (5.27).

The full solution space is then

$$\mathcal{V}^\eta = \{(\boldsymbol{u}, \eta) \in V_f^\eta \times V_s : \boldsymbol{u}|_\Gamma = \partial_t \eta \boldsymbol{e}_r.\}.$$

Similarly, the test space using the fixed reference fluid domain formulation is defined as

$$\mathcal{Q}^\eta = \{(\boldsymbol{q}, \psi) \in C_c^\infty([0, T); V_f^\eta \times V_s) : \boldsymbol{q}|_\Gamma = \psi \boldsymbol{e}_r\}. \tag{5.38}$$

The weak formulation of the nonlinearly coupled FSI problem defined on the fixed reference domain is as follows.

Definition 5.2.2 (Fixed Domain Weak Formulation) An ordered pair $(\boldsymbol{u}, \eta) \in \mathcal{V}^\eta$ with $\boldsymbol{u}$ defined on the fixed reference fluid domain Ω_f of the nonlinearly coupled FSI problem (5.2)–(5.9) if the following holds all test functions $(\boldsymbol{q}, \psi) \in \mathcal{Q}^\eta$:

$$\begin{aligned}
&-\int_0^T \int_{\Omega_f} \left(1 + \frac{\eta}{R}\right) \boldsymbol{u} \cdot \partial_t \boldsymbol{q} \\
&+ \int_0^T \int_{\Omega_f} \left(1 + \frac{\eta}{R}\right) \Big([((\boldsymbol{u} - \boldsymbol{w}^\eta) \cdot \nabla^\eta)\boldsymbol{u}] \cdot \boldsymbol{q} - [((\boldsymbol{u} - \boldsymbol{w}^\eta) \cdot \nabla^\eta)\boldsymbol{q}] \cdot \boldsymbol{u}\Big) \\
&- \int_0^T \int_{\Omega_f} \frac{1}{2R}(\partial_t \eta)\boldsymbol{u} \cdot \boldsymbol{q} + 2\mu_f \int_0^T \int_{\Omega_f} \left(1 + \frac{\eta}{R}\right) \boldsymbol{D}^\eta(\boldsymbol{u}) : \boldsymbol{D}^\eta(\boldsymbol{q}) \\
&- \int_0^T \int_\Gamma \partial_t \eta \cdot \partial_t \psi \\
&+ C_1 \int_\Gamma \partial_z \eta \cdot \partial_z \psi + C_2 \int_\Gamma \partial_z^2 \eta \cdot \partial_z^2 \psi = \int_0^T \int_{\Gamma_{in}} P_{in}(t) \cdot q_z - \int_0^T \int_{\Gamma_{out}} P_{out}(t) \cdot q_z \\
&+ \int_{\Omega_f} (1 + \frac{\eta_0}{R})\boldsymbol{u}_0 \cdot \boldsymbol{q}(0) + \int_\Gamma v_0 \cdot \psi(0).
\end{aligned} \tag{5.39}$$

To obtain this weak formulation, we begin by transforming the first integral on the left-hand side of (5.30) in Definition 5.2.1, describing the fluid kinetic energy. Namely, by recalling the definition of the Jacobian of the ALE transformation (5.31), which is equal to $1 + \eta/R$, we formally calculate

$$-\int_{\Omega_f(t)} \boldsymbol{u} \cdot \partial_t \boldsymbol{q} d\boldsymbol{x} = -\int_{\Omega_f} \boldsymbol{u}^\eta \cdot (\partial_t \boldsymbol{q} - (\mathbf{w}^\eta \cdot \nabla^\eta)\boldsymbol{q})(1 + \frac{\eta}{R}) d\tilde{\boldsymbol{x}}$$

$$= -\int_{\Omega_f} (1 + \frac{\eta}{R}) \boldsymbol{u}^\eta \cdot \partial_t \boldsymbol{q} d\tilde{\boldsymbol{x}}$$

$$+ \frac{1}{2} \int_{\Omega_f} (1 + \frac{\eta}{R})(\mathbf{w}^\eta \cdot \nabla^\eta)\boldsymbol{q} \cdot \boldsymbol{u}^\eta d\tilde{\boldsymbol{x}} + \frac{1}{2} \int_{\Omega_f} (1 + \frac{\eta}{R})(\mathbf{w}^\eta \cdot \nabla^\eta)\boldsymbol{q} \cdot \boldsymbol{u}^\eta d\tilde{\boldsymbol{x}}.$$

In the last integral on the right-hand side, we use the definition of $\mathbf{w}^\eta$ in (5.34) and of ∇^η in (5.32), to obtain

$$\int_{\Omega_f} (1 + \frac{\eta}{R})(\mathbf{w}^\eta \cdot \nabla^\eta)\boldsymbol{q} \cdot \boldsymbol{u}^\eta d\tilde{\boldsymbol{x}} = \frac{1}{R} \int_{\Omega_f} \partial_t \eta \, \tilde{r} \, \partial_{\tilde{r}} \boldsymbol{q} \cdot \boldsymbol{u}^\eta d\tilde{\boldsymbol{x}}.$$

Using integration by parts with respect to $\tilde{r}$, and keeping in mind that η does not depend on $\tilde{r}$, we obtain

$$-\int_{\Omega_f(t)} \boldsymbol{u} \cdot \partial_t \boldsymbol{q} = -\int_{\Omega_f} (1 + \frac{\eta}{R}) \boldsymbol{u}^\eta \cdot \partial_t \boldsymbol{q} + \frac{1}{2} \int_{\Omega_f} (1 + \frac{\eta}{R})(\mathbf{w}^\eta \cdot \nabla^\eta)\boldsymbol{q} \cdot \boldsymbol{u}^\eta$$

$$-\frac{1}{2} \int_{\Omega_f} (1 + \frac{\eta}{R})(\mathbf{w}^\eta \cdot \nabla^\eta)\boldsymbol{u}^\eta \cdot \boldsymbol{q} - \frac{1}{2R} \int_{\Omega_f} (\partial_t \eta) \boldsymbol{u}^\eta \cdot \boldsymbol{q} + \frac{R}{2R} \int_\Gamma (\partial_t \eta)^2 \psi,$$

By using this calculation and the definition of weak solutions Definition 5.2.1, we recover (5.39).

Notice that in the weak formulation (5.39) above, there is an extra (yet important) term that arises due to the motion of the fluid domain: $\int_0^T \int_{\Omega_f} \frac{1}{2R} (\partial_t \eta) \boldsymbol{u} \cdot \boldsymbol{q}$. This term is not present in the weak formulation of the linearly coupled problem.

5.3 Statement of the Main Result

Now that we have defined an appropriate weak formulation for the nonlinearly coupled prototypical FSI problem, we can state the main existence result of this chapter as follows.

Theorem 5.3.1 (Main Existence Result) *Let the fluid viscosity μ and the elasticity coefficients C_1 and C_2 be positive. Furthermore, suppose that the initial structure displacement $\eta_0 \in H_0^2(\Gamma)$, initial structure velocity $v_0 \in L^2(\Gamma)$, and initial fluid velocity $\boldsymbol{u}_0 \in L^2(\Omega_f(0))$ be such that $(R + \eta_0(z)) > 0$, $z \in \Gamma$, and let the inlet and* *outlet pressure data P_{in}, $P_{out} \in L^2_{loc}(0, \infty)$. Then there exists a time $T > 0$ and a weak solution $(\boldsymbol{u}, \eta) \in \mathcal{V}$ on $(0, T)$ of problem (5.2)–(5.9) that satisfies the weak formulation posed on the physical moving fluid domain given by Definition 5.2.1 (or, equivalently, Definition 5.2.2), which satisfies the following energy estimate:*

$$E(t)+\int_0^t D(\tau)d\tau \leq E_0+C(\|P_{in}\|^2_{L^2(0,t)}+\|P_{out}\|^2_{L^2(0,t)}), \quad t \in [0, T], \tag{5.40}$$

where C depends only on the coefficients in the problem, E_0 is the kinetic energy of the initial data, and $E(t)$ and $D(t)$ are given by

$$E(t) = \frac{1}{2}\|\mathbf{u}\|^2_{L^2(\Omega^\eta(t))} + \frac{1}{2}\|\partial_t \eta\|^2_{L^2(\Gamma))} + \frac{1}{2}\big(C_1\|\partial_z \eta\|^2_{L^2(\Gamma)} + C_2\|\partial_z^2 \eta\|^2_{L^2(\Gamma)}\big),$$

$$D(t) = \mu\|\mathbf{D}(\mathbf{u})\|^2_{L^2(\Omega^\eta(t)))}.$$

Furthermore, the solution can be maximally extended in time in the sense that T can be taken to be infinite, or else, $T = \sup\{t \geq 0 : R + \eta(t, z) > 0$ for all $z \in \Gamma\}$ so that T is the first instance of domain degeneracy.

In the remainder of this chapter, we present the main steps of the proof of the main result. The proof will be based on a Lie operator splitting scheme, specified in terms of the weak formulation of the coupled problem defined on the fixed reference domain.

5.4 The Lie Operator Splitting Scheme

The core concept behind the splitting methodology remains the same as in the linear case (see (4.30) and (4.31)). However, ensuring that this approach yields a weak solution in the nonlinearly coupled case presents significant challenges. These arise from the fact that the fluid equations are now nonlinear and are defined on moving, time-dependent domains that are not known a priori.

As in the linearly coupled case, the goal is to develop a semidiscretized scheme, discretized in time by the time step $\Delta t = T/N$, where N is the number of discretized time subintervals. On each time subinterval $[t_n, t_{n+1}]$ for $t_n = n\Delta t$, two subproblems are "solved," which define an approximate solution vector at the nth time step for the time step $\Delta t = T/N$, denoted by

$$X_N^{n+\frac{i}{2}} = \begin{pmatrix} \boldsymbol{u}_N^{n+\frac{i}{2}} \\ \eta_N^{n+\frac{i}{2}} \\ v_N^{n+\frac{i}{2}} \end{pmatrix} \qquad \text{for } i = 1, 2 \text{ and } n = 0, 1, \ldots, N-1,$$

where $\boldsymbol{u}_N^{n+\frac{i}{2}}$ is the approximate fluid velocity defined on the *fixed reference domain* Ω_f, $\eta_N^{n+\frac{i}{2}}$ is the approximate structure displacement, and $v_N^{n+\frac{i}{2}}$ is the approximate structure velocity.

Before we specify the structure and fluid subproblems in weak formulation, we introduce the following notation for the semidiscrete versions of the kinetic and elastic energy and for dissipation

$$\begin{aligned} E_N^{n+\frac{i}{2}} &= \frac{1}{2}\int_{\Omega_f}\left(1+\frac{\eta_N^{n+\frac{i}{2}}}{R}\right)|\boldsymbol{u}_N^{n+\frac{i}{2}}|^2 + \frac{1}{2}\int_\Gamma |v_N^{n+\frac{i}{2}}|^2 \\ &+ \frac{1}{2}C_1\int_\Gamma |\partial_z \eta_N^{n+\frac{i}{2}}|^2 + \frac{1}{2}C_2\int_\Gamma |\partial_z^2 \eta_N^{n+\frac{i}{2}}|^2, \end{aligned} \tag{5.41}$$

and the total fluid dissipation

$$D_N^n = \mu_f \int_{\Omega_f}\left(1+\frac{\eta_N^n}{R}\right)|\boldsymbol{D}^{\eta_N^n}(\boldsymbol{u}_N^n)|^2. \tag{5.42}$$

We remark that these are the same expressions as for the linearly coupled case, except for the integrals over the fluid domain, which now have the Jacobian of the transformation from $\Omega_{f,N}^n$ to Ω_f as an additional factor, which corresponds to integrating in the physical space over $\Omega_{f,N}^n$, defined by

$$\Omega_{f,N}^n(t) = \{(z,r)\in\mathbb{R}^2 : 0\le z\le R, 0\le r\le R+\eta_N^n(t,z)\}. \tag{5.43}$$

The Structure Subproblem In this subproblem, the fluid velocity remains same so that

$$\boldsymbol{u}_N^{n+\frac{1}{2}} = \boldsymbol{u}_N^n,$$

and we update the structure displacement and velocity by letting $(\eta_N^{n+\frac{1}{2}}, v_N^{n+\frac{1}{2}})$ be the unique solution that satisfies the weak following formulation:

$$\begin{cases} \displaystyle\int_\Gamma \frac{\eta_N^{n+\frac{1}{2}}-\eta_N^n}{\Delta t}\varphi dz = \int_\Gamma v_N^{n+\frac{1}{2}}\varphi dz, \quad \text{for all } \varphi\in L^2(\Gamma), \\ \displaystyle\int_\Gamma \frac{v_N^{n+\frac{1}{2}}-v_N^n}{\Delta t}\psi + C_1\int_\Gamma \partial_z\eta_N^{n+\frac{1}{2}}\partial_z\psi \\ \displaystyle\quad +C_2\int_\Gamma \partial_z^2\eta_N^{n+\frac{1}{2}}\partial_z^2\psi = 0, \text{ for all } \psi\in H_0^2(\Gamma). \end{cases} \tag{5.44}$$

We remark that this weak formulation is identical to that in the structure subproblem for the prototypical linearly coupled FSI problem in (4.34).

Proposition 5.4.1 *For each fixed* $\Delta t > 0$, *problem* (5.44) *has a unique solution* $(v_N^{n+\frac{1}{2}}, \eta_N^{n+\frac{1}{2}}) \in H_0^2(\Gamma) \times H_0^2(\Gamma)$.

Proof The proof is a direct consequence of the Lax-Milgram lemma as in the linearly coupled case. See Proposition 4.3.1. □

Proposition 5.4.2 *For each fixed* $\Delta t > 0$, *the solution of problem* (5.44) *satisfies the following discrete energy equality:*

$$E_N^{n+\frac{1}{2}} + \frac{1}{2}\left(\|v_N^{n+\frac{1}{2}} - v_N^n\|^2 + C_1\|\partial_z(\eta_N^{n+\frac{1}{2}} - \eta_N^n)\|^2 + C_2\|\partial_z^2(\eta_N^{n+\frac{1}{2}} - \eta^n)\|^2\right) = E_N^n, \tag{5.45}$$

where the kinetic energy E_N^n *is defined in* (5.41).

Proof The proof is the same as in the linear case. See Proposition 4.3.2. □

The Fluid Subproblem For the prototypical nonlinearly coupled FSI problem, the fluid subproblem is more complex than the corresponding fluid subproblem for the prototypical linearly coupled FSI model, discussed in the previous chapter. This is because nonlinearly coupled FSI models have weak formulations, which include additional terms that take into account the time-dependent geometry of the fluid domain, which must be handled properly to ensure that the fluid subproblem indeed has a unique solution, and so that the fluid subproblem has an appropriate semidiscrete energy estimate.

In this subproblem, the structure displacement remains the same so that

$$\eta_N^{n+1} = \eta_N^{n+\frac{1}{2}},$$

and we update the coupled fluid/structure velocities $\boldsymbol{u}_N^{n+1}$ and v_N^{n+1}, which belong to the following solution space:

$$\tilde{Q}_N^n = \{(\boldsymbol{u}, v) \in V_f^{\eta_N^n} \times L^2(\Gamma) : \boldsymbol{u}|_\Gamma = v\boldsymbol{e}_r\}, \tag{5.46}$$

where we recall the definition of V_f^η for a structure displacement η from (5.36). The functions $\boldsymbol{u}_N^{n+1}$ and v_N^{n+1} are updated so that they satisfy the following weak formulation, which holds for all test functions $(\boldsymbol{q}, \psi) \in \tilde{Q}_N^n$:

$$\int_{\Omega_f}\left(1+\frac{\eta_N^n}{R}\right)\frac{\boldsymbol{u}_N^{n+1}-\boldsymbol{u}_N^n}{\Delta t}\cdot\boldsymbol{q}$$
$$+\frac{1}{2}\int_{\Omega_f}\left(1+\frac{\eta_N^n}{R}\right)\left(\left(\boldsymbol{u}_N^n-v_N^{n+\frac{1}{2}}\frac{r}{R}\boldsymbol{e}_y\right)\cdot\nabla^{\eta_N^n}\boldsymbol{u}_N^{n+1}\right)\cdot\boldsymbol{q}$$
$$-\frac{1}{2}\int_{\Omega_f}\left(1+\frac{\eta_N^n}{R}\right)\left(\left(\boldsymbol{u}_N^n-v_N^{n+\frac{1}{2}}\frac{r}{R}\boldsymbol{e}_y\right)\cdot\nabla^{\eta_N^n}\boldsymbol{q}\right)\cdot\boldsymbol{u}_N^{n+1}$$
$$+\int_{\Omega_f}\frac{1}{2R}v_N^{n+\frac{1}{2}}\boldsymbol{u}_N^{n+1}\cdot\boldsymbol{q}+2\mu_f\int_{\Omega_f}\left(1+\frac{\eta_N^n}{R}\right)\boldsymbol{D}^{\eta_N^n}(\boldsymbol{u}_N^{n+1}):\boldsymbol{D}^{\eta_N^n}(\boldsymbol{q})\psi$$
$$+\int_{\Gamma}\frac{v_N^{n+1}-v_N^{n+\frac{1}{2}}}{\Delta t}=\int_{\Gamma_{in}}P_{N,in}^n(t)q_z-\int_{\Gamma_{out}}P_{N,out}^n(t)q_z, \tag{5.47}$$

where $P_{N,in/out}^n=\frac{1}{\Delta t}\int_{n\Delta t}^{(n+1)\Delta t}P_{in/out}(t)dt$ is the discretized inlet/outlet pressure.

Remark 5.4.1 Note that in the original weak formulation for the problem defined on a fixed reference domain, we have the following terms:

$$-\int_0^T\int_{\Omega_f}\left(1+\frac{\eta}{R}\right)\boldsymbol{u}\cdot\partial_t\boldsymbol{q}-\frac{1}{2R}\int_0^T\int_{\Omega_f}(\partial_t\eta)\boldsymbol{u}\cdot\boldsymbol{q}.$$

While the signs in the above weak formulation (5.47) may seem incongruous with these terms at first, integration by parts of the above expression gives

$$\int_0^T\int_{\Omega_f}\left(1+\frac{\eta}{R}\right)\partial_t\boldsymbol{u}\cdot\boldsymbol{q}+\frac{1}{2R}\int_0^T\int_{\Omega_f}(\partial_t\eta)\boldsymbol{u}\cdot\boldsymbol{q},$$

since the time derivative applies to both the term $\left(1+\frac{\eta}{R}\right)$ and $\boldsymbol{u}$.

Remark 5.4.2 Problem (5.47) is a linear problem since η_N^n and $v_N^{n+\frac{1}{2}}$ have already been calculated in the structure subproblem, and the quadratic advection term is linearized around the previously calculated velocity $\boldsymbol{u}_N^n$. This is crucial for allowing us to find a unique solution, which will be done using a standard Lax-Milgram lemma argument, as we show below.

Remark 5.4.3 Notice that the Jacobian $\left(1+\frac{\eta_N^n}{R}\right)$ lags the "current" fluid domain $\Omega_{f,N}^{n+1}$ associated with the current interface η_N^{n+1}. This means, for example, that the first term in (5.47) is integrated over $\Omega_{f,N}^n$, and not $\Omega_{f,N}^{n+1}$, since

$$\int_{\Omega_f}\left(1+\frac{\eta_N^n}{R}\right)\frac{\boldsymbol{u}_N^{n+1}-\boldsymbol{u}_N^n}{\Delta t}\cdot\boldsymbol{q}=\int_{\Omega_{f,N}^n}\frac{\boldsymbol{u}_N^{n+1}-\boldsymbol{u}_N^n}{\Delta t}\cdot\boldsymbol{q},$$

where $\Omega^n_{f,N} = \{(z,r) \in \mathbb{R}^2 : 0 \le z \le L, 0 \le r \le R + \eta^n_N(z)\}$. The main reason for the choice of the "previous" displacement η^n_N is related to energy considerations, as we explain in the proof of the energy estimate presented in Proposition 5.4.4.

Proposition 5.4.3 *Assuming that $\eta^n_N > -R$, the fluid subproblem defined by (5.47) has a unique weak solution $(\boldsymbol{u}^{n+1}_N, v^{n+1}_N) \in \tilde{Q}^{n+1}_N$ satisfying $\boldsymbol{u}^{n+1}_N|_\Gamma = v^{n+1}_N$.*

Proof The proof is again a consequence of the Lax-Milgram lemma (see Lemma 2.2.1). More precisely, for $(\boldsymbol{u}, v), (\boldsymbol{q}, \psi) \in \tilde{Q}^{n+1}_N$ define the bilinear form associated with (5.47):

$$\mathcal{B}((\boldsymbol{u}, v), (\boldsymbol{q}, \psi)) := \int_{\Omega_f} (1 + \frac{\eta^n_N}{R}) \Big(\boldsymbol{u} \cdot \boldsymbol{q} + \frac{\Delta t}{2R} \Big[(\boldsymbol{u}^n_N - v_N^{n+\frac{1}{2}} \frac{r}{R} \boldsymbol{e}_r) \cdot \nabla^{\eta^n_N} \Big] \boldsymbol{u} \cdot \boldsymbol{q}$$

$$- \frac{\Delta t}{2} \Big[(\boldsymbol{u}^n_N - v_N^{n+\frac{1}{2}} \frac{r}{R} \boldsymbol{e}_r) \cdot \nabla^{\eta^n_N} \Big] \boldsymbol{q} \cdot \boldsymbol{u} \Big) + \int_\Gamma v\psi$$

$$+ \Delta t \frac{1}{2R} \int_{\Omega_f} v_N^{n+\frac{1}{2}} \boldsymbol{u} \cdot \boldsymbol{q} + \Delta t 2\mu_f \int_{\Omega_f} (1 + \frac{\eta^n_N}{R}) \boldsymbol{D}^{\eta^n_N}(\boldsymbol{u}) : \boldsymbol{D}^{\eta^n_N}(\boldsymbol{q}).$$

We need to prove that this bilinear form $\mathcal{B}$ is coercive and continuous on $\tilde{Q}^{n+1}_N$. To see that $\mathcal{B}$ is coercive, we write

$$\mathcal{B}((\boldsymbol{u}, v), (\boldsymbol{u}, v)) = \int_{\Omega_f} (1 + \frac{\eta^n_N}{R} + \frac{\Delta t}{2R} v_N^{n+\frac{1}{2}}) |\boldsymbol{u}|^2 + \int_\Gamma v^2 + \Delta t 2\mu_f$$

$$\times \int_{\Omega_f} (1 + \frac{\eta^n_N}{R}) |\mathbf{D}^{\eta^n_N}(\boldsymbol{u})|^2.$$

It is easy to see now that coercivity follows immediately, once we recall that $\eta^n_N > -R$, which implies that $1 + \frac{\eta^n_N}{R} + \frac{\Delta t}{2} v_N^{n+\frac{1}{2}} = 1 + \frac{1}{2R}(\eta^n_N + \eta_N^{n+\frac{1}{2}}) > 0$.

To prove continuity of $\mathcal{B}$, we first notice that from (5.32) we have

$$\|\nabla^{\eta^n_N} \boldsymbol{u}\|_{L^2(\Omega_f)} \le C(1 + \|\eta^n_N\|_{H^2(\Gamma)}) \|\boldsymbol{u}\|_{H^1(\Omega_f)}.$$

By applying the generalized Hölder inequality and the continuity of the embedding of H^1 into L^4, we obtain

$$\mathcal{B}((\boldsymbol{u}, v), (\boldsymbol{q}, \psi)) \le C \Big(\|\boldsymbol{u}\|_{L^2(\Omega_f)} \|\boldsymbol{q}\|_{L^2(\Omega_f)} + \|v\|_{L^2(\Gamma)} \|\psi\|_{L^2(\Gamma)}$$

$$+ \Delta t \|\eta^n_N\|_{H^2(\Gamma)} (\|\boldsymbol{u}\|_{H^1(\Omega_f)} + \|v_N^{n+\frac{1}{2}}\|_{H^1(\Gamma)}) \|\boldsymbol{u}\|_{H^1(\Omega_f)} \|\boldsymbol{q}\|_{H^1(\Omega_f)}$$

$$+ \Delta t \mu_f \|\eta^n_N\|^2_{H^2(\Gamma)} \|\boldsymbol{u}\|_{H^1(\Omega_f)} \|\boldsymbol{q}\|_{H^1(\Omega_f)} \Big).$$

This shows that $\mathcal{B}$ is continuous. The Lax-Milgram lemma now implies the existence of a unique solution $(\boldsymbol{u}_N^{n+1}, v_N^{n+1})$ of problem (5.47). □

Proposition 5.4.4 *For each fixed $\Delta t > 0$, the solution of problem (5.47) satisfies the following discrete energy inequality:*

$$\begin{aligned} &E_N^{n+1} + \|\boldsymbol{u}_N^{n+1} - \boldsymbol{u}_N^n\|^2_{L^2(\Omega_f)} + \|v_N^{n+1} - v_N^{n+\frac{1}{2}}\|^2_{L^2(\Gamma)} \\ &+D_N^{n+1} \le E_N^{n+\frac{1}{2}} + C\left(\|P_{in}(t)\|^2_{L^2(n\Delta t,(n+1)\Delta t)} + \|P_{out}(t)\|^2_{n\Delta t,(n+1)\Delta t)}\right), \end{aligned} \tag{5.48}$$

where the kinetic energy E_N^n and dissipation D_N^n are defined in (5.41) *and* (5.42), *and the constant C depends only on the parameters in the problem, and not on Δt (or N).*

Proof As in the linearly coupled FSI problem, we begin by replacing the test functions $\boldsymbol{q}$ and ψ in the weak formulation (5.47) by the fluid velocity $\boldsymbol{u}_N^{n+1}$ and structure velocity v_N^{n+1}, respectively, and multiply the resulting equation by Δt. We notice that the first term on the left-hand side becomes

$$\frac{1}{2}\int_{\Omega_f}(1+\frac{\eta_N^n}{R})|\boldsymbol{u}_N^{n+1}|^2.$$

This is the term that contributes to the discrete kinetic energy at the time step $n+1$, but it does not have the correct form, since the discrete kinetic energy at $n+1$ is given in terms of the structure location at $n+1$, not n. To get around this difficulty, we notice the term associated with the fluid advection:

$$\frac{1}{2R}\int_{\Omega_f}\Delta t v_N^{n+\frac{1}{2}}|\boldsymbol{u}_N^{n+1}|^2$$

and express $\Delta t v_N^{n+\frac{1}{2}}$ in terms of the displacement η as follows: $\Delta t v_N^{n+\frac{1}{2}} := (\eta_N^{n+1/2} - \eta_N^n)$, which is equal to $(\eta_N^{n+1} - \eta_N^n)$ since in this subproblem $\eta_N^{n+1} = \eta_N^{n+1/2}$. A combination of these two terms then gives the "correct" discrete kinetic energy of the fluid at the time step $n+1$, namely, we get

$$\frac{1}{2}\Big(\int_{\Omega_f}(1+\frac{\eta_N^n}{R})|\boldsymbol{u}_N^{n+1}|^2 + \Delta t v_N^{n+\frac{1}{2}}|\boldsymbol{u}_N^{n+1}|^2\Big) = \frac{1}{2}\int_{\Omega_f}(1+\frac{\eta_N^{n+1}}{R})|\boldsymbol{u}_N^{n+1}|^2.$$

To complete the proof, one simply uses the algebraic identity $(a-b)\cdot a = \frac{1}{2}(|a|^2 + |a-b|^2 - |b|^2)$ in the same way as in the proof of Proposition 4.3.2. □

Remark 5.4.4 Notice how the nonlinearity of the coupling at the deformed interface requires the presence of nonlinear fluid advection in order for the discrete kinetic energy of the fluid subproblem to be bounded and decreasing in time.

Remark 5.4.5 Recall that the splitting scheme holds as long as $\eta_N^n > -R$, and hence the uniform energy estimates for the structure and fluid subproblems hold until the time step for which this condition no longer holds.

5.4.1 Semidiscrete Coupled Problem: Weak Formulation and Uniform Energy Estimates

Similarly as in the linearly coupled case, we combine the semidiscrete versions of the structure and fluid subproblems and obtain the following **semidiscrete weak formulation** of the coupled problem on a **fixed, reference domain**:

$$
\begin{aligned}
&\int_{\Omega_f}\left(1+\frac{\eta_N^n}{R}\right)\frac{\boldsymbol{u}_N^{n+1}-\boldsymbol{u}_N^n}{\Delta t}\cdot\boldsymbol{q}\\
&+\frac{1}{2}\int_{\Omega_f}\left(1+\frac{\eta_N^n}{R}\right)\left(\left(\boldsymbol{u}_N^n-\frac{r}{R}v_N^{n+\frac{1}{2}}\boldsymbol{e}_y\right)\cdot\nabla^{\eta_N^n}\boldsymbol{u}_N^{n+1}\right)\cdot\boldsymbol{q}\\
&-\frac{1}{2}\int_{\Omega_f}\left(1+\frac{\eta_N^n}{R}\right)\left(\left(\boldsymbol{u}_N^n-\frac{r}{R}v_N^{n+\frac{1}{2}}\boldsymbol{e}_y\right)\cdot\nabla^{\eta_N^n}\boldsymbol{q}\right)\cdot\boldsymbol{u}_N^{n+1}\\
&+\int_{\Omega_f}\frac{1}{2R}v_N^{n+\frac{1}{2}}\boldsymbol{u}_N^{n+1}\cdot\boldsymbol{q}+2\mu_f\int_{\Omega_f}\left(1+\frac{\eta_N^n}{R}\right)\boldsymbol{D}^{\eta_N^n}(\boldsymbol{u}_N^{n+1}):\boldsymbol{D}^{\eta_N^n}(\boldsymbol{q})\\
&+\int_\Gamma\frac{v_N^{n+1}-v_N^n}{\Delta t}\psi\\
&+C_1\int_\Gamma\partial_z\eta_N^{n+\frac{1}{2}}\partial_z\psi+C_2\int_\Gamma\partial_z^2\eta_N^{n+\frac{1}{2}}\partial_z^2\psi=\int_{\Gamma_{in}}P_{N,in}^n(t)q_z-\int_{\Gamma_{out}}P_{N,out}^n(t)q_z,
\end{aligned}
\tag{5.49}
$$

which holds for all the test functions $(\boldsymbol{q},\psi)$ that belong to the test space Q_N^n:

$$
Q_N^n=\{(\boldsymbol{q},\psi)\in\mathcal{V}_f^{\eta_N^n}\times H_0^2(\Gamma):\boldsymbol{q}|_\Gamma=\psi\boldsymbol{e}_r\}. \tag{5.50}
$$

Our goal is to show that approximations constructed above, which satisfy the semidiscrete weak formulation of the coupled problem (5.49), are uniformly bounded. Uniform boundedness will help us construct subsequences of approximate solutions that converge weakly and/or weakly* to the limiting function as the time step $\Delta t\to 0$ (or $N\to\infty$). Our ultimate goal is to show that these limiting functions satisfy the weak formulation stated in Definition 5.2.1 of the continuous coupled problem, which means that they are weak solutions of the original problem. As we shall see later when we construct the compactness arguments to pass to the limit in the approximate weak formulations, it will be easier to work on physical domains, rather than the reference, fixed domain.

For this purpose, we introduce here the **semidiscrete weak formulation** of the coupled problem defined on approximate **moving domains**. We introduce the following function spaces defined on domains $\Omega_f^{\eta_N^n}$:

$$(\mathcal{V}_F^\eta)_N^{n+1}(\Omega_f^{\eta_N^n}) = \{(\boldsymbol{u}^{n+1}, v^{n+1}) \in \mathcal{V}_F(\Omega_f^{\eta_N^n}) \times L^2(\Gamma) \; : \; \boldsymbol{u}^{n+1}_{|\Gamma^{\eta_N^n}} = v^{n+1}\mathbf{e}_r\},$$
$$(\mathcal{Q}^\eta)_N^n(\Omega_f^{\eta_N^n}) = \{(\boldsymbol{q}, \psi) \in \mathcal{V}_F(\Omega_f^{\eta_N^n}) \times H_0^2(\Gamma) \; : \; \boldsymbol{q}_{|\Gamma^{\eta_N^n}} = \psi\mathbf{e}_r\}. \tag{5.51}$$

To state the weak formulation of the semidiscretized problem on moving domains, we introduce the notation $\tilde{\mathbf{u}}_N^n$ for the fluid velocity defined on $\Omega_f^{\eta_N^n}$, which is obtained by taking the function $\mathbf{u}_N^n$, which is defined on $\Omega_f^{\eta_N^{n-1}}$, and then mapping $\Omega_f^{\eta_N^n}$ onto the fixed, reference domain Ω_f via $\boldsymbol{\Phi}^{\eta_N^{n-1}}$, and then composing it with $(\boldsymbol{\Phi}^{\eta_N^n})^{-1}$ to get the function defined on $\Omega_f^{\eta_N^n}$. Thus,

$$\tilde{\mathbf{u}}_N^n = \boldsymbol{u}_N^n \circ \boldsymbol{\Phi}^{\eta_N^{n-1}} \circ (\boldsymbol{\Phi}^{\eta_N^n})^{-1}. \tag{5.52}$$

The weak formulation of the **semi-discretized, coupled** FSI problem on **moving domains** $\Omega_f^{\eta_N^n}$ now reads find $(\boldsymbol{u}_N^{n+1}, v_N^{n+1}) \in (\mathcal{V}_F^\eta)_N^{n+1}(\Omega_f^{\eta_N^n})$ and $(v_N^{n+\frac{1}{2}}, \eta_N^{n+\frac{1}{2}}) \in H_0^2(\Gamma) \times H_0^2(\Gamma)$ such that

$$\int_{\Omega_f^{\eta_N^n}} \frac{\boldsymbol{u}_N^{n+1} - \tilde{\boldsymbol{u}}_N^n}{\Delta t} \cdot \boldsymbol{q} + \frac{1}{2}\int_{\Omega_f^{\eta_N^n}} \frac{v_N^{n+1/2}}{R + \eta_N^n} \boldsymbol{u}_N^{n+1} \cdot \boldsymbol{q} + 2\mu_f \int_{\Omega_f^{\eta_N^n}} \mathbf{D}(\boldsymbol{u}_N^{n+1}) : \mathbf{D}(\boldsymbol{q})$$
$$+ \int_{\Omega_f^{\eta_N^n}} \frac{1}{2}\left[(\tilde{\boldsymbol{u}}_N^n - \frac{v_N^{n+1/2} r}{R + \eta_N^n}\mathbf{e}_r) \cdot \nabla\, \boldsymbol{u}_N^{n+1} \cdot \boldsymbol{q} - (\tilde{\boldsymbol{u}}_N^n - \frac{v_N^{n+1/2}}{R + \eta_N^n}\mathbf{e}_r) \cdot \nabla \right] \boldsymbol{q} \cdot \boldsymbol{u}_N^{n+1}$$
$$+ \int_\Gamma \frac{v_N^{n+1} - v_N^n}{\Delta t}\psi + a_e(\eta_N^{n+1}, \psi) = F^n(\boldsymbol{q}),$$

$$\int_\Gamma \frac{\eta_N^{n+1} - \eta_N^n}{\Delta t}\psi = \int_\Gamma v_N^{n+1/2}\psi, \quad \forall (\boldsymbol{q}\psi) \in \mathcal{Q}_N^n(\Omega_f^{\eta_N^n}),$$
$$\text{with } \nabla \cdot \boldsymbol{u}_N^{n+1} = 0. \tag{5.53}$$

We are now ready to prove the following **uniform boundedness result** for the approximations constructed using the Lie operator splitting scheme.

Lemma 5.4.1 (Uniform Energy Estimates) *Let $\Delta t > 0$ and $N = T/\Delta t > 0$. Furthermore, let $E_N^{n+\frac{1}{2}}$, E_N^{n+1}, and D_N^j be the kinetic energy and dissipation given by* (5.41) *and* (5.42), *respectively. There exists a constant $C > 0$ independent of*

Δt (and N), which depends only on the parameters in the problem, on the kinetic energy of the initial data E_0, and on the energy norm of the inlet and outlet data $\|P_{in/out}\|^2_{L^2(0,T)}$, such that:

1. *The kinetic and elastic energy are uniformly bounded:*

$$E_N^{n+\frac{1}{2}} \leq C,\ E_N^{n+1} \leq C, n = 0, \ldots, N-1;$$

2. *Viscous fluid dissipation is uniformly bounded:*

$$\sum_{j=1}^{N} D_N^j \leq C;$$

3. *Numerical dissipation is uniformly bounded:*

$$\sum_{n=0}^{N-1}\left(\int_{\Omega_f}(R+\eta_N^n)|\boldsymbol{u}_N^{n+1}-\boldsymbol{u}_N^n|^2 + \|v_N^{n+1}-v_N^{n+\frac{1}{2}}\|^2_{L^2(\Gamma)}\right.$$
$$\left. +\|v_N^{n+\frac{1}{2}}-v_N^n\|^2_{L^2(\Gamma)}\right) \leq C$$

$$\sum_{n=0}^{N-1}\left(C_0\|\eta_N^{n+1}-\eta_N^n\|^2_{L^2(\Gamma)} + C_1\|\partial_z(\eta_N^{n+1}-\eta_N^n)\|^2_{L^2(\Gamma)}\right.$$
$$\left. +C_2\|\partial_z^2(\eta_N^{n+1}-\eta_N^n)\|^2_{L^2(\Gamma)}\right) \leq C.$$

In fact, $C = E_0 + \tilde{C}\left(\|P_{in}\|^2_{L^2(0,T)} + \|P_{out}\|^2_{L^2(0,T)}\right)$, where $\tilde{C}$ is the constant from (5.48), *which depends only on the parameters in the problem and T.*

Proof The proof follows by adding the energy estimates (5.45) and (5.48) and taking the sum, on both sides, over n from 0 to $N-1$. After canceling the same terms in the kinetic energy that appear on both sides of the inequality, we obtain

$$E_N^N + \sum_{n=0}^{N-1} D_N^{n+1} + \frac{1}{2}\sum_{n=0}^{N-1}\left(\int_{\Omega_f}(R+\eta_N^n)|\boldsymbol{u}_N^{n+1}-\boldsymbol{u}_N^n|^2 + \|v_N^{n+1}-v_N^{n+\frac{1}{2}}\|^2_{L^2(\Gamma)}+\right.$$
$$+\rho_s h\|v_N^{n+\frac{1}{2}}-v_N^n\|^2_{L^2(\Gamma)} + C_0\|\eta_N^{n+\frac{1}{2}}-\eta_N^n\|^2_{L^2(\Gamma)} + C_1\|\partial_z(\eta_N^{n+\frac{1}{2}}-\eta_N^n)\|^2_{L^2(\Gamma)}+$$
$$\left.+C_2\|\partial_z^2(\eta_N^{n+\frac{1}{2}}-\eta_N^n)\|^2_{L^2(\Gamma)}\right) \leq E_0 + \tilde{C}\Delta t\sum_{n=0}^{N-1}((P_{in}^n)^2 + (P_{out}^n)^2).$$

We estimate the term on the right-hand side involving the inlet and outlet pressure. For this purpose, we recall that on every subinterval (t^n, t^{n+1}), the pressure data is approximated by a constant that is equal to the average value of the pressure over that time interval. Therefore, after using Hölder's inequality, we obtain

$$\Delta t \sum_{n=0}^{N-1} (P_{in}^n)^2 = \Delta t \sum_{n=0}^{N-1} \left(\frac{1}{\Delta t} \int_{n\Delta t}^{(n+1)\Delta t} P_{in}(t)dt \right)^2 \leq \| P_{in} \|^2_{L^2(0,T)}.$$

By using the pressure estimate to bound the right-hand side in the above energy estimate, we have obtained all the statements in the lemma, with the constant C given by $C = E_0 + \tilde{C} \| P_{in/out} \|^2_{L^2(0,T)}$.

Notice that Statement 1 can be obtained in the same way by summing from 0 to $n-1$, for each n, instead of from 0 to $N-1$. □

Remark 5.4.6 The numerical dissipation terms above are associated with the time-splitting scheme.

This result will be crucial to show uniform boundedness of the approximate functions, which are defined on the entire time interval, and not only at the discrete time points. A definition of the approximate functions and their properties is studied next.

5.5 Approximate Functions

To study convergence of approximate solutions to a weak solution, we need to extend the approximate functions defined by the Lie operator splitting scheme, which are defined at discrete time points, to the entire time interval. We first introduce the piecewise constant extensions as follows: For any $t \in (n\Delta t, (n+1)\Delta t]$, we set

$$\eta_N(t) = \eta_N^{n+1}, \quad v_N(t) = v_N^{n+1}, \quad v_N^*(t) = v_N^{n+\frac{1}{2}}, \quad \boldsymbol{u}_N = \boldsymbol{u}_N^{n+1}. \tag{5.54}$$

We also define linear interpolations, which will be useful in analyzing the time derivatives. In particular, we define the linearly interpolated approximate solutions as follows:

$$\overline{\eta}_N(n\Delta t) = \eta_N^n, \quad \overline{v}_N(n\Delta t) = v_N^n, \quad \overline{\boldsymbol{u}}_N(n\Delta t) = \boldsymbol{u}_N^n, \qquad \text{for } n = 0, 1, \ldots, N, \tag{5.55}$$

and note that $\partial_t \overline{\eta}_N = v_N^*$.

Remark 5.5.1 Notice that functions $v_N^* = v_N^{n-1/2}$ are determined by the structure elastodynamics subproblem, while functions $v_N = v_N^n$ are determined by the fluid subproblem. As a consequence, only the functions v_N are equal to the trace of the fluid velocity on Γ, i.e., $\boldsymbol{u}_N = v_N \mathbf{e_r}$. This is not necessarily the case for the functions v_N^*. However, we will show later that the difference between the two sequences converges to zero in $L^2(\Gamma)$.

Now that the approximate functions are defined on the entire time interval $(0, T)$, we study their convergence as $\Delta t \to 0$.

5.5.1 Uniform Boundedness of Approximate Functions

The uniform energy estimates for the approximate functions will be crucial to obtain weak and weak* convergence of approximate subsequences. In this subsection, we show that the approximate functions specified above are uniformly bounded in Δt (or N), and that the approximate displacement η_N is uniformly bounded in such a way that the Jacobian $1 + \eta_N/R$ is bounded uniformly and away from zero, i.e., $0 < R_{min}/R \le 1 + \eta_N/R \le R_{max}/R$, for all Δt (or N) so that the Lie operator splitting scheme is well-defined for some $T > 0$.

Proposition 5.5.1 *The following uniform boundedness results hold for the approximate solutions $\eta_N, \overline{\eta}_N, \boldsymbol{u}_N, v_N^*$ and v_N, mapped onto the fixed domain Ω_f, uniformly in N:*

1. ***Structure displacement** η_N is uniformly bounded in $L^\infty(0, T; H_0^2(\Gamma))$ and $\overline{\eta}_N$ is uniformly bounded in $W^{1,\infty}(0, T; L^2(\Gamma)) \cap L^\infty(0, T; H_0^2(\Gamma))$.*
2. ***Jacobian** $1 + \frac{\eta_N}{R}$ **of the transformation** $\boldsymbol{\Phi}^{\eta_N}$, defined in (5.31), is uniformly bounded. Namely, there exists a $T > 0$ small enough, such that*

$$0 < \frac{R_{min}}{R} \le 1 + \frac{\eta_N}{R} \le \frac{R_{max}}{R}, \forall N \in \mathbb{N}, t \in (0, T).$$

3. ***Fluid velocity** $\boldsymbol{u}_N$ is uniformly bounded in $L^\infty(0, T; L^2(\Omega_f)) \cap L^2(0, T; H^1(\Omega_f))$.*
4. ***Structure velocity** v_N^* is uniformly bounded in $L^\infty(0, T; L^2(\Gamma))$, and $v_N = \boldsymbol{u}_N|_\Gamma \cdot \boldsymbol{e}_r$ is uniformly bounded in $L^\infty(0, T; L^2(\Gamma)) \cap L^2(0, T; H^{1/2}(\Gamma))$.*

Proof We begin by proving the first two statements about the uniform boundedness of the structure displacement. For this purpose, first notice that the uniform energy estimate $E_N^n \le C$, where C is independent of N, established in Lemma 5.4.1, implies

$$\|\eta_N(t)\|^2_{L^2(\Gamma)}, \|\partial_z \eta_N(t)\|^2_{L^2(\Gamma)}, \|\partial^2_{zz} \eta_N(t)\|^2_{L^2(\Gamma)} \le C, \ \forall t \in [0, T],$$

which immediately gives

$$\|\eta_N\|_{L^\infty(0,T;H_0^2(\Gamma))} \le C. \tag{5.56}$$

Next, we address the uniform boundedness of $\bar{\eta}_N(t)$. From the definition of the linear interpolant $\bar{\eta}_N(t)$ (see (5.55)), we observe that

$$\partial_t \bar{\eta}_N(t) = \frac{\eta_N^{n+1} - \eta_N^n}{\Delta t} = \frac{\eta_N^{n+1/2} - \eta_N^n}{\Delta t} = v_N^{n+\frac{1}{2}}, \ t \in (n\Delta t, (n+1)\Delta t),$$

and so, since v_N^* was defined in (5.54) as a piecewise constant function defined via $v_N^*(t, \cdot) = v^{n+\frac{1}{2}}$, for $t \in (n\Delta t, (n+1)\Delta t]$, we see that

$$\partial_t \bar{\eta}_N = v_N^* \ a.e. \text{ on } (0, T). \tag{5.57}$$

We can now use the boundedness of $E_N^{n+\frac{i}{2}}$, stated in Lemma 5.4.1 to conclude

$$(\bar{\eta}_N)_{N\in\mathbb{N}} \text{ is bounded in } L^\infty(0, T; H_0^2(\Gamma)) \cap W^{1,\infty}(0, T; L^2(\Gamma)).$$

This finishes the proof of Statement 1.

To prove Statement 2, we first notice that inequality (5.56) above implies

$$\|\eta_N^n - \eta_0\|_{H_0^2(\Gamma)} \leq 2C, \ n = 1, \ldots, N, \ N \in \mathbb{N},$$

where we recall that $\eta_N^0 = \eta_0$. We calculate

$$\|\eta_N^n - \eta_0\|_{L^2(\Gamma)} \leq \sum_{i=0}^{n-1} \|\eta_N^{i+1} - \eta_N^i\|_{L^2(\Gamma)} = \Delta t \sum_{i=0}^{n-1} \|v_N^{i+\frac{1}{2}}\|_{L^2(\Gamma)}.$$

To bound the right-hand side, we use the uniform bound $E_N^{n+\frac{1}{2}} \leq C$, where C is independent of N, from Lemma 5.4.1 to obtain

$$\|\eta_N^n - \eta_0\|_{L^2(\Gamma)} \leq Cn\Delta t \leq CT, \ n = 1, \ldots, N, \ N \in \mathbb{N}.$$

Now that we have uniform bounds for $\|\eta_N^n - \eta_0\|_{L^2(\Gamma)}$ and $\|\eta_N^n - \eta_0\|_{H_0^2(\Gamma)}$, we can use the interpolation inequality for Sobolev spaces (see Theorem 2.5.3 in Chap. 2, see also [1], Thm. 4.17, p. 79), to get

$$\|\eta_N^n - \eta_0\|_{H^1(\Gamma)} \leq 2C\sqrt{T}, \ n = 1, \ldots, N, \ N \in \mathbb{N}.$$

From Lemma 5.4.1, we see that $C = C(T)$ depends on T through the norms of the inlet and outlet data in such a way that C is an increasing function of T. Therefore, by choosing T small, we can make $\|\eta_N^n - \eta_0\|_{H^1(\Gamma)}$ arbitrary small for $n = 1, \ldots, N$, $N \in \mathbb{N}$. Because of the Sobolev embedding of $H^1(\Gamma)$ into $C[0, L]$, we can also make $\|\eta_N^n - \eta_0\|_{C[0,L]}$ arbitrary small. Since the initial data η_0 is such that $R + \eta_0(z) > 0$ (due to the conditions listed in (5.11)), we see that for a $T > 0$ small enough, there exist $R_{\min}, R_{\max} > 0$, such that

$$0 < R_{\min} \leq R + \eta_N(t, z) \leq R_{\max}, \ \forall N \in \mathbb{N}, z \in (\Gamma), t \in (0, T),$$

which finishes the proof of Statement 2.

To prove Statement 3, namely, to show uniform boundedness of $(\boldsymbol{u}_N)_{N\in\mathbb{N}}$ in $L^\infty(0, T, L^2(\Omega_f) \cap L^2(0, T; H^1(\Omega_f))$, we first notice that the uniform boundedness of $(\boldsymbol{u}_N)_{N\in\mathbb{N}}$ in $L^\infty(0, T, L^2(\Omega_f)$ follows directly from Statements 1 and 2 in Lemma 5.4.1. What is left to show is that $(\boldsymbol{u}_N)_{N\in\mathbb{N}}$ is uniformly bounded in $L^2(0, T; H^1(\Omega_f))$. Notice that Lemma 5.4.1 implies that only the symmetrized gradient is bounded and not the gradient itself. More precisely, Lemma 5.4.1 implies that

$$\sum_{n=1}^{N} \int_{\Omega_f} (R + \eta_N^{n-1}) |\mathbf{D}^{\eta_N^{n-1}}(\boldsymbol{u}_N^n)|^2 d\tilde{\boldsymbol{x}} \Delta t \leq C. \tag{5.58}$$

We would like to use Korn's inequality here to estimate the gradient in terms of the symmetrized gradient, which is bounded, but we cannot immediately do this since estimate (5.58) is given in terms of the **transformed** symmetrized gradient. Moreover, Korn's constant depends on the fluid domain, which is moving in our case. Therefore, we need a uniform Korn's constant, independent of the family of moving domains to get a uniform bound on the gradient $(\nabla \boldsymbol{u}_N)_{N\in\mathbb{N}}$, independent of N.

To get around this difficulty, we take the following approach. We first transform the problem back into the original domain $\Omega_f^{\eta_N^{n-1}}$ on which $\boldsymbol{u}_N$ is defined and work with the original, physical symmetrized gradient, not the transformed one. We then apply Korn's inequality in the usual way. However, since Korn's constant depends on the domain, we will need a result that provides a universal Korn's constant, independent of the family of domains under consideration. We address this below.

We first transform the problem onto the physical domain. For each fixed $N \in \mathbb{N}$, and for all $n = 1, \ldots, N$, we map the fluid velocity $\boldsymbol{u}_N^n$ to the physical domain, which is determined by the location of η_N^{n-1}. In this proof ,we will be using N as a *superscript* to denote the corresponding functions defined on physical domains:

$$\boldsymbol{u}^{N,n} := \boldsymbol{u}_N^n \circ (\boldsymbol{\Phi}_f^{\eta_N^{n-1}})^{-1}, \ n = 1, \ldots, N, \ N \in \mathbb{N}.$$

By using formula (11.24), we get

$$\int_{\Omega_f} (1 + \eta_N^{n-1}) |\mathbf{D}^{\eta_N^{n-1}}(\boldsymbol{u}_N^n)|^2 = \int_{\Omega_f^{\eta_N^{n-1}}} |\mathbf{D}(\boldsymbol{u}^{N,n})|^2 = \|\mathbf{D}(\boldsymbol{u}^{N,n})\|^2_{L^2(\Omega_f^{\eta_N^{n-1}})}. \tag{5.59}$$

Now, we want to design a Korn's inequality in such a way that Korn's constant is independent of N.

There are two ways to do it. One is the approach from [81, 265, 342], which assumes certain domain regularity. In our case, the conditions of [342] translate into requiring that the family of fluid domains $\{\Omega_f^{\eta_N^{n-1}}\}_{N\in\mathbb{N}}$ has a uniform Lipschitz constant, which is indeed the case, and the proof using the result from [342] can be found in [265].

Another approach, which is more straightforward, and does not require the uniform Lipschitz property of the family of fluid domains, is an approach based on directly showing that under the following two conditions:

- The fluid velocity is divergence-free
- The tangential component of the trace of the fluid velocity on the moving boundary is zero,

the L^2-norm of the velocity gradient is equal to two times the L^2-norm of the symmetrized gradient of velocity. This is similar to the result by Chambolle et al. in [81], Lemma 6, pg. 377. See also Lemma 2.8.4 in Chap. 2. We present it below. □

Proposition 5.5.2 (Uniform Korn's Equality) *Let $u \in \mathcal{V}_F^\eta(t)$ where $\mathcal{V}_F^\eta(t)$ is defined in* (5.37). *Then, the following Korn's equality holds:*

$$\|\nabla \boldsymbol{u}\|^2_{L^2(\Omega_f^\eta)} = 2\|\mathbf{D}(\boldsymbol{u})\|^2_{L^2(\Omega_f^\eta)}. \tag{5.60}$$

Proof We rewrite the right-hand side of (5.60) in terms of the gradient and calculate the square of the norms on both sides. From that calculation, it is then easy to see that it is sufficient to show that

$$\int_{\Omega_f^\eta} \nabla \boldsymbol{u} : \nabla^t \boldsymbol{u} = 0.$$

We prove that this equality holds true for smooth functions $\boldsymbol{u}$, and then the conclusion follows by density arguments. From integration by parts and $\nabla\cdot\boldsymbol{u} = 0$, we get

$$\int_{\Omega_f^\eta} \nabla \boldsymbol{u} : \nabla^t \boldsymbol{u} = -\int_{\Omega_f^\eta} \boldsymbol{u}\cdot\nabla(\nabla\cdot\boldsymbol{u}) + \int_{\partial\Omega_f^\eta} (\nabla^t \boldsymbol{u})\boldsymbol{n}\cdot\boldsymbol{u} = \int_{\partial\Omega_f^\eta} (\nabla^t \boldsymbol{u})\boldsymbol{n}\cdot\boldsymbol{u},$$

where $\boldsymbol{n} = (-\eta_z, 1)^t$. We now show that $(\nabla^t \boldsymbol{u})\boldsymbol{n}\cdot\boldsymbol{u} = 0$ on $\partial\Omega_f^\eta$. Since $\partial\Omega_f^\eta = \Gamma^\eta \cup \Gamma_{in/out} \cup \Gamma_b$, we consider each part of the boundary separately:

1. On Γ^η we have $\boldsymbol{u} = (0, u_r)$, i.e., we have $u_z(z, R+\eta(z)) = 0$. Since $\boldsymbol{u}$ is smooth, we can differentiate this equality w.r.t. z to get $\partial_z u_z + \partial_r u_z \eta_z = 0$ on Γ, i.e., for $z \in (\Gamma)$. By using $\nabla\cdot\boldsymbol{u} = 0$, we get $-\partial_r u_z \eta_z = \partial_z u_z = -\partial_r u_r$. Since $\boldsymbol{n} = (-\eta_z, 1)^\tau$, we get

$$(\nabla^\tau \boldsymbol{u})\boldsymbol{n} \cdot \boldsymbol{u} = ((\nabla^\tau \boldsymbol{u})\boldsymbol{n})_r u_r = (-\partial_r u_z \eta_z + \partial_r u_r) u_r = 0.$$

2. On $\Gamma_{in/out}$, we have $\boldsymbol{u} = (u_z, 0)$ and $\boldsymbol{n} = (\pm 1, 0)$. Hence,

$$(\nabla^\tau \boldsymbol{u})\boldsymbol{n} \cdot \boldsymbol{u} = u_z((\nabla^\tau \boldsymbol{u})\boldsymbol{n})_z = u_z(\partial_z u_z) = -u_z \partial_r u_r = 0.$$

3. On Γ_b, we have $\boldsymbol{u} = (u_z, 0)$, $\partial_r u_z = 0$, and $\boldsymbol{n} = (0, -1)$. Hence,

$$(\nabla^\tau \boldsymbol{u})\boldsymbol{n} \cdot \boldsymbol{u} = u_z((\nabla^\tau \boldsymbol{u})\boldsymbol{n})_z = u_z(-\partial_z u_r) = 0.$$

This concludes the proof of Korn's equality (5.60). □

We can now continue estimating (5.59) on the physical domain by using the uniform Korn's equality (5.60) to obtain

$$\int_{\Omega_f} (1 + \eta_N^{n-1}) |\nabla^{\eta_N^{n-1}} (\boldsymbol{u}_N^n)|^2 = 2 \int_{\Omega_f} (1 + \eta_N^{n-1}) |\mathbf{D}^{\eta_N^{n-1}} (\boldsymbol{u}_N^n)|^2. \tag{5.61}$$

Notice that on the left-hand side, we still have the transformed gradient $\nabla_N^{\eta^{n-1}}$ and not ∇, and so we employ (5.33) to calculate the relationship between the two:

$$\nabla \boldsymbol{u}_N^n = \left(\nabla^{\eta_N^{n-1}} \boldsymbol{u}_N^n\right) \left(\nabla \boldsymbol{\Phi}^{\eta_N^{n-1}}\right), \quad n = 1, \dots, N, \; N \in \mathbb{N}. \tag{5.62}$$

Since η_N are bounded in $L^\infty(0, T; H^2(\Gamma))$, the gradient of the ALE mapping is bounded in $L^\infty(\Omega_f)$:

$$\|\nabla \boldsymbol{\Phi}^{\eta_N^{n-1}}\|_{L^\infty(\Omega_f)} \leq C, \quad n = 1, \dots, N, \; N \in \mathbb{N}.$$

Using this estimate in (5.62), and by summing from $n = 1, \dots, N$ in (5.62), we obtain the following estimate for $\nabla \boldsymbol{u}_N^n$:

$$\sum_{n=1}^N \|\nabla \boldsymbol{u}_N^n\|_{L^2(\Omega_f)}^2 \Delta t \leq C \sum_{n=1}^N \int_{\Omega_f} (R + \eta_N^{n-1}) |\mathbf{D}^{\eta_N^{n-1}} (\boldsymbol{u}_N^n)|^2 d\tilde{\boldsymbol{x}} \Delta t,$$

where C is independent of N. This implies that the sequence $(\nabla \boldsymbol{u}_N)_{N \in \mathbb{N}}$ is uniformly bounded in $L^2((0, T) \times \Omega_f)$, and so the sequence $(\boldsymbol{u}_N)_{N \in \mathbb{N}}$ is uniformly bounded in $L^2(0, T; H^1(\Omega_f))$.

Since Lemma 5.4.1 implies uniform boundedness of $(\boldsymbol{u}_N)_{N \in \mathbb{N}}$ in $L^\infty(0, T, L^2(\Omega_f))$, as mentioned before, we now have that the sequence $(\boldsymbol{u}_N)_{N \in \mathbb{N}}$ is uniformly bounded in $L^\infty(0, T, L^2(\Omega_f) \cap L^2(0, T; H^1(\Omega_f))$. This completes the proof of Statement 3.

Statement 4 follows directly from Statements 1 and 2 of Lemma 5.4.1.

This completes the proof of Proposition 5.5.1. □

The uniform boundedness results above will allow us to obtain weakly and weakly* convergent subsequences, as we show in the next subsection. Our ultimate goal is to show that the limits of these subsequences converge to a weak solution of the coupled problem. Namely, we would like to show that the limits of these approximate solutions that satisfy the weak formulations (5.49) (or alternatively (5.53)) satisfy the weak formulation of the coupled, continuous problem (5.39) (or alternatively (5.30)). Since the problem is nonlinear, weak convergence of velocities needs to be upgraded to strong convergence in order to be able to pass to the limit in the semidiscrete weak formulation of the coupled problem (5.53) and recover (5.30). This will be done using certain compactness arguments, as we show in Sect. 5.7.

Furthermore, to pass to the limit in the weak formulations of approximate solutions, we need additional regularity in time of the structure displacement. To get this additional regularity of the structure displacement, we need the following uniform boundedness result for the linear interpolants $\bar{\eta}$. We note that the final result is independent of the type of extrapolation used.

Proposition 5.5.3 *The sequence* $(\bar{\eta}_N)_{N\in\mathbb{N}}$ *is uniformly bounded in* $L^\infty(0,T;H_0^2(\Gamma))\cap W^{1,\infty}(0,T;L^2(\Gamma))$.

Proof We now observe that

$$\partial_t\bar{\eta}_N(t)=\frac{\eta_N^{n+1}-\eta_N^n}{\Delta t}=\frac{\eta_N^{n+1/2}-\eta_N^n}{\Delta t}=v_N^{n+\frac{1}{2}},\quad t\in(n\Delta t,(n+1)\Delta t),$$

and so, since v_N^* was defined in (5.54) as a piecewise constant function defined via $v_N^*(t,\cdot)=v^{n+\frac{1}{2}}$, for $t\in(n\Delta t,(n+1)\Delta t]$, we see that

$$\partial_t\bar{\eta}_N=v_N^*\ a.e.\ \text{on}\ (0,T). \tag{5.63}$$

By using Lemma 5.4.1 (the boundedness of $E_N^{n+\frac{i}{2}}$), we get

$$(\bar{\eta}_N)_{N\in\mathbb{N}}\ \text{is bounded in}\ L^\infty(0,T;H_0^2(\Gamma))\cap W^{1,\infty}(0,T;L^2(\Gamma)).$$

□

We will use this proposition to show precompactness of $(\eta_N)_{N\in\mathbb{N}}$ in $L^\infty(0,T;H^s(\Gamma))$, $0<s<2$, in Sect. 5.7.

5.6 Weak Convergence of Approximate Functions

The uniform boundedness results presented in Proposition 5.5.1 give us the following weak convergence results.

Lemma 5.6.1 *There exist subsequences* $(\boldsymbol{u}_N)_{N\in\mathbb{N}}$, $(\overline{\eta}_N)_{N\in\mathbb{N}}$, $(\eta_N)_{N\in\mathbb{N}}$, $(v_N^*)_{N\in\mathbb{N}}$, $(v_N)_{N\in\mathbb{N}}$ *and the functions* $\eta \in L^\infty(0,T;H_0^2(\Gamma))$, $v \in L^\infty(0,T;L^2(\Gamma)) \cap L^2(0,T;H_0^2(\Gamma))$, $v^* \in L^\infty(0,T;L^2(\Gamma))$, *and* $\boldsymbol{u} \in L^\infty(0,T;L^2(\Omega_f)) \cap L^2(0,T;H^1(\Omega_f))$, *such that*

$$\boldsymbol{u}_N \rightharpoonup \boldsymbol{u} \text{ weakly-star in } L^\infty(0,T;L^2(\Omega_f)) \text{ and weakly in } L^2(0,T;H^1(\Omega_f)),$$

$$\overline{\eta}_N \rightharpoonup \overline{\eta} \text{ weakly-star in } W^{1,\infty}(0,T;L^2(\Gamma)) \text{ and weakly-star in } L^\infty(0,T;H_0^2(\Gamma)),$$

$$\eta_N \rightharpoonup \eta \text{ weakly-star in } L^\infty(0,T;H_0^2(\Gamma)),$$

$$v_N^* \rightharpoonup v^* \text{ weakly-star in } L^\infty(0,T;L^2(\Gamma)),$$

$$v_N \rightharpoonup v \text{ weakly-star in } L^\infty(0,T;L^2(\Gamma)) \text{ and weakly in } L^2(0,T;H^{1/2}(\Gamma)).$$

Furthermore,

$$\bar{\eta} = \eta \text{ and } v = v^*. \tag{5.64}$$

Proof The statement $\bar{\eta} = \eta$ is a trivial consequence of the uniqueness of the distributional limit. As a consequence, we get that the weak* limit of η_N is, in fact, in $W^{1,\infty}(0,T;L^2(\Gamma))$.

To show that $v = v^*$, we multiply the uniform energy estimates by Δt, and notice again that $\|v_N\|^2_{L^2((0,T)\times(\Gamma))} = \Delta t \sum_{n=1}^N \|v_N^n\|^2_{L^2(\Gamma)}$. This implies $\|v_N - v_N^*\|_{L^2((0,T)\times(\Gamma))} \le C\sqrt{\Delta t}$, and so in the limit as $\Delta t \to 0$, we get $v = v^*$. □

In the linearly coupled case, discussed in Chap. 4, weak and weak* convergence were sufficient to pass to the limit in the weak formulation to show that the limits satisfy the weak formulation and are, therefore, weak solutions. However, in the nonlinearly coupled FSI problem, weak convergence is not strong enough to pass to the limit in the semidiscrete formulation (5.49). This is why we need compactness arguments to upgrade the weak convergence of fluid and structure velocities and of structure displacement to a strong convergence. The next section addresses strong convergence of approximate functions and discusses the compactness arguments that we will use to obtain strong convergence of approximating fluid and structure velocities. The compactness arguments used to obtain strong convergence for the fluid and structure velocities will rely on the use of a recent generalization of the Aubin-Lions-Simon compactness lemma to problems on moving domains [270], while the compactness arguments for the structure displacement will rely on the Arzelà-Ascoli theorem. See Sect. 5.7 in the Preliminaries chapter for the classical compactness results for function spaces.

5.7 Strong Convergence of Approximate Functions

We start by discussing strong convergence of structure displacements and then move onto discussing compactness arguments to deduce strong convergence of fluid and structure velocities.

5.7.1 Strong Convergence of Approximate Structure Displacements

Our ultimate goal in this subsection is to show that $\eta_N \to \eta$ in $L^\infty(0, T; C^1[0, L])$, and that the time-shifts by Δt, denoted by $\tau_{\Delta t}\eta_N$, converge strongly in $L^\infty(0, T; C^1[0, L])$.

Definition 5.7.1 For an arbitrary function $\mathbf{f}$, the time-shift $\tau_h\mathbf{f}$ is defined by

$$\tau_h\mathbf{f}(t, .) = \mathbf{f}(t - h, .), \; h \in \mathbb{R}. \tag{5.65}$$

To show strong convergence of structure displacements, we first need the following result on strong convergence of linear interpolants of structure displacements $\bar{\eta}_N$.

Proposition 5.7.1 *There exists a subsequence of linear interpolants* $(\bar{\eta})_{N\in\mathbb{N}}$, *defined in* (5.55), *such that the following convergence result holds:*

$$\bar{\eta}_N \to \eta \text{ in } C([0, T]; H^s(\Gamma)), \; 0 < s < 2.$$

Proof We start by recalling the uniform boundedness result of Proposition 5.5.3, which states that the sequence $(\bar{\eta}_N)_{N\in\mathbb{N}}$ is uniformly bounded in $L^\infty(0, T; H_0^2(\Gamma)) \cap W^{1,\infty}(0, T; L^2(\Gamma))$. Now, by the interpolation estimate (see Theorem 5.2, [1]), there exists a constant $C > 0$ independent of N, such that for $0 < \alpha < 1$, the following estimate holds:

$$\|\bar{\eta}_N(t + \Delta t) - \bar{\eta}_N(t)\|_{H^{2\alpha}(\Gamma)} \leq C\|\bar{\eta}_N(t + \Delta t) - \bar{\eta}_N(t)\|_{L^2(\Gamma)}^{1-\alpha}\|\bar{\eta}_N(t + \Delta t) - \bar{\eta}_N(t)\|_{H^2(\Gamma)}^{\alpha}.$$

By multiplying and dividing the right-hand side by $(\Delta t)^{1-\alpha}$, and using uniform boundedness of v_N, we get

$$\|\bar{\eta}_N(t + \Delta t) - \bar{\eta}_N(t)\|_{H^{2\alpha}(\Gamma)} \leq C(\Delta t)^{1-\alpha}, \text{ where } 0 < \alpha < 1. \tag{5.66}$$

Therefore, $(\bar{\eta}_N)_{N\in\mathbb{N}}$ is uniformly bounded in $C^{0,1-\alpha}([0, T]; H^{2\alpha}(\Gamma))$, $0 < \alpha < 1$. Now, from the continuous embedding of $H^{2\alpha}$ into $H^{2\alpha-\varepsilon}$, and by applying the Arzelà-Ascoli theorem, we conclude that $(\bar{\eta}_N)_{N\in\mathbb{N}}$ has a convergent subsequence, which we again denote by $(\bar{\eta}_N)_{N\in\mathbb{N}}$, such that

$$\bar{\eta}_N \to \eta \text{ in } C([0,T]; H^s(\Gamma)), \ 0 < s < 2.$$

Here, we used the fact that the sequences $(\bar{\eta}_N)_{N\in\mathbb{N}}$ and $(\eta_N)_{N\in\mathbb{N}}$ have the same limit. □

Using this result, we can now prove the following lemma:

Lemma 5.7.1 *$\eta_N \to \eta$ in $L^\infty(0,T; H^s(\Gamma))$, $0 < s < 2$.*

Proof The proof follows from the continuity in time of η, and from the fact that $\bar{\eta}_N \to \eta$ in $C([0,T]; H^s(\Gamma))$, $0 < s < 2$. Namely, let $\varepsilon > 0$. Then, from the continuity of η in time, we have that there exists a $\delta t > 0$ such that

$$\|\eta(t_1) - \eta(t_2)\|_{H^s(\Gamma)} < \frac{\varepsilon}{2}, \text{ for } t_1, t_2 \in [0,T], \text{ and } |t_1 - t_2| \le \delta t.$$

Furthermore, from the convergence $\bar{\eta}_N \to \eta$ in $C([0,T]; H^s(\Gamma))$, $0 < s < 2$, we know that there exists an $N^* \in \mathbb{N}$ such that

$$\|\bar{\eta}_N - \eta\|_{C([0,T]; H^s(\Gamma))} < \frac{\varepsilon}{2}, \ \forall N \ge N^*.$$

Now, let N be any natural number such that $N > \max\{N^*, T/\delta t\}$. Denote by $\Delta t = T/N$, and let $t \in [0,T]$. Furthermore, let $n \in \mathbb{N}$ be such that $(n-1)\Delta t < t \le n\Delta t$. Recall that $\bar{\eta}_N(n\Delta t) = \eta_N(n\Delta t) = \eta_N(t)$ from the definition of $\bar{\eta}_N$ and η_N. By using this, and by combining the two estimates above, we get that for $(n-1)\Delta t < t \le n\Delta t$,

$$\begin{aligned}
\|\eta_N(t) - \eta(t)\|_{H^s(\Gamma)} &= \|\eta_N(t) - \eta(n\Delta t) + \eta(n\Delta t) - \eta(t)\|_{H^s(\Gamma)} \\
&= \|\eta_N(n\Delta t) - \eta(n\Delta t) + \eta(n\Delta t) - \eta(t)\|_{H^s(\Gamma)} \\
&\le \|\eta_N(n\Delta t) - \eta(n\Delta t)\| + \|\eta(n\Delta t) - \eta(t)\|_{H^s(\Gamma)} \\
&= \|\bar{\eta}_N(n\Delta t) - \eta(n\Delta t)\|_{H^s(\Gamma)} + \|\eta(n\Delta t) - \eta(t)\|_{H^s(\Gamma)} < \varepsilon.
\end{aligned}$$

Here, the first term is bounded by $\varepsilon/2$ due to the convergence $\bar{\eta}_N \to \eta$, while the second term is bounded by $\varepsilon/2$ due to the continuity of η. Since the obtained estimate is uniform in N and t, the statement of the lemma is proved. □

Finally, we can now prove the following strong convergence results for structure displacements.

Lemma 5.7.2 *The following strong convergence results hold for structure displacements $(\eta_N)_{N\in\mathbb{N}}$ and for the time-shifts $(\tau_{\Delta t}\eta_N)_{N\in\mathbb{N}}$:*

$$\begin{aligned}
\eta_N &\to \eta \text{ in } L^\infty(0,T; C^1[0,L]), \\
\tau_{\Delta t}\eta_N &\to \eta \text{ in } L^\infty(0,T; C^1[0,L]).
\end{aligned} \tag{5.67}$$

Proof The first statement is a direct consequence of Lemma 5.7.1 in which we proved that $\eta_N \to \eta$ in $L^\infty(0, T; H^s(0, L))$, $0 < s < 2$. This means that for $s > \frac{3}{2}$, we immediately have

$$\eta_N \to \eta \text{ in } L^\infty(0, T; C^1[0, L]). \tag{5.68}$$

To show convergence of the shifted displacements $\tau_{\Delta t}\eta_N$ to the same limiting function η, we recall that

$$\bar{\eta}_N \to \eta \text{ in } C([0, T]; H^s[0, L]), \ 0 < s < 2,$$

and that $(\bar{\eta}_N)_{N\in\mathbb{N}}$ is uniformly bounded in $C^{0,1-\alpha}([0, T]; H^{2\alpha}(0, L))$, $0 < \alpha < 1$. Uniform boundedness of $(\bar{\eta}_N)_{N\in\mathbb{N}}$ in $C^{0,1-\alpha}([0, T]; H^{2\alpha}(0, L))$ implies that there exists a constant $C > 0$, independent of N, such that

$$\|\bar{\eta}_N((n-1)\Delta t) - \bar{\eta}_N(n\Delta t)\|_{H^{2\alpha}(0,L)} \leq C|\Delta t|^{1-\alpha}.$$

This means that for each $\varepsilon > 0$, there exists an $N_1 > 0$ such that

$$\|\bar{\eta}_N((n-1)\Delta t) - \bar{\eta}_N(n\Delta t)\|_{H^{2\alpha}(0,L)} \leq \frac{\varepsilon}{2}, \text{ for all } N \geq N_1.$$

Here, N_1 is chosen by recalling that $\Delta t = T/N$, and so the right-hand side implies that we want an N_1 such that

$$C\left(\frac{T}{N}\right)^{1-\alpha} < \frac{\varepsilon}{2} \text{ for all } N \geq N_1.$$

Now, convergence $\bar{\eta}_N \to \eta$ in $C([0, T]; H^s[0, L])$, $0 < s < 2$, implies that for each $\varepsilon > 0$, there exists an $N_2 > 0$ such that

$$\|\bar{\eta}_N(n\Delta t) - \eta(t)\|_{H^s(0,L)} < \frac{\varepsilon}{2}, \text{ for all } N \geq N_2.$$

We will use this to show that for each $\varepsilon > 0$, there exists an $N^* \geq \max\{N_1, N_2\}$, such that

$$\|\tau_{\Delta t}\bar{\eta}_N(t) - \eta(t)\|_{H^s(0,L)} < \varepsilon, \text{ for all } N \geq N^*.$$

Let $t \in (0, T)$. Then there exists an n such that $t \in ((n-1)\Delta t, n\Delta t]$. We calculate

$$\begin{aligned}
\|\tau_{\Delta t}\bar{\eta}_N(t) - \eta(t)\|_{H^s(0,L)} &= \|\tau_{\Delta t}\bar{\eta}_N(t) - \bar{\eta}_N(n\Delta t) + \bar{\eta}_N(n\Delta t) - \eta(t)\|_{H^s(0,L)} \\
&= \|\bar{\eta}_N((n-1)\Delta t) - \bar{\eta}_N(n\Delta t) + \bar{\eta}_N(n\Delta t) - \eta(t)\|_{H^s(0,L)} \\
&\leq \|\bar{\eta}_N((n-1)\Delta t) - \bar{\eta}_N(n\Delta t)\|_{H^s(0,L)} + \|\bar{\eta}_N(n\Delta t) - \eta(t)\|_{H^s(0,L)}.
\end{aligned}$$

The first term is less than ε for all $N > N^*$ by the uniform boundedness of $(\bar{\eta}_N)_{N\in\mathbb{N}}$ in $C^{0,1-\alpha}([0,T]; H^{2\alpha}(0,L))$, while the second term is less than ε for all $N > N^*$ by the convergence of $\bar{\eta}_N$ to η in $C([0,T]; H^s[0,L])$, $0 < s < 2$.

Now, by noticing that $\tau_{\Delta t}\bar{\eta}_N = \overline{(\tau_{\Delta t}\eta_N)}$, we can use the same argument as in Lemma 5.6.1 to show that sequences $(\overline{\tau_{\Delta t}\eta_N})_{N\in\mathbb{N}}$ and $(\tau_{\Delta t}\eta_N)_{N\in\mathbb{N}}$ both converge to the same limit η in $L^\infty(0,T; H^s(0,L))$, for $s < 2$. □

To pass to the limit in semidiscrete weak formulation of the coupled FSI problem, what is left to show is that the sequences of fluid and structure velocities, $(\boldsymbol{u}_N)_{N\in\mathbb{N}}$, and $(v_N)_{N\in\mathbb{N}}$, constructed above, are relatively compact in $L^2(0,T; L^2(\Omega_f))$ and $L^2(0,T; L^2(\Gamma))$. This implies the existence of subsequences that converge strongly in $L^2(0,T; L^2(\Omega_f))$ and $L^2(0,T; L^2(\Gamma))$, respectively, which will be sufficient to pass to the limit. The strong convergence of fluid and structure velocities is discussed next.

5.7.2 *Strong Convergence of Fluid and Structure Velocities*

To show strong convergence of fluid and structure velocities, we will have to use compactness arguments. The general principle behind compactness arguments is that by obtaining uniform boundedness of spacetime functions in a particular function space and by obtaining additional uniform boundedness of the *time derivatives* of these functions in a weaker function space, we can obtain strong convergence of these functions to a limiting function. This is the fundamental principle behind the *Aubin-Lions compactness lemma* [11, 240]. See also Theorem 2.9.4 in Sect. 2. The Aubin-Lions compactness lemma relies on a chain of embeddings of function spaces in spatial variables:

$$Z \subset\subset Y \subset X,$$

where the embedding of Z into Y is a compact embedding. Then, for any $1 < p \leq \infty$, showing that the functions $\{f_n\}_{n=1}^{\infty}$ are uniformly bounded in $L^p(0,T; Z)$ and showing that their time derivatives are uniformly bounded in $L^p(0,T; X)$ implies that the functions f_n converge strongly along a subsequence in $L^p(0,T; Y)$ to a limiting function f.

Note that the Aubin-Lions lemma immediately gives strong convergence of the structure displacements $\overline{\eta}_N$, via the chain of embeddings:

$$H_0^2(\Gamma) \subset\subset L^2(\Gamma) \subset L^2(\Gamma).$$

Since $\overline{\eta}_N$ are uniformly bounded in $L^\infty(0,T; H_0^2(\Gamma))$ and $\partial_t\overline{\eta}_N = v_N^*$ is uniformly bounded in $L^\infty(0,T; L^2(\Gamma))$, the Aubin-Lions compactness lemma (Theorem 2.9.4) implies that η_N converges along a subsequence strongly to a limiting function η in $L^\infty(0,T; L^2(\Gamma))$.

We will need a more involved argument to show the convergence of the fluid velocities and the structure velocities $(\boldsymbol{u}_N, v_N)$. In principle, we should be able to use an Aubin-Lions-type argument to show strong convergence of $(\boldsymbol{u}_N, v_N)$ in an appropriate function space along a subsequence. To see this, let us return to the linearly coupled case for a brief moment and consider what would happen in that case. In the linearly coupled case, the semidiscrete formulation reads

$$\int_{\Omega_f} \frac{\boldsymbol{u}_N^{n+1} - \boldsymbol{u}_N^n}{\Delta t} \cdot \boldsymbol{q} + 2\mu_f \int_{\Omega_f} \boldsymbol{D}(\boldsymbol{u}_N^{n+1}) : \boldsymbol{D}(\boldsymbol{q}) + \int_\Gamma \frac{v_N^{n+1} - v_N^n}{\Delta t} \psi$$

$$+ C_1 \int_\Gamma \partial_z v_N^{n+\frac{1}{2}} \cdot \partial_z \psi + C_2 \int_\Gamma \partial_z^2 v^{n+\frac{1}{2}} \cdot \partial_z^2 \psi = \int_{\Gamma_{in}} P_{N,in}^n q_z - \int_{\Gamma_{out}} P_{N,out}^n q_z,$$

which we have reproduced from (4.39) for convenience, where $(\boldsymbol{q}, \psi)$ is a test function in a fixed test function space $\mathcal{Q} := \mathcal{V}_f \times \mathcal{V}_s$, where $\mathcal{V}_f$ and $\mathcal{V}_s$ are defined by (4.22) and (4.21), respectively. We can formally think of $\dfrac{\boldsymbol{u}_N^{n+1} - \boldsymbol{u}_N^n}{\Delta t}$ and $\dfrac{v_N^{n+1} - v_N^n}{\Delta t}$ as a discretized form of the time derivative of $(\boldsymbol{u}_N, v_N)$, in which case we can use the weak formulation for fixed but arbitrary test functions $(\boldsymbol{q}, \psi)$ satisfying $\|(\boldsymbol{q}, \psi)\|_{\mathcal{Q}} \le 1$ to show that the discretized time derivative of $(\boldsymbol{u}_N, v_N)$ is at least formally uniformly bounded in $L^2(0, T; \mathcal{Q}')$. In addition, $(\boldsymbol{u}_N, v_N)$ itself is uniformly bounded in $L^2(0, T; H^1(\Omega_f) \times H^{1/2}(\Gamma))$, and hence we conclude that the fluid velocities and structure velocities $(\boldsymbol{u}_N, v_N)$ are strongly precompact in $L^2(0, T; L^2(\Omega_f) \times L^2(\Gamma))$ via the chain of embeddings:

$$H^1(\Omega_f) \times H^{1/2}(\Gamma) \subset\subset L^2(\Omega_f) \times L^2(\Gamma) \subset \mathcal{Q}'.$$

Even though in principle, we should expect there to be a similar convergence of $(\boldsymbol{u}_N, v_N)$ in the nonlinearly coupled case, the same standard Aubin-Lions argument will not apply here. This is because the approximate fluid velocities $\boldsymbol{u}_N^n$ are defined on **moving fluid domains** $\Omega_{f,N}^n$. Therefore, all of the approximate fluid velocities $\boldsymbol{u}_N^n$ are **defined on different fluid domains**, and hence, we cannot create a uniform chain of embeddings needed for the standard Aubin-Lions compactness lemma.

We could try to pull back all of the approximate fluid velocities $\boldsymbol{u}_N^n$ to the fixed reference domain Ω_f so that all of the resulting approximate fluid velocities will be defined on the same domain. However, the difficulty here is that **the divergence-free condition** $\nabla \cdot \boldsymbol{u}_N^n = 0$ on the physical domain will transform nonlinearly as $\nabla^{\eta_N^n} \cdot (\boldsymbol{u}_N^{n+1} \circ \boldsymbol{\Phi}_f^{\eta_N^n}) = 0$, so that the divergence-free condition on the reference domain will change depending on n and N. This will make it hard to properly identify an appropriate test space for the semidiscrete formulations on the fixed reference domain, since the divergence-free condition on the fixed reference domain becomes a nonlinear condition.

Instead, we will take the approach of **working on the physical fluid domains** $\Omega^n_{f,N}$ so that the divergence-free condition remains the same for all of the approximate fluid velocities $\boldsymbol{u}^n_N$. However, the difficulty here is that the approximate fluid velocities $\boldsymbol{u}^n_N$ are still all defined on different domains $\Omega^n_{f,N}$. To address this difficulty, we will define the approximate fluid velocities $\boldsymbol{u}^n_N$ on a common maximal fluid domain Ω^M_f that contains all of the approximate fluid domains $\Omega^n_{f,N}$ by extending $\boldsymbol{u}^n_N$ by zero in $\Omega^M_f \setminus \Omega^n_{f,N}$. The existence of such a maximal fluid domain Ω^M_f can be shown by establishing the existence of a continuous function $M(z)$ satisfying $M(0) = M(L) = 0$ such that

$$\eta^n_N(z) \le M(z), \qquad \text{for all } N \text{ and } n = 0, 1, \dots, N,$$

which follows from the uniform boundedness properties of the structure displacements. More precisely, the following result guarantees the desired uniform boundedness.

Lemma 5.7.3 *There exist smooth functions $m(z)$ and $M(z)$, $z \in \overline{\Gamma} = [0, L]$, such that the sequence $\{\eta^n_N(z)\}$ corresponding to the semi-discretization of $(0, T)$ can be uniformly bounded as follows:*

- $m(z) \le \eta^n_N(z) \le M(z),\ z \in [0, L],\ n = 0, \dots, N,\ N \in \mathbb{N},\ \Delta t = T/N,$ *and*
- $M(z) - m(z) \le C(T),\ z \in [0, L],$ *where* $C(T) \to 0,\ T \to 0.$

Moreover, for each fixed N, $l \in \{1, \dots, N\}$, and $n \in \{1, \dots, N - l\}$, there exist smooth functions $m^{n,l}_N(z)$, $M^{n,l}_N(z)$ such that the functions $\{\eta^n_N(z)\}$ corresponding to the semi-discretization of $(t, t+h) = (n\Delta t, (n+l)\Delta t)$ can be bounded as follows:

- $m^{n,l}_N(z) \le \eta^{n+i}_N(z) \le M^{n,l}_N(z),\ z \in [0, L],\ i = 0, \dots, l,$ *and*
- $M^{n,l}_N(z) - m^{n,l}_N(z) \le C\sqrt{l\Delta t},\ z \in [0, L],$ and $\|M^{n,l}_N - m^{n,l}_N\|_{L^2(\Gamma)} \le Cl\Delta t.$

To prove this lemma, we first show the following uniform boundedness result.

Proposition 5.7.2 *Sequence $(\eta_N)_{N\in\mathbb{N}}$ is uniformly bounded in*

$$L^\infty(0, T; H^2_0(0, L)).$$

Moreover, for T small enough, we have

$$0 < R_{\min} \le R + \eta_N(t, z) \le R_{\max},\ \forall N \in \mathbb{N}, z \in (0, L), t \in (0, T). \tag{5.69}$$

Proof of Proposition 5.7.2 The argument proceeds analogously to the proof of Proposition 5.5.1, Claim 2.

Proof of Lemma 5.7.3 The existence and properties of the functions M and m are a direct consequence of Proposition 5.7.2.

To prove the second statement of Lemma 5.7.3, we fix $N \in \mathbb{N}$ and consider finitely many functions $\eta^{n+i}_N(z)$ for $i = 0, \dots, l$, which are defined on the time interval from $t = n\Delta t$ to $t + h = n\Delta t + l\Delta t$. Define $m_N^{n,l}(z)$ and $M_N^{n,l}(z)$, $z \in [0, L]$ to be the functions obtained by considering the minimum and maximum of the finitely many functions $\eta_N^{n+i}(z)$ for $i = 0, \ldots, l$, mollified if necessary to get the smooth functions. The properties of the functions $m_N^{n,l}$, $M_N^{n,l}$ then follow from the proof of Proposition 5.7.2. Namely, from the proof of Proposition 5.7.2, it follows that the upper bound on $\|\eta_N^{n+i} - \eta_N^n\|_{H^1}$ only depends on the width of the time interval, which is h, namely, $\|\eta_N^{n+i} - \eta_N^n\|_{H^1} \leq C\sqrt{h}$, for all $i = 1, \ldots, l$. Since η_N^n are defined on $\overline{\Gamma} = [0, L]$, then the estimate also holds point-wise (because of the Sobolev embedding of $H^1(\Gamma)$ into $C[0, L]$). Thus, the first inequality is proven since $h = l\Delta t$. The second inequality, namely, the L^2-bound on the difference between $M_N^{n,l}$ and $m_N^{n,l}$, follows from

$$\|\eta_N^n - \eta_0\|_{L^2(\Gamma)} \leq \sum_{i=0}^{n-1} \|\eta_N^{i+1} - \eta_N^i\|_{L^2(\Gamma)} = \Delta t \sum_{i=0}^{n-1} \|v_N^{i+\frac{1}{2}}\|_{L^2(\Gamma)} \leq \|v_N\|_{L^2(0,T,L^2(\Gamma))},$$

where $\eta_N^0 = \eta_0$, and from the uniform energy bound of the structure velocity v_N in $L^2(0, T; L^2(\Gamma))$, which follows from the uniform energy bounds in Lemma 5.4.1.

□

Now we can define the **maximal fluid domain** by

$$\Omega_f^M = \{(z, r) \in \mathbb{R}^2 : 0 \leq z \leq L, 0 \leq r \leq R + M(z)\},$$

and we can extend the fluid velocities $\boldsymbol{u}_N^n$ to Ω_f^M by keeping these fluid velocities the same on $\Omega_{f,N}^n \subset \Omega_f^M$ and then using an extension by zero outside of $\Omega_{f,N}^n$ so that

$$\boldsymbol{u}_N^n = 0, \qquad \text{on } \Omega_f^M \setminus \Omega_{f,N}^n.$$

This gives us approximate fluid velocities that still satisfy the divergence-free condition, but now all of the fluid velocities are defined on a common fluid domain Ω_f^M.

While this extension by zero may introduce discontinuities into the functions $\boldsymbol{u}_N^n$, because of the fact that the structure displacements η_N^n are uniformly Lipschitz and the fact that the fluid velocities $\boldsymbol{u}_N$ are uniformly bounded in $L^2(0, T; H^1(\Omega_f(t)))$, we then have that the extended fluid velocities $\boldsymbol{u}_N$ are uniformly bounded in $L^2(0, T; H^s(\Omega_f^M))$ for $0 \leq s < 1/2$. Notice that we use the same notation for the extended velocities. See Theorem 2.8.2 in the Preliminaries chapter for the precise result regarding the zero extension across a Lipschitz boundary of an H^s-function.

More precisely, we have the following lemma.

Lemma 5.7.4 *The extensions by* 0 *to* Ω^M *of* $\boldsymbol{u}_N^n$ *are such that*

$$(\boldsymbol{u}_N^n, v_N^n) \in H^s(\Omega^M) \times H^s(\Gamma), 0 < s < 1/2.$$

Proof of Proposition 5.7.2 The proof is a consequence of the extension theorem Theorem 2.8.2 in Chap. 2. See also Theorem 2.7 in [257]. More precisely, to apply Theorem 2.8.2, we first notice that the functions η_N^n are uniformly Lipschitz on Γ. This follows from the uniform energy estimates presented in Lemma 5.4.1, which imply that there exists a constant $C > 0$, independent of Δt, such that the $H^2(\Gamma)$-norms of η_N^n are uniformly bounded by C. Since $\Gamma \subset \mathbb{R}$, this implies that η_N^n are uniformly Lipschitz, where the Lipschitz constant is independent of Δt. Now we can apply Theorem 2.8.2 to the functions $\boldsymbol{u}_N^n$. First, the uniform energy estimates from Lemma 5.4.1 imply that $\boldsymbol{u}_N^n$ are in $H^1(\Omega^{\eta_N^n})$. Then, Theorem 2.8.2 implies that the extensions by 0 to Ω^M belong to $H^s(\Omega^M)$, $0 < s < 1/2$. □

Moreover, there exists an absolute constant $C > 0$, which depends only on the maximum of the Lipschitz constants for η_N^n, such that

$$\|\boldsymbol{u}_N^n\|_{H^s(\Omega^M)} \leq C\|\nabla \boldsymbol{u}_N^n\|_{L^2(\Omega^{\eta_N^n})}, \, 0 < s < 1/2. \tag{5.70}$$

In 1D the constant C "measures" the jump across Γ.

Similarly, we extend the test functions $\boldsymbol{q} \in Q_N^n$ to the maximal domain Ω^M by 0.

Remark 5.7.1 (Extensions of Test Functions) Notice that the extended test functions are no longer smooth. However, as we shall see below, we will work with "admissible" test functions whose smoothness will depend only on the smoothness of $\boldsymbol{q}$ within the domains $\Omega^{\eta_N^n}$.

Now that we have obtained that the approximate fluid velocities $\boldsymbol{u}_N$ defined on the common maximal fluid domain Ω_f^M are uniformly bounded in $L^2(0, T; H^s(\Omega_f^M))$ for $0 \leq s < 1/2$, we would like to apply a compactness argument alla Aubin-Lions. However, for an Aubin-Lions-type compactness argument to work, we need further bounds on the "time derivatives" of $\boldsymbol{u}_N$. We get estimates on the "time derivatives" of $\boldsymbol{u}_N$ by using the semidiscretized form of the weak formulation, but since we are working with fluid velocities that are defined on the physical domains $\Omega_{f,N}^n$, we recall the weak formulation of the semidiscretized coupled problem (5.53) defined on moving domains:

$$\int_{\Omega_{f,N}^n} \frac{\boldsymbol{u}_N^{n+1} - \tilde{\boldsymbol{u}}_N^n}{\Delta t} \cdot \boldsymbol{q} + \frac{1}{2}\int_{\Omega_{f,N}^n} \left[\left(\left(\tilde{\boldsymbol{u}}_N^n - \frac{r}{R+\eta_N^n} v_N^{n+\frac{1}{2}} \boldsymbol{e}_r\right) \cdot \nabla \boldsymbol{u}_N^{n+1}\right) \cdot \boldsymbol{q}\right]$$
$$- \left[\left(\left(\tilde{\boldsymbol{u}}_N^n - \frac{r}{R+\eta_N^n} v_N^{n+\frac{1}{2}} \boldsymbol{e}_r\right) \cdot \nabla \boldsymbol{q}\right) \cdot \boldsymbol{u}_N^{n+1}\right]$$
$$+ \int_{\Omega_{f,N}^n} \frac{1}{2(R+\eta_N^n)} v_N^{n+\frac{1}{2}} \boldsymbol{u}_N^{n+1} \cdot \boldsymbol{q} + 2\mu_f \int_{\Omega_{f,N}^n} \boldsymbol{D}(\boldsymbol{u}_N^{n+1}) : \boldsymbol{D}(\boldsymbol{q})$$

$$+\int_\Gamma \frac{v_N^{n+1} - v_N^{n+\frac{1}{2}}}{\Delta t}\psi$$
$$= \int_{\Gamma_{in}} P^n_{N,in}(t)q_z - \int_{\Gamma_{out}} P^n_{N,out}(t)q_z,$$

where $\tilde{\boldsymbol{u}}^n_N = \boldsymbol{u}^n_N \circ \boldsymbol{\Phi}_f^{\eta^n_N} \circ \left(\boldsymbol{\Phi}_f^{\eta^{n-1}_N}\right)^{-1}$ is defined on $\Omega^n_{f,N}$. Thus, at least on a formal level, we can estimate the discretized time derivative in a dual space, where the test functions are test functions defined on $\Omega^n_{f,N}$. However, because the solution space and test space for this semidiscretized weak formulation on the physical domains depend on n and N, we will have to use a *generalization of the Aubin-Lions-Simon compactness lemma to problems on moving domains*, introduced by Muha and Čanić in [270], and stated in the Preliminaries of this book, Chap. 2, as Theorem 2.10.1. To handle the fact that the physical domains are changing, see Fig. 5.2, where we will have to verify additional conditions to show that there exist mappings between the different test and solution spaces that are uniformly bounded in an appropriate sense, which will allow us to be able to properly compare functions defined on different physical domains with each other in a uniform way.

To set the framework for the application of the generalized Aubin-Lions-Simon compactness result by Muha and Čanić, we will consider the integral semidiscrete formulation of the coupled FSI problem above, first specified in (5.53), defined on slightly different, "stronger" function spaces. Namely, for each fixed N, we introduce

$$(V_F^\eta)_N^{n+1}(\Omega^{\eta^n_N}) = \{(\boldsymbol{u}, v) \in V_F(\Omega^{\eta^n_N}) \times H^{1/2}(\Gamma) : \boldsymbol{u}_{|\Gamma^{\eta^n_N}} = v\}, \tag{5.71}$$

which is a solution space that utilizes the fact that our weak solutions have the trace of the fluid velocity not only in $L^2(\Gamma)$ but also in $H^{1/2}(\Gamma)$. Similarly, we introduce the test space

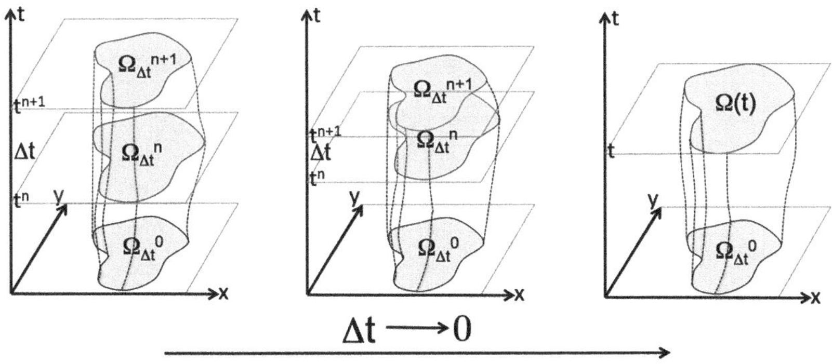

Fig. 5.2 A sketch of moving domains $\Omega_N^{\eta^n}$ as they change with $\Delta t \to 0$, i.e., with $N \to \infty$

$$(Q^{\eta})^n_N(\Omega^{\eta^n_N}) = \{(\boldsymbol{q}, \psi) \in \left(V_F(\Omega^{\eta^n_N}) \cap H^3(\Omega^{\eta^n_N})\right) \times H^2_0(\Gamma) : \boldsymbol{q}_{|\Gamma^{\eta^n_N}} = \psi\} \tag{5.72}$$

by requiring that the fluid velocity test functions additionally belong to $H^3(\Omega^{\eta^n_N})$. The resulting test space is dense in $(Q^{\eta})^n_N(\Omega^{\eta^n_N})$. This additional regularity simplifies some calculations in the compactness proof (although it is not optimal in the sense that weaker assumptions could be used in exchange for a slightly more complicated estimates).

We will now apply the generalized compactness result Theorem 2.10.1 in Chap. 2 to the sequences of approximate fluid and structure velocities. We first define the overarching Hilbert spaces H and V, which do not depend on N, and satisfy the properties listed in Theorem 2.10.1. For this purpose, we recall the definition of the maximal fluid domain Ω_M and consider the fluid velocity functions $\boldsymbol{u}^n_N$ extended to Ω_M by zero. By Lemma 5.7.4, the functions $(\boldsymbol{u}^n_N, v^n_N)$ belong to the space $H^s(\Omega^M) \times H^s(\Gamma)$, $0 < s < 1/2$. Keeping this result in mind, we define the spaces H and V of functions $(\boldsymbol{u}^n_N, v^n_N)$ as follows:

$$H = L^2(\Omega^M) \times L^2(\Gamma), \quad V = H^s(\Omega^M) \times H^s(\Gamma), 0 < s < 1/2. \tag{5.73}$$

Obviously, H and V are Hilbert spaces and $V \subset\subset H$.

Furthermore, let V^n_N and Q^n_N from Theorem 2.10.1 be given by the spaces (5.71) and (5.72), respectively:

$$V^n_N := (V^{\eta}_F)^{n+1}_N(\Omega^{\eta^n_N}), \quad Q^n_N := (Q^{\eta})^{n+1}_N(\Omega^{\eta^n_N}).$$

Then, from inequality (5.70), we see that V^n_N is continuously embedded in V, uniformly in N and n. Furthermore, $V^n_N \subset\subset \overline{Q^n_N}^H \hookrightarrow (Q^n_N)'$.

Now we recall the definition of the velocity functions $(\boldsymbol{u}_N, v_N) \subset L^2(0, T; H)$, which depend on both time and space and which are piecewise constant in t:

$$\left.\begin{aligned} \boldsymbol{u}_N &= \boldsymbol{u}^n_N \\ v_N &= v^n_N \end{aligned}\right\} \quad \text{on } ((n-1)\Delta t, n\Delta t],\ n = 1, \ldots, N,$$

as well as the corresponding time-shifts,

$$\tau_h \boldsymbol{u}_N(t, .) = \boldsymbol{u}_N(t - h, .),\ h \in \mathbb{R}.$$

Theorem 5.7.1 (Compactness of $(\boldsymbol{u}_N, v_N)_{N\in\mathbb{N}}$) *The sequence $(\boldsymbol{u}_N, v_N)_{N\in\mathbb{N}}$, satisfying the weak formulation* (5.53) *and energy estimates from Lemma 5.4.1, is relatively compact in $L^2(0, T; H)$, where $H = L^2(\Omega^M) \times L^2(\Gamma)$.*

Proof of Proposition 5.7.2 We would like to show that assumptions (A)–(C) from Theorem 2.10.1 hold true.

Property (A): Strong Bounds. The strong bounds (A1) and (A2) follow directly from the uniform energy bounds in Lemma 5.4.1 and from Lemma 5.7.4. As stated before, property (A3) is not needed for the proof.

Property (B): The Time Derivative Bound. We want to estimate the following norm:

$$\|P_N^n \frac{\boldsymbol{u}_N^{n+1}-\boldsymbol{u}_N^n}{\Delta t}\|_{(Q_N^n)'} = \sup_{\|(\boldsymbol{q},\psi)\|_{Q_N^n}=1} \Big| \int_{\Omega^{\eta_N^n}} \frac{\boldsymbol{u}_N^{n+1}-\boldsymbol{u}_N^n}{\Delta t}\cdot \boldsymbol{q} + \int_0^L \frac{v_N^{n+1}-v_N^n}{\Delta t}\psi \Big|.$$

For this purpose, we add and subtract the function $\tilde{\boldsymbol{u}}_N^n$, which is defined in (5.52):

$$\begin{aligned} &\Big| \int_{\Omega^{\eta_N^n}} \frac{\boldsymbol{u}_N^{n+1}-\boldsymbol{u}_N^n \pm \tilde{\boldsymbol{u}}_N^n}{\Delta t}\cdot \boldsymbol{q} + \int_0^L \frac{v_N^{n+1}-v_N^n}{\Delta t}\psi \Big| \\ &\le \Big| \int_{\Omega^{\eta_N^n}} \frac{\boldsymbol{u}_N^{n+1}-\tilde{\boldsymbol{u}}_N^n}{\Delta t}\cdot \boldsymbol{q} + \int_0^L \frac{v_N^{n+1}-v_N^n}{\Delta t}\psi \Big| + \Big| \int_{\Omega^{\eta_N^n}} \frac{\tilde{\boldsymbol{u}}_N^n-\boldsymbol{u}_N^n}{\Delta t}\cdot \boldsymbol{q} \Big|, \end{aligned} \tag{5.74}$$

and estimate the two terms on the right-hand side.

The first term is estimated by using the weak form of the problem given by Eq. (5.53):

$$\Big| \int_{\Omega^{\eta_N^n}} \frac{\boldsymbol{u}_N^{n+1}-\hat{\boldsymbol{u}}_N^n}{\Delta t}\cdot \boldsymbol{q} + \int_0^L \frac{v_N^{n+1}-v_N^n}{\Delta t}\psi \Big|$$

$$\le C\frac{R+M}{R+m}\|\nabla \boldsymbol{q}\|_{L^\infty}(\|v_N^{n+1/2}\|_{L^2}+\|\boldsymbol{u}_N^n\|_{L^2})\|\nabla \boldsymbol{u}_N^{n+1}\|_{L^2}+C_1\|\nabla \boldsymbol{u}_N^{n+1}\|_{L^2}\|\nabla \boldsymbol{q}\|_{L^2}$$

$$+\|\eta_N^{n+1}\|_{H^2}\|\psi\|_{H^2}+C\|\boldsymbol{q}\|_{H^1} \le C(\|\nabla \boldsymbol{u}_N^{n+1}\|_{L^2}+\|\eta_N^{n+1}\|_{H^2}+1)\|(\boldsymbol{q},\psi)\|_{(Q_N^{n,l})}.$$

Here we used the energy estimates from Lemma 5.4.1 from where we concluded that $\|\boldsymbol{u}_N^n\|_{L^2}$, $\|v_N^n\|_{L^2}$ are uniformly bounded by C. □

Remark 5.7.2 Notice how the choice of the space Q_N^n, which includes high regularity Sobolev spaces, is useful in the last inequality to provide the upper bound in terms of $\|(\boldsymbol{q},\psi)\|_{(Q_N^n)}$.

To estimate the second term, we first notice that function $\tilde{\boldsymbol{u}}_N^n$ is nonzero on $\Omega^{\eta_N^n}$, while function $\boldsymbol{u}_N^n$ is nonzero on $\Omega^{\eta_N^{n-1}}$. This is why we introduce $A=\Omega^{\eta_N^{n-1}}\cap\Omega^{\eta^n}$, $B_1=\Omega^{\eta^n}\setminus\Omega^{\eta_N^{n-1}}$, and $B_2=\Omega^{\eta_N^{n-1}}\setminus\Omega^{\eta^n}$ (see Fig. 5.3) and estimate the integrals over A, B_1, and B_2 separately:

$$\begin{aligned} \Big|\int_A(\tilde{\mathbf{u}}_N^n-\boldsymbol{u}_N^n)\cdot\boldsymbol{q}\Big| =& \Big|\int_A\Big(\boldsymbol{u}_N^n(z,r)-\boldsymbol{u}_N^n(z,\frac{R+\eta_N^n}{R+\eta_N^{n-1}}r)\Big)\cdot\boldsymbol{q}(z,r)dzdr\Big| \\ \le& C\Delta t\|v^{n-1/2}\|_{L^2}\|\nabla\boldsymbol{u}_N^n\|_{L^2(A)}\|\boldsymbol{q}\|_{L^\infty(A)}. \end{aligned}$$

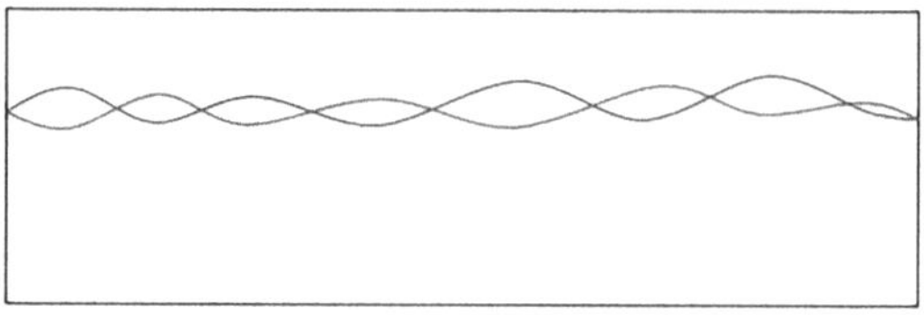

Fig. 5.3 The 2D fluid domains at time steps t_n and t_{n+1}

To estimate the integral over B_1, we use the fact that $\boldsymbol{u}^n_N = 0$ on B_1 to obtain:

$$
\begin{aligned}
|\int_{B_1} (\tilde{\mathbf{u}}^n_N - \boldsymbol{u}^n_N)\cdot \boldsymbol{q}| \leq& |\int_0^L dz \int_{R+\eta_N^{n-1}(z)}^{R+\eta_N^n(z)} (\tilde{\mathbf{u}}^n_N - \boldsymbol{u}^n_N)(z,r)\cdot \boldsymbol{q} dr| \\
=& \int_0^L dz \int_{R+\eta_N^{n-1}(z)}^{R+\eta_N^n(z)} \tilde{\mathbf{u}}^n_N(z,r)\cdot \boldsymbol{q} dr| \\
\leq& \|\boldsymbol{q}\|_{L^\infty} \int_0^L \max_r(\tilde{\boldsymbol{u}}^n_N(z,r)) dz \int_{R+\eta_N^{n-i}(z)}^{R+\eta_N^n(z)} dr \\
\leq& C\|\boldsymbol{q}\|_{L^\infty} \int_0^L \|\partial_r u^n_{N,r}(.,z)\|_{L^2_r} |\Delta t v^{n-1/2}(z)| dz \\
\leq& C\Delta t \|\boldsymbol{q}\|_{L^\infty} \|\nabla \boldsymbol{u}^n_N\|_{L^2(\Omega^n_N)}.
\end{aligned}
$$

Here, we used the fact that for $f \in H^1(0,1)$, we can estimate $\|f\|_{L^\infty} \leq C\|f'\|_{L^2}$, and applied this to the function $\mathbf{u}^n_N$ above, viewed as a function of only one variable, r, satisfying the condition that the r-component $u_r(z,0) = 0$ is equal to zero at $r = 0$, and the z-component u_z is equal to zero on the boundary $r = R + \eta_N^{n-1}(z)$. This is a formal estimate that can be justified by density arguments.

The integral over B_2 is equal to zero.

Property (C): Smooth Dependence of Function Spaces on Time. We want to define common function spaces for test functions and for approximate solutions, which will help with estimating the time shifts. For this purpose, we recall that our scheme is designed in such a way that the functions $\boldsymbol{u}^{n+1}_N$ are defined on the "previous" domain $\Omega^{\eta^n_N}$.

Property C1: The Common Test Space. Consider the maximal and minimal domains $\Omega^{M_N^{n,l-1}}$ and $\Omega^{m_N^{n,l-1}}$ determined by the maximum and minimum functions $M_N^{n,l-1}$ and $m_N^{n,l-1}$ given by Lemma 5.7.3. We define the "common" test space required by the general property *(C1)* to be the space consisting of all the (smooth) functions $(\boldsymbol{q},\psi)$ such that $\boldsymbol{q}$ is defined on $\Omega^{m_N^{n,l-1}}$ and then extended to the maximal domain $\Omega^{M_N^{n,l-1}}$ by the trace $\psi\mathbf{e}_r$ (which is constant in the $\mathbf{e}_r$ direction):

$$
\begin{aligned}
Q_N^{n,l} =& \{(\boldsymbol{q},\psi) \in \left(V_F(\Omega^{M_N^{n,l-1}}) \cap H^3(\Omega^{M_N^{n,l-1}})\right) \times H^3(\Gamma) : \boldsymbol{q} \\
=& \psi\mathbf{e}_r \text{ on } \Omega^{M_N^{n,l-1}} \setminus \Omega^{m_N^{n,l-1}}\}.
\end{aligned}
$$

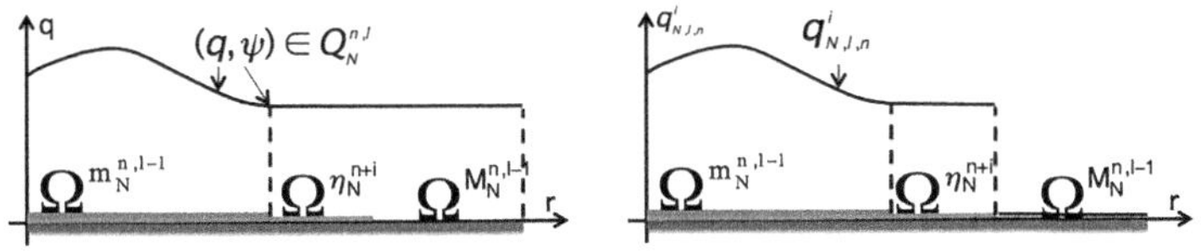

Fig. 5.4 A sketch of the test functions $\boldsymbol{q} \in Q_N^{n,l}$ (left) and $\boldsymbol{q}^i_{N,l,n}$ (right) as functions of r only

See Fig. 5.4.

Remark 5.7.3 Notice that $Q_N^{n,l} \subset Q_N^{n+i}$, for all $i = 0, \ldots, l$, and the functions from $Q_N^{n,l}$ are admissible for problem (5.53) for all domains $\Omega^{\eta_N^{n+i}}$, $i = 0, \ldots, l$. In particular, they are divergence-free for all $i = 0, \ldots, l$.

We can now define the common test space for all the time shifts by $i\Delta t$, where $i = 0, \ldots, l$, in the following way. First notice that the test functions from $Q_N^{n,l}$ are nonzero on $\Omega_N^{M^{n,l}}$. Since $\Omega_N^{M^{n,l}} \supset \Omega_N^{\eta_N^{n+i}}$, we can define the test functions belonging to Q_N^{n+i} to be the **restrictions** onto $\Omega^{\eta_N^{n+i}}$ of the test functions from $Q_N^{n,l}$. More precisely, we define

$$\boldsymbol{q}^i_{N,l,n} = \begin{cases} \boldsymbol{q}, & \text{in } \Omega^{\eta_N^{n+i}} \\ 0, & \text{elsewhere} \end{cases}, \quad \boldsymbol{q} \in Q_N^{n,l}. \tag{5.75}$$

Now, the mappings $J^i_{N,l,n} : Q_N^{n,l} \to Q_N^{n+i}$ are defined by

$$J^i_{N,l,n}(\boldsymbol{q}, \psi) = (\boldsymbol{q}^i_{N,l,n}, \psi).$$

Indeed, from the definition of the test space $Q_N^{n,l}$, we can see that $J^i_{N,l,n}(\boldsymbol{q}, \psi) \in Q_N^{n+i}$. Moreover, we immediately see that

$$\|J^i_{N,l,n}\boldsymbol{q}\|_{Q_N^{n+i}} \leq C\|\boldsymbol{q}\|_{Q_N^{n,l}}.$$

To check that inequality (2.21) holds, we use Lemma 5.7.3 and compute

$$\Big((J^{j+1}_{N,l,n}(\boldsymbol{q}, \psi) - J^j_{N,l,n}(\boldsymbol{q}, \psi)), (\boldsymbol{u}_N^{n+j+1}, v_N^{n+j+1})\Big)_H$$

$$= \Big|\int_0^L \int_{\eta_N^{n+j}(z)}^{\eta_N^{n+j+1}(z)} \psi(z)\boldsymbol{u}_N^{n+j+1}(z, r)drdz\Big|$$

$$\leq \|\psi\|_{L^\infty}\|\nabla \boldsymbol{u}_N^{n+j+1}\|_{L^2}\|\eta_N^{n+j+1} - \eta_N^{n+j}\|_{L^1(\Gamma)}$$

$$\leq C\|(\boldsymbol{q}, \psi)\|_{Q_N^{n,l}}\|\boldsymbol{u}_N^{n+j+1}\|_{V_N^{n+j+1}}N, \quad j \in \{0, \ldots, l-1\}.$$

The penultimate inequality is proved analogously as in the proof of property B.

To check that inequality (2.22) holds, we recall that $J^i_{N,l,n}\boldsymbol{q}$ and $\boldsymbol{q}$ differ only in the region $\Omega^M \setminus \Omega^{\eta_N^{n+i}}$, and so the H-norm difference between the two functions can be bounded by the $Q_N^{n,l}$-norm of $\boldsymbol{q}$ and the difference between $M_N^{n,l}$ and $m_N^{n,l}$, which is bounded by $C\sqrt{l\Delta t}$ according to Lemma 5.7.3. This completes the construction of a common test space for the time shifts, as specified by Property C1.

Property C2: Approximation Property of Solution Spaces. We would like to be able to compare and estimate the time shifts by h of the fluid velocity function $\boldsymbol{u}$, given at time $t+h$, with the function $\boldsymbol{u}$ given at time t. Again, as we mentioned earlier, the time-shift and the function itself are defined on different physical domains, since they also depend on time. We would like to define a common solution space on which we can make the function comparisons, where the common function space needs to be constructed in such a way that the functions from that space approximate well the original functions defined on different domains. The common space will be defined on a domain that is contained in all the other domains (we can do this because our approximate domains are close). The functions corresponding to the time-shifts are going to be defined on this common domain by a "squeezing" procedure, also used in [81], and then extended to the largest domain by the trace on the boundary in a way that keeps the divergence-free condition satisfied. Keeping in mind that $h = l\Delta t$, we introduce the space $V_N^{n,l}$ to be the closure of $Q_N^{n,l}$ in V. We would like to define the mappings $I^i_{N,l,n}$:

$$I^i_{N,l,n} : V_N^{n+i} \to V_N^{n,l},\ i = 0, 1, \ldots, l, \text{ where } V_N^{n,l} = \overline{Q_N^{n,l}}^V,$$

to approximate the functions in V_N^{n+i}, by the functions from $V_N^{n,l}$, which are all defined on the same domain. This will be done via a "squeezing" operator, defined below (see also [81]), and then extending the squeezed functions onto the larger, common domain, by the trace v on the boundary.

Definition 5.7.2 (The Squeezing Operator) Let η_m, η, and η_M be three functions defined on $[0, L]$ such that $-R < \eta_m(z) \le \eta(z) \le \eta_M(z)$, $z \in [0, L]$, so that $R + \eta_m$ defines Ω^{η_m} with $R + \eta_m > 0$. Let $\boldsymbol{u}$ be a divergence-free function defined on Ω^η such that $\boldsymbol{u} = v\mathbf{e}_r$ on Γ^η. For a given $\sigma \ge 1$, such that $\sigma(R + \eta_m) \ge R + \eta$, we define $\boldsymbol{u}_\sigma \in H^1(\Omega^{\eta_M})$ in the following way:

$$\boldsymbol{u}_\sigma(z, r) = \begin{cases} (\sigma u_z(z, \sigma r), u_r(z, \sigma r)), & \sigma r \le R + \eta(z), \\ v\mathbf{e}_r, & \text{elsewhere}. \end{cases} \tag{5.76}$$

This "squeezing" operator associates with any function $\boldsymbol{u}$ defined on Ω^η a function $\boldsymbol{u}_\sigma$ defined on the large domain Ω^{η_M}, containing all the important information about the original function $\boldsymbol{u}$ squeezed within the minimal domain Ω^{η_m}. See Fig. 5.5. The operator is designed by (1) first squeezing the function $\boldsymbol{u}$ from domain Ω^η into Ω^{η_m} and rescaling the function u_z by σ so that the divergence-free condition remains

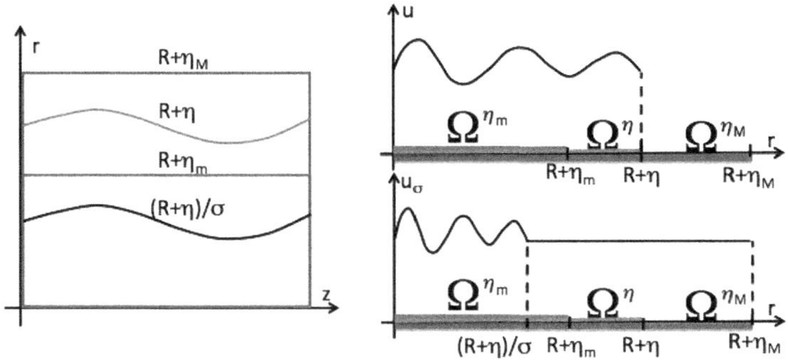

Fig. 5.5 The squeezing operator. Left: a sketch of the fluid domains. Right: a sketch of an r-cut of $\mathbf{u}$ and $\mathbf{u}_\sigma$

to be satisfied and (2) extending the squeezed $\boldsymbol{u}$ to the remainder of the maximal domain Ω^{η_M} by the values of its trace on η, where the extension is constant in the $\mathbf{e}_r$ direction.

Remark 5.7.4 Notice that $\nabla \cdot \boldsymbol{u} = 0$ implies that also $\nabla \cdot \boldsymbol{u}_\sigma = 0$. Moreover, if $(\boldsymbol{u}, v) \in H^s(\Omega^\eta) \times H^s(0, L)$, then $\boldsymbol{u}_\sigma \in H^s(\Omega^{\eta_M})$, $0 \le s \le 1$.

We now define the mappings $I^i_{N,l,n} : V_N^{n+i} \to V_N^{n,l}$ by

$$I^i_{N,l,n} : \boldsymbol{u}_N^{n+i} \mapsto (\boldsymbol{u}_N^{n+i})_{\sigma^i_{n,l}} \in V_N^{n,l}, \tag{5.77}$$

where $(\boldsymbol{u}_N^{n+i})_{\sigma^i_{n,l}}$ is obtained from Definition 5.7.2 by setting the following:

- $\eta_M = M_N^{n-1,l+1}$, $\eta = \eta_N^{n+i-1}$, $\eta_m = m_N^{n,l}$
- $\sigma^i_{n,l} > 0$ is such that $\sigma^i_{n,l} \ge \max\limits_{z\in[0,L]} \dfrac{R + M^{n-1,l+1}(z)}{R + \eta_N^{n+i-1}(z)}$.

Thus, the mappings $I^i_{N,l,n}$ take the functions $\boldsymbol{u}_N^{n+i}$ defined by the time-shifts $i\Delta t$, $i = 0, \ldots, l$, and associate with those functions the functions in the common velocity space $V_N^{n,l}$ that are all defined on the same domain $\Omega^{M_N^{n,l}}$ and are obtained by squeezing the functions into a domain that is contained in the minimal domain $\Omega^{m_N^{n,l}}$, and then extending them to $\Omega^{M_N^{n,l}}$ by their trace v_N^{n+1} on the moving boundary.

We need to prove that there exists a universal constant $C > 0$ such that

$$\|I^i_{N,l,n}(\boldsymbol{u}_N^{n+i})\|_{V_N^{n,l}} \le C\|\boldsymbol{u}_N^{n+i}\|_{V_N^{n+i}}, \quad i = 0, \ldots, l \tag{5.78}$$

and a universal, monotonically increasing function g, which converges to 0 as $h \to 0$, where $h = l\Delta t$, such that

$$\|I^i_{N,l,n}(\boldsymbol{u}_N^{n+i}) - \boldsymbol{u}_N^{n+i}\|_{L^2(\Omega^M)} \leq g(l\Delta t)\|\boldsymbol{u}_N^{n+i}\|_{V_N^{n+i}}, \quad i = 0, \ldots, l. \tag{5.79}$$

The first inequality follows from Lemma 2.7 in [257] and the fact that the squeezed functions remain in H^1, with the uniformly bounded norm provided by the energy estimates in Lemma 5.4.1.

To show (5.79), we first prove the following lemma, which states that the difference between $\boldsymbol{u}$ and $\boldsymbol{u}_\sigma$ on Ω^M can be estimated by the L^2-norms of the gradient of $\boldsymbol{u}$ on Ω^η and the L^2-norm of the velocity on the boundary, where the constant in the estimate depends only on the Lipschitz constant of the fluid domain boundary.

Lemma 5.7.5 *Let η_m, η, and η_M be as in Definition 5.7.2, $\boldsymbol{u} \in H^1(\Omega^\eta)$, $\boldsymbol{u} = v\mathbf{e}_2$ on Γ^η, and let σ be such that $\sigma \geq \max_{z\in[0,L]} \dfrac{R+\eta_M(z)}{R+\eta(z)}$. Then the following estimate holds:*

$$\|\boldsymbol{u} - \boldsymbol{u}_\sigma\|_{L^2(\Omega^{\eta_M})} \leq C\sqrt{\sigma - 1}(\|\nabla \boldsymbol{u}\|_{L^2(\Omega^\eta)} + \|v\|_{L^2(0,L)}),$$

where C depends only on $\|R + \eta(z))\|_{L^\infty}$.

Proof of Proposition 5.7.2 We will prove this estimate in two steps. First, we obtain an $L^2(\Omega^{\eta_M})$ estimate of the difference between $\boldsymbol{u}$ and a slightly modified function $\tilde{\boldsymbol{u}}_\sigma$, which is *not divergence-free*, defined by

$$\tilde{\boldsymbol{u}}_\sigma(z, r) = \begin{cases} (u_z(z, \sigma r), u_r(z, \sigma r)), & \sigma r \leq R + \eta(z), \\ v\mathbf{e}_r, & \text{elsewhere}, \end{cases},$$

and then estimate the difference between $\tilde{\boldsymbol{u}}_\sigma$ and $\boldsymbol{u}_\sigma$. Notice that the only difference between $\tilde{\boldsymbol{u}}_\sigma$ and $\boldsymbol{u}_\sigma$ is the factor σ in front of u_z.

The L^2-norm of the difference between $\boldsymbol{u}$ and $\tilde{\boldsymbol{u}}_\sigma$ can be broken into three parts:

$$\begin{aligned}\int_{\Omega^{\eta_M}} |\boldsymbol{u}(z, r) - \tilde{\boldsymbol{u}}_\sigma(z, r)|^2 drdz &= \int_0^L \int_0^{\frac{R+\eta(z)}{\sigma}} |\boldsymbol{u}(z, r) - \tilde{\boldsymbol{u}}_\sigma(z, r)|^2 drdz \\ + \int_0^L \int_{\frac{R+\eta(z)}{\sigma}}^{R+\eta} |\boldsymbol{u}(z, r) - \tilde{\boldsymbol{u}}_\sigma(z, r)|^2 drdz &+ \int_0^L \int_{R+\eta}^{R+\eta_M} |\boldsymbol{u}(z, r) - \tilde{\boldsymbol{u}}_\sigma(z, r)|^2 drdz \\ &= I_1 + I_2 + I_3.\end{aligned}$$

The first part, I_1, can be estimated as follows:

$$\begin{aligned} I_1 &= \int_0^L \int_0^{\frac{R+\eta(z)}{\sigma}} |\boldsymbol{u}(z, r) - \boldsymbol{u}(z, \sigma r)|^2 drdz \\ &= \int_0^L \int_0^{\frac{R+\eta(z)}{\sigma}} |\partial_r \boldsymbol{u}(z, \xi)|^2 |r(1 - \sigma)|^2 drdz \quad \leq C\|\nabla \boldsymbol{u}\|^2_{L^2(\Omega^{\eta_m})}(1 - \sigma)^2. \end{aligned}$$

Notice that the constant C depends on the size of Ω^{η_m}, which can be estimated from above by the size of the maximal domain.

To estimate I_2, we recall that $\tilde{\boldsymbol{u}}_\sigma$ is defined to be $v\mathbf{e}_r$ outside the squeezed domain bounded by $(R+\eta(z))/\sigma$, which is exactly the trace of $\boldsymbol{u}$ on $R+\eta(z)$. Therefore, we get

$$I_2 = \int_0^L \int_{\frac{R+\eta}{\sigma}}^{R+\eta} |\mathbf{u}(z,r) - \boldsymbol{u}(z, R+\eta(z))|^2 dr\, dz$$

$$\leq \int_0^L \int_{\frac{R+\eta}{\sigma}}^{R+\eta} |\partial_r \boldsymbol{u}(z,\xi)|^2 (R+\eta(z))^2 (1-\frac{1}{\sigma})^2 \leq C(\sigma-1)^2 \|\nabla \boldsymbol{u}\|^2_{L^2(\Omega^{\eta_m})}.$$

Finally, to estimate I_3, we recall that $\boldsymbol{u}$ is extended by zero outside Ω^η, and that, as before, $\tilde{\boldsymbol{u}}_\sigma$ is equal to $v\mathbf{e}_r$ in the same region. Therefore, we obtain

$$I_3 = \int_0^L \int_{R+\eta}^{R+\eta_M} |v(z)|^2 dr dz \leq \int_0^L |v(z)|^2 (\eta_M - \eta)$$
$$\leq \|v\|^2_{L^2(0,L)} (\sigma - 1) \|R+\eta(z))\|_{L^\infty},$$

where the last inequality holds due to the assumption $\sigma \geq \max_{z\in[0,L]} \dfrac{R+\eta_M(z)}{R+\eta(z)}$.

In order to complete the proof and take into account the multiplication of u_z by σ, we use the triangle inequality:

$$\|u_z - (u_\sigma)_z\|_{L^2} \leq \|u_z - (u_\sigma)_z \pm (\tilde{u}_\sigma)_z\|_{L^2} = \|u_z - (\tilde{u}_\sigma)_z\|_{L^2} + \|u_z\|_{L^2}(\sigma - 1)$$
$$\leq C\sqrt{\sigma - 1}(\|\nabla \boldsymbol{u}\|_{L^2} + \|v\|_{L^2}).$$

This proves the lemma. □

To show (5.79), we use this lemma to get

$$\|I^i_{N,n,l}\boldsymbol{u}^{n+i}_N - \boldsymbol{u}^{n+i}_N\|_{L^2(\Omega^{\eta_M})} \leq C\sqrt{\sigma^i_{n,l} - 1}\|\boldsymbol{u}^{n+i}_N\|_{V^{n+i}_N} \leq C\sqrt{\sigma_{n,l} - 1}\|\boldsymbol{u}^{n+i}_N\|_{V^{n+i}_N},$$

where

$$\sigma_{n,l} := \max_{z\in[0,L]} \frac{R + M^{n-1,l+1}_N}{R + m^{n,l}_N}.$$

It is now crucial to show that we can estimate $(\sigma_{n,l} - 1)$ appearing on the right-hand side of the above inequality by a function $g(h)$ such that $g \to 0$ as $h \to 0$, where $h = l\Delta t$. Indeed, we have

$$\sigma_{n,l} - 1 = \frac{R + M_N^{n-1,l+1}}{R + m_N^{n,l}} - \frac{R + m_N^{n,l}}{R + m_N^{n,l}} = \frac{M_N^{n-1,l+1} - m_N^{n,l}}{R + m_N^{n,l}}$$

$$\leq C(M_N^{n-1,l+1} - m_N^{n,l}) \leq C\sqrt{l\Delta t},$$

where the last inequality follows from Lemma 5.7.3. Thus, we have shown that $g(h) = Ch^{1/4}$, and that the following density results hold.

Corollary 5.7.1 (Density) *Let $n, l, N \in \mathbb{N}$ be such that $n + l \leq N$, $h = l\Delta t$ and $i \in \{0, \ldots, l\}$. Let $(\boldsymbol{u}, v) \in V_N^{n+i}$. Then there exist a constant C (independent of N, n, and l) and the operators $I_{N,n,l}^i$ defined by* (5.77) *such that $I_{N,n,l}^i(\boldsymbol{u}, v) \in V_N^{n,l}$ and*

$$\|I_{N,n,l}^i \boldsymbol{u} - \boldsymbol{u}\|_{L^2} \leq Ch^{1/4}\|\boldsymbol{u}_N^{n+i}\|_{V_N^{n+i}}, \ i \in \{0, \ldots, l\}.$$

Property C3: Uniform Ehrling Property. We need to prove the uniform Ehrling property, stated in (2.25). The main difficulty comes, again, from the fact that we have to work with moving domains, which are parameterized by N, n and l.

Remark 5.7.5 (Notation) To show that the uniform Ehrling estimate (2.25) holds, independently of all three parameters, we simplify the notation (only in this proof) and replace the indices N, n, l with only one index, n, so that, e.g., the minimum and maximum functions $m_N^{n,l}$, $M_N^{n,l}$ are re-enumerated as M_n, m_n, with the corresponding maximal fluid domains Ω^{M_n} and the function spaces H_n, V_n, and Q_n' all defined on Ω^{M_n}.

We prove the uniform Ehrling property by contradiction and by using the "uniform squeezing property" (see Lemma 5.7 in [275]; for the classical Ehrling property, see Lemma 2.9.1 in Chap. 2). We start by assuming that the statement of the uniform Ehrling property (2.25) is false. More precisely, we assume that there exist a $\delta_0 > 0$ and a sequence $(\mathbf{f}_n, g_n) \in H_n$ such that

$$\|(\mathbf{f}_n, g_n)\|_H = \|(\mathbf{f}_n, g_n)\|_{H_n} > \delta_0\|(\mathbf{f}_n, g_n)\|_{V_n} + n\|(\mathbf{f}_n, g_n)\|_{Q_n'}.$$

Here, as before, we have extended the functions f_n onto the entire domain Ω^M, which is determined by the maximal function $M(z)$ defined in Lemma 5.7.3, so that $\|(\mathbf{f}_n, g_n)\|_H = \|(\mathbf{f}_n, g_n)\|_{H_n}$. Recall that H is defined as the L^2 product space, so extensions by 0 do not change the norm.

It will be convenient to also replace the V_n norm on the right-hand side by the norm on V. Here, however, since V is defined as the H^s product space, the norm on V_n is bounded from below by a constant times the norm on V; see Lemma 5.7.4. Thus, we have

$$\|(\mathbf{f}_n, g_n)\|_H > \delta_0\|(\mathbf{f}_n, g_n)\|_{V_n} + n\|(\mathbf{f}_n, g_n)\|_{Q_n'} \geq C\delta_0\|(\mathbf{f}_n, g_n)\|_V + n\|(\mathbf{f}_n, g_n)\|_{Q_n'}.$$

Without the loss of generality, we can assume that our sequence $(\mathbf{f}_n, g_n)$ is such that $\|(\mathbf{f}_n, g_n)\|_H = 1$. For example, we could consider $\frac{1}{\|(\mathbf{f}_n, g_n)\|_H}(\mathbf{f}_n, g_n)$ instead of $(\mathbf{f}_n, g_n)$. Notice that we now have a sequence such that the two terms on the right-hand side are uniformly bounded in n, which implies that there exists a subsequence, which we again denote by $(\mathbf{f}_n, g_n)$, such that

$$\|(\mathbf{f}_n, g_n)\|_H = 1, \quad \|(\mathbf{f}_n, g_n)\|_V \le \frac{1}{C\delta_0}, \quad \|(\mathbf{f}_n, g_n)\|_{Q_n'} \to 0. \tag{5.80}$$

Since $(\mathbf{f}_n, g_n)$ is uniformly bounded in V, and by the compactness of the embedding of V into H, we conclude that there exists a subsequence $(\mathbf{f}_{n_k}, g_{n_k}) \to (\mathbf{f}, g)$ strongly in H.

Now, from the energy estimates, we recall that sequences $\{M_n\}$, $\{m_n\}$ are uniformly bounded in $H^2(\Gamma)$, and so by the compactness of the embedding of $H^2(\Gamma)$ into $C(\bar{\Gamma})$, there exist subsequences $M_{n_k} \to \tilde{M}$, $m_{n_k} \to \tilde{m}$. With a slight abuse of notation, we denote the convergent subsequences by index n. Therefore, we are now working with the convergent subsequences:

$$M_n \to \tilde{M},\ m_n \to \tilde{m},\ (\mathbf{f}_n, g_n) \to (\mathbf{f}, g) \text{ in } H.$$

These functions define the maximal domain $\Omega^{\tilde{M}}$, the minimal domain $\Omega^{\tilde{m}}$, and the corresponding minimal and maximal domains that depend on n: Ω^{m_n} and Ω^{M_n}.

Let $\tilde{S}$ be the strip between the minimal and maximal domains $\Omega^{\tilde{m}}$ and $\Omega^{\tilde{M}}$, and S_n be the strip between the minimal and maximal domains that depend on n:

$$S_n = \Omega^{M_n} \setminus \Omega^{m_n}, \quad \tilde{S} = \Omega^{\tilde{M}} \setminus \Omega^{\tilde{m}}.$$

We now want to show that $(\mathbf{f}, g) = 0$ first outside $\Omega^{\tilde{M}}$, i.e., in $\Omega^M \setminus \Omega^{\tilde{M}}$, then in $\Omega^{\tilde{M}}$. (Recall that Ω^M was defined to contain all the approximate domains.) This will contradict the assumption that $\|(\mathbf{f}_n, g_n)\|_H = 1$ and complete the proof.

First, we show that $\mathbf{f} = 0$ outside of $\Omega^{\tilde{M}}$, and inside Ω^M. This will follow simply because $\mathbf{f}$ is the limit of a sequence of functions that are zero outside Ω^{M_n}, which converge to $\Omega^{\tilde{M}}$. More precisely, we introduce the characteristic functions $\chi_{\Omega^{M_n}}$ of the sets Ω^{M_n} and recall that $\chi_{\Omega^{M_n}}$ converge uniformly to $\chi_{\Omega^{\tilde{M}}}$. Therefore,

$$\mathbf{f}(1 - \chi_{\Omega^{\tilde{M}}}) = \lim_n \mathbf{f}_n(1 - \chi_{\Omega^{M_n}}) = 0,$$

since $\mathbf{f}_n$ are extended by 0 to Ω^M.

Next, we show that $\mathbf{f} = 0$ inside $\Omega^{\tilde{M}}$. We start by showing that $\mathbf{f} = g\mathbf{e}_r$ in $\tilde{S}$. This follows immediately from

$$(\mathbf{f} - g\mathbf{e}_r)\chi_{\tilde{S}} = \lim_n (\mathbf{f}_n - g_n\mathbf{e}_r)\chi_{S_n} = 0,$$

because $(\mathbf{f}_n, g_n) \in V_n = \overline{Q_n}^V$, which implies $\mathbf{f}_n = g_n\mathbf{e}_r$ in S_n.

We finish the proof by using $\|(\mathbf{f}_n, g_n)\|_{Q'_n} \to 0$ to show $(\mathbf{f}, g) = (\mathbf{0}, 0)$. Recall that

$$Q_n = \{(\boldsymbol{q}, \psi) \in \left(V_F(\Omega^{M_n}) \cap H^4(\Omega^{M_n})\right) \times H_0^2(\Gamma) : \boldsymbol{q} = \psi \mathbf{e}_3 \text{ on } S_n\}.$$

Let $\varepsilon > 0$. By the density of Q_n in H, and by the uniform convergence of M_n and m_n, combined with the "squeezing procedure" from the proof of Property C2, we can construct a test function $(\boldsymbol{q}, \psi)$ and $n_0 \in \mathbb{N}$ such that

$$\|(\mathbf{f}, g) - (\boldsymbol{q}, \psi)\|_H \leq \varepsilon, \quad (\boldsymbol{q}, \psi) \in Q_n, \ n \geq n_0. \tag{5.81}$$

Therefore, we have

$$\begin{aligned} \langle(\mathbf{f}, g), (\boldsymbol{q}, \psi)\rangle_H &= \lim_n \langle(\mathbf{f}_n, g_n), (\boldsymbol{q}, \psi)\rangle_H = \lim_n {}_{Q'_n}\langle(\mathbf{f}_n, g_n), (\boldsymbol{q}, \psi)\rangle_{Q_n} \\ &\leq \|(\mathbf{f}_n, g_n)\|_{Q'_n} \underbrace{\|(\boldsymbol{q}, \psi)\|_{Q_n}}_{\leq C} \to 0. \end{aligned}$$

Here, the uniform boundedness of $\|(\boldsymbol{q}, \psi)\|_{Q_n}$ follows from the uniform convergence of M_n and m_n. Hence,

$$\|(\mathbf{f}, g)\|_H^2 = \langle(\mathbf{f}, g), (\mathbf{f}, g) \pm (\boldsymbol{q}, \psi)\rangle_H = \langle(\mathbf{f}, g), (\mathbf{f} - \boldsymbol{q}, g - \psi)\rangle_H \leq \varepsilon\|(\mathbf{f}, g)\|_H.$$

Therefore, $\|(\mathbf{f}, g)\|_H < \varepsilon$. Since ε is arbitrary, $(\mathbf{f}, g) = (\mathbf{0}, 0)$, and this completes the proof of the uniform Ehrling property.

Since all the assumptions of Theorem 2.10.1 have been verified, this concludes the proof of Theorem 5.7.1. □

In contrast to the linearly coupled prototypical model, we have seen that for the nonlinearly coupled prototypical model, we had to show strong convergence of the approximate solutions, as weak and weak-star convergence are not sufficient to handle the nonlinear terms in the weak formulation. Even upon obtaining strong convergence of the approximate solutions, there are still additional difficulties in passing to the limit that arise from the time-dependent nature of the fluid domain. Those are related to the problems with test functions defined on moving domains, as we explain next.

5.7.3 Convergence of Test Functions

Recall the semidiscrete formulation for the approximate solutions, which was stated in (5.49). The approximate solutions constructed via the splitting scheme, for each N and corresponding time step Δt, satisfy the semidiscrete formulation in (5.49) for all appropriate test functions $(\boldsymbol{q}, \psi) \in Q_N^n$. Note that the test space Q_N^n depends on

n and N, however, because the test function $\boldsymbol{q} \in H^1(\Omega_f)$ must satisfy $\nabla^{\eta_N^n} \cdot \boldsymbol{q} = 0$ on Ω_f. Therefore, even though the test functions in Q_N^n are all defined on the same domain $\Omega_f \times \Gamma$, the divergence-free condition transforms differently back to the reference domain depending on η_N^n and hence, depending on n and N. The fact that the divergence-free condition for the test space transforms differently depending on n and N is a problem that directly arises from the fact that the moving domain $\Omega_f(t)$ depends on the displacement η of the structure.

Therefore, test functions that are admissible in the test space for a certain choice of n and N will not generally be admissible for all of the semidiscrete formulations for all n and N. This makes it difficult to compare the semidiscrete formulations for different values of n and N, and hence, we have to find a way to transfer test functions between the various semidiscrete formulations, in such a way that these test functions become close to each other in the limit as N goes to infinity. While it is nice to have test functions $(\boldsymbol{q}, \psi)$ that are defined on a common reference domain $\Omega_f \times \Gamma$, the divergence-free condition $\nabla^{\eta_N^n} \cdot \boldsymbol{q} = 0$ on Ω_f is a nonlinear condition that is highly nontrivial to work with. Hence, we will instead consider test functions on the physical domain, where the divergence-free condition always reads $\nabla \cdot \boldsymbol{q} = 0$ (albeit on a time-dependent domain), and then we will pull test functions back to the reference domain using the ALE mapping for the structure displacement η_N^n for a given value of n and N.

To do this, recall that we have defined a minimal and maximal domain by showing the existence of a function $m(z)$, $M(z)$ for $z \in [0, L]$, satisfying $M(0) = M(L) = 0 = m(0) = m(L)$ and

$$m(z) \leq \eta_N^n(z) \leq M(z), \quad \text{for all } z \in [0, L], \text{ positive integers } N, \text{ and}$$
$$n = 0, 1, 2, \dots, N.$$

We then defined the associated minimal and maximal fluid domain in the physical space, given by

$$\Omega_f^m = \{(z, r) \in \mathbb{R}^2 : 0 \leq z \leq L, 0 \leq r \leq R + m(z)\},$$
$$\Omega_f^M = \{(z, r) \in \mathbb{R}^2 : 0 \leq z \leq L, 0 \leq r \leq R + M(z)\}.$$

Note that for all n and N, $\Omega_f^m \subset \Omega_{f,N}^n \subset \Omega_f^M$ and furthermore, $\Omega_f^m \subset \Omega_f^\eta(t) \subset \Omega_f^M$ for all $t \in [0, T]$ by the convergence of the approximate structure displacements η_N to the limiting structure displacement η, where

$$\Omega_f^\eta(t) = \{(z, r) \in \mathbb{R}^2 : 0 \leq z \leq L, 0 \leq r \leq R + \eta(t, z)\}.$$

Recall that the test space for the problem on the fixed domain is given by $\mathcal{Q}^\eta$, defined in (5.38). Our approach to constructing suitable test functions will involve constructing an appropriate dense subset of test functions in $\mathcal{Q}^\eta$ defined on the fixed reference domain, for which we can find corresponding test functions for the semidiscrete weak formulations that converge appropriately to the limiting test function as N goes to infinity.

We emphasize that we have already extracted the limiting structure displacement η as a limit of the approximate structure displacements η_N so that η is a known function.

Before proceeding with the construction of the test functions, we make the following observation. Let $(\boldsymbol{q}, \psi) \in \mathcal{Q}^\eta$ be a test function. Recall that the test function $\boldsymbol{q}$ is defined on the fixed reference domain. However, we can transform the test function $\boldsymbol{q}$ to the physical domain using the ALE mapping $(\boldsymbol{\Phi}_f^\eta)^{-1}$. With a slight abuse of notation, we will continue to denote the transformed test function by $\boldsymbol{q}$. Moreover, we can extend $\boldsymbol{q}$ to be a divergence-free function on the entire maximal fluid domain Ω_f^M by defining $\boldsymbol{q}(t, \cdot) = \psi(t, \cdot)\boldsymbol{e}_r$ in the region $\Omega_f^M \setminus \Omega_f^\eta(t)$. In this way, we can consider the test functions $\boldsymbol{q}$ as being defined on the entire maximal fluid domain.

Moreover, we define $\boldsymbol{q}_\psi$ as follows:

$$\boldsymbol{q}_\psi(t, \cdot) = \psi(t, \cdot)\boldsymbol{e}_r \qquad \text{in } \Omega_f^M \setminus \Omega_f^m,$$

where $\boldsymbol{q}_\psi$ is a divergence-free extension of the function $\psi(t, \cdot)\boldsymbol{e}_r$ to Ω_f^m, satisfying the following boundary conditions:

$$\begin{aligned} (q_\psi)_r &= 0, \qquad && \text{on } \Gamma_{in/out}, \\ (q_\psi)_r &= 0, && \text{on } \Gamma_b. \end{aligned}$$

Notice that $(\boldsymbol{q}_\psi, \psi)$ is an admissible test function in the test space $\mathcal{Q}^\eta$. Moreover, $\boldsymbol{q} - \boldsymbol{q}_\psi$ has zero trace on $\Gamma(t)$ and can therefore be approximated by smooth divergence-free functions with compact support in $\Omega_f^\eta(t) \cup \Gamma_{in/out} \cup \Gamma_b$ (see [145, Theorem III.4.1]). This observation leads to the following definition of a dense subset $\hat{\mathcal{Q}}$ of the test function space $\mathcal{Q}^\eta$.

Definition 5.7.3 The set $\hat{\mathcal{Q}}$ consists of all test functions of the form $(\boldsymbol{q}_0 + \boldsymbol{q}_\psi, \psi)$, where:

- $\psi \in C_c^\infty([0, T); H_0^2(\Gamma))$,
- $\boldsymbol{q}_0$ is obtained by extending a smooth function $\boldsymbol{q}$, with compact support in $\Omega_f^\eta \cup \Gamma_{in/out} \cup \Gamma_b$ and satisfying $\text{div}\boldsymbol{q} = 0$, to $(0, T) \times \Omega_f^M$ by zero. Note that $\boldsymbol{q}_0$ remains divergence-free on Ω_f^M.

Now we use test functions from $\hat{\mathcal{Q}}$ to construct admissible test functions for the semidiscrete weak formulations.

Step 1. Given a function $(\boldsymbol{q}, \psi) \in \hat{\mathcal{Q}}$, we define test functions $(\tilde{\boldsymbol{q}}_N, \psi)$ for the semidiscrete weak formulations and the test function $(\tilde{\boldsymbol{q}}, \psi)$ for the limiting weak formulation by pulling the divergence-free fluid functions on the maximal fluid domain back to the reference fluid domain using the appropriate ALE mapping:

$$\tilde{q}_N(t,\cdot) = q|_{\Omega_N^{\eta_N}(t)} \circ \Phi_f^{\eta_N}(t,\cdot), \quad \tilde{q}(t,\cdot) = q|_{\Omega_f^{\eta}(t)} \circ \Phi_f^{\eta}(t,\cdot). \tag{5.82}$$

Note that $\nabla^{\eta_N} \cdot \tilde{q}_N = 0$ and $\nabla^{\eta} \cdot \tilde{q} = 0$ on Ω_f, and we also have that $\tilde{q}_N|_\Gamma = \psi e_r$ and $\tilde{q}|_\Gamma = \psi e_r$ as desired for all sufficiently large N, by the strong convergence of η_N to η in $C(0,T;C(\Gamma))$. We also remark that the test function $\tilde{q}_N$ is actually discontinuous because of the fact that η_N itself is piecewise constant on the subintervals $[n\Delta t, (n+1)\Delta t)$ and hence, (potentially) discontinuous at the endpoints of each time step.

Step 2. We then use the test functions $(\tilde{q}_N, \psi_N)$ restricted to the subintervals $[n\Delta t, (n+1)\Delta t)$ as the test functions in the semidiscrete weak formulations, and we use $(\tilde{q}, \psi)$ as the test functions for the limiting weak formulation, whenever $(q, \psi) \in \hat{\mathcal{Q}}$.

The test functions $(\tilde{q}, \psi)$ on $\Omega_f^{\eta}(t)$ that can be generated in this way are dense in the test space $\mathcal{Q}^{\eta}$.

In general, the test functions $\tilde{q}_N$ and $\tilde{q}$ are different. However, we can show the following uniform convergence result.

Lemma 5.7.6 *The test functions constructed above in **STEPs 1–2** have the following convergence properties:*

$$\tilde{q}_N \to \tilde{q} \text{ and } \nabla\tilde{q}_N \to \nabla\tilde{q} \quad \text{pointwise uniformly in } [0,T] \times \overline{\Omega}_f.$$

Proof of Proposition 5.7.2 The uniform convergence of $\tilde{\mathbf{q}}_N$ follows from the uniform convergence of η_N, and from the fact that $\mathbf{q}$ are smooth, once the mean value theorem is used to express

$$\begin{aligned} |\tilde{\mathbf{q}}_N(t,z,r) - \tilde{q}(t,z,r)| &= |q(t,z,(R+\tau_{\Delta t}\eta_N)r) - q(t,z,(R+\eta)r)| \\ &= |\partial_r q(t,z,\zeta)r|\,|\eta(t,z) - \eta_N(t-\Delta t,z)|. \end{aligned}$$

To show the uniform convergence of the gradients, one can use the chain rule to calculate

$$\begin{aligned} \partial_z\tilde{\mathbf{q}}_N(t,z,r) = \partial_z q(t,z,(R+\tau_{\Delta t}\eta_N)r) &+ [\partial_z\tau_{\Delta t}\eta_N(t,z)r] \\ &[\partial_r q(t,z,(R+\tau_{\Delta t}\eta_N)r)]\,. \end{aligned}$$

The uniform convergence of $\partial_z\tilde{\mathbf{q}}_N(t,z,r)$ follows from the uniform convergence of $\partial_z\tau_{\Delta t}\eta_N$. Combined with the first part of the proof, we get $\partial_z\tilde{\mathbf{q}}_N \to \partial_z\tilde{q}$ uniformly on $[0,T] \times \Omega_f$. The uniform convergence of $\partial_r\tilde{\mathbf{q}}_N$ can be shown in a similar way.

□

This convergence result for the test functions combined with the strong convergence results for the approximate solutions will allow us to pass to the limit in the semidiscrete weak formulations to show that the limiting fluid velocity and structure displacement $(\boldsymbol{u}, \eta)$ satisfy the limiting weak formulation.

Remark 5.7.6 Notice that the "corresponding" test functions ψ_N are independent of N on Γ and are equal to ψ there. The particular "constant" extension of ψ is crucial in avoiding the dependence of ψ_N on the regularity of η_N (via the ALE mapping).

We are now ready to pass to the limit in semidiscrete weak formulations as $N \to \infty$ to show that the limits satisfy the weak formulation of the continuous problem.

5.8 Convergence to a Weak Solution

We start by writing the weak formulation of the coupled, semidiscretized problem (5.49) by using the test functions $(\tilde{\boldsymbol{q}}_N, \psi)$ constructed above in Sect. 5.7.3, where $\tilde{\boldsymbol{q}}_N$ are defined in (5.82).

Remark 5.8.1 In this section, to simplify the notation, we will be using $\boldsymbol{q}$ in place of $\tilde{\boldsymbol{q}}$, and keep in mind that the test functions $(\tilde{\boldsymbol{q}}, \psi)$ constructed above are dense in the test space $\mathcal{Q}^\eta$.

We integrate (5.49) with respect to t from $n\Delta t$ to $(n+1)\Delta t$ and then take the sum from $n = 0, \dots, N-1$ to get the time integrals over $(0, T)$. Furthermore, we use the fact that the following is true:

$$\sum_{n=0}^{N-1} \int_{\Omega_f} (1 + \frac{\tau_{\Delta t}\, \eta_N}{R}) \frac{\boldsymbol{u}_N^{n+1} - \boldsymbol{u}_N^n}{\Delta t} \cdot \boldsymbol{q}_N = \int_0^T \int_{\Omega_f} (1 + \frac{\tau_{\Delta t}\, \eta_N}{R}) \partial_t \bar{\boldsymbol{u}}_N \cdot \boldsymbol{q}_N$$

where $\bar{\boldsymbol{u}}_N$ is the *piecewise linear extrapolation*, given in (5.55). Now, for each fixed N, i.e., fixed Δt, the **weak formulation satisfied by the approximate functions, defined on the fixed domain** Ω_f, is given by the following:

$$\begin{aligned}
&\int_0^T \int_{\Omega_f} (1 + \frac{\tau_{\Delta t}\, \eta_N}{R}) \Big(\partial_t \bar{\boldsymbol{u}}_N \cdot \boldsymbol{q}_N + \frac{1}{2} (\tau_{\Delta t} \boldsymbol{u}_N - \mathbf{w}_N) \cdot \nabla^{\tau_{\Delta t}\eta_N} \boldsymbol{u}_N \cdot \boldsymbol{q}_N \\
&\quad - \frac{1}{2} (\tau_{\Delta t} \boldsymbol{u}_N - \mathbf{w}_N) \cdot \nabla^{\tau_{\Delta t}\eta_N} \boldsymbol{q}_N \cdot \boldsymbol{u}_N \Big) + \frac{1}{2R} \int_0^T \int_{\Omega_f} v_N^* \boldsymbol{u}_N \cdot \boldsymbol{q}_N \\
&+ 2\mu \int_0^T \int_{\Omega_f} (1 + \frac{\tau_{\Delta t}\eta_N}{R}) \mathbf{D}^{\tau_{\Delta t}\eta_N}(\boldsymbol{u}_N) : \mathbf{D}^{\tau_{\Delta t}\eta_N}(\boldsymbol{q}_N) + \int_0^T \int_\Gamma \partial_t \bar{v}_N \psi \\
&\qquad + C_1 \int_0^T \int_\Gamma \partial_z \eta_N\, \partial_z \psi + C_2 \int_0^T \int_\Gamma \partial_z^2 \eta_N\, \partial_z^2 \psi \\
&= \Big(\int_0^T P_{in}^N dt \int_0^R q_z(t, 0, r) dr - \int_0^T P_{out}^N dt \int_0^R q_z(t, L, r) dr \Big),
\end{aligned} \tag{5.83}$$

with

$$\nabla^{\tau_{\Delta t}\eta} \cdot \boldsymbol{u}_N = 0, \quad v_N = ((u_r)_N)_{|\Gamma},$$
$$\boldsymbol{u}_N(0,.) = \boldsymbol{u}_0,\ \eta(0,.)_N = \eta_0,\ v_N(0,.) = v_0. \tag{5.84}$$

Here $\bar{\mathbf{u}}_N$ and $\bar{v}_N$ are the piecewise linear functions defined in (5.55); $\tau_{\Delta t}$ is the shift in time by Δt to the left, defined in (5.65); $\nabla^{\tau_{\Delta t}\eta_N}$ is the transformed gradient via the ALE mapping $\boldsymbol{\Phi}^{\tau_{\Delta t}\eta_N}$, defined in (5.32); and v_N^*, $\mathbf{u}_N$, v_N, and η_N are defined in (5.54).

We would like to pass to the limit in (5.83) as $N \to \infty$ to show that the limits satisfy the weak formulation of the continuous problem, given by (5.39). To pass to the limit, we recall the weak- and weak*-convergence results presented in Lemma 5.6.1, the convergence of test functions discussed in Sect. 5.7.3, and the strong convergence results obtained in Sect. 5.7.

We summarize the **strong convergence** results here. We have shown that there exist subsequences $(\boldsymbol{u}_N)_{N\in\mathbb{N}}$, $(\eta_N)_{N\in\mathbb{N}}$, $(\bar{\eta}_N)_{N\in\mathbb{N}}$, $(v_N)_{N\in\mathbb{N}}$ such that

$$\begin{aligned}
\boldsymbol{u}_N &\to \boldsymbol{u} \text{ in } L^2(0,T;L^2(\Omega_M)),\\
\tau_{\Delta t}\boldsymbol{u}_N &\to \boldsymbol{u} \text{ in } L^2(0,T;L^2(\Omega_M)),\\
\eta_N &\to \eta \text{ in } L^\infty(0,T;C^1(\bar{\Gamma})),\\
\bar{\eta}_N &\to \bar{\eta} \text{ in } L^\infty(0,T;C^1(\bar{\Gamma})),\\
v_N &\to v \text{ in } L^2(0,T;L^2(\Gamma)),\\
\tau_{\Delta t} v_N &\to v \text{ in } L^2(0,T;L^2(\Gamma)).
\end{aligned} \tag{5.85}$$

The statements about the convergence of the time-shifts $(\tau_{\Delta t}\boldsymbol{u}_N)_{N\in\mathbb{N}}$ and $(\tau_{\Delta t} v_N)_{N\in\mathbb{N}}$ follow directly from Statement 3 of Theorem 5.4.1, i.e., from the numerical dissipation estimate for the splitting scheme. Namely, if we multiply the third equality of Lemma 5.4.1 by Δt, we get

$$\|\tau_{\Delta t}\boldsymbol{u}_N - \boldsymbol{u}_N\|^2_{L^2((0,T)\times\Omega)} + \|\tau_{\Delta t} v_N - v_N\|^2_{L^2((0,T)\times(0,L))} \le C\Delta t. \tag{5.86}$$

We can now **pass to the limit** directly in all the terms except in the term that contains $\partial_t \bar{\mathbf{u}}_N$. To deal with this term, we notice that, since $\boldsymbol{q}_N$ are smooth on subintervals $(n\Delta t, (n+1)\Delta t)$, we can use integration by parts on these subintervals to obtain

$$\int_0^T \int_{\Omega_f} (1 + \frac{\tau_{\Delta t}\,\eta_N}{R})\partial_t \bar{\boldsymbol{u}}_N \cdot \boldsymbol{q}_N = \sum_{n=0}^{N-1} \int_{n\Delta t}^{(n+1)\Delta t} \int_{\Omega_f} (1 + \frac{\eta_N^n}{R})\partial_t \bar{\boldsymbol{u}}_N \cdot \boldsymbol{q}_N$$
$$= -\sum_{n=0}^{N-1} \int_{n\Delta t}^{(n+1)\Delta t} \int_{\Omega_f} (1 + \frac{\tau_{\Delta t}\,\eta_N}{R})\bar{\boldsymbol{u}}_N \cdot \partial_t \boldsymbol{q}_N$$

$$+\sum_{n=0}^{N-1}\Big[\int_{\Omega_f}\left(1+\frac{\eta_N^{n+1}-\eta_N^{n+1}+\eta_N^n}{R}\right)\boldsymbol{u}_N^{n+1}\cdot\boldsymbol{q}_N((n+1)\Delta t-)$$

$$-\int_{\Omega_f}(1+\frac{\eta_N^n}{R})\boldsymbol{u}_N^n\cdot\boldsymbol{q}_N(n\Delta t+)\Big]. \tag{5.87}$$

Here, we have denoted by $\boldsymbol{q}_N((n+1)\Delta t-)$ and $\boldsymbol{q}_N(n\Delta t+)$ the limits from the left and right, respectively, of $\boldsymbol{q}_N$ at the appropriate points.

The integral involving $\partial_t\boldsymbol{q}_N$ can be simplified by recalling that $\mathbf{q}_N(t,\tilde{z},\tilde{r}) = \mathbf{q}(t,\boldsymbol{\Phi}_f^{\eta_N}(t)(\tilde{z},\tilde{r})) = \mathbf{q}(t,\tilde{z},(1+\frac{\eta_N}{R}(t,\tilde{z},\tilde{r})))$, where η_N are constant on each subinterval $(n\Delta t,(n+1)\Delta t)$. Thus, by the chain rule, we see that $\partial_t\boldsymbol{q}_N = \partial_t\boldsymbol{q}$ on $(n\Delta t,(n+1)\Delta t)$. After summing over all $n=0,\dots,N-1$, we obtain

$$-\sum_{n=0}^{N-1}\int_{n\Delta t}^{(n+1)\Delta t}\int_{\Omega_f}(1+\frac{\tau_{\Delta t}\eta_N}{R})\bar{\boldsymbol{u}}_N\cdot\partial_t\boldsymbol{q}_N = -\int_0^T\int_{\Omega_f}(1+\frac{\tau_{\Delta t}\eta_N}{R})\bar{\boldsymbol{u}}_N\cdot\partial_t\boldsymbol{q}.$$

To deal with the last two terms in (5.87), we calculate

$$\sum_{n=0}^{N-1}\Big(\int_{\Omega_f}(1+\frac{\eta_N^{n+1}-\eta_N^{n+1}+\eta_N^n}{R})\boldsymbol{u}_N^{n+1}\cdot\boldsymbol{q}_N((n+1)\Delta t-)$$

$$-\int_{\Omega}(1+\frac{\eta_N^n}{R})\boldsymbol{u}_N^n\cdot\boldsymbol{q}_N(n\Delta t+)\Big)$$

$$=\sum_{n=0}^{N-1}\int_{\Omega_f}\left((1+\frac{\eta_N^{n+1}}{R})\boldsymbol{u}_N^{n+1}\cdot\boldsymbol{q}_N((n+1)\Delta t-)\right.$$

$$\left.-\frac{(\eta_N^{n+1}-\eta_N^n)}{R}\boldsymbol{u}_N^{n+1}\cdot\boldsymbol{q}_N((n+1)\Delta t-)\right)$$

$$-\int_{\Omega_f}(1+\frac{\eta_0}{R})\boldsymbol{u}_0\cdot\mathbf{q}(0)-\sum_{n=1}^{N-1}\int_{\Omega_f}(1+\frac{\eta_N^n}{R})\boldsymbol{u}_N^n\cdot\boldsymbol{q}_N(n\Delta t+)$$

After a small manipulation involving rewriting $(\eta_N^{n+1}-\eta_N^n)$ as $v_N^{n+\frac{1}{2}}\Delta t$, and rewriting the summation indices in the first term, we obtain that the above expression is equal to

$$=\sum_{n=1}^{N}\int_{\Omega_f}(1+\frac{\eta_N^n}{R})\boldsymbol{u}_N^n\cdot\boldsymbol{q}_N(n\Delta t-)-\frac{1}{R}\int_0^T\int_{\Omega_f}v_N^*\boldsymbol{u}_N\cdot\boldsymbol{q}_N$$

$$-\int_{\Omega_f}(1+\frac{\eta_0}{R})\boldsymbol{u}_0\cdot\mathbf{q}(0)-\sum_{n=1}^{N-1}\int_{\Omega_f}(1+\frac{\eta_N^n}{R})\boldsymbol{u}_N^n\cdot\boldsymbol{q}_N(n\Delta t+).$$

Since the test functions have compact support in $[0, T)$, the value of the first term at $n = N$ is zero, and so we can combine the two sums to obtain

$$= \sum_{n=1}^{N} \int_{\Omega_f} (1 + \frac{\eta_N^n}{R}) \boldsymbol{u}_N^n \cdot \big(\boldsymbol{q}_N(n\Delta t-) - \boldsymbol{q}_N(n\Delta t+)\big)$$

$$- \int_{\Omega_f} (1 + \frac{\eta_0}{R}) \boldsymbol{u}_0 \cdot \mathbf{q}(0) - \frac{1}{R} \int_0^T \int_{\Omega_f} v_N^* \boldsymbol{u}_N \cdot \boldsymbol{q}_N.$$

Therefore, we have obtained that

$$\int_0^T \int_{\Omega_f} (1 + \frac{\tau_{\Delta t}\, \eta_N}{R}) \partial_t \bar{\boldsymbol{u}}_N \cdot \boldsymbol{q}_N = \sum_{n=1}^{N} \int_{\Omega_f} (1 + \frac{\eta_N^n}{R}) \boldsymbol{u}_N^n \cdot \big(\boldsymbol{q}_N(n\Delta t-) - \boldsymbol{q}_N(n\Delta t+)\big)$$

$$- \int_{\Omega_f} (1 + \frac{\eta_0}{R}) \boldsymbol{u}_0 \cdot \mathbf{q}(0) - \frac{1}{R} \int_0^T \int_{\Omega_f} v_N^* \boldsymbol{u}_N \cdot \boldsymbol{q}_N.$$

We know how to pass to the limit in all the terms on the right-hand side except the first one. To deal with this term, we use the mean value theorem to rewrite

$$\boldsymbol{q}_N(n\Delta t-, z, r) - \boldsymbol{q}_N(n\Delta t+, z, r) = \boldsymbol{q}(n\Delta t, z, (1 + \frac{\eta_N^n}{R})r)$$

$$- \boldsymbol{q}(n\Delta t, z, (1 + \frac{\eta_N^{n+1}}{R})r) =$$

$$= \frac{1}{R} \partial_r \boldsymbol{q}(n\Delta t, z, \zeta) r (\eta_N^n - \eta_N^{n+1}) = -\frac{1}{R} \Delta t \partial_r \boldsymbol{q}(n\Delta t, z, \zeta) v_N^{n+\frac{1}{2}} r$$

$$= -\frac{1}{R} \Delta t \partial_r \boldsymbol{q}(n\Delta t, z, \zeta)(\tau_{-\Delta t} v_N^*) r$$

Now, the integral containing the difference between the $\boldsymbol{q}_N$'s becomes

$$\sum_{n=1}^{N-1} \int_{\Omega_f} (1 + \frac{\eta_N^n}{R}) \boldsymbol{u}_N^n \big(\boldsymbol{q}_N(n\Delta t-) - \boldsymbol{q}_N(n\Delta t+)\big)$$

$$= - \int_0^{T-\Delta t} \int_{\Omega} (1 + \frac{\eta_N^n}{R}) \boldsymbol{u}_N \frac{r}{R} (\tau_{-\Delta t} v_N^*) \partial_r \boldsymbol{q}.$$

Now, we know how to pass to the limit in the expression on the right-hand side. Namely, we get

$$\int_0^{T-\Delta t}\int_{\Omega_f}(1+\frac{\eta_N^n}{R})\boldsymbol{u}_N\frac{r}{R}(\tau_{-\Delta t}v_N^*)\partial_r\boldsymbol{q} \to \int_0^T\int_{\Omega_f}(1+\frac{\eta_N}{R})\boldsymbol{u}\frac{r}{R}\partial_t\eta\partial_r\boldsymbol{q}$$

as $N \to \infty$.

Therefore, we now have that as $N \to \infty$,

$$\int_0^T\int_{\Omega_f}(1+\frac{\tau_{\Delta t}\,\eta_N}{R})\partial_t\bar{\boldsymbol{u}}_N\cdot\boldsymbol{q}_N \to$$

$$-\int_0^T\int_{\Omega_f}(1+\frac{\eta}{R})\boldsymbol{u}\frac{r}{R}\partial_t\eta\partial_r\boldsymbol{q} - \int_0^T\int_{\Omega_f}(1+\frac{\eta}{R})\boldsymbol{u}\cdot\partial_t\boldsymbol{q}$$

$$-\int_{\Omega_f}(1+\frac{\eta_0}{R})\boldsymbol{u}_0\cdot\mathbf{q}(0) - \frac{1}{R}\int_0^T\int_{\Omega_f}\partial_t\eta\boldsymbol{u}\cdot\boldsymbol{q}$$

Therefore, by recalling that $\tilde{\boldsymbol{q}} = \boldsymbol{q}\circ\boldsymbol{\Phi}^\eta$ and noticing that $\partial_t\tilde{\boldsymbol{q}} = \partial_t\boldsymbol{q} + \frac{r}{R}(\partial_t\eta)\partial_r\boldsymbol{q}$, we have finally obtained

$$\int_0^T\int_{\Omega_f}(1+\frac{\tau_{\Delta t}\,\eta_N}{R})\partial_t\bar{\boldsymbol{u}}_N\cdot\boldsymbol{q}_N \to -\int_0^T\int_{\Omega_f}(1+\frac{\eta}{R})\boldsymbol{u}\cdot\partial_t\tilde{\boldsymbol{q}} - \frac{1}{R}\int_0^T\int_{\Omega_f}\partial_t\eta\boldsymbol{u}\cdot\tilde{\boldsymbol{q}}$$

$$-\int_{\Omega}(1+\frac{\eta_0}{R})\boldsymbol{u}_0\cdot\tilde{\boldsymbol{q}}(0).$$

Now we can go back to the semidiscrete weak formulation (5.83) and take the limits in all the terms. Notice that the term $-\frac{1}{R}\int_0^T\int_{\Omega_f}\partial_t\eta\boldsymbol{u}\cdot\tilde{\boldsymbol{q}}$ obtained above combines with the alike term in (5.83) to give the term $-\frac{1}{2R}\int_0^T\int_{\Omega_f}\partial_t\eta\boldsymbol{u}\cdot\tilde{\boldsymbol{q}}$ in the limit. We emphasize one more time that this is an extra term, which exists in this weak formulation due to the motion of the fluid domain.

After taking the limits and combining the alike terms, one recovers the weak formulation (5.39), which holds for all the test functions $(\boldsymbol{q},\psi)$ that are dense in the test space $\mathcal{Q}^\eta$. This proves the first part of Theorem 5.3.1 (**Main existence theorem**).

To complete the proof of Theorem 5.3.1, we need to show that either our result is **global in time or the result holds until the time of degeneracy** when, e.g., the walls of the cylinder touch each other. This can be proved by using similar argument as in [81] p. 397-398; see also [333]. More precisely, let $(0, T_1)$, $T_1 > 0$, be the interval on which we have constructed our solution $(\mathbf{u}, \eta)$, and let $m_1 = \min_{(0,T_1)\times(0,L)}(R+\eta)$. From Lemma 5.7.2, we know that $m_1 > 0$. Furthermore, since $\eta \in W^{1,\infty}(0, T; L^2(0, L)) \cap L^\infty(0, T; H_0^2(0, L))$ and $\mathbf{u} \in L^\infty(0, T; L^2(\Omega))$, we can take T_1 such that $\eta(T_1) \in H_0^2(0, L)$, $\partial_t \eta(T_1) \in L^2(0, L)$, and $\mathbf{u}(T_1) \in L^2(\Omega)$. We can now use the first part of Theorem 5.3.1 to prolong the solution $(\mathbf{u}, \eta)$ to the interval $(0, T_2)$, $T_2 > T_1$. By iteration, we can continue the construction of our solution to the interval $(0, T_k)$, $k \in \mathbb{N}$, where $(T_k)_{k \in \mathbb{N}}$ is an increasing sequence. We set $m_k = \min_{(0,T_k)\times(0,L)}(R + \eta) > 0$.

Since $m_k > 0$, we can continue the construction further. Without loss of generality, we could choose a $T_{k+1} > T_k$ so that $m_{k+1} \geq \frac{m_k}{2}$. From (5.40) and from the embedding

$$L^\infty(0, T; H_0^2(0, L)) \cap W^{1,\infty}(0, T; L^2(0, L)) \hookrightarrow C^{0,1-\alpha}([0, T]; H^{2\alpha}(0, L)),$$

by taking $\alpha = 1/2$, we have that the displacement η is Hölder continuous in time, namely,

$$\|\eta\|_{C^{0,1/2}(0,T_{k+1};C[0,L])} \leq C(T_{k+1}).$$

Therefore, the following estimate holds:

$$\begin{aligned} R + \eta(T_{k+1}, z) &\geq R + \eta(T_k, z) - C(T_{k+1})(T_{k+1} - T_k)^{1/2} \\ &\geq m_k - C(T_{k+1})(T_{k+1} - T_k)^{1/2}. \end{aligned}$$

For a T_{k+1} chosen so that $m_{k+1} \geq \frac{m_k}{2}$, this estimate implies

$$T_{k+1} - T_k \geq \frac{m_k^2}{4C(T_{k+1})^2}, \quad k \in \mathbb{N}. \tag{5.88}$$

Now, let us take $T^* = \sup_{k \in \mathbb{N}} T_k$ and set $m^* = \min_{(0,T^*)\times(0,L)}(R + \eta)$. Obviously, $m_k \geq m^*$, $k \in \mathbb{N}$. There are two possibilities. Either $m^* = 0$ or $m^* > 0$. If $m^* = 0$, this means that $\lim_{t \to T} \min_{z \in [0,L]} (R + \eta(z)) = 0$, and the second statement in the theorem is proved. If $m^* > 0$, we need to show that $T^* = \infty$. To do that, notice that (5.40) gives the form of the constant $C(T)$, which is a nondecreasing function of T. Therefore, we have $C(T_k) \leq C(T^*)$, $\forall k \in \mathbb{N}$. Using this observation and the fact that $m_k \geq m^*$, $k \in \mathbb{N}$, estimate (5.88) implies

$$T_{k+1} - T_k \geq \frac{(m^*)^2}{2C(T^*)^2}, \quad \forall k \in \mathbb{N}.$$

Since this holds for all $k \in \mathbb{N}$, we have that $T^* = \infty$.

This concludes the final step of the proof of Theorem 5.3.1 stating the existence of a weak solution to the nonlinearly coupled prototypical FSI model. □

5.9 Bibliographical Notes

We conclude this chapter on nonlinearly coupled FSI problem by making some final bibliographical notes. We concentrate on FSI problems where the fluid is incompressible, as this is the primary focus of this book and the area where the theory is most well-developed. Toward the end, we will briefly discuss some recent advances involving more complex fluid models.

First, we note that, depending on the specific physical FSI problem, the structure can be modeled as a rigid body, a bulk elastic body (with the same dimension as the fluid), or a lower-dimensional model such as a plate, shell, beam, or string.

Historically, the first well-posedness results for nonlinearly coupled FSI problems focused on the interaction of a rigid body with a viscous, incompressible, or compressible Newtonian fluid. The existence theory for both weak and strong solutions involving FSI with rigid bodies has been extensively studied in various contexts, including bounded and unbounded domains, slip and no-slip kinematic coupling conditions, and scenarios involving one or multiple rigid bodies. For a comprehensive overview, we refer the reader to the extensive literature on this topic, including [4, 96, 102, 116, 117, 138, 157, 159, 171, 176, 184, 248, 301, 317–319, 344].

Strong solutions are typically local-in-time or apply to small data, while weak solutions exist until contact occurs—that is, until the moment when the rigid body touches the boundary or another rigid body, causing the fluid domain to degenerate. In [82], however, the authors prove the existence of global-in-time weak solutions regardless of contact. Nevertheless, this result does not provide information about the system's behavior after contact. In fact, it is expected that solutions may not be unique post-contact, with some being nonphysical [137, 311]. There is still no consensus on how to model contact in these systems. This issue is illustrated by the so-called no contact paradox, which asserts that under certain assumptions, contact will not occur in finite time [181, 182, 283], a clearly nonphysical outcome. Several explanations for this paradox have been proposed in recent years [158, 160, 172, 343].

To conclude our review of the results for fluid-rigid body systems, we turn to several related topics that have garnered significant attention. Weak-strong uniqueness results, which assert that strong solutions (defined in an appropriate sense) are unique within a broader class of finite energy weak solutions, were established in [83, 123, 150, 271]. Uniqueness results in the 2D case were proved in [50, 163]. Recently, conditional regularity results in $3D$ case have shown that if a solution satisfies additional integrability conditions (the so-called Prodi-Serrin conditions), then the solution is smooth [246, 247, 272]. There is also an extensive body of literature on the stability and long-term behavior of fluid-rigid body systems; see, for example, [46, 144, 148, 151, 183]. Finally, we mention results on time-periodic solutions [45, 147, 152] and the self-propulsion of rigid bodies [143, 146, 251, 306].

In many fluid-structure interaction (FSI) applications where the fluid occupies a time-dependent domain that evolves according to the movement of a structure, the structure is often elastic or deformable rather than rigid. This is the case, for example, in the study of blood flow through the heart or arteries, where the arterial or cardiac walls are elastic or viscoelastic structures. Models describing biological fluid-structure interaction are common in the literature; see, for example, [141, 282, 286, 340]. Given their importance in practical applications, the development of rigorous mathematical techniques to analyze nonlinearly coupled fluid-structure dynamics is essential.

An early work in this area, published in [118], considered only a finite number of eigenmodes to update the structure's displacement. However, our focus is on elastic structures modeled by the full equations of elastodynamics, which govern the unknown configuration of a time-dependent fluid domain in nonlinearly coupled models. In the literature on well-posedness for moving boundary problems, two main types of prototypical nonlinearly coupled FSI problems are prevalent: (1) problems where the structure domain is of lower dimension than the fluid domain (see, e.g., [31, 81, 167, 168, 233, 234, 265, 273, 330, 333]) and (2) problems where the structure domain has the same dimension as the fluid domain (see, e.g., [32, 84, 85, 99, 100, 192, 193, 205, 221, 287]).

The first results for nonlinearly coupled FSI were obtained for fluid-structure systems where the *structure domain is of lower dimension than the fluid domain.* One of the earliest well-posedness results for such problems was presented in [31], where the existence of a local strong solution for sufficiently smooth and small initial data, under an additional smallness condition on the coupling parameter, was proved. This result concerned a nonlinearly coupled FSI model involving the interaction between a 1D viscoelastic structure, described by a generalized string model, and the Navier-Stokes equations in a 2D domain. The local existence result was shown using a fixed-point argument, after mapping the problem from a moving domain to a fixed one and linearizing the resulting system appropriately. This local existence result for strong solutions was extended in [234], where a similar model was considered involving a 1D structure modeled by the damped beam equation interacting with a fluid modeled by the Navier-Stokes equations. The smallness condition on the initial data and smallness parameter from [31] are removed, so that in [234], local existence of strong solutions is established for general sufficiently regular initial data, under the additional assumption that the structure is a viscoelastic plate (so that it has fourth-order spatial derivatives in the defining equations). Furthermore, if the initial data is sufficiently small, the result in [234] establishes global existence for small data. These past results for strong solutions [31, 234] involve local existence of strong solutions and were recently improved in [168], where it is established that there is global existence of strong solutions to an FSI problem involving a viscoelastic plate interacting with a fluid modeled by the Navier-Stokes equations, assuming the initial data is sufficiently regular and it does not touch the bottom boundary of the fluid domain at the initial time. In particular, the work in [168] shows that if the structure is not touching the bottom of the fluid domain at the initial time, then the structure will never contact the bottom of the fluid domain, and the strong solution will exist globally in time. In the case where the structure is a purely elastic rod or beam (i.e., without structural dissipation), the existence of a unique local-in-time strong solution was established in [169]. It is important to note that in this work, fluid dissipation is crucial for controlling the structure's velocity in the appropriate norms, and there is no gap between the regularity of the solution and the regularity of the initial data. All the aforementioned results apply to the Cauchy problem. The existence of time-periodic strong solutions for an FSI problem involving a damped beam was established in [76].

For 3D fluids interacting with 2D structures, several results concerning strong solutions have been obtained. In [84], the local-in-time existence for an FSI problem was established, where the structure was a biofluid shell modeled by nonlinear equations. Similarly, in [85], the structure was an elastic shell whose dynamics were influenced by inertial, membrane, and bending effects, and local-in-time existence for strong solutions was also proved. In [235], the structure is modeled by the damped wave equation, and local-in-time as well as global-in-time small-data existence results for strong solutions were proved. A cylindrical viscoelastic shell model was considered in [243], where the existence of a maximal-in-time strong solution for the corresponding 3D-2D FSI problem was established. The development of L^p theory for strong solutions to a 3D-2D FSI problem involving a damped plate was presented in [245]. Finally, recent results on strong solutions for FSI problems involving inviscid fluids can be found in [23, 222].

The previously described results for nonlinearly coupled models involving elastic structures of lower dimension than the fluid are results involving strong solutions. Typically, strong solution results for FSI problems are local in time and/or valid for small initial data. Moreover, they are often obtained using mathematical techniques that are not constructive. To gain additional insights into the behavior of solutions for this class of problems, it is important to study weak solutions in finite energy spaces.

One of the earliest results in the area of weak solutions for nonlinearly coupled FSI problems was presented in [81], where the existence of a weak solution to a 3D fluid-structure interaction problem—featuring a fluid governed by the Navier-Stokes equations and a viscoelastic plate—was proved until the time of self-contact of the structure. This result was extended in [167], where the existence of weak solutions was proved for a similar model without the viscoelastic regularizing term, involving an elastic plate instead. Further development in the study of weak solutions for nonlinearly coupled FSI problems was achieved in [265], which considered a 2D fluid modeled by the Navier-Stokes equations interacting with a Koiter shell. In this case, the fluid flow was driven by inlet and outlet dynamic pressure data. This work introduced a new methodology for proving the existence of weak solutions, featuring a constructive proof based on an operator splitting scheme that explicitly constructed approximate solutions.

It is worth noting that the first constructive existence proof for a moving boundary problem, though with a prescribed motion of the boundary, was presented by Ladyzhenskaya in 1970 (see [231]). This work dealt with the interaction between an incompressible, viscous fluid and a given moving boundary, using a time-discretization approach known as Rothe's method, under the assumption of high regularity of the interface.

The method introduced by Muha and Čanić in [265] was extended to a broader class of problems, including FSI models with the Navier-slip condition [269], nonlinear Koiter shell equations [268], multilayered structures [267], and elastic structures supported by thin mesh-like tubes [334, 335]. We discuss some more details on these extensions in the next chapter, Chap. 6.

These results on weak solutions were further expanded to nonlinearly coupled FSI models with curved structures in [233], where a Koiter shell with a curved reference configuration was studied, and to models involving nonlinear plates [330]. These advancements prompted further developments in the theory of weak solutions to FSI problems. A functional framework allowing for the definition of global-in-time weak solutions, including the possibility of contact, was introduced in [77]. Additionally, weak-strong uniqueness results for FSI problems were established in [298].

In [273], fluid dissipation was employed to prove additional regularity for the structure displacement, which enabled the existence proof for weak solutions in FSI problems involving nonlinear Koiter shells. This technique of using fluid dissipation to derive extra regularity for the structure has since been applied in a range of subsequent works, proving the existence of weak solutions for more complex FSI models, such as compressible fluids [212, 329], polymer fluids [54], heat-conducting fluids [58], and multicomponent fluids [202].

The regularity of weak solutions to 2D-1D FSI problems was proved in [299], while conditional regularity of weak solutions to 3D-2D FSI problems was analyzed in [56]. Additionally, time-periodic weak solutions for FSI problems involving lower-dimensional structures were studied in [212, 259]. Recently, a singular limit problem involving fluid-structure interaction (FSI) was analyzed, leading to the rigorous derivation of a sixth-order thin film equation as a simplified model of the FSI system under specific scaling regimes [64, 65]. All these results were derived under the assumption that the structure deforms in a predefined direction, typically normal to the reference configuration. However, the case where the structure can deform in all three spatial directions presents significant new challenges, and the first existence results for such models have only recently been obtained [206, 269, 335, 338].

For FSI models in which *the structure is of the same dimension as the fluid domain*, the first well-posedness result was established in [99], where it was shown that for a nonlinearly coupled model involving a fluid governed by the Navier-Stokes equations interacting with a linearly elastic solid, there is local existence of strong solutions provided the initial fluid and structure velocities are in H^5 and H^2, respectively.

This local existence result for strong solutions was extended in [100], where the elastic solid was modeled by quasilinear equations of elasticity. The regularity assumptions on the initial data were later improved in [221], where local existence of strong solutions was proven with lower regularity, requiring only that the fluid and structure velocities be in H^3 and H^2. Further analysis of this model, focusing on the a priori energy estimates necessary for studying well-posedness, was conducted in [192].

A variant of the original model from [99], involving an additional transmission boundary condition that stabilizes the dynamics, was studied in [193]. In this case, it was shown that the damping and transmission boundary conditions lead to a global existence result for smooth solutions, provided the initial data is sufficiently small in higher-order Sobolev norms. Finally, the well-posedness of a model where the Navier-Stokes equations interact with a structure governed by the Lamé equations of elasticity was considered in [287], which established local existence of strong solutions with less stringent regularity assumptions on the initial data compared to previous works like [99] and [221]. Additionally, recent work in [220] proved small-data global existence and exponential decay for a 3D-3D FSI problem where the structure is modeled by a damped wave equation.

The theory of weak solutions for 3D-3D FSI problems is less developed, as finite energy weak solutions generally lack the regularity required to properly define all terms in the weak formulation. However, in cases where the structure equation includes a higher-order regularizing term, the existence of weak solutions was proven in [47]. Recently, the existence of weak solutions for a 3D-3D FSI problem was shown when the structure is described by a second-grade nonlinear viscoelastic model, where the elastic energy is coercive in $W^{1,p}$, $p > 3$ [32]. A similar result was obtained for a model involving a compressible fluid, where global-in-time existence for weak solutions was proven, even allowing for self-contact of the solid structure [205].

At the end of this literature review, let us mention that in recent years, there has been significant development in the analysis of FSI problems involving various fluid models. These include compressible fluids, e.g., [48, 49, 57, 329]; heat-conducting fluids, e.g., [58, 242, 244]; electrically conducting fluids, e.g., [33, 297]; non-Newtonian fluids, e.g., [190, 293]; and magnetohydrodynamic fluids, e.g., [188]. However, since this is not the main topic of this book, we will not delve into the details of these results.

For a more in-depth discussion of nonlinearly coupled FSI models, including both analytical results and numerical methods for weak and strong solutions, we refer the reader to the references [170, 203, 291, 333].

In the next chapter, we provide some more details on the works where the splitting scheme was extended to prove existence of weak solutions to FSI problems with the Navier-slip condition, multilayered structures, nonlinear structures, FSI with mesh-supported shells, and 3D-2D coupled systems.

Chapter 6
Extensions of the Splitting Scheme

The splitting scheme approach to fluid-structure interaction, demonstrated in the previous two chapters, has proven to be robust and adaptable, having been extended to a variety of physically significant contexts. The mathematical framework developed through this approach has been instrumental in establishing the constructive existence of weak solutions for complex fluid-structure interaction models. It has also been successfully applied to analyze coupled fluid-structure systems of practical importance. In this section, we outline the more intricate FSI models that have been studied using extensions of the Lie operator splitting method, highlighting the versatility of this approach and laying the groundwork for the work presented in the rest of this book.

6.1 3D-2D Fluid-Structure Systems

The prototypical FSI models discussed above involved a 2D incompressible viscous fluid interacting with a 1D elastic structure, both in the context of linear coupling (see Chap. 4) and nonlinear coupling (see Chap. 5). However, for practical applications, it is crucial to extend constructive existence results for weak solutions to systems with more physically relevant dimensions. Specifically, we aim to achieve analogous results for incompressible viscous fluids in three dimensions interacting with two-dimensional elastic structures, such as plates, membranes, or shells. Similar to the 2D-1D case, our fluid domain boundary in the 3D-2D prototypical FSI problem consists of the inlet and outlet portions Γ_{in} and Γ_{out}, the bottom boundary Γ_b, and the time-dependent moving 2D structure $\Gamma(t)$.

In this context, it is worth noting that in the linearly coupled prototypical model of FSI, the analysis would remain largely unchanged if we considered a three-dimensional Stokes flow interacting with a two-dimensional elastic structure. This is because, in the case of linear coupling, the time evolution of the fluid domain's

S. Čanić et al., *Deterministic and Stochastic Fluid-Structure Interaction*, Advances in Mathematical Fluid Mechanics, https://doi.org/10.1007/978-3-032-00898-5_6

geometry is not a factor. As a result, the fluid equations, even in three dimensions, are defined on a fixed reference domain Ω_f, which maintains a regular and well-behaved geometry despite being a three-dimensional spatial domain.

In contrast, there are additional challenges that appear when considering **three-dimensional incompressible Navier-Stokes equations** coupled to two-dimensional elastic structures, in the **nonlinearly coupled** prototypical model of FSI. The main difficulty lies in the fact that when considering, e.g., a plate equation for the structure, the structure displacement η is in the finite energy space $W^{1,\infty}(0, T; L^2(\Gamma)) \cap L^\infty(0, T; H_0^2(\Gamma))$. Because the problem is nonlinearly coupled, the regularity of the structure displacement η is especially important, as the structure displacement determines the time-dependent location of the structure $\Gamma(t)$, which in turn determines the three-dimensional time-dependent fluid domain $\Omega_f(t)$.

Because we are considering three spatial dimensions for the fluid and two spatial dimensions for the elastic structure, we have different Sobolev embeddings. For a one-dimensional structure, which is what we have in the 2D-1D prototypical fluid-structure model described before, the structure displacement $\eta \in W^{1,\infty}(0, T; L^2(\Gamma)) \cap L^\infty(0, T; H_0^2(\Gamma))$ defines a function that is uniformly Lipschitz continuous for all $t \in [0, T]$. However, for a two-dimensional structure, a function $\eta \in H_0^2(\Gamma)$ is not a Lipschitz continuous function, so **the domain $\Omega_f(t)$ is not a Lipschitz domain**. This has several important ramifications that complicate the analysis of three-dimensional nonlinearly coupled FSI, which was carried out in [266] and was later extended to three-dimensional FSI involving nonlinear elasticity in [268].

First, the standard *trace theorem* from the classical theory of Sobolev spaces states that functions in $H^1(\Omega)$ have well-defined traces in $L^2(\partial\Omega)$ and even $H^{1/2}(\partial\Omega)$ under the condition that Ω is a Lipschitz domain. However, because $\Omega_f(t)$ in the three-dimensional case of nonlinearly coupled FSI is no longer necessarily a Lipschitz domain due to the limited regularity of the structure displacement η, we must use **an extension of the trace theory for Sobolev spaces**. This extension of the Sobolev trace theory was developed in a note [263], which states an extension of the Sobolev trace theorem to functions defined on domains which are locally **subgraphs of Hölder continuous functions**. This is exactly what is required in the current scenario of three-dimensional nonlinearly coupled FSI, since structure displacements η which are in $H_0^2(\Gamma)$ are not Lipschitz continuous, but are α-Hölder continuous for $\alpha \in (0, 1)$. The result from [263] states that the trace of an $H^1(\Omega)$ function on a domain Ω that is locally a subgraph of an α-Hölder continuous function belongs to $H^s(\partial\Omega)$ for any $s \in (0, \alpha/2)$.

This result on the extension of the Sobolev trace theory, described in [263], is exactly what is needed to properly define the trace of the fluid velocity $\boldsymbol{u} \in L^2(0, T; H^1(\Omega_f(t)))$. This allows us to impose the boundary conditions on Γ_{in}, Γ_{out}, and Γ_b on $\boldsymbol{u}$, as the trace of $\boldsymbol{u}$ along the inlet, outlet, and bottom boundary are well-defined. The fact that we can define the trace of $\boldsymbol{u}$ in a well-defined manner allows us to also precisely make sense of the kinematic coupling condition

$\boldsymbol{u}|_{\Gamma(t)} = \partial_t \eta \boldsymbol{e}_r$, which involves the trace of $\boldsymbol{u}$ along the time-dependent moving interface $\Gamma(t)$.

For more details on this FSI topic, the interested reader can visit [266, 268]

6.2 Nonlinear Structures

The study of nonlinearly elastic structures , for example, St. Venant-Kirchhoff elastic structures, nonlinearly elastic Koiter shells, and nonlinear biomembranes, has also been considered in the literature on FSI [84, 85, 100, 149, 166, 268, 273, 288]. The splitting scheme methodology of establishing existence of weak solutions to fluid-structure interaction has been successfully extended from the prototypical model of FSI to fluid-structure systems with nonlinear elastic structures in [268, 273]. These elastic structures with nonlinear behavior appear in many applications of real-life significance. For example, a computational study of fluid-structure interaction in a curved coronary artery [70] involves considering nonlinear equations of elasticity for the displacement of the curved cylindrical arterial walls from its reference configuration, where a cubic-type nonlinearity appears in the elasticity equations as a result of the curved cylindrical geometry of the coronary artery. The extension of the splitting scheme to FSI systems involving nonlinear structures was done first in [268], and splitting scheme methods have been used in FSI systems of practical significance involving nonlinearly elastic structures; see, for example, [70, 273].

The main difference in the splitting scheme for the problem with nonlinear Koiter shell or plate is the structure subproblem. Namely, recall that in the linearly elastic Koiter shell case, the structure subproblem involves solving a time-discretized form of the linear Koiter shell with zero forcing, where the time derivative is discretized by using a backward Euler scheme. This gives rise to a linear elliptic problem for the new structure displacement $\eta_N^{n+\frac{1}{2}}$ written in weak formulation; see (4.34), where the existence and uniqueness of the new updated structure displacement $\eta_N^{n+\frac{1}{2}}$ can be obtained using standard tools from the theory of linear elliptic PDEs, namely, the Lax-Milgram theorem.

However, in the case of a nonlinearly elastic Koiter shell, there are notable differences. First, one cannot use just a backward Euler time discretization to discretize the structure subproblem, as a more involved time discretization is required in order to obtain a **semidiscrete energy estimate which mirrors the corresponding continuous energy estimates** for the nonlinear Koiter shell PDE. In addition, because the resulting problem that arises after time discretization is no longer a linear problem, the Lax-Milgram theorem is not sufficient for obtaining existence and uniqueness of a solution to the (nonlinear) weak formulation of the structure subproblem, and instead, one must invoke a fixed point argument using the Schaefer Fixed Point Theorem in order to obtain existence of an updated structure displacement $\eta_N^{n+\frac{1}{2}}$ from the structure subproblem. However, if one is able to obtain existence of a solution to the structure subproblem via a fixed point argument and if one is able to discretize time derivatives in an appropriate way that preserves the balance of energy of the fully coupled FSI dynamics, then the splitting scheme remains an effective method for studying well-posedness of FSI systems containing structures with nonlinear effects.

The consideration of splitting scheme methods for studying FSI with nonlinearly elastic structures was extended in recent work [273], in which a more general nonlinear Koiter shell considering contributions from both membrane and bending energy is considered, whereas the past work in [268] does not include the fully nonlinear terms arising from the bending energy. The existence of a weak solution is established for this problem in [273] by using a splitting scheme approach to show existence of a weak solution to a form of the weak formulation containing a linear sixth order regularization of the structure dynamics and then obtaining uniform estimates that are independent of the regularization parameter. In addition to showing well-posedness for this nonlinearly coupled FSI problem with a nonlinear Koiter shell, this work also establishes regularity results for the structure displacement η and the structure velocity $\partial_t \eta$.

6.3 Fluid-Structure Interaction with the Navier Slip Condition

In the prototypical model of fluid-structure interaction discussed in Chaps. 4 and 5, we use the no-slip condition as a kinematic coupling condition to couple the fluid and structure dynamics, thereby imposing continuity of velocities at the fluid-structure interface:

$$\boldsymbol{u}|_{\Gamma} = (\partial_t \eta)\boldsymbol{e}_r \text{ for linear coupling, and } \boldsymbol{u}|_{\Gamma(t)} = (\partial_t \eta)\boldsymbol{e}_r \text{ for nonlinear coupling.}$$

In both the linearly coupled and nonlinearly coupled models studied in Chaps. 4 and 5, the no-slip condition, as stated above, implies that the fluid velocity along the fluid-structure interface is purely transversal (radial), namely, that its dominant component is in the transverse direction, since one of the assumptions in the prototypical model is that the structure displaces in only the transverse (radial) direction. This is not always the best representation of reality. In particular, if one can assume that the structure displacement is a vector function $\boldsymbol{\eta}$ with components in all $\mathbb{R}^n$, $n = 2, 3$ directions, then the no-slip condition reads

$$\boldsymbol{u}|_{\Gamma} = \partial_t \boldsymbol{\eta} \text{ for linear coupling, and } \boldsymbol{u}|_{\Gamma(t)} = \partial_t \boldsymbol{\eta} \text{ for nonlinear coupling.}$$

One example where it is imperative to assume that the structure displacement is a vector function is the case of the **Navier slip condition**, which can be used in place of the no-slip condition. The slip boundary condition might be more realistic than the classical no-slip boundary condition in situations, e.g., when the structure is "rough," and in modeling FSI dynamics near, or at a contact. See [158, 160, 172, 181, 182, 283, 343]. Cardiovascular tissue and cell-seeded tissue constructs, which consist of grooves in tissue scaffolds that are lined with cells, are examples of "rough" elastic interfaces interacting with an incompressible, viscous fluid.

The Navier-slip condition allows for a mismatch between the tangential components of the structure and fluid velocities in the tangential direction, giving rise to a new set of kinematic coupling conditions. In the linearly coupled problem, the kinematic coupling involving the Navier-slip condition reads:

$$\begin{cases} \boldsymbol{u}|_\Gamma \cdot \boldsymbol{n} = \partial_t \boldsymbol{\eta} \cdot \boldsymbol{n}, \\ \beta(\partial_t \boldsymbol{\eta} - \boldsymbol{u}|_\Gamma) \cdot \boldsymbol{\tau} = \boldsymbol{\sigma}(\boldsymbol{u}, \pi)\boldsymbol{n} \cdot \boldsymbol{\tau}|_\Gamma, \end{cases}$$

where $\boldsymbol{n}$ is the unit outward pointing normal and $\boldsymbol{\tau}$ is/are unit tangent(s) to the reference configuration of the fluid-structure interface Γ. Coefficient $\beta \geq 0$ is a nonnegative coefficient with the units of frictions, the inverse of which is related to slip length. The first equation, which describes continuity of normal velocities, is also known as the *no-penetration condition.*

In the nonlinearly coupled problem, the kinematic coupling involving the Navier-slip condition reads:

$$\begin{cases} \boldsymbol{u}|_{\Gamma(t)} \cdot \boldsymbol{n}(t) = \partial_t \boldsymbol{\eta} \cdot \boldsymbol{n}(t), \\ \beta(\partial_t \boldsymbol{\eta} - \boldsymbol{u}|_{\Gamma(t)}) \cdot \boldsymbol{\tau}(t) = \boldsymbol{\sigma}(\boldsymbol{u}, \pi)\boldsymbol{n}(t) \cdot \boldsymbol{\tau}(t)|_{\Gamma(t)}, \end{cases} \tag{6.1}$$

where $\boldsymbol{n}(t)$ is the outward pointing unit normal and $\boldsymbol{\tau}(t)$ is/are the unit tangent vector(s) to the moving fluid-structure interface $\Gamma(t)$. As before, $\beta \geq 0$ is a nonnegative friction/slip coefficient.

To provide more details about the extension of the Lie operator splitting method to study the Navier-slip FSI case, we will follow the approach first presented in [269]. To deal with the nonlinearity, we will use the compactness arguments for the Navier slip FSI model carried out in [270].

The FSI problem studied in [269] consists of the nonlinear Navier-Stokes equations describing an incompressible, viscous Newtonian fluid, stated previously in (5.2) and (5.3), where the Navier-Stokes equations are posed on a time-dependent and a priori unknown fluid domain $\Omega_f(t)$, which is two-dimensional. However, for the elastic structure, we will allow for *arbitrary vector-valued displacements* rather than considering only transverse (radial) structure displacements, as was done in the prototypical model of nonlinearly coupled FSI. Therefore, we will describe the structure displacement $\boldsymbol{\eta} : \Gamma \to \mathbb{R}^2$ from the reference configuration Γ by an elastodynamics equation governed by an elasticity operator $\mathcal{L}_e$:

$$\partial_{tt}\boldsymbol{\eta} + \mathcal{L}_e \boldsymbol{\eta} = F, \qquad \text{on } \Gamma.$$

Operator $\mathcal{L}_e$ has an associated elastic energy, which defines the norm $\|\boldsymbol{\eta}\|_E^2 = (\mathcal{L}_e\boldsymbol{\eta}, \boldsymbol{\eta})_{L^2(\Gamma)}$. The elastodynamics equation is supplemented by the appropriate clamped boundary conditions. The time-dependent location of the structure $\Gamma(t)$ will be described by

$$\Gamma(t) = \{(z, R) + \boldsymbol{\eta}(t, z) : z \in [0, L]\}$$

and since $\boldsymbol{\eta}(0) = \boldsymbol{\eta}(L)$, we can define the time-dependent fluid domain $\Omega_f(t)$ to be the two-dimensional region that is bounded by Γ_{in}, Γ_{out}, Γ_b, and $\Gamma(t)$.

The kinematic and dynamic coupling conditions describing two-way coupling between the fluid and structure are now given by:

- **Kinematic coupling condition:**

$$\begin{cases} \boldsymbol{u}|_{\Gamma(t)} \cdot \boldsymbol{n}(t) = \partial_t \boldsymbol{\eta} \cdot \boldsymbol{n}(t), \\ \beta(\partial_t \boldsymbol{\eta} - \boldsymbol{u}|_{\Gamma(t)}) \cdot \boldsymbol{\tau}(t) = \boldsymbol{\sigma}(\boldsymbol{u}, \pi)\boldsymbol{n}(t) \cdot \boldsymbol{\tau}(t)|_{\Gamma(t)}, \end{cases}$$

  where $\boldsymbol{n}(t)$ is the outward pointing unit normal and $\boldsymbol{\tau}(t)$ is the unit tangent vector to the moving fluid-structure interface $\Gamma(t)$, and $\beta \geq 0$ is a nonnegative friction/slip coefficient.
- **Dynamic coupling condition:**

$$\partial_{tt}\boldsymbol{\eta} + \mathcal{L}_e\boldsymbol{\eta} = -\mathcal{J}^\eta \boldsymbol{\sigma}(\boldsymbol{u}, \pi)\boldsymbol{n}(t)|_{\Gamma(t)}.$$

  This is similar to (5.9) in Chap. 5, except that now we have a vector equation. Furthermore, the arc length measure $\mathcal{J}^\eta$ in the case of vector displacements $\boldsymbol{\eta} = (\eta_z, \eta_r)$, is now given by

$$\mathcal{J}^\eta = \sqrt{(1 + \partial_z \eta_z(t, z))^2 + (\partial_r \eta_r(t, z))^2}.$$

To prove the existence of a weak solution to this nonlinearly coupled FSI problem with the Navier-slip condition, one can again use the Lie splitting scheme. However, there are some key differences that we summarize below. For the full proof, we refer the reader to [269, 270].

The first key difference is in the energy estimate. The Navier-slip condition brings an additional term in the energy estimate, which captures the dissipation due to friction in the Navier-slip condition. In particular, the energy estimate for the nonlinearly coupled Navier slip FSI model is given by:

$$\frac{1}{2}\|\partial_t \boldsymbol{\eta}(t)\|^2_{L^2(\Gamma)} + \|\boldsymbol{\eta}(t)\|^2_E + \frac{1}{2}\|\boldsymbol{u}(t)\|^2_{L^2(\Omega_f(t))}$$
$$+ \mu_f \int_0^t \|\boldsymbol{D}(\boldsymbol{u})(s)\|^2_{L^2(\Omega_f(s))} ds + \beta \int_0^t \|(\partial_t \boldsymbol{\eta} - \boldsymbol{u})|_{\Gamma(t)} \cdot \boldsymbol{\tau}(s)\|_{L^2(\Gamma(s))} ds$$

$$\leq \frac{1}{2}\|v_0\|^2_{L^2(\Gamma)}+\|\boldsymbol{\eta}_0\|^2_E+\frac{1}{2}\|\boldsymbol{u}_0\|^2_{L^2(\Omega_f(0))}+C\left(\|P_{in}(t)\|^2_{L^2(0,T)}+\|P_{out}(t)\|^2_{L^2(0,T)}\right),$$

where the term involving β represents the frictional dissipation due to the tangential slip of the fluid along the structure interface.

The second key difference is in the weak formulation. In the prototypical nonlinearly coupled no-slip model, studied in Chap. 5, the solution space incorporates a no-slip condition $\boldsymbol{u}|_{\Gamma(t)} = (\partial_t \eta)\boldsymbol{e}_r$, and the test space for the test functions $(\boldsymbol{q}, \psi)$ incorporates this condition also as $\boldsymbol{q}|_{\Gamma(t)} = \psi \boldsymbol{e}_r$. However, in the Navier slip case, the Navier-slip condition is enforced weakly via integration by parts in the weak formulation, and only the no-penetration condition is enforced strongly in the solution and test spaces. Therefore, we instead define the coupled solution space to be

$$\mathcal{V} = \{(\boldsymbol{u}, \boldsymbol{\eta}) \in V_f \times V_s : \boldsymbol{u} \cdot \boldsymbol{n}(t)|_{\Gamma(t)} = \partial_t \boldsymbol{\eta} \cdot \boldsymbol{n}(t)\}$$

for appropriate finite energy spaces V_f and V_s, and we define the test space to be

$$\mathcal{Q} = \{(\boldsymbol{q}, \boldsymbol{\psi}) \in C_c^\infty([0, T); \mathcal{V}_f(t) \times \mathcal{V}_s) : \boldsymbol{q} \cdot \boldsymbol{n}(t)|_{\Gamma(t)} = \boldsymbol{\psi} \cdot \boldsymbol{n}(t)\},$$

where $\mathcal{V}_f$ and $\mathcal{V}_s$ are defined similarly to (5.26) and (5.27) as in the prototypical nonlinearly coupled FSI model, but in the current context, the trace of the fluid velocity along $\Gamma(t)$ is no longer restricted to be purely radial, and the structure displacement can more generally be vector-valued.

In this case, the weak formulation involves finding a solution $(\boldsymbol{u}, \boldsymbol{\eta}) \in \mathcal{V}$ such that for all test functions $(\boldsymbol{q}, \boldsymbol{\psi}) \in \mathcal{Q}$,

$$\begin{aligned}
&-\int_0^T\int_{\Omega_f(t)} \boldsymbol{u}\cdot\partial_t\boldsymbol{q}+\frac{1}{2}\int_0^T\int_{\Omega_f(t)}\Big([(\boldsymbol{u}\cdot\nabla)\boldsymbol{u}]\cdot\boldsymbol{q}-[(\boldsymbol{u}\cdot\nabla)\boldsymbol{q})\boldsymbol{u}]\Big)-\int_0^T\int_{\Gamma(t)}(\partial_t\boldsymbol{\eta}\cdot\boldsymbol{n}(t))(\boldsymbol{u}\cdot\boldsymbol{q})\\
&\quad+\frac{1}{2}\int_0^T\int_{\Gamma(t)}(\boldsymbol{u}\cdot\boldsymbol{n}(t))(\boldsymbol{u}\cdot\boldsymbol{q})-\beta\int_{\Gamma(t)}[(\partial_t\boldsymbol{\psi}-\boldsymbol{q})\cdot\tau(t)]\cdot[(\partial_t\boldsymbol{\eta}-\boldsymbol{u})\cdot\tau(t)]\\
&\quad+2\mu_f\int_0^T\int_{\Omega_f(s)}\boldsymbol{D}(\boldsymbol{u}):\boldsymbol{D}(\boldsymbol{q})-\int_0^T\int_\Gamma\partial_t\boldsymbol{\eta}\cdot\partial_t\boldsymbol{\psi}+\int_0^T(\boldsymbol{\eta},\boldsymbol{\psi})_E\\
&=\int_0^T\int_{\Gamma_{in}}P_{in}(t)\cdot q_z-\int_0^T\int_{\Gamma_{out}}P_{out}(t)\cdot q_z+\int_{\Omega_f(0)}\boldsymbol{u}_0\cdot\boldsymbol{q}(0)+\int_\Gamma v_0\cdot\psi(0).
\end{aligned}$$

We conclude by making a few observations about this weak formulation. Rather than having a single term

$$-\int_0^T\int_\Gamma(\partial_t\eta)^2\cdot\psi$$

in the weak formulation as in the no-slip prototypical model, the weak formulation instead has the terms

$$-\int_0^T \int_{\Gamma(t)} (\partial_t \boldsymbol{\eta} \cdot \boldsymbol{n}(t))(\boldsymbol{u} \cdot \boldsymbol{q}) + \frac{1}{2} \int_0^T \int_{\Gamma(t)} (\boldsymbol{u} \cdot \boldsymbol{n}(t))(\boldsymbol{u} \cdot \boldsymbol{q}),$$

which can also be expressed as the single term $-\frac{1}{2} \int_0^T \int_{\Gamma(t)} (\partial_t \boldsymbol{\eta} \cdot \boldsymbol{n}(t))(\boldsymbol{u} \cdot \boldsymbol{q})$ by the no penetration condition. Compared to the weak formulation for the prototypical model, there is also an additional dissipation term in the weak formulation, which is the term involving the Navier slip coefficient β.

A similar fluid-structure splitting scheme can be developed for this Navier slip model; see [269]. However, in contrast with the no-slip boundary condition which "transmits" the regularizing mechanism by the viscous fluid dissipation onto the fluid-structure interface, in the Navier-slip condition, the tangential components of the fluid and structure velocities are no longer continuous, and thus information is lost in the tangential direction. The additional estimate that comes from the frictional dissipation provides only an L^2 estimate for the tangential components of the structure velocity, which is insufficient for the analogous compactness argument as in the no-slip case. However, we can still use the generalized Aubin-Lions-Simon theorem (Theorem 2.10.1) to prove the compactness of the sequence of approximate solutions, but we must use a different function space for the structure velocity; namely, we must use $H^{-s}(\Gamma)$, $0 < s < 1/2$, instead of $L^2(\Gamma)$. The function space for the fluid velocity remains the same as in the no-slip case. The L^2 convergence of the normal component of the structure displacement can be recovered a posteriori. For more details about the compactness argument, we refer the reader to [270, Section 4.3].

6.4 Multilayered Structures in FSI

Another important extension of the prototypical FSI model is the extension to systems involving *multilayered (composite) structures*, which are made up of different layers with possibly different mechanical properties. This extension is highly relevant to many real-world FSI applications, particularly in biomedical contexts where structures often comprise multiple layers with distinct physical characteristics. For example, arterial walls are made up of layers of tissue with different thicknesses and elastic properties, and bioartificial organs are typically encapsulated within an elastic or poroelastic thin layer [267]. The extension of the splitting scheme approach for FSI to the interaction between a fluid and a multilayered structure was first carried out in [267]. In this model, the fluid is an incompressible, viscous Newtonian fluid which is flowing in an elastic channel where the walls of the channel are elastic and are made up of multiple layers. The structure now consists of an external thick layer and an internal thin layer that is in contact with the fluid. The problem is posed on the following geometry: The fluid domain has a reference configuration of $\Omega_f = [0, L] \times [0, R]$, the thick structure has a reference configuration of $\Omega_s = [0, L] \times [R, R + h]$ where h is the thickness of the thick structure, and the thin structure is a one-dimensional elastic structure with a reference configuration of $\Gamma = [0, L] \times \{R\}$.

Let us define $\boldsymbol{d}$ to be the vector displacement of the two-dimensional thick structure from its reference configuration Ω_s, and let η be the transverse (vertical) displacement of the thin structure from its one-dimensional reference configuration Γ. The incompressible viscous Newtonian fluid is described by the Navier-Stokes equations (5.2) posed on a moving time-dependent fluid domain, where the moving fluid domain $\Omega_f(t)$ is determined by the position $\Gamma(t)$ of the inner thin structure layer in direct contact with the fluid so that

$$\Omega_f(t) = \{(z, r) \in \mathbb{R}^2 : 0 \leq z \leq L, 0 \leq r \leq R + \eta(t, z)\},$$

where we assume that the thin layer displaces in only the transverse (radial) direction so that

$$\Gamma(t) = \{(z, r) \in \mathbb{R}^2 : 0 \leq z \leq L, r = R + \eta(t, z)\}.$$

The elastodynamics of the **thin structure** is described by the wave equation, where the scalar transverse displacement η of the one-dimensional thin structure from its reference configuration Γ is given by

$$\partial_{tt}\eta - \Delta\eta = F, \qquad \text{on } \Gamma,$$

where the external load F on the thin structure will be described later in the dynamic coupling condition. We refer the reader to [67] for a FSI problem with multilayered structures where the thin structure is modeled by the Koiter membrane model allowing *vector displacement* in both r and z directions.

The elastodynamics of the thick structure is described by the equations of elasticity for the (two-dimensional) vector-valued displacement $\boldsymbol{d}$ of the thick structure from its reference configuration Ω_s so that

$$\partial_{tt}\boldsymbol{d} - \nabla \cdot \boldsymbol{S}(\boldsymbol{d}) = 0, \qquad \text{on } \Omega_s.$$

The elasticity stress tensor $\boldsymbol{S}$ for the thick structure is given by the Piola-Kirchhoff stress tensor

$$\boldsymbol{S}(d) = 2\mu \boldsymbol{D}(\boldsymbol{d}) + \lambda(\nabla \cdot \boldsymbol{d}),$$

where $\boldsymbol{D}(\boldsymbol{d})$ is the symmetrized gradient.

The kinematic and dynamic coupling conditions must now account for the fact that there are three subproblems that must be coupled together.

- The kinematic coupling conditions now describe the kinematic coupling between the fluid and the thin structure (interface), which is given by a no-slip condition

$$\boldsymbol{u}|_{\Gamma(t)} = (\partial_t \eta)\boldsymbol{e}_r,$$

and a coupling between the thin and thick structure, which is given by the continuity of displacements:

$$\boldsymbol{d}|_\Gamma = \eta \boldsymbol{e}_r, \qquad \text{on } \Gamma.$$

- The dynamic coupling condition specifies the load on the interface, which drives the interface elastodynamics. The load is given by the jump in the contact force, i.e., normal stress, across the interface, coming from the fluid Cauchy normal stress on one side, and the thick structure Piola-Kirchoff normal stress on the other:

$$\partial_{tt}\eta - \Delta\eta = -\mathcal{J}^\eta \boldsymbol{\sigma}(\boldsymbol{u}, p)\boldsymbol{n} \cdot \boldsymbol{e}_r|_{\Gamma(t)} + \boldsymbol{S}(\boldsymbol{\eta})\boldsymbol{e}_r \cdot \boldsymbol{e}_r|_\Gamma,$$

where $\mathcal{J}^\eta = \sqrt{1 + (\partial_z \eta)^2}$.

For the multilayered structure FSI problem, we obtain an energy estimate of the following form for the nonlinearly coupled multilayered structure FSI problem described above:

$$\frac{1}{2}\|\partial_t \eta(t)\|^2_{L^2(\Gamma)} + \|\eta(t)\|^2_{H^1(\Gamma)} + \frac{1}{2}\|\boldsymbol{u}(t)\|^2_{L^2(\Omega_f(t))} + \mu_f \int_0^t \|\boldsymbol{D}(\boldsymbol{u})(s)\|^2_{L^2(\Omega_f(s))} ds$$
$$+ \frac{1}{2}\|\partial_t \boldsymbol{d}(t)\|^2_{L^2(\Omega_s)} + \mu\|\boldsymbol{D}(\boldsymbol{d})(t)\|^2_{L^2(\Omega_s)} + \frac{\lambda}{2}\|\nabla \cdot \boldsymbol{d}(t)\|^2_{L^2(\Omega_s)}$$
$$\leq \frac{1}{2}\|v_0\|^2_{L^2(\Gamma)} + \|\eta_0\|^2_{H^1(\Gamma)} + \frac{1}{2}\|\boldsymbol{u}_0\|^2_{L^2(\Omega_f(0))} + \frac{1}{2}\|\partial_t \boldsymbol{d}(0)\|^2_{L^2(\Omega_s)}$$
$$+ \mu\|\boldsymbol{D}(\boldsymbol{d}_0)\|^2_{L^2(\Omega_s)} + \frac{\lambda}{2}\|\nabla \cdot \boldsymbol{d}_0\|^2_{L^2(\Omega_s)} + C\left(\|P_{in}(t)\|^2_{L^2(0,T)} + \|P_{out}(t)\|^2_{L^2(0,T)}\right).$$

As emphasized in the original manuscript [267], this energy estimate reveals one of the primary reasons why considering multilayered structures in FSI is mathematically advantageous, especially for problems involving thick structures that are of the same dimension as the fluid domain. Having a multilayered structure where there is a thin layer that has a lower dimension than the full composite structure, **regularizes the geometry** of the problem in the following way. For a thick elastic structure, the structure displacement $\boldsymbol{d}$ belongs to the finite energy space $L^\infty(0, T; H^1(\Omega_s))$ where Ω_s is two-dimensional. If the FSI model just involved a two-dimensional fluid in direct contact with a two-dimensional thick structure, then the time-dependent location of the interface between the fluid and the thick structure would be determined by the trace of $\boldsymbol{d}$ along the bottom boundary of Ω_s, which would only be in $L^\infty(0, T; H^{1/2}(\Gamma))$ in the finite energy space. Even with the interface Γ between the thick structure and the fluid being one-dimensional, $H^{1/2}(\Gamma)$ does not provide sufficient Sobolev regularity to ensure that we even have a continuous moving interface, and this lack of boundary regularity of the structure displacement for thick structures is one of the predominant mathematical obstacles in the analysis of FSI models between fluids and thick structures of the same dimension.

If we instead consider a composite multilayered structure interacting with a fluid as in the model described above, we see that the elastodynamics of the thin membrane separating the thick structure and the fluid regularize the time-dependent fluid-structure interface. In particular, even though $\boldsymbol{d}$ is still in $L^{\infty}(0, T; H^1(\Omega_s))$, we have by the kinematic coupling condition that the trace of $\boldsymbol{d}$ along Γ is equal to $\eta \boldsymbol{e}_r$, where by the energy estimates, $\eta \in L^{\infty}(0, T; H_0^1(\Gamma))$. Thus, the elastic energy of the interface $\Gamma(t)$ places its displacement η in a higher regularity space $H^1(\Gamma)$ in contrast to the space $H^{1/2}(\Gamma)$ that would arise from the Sobolev trace theorem applied to $\boldsymbol{d}$. Since we have that $\boldsymbol{d}|_{\Gamma} = \eta \boldsymbol{e}_r$ by the kinematic coupling condition, we see that the higher $H^1(\Gamma)$ regularity of the interface displacement (which is now the thin structure displacement) transfers to the trace of the thick structure displacement along the bottom boundary Γ of Ω_s. This allows us to define the fluid-structure interface $\Gamma(t)$ to be a continuous curve, which is a significant advantage for the mathematical analysis of the problem.

One can then extend the splitting scheme for the nonlinearly coupled prototypical model of FSI to this multilayered model by developing a similar fluid-structure splitting, where the structure subproblem involves updating both the thick layer and the thin layer elastodynamics, and the thin layer structure velocity is updated in both the fluid and structure subproblem, in order to reflect the fact that the thin layer structure velocity must match the fluid velocity along the moving fluid-structure interface $\Gamma(t)$. For full details on the extension of the Lie operator splitting scheme to the case of multilayered structures interacting with incompressible fluids, we refer the reader to the exposition in the manuscript [267].

6.5 FSI with Mesh-Supported Elastic Structures

We conclude this literature review by discussing a recent extension of the prototypical FSI model, motivated by advancements in biomedical engineering. Atherosclerosis, a cardiovascular condition caused by plaque buildup on arterial walls, restricts blood flow through the coronary arteries [208, 286]. Recent biomedical advances have significantly improved outcomes for patients affected by this condition. One of the most successful treatments is the use of stents—mesh-like prosthetic structures inserted into plaque-narrowed arteries to restore healthy blood circulation. The specific geometry and arrangement of the metallic rods in a stent are critical to its effectiveness. However, constructing physical stents for real-world testing can be costly and time-consuming, underscoring the importance of developing numerical methods and mathematical analysis to analyze stent designs and assess their performance computationally.

In past works, these thin mesh-like structures have been modeled as three-dimensional elastic bodies, which make finite element discretizations of stents costly and memory consuming. See the discussions, for example, in [335, 339]. To rectify these challenges, a new approach was introduced in [320], where stents are modeled as one-dimensional nets/networks of curved rods defined on a given mesh or graph topology. Namely, the stent is modeled as a network of one-dimensional hyperbolic equations which are coupled to each other at junction points where curved rods (i.e., stent struts) meet. This gives rise to a set of equations that can be solved efficiently and which accurately model stent dynamics. The rigorous justification of coupled curved rod models can be found in, e.g., [173, 198–200, 294].

Due to interest in simulating the behavior of stents implanted in arterial walls, the work presented in [336] addresses the coupled stent-shell problem where the shell simulates the blood vessel wall. In particular, the work in [336] considers Naghdi type shells coupled to the one-dimensional stent net equations and addresses issues such as existence of a unique solution to the coupled problem. The work in [336] does not consider the blood flow through the arteries. This was, however, done in [334, 335] where a coupled fluid-stent-shell interaction model was considered. In [334], the fluid was coupled to a mesh-reinforced shell via linear coupling, and the time-dependent Stokes equations were used to model the blood flow. In [335], the fluid and mesh-reinforced shell were nonlinearly coupled, and the fluid was modeled using the Navier-Stokes equations for an incompressible, viscous fluid. In both works, a time-splitting scheme was used to prove the existence of a weak solution.

A FSI problem involving mesh-supported shells has also been studied in [68] where the authors modeled the presence of a stent by simply changing the elasticity properties of the vascular wall. The vascular wall was modeled by a Koiter shell with higher elasticity and thickness coefficients at the location where the stent struts are located, making the shell stiffer at those places. In that work, the elasticity parameters and thickness of the elastic shell are allowed to have jump discontinuities and are hence only prescribed to be in L^∞. Again, the splitting scheme was used to prove the existence of weak solutions to this nonlinearly coupled FSI problem.

Since these first existence results via Lie operator splitting for mesh-supported structure FSI models, there has been additional analysis of such models. Analysis of the regularity of weak solutions to the linearly coupled model was carried out in recent work in [153]. In addition, the mathematical analysis of mesh-supported structures in FSI has set the foundation for the development of accurate numerical schemes which can be used to model real-life stent dynamics and assess the performance of various stent designs through computational simulations. This is done in [70] for example, where four different stent geometries are assessed through computations that are performed using a numerical method, where the full dynamics of blood flow interacting with arterial walls supported by different stent geometries are computationally analyzed. We remark that the study of stents is now an active area of research with many important practical implications for biomedicine. As a further example, we refer the reader to work in [341], where a numerical solver is used to model drug-eluting stents, which are stents which release anti-inflammatory agents into surrounding tissue in order to reduce the risk of arterial reclosure after stent implantation. In addition, studies of optimal stent design have been carried out in order to inform decisions about the design of stents in bioengineering, using mathematical techniques for constrained optimization problems; see [337].

Part III
Stochastic Fluid-Structure Interaction (SFSI)

Chapter 7
Stochastic FSI: A Reduced Model

In this chapter, we begin the study of stochastic fluid-structure interaction by considering a *reduced* model of stochastic FSI. As seen in Chaps. 4 and 5, even the analysis of fully coupled deterministic models of FSI, where *fully coupled* refers to FSI systems modeled by a coupled system of PDEs, is already challenging due to the two-way coupling between the fluid and the structure. Hence, in this chapter, we begin the study of FSI by studying a *reduced model*, in which the full stochastic dynamics of both the fluid and the structure under the influence of external stochastic perturbations can be modeled by a single self-contained equation for the structure displacement, known as the *stochastic viscous wave equation*. We begin with this reduced model and note that because we are dealing with a single stochastic equation, we can use mathematical tools, such as an explicit fundamental solution, in order to analyze the equation. In particular, the study of this reduced model, as a result of having a single stochastic equation describing the full system, allows for the use of explicit mathematical tools that are unavailable in the case of fully coupled stochastic models, such as those introduced in the later chapters, Chaps. 8 and 9. The availability of mathematical methods that can be used for studying the single reduced stochastic equation rather than a fully coupled stochastic system allows us to obtain not just well-posedness results (Theorem 7.3.1) but also very explicit information about the regularity of sample paths of solutions; see Theorem 7.4.1.

The model under consideration in this chapter, from which a reduced stochastic equation for FSI arises, is a fully coupled fluid-structure interaction model involving a viscous incompressible Newtonian fluid modeled by the stationary Stokes equations in the lower half space $(z < 0) \subset \mathbb{R}^3$ interacting dynamically with an elastic structure which displaces transversally from its reference configuration $(z = 0)$. We assume two-way coupling between the fluid and structure, and the model is *linearly coupled*, so that the stationary Stokes equations for the fluid are posed on the fixed reference domain for the fluid $\Omega_f = (z < 0)$. In order to consider the effects of randomness on the coupled fluid-structure dynamics, we consider space-time white noise forcing, scaled by a nonlinear Lipschitz function of the structure

S. Čanić et al., *Deterministic and Stochastic Fluid-Structure Interaction*, Advances in Mathematical Fluid Mechanics, https://doi.org/10.1007/978-3-032-00898-5_7

displacement, acting on the elastic membrane, where the stochastic forcing on the membrane perturbs the coupled dynamics of the fluid and structure, as a result of the two-way coupling between the fluid and structure.

This model is a *reduced model*, because even though the original model is described in terms of a fluid subproblem and a structure subproblem that are appropriately coupled through a kinematic and dynamic coupling condition, one can show that the full dynamics of the structure and fluid can be described by a single self-contained equation for the structure displacement, known as the stochastic viscous wave equation. This is a wave equation perturbed by stochastic space-time white noise forcing, with an additional term $\sqrt{-\Delta}\eta_t$ arising from the Dirichlet-to-Neumann operator for the lower half space, which accounts for the regularizing effects of the fluid viscosity on the structure dynamics. The original fully coupled FSI model, involving a coupled system of PDEs, can be reduced to a single self-contained stochastic equation for the structure displacement, since the fluid normal stress on the structure can be expressed purely in terms of the structure displacement with the help of the dynamic coupling condition. This allows us to consider a single stochastic equation for the whole stochastic FSI model, and the specific geometry for this model allows us to consider a stochastic equation on $\mathbb{R}^2$ for the physical model, called the **stochastic viscous wave equation**. We can more generally consider this stochastic viscous wave equation for general spatial dimensions n, where we emphasize that $n = 2$ is the dimension of the physical FSI model from which we derived this equation.

Having a single self-contained stochastic equation on $\mathbb{R}^n$ allows us to analyze this fluid-structure interaction model by using classical methods from stochastic PDEs that have been used to study stochastic heat and wave equations, for example. This involves considering *mild solutions*, which are solutions that are defined via convolution of the fundamental solution with the space-time white noise via an appropriately defined stochastic integral. In this chapter, we will derive the stochastic viscous wave equation and analyze its well-posedness in various dimensions. The important result here will be well-posedness in terms of mild solutions in spatial dimensions $n = 1$ and $n = 2$, which improves upon classically known results for the stochastic heat and wave equations. We will also show that the stochastic viscous wave equation has better Hölder regularity of mild solutions in both spatial dimensions $n = 1$ and $n = 2$ than the stochastic heat and wave equations. The improvements in terms of well-posedness and regularity of mild solutions to the stochastic viscous wave equation when compared to the classical stochastic heat and wave equations with space-time white noise arise from a combination of a favorable space-time scaling in the equation, combined with the regularizing effects of fluid viscosity on the structure. While the methods we use will be based on fundamental solutions, Fourier analysis, and stochastic integration rather than the splitting scheme approaches found in previous chapters, we discuss this reduced model, since the first well-posedness results for stochastic FSI were developed in the context of this reduced model and led the way for further analysis of such stochastic fluid-structure systems. In addition, the methods used in the analysis of this reduced model and the stochastic viscous wave equation provide a nice application of the theory of space-time white noise and stochastic integration.

The reduced stochastic FSI model considered in this chapter was studied in [216].

7.1 Introduction

We focus on a stochastic model for fluid-structure interaction given by a stochastic wave equation augmented by dissipation associated with the effects of an incompressible, viscous fluid:

$$\eta_{tt} + 2\mu\sqrt{-\Delta}\eta_t - \Delta\eta = f(\eta)W(dt, dx), \qquad \text{in } \mathbb{R}^n. \tag{7.1}$$

The wave operator models the elastodynamics of a linearly elastic membrane, where η denotes membrane displacement, while the dissipative part, which is in the form of the Dirichlet-to-Neumann operator applied to the time derivative of displacement, accounts for dissipation due to fluid viscosity, where μ denotes the fluid viscosity coefficient. The equation is forced by space-time white noise $W(dt, dx)$, which accounts for stochastic effects in real-life problems; see Sect. 3.7 for a discussion of space-time white noise. The space-time white noise is scaled by a nonlinear, Lipschitz function $f(\eta)$. We show below how this equation is derived from a coupled fluid-structure interaction problem involving the Stokes equations describing the flow of an incompressible, viscous fluid, and the wave equation modeling the elastodynamics of a (stretched) linearly elastic membrane. We consider Eq. (7.1) in $\mathbb{R}^n$ with $n = 1$ and $n = 2$, focusing primarily on $n = 2$, which is the physical dimension.

We prove the existence of a function-valued *mild solution* to a Cauchy problem for Eq. (7.1), which holds both in dimensions 1 and 2. Here, by "mild solution," we refer to a stochastic mild solution defined via stochastic integration involving the Green's function, specified below in Definition 7.3.1. This is interesting, because our result contrasts the results that hold for the stochastic heat and wave equations: the stochastic heat and the stochastic wave equations do not have function-valued mild solutions in spatial dimension 2 or higher. Additionally, we prove that sample paths of the stochastic mild solution for the stochastic viscous wave equation are Hölder continuous with Hölder exponents $\alpha \in [0, 1)$ for $n = 1$, and $\alpha \in [0, 1/2)$ for $n = 2$.

These well-posedness results show that the viscous fluid dissipation in fluid-structure interaction is sufficient to smooth out the rough stochastic nature of the real-life data in the problem modeled by the space-time white noise. In particular, the Dirichlet-to-Neumann operator controls the high frequencies in the structure (membrane) displacement that are driven by the space-time white noise. The work done on this reduced model represents the first rigorous study of well-posedness for stochastic fluid-structure interaction systems.

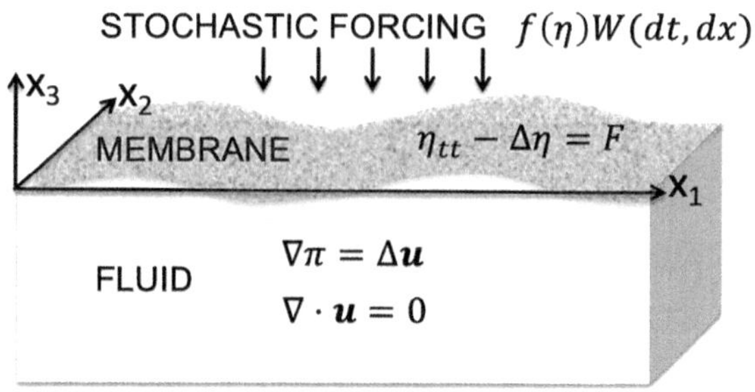

Fig. 7.1 A sketch of the fluid and structure domains

The Fluid-Structure Interaction Model We begin by describing the fluid-structure interaction model from which the Eq. (7.1) arises. Consider a prestressed infinite elastic membrane surface, which is modeled by the linear wave equation

$$\eta_{tt} - \Delta\eta = F_s, \quad \text{on } \Gamma := \{(x_1, x_2, 0) \in \mathbb{R}^3 : (x_1, x_2) \in \mathbb{R}^2\}, \tag{7.2}$$

where $\eta(x_1, x_2)$ denotes the transverse displacement (in the x_3 direction) of the elastic surface from its reference configuration Γ and the external load F_s will be specified later in the dynamic coupling condition. See Fig. 7.1.

Beneath this elastic drum surface, we consider a viscous, incompressible fluid, which resides in the lower half-space in $\mathbb{R}^3$,

$$\mathcal{O}_f = \{(x_1, x_2, x_3) \in \mathbb{R}^3 : x_3 < 0\}, \tag{7.3}$$

modeled by the stationary Stokes equations for an incompressible, viscous fluid:

$$\left.\begin{aligned} \nabla\cdot\boldsymbol{\sigma}(\pi, \boldsymbol{u}) = 0, \\ \nabla\cdot\boldsymbol{u} = 0, \end{aligned}\right\} \quad \text{in } \mathcal{O}_f = \{(x_1, x_2, x_3) \in \mathbb{R}^3 : x_3 < 0\}, \tag{7.4}$$

where $\boldsymbol{\sigma}$ is the Cauchy stress tensor, and the unknown quantities are the fluid pressure $\pi : \mathcal{O}_f \to \mathbb{R}$ and the fluid velocity $\boldsymbol{v} : \mathcal{O}_f \to \mathbb{R}^3$. See Sect. 2.13.2 for more information about the stationary Stokes equations. We will be assuming that the fluid is Newtonian, so that

$$\boldsymbol{\sigma} = -\pi\boldsymbol{I} + 2\mu\boldsymbol{D}(\boldsymbol{u}),$$

where μ denotes the fluid viscosity coefficient, $\boldsymbol{I}$ is the three by three identity matrix, and $\boldsymbol{D}(\boldsymbol{u})$ is the symmetrized gradient of fluid velocity $\boldsymbol{D}(\boldsymbol{u}) = (\nabla\boldsymbol{u} + (\nabla\boldsymbol{u})^T)/2$. Therefore, the Stokes equations now read

$$\left.\begin{aligned} \nabla\pi &= \mu\Delta\boldsymbol{u}, \\ \nabla\cdot\boldsymbol{u} &= 0, \end{aligned}\right\} \quad \text{in } \mathcal{O}_f = \{(x_1, x_2, x_3) \in \mathbb{R}^3 : x_3 < 0\}, \tag{7.5}$$

where we require that the fluid velocity is bounded in the lower half space and the pressure $\pi \to 0$ as $|x| \to \infty$.

We consider the problem in which the elastic surface is displaced from its reference configuration Γ with some given initial displacement and velocity, allowing only vertical displacement, where the elastodynamics of the elastic surface is driven by the total force exerted onto the membrane, which comes from the fluid on one side, and an external stochastic forcing on the other. See Fig. 7.1.

Infinite domains are considered to simplify the analysis, since the main purpose of this work is to understand the interplay between the dispersion effects in the 2D wave equation, dissipation due to fluid viscosity, and stochasticity imposed by the external forcing, which can be related to the stochasticity of not only the external forcing, but also to the stochasticity of data (e.g., inlet/outlet data) in real-life applications, such as blood flow through arteries.

The fluid and the structure are coupled via two coupling conditions, the kinematic and dynamic coupling conditions, giving rise to the so-called two-way coupled fluid-structure interaction problem. The coupling conditions in the present study are evaluated at a *fixed (linearized) fluid-structure interface* corresponding to the structure's reference configuration Γ. This is known as *linear coupling* (see the discussion of linearly coupled FSI models, e.g., in Chap. 4). The two conditions read:

Kinematic Coupling Condition The kinematic coupling condition describes the coupling between the kinematic quantities such as velocity. We will be assuming the no-slip condition, meaning that the fluid and structure velocities are continuous at the interface (there is no slip between the two):

$$(\partial_t \eta)\boldsymbol{e}_{x_3} = \boldsymbol{u}|_\Gamma, \quad \text{for } x = (x_1, x_2) \in \mathbb{R}^2, t \geq 0. \tag{7.6}$$

Dynamic Coupling Condition The dynamic coupling condition describes the balance of forces at the fluid-structure interface Γ; namely, it states that the elastodynamics of the membrane is driven by the force corresponding to the jump in traction (normal stress) across the membrane. This specifies the external load F_s on the structure elastodynamics in the structure equation (7.2). On the fluid side, the traction (normal stress) at the interface is given by $-\boldsymbol{\sigma}\boldsymbol{e}_{x_3}$, where $\boldsymbol{e}_{x_3}$ is the normal vector to Γ, while on the other side, we are assuming a given loading $F_{ext}(\eta)$ to be a stochastic process $f(\eta)W(dt, dx)$ in the $\boldsymbol{e}_{x_3}$ direction. Examples of such a loading can be found in cardiovascular applications; see, e.g., [260]. Since we assume that the structure only has transversal displacement, the dynamic coupling condition reads

$$\eta_{tt} - \Delta\eta = -\boldsymbol{\sigma}\boldsymbol{e}_{x_3} \cdot \boldsymbol{e}_{x_3} + f(\eta(t, x))W(dt, dx) \text{ where } x = (x_1, x_2) \in \mathbb{R}^2, t \geq 0. \tag{7.7}$$

In fluid-structure interaction problems and physical problems in general, physical phenomena are subject to small random deviations that cause deviations from deterministic behavior. The consideration of such stochastic effects in partial differential equations can give rise to new phenomena and is an area of active research. Furthermore, in real-life data, one observes such stochastic noise both in terms of the force exerted onto the structure, as well as in the data that drives the problem. For example, the measured inlet/outlet pressure data in a fluid-structure interaction problem describing arterial blood flow has similar stochastic noise deviations to $F_{ext}(\eta) = f(\eta(t,x))W(dt,dx)$. Here $W(dt,dx)$ is space-time white noise in $(t,x) \in \mathbb{R}^+ \times \mathbb{R}^2$, which was discussed previously in Sect. 3.7. We will assume that $f : \mathbb{R} \to \mathbb{R}$ is a Lipschitz continuous function. In particular, the case $f \not\equiv 1$ allows dependence of the magnitude of the stochastic noise at each point on the structure displacement $\eta(t,x)$ itself.

To derive equation (7.1) as a model which describes the fluid-structure interaction problem (7.2)–(7.7), we focus on the dynamic coupling condition (7.7). The goal is to try to express the effects of fluid normal stress via the Dirichlet-to-Neumann operator defined entirely in terms of η and/or its derivatives. In this derivation, we also use the kinematic coupling condition (7.6) as explained below. We recall the following derivation of the viscous wave operator in (7.1) from [213].

First notice that the right-hand side of (7.7) is given by

$$-\boldsymbol{\sigma}\boldsymbol{e}_{x_3} \cdot \boldsymbol{e}_{x_3} = \pi - 2\mu \frac{\partial u_{x_3}}{\partial x_3} \quad \text{on } \Gamma.$$

Since the tangential displacements are assumed to be zero, the kinematic coupling condition (7.6) implies that the x_1 and x_2 components of the fluid velocity $\boldsymbol{u}$ are zero on Γ, namely, $u_{x_1} = u_{x_2} = 0$ on Γ. By the divergence free condition, one immediately gets that $\partial u_{x_3}/\partial x_3 = 0$ on Γ. Therefore,

$$-\boldsymbol{\sigma}\boldsymbol{e}_{x_3} \cdot \boldsymbol{e}_{x_3} = \pi \quad \text{on } \Gamma, \tag{7.8}$$

where π is the fluid pressure given as a solution to the Stokes equations (2.30). So it remains to find an appropriate expression for π on Γ in terms of the structure displacement η. In fact, we will show that the following formula holds

$$\pi = -2\mu\sqrt{-\Delta}\partial_t \eta \quad \text{on } \Gamma, \tag{7.9}$$

under the assumption that η and $\partial_t \eta$, along with their spatial derivatives, are smooth functions that are rapidly decreasing at infinity. We will also impose the boundary conditions on (2.30), stating that the fluid velocity is bounded on the lower half space, and the pressure π has a limit equal to zero as $|x| \to \infty$ in the lower half space.

To derive the formula (7.9), we note that by taking the inner product of the first equation in (2.30) with $\boldsymbol{e}_{x_3}$, we obtain

$$\frac{\partial \pi}{\partial x_3} = \mu \Delta_{x_1,x_2} u_{x_3} + \mu \frac{\partial^2 u_{x_3}}{\partial x_3^2} = \mu \Delta u_{x_3}, \tag{7.10}$$

where $\Delta_{x_1,x_2} := \frac{\partial^2}{\partial x_1^2} + \frac{\partial^2}{\partial x_2^2}$. Furthermore, by taking the divergence of the first equation in (2.30), and by using the divergence-free condition, we get that the pressure π is harmonic. Thus, if we can compute the right-hand side of (7.10) on Γ, we can recover π as the solution to a Neumann boundary value problem for Laplace's equation in the lower half space, with the boundary condition requiring that π goes to zero at infinity.

To compute the right-hand side of (7.10), we need to compute u_{x_3}. Taking the Laplacian on both sides of (7.10), and using the fact that π is harmonic, we obtain

$$\Delta^2 u_{x_3} = 0, \qquad \text{on } \mathcal{O}_f = \{(x_1, x_2, x_3) \in \mathbb{R}^3 : x_3 < 0\}. \tag{7.11}$$

Thus, u_{x_3} satisfies the biharmonic equation with the following two boundary conditions: (1) from the kinematic coupling condition, we get

$$u_{x_3}(x_1, x_2, 0) = \eta_t(x_1, x_2, 0), \qquad \text{on } \Gamma = \{(x_1, x_2, x_3) \in \mathbb{R}^3 : x_3 = 0\}, \tag{7.12}$$

and (2) from the fact that $u_{x_1} = u_{x_2} = 0$ on Γ, by the kinematic coupling condition and the fact that $\boldsymbol{u}$ is divergence free, we get

$$\frac{\partial u_{x_3}}{\partial x_3}(x_1, x_2, 0) = 0, \qquad \text{on } \Gamma = \{(x_1, x_2, x_3) \in \mathbb{R}^3 : x_3 = 0\}. \tag{7.13}$$

We solve (7.11) with boundary conditions (7.12) and (7.13) by taking a Fourier transform in the variables x_1 and x_2 but not in x_3. We will denote the Fourier variables associated with x_1 and x_2 by ξ_1 and ξ_2, and we will denote $\xi = (\xi_1, \xi_2)$, $|\xi|^2 = \xi_1^2 + \xi_2^2$. The Fourier transform equation then reads

$$|\xi|^4 \widehat{u_{x_3}}(\xi, x_3) - 2|\xi|^2 \frac{\partial^2}{\partial x_3^2} \widehat{u_{x_3}}(\xi, x_3) + \frac{\partial^4}{\partial x_3^4} \widehat{u_{x_3}}(\xi, x_3) = 0. \tag{7.14}$$

We want to solve for $\widehat{u_{x_3}}(\xi, x_3)$ by regarding this Eq. (7.14) as a differential equation in x_3 for each $\xi \in \mathbb{R}^2$, supplemented with the boundary conditions

$$\widehat{u_{x_3}}(\xi, 0) = \widehat{\eta_t}(\xi), \qquad \frac{\partial \widehat{u_{x_3}}}{\partial x_3}(\xi, 0) = 0, \tag{7.15}$$

which are the Fourier transforms of the respective boundary conditions (7.12) and (7.13). For each $\xi \in \mathbb{R}^2$, the general solution to (7.14) is a linear combination (with coefficients depending on $\xi \in \mathbb{R}^2$) of the four solutions: $e^{|\xi|x_3}, x_3 e^{|\xi|x_3}, e^{-|\xi|x_3}, x_3 e^{-|\xi|x_3}$, but taking into account the desired decay of the resulting solution and its derivatives at infinity, we exclude the last two of these $e^{-|\xi|x_3}$ and $x_3e^{-|\xi|x_3}$, which grow exponentially as $x_3 \to -\infty$. Thus, taking into account the boundary conditions stated in (7.15), we conclude that the solution to (7.14) is:

$$\widehat{u_{x_3}}(\xi, x_3) = \widehat{\eta_t}(\xi)e^{|\xi|x_3} - |\xi|\widehat{\eta_t}(\xi)x_3e^{|\xi|x_3}. \tag{7.16}$$

We can now compute the right-hand side of (7.10). Taking the Fourier transform of (7.10) in the x_1 and x_2 variables, and evaluating the equation on Γ by using the kinematic coupling condition (7.6), we get

$$\frac{\partial\widehat{\pi}}{\partial x_3}(\xi, 0) = -\mu|\xi|^2\widehat{\eta_t}(\xi) + \mu\frac{\partial^2\widehat{u_{x_3}}}{\partial x_3^2}(\xi, 0) = -2\mu|\xi|^2\widehat{\eta_t}(\xi), \tag{7.17}$$

where the last equality follows by using the explicit formula for $\widehat{u_{x_3}}(\xi, x_3)$ in (7.16).

We now know that the pressure π is a harmonic function in the lower half space, satisfying the Neumann boundary condition (7.17) given in Fourier space. To recover formula (7.9), we want π on Γ. This can be obtained via the Neumann to Dirichlet operator. It is well known that the *Dirichlet to Neumann* operator for Laplace's equation in the lower half space is given by $\sqrt{-\Delta}$, thereby having a Fourier multiplier $|\xi|$; see e.g., [72]. Therefore, the Neumann to Dirichlet operator for Laplace's equation in the lower half space (with the solution to Laplace's equation having a limit of zero at infinity) is a Fourier multiplier of the form $\frac{1}{|\xi|}$. Thus, the Neumann to Dirichlet operator applied to the Neumann data (7.17) gives the pressure as Dirichlet data:

$$\widehat{\pi}(\xi) = -2\mu|\xi|\widehat{\eta_t}(\xi) \qquad \text{on } \Gamma,$$

which establishes the desired formula (7.9).

The dynamic coupling condition (7.7), together with (7.8) and (7.9) gives the stochastic model

$$\eta_{tt} + \sqrt{-\Delta}\eta_t - \Delta\eta = f(\eta)W(dt, dx) \quad \text{on } \mathbb{R}^2,$$

where we have set the fluid viscosity $\mu = 1/2$. We will refer to this model as the *stochastic viscous wave equation*, as it is a stochastic wave equation augmented by the viscous effects of the fluid, which are captured by the Dirichlet to Neumann operator acting on the structure velocity $\partial_t\eta$.

On the other hand, the study of stochasticity in PDEs has been of recent interest. Most physical phenomena occurring in real-life applications feature the presence of some sort of random noise that perturbs the system from what may be deterministically expected. The types of noise added to the equation can vary, from the simplest space-time white noise, which intuitively is noise that is "independent at each time and space," to forms of noise with smoother spatial correlation, so that the noise is still independent at separate times, but the noise in space allows for correlation between points and is hence "smoother" than white noise.

Many classical partial differential equations, such as the heat and wave equations, have been studied with the addition of stochastic random forcing. There are many approaches to the study of such stochastic PDEs. One approach uses Walsh's theory of martingale measures of which white noise is an example. The theory of integration against such martingale measures can be found in Walsh's work [346]. Upon defining such an appropriate theory of stochastic integration, one can define what is called a *mild solution* to a given stochastic partial differential equation by means of the Green's function.

The choice of the stochastic noise used in the PDE being studied is of utmost importance. White noise, which is noise that is intuitively "independent at all spaces and times," is a starting point for many studies, and we refer the reader to the previous discussion in Sect. 3.7 for details about its properties and stochastic integration against white noise. From Sect. 3.7, we recall that in formal mathematical notation, the time and space independence property of white noise is expressed via expectation as

$$\mathbb{E}(\dot{W}(t,x)\dot{W}(s,y)) = \delta_0(t-s)\delta_0(x-y),$$

where δ_0 is the Dirac delta function, and $\dot{W}(t,x)$, or $W(dt,dx)$, denotes space-time white noise.

The white noise perturbed heat and wave equations have interesting properties. When perturbed by white noise, the two equations:

$$\eta_t - \Delta\eta = W(dt,dx) \quad \text{and} \quad \eta_{tt} - \Delta\eta = W(dt,dx) \qquad \text{in } \mathbb{R}^n,$$

do not allow function-valued mild solutions in spatial dimensions two and higher, while they do in spatial dimension one. See, for example, [107] and [210], where questions of existence and uniqueness of mild solutions are addressed. This interesting property related to spatial dimensions two and higher is due to the lack of square integrability of the Green's function in time and space, as we discuss later. Because of this property, "smoother" types of stochastic noise, such as spatially homogeneous Gaussian noise, are used to perturb the stochastic heat and wave equations in higher dimensions in order to yield function-valued mild solutions. In formal mathematical notation, such spatially homogeneous Gaussian noise $\dot{F}(t,x)$ has a covariance structure

$$\mathbb{E}(\dot{F}(t,x)\dot{F}(s,y)) = \delta_0(t-s)k(x-y), \tag{7.18}$$

where $k : \mathbb{R}^n \to \mathbb{R}$. Note that the formal case of setting k to be the Dirac delta "function" recovers the previous white noise case, though choosing smoother functions k allows us to formulate "smoother" types of noise. See Sec. 3 of [107] and the work in [108] for more information about spatially homogeneous Gaussian noise. One of the key questions in studying stochastically perturbed PDEs is: What conditions on k need to be imposed, so that the resulting equation with the spatially homogeneous Gaussian noise with covariance structure (7.18) has function-valued mild solutions? See, for example, [108], and for more general contexts [106] and [207].

In the current chapter, we will not take the approach of considering the stochastic viscous wave equation perturbed by general spatially homogeneous Gaussian noise. This is because, as we shall see below, the viscous wave equation

$$\eta_{tt} + \sqrt{-\Delta}\eta_t - \Delta\eta = F, \qquad \text{on } \mathbb{R}^n, \tag{7.19}$$

considered in [213, 214] as a model for fluid-structure interaction, combines the following two desirable properties: the "right" space-time scaling (c.f. wave equation) and adequate dissipative effects. The resulting behavior is "in between" the wave and heat equations. The viscous wave equation (7.19) turns out to have just the right scaling and dissipation to allow function-valued mild solutions even in spatial dimension two for the white noise perturbed equation

$$\eta_{tt} + \sqrt{-\Delta}\eta_t - \Delta\eta = W(dt, dx) \qquad \text{in } \mathbb{R}^n. \tag{7.20}$$

This is of great interest, since Eqs. (7.1) and (7.20) in two spatial dimensions correspond exactly to the physical fluid-structure interaction model we are considering and hence have direct physical significance.

The main results of this Chapter are (1) the existence of a function-valued mild solution for the space-time white noise perturbed viscous wave equation (7.1) (and (7.20)) for dimensions $n = 1$ and $n = 2$, and (2) Hölder continuity $C^{0,\alpha}$ with $\alpha \in [0, 1)$ for $n = 1$, and $\alpha \in [0, 1/2)$ for $n = 2$, of "every" realization of the displacement η, obtained as a mild solution to the randomly perturbed viscous wave equation.

In particular, in terms of Hölder continuity, our results imply that the stochastic mild solution to Eq. (7.1) with zero initial data has a continuous *modification* that is α-Hölder continuous in time and space, with $\alpha \in [0, 1)$ for $n = 1$, and $\alpha \in [0, 1/2)$ for $n = 2$. Here, a *modification* of a stochastic process $\{X_i\}_{i\in I}$ is defined to be a stochastic process $\{\tilde{X}_i\}_{i\in I}$ such that the probability $\mathbb{P}(\tilde{X}_i = X_i) = 1$, for all $i \in I$. Thus, we show that the stochastic function-valued mild solution has a modification that is a Hölder continuous function for every realization η of the displacement, obtained as a mild solution to the randomly perturbed viscous wave equation (7.1).

Even in dimension $n = 1$, this contrasts the results for the stochastically perturbed heat and wave equations:

$$\eta_t - \Delta\eta = f(\eta)W(dt, dx) \quad \text{and} \quad \eta_{tt} - \Delta\eta = f(\eta)W(dt, dx) \qquad \text{on } \mathbb{R}. \tag{7.21}$$

Namely, in $n = 1$, the function-valued mild solutions (up to modification) for zero initial data are α-Hölder continuous in time and β-Hölder continuous in space, where $\alpha \in [0, 1/4)$, $\beta \in [0, 1/2)$ for the stochastic heat equation, and $\alpha, \beta \in$

[0, 1/2) for the stochastic wave equation; see [107, 210], and [178]. The difference in space and time Hölder regularity between the heat and wave equations is due to the scaling of space and time, where for the heat equation, one-time derivative "corresponds" to two spatial derivatives, while for the wave equation, one-time derivative "corresponds" to one spatial derivative. In the stochastic viscous wave equation, the additional regularizing effect of the fluid viscosity implies improved Hölder regularity. In spatial dimension one, the solution is Hölder continuous of order $\alpha \in [0, 1)$ in space and time, which is an improvement over the results for both the stochastic heat and wave equations. In spatial dimension two, the solution is Hölder continuous of order $\alpha \in [0, 1/2)$ in space and time, whereas the stochastic heat and wave equations do not have function-valued mild solutions in spatial dimension two. We summarize these Hölder continuity results for mild solutions with appropriate initial data in Table 7.1, for the three stochastic equations with space-time white noise stochasticity that we have previously mentioned.

The literature on the Hölder continuity properties of the solutions to the heat equation and the wave equation with random noise in spatial dimensions two and higher is an area of extensive study. However, we emphasize again that for these equations, the stochastic noise is not white noise but something smoother, such as, e.g., spatially homogeneous Gaussian noise, as a function-valued mild solution does not exist with space-time white noise in spatial dimensions two and higher for these equations. We refer the reader to [98, 110, 295] for more details.

This chapter is organized as follows. In Sect. 7.2, we recall the properties of the deterministic forms of the heat, wave, and viscous wave equation solutions that will be necessary to show the main result. In Sect. 7.3, we show the existence and uniqueness of a stochastic function-valued mild solution for Eq. (7.1), in dimensions $n = 1, 2$, where $f : \mathbb{R}^n \to \mathbb{R}$ is a Lipschitz continuous function, and in Sect. 7.4, we study the Hölder continuity of sample paths of solutions to (7.20).

Table 7.1 This table shows the α-Hölder regularity in time and the β-Hölder regularity in space for the mild solutions of the various stochastic equations with space-time white noise in spatial dimensions $n = 1$ and $n = 2$. Note that the regularity of the mild solutions is improved for the stochastic viscous wave equation over the classical stochastic heat and wave equations, where these equations are all considered with space-time white noise forcing

Stochastic equation with $W(dt, dx)$	Spatial dimension n	Regularity of mild solution
Stochastic heat equation	$n = 1$	$\alpha \in [0, 1/4)$, $\beta \in [0, 1/2)$
Stochastic wave equation	$n = 1$	$\alpha \in [0, 1/2)$, $\beta \in [0, 1/2)$
Stochastic viscous wave equation	$n = 1$	$\alpha \in [0, 1)$, $\beta \in [0, 1)$
Stochastic heat equation	$n = 2$	Mild solution does not exist
Stochastic wave equation	$n = 2$	Mild solution does not exist
Stochastic viscous wave equation	$n = 2$	$\alpha \in [0, 1/2)$, $\beta \in [0, 1/2)$

7.2 Deterministic Linear Equations

In this section, we recall some basic facts about the deterministic linear heat, wave, and viscous wave equations that will be needed later in the analysis of the stochastic viscous wave equation.

7.2.1 *The Viscous Wave Equation*

We begin by considering the (linear) viscous wave equation

$$\eta_{tt} + \sqrt{-\Delta}\eta_t - \Delta\eta = 0, \tag{7.22}$$

with initial data

$$\eta(0, x) = g(x), \qquad \partial_t \eta(0, x) = h(x).$$

We recall some basic properties related to the analysis of this equation here and recommend [213] for more information. Assuming that the initial data g, h are regular enough, for example, $g, h \in \mathcal{S}(\mathbb{R}^n)$, we can explicitly solve this equation using the Fourier transform (see Sect. 2.6) to obtain

$$\begin{aligned}\widehat{\eta}(t,\xi) &= \widehat{g}(\xi)e^{-\frac{|\xi|}{2}t}\left(\cos\left(\frac{\sqrt{3}}{2}|\xi|t\right) + \frac{1}{\sqrt{3}}\sin\left(\frac{\sqrt{3}}{2}|\xi|t\right)\right) \\ &\quad + \widehat{h}(\xi)e^{-\frac{|\xi|}{2}t}\frac{\sin\left(\frac{\sqrt{3}}{2}|\xi|t\right)}{\frac{\sqrt{3}}{2}|\xi|}.\end{aligned} \tag{7.23}$$

From the Fourier representation, one can see that this equation has both parabolic and wave-like properties. The wavelike behavior is represented by the presence of cosine and sine, and the strong parabolic dissipation is given by the exponential factor $e^{-\frac{|\xi|}{2}t}$, which causes damping of frequencies over time.

Of particular interest to this work is the solution to the general inhomogeneous problem

$$\eta_{tt} + \sqrt{-\Delta}\eta_t - \Delta\eta = F, \tag{7.24}$$

with initial data $u(0, x) = g(x)$, $\partial_t u(0, x) = h(x)$, which can be obtained using Duhamel's principle:

$$\widehat{\eta}(t,\xi) = \widehat{g}(\xi)e^{-\frac{|\xi|}{2}t}\left(\cos\left(\tfrac{\sqrt{3}}{2}|\xi|t\right) + \tfrac{1}{\sqrt{3}}\sin\left(\tfrac{\sqrt{3}}{2}|\xi|t\right)\right) + \widehat{h}(\xi)e^{-\frac{|\xi|}{2}t}\frac{\sin\left(\frac{\sqrt{3}}{2}|\xi|t\right)}{\frac{\sqrt{3}}{2}|\xi|}$$

$$+\int_0^t \widehat{F}(\tau,\xi)e^{-\frac{|\xi|}{2}(t-\tau)}\frac{\sin\left(\frac{\sqrt{3}}{2}|\xi|(t-\tau)\right)}{\frac{\sqrt{3}}{2}|\xi|}d\tau. \tag{7.25}$$

The inverse Fourier transform gives the solution u in physical space:

$$\begin{aligned}\eta(t,\cdot) &= e^{-\frac{\sqrt{-\Delta}}{2}t}\left(\cos\left(\frac{\sqrt{3}}{2}\sqrt{-\Delta t}\right)+\frac{1}{\sqrt{3}}\sin\left(\frac{\sqrt{3}}{2}\sqrt{-\Delta t}\right)\right)g\\ &+e^{-\frac{\sqrt{-\Delta}}{2}t}\frac{\sin\left(\frac{\sqrt{3}}{2}\sqrt{-\Delta t}\right)}{\frac{\sqrt{3}}{2}\sqrt{-\Delta}}h\\ &+\int_0^t e^{-\frac{\sqrt{-\Delta}}{2}(t-\tau)}\frac{\sin\left(\frac{\sqrt{3}}{2}\sqrt{-\Delta}(t-\tau)\right)}{\frac{\sqrt{3}}{2}\sqrt{-\Delta}}F(\tau,\cdot)d\tau.\end{aligned} \tag{7.26}$$

Of considerable importance in future sections will be the effect of an inhomogeneous source term on the linear operator. In particular, the solution to (7.24) with zero initial data is given by the formula:

$$\eta(t,\cdot)=\int_0^t e^{-\frac{\sqrt{-\Delta}}{2}(t-\tau)}\frac{\sin\left(\frac{\sqrt{3}}{2}\sqrt{-\Delta}(t-\tau)\right)}{\frac{\sqrt{3}}{2}\sqrt{-\Delta}}F(\tau,\cdot)d\tau. \tag{7.27}$$

By recalling that the Fourier transform interchanges multiplication of functions and convolution (Proposition 2.6.2), we can rewrite the formula (7.27) in a more explicit manner. Let us define the kernel $K_t(x)$ by the inverse Fourier transform,

$$K_t(x)=\frac{1}{(2\pi)^n}\int_{\mathbb{R}^n} e^{ix\cdot\xi}e^{-\frac{|\xi|}{2}t}\frac{\sin\left(\frac{\sqrt{3}}{2}|\xi|t\right)}{\frac{\sqrt{3}}{2}|\xi|}d\xi. \tag{7.28}$$

To take advantage of the scaling of this PDE, we introduce the unit scale kernel $K(x)$, defined by

$$K(x)=\frac{1}{(2\pi)^n}\int_{\mathbb{R}^n} e^{ix\cdot\xi}e^{-\frac{|\xi|}{2}}\frac{\sin\left(\frac{\sqrt{3}}{2}|\xi|\right)}{\frac{\sqrt{3}}{2}|\xi|}d\xi, \tag{7.29}$$

which is just the kernel $K_t(x)$ at unit time $t=1$. A simple change of variables shows the following crucial scaling relation:

$$K_t(x)=t^{1-n}K\left(\frac{x}{t}\right). \tag{7.30}$$

Equipped with the notation above, we can rewrite (7.27) in physical spatial variables as

$$\eta(t,\cdot) = \int_0^t K_{t-s}(\cdot) * F(s,\cdot)ds = \int_0^t \int_{\mathbb{R}^n} K_{t-s}(x-y)F(s,y)dyds, \tag{7.31}$$

where the convolution operator $*$ denotes a convolution only in the spatial variables; see Definition 2.6.2 and Proposition 2.6.2.

The importance of using the kernel $K_t(x)$ to express the solution to the viscous wave equation explicitly as (7.31) lies in the fact that we have the following strong estimate for the unit-scale kernel $K(x)$, which carries over to the general kernel $K_t(x)$ by the scaling relation (7.30). The following lemma reflects the strong dissipative effects of the fluid viscosity, represented by the presence of the Dirichlet to Neumann operator.

Lemma 7.2.1 *For all dimensions n, the kernel $K(x)$ is in $L^q(\mathbb{R}^n)$ for all $1 \leq q \leq \infty$.*

Proof The proof of this lemma is by estimates using a repeated integration by parts. We refer the reader to the proof of Lemma 3.3 in [213]. □

This representation of solution will be important later in Sect. 7.3.2 when we discuss well-posedness of the *stochastic* viscous wave equation. To compare the stochastic viscous wave equation with the stochastic heat and the stochastic wave equations as will be done in Sect. 7.3.2, we now give the analogue of the above analysis using a convolution kernel for the heat and wave equations, focusing on the inhomogeneous forms of these equations with zero initial data.

First, we consider the inhomogeneous heat equation. Define the heat equation kernel and the corresponding unit scale kernel:

$$K_t^H(x) = \frac{1}{(2\pi)^n}\int_{\mathbb{R}^n} e^{ix\cdot\xi}e^{-|\xi|^2 t}d\xi = \frac{1}{(4\pi t)^{n/2}}e^{-\frac{x^2}{4t}}, \tag{7.32}$$

$$K^H(x) = \frac{1}{(2\pi)^n}\int_{\mathbb{R}^n} e^{ix\cdot\xi}e^{-|\xi|^2}d\xi = \frac{1}{(4\pi)^{n/2}}e^{-\frac{x^2}{4}}. \tag{7.33}$$

A simple change of variables shows the following scaling relation for the heat equation kernel:

$$K_t^H(x) = t^{-\frac{n}{2}}K^H\left(\frac{x}{t^{1/2}}\right). \tag{7.34}$$

In terms of the heat equation kernel, the solution to the inhomogeneous heat equation

$$\eta_t - \Delta\eta = F$$

with zero initial data is given by the formula:

$$
\begin{aligned}
\eta(t,x) &= \int_0^t e^{-(t-s)\Delta} F(s,\cdot)ds = \int_0^t K_{t-s}^H(\cdot) * F(s,\cdot)ds \\
&= \int_0^t \int_{\mathbb{R}^n} K_{t-s}^H(x-y)F(s,y)dyds.
\end{aligned}
\tag{7.35}
$$

Note that in all dimensions n, and for all times $t > 0$, the kernel $K_t^H(x)$ defined in (7.32), is function-valued and is, in fact, a Schwartz function.

Next, we carry out the same analysis for the inhomogeneous wave equation. The wave equation kernel and the corresponding unit scale kernel can be defined similarly as

$$
K_t^W(x) = \frac{1}{(2\pi)^n} \int_{\mathbb{R}^n} e^{ix\cdot\xi} \frac{\sin(|\xi|t)}{|\xi|} d\xi, \; K^W(x) = \frac{1}{(2\pi)^n} \int_{\mathbb{R}^n} e^{ix\cdot\xi} \frac{\sin(|\xi|)}{|\xi|} d\xi.
\tag{7.36}
$$

The corresponding scaling relation is

$$
K_t^W(x) = t^{1-n} K^W\left(\frac{x}{t}\right),
\tag{7.37}
$$

which is the same as the scaling (7.30) of the kernel for the viscous wave equation. The solution to the inhomogeneous wave equation

$$
\eta_{tt} - \Delta\eta = F
$$

with zero initial data is then given by the formula:

$$
\begin{aligned}
\eta(t,x) &= \int_0^t \frac{\sin((t-s)\sqrt{-\Delta})}{\sqrt{-\Delta}} F(s,\cdot)ds = \int_0^t K_{t-s}^W(\cdot) * F(s,\cdot)ds \\
&= \int_0^t \int_{\mathbb{R}^n} K_{t-s}^W(x-y)F(s,y)dyds.
\end{aligned}
\tag{7.38}
$$

It is important to note that unlike the case for the viscous wave and heat equation, the kernel $K_t^W(x)$ is no longer necessarily function-valued. In fact, we have, for example, the following well-known formulas for the kernel $K_t^W(x)$, giving the fundamental solution for the wave equation:

$$
K_t^W(x) = \begin{cases}
\frac{1}{2} 1_{|x|<t} & \text{for } n = 1, \\
\frac{1}{\sqrt{2\pi}} \frac{1}{\sqrt{t^2 - |x|^2}} 1_{|x|<t} & \text{for } n = 2, \\
\frac{1}{4\pi} \sigma_t(dx) & \text{for } n = 3,
\end{cases}
\tag{7.39}
$$

where $\sigma_t(dx)$ in the last expression denotes the surface measure on the sphere of radius t centered at the origin. There are more complicated formulas for higher dimensions also, but $K_t^W(x)$ is function-valued only in dimensions one and two. In fact, $K_t^W(x)$ becomes increasingly singular as the dimension increases.

It is interesting to note that the kernel for the wave equation K_t^W can be tied to the kernel for the viscous wave equation K_t by the following result.

Proposition 7.2.1 *The kernel $K_t(x)$ for the viscous wave equation in dimension n, defined by* (7.28)*, is given by the convolution*

$$K_t(x) = c_n \int_{\mathbb{R}^n} \frac{t}{(t^2 + 4|x-y|^2)^{\frac{n+1}{2}}} K^W_{\frac{\sqrt{3}}{2}t}(y)dy,$$

where c_n is a constant depending only on the dimension n.

Proof We use formula (7.28) and recall that the Fourier transform interchanges multiplication and convolutions. The inverse Fourier transform of $e^{-\frac{|\xi|}{2}t}$ is

$$\frac{1}{(2\pi)^n} \int_{\mathbb{R}^n} e^{ix\cdot\xi} e^{-\frac{|\xi|}{2}t} d\xi = \tilde{c}_n \frac{t}{(t^2 + 4|x|^2)^{\frac{n+1}{2}}},$$

where $\tilde{c}_n$ depends only on n. From the definition of K_t^W, we get

$$\frac{1}{(2\pi)^n} \int_{\mathbb{R}^n} e^{ix\cdot\xi} \frac{\sin\left(\frac{\sqrt{3}}{2}|\xi|t\right)}{\frac{\sqrt{3}}{2}|\xi|} d\xi = \frac{2}{\sqrt{3}} K^W_{\frac{\sqrt{3}}{2}t}(x). \tag{7.40}$$

The result then follows by using the fact that the Fourier transform interchanges multiplication and convolution, where we replaced the constant $\frac{2}{\sqrt{3}}\tilde{c}_n$ by c_n; see Proposition 2.6.2. □

We have so far considered the inhomogeneous viscous wave equation with zero initial data. However, we will consider eventually the stochastic form of this equation with continuous bounded initial data and will hence have to consider the full form of the solution given in (7.26), which takes into account the possibility of nonzero initial displacement and velocity. Note that the convolution kernel $K(x)$ defined in (7.28) can be used to describe the effect of an inhomogeneous source term and an initial velocity, as seen in (7.26). However, the kernel $K(x)$ does not describe the effect of an initial displacement g on the solution. For this reason, we introduce the corresponding convolution kernel $J_t(x)$ and the respective unit scale kernel $J(x)$ associated with the propagation of $g(x)$ in (7.26):

$$J_t(x) = \frac{1}{(2\pi)^n} \int_{\mathbb{R}^n} e^{ix\cdot\xi} e^{-\frac{|\xi|}{2}t} \left(\cos\left(\frac{\sqrt{3}}{2}|\xi|t\right) + \frac{1}{\sqrt{3}} \sin\left(\frac{\sqrt{3}}{2}|\xi|t\right)\right) d\xi, \tag{7.41}$$

$$J(x) = \frac{1}{(2\pi)^n} \int_{\mathbb{R}^n} e^{ix\cdot\xi} e^{-\frac{|\xi|}{2}} \left(\cos\left(\frac{\sqrt{3}}{2}|\xi| \right) + \frac{1}{\sqrt{3}} \sin\left(\frac{\sqrt{3}}{2}|\xi| \right) \right) d\xi. \tag{7.42}$$

A change of variables shows that

$$J_t(x) = t^{-n} J\left(\frac{x}{t}\right). \tag{7.43}$$

We can then write the representation formula (7.26) for the solution of the general viscous wave equation with nonzero initial data $\eta(0, x) = g(x)$ and $\partial_t \eta(0, x) = h(x)$ and inhomogeneous source term F, as

$$\begin{aligned} \eta(t, x) = & \int_{\mathbb{R}^n} J_t(x-y)g(y)dy + \int_{\mathbb{R}^n} K_t(x-y)h(y)dy \\ & + \int_0^t \int_{\mathbb{R}^n} K_{t-s}(x-y)F(s, y)dsdy. \end{aligned} \tag{7.44}$$

In analogy to Lemma 7.2.1, one can show the following lemma, which shows that the unit scale kernel $J(x)$ has strong integrability properties.

Lemma 7.2.2 *For all dimensions n, the kernel $J(x)$ is in $L^q(\mathbb{R}^n)$ for all $1 \leq q \leq \infty$. Furthermore, we have the estimate*

$$|J(x)| \leq C_N |x|^{-1-n\left(\frac{N-1}{N}\right)},$$

for any $N \geq n+1$, where C_N is a constant depending on N.

Proof We refer the reader to the proof of Lemma 3.3 in [213]. While the proof there is for the slightly different unit kernel $K(x)$, the corresponding proof for $J(x)$ is just a slight modification of the proof given there. □

Finally, we establish a final lemma in this section, which shows the effect of the viscous wave operator on initial data $\eta(0, x) = g(x)$ and $\partial_t \eta(0, x) = h(x)$, where g and h are the continuous versions of functions in $H^2(\mathbb{R}^n)$. This will be useful when showing existence and uniqueness of a mild solution to the stochastic viscous wave equation with initial data g and h, which are the continuous versions of functions in $H^2(\mathbb{R}^n)$; see Sect. 7.3.2.

Lemma 7.2.3 *Let $n = 1$ or $n = 2$, and let g and h be the continuous versions of functions in $H^2(\mathbb{R}^n)$. Then, for any positive time $T > 0$, the solution to*

$$\eta_{tt} + \sqrt{-\Delta}\eta_t - \Delta\eta = 0,$$

with initial data

$$\eta(0, x) = g(x), \qquad \partial_t \eta(0, x) = h(x),$$

is a bounded, continuous function on $[0, T] \times \mathbb{R}^n$. *Furthermore, the solution* $\eta(t, x)$ *has the following Hölder continuity properties depending on the dimension* n*:*

- *If* $n = 1$, *then for every* $\rho \in [0, 1]$, *there exists a constant* $C_{\rho,T}$ *depending only on* ρ *and* T *such that for all* $t, t' \in [0, T]$, $x, x' \in \mathbb{R}$,

$$|\eta(t, x) - \eta(t, x')| \leq C_{\rho,T}|x - x'|^{\rho}, \qquad |\eta(t, x) - \eta(t', x)| \leq C_{\rho,T}|t - t'|^{\rho}.$$

- *If* $n = 2$, *then for every* $\rho \in [0, 1)$, *there exists a constant* $C_{\rho,T}$ *depending only on* ρ *and* T *such that for all* $t, t' \in [0, T]$, $x, x' \in \mathbb{R}^2$,

$$|\eta(t, x) - \eta(t, x')| \leq C_{\rho,T}|x - x'|^{\rho}, \qquad |\eta(t, x) - \eta(t', x)| \leq C_{\rho,T}|t - t'|^{\rho/2}.$$

Remark 7.2.1 By Sobolev embedding (see Corollary 2.5.1), every function in $H^2(\mathbb{R}^n)$ for $n = 1, 2$ is a continuous function, up to a set of measure zero. Because we need the initial data g, h to be continuous for our arguments, we state that g, h are the *continuous versions* of functions in $H^2(\mathbb{R}^n)$ in order to indicate that $g, h \in H^2(\mathbb{R}^n)$ are potentially redefined on a set of measure zero, so that they are also continuous functions. We will use this terminology for the remainder of the chapter.

Proof First we show that $\eta(t, x)$ is bounded. By Lemmas 7.2.1 and 7.2.2, $J, K \in L^1(\mathbb{R}^n)$. Therefore, by using the scaling relations (7.43) and (7.30), we have that for $t \in [0, T]$,

$$\int_{\mathbb{R}^n} |J_t(x)| dx = t^{-n} \int_{\mathbb{R}^n} \left| J\left(\frac{x}{t}\right) \right| dx = ||J||_{L^1} < \infty, \tag{7.45}$$

$$\int_{\mathbb{R}^n} |K_t(x)| dx = t^{1-n} \int_{\mathbb{R}^n} \left| K\left(\frac{x}{t}\right) \right| dx = t||K||_{L^1} \leq T||K||_{L^1} < \infty. \tag{7.46}$$

Using these facts along with the fact that g, h are bounded (by Sobolev embedding since they are in $H^2(\mathbb{R}^n)$), the explicit formula

$$\eta(t, x) = \int_{\mathbb{R}^n} J_t(y) g(x - y) dy + \int_{\mathbb{R}^n} K_t(y) h(x - y) dy \tag{7.47}$$

implies that $\eta(t, x)$ is bounded on $[0, T] \times \mathbb{R}^n$.

Next, we establish continuity. First, we consider spatial increments. Since $g, h \in H^2(\mathbb{R}^n)$ are continuous, they are ρ-Hölder continuous for $\rho \in [0, 1)$ by Sobolev embedding, and in fact also Lipschitz continuous in dimension one. Then, for $x, x' \in \mathbb{R}^n$ and $t > 0$,

$$
\begin{aligned}
|\eta(t,x)-\eta(t,x')| &\le \int_{\mathbb{R}^n} |J_t(y)|\cdot|g(x-y)-g(x'-y)|dy \\
&+\int_{\mathbb{R}^n} |K_t(y)|\cdot|h(x-y)-h(x'-y)|dy \\
&\le C_\rho|x-x'|^\rho\left(\int_{\mathbb{R}^n}|J_t(y)|dy+\int_{\mathbb{R}^n}|K_t(y)|dy\right)\le C_{\rho,T}|x-x'|^\rho,
\end{aligned} \tag{7.48}
$$

where $\rho = [0, 1]$ if $n = 1$, and $\rho \in [0, 1)$ if $n = 2$, with C_ρ depending on ρ. In particular, the Lipschitz or Hölder continuity of the initial data is propagated in time on a finite time interval. We have also shown that $\eta(t, x)$ at each fixed time is a continuous function.

Next, we consider time increments. Consider $0 \le t' < t \le T$. We want to estimate the quantity $|\eta(t, x) - \eta(t', x)|$ for arbitrary $x \in \mathbb{R}^n$. We consider the two cases of $n = 1$ and $n = 2$ separately.

Case 1: If $n = 1$, then we have that $(\eta, \eta_t) \in C([0, T]; H^2(\mathbb{R}))\times C([0, T]; H^1(\mathbb{R}))$. Since $H^1(\mathbb{R})$ embeds continuously into the bounded continuous functions on $\mathbb{R}$, η_t is bounded and continuous on $[0, T] \times \mathbb{R}$. Hence, $|\eta_t(t, x)| \le C_T$, $\forall x \in \mathbb{R}$ and $t \in [0, T]$. By the fundamental theorem of calculus,

$$
|\eta(t,x)-\eta(t',x)| \le \int_{t'}^{t} |\eta_t(s,x)|ds \le C_T|t-t'|. \tag{7.49}
$$

Case 2: If $n = 2$, by uniqueness of the solution in the space $C([0, T]; H^2(\mathbb{R}^2)) \times C([0, T]; H^1(\mathbb{R}^2))$, we can consider $\eta(t', x)$ and $\partial_t\eta(t', x)$ as initial data at time t', to get

$$
|\eta(t,x)-\eta(t',x)| =
$$

$$
\left|\int_{\mathbb{R}^n} J_{t-t'}(y)\eta(t',x-y)dy-\eta(t',x)+\int_{\mathbb{R}^n} K_{t-t'}(y)\partial_t\eta(t',x-y)dy\right|.
$$

Since $\int_{\mathbb{R}^n} J_t(y)dy = \widehat{J}_t(\xi = 0) = 1$, the following estimate holds:

$$
\begin{aligned}
|\eta(t,x)-\eta(t',x)| &\le \int_{\mathbb{R}^n} |J_{t-t'}(y)||\eta(t',x-y)-\eta(t',x)|dy \\
&+\int_{\mathbb{R}^n} |K_{t-t'}(y)||\partial_t\eta(t',x-y)|dy \\
&:= I_1 + I_2.
\end{aligned} \tag{7.50}
$$

To complete the estimate, we first consider integral I_1. We break up the integral I_1 into two parts,

$$
\begin{aligned}
I_1 &= \int_{|y| \leq |t-t'|^{1/2}} |J_{t-t'}(y)| \cdot |\eta(t', x-y) - \eta(t', x)| dy \\
&+ \int_{|y| > |t-t'|^{1/2}} |J_{t-t'}(y)| \cdot |\eta(t', x-y) - \eta(t', x)| dy \\
&= I_{1,1} + I_{1,2}.
\end{aligned} \tag{7.51}
$$

Using the Hölder continuity in space from (7.48), and using the estimate (7.45), we get for $\rho \in [0, 1)$,

$$
\begin{aligned}
I_{1,1} &\leq C_{\rho,T} \int_{|y| \leq |t-t'|^{1/2}} |J_{t-t'}(y)| \cdot |y|^{\rho} dy \\
&\leq C_{\rho,T} |t-t'|^{\rho/2} \int_{|y| \leq |t-t'|^{1/2}} |J_{t-t'}(y)| dy \leq C_{\rho,T} |t-t'|^{\rho/2}.
\end{aligned} \tag{7.52}
$$

To estimate $I_{1,2}$, we recall that we already showed that $\eta(t, x)$ is bounded on $[0, T] \times \mathbb{R}^n$ by some constant M_T. Therefore, by the scaling relation (7.43), and by using a change of variables, we get

$$
\begin{aligned}
I_{1,2} &\leq 2M_T \int_{|y| > |t-t'|^{1/2}} |J_{t-t'}(y)| dy \\
&= 2M_T (t-t')^{-n} \int_{|y| > |t-t'|^{1/2}} \left| J\left(\frac{y}{t-t'}\right)\right| dy \\
&= 2M_T \int_{|z| > |t-t'|^{-1/2}} |J(z)| dz.
\end{aligned} \tag{7.53}
$$

To estimate the last integral, we recall the estimate stated in Lemma 7.2.2 and choose an N in that estimate (which depends on ρ) $N_\rho \geq n+1$, sufficiently large so that

$$
1 + n\left(\frac{N_\rho - 1}{N_\rho}\right) \geq n + \rho, \quad \text{or equivalently} \quad \rho \leq 1 - \frac{n}{N_\rho}, \tag{7.54}
$$

for arbitrary $\rho \in [0, 1)$. Then, continuing from (7.53) and switching to polar coordinates, the estimate from Lemma 7.2.2, together with the inequality (7.54), imply

$$
\begin{aligned}
I_{1,2} &\leq 2M_T C_{N_\rho} \int_{|z| > |t-t'|^{-1/2}} \frac{1}{|z|^{1+n\left(\frac{N_\rho - 1}{N_\rho}\right)}} dz \\
&= 2M_T C_{N_\rho} |\mathbb{S}^{n-1}| \int_{|t-t'|^{-1/2}}^{\infty} r^{-1-n\left(\frac{N_\rho - 1}{N_\rho}\right)} r^{n-1} dr
\end{aligned}
$$

$$= 2M_T C_{N_\rho}|\mathbb{S}^{n-1}| \int_{|t-t'|^{-1/2}}^{\infty} r^{-2+\frac{n}{N_\rho}} dr$$

$$= C_{\rho,T}|t-t'|^{\frac{1}{2}\left(1-\frac{n}{N_\rho}\right)} \leq C_{\rho,T}|t-t'|^{\rho/2}, \tag{7.55}$$

for $\rho \in [0, 1)$, where C_{N_ρ} denotes the constant for $N = N_\rho$ in the inequality in Lemma 7.2.2. In the last inequality, we used the fact that t', t belong to a bounded interval $[0, T]$, and in the last step, we reused the notation for the constant $C_{\rho,T}$, though the value may change.

Finally, we estimate

$$I_2 := \int_{\mathbb{R}^n} |K_{t-t'}(y)| \cdot |\partial_t \eta(t', x-y)| dy.$$

Since $(\eta, \partial_t\eta) \in C([0, T]; H^2(\mathbb{R}^n)) \times C([0, T]; H^1(\mathbb{R}^n))$, we have that $\partial_t\eta(t, \cdot)$ is uniformly bounded in $H^1(\mathbb{R}^n)$ for $t \in [0, T]$. We note that for $n = 2$, $H^1(\mathbb{R}^n)$ embeds into $L^q(\mathbb{R}^n)$ for all $2 \leq q < \infty$. This is because for general dimension n, if a function $f \in H^{n/2}(\mathbb{R}^n)$, we can show that for all $1 < p \leq 2$, $\widehat{f} \in L^p(\mathbb{R}^n)$, which implies the result by the Hausdorff-Young inequality. Using Hölder's inequality with the conjugate exponents $2/p$ and $2/(2-p)$, one can compute:

$$\int_{\mathbb{R}^n} |\widehat{f}(\xi)|^p d\xi = \int_{\mathbb{R}^n} (1+|\xi|^2)^{-np/4}(1+|\xi|^2)^{np/4}|\widehat{f}(\xi)|^p d\xi$$

$$\leq \left(\int_{\mathbb{R}^n} (1+|\xi|^2)^{-\frac{np}{2(2-p)}} d\xi\right)^{\frac{2-p}{2}} \cdot ||f||^p_{H^{n/2}(\mathbb{R}^n)} = C_p||f||^p_{H^{n/2}(\mathbb{R}^n)},$$

since $\frac{p}{2-p} > 1$ for $1 < p \leq 2$. Hence, for q such that $2 \leq q < \infty$, by the Hausdorff-Young inequality (Proposition 2.6.4), we have that for conjugate exponents p and q,

$$||f||_{L^q(\mathbb{R}^n)} \leq ||\widehat{f}||_{L^p(\mathbb{R}^n)} \leq C_p||f||_{H^{n/2}(\mathbb{R}^n)},$$

so that $H^1(\mathbb{R}^2)$ embeds continuously into $L^q(\mathbb{R}^2)$ for $2 \leq q < \infty$.

Hence, since $\partial_t\eta(t, \cdot)$ is uniformly bounded in $H^1(\mathbb{R}^n)$ on $t \in [0, T]$, we have that

$$\sup_{0\leq t\leq T} ||\partial_t\eta(t, \cdot)||_{L^q(\mathbb{R}^2)} \leq C_{T,q}, \qquad \text{for } 2 \leq q < \infty. \tag{7.56}$$

In addition, we use the scaling property for the kernel $K_t(x)$ given by (7.30) to deduce that

$$||K_t(x)||_{L^p(\mathbb{R}^n)} = t^{1-n}\left(\int_{\mathbb{R}^n} \left|K\left(\frac{x}{t}\right)\right|^p dx\right)^{1/p} = t^{1-n+\frac{n}{p}}\left(\int_{\mathbb{R}^n} |K(y)|^p dy\right)^{1/p}$$

$$= t^{1-\frac{n}{q}}||K||_{L^p(\mathbb{R}^n)} = C_p t^{1-\frac{n}{q}}, \tag{7.57}$$

where p and q are conjugate exponents. Therefore, by (7.56) and (7.57),

$$I_2 \leq \tilde{C}_{T,q}|t-t'|^{1-\frac{n}{q}}, \tag{7.58}$$

for $2 \leq q < \infty$, where $n = 2$. By choosing $2 \leq q < \infty$ appropriately, this implies the desired estimate

$$I_2 \leq C_{T,\rho}|t-t'|^{\rho/2}$$

for $\rho \in [0, 1)$.

By combining (7.50), (7.51), (7.52), (7.55), and (7.58), we get the desired result:

$$|\eta(t,x) - \eta(t',x)| \leq C_{\rho,T}|t-t'|^{\rho/2}, \tag{7.59}$$

for $\rho \in [0, 1)$ in the case $n = 2$.

The spatial estimate (7.48) and the time estimates (7.49) for $n = 1$ and (7.59) for $n = 2$, which are *uniform on* $[0, T] \times \mathbb{R}^n$, establish the continuity of $\eta(t, x)$ on $[0, T] \times \mathbb{R}^n$. □

7.3 The Stochastic Viscous Wave Equation in Dimensions $n = 1, 2$

We are now in the position to study the stochastic viscous wave equation:

$$\eta_{tt} + \sqrt{-\Delta}\eta_t - \Delta\eta = f(\eta)W(dt, dx), \qquad \text{in } \mathbb{R}^n, \tag{7.60}$$

with initial data:

$$\eta(0, x) = g(x), \qquad \partial_t\eta(0, x) = h(x). \tag{7.61}$$

For simplicity, we assume that g and h are the continuous versions of functions in $H^2(\mathbb{R}^n)$, $f : \mathbb{R} \to \mathbb{R}$ is Lipschitz continuous, and $W(dt, dx)$ is space-time white noise. In particular, since f is Lipschitz continuous, there exists a constant $L > 0$ such that

$$|f(x) - f(y)| \leq L|x - y|, \qquad \text{for all } x, y \in \mathbb{R},$$

$$|f(x)| \leq L(1 + |x|), \qquad \text{for all } x \in \mathbb{R}. \tag{7.62}$$

We will show that the Cauchy problem (7.60), (7.61) has a *mild solution* in the sense of a stochastic process satisfying a stochastic integral equation (see Definition 7.3.1

below), which is function-valued in dimensions $n = 1, 2$. This is in contrast to the corresponding stochastic heat and wave equations,

$$\eta_t - \Delta\eta = f(\eta)W(dt, dx), \qquad \text{in } \mathbb{R}^n, \tag{7.63}$$

$$\eta_{tt} - \Delta\eta = f(\eta)W(dt, dx), \qquad \text{in } \mathbb{R}^n, \tag{7.64}$$

which have function-valued mild solutions only in dimension $n = 1$.

In the next section, we review the concept of mild solution for the stochastic heat and wave equations and demonstrate the well-known fact that there are function-valued mild solutions only in dimension one. We then consider the concept of mild solutions for the stochastic viscous wave equation, showing heuristically why we will be able to consider such solutions in dimension two. In Sect. 7.3.2, we rigorously prove existence and uniqueness of a mild solution to (7.60), (7.61) using a Picard iteration argument to deal with the nonlinearity $f(\eta)$.

7.3.1 The Concept of Mild Solution

To define the concept of mild solution for the *stochastic* viscous wave equation (7.60), we first recall the solution for the *deterministic* inhomogeneous problem (7.24). Namely, as shown earlier, the solution to the *deterministic* inhomogeneous problem (7.24) with initial data $\eta(0, x) = g(x)$ and $\partial_t\eta(0, x) = h(x)$ is given by the formula

$$\begin{aligned}\eta(t, x) = &\int_{\mathbb{R}^n} J_t(x - y)g(y)dy + \int_{\mathbb{R}^n} K_t(x - y)h(y)dy \\ &+ \int_0^t \int_{\mathbb{R}^n} K_{t-s}(x - y)F(s, y)dyds,\end{aligned} \tag{7.65}$$

where $J_t(x)$ is defined by (7.41) and $K_t(x)$ by (7.28). For the general *stochastic* case (7.60) with initial data (7.61), we can formally regard the stochastic forcing $f(\eta)W(dt, dx)$ as the forcing term F in the deterministic equation and formally require that the solution η to the stochastic viscous wave equation satisfies the *stochastic integral equation* alla (7.65):

$$\begin{aligned}\eta(t, x, \omega) = &\int_{\mathbb{R}^n} J_t(x - y)g(y)dy + \int_{\mathbb{R}^n} K_t(x - y)h(y)dy \\ &+ \int_0^t \int_{\mathbb{R}^n} K_{t-s}(x - y)f(\eta(s, y, \omega))W(ds, dy).\end{aligned}$$

The result of this formal argument gives rise to the concept of a mild solution.

Definition 7.3.1 A stochastic process $\eta(t, x)$ is a *mild solution* to the stochastic viscous wave equation (7.60) with initial data (7.61) if $\eta(t, x)$ is jointly measurable and adapted to the filtration $\{\mathcal{F}_t\}_{t\geq 0}$ with

$$\begin{aligned}\eta(t, x, \omega) = &\int_{\mathbb{R}^n} J_t(x-y)g(y)dy + \int_{\mathbb{R}^n} K_t(x-y)h(y)dy \\ &+ \int_0^t \int_{\mathbb{R}^n} K_{t-s}(x-y)f(\eta(s, y, \omega))W(ds, dy),\end{aligned} \tag{7.66}$$

and the stochastic integral on the right-hand side of (7.66) is defined.

Remark 7.3.1 (Probabilistic Notation) In the remainder of this discussion on stochastic fluid-structure interaction, we will generally follow the probabilistic convention of not writing the explicit ω dependence of random variables and stochastic processes. In particular, while we wrote out the explicit ω dependence in the stochastic process $\eta(t, x, \omega)$ in (7.66), we will henceforth omit the explicit ω dependence when it is clear from context that the mathematical quantity involved is a random variable. For example, we would write

$$\begin{aligned}\eta(t, x) = &\int_{\mathbb{R}^n} J_t(x-y)g(y)dy + \int_{\mathbb{R}^n} K_t(x-y)h(y)dy \\ &+ \int_0^t \int_{\mathbb{R}^n} K_{t-s}(x-y)f(\eta(s, y))W(dy, ds),\end{aligned}$$

for the full expression in (7.66).

We can define the concept of a mild solution to the stochastic heat and the stochastic wave equations (7.63) and (7.64) in the same way using the deterministic heat and wave equation representation formulas for the solutions of the corresponding inhomogeneous equation, given in (7.35) and (7.38).

As mentioned earlier, the existence of a function-valued mild solution to the stochastic heat and wave equations (7.63) and (7.64), defined this way, can be obtained only in dimension $n = 1$, as was discussed in [107] and [210]. However, we will be able to prove the existence of a function-valued mild solution to the stochastic viscous wave equation (7.60) in both dimensions $n = 1, 2$. To give an idea of why we might expect this to be true, we present the following heuristic argument.

A heuristic argument for the existence of a mild solution to (7.60)**,** (7.61) **in** $n = 2$. For simplicity, let $f : \mathbb{R} \to \mathbb{R}$ on the right-hand side in (7.60), (7.63) and (7.64) be identically equal to 1, so that we can just consider the case of additive noise. Therefore, we consider the equations

$$\eta_t - \Delta\eta = W(dt, dx), \qquad \text{on } \mathbb{R}^n, \tag{7.67}$$

$$\eta_{tt} - \Delta\eta = W(dt, dx), \qquad \text{on } \mathbb{R}^n, \tag{7.68}$$

$$\eta_{tt} + \sqrt{-\Delta}\eta_t - \Delta\eta = W(dt, dx), \qquad \text{on } \mathbb{R}^n. \tag{7.69}$$

Furthermore, for simplicity, we consider *zero initial data* for the purposes of this heuristic argument.

For the stochastic heat equation with additive noise (7.67), we have an explicit formula for the solution as a stochastic integral

$$\eta(t, x) = \int_0^t \int_{\mathbb{R}^n} K_{t-s}^H(x - y) W(ds, dy),$$

where the kernel K_t^H is defined by (7.32). For this stochastic integral to make sense, we must have

$$\int_0^t \int_{\mathbb{R}^n} |K_{t-s}^H(x - y)|^2 dy ds < \infty. \tag{7.70}$$

Using the scaling relation (7.34), we can rewrite condition (7.70) in terms of the unit kernel as

$$\int_0^t \int_{\mathbb{R}^n} |K_{t-s}^H(x - y)|^2 dy ds = \int_0^t \int_{\mathbb{R}^n} (t - s)^{-n} \left| K^H\left(\frac{x - y}{(t - s)^{1/2}}\right)\right|^2 dy ds$$
$$= \int_0^t \int_{\mathbb{R}^n} (t - s)^{-n/2} |K^H(y)|^2 dy ds = \left(\int_0^t (t - s)^{-n/2} ds\right) ||K^H||_{L^2(\mathbb{R}^n)}^2 < \infty. \tag{7.71}$$

Because K^H is a Gaussian in all dimensions n, we have that $||K^H||_{L^2(\mathbb{R}^n)}^2 < \infty$ for all n. However, the time integral $\int_0^t (t - s)^{-n/2} ds$ only converges in dimension $n = 1$. This is the reason why a function-valued mild solution to the stochastic heat equation with additive noise (7.67) exists only in dimension 1.

Let us carry out a similar analysis for the stochastic wave equation with additive noise (7.68). A mild solution, if it exists, must be given by the stochastic integral

$$\eta(t, x) = \int_0^t \int_{\mathbb{R}^n} K_{t-s}^W(x - y) W(ds, dy),$$

where K_t^W is defined by (7.36). This stochastic integral exists only if the following integrability condition is satisfied:

$$\int_0^t \int_{\mathbb{R}^n} |K_{t-s}^W(x - y)|^2 dy ds < \infty. \tag{7.72}$$

By using the scaling relation (7.37), this condition can be rewritten as

$$\int_0^t \int_{\mathbb{R}^n} |K_{t-s}^W(x-y)|^2 dyds = \int_0^t \int_{\mathbb{R}^n} (t-s)^{2-2n} \left| K^W\left(\frac{x-y}{t-s}\right)\right|^2 dyds$$
$$= \int_0^t \int_{\mathbb{R}^n} (t-s)^{2-n} |K^W(y)|^2 dyds = \left(\int_0^t (t-s)^{2-n} ds\right) ||K^W||^2_{L^2(\mathbb{R}^n)} < \infty. \tag{7.73}$$

Note here that the time integral $\int_0^t (t-s)^{2-n}$ converges for $n = 1, 2$. However, it is easy to check from the explicit form of the fundamental solution in (7.39) that K^W is in $L^2(\mathbb{R}^n)$ only for dimension $n = 1$. This is the reason why a function-valued mild solution to the stochastic wave equation with additive noise (7.68) exists only in dimension one.

The stochastic viscous wave equation with additive noise (7.69) is exactly in between these two cases, with both the time and space integral arising from the integral of the square of the kernel being integrable in both $n = 1$ and $n = 2$. Namely, a function-valued mild solution to (7.69) would be defined by

$$\eta(t,x) = \int_0^t \int_{\mathbb{R}^n} K_{t-s}(x-y) W(ds, dy),$$

where $K_t(x)$ is the kernel given by (7.28). Using the scaling relation (7.30) involving the unit kernel (7.29), we compute, similarly as in the previous examples, that this integral exists only if the following integrability condition is satisfied:

$$\int_0^t \int_{\mathbb{R}^n} |K_{t-s}(x-y)|^2 dyds = \left(\int_0^t (t-s)^{2-n} ds\right) ||K||^2_{L^2(\mathbb{R}^n)} < \infty. \tag{7.74}$$

However, by Lemma 7.2.1, $K \in L^2(\mathbb{R}^n)$ for all n. Therefore, this integrability condition is satisfied in both dimensions one and two. Thus, the stochastic viscous wave equation with additive noise (7.69) has a function-valued mild solution *both in dimensions one and two*.

Finally, we note that the nature of the parabolic damping is essential for the stochastic viscous wave equation to have a function-valued mild solution in dimension two also. In particular, the stochastic damped wave equation

$$\eta_{tt} + 2c\eta_t - \Delta\eta = W(dx, dt), \qquad \text{on } \mathbb{R}^n \tag{7.75}$$

with $c > 0$ and zero initial data has a function-valued mild solution only in dimension one. To see this, we use the explicit formula for the fundamental solution from [106] in frequency space,

$$\widehat{K}_t^{DW}(\xi) = \begin{cases} (c^2 - |\xi|^2)^{-1/2} e^{-ct} \sinh\left(t\sqrt{c^2 - |\xi|^2}\right) & \text{if } |\xi| < c, \\ (|\xi|^2 - c^2)^{-1/2} e^{-ct} \sin\left(t\sqrt{|\xi|^2 - c^2}\right) & \text{if } |\xi| > c, \end{cases} \tag{7.76}$$

so that the mild solution if it exists for (7.75) must be given by

$$\eta(t, x) = \int_0^t \int_{\mathbb{R}^n} K_{t-s}^{DW}(x - y) W(ds, dy).$$

Here, the superscript DW indicates that we are considering the fundamental solution for the damped wave equation. Thus, the integrability condition for the mild solution to exist is that

$$\int_0^t \int_{\mathbb{R}^n} |K_{t-s}^{DW}(x - y)|^2 dy ds < \infty.$$

This is equivalent, by Plancherel's theorem, to

$$\begin{aligned} \int_0^t \int_{\mathbb{R}^n} |K_{t-s}^{DW}(x - y)|^2 dy ds &= \int_0^t \int_{\mathbb{R}^n} |K_{t-s}^{DW}(y)|^2 dy ds \\ &= \int_0^t \int_{\mathbb{R}^n} |\widehat{K}_{t-s}^{DW}(\xi)|^2 d\xi ds < \infty. \end{aligned} \tag{7.77}$$

However, the condition (7.77) holds only in dimension $n = 1$. This is because $\widehat{K}_t(\xi) \notin L^2(\mathbb{R}^n)$ for $n \geq 2$ for all $t > 0$. To see this, note that for $|\xi| > c$, the explicit formula (7.76) gives that

$$\widehat{K}_t^{DW}(\xi) = (|\xi|^2 - c^2)^{-1/2} e^{-ct} \sin\left(t\sqrt{|\xi|^2 - c^2}\right), \qquad \text{for } |\xi| > c.$$

We then compute that

$$\begin{aligned} ||\widehat{K}_t^{DW}||^2_{L^2(\mathbb{R}^n)} &\geq e^{-2ct} \int_{|\xi|>c} (|\xi|^2 - c^2)^{-1} \sin^2\left(t\sqrt{|\xi|^2 - c^2}\right) d\xi \\ &= \alpha_{n-1} e^{-2ct} \int_c^\infty \frac{r^{n-1}}{r^2 - c^2} \sin^2\left(t\sqrt{r^2 - c^2}\right) dr, \end{aligned}$$

where α_{n-1} denotes the surface area of the sphere $S^{n-1} \subset \mathbb{R}^n$. We use a change of variables,

$$\rho = \sqrt{r^2 - c^2} \iff r = \sqrt{\rho^2 + c^2} \qquad d\rho = \frac{r}{\sqrt{r^2 - c^2}} dr.$$

Then,

$$||\widehat{K}_t^{DW}||^2_{L^2(\mathbb{R}^n)} \geq \alpha_{n-1} e^{-2ct} \int_0^\infty \frac{\sin^2(t\rho)}{\rho} (\rho^2 + c^2)^{\frac{n}{2}-1} d\rho.$$

So for $n \geq 2$, we have that $\frac{n}{2} - 1 \geq 0$ and hence for any $t > 0$,

$$||\widehat{K}_t^{DW}||^2_{L^2(\mathbb{R}^n)} \geq \alpha_{n-1}e^{-2ct}c^{n-2}\int_0^\infty \frac{\sin^2(t\rho)}{\rho}d\rho = \infty \qquad \text{for } n \geq 2,$$

since this integral diverges. Thus, the integrability condition (7.77) does not hold in dimensions two and higher and holds only in dimension one. So the stochastic damped wave equation has a function-valued mild solution only in dimension one.

7.3.2 *Existence and Uniqueness for the Stochastic Viscous Wave Equation*

While the heuristic argument above was done for a simpler case of additive white noise when $f(\eta) = 1$, we can get an existence and uniqueness result for the more general equation (7.60) with a general, Lipschitz $f(\eta)$ in dimensions one and two, by a standard Picard iteration procedure, and by estimates of the kernel. We then obtain estimates on the higher moments of the solution for later use in Sect. 7.4. Such a Picard iteration and higher moment bound procedure is standard in the stochastic PDE literature [106, 107, 210, 346]. More precisely, we have the following main result.

Theorem 7.3.1 (Existence and Uniqueness) *Let $n = 1$ or $n = 2$, and let g and h be the continuous versions of functions in $H^2(\mathbb{R}^n)$. Suppose $f : \mathbb{R} \to \mathbb{R}$ is a Lipschitz continuous function. Then, there exists a function-valued mild solution to the equation*

$$\eta_{tt} + \sqrt{-\Delta}\eta_t - \Delta\eta = f(\eta)W(dt, dx) \qquad \text{on } \mathbb{R}^n \tag{7.78}$$

with initial data $\eta(0, x) = g(x)$, $\partial_t\eta(0, x) = h(x)$, which is unique up to stochastic modification.

Proof To establish existence, we use Picard iterations. We begin by setting the first iterate η_0 to be the deterministic function

$$\eta_0(t, x) = \int_{\mathbb{R}^n} J_t(x - y)g(y)dy + \int_{\mathbb{R}^n} K_t(x - y)h(y)dy, \tag{7.79}$$

which is the solution to the deterministic linear homogeneous viscous wave equation with initial data given by g and h. By Lemma 7.2.3, $\eta_0(t, x)$ is a bounded, continuous function on $[0, T] \times \mathbb{R}^n$.

Then, define the Picard iterates η_k for $k \geq 1$ inductively by

$$\eta_k(t, x) = \eta_0(t, x) + \int_0^t \int_{\mathbb{R}^n} K_{t-s}(x - y)f(\eta_{k-1}(s, y))W(ds, dy), \tag{7.80}$$

where η_0 captures the deterministic evolution of the initial data g and h. However, we must check that the stochastic integral on the right-hand side makes sense. This is the content of the following lemma.

Lemma 7.3.1 *The Picard iteration procedure* (7.80) *is well-defined at each step. Furthermore,*

$$\sup_{t\in[0,T]}\sup_{x\in\mathbb{R}^n}\mathbb{E}\left(|\eta_k(t,x)|^2\right) < \infty, \qquad \textit{for all } k \geq 0,\ T \geq 0.$$

The proof of Lemma 7.3.1 is given in the Appendix.

Remark 7.3.2 Note that because random variables are only defined up to a measure zero set, the kth Picard iterate $\{\eta_k(t,x)\}_{t\in[0,T],x\in\mathbb{R}^n}$ is defined only up to stochastic modification. However, as we show in the proof of Lemma 7.3.1, there exists a modification of $\{\eta_k(t,x)\}_{t\in[0,T],x\in\mathbb{R}^n}$ for which the stochastic integral

$$\int_0^t\int_{\mathbb{R}^n} K_{t-s}(x-y) f(\eta_k(s,y)) W(dy,ds)$$

is defined, where this stochastic integral is needed to obtain the next iterate η_{k+1}. Hence, when defining η_k at each point $(t,x) \in [0,T]\times\mathbb{R}^n$ in (7.80), we choose the modification that allows the stochastic integral needed for the next step of the Picard iteration to be defined. Note that all of the arguments that follow are well suited to the fact that the Picard iterates η_k are defined only up to stochastic modification. For example, we will later consider the quantity for each k and $t \in [0,T]$,

$$\sup_{0\leq s\leq t, x\in\mathbb{R}^n} \mathbb{E}[(\eta_k - \eta_{k-1})^2(s,x)],$$

when studying convergence of the iterates, and this quantity is unchanged by stochastic modification of any of the individual iterates.

The next step is to show that $\eta_k(t,x)$ converge in an appropriate sense as $k \to \infty$ and that the limit is a unique mild solution to the stochastic viscous wave equation (7.78).

Convergence We start by considering the difference between consecutive iterates: for any $k \geq 2$,

$$\begin{aligned}\eta_k(t,x) - \eta_{k-1}(t,x) = &\int_0^t\int_{\mathbb{R}^n} K_{t-s}(x-y)[f(\eta_{k-1}(s,y)) \\ &- f(\eta_{k-2}(s,y))]W(ds,dy).\end{aligned} \tag{7.81}$$

Using the Itö isometry (3.25) and the fact that f is Lipschitz continuous with a global Lipschitz constant L, we obtain

$$
\begin{aligned}
&\mathbb{E}[(\eta_k - \eta_{k-1})^2(t,x)] \\
&\leq L^2\mathbb{E}\left(\int_0^t \int_{\mathbb{R}^n} K_{t-s}^2(x-y)|\eta_{k-1}(s,y) - \eta_{k-2}(s,y)|^2 dyds\right) \\
&= L^2 \int_0^t \int_{\mathbb{R}^n} K_{t-s}^2(x-y)\mathbb{E}[(\eta_{k-1} - \eta_{k-2})^2(s,y)]dyds,
\end{aligned}
$$

where we used Fubini's theorem in the last step.

Let

$$
H_k^2(t) := \sup_{0\leq s\leq t, x\in\mathbb{R}^n} \mathbb{E}[(\eta_k - \eta_{k-1})^2(s,x)]. \tag{7.82}
$$

We want to show that for every $t \geq 0$, $\sum_{k=1}^{\infty} H_k(t) < \infty$, as this would imply that $\{\eta_k(t,x)\}_{k=0}^{\infty}$ is a Cauchy sequence in $L^2(\Omega)$ for each $t \geq 0$, $x \in \mathbb{R}^n$. Indeed, first notice that the following inequality holds:

$$
H_k^2(t) \leq L^2 \int_0^t \int_{\mathbb{R}^n} K_{t-s}^2(x-y)H_{k-1}^2(s)dyds. \tag{7.83}
$$

To further estimate the right-hand side, we estimate the kernel using a calculation as in (7.74) for dimensions $n = 1$ and 2 to obtain

$$
\int_{\mathbb{R}^n} K_{t-s}^2(x-y)dy = (t-s)^{2-n}||K||_{L^2(\mathbb{R}^n)}^2 = c_n(t-s)^{2-n} \leq c_n t^{2-n}, \tag{7.84}
$$

for some constant c_n depending only on n, for $s \in [0,t]$. Combining this estimate with (7.83), one obtains

$$
H_k^2(t) \leq c_{n,t} \int_0^t H_{k-1}^2(s)ds, \ k = 2, 3, \ldots \tag{7.85}
$$

for a finite constant $c_{n,t}$ that depends only on t and the dimension $n = 1, 2$. We will use this inequality inductively for $k = 2, 3, \ldots$ to obtain the desired result. For this purpose, we must first show that $H_1(t)$ is finite. In particular, recalling (7.82) and using the result in Lemma 7.3.1, we have

$$
H_1^2(t) \leq 2\left(\sup_{0\leq s\leq t, x\in\mathbb{R}^n} \mathbb{E}\left(|\eta_1(s,x)|^2\right) + \sup_{0\leq s\leq t, x\in\mathbb{R}^n} \mathbb{E}\left(|\eta_0(s,x)|^2\right)\right) = A_t < \infty,
$$

where A_t is a constant depending only on t. Hence, by inductively using (7.85), we have that

$$
H_{k+1}^2(t) \leq \frac{A_t \cdot (c_{n,t})^k t^k}{k!}. \tag{7.86}
$$

Thus,

$$\sum_{k=1}^{\infty} H_k(t) \leq A_t^{1/2} \sum_{k=0}^{\infty} \frac{(c_{n,t})^{k/2} t^{k/2}}{(k!)^{1/2}} < \infty, \tag{7.87}$$

as this series converges. Recalling the definition of $H_k^2(t)$ in (7.82), we conclude that $\{\eta_k(t, x)\}_{k=0}^{\infty}$ for each $t \geq 0, x \in \mathbb{R}^n$ is a Cauchy sequence in $L^2(\Omega)$. Hence, $\eta_k(t, x)$ converges in $L^2(\Omega)$ to some $\eta(t, x)$ for each $t \geq 0, x \in \mathbb{R}^n$.

Existence of a Mild Solution We now show that the limit $\eta(t, x)$ is a mild solution to (7.78). Indeed, after passing to the limit on both sides of (7.80), we immediately see that the left hand side of (7.80) converges to $\eta(t, x)$ in $L^2(\Omega)$. To deal with the limit on the right-hand side of (7.80), we first calculate the following estimate: by the Lipschitz property of f and the Itô isometry (3.25), we have

$$\begin{aligned} &\left\| \int_0^t \int_{\mathbb{R}^n} K_{t-s}(x-y)[f(\eta_{k-1}(s, y)) - f(\eta(s, y))]W(ds, dy) \right\|_{L^2(\Omega)}^2 \\ &\quad \leq L^2 \int_0^t \int_{\mathbb{R}^n} K_{t-s}^2(x-y)\mathbb{E}\left(|\eta_{k-1}(y, s) - \eta(y, s)|^2\right) dyds. \end{aligned} \tag{7.88}$$

To further estimate the right-hand side, we recall the convergence of the series (7.87) and the definition (7.82) of $H_k^2(t)$ to conclude:

$$\mathbb{E}(|\eta_{k-1}(s, y) - \eta(s, y)|^2) \to 0, \text{ as } k \to \infty, \text{ uniformly for } 0 \leq s \leq t \text{ and } y \in \mathbb{R}^n. \tag{7.89}$$

Additionally, by recalling (7.84), we get

$$\int_0^t \int_{\mathbb{R}^n} K_{t-s}^2(x-y) dyds = \frac{c_n}{3-n} t^{3-n}. \tag{7.90}$$

Therefore, combining this equality with (7.89), and using it in the right-hand side of (7.88), we obtain that for every fixed $t \geq 0$, the following convergence result holds:

$$\begin{aligned} &\int_0^t \int_{\mathbb{R}^n} K_{t-s}(x, y) f(\eta_{k-1}(s, y)) W(ds, dy) \\ &\quad \to \int_0^t \int_{\mathbb{R}^n} K_{t-s}(x, y) f(\eta(s, y)) W(ds, dy) \text{ in } L^2(\Omega). \end{aligned}$$

This shows that η satisfies (7.66).

To complete the proof that η is a mild solution, we must show according to Definition 7.3.1 that $\{\eta(t, x)\}_{t \in \mathbb{R}^+ \times \mathbb{R}^n}$ is jointly measurable and adapted to $\mathcal{F}_t$. Since each η_k is adapted to $\mathcal{F}_t$, so is the limit u. In addition, by the uniform convergence

(7.89), η is continuous in $L^2(\Omega)$ on $\mathbb{R}^+ \times \mathbb{R}^n$ since each η_k has this property, by the proof of Lemma 7.3.1 in the Appendix. Hence, by Remark 3.7.2, η has a stochastic modification that is jointly measurable. This completes the proof that η is a mild solution.

Uniqueness Uniqueness follows from Gronwall's inequality, Theorem 2.12.1. More precisely, suppose that η and ϕ are both mild solutions with the same initial data (7.61). Then, their difference $\psi := \eta - \phi$ satisfies the following stochastic integral equation:

$$\psi(t,x) = \int_0^t \int_{\mathbb{R}^n} K_{t-s}(x-y)[f(\eta(s,y)) - f(\phi(s,y))]W(dy,ds).$$

Taking the $L^2(\Omega)$ norm of both sides, we get that

$$\mathbb{E}[\psi^2(t,x)] \leq L^2 \int_0^t \int_{\mathbb{R}^n} K_{t-s}^2(x-y)\mathbb{E}\left(|\eta(s,y) - \phi(s,y)|^2\right) dyds.$$

So defining

$$H(t) := \sup_{0 \leq s \leq t, x \in \mathbb{R}^n} \mathbb{E}[\psi^2(s,x)],$$

we get after using (7.84), the following inequality:

$$H(t) \leq L^2 \int_0^t H(s) \left(\int_{\mathbb{R}^n} K_{t-s}^2(x-y)dy \right) ds = c_{n,t} L^2 \int_0^t H(s)ds.$$

Since $H(0) = 0$, using Gronwall's inequality then implies that $H(t)$ is identically zero for all t. In particular, η is unique up to stochastic modification, since the expectation $\mathbb{E}[\psi^2(t,x)] = 0$ for all $t \geq 0$ and $x \in \mathbb{R}^n$. This completes the uniqueness proof and the proof of Theorem 7.3.1. □

Now that we have shown an appropriate notion of existence and uniqueness of a mild solution for (7.60) in dimensions one and two, we would like to understand more details of the solution behavior. In particular, we would like to study the *Hölder continuity of the sample paths*, defined below in Sect. 7.4. In order to do that, it is useful to obtain uniform boundedness of L^p moments of the unique mild solution, for $p \geq 2$, uniformly in space and time on a bounded time interval. The proof of this result will rely on the BDG inequality for integration against space-time white noise, which is stated in Theorem 3.7.1.

Theorem 7.3.2 *Let $n = 1$ or 2, and let g and h be the continuous versions of functions in $H^2(\mathbb{R}^n)$. Let $\eta(t,x)$ be the unique function-valued mild solution to* (7.60) *with initial data* (7.61). *Then, for each $T > 0$ and $p \geq 2$,*

$$\sup_{0 \leq t \leq T} \sup_{x \in \mathbb{R}^n} \mathbb{E}(|\eta(t,x)|^p) < \infty. \tag{7.91}$$

Proof Note that we have already established this result for $p = 2$ by using Lemma 7.3.1 and the uniform convergence in $L^2(\Omega)$ given by (7.89). To prove the higher moment bound (7.91), we reexamine our Picard iterates (7.80):

$$\eta_k(t,x) - \eta_{k-1}(t,x) = \int_0^t \int_{\mathbb{R}^n} K_{t-s}(x-y)[f(\eta_{k-1}(s,y)) \\ - f(\eta_{k-2}(s,y))]W(ds,dy).$$

Using the BDG inequality stated in Theorem 3.7.1, for $k \geq 2$, we get

$$\mathbb{E}(|\eta_k(t,x) - \eta_{k-1}(t,x)|^p) \\ \leq c_p \mathbb{E}\left[\left(\int_0^t \int_{\mathbb{R}^n} K_{t-s}^2(x-y)|f(\eta_{k-1}(s,y)) - f(\eta_{k-2}(s,y))|^2 dyds\right)^{p/2}\right].$$

Since f is Lipschitz, we can further estimate the right-hand side to obtain

$$\mathbb{E}(|\eta_k(t,x) - \eta_{k-1}(t,x)|^p) \\ \leq c_p L^p \mathbb{E}\left[\left(\int_0^t \int_{\mathbb{R}^n} K_{t-s}^2(x-y)|\eta_{k-1}(s,y) - \eta_{k-2}(s,y)|^2 dyds\right)^{p/2}\right]. \tag{7.92}$$

We would like to move the expectation inside the integral sign on the right-hand side, but we cannot do this yet because of the exponent of $p/2$. To handle this, we will separate $K_{t-s}^2(x-y)$ into

$$K_{t-s}^2(x-y) = |K_{t-s}(x-y)|^{\frac{2p-4}{p}} \cdot |K_{t-s}(x-y)|^{\frac{4}{p}}. \tag{7.93}$$

We then apply Hölder's inequality with the conjugate exponents $p/2$ and $p/(p-2)$ in (7.92) to obtain

$$\mathbb{E}(|\eta_k(t,x) - \eta_{k-1}(t,x)|^p) \\ \leq c_p L^p \left(\int_0^t \int_{\mathbb{R}^n} K_{t-s}^2(x-y)dyds\right)^{\frac{p}{2}-1} \\ \times \mathbb{E}\left(\int_0^t \int_{\mathbb{R}^n} K_{t-s}^2(x-y)|\eta_{k-1}(s,y) - \eta_{k-2}(s,y)|^p dyds\right) \\ = c_p L^p \left(\int_0^t \int_{\mathbb{R}^n} K_{t-s}^2(x-y)dyds\right)^{\frac{p}{2}-1} \\ \times \left(\int_0^t \int_{\mathbb{R}^n} K_{t-s}^2(x-y)\mathbb{E}\left(|\eta_{k-1}(s,y) - \eta_{k-2}(s,y)|^p\right) dyds\right). \tag{7.94}$$

Therefore, defining

$$J_k^p(t) := \sup_{0 \le s \le t, x \in \mathbb{R}^n} \mathbb{E}(|\eta_k(s,x) - \eta_{k-1}(s,x)|^p),$$

we get that

$$J_k^p(t) \le c_p L^p \left(\int_0^t \int_{\mathbb{R}^n} K_{t-s}^2(x-y) dy ds \right)^{\frac{p}{2}-1} \\ \times \int_0^t J_{k-1}^p(s) \left(\int_{\mathbb{R}^n} K_{t-s}^2(x-y) dy \right) ds. \tag{7.95}$$

Using (7.84) and (7.90), we obtain the following recursive inequality:

$$J_k^p(t) \le c_{p,n} t^{(3-n)(\frac{p}{2}-1)} \int_0^t J_{k-1}^p(s)(t-s)^{2-n} ds \le c_{p,n} t^{(3-n)\frac{p}{2}-1} \int_0^t J_{k-1}^p(s) ds. \tag{7.96}$$

Note that $J_1^p(t)$ is finite. Namely, by using the BDG inequality from Theorem 3.7.1, one obtains

$$\begin{aligned}
&\mathbb{E}(|\eta_1(t,x) - \eta_0(t,x)|^p) \\
&= \mathbb{E}\left(\left| \int_0^t \int_{\mathbb{R}^n} K_{t-s}(x-y) f(\eta_0(s,y)) W(ds,dy) \right|^p \right) \\
&\le c_p \mathbb{E}\left[\left(\int_0^t \int_{\mathbb{R}^n} K_{t-s}^2(x-y) |f(\eta_0(s,y))|^2 dy ds \right)^{p/2} \right] \\
&\le c_p L^p \cdot \left[\left(\int_0^t \int_{\mathbb{R}^n} K_{t-s}^2(x-y)(1 + |\eta_0(s,y)|)^2 dy ds \right)^{p/2} \right],
\end{aligned}$$

where we eliminated the expectation because $\eta_0(t,x)$ is deterministic. We then use the splitting from (7.93) above, and the same Hölder inequality argument as before, to obtain

$$\begin{aligned}
&\mathbb{E}(|\eta_1(t,x) - \eta_0(t,x)|^p) \\
&\le c_p L^p \left(\int_0^t \int_{\mathbb{R}^n} K_{t-s}^2(x-y) dy ds \right)^{\frac{p-2}{2}} \\
&\quad \times \left(\int_0^t \int_{\mathbb{R}^n} K_{t-s}^2(x-y)(1 + |\eta_0(s,y)|)^p dy ds \right).
\end{aligned}$$

The right-hand side is uniformly bounded for $t \in [0,T]$ and $x \in \mathbb{R}^n$ by Lemma 7.2.3 and (7.74). So $J_1^p(t)$ is finite for each $t \ge 0$.

The recursive inequality (7.96) implies that for any fixed $T > 0$, we have that for all $0 \leq t \leq T$ and $k \geq 2$,

$$J_k^p(t) \leq C_{T,p,n} \int_0^t J_{k-1}^p(s)ds. \tag{7.97}$$

Since $J_1^p(t)$ is finite and bounded by a constant A_T for all $0 \leq t \leq T$, we then apply (7.97) inductively to conclude that

$$\sum_{k=1}^{\infty} J_k(t) \leq A_T^{1/p} \sum_{k=0}^{\infty} \frac{(C_{T,p,n})^{k/p} t^{k/p}}{(k!)^{1/p}} < \infty, \qquad \text{for all } t \in [0, T]. \tag{7.98}$$

Since η_0 is bounded on $[0, T] \times \mathbb{R}^n$ for all $T > 0$, deterministic, and continuous by Lemma 7.2.3, we have that $\eta_0 \in C([0, T] \times \mathbb{R}^n; L^p(\Omega))$. Since $\sum_{k=1}^{\infty} J_k(T) < \infty$, we have that the sequence η_k, $k = 1, 2, \ldots$ is a Cauchy sequence in the complete space of bounded functions on $[0, T] \times \mathbb{R}^n$, taking values in $L^p(\Omega)$, equipped with the appropriate supremum norm:

$$||f|| = \sup_{0 \leq t \leq T} \sup_{x \in \mathbb{R}^n} \left(\mathbb{E}(|f(t, x)|^p)\right)^{1/p}.$$

Hence, the sequence η_k converges in this space as $k \to \infty$, and the limit must be η. Thus, $||\eta|| = \sup_{0 \leq t \leq T} \sup_{x \in \mathbb{R}^n} (\mathbb{E}(|\eta(t, x)|^p))^{1/p}$ is bounded. □

7.4 Hölder Continuity of Sample Paths

In this section, we investigate additional properties of our unique mild solution by focusing on what the sample paths of the solution look like. In particular, we study Hölder continuity of sample paths. Because we are working with a stochastic process, we have to precisely define what we mean by Hölder continuity of the sample paths.

For this purpose, we recall the notion of a *modification* from Definition 3.4.1. If $\{X_i\}_{i \in I}$ where I is an index set is a stochastic process on a complete probability space $(\Omega, \mathcal{F}, \mathbb{P})$, then $\{\tilde{X}_i\}$ is a modification if $\mathbb{P}(\tilde{X}_i = X_i) = 1, \forall i \in I$.

We also recall that, given a stochastic process $\{X_i\}_{i \in I}$, the *finite dimensional distributions* are the distributions of the random vectors $(X_{i_1}, X_{i_2}, ..., X_{i_k})$ for all finite collections $(i_1, i_2, ..., i_k)$ of indices in I.

Note that $\{X_i\}_{i \in I}$ and a modification have the same finite dimensional distributions. Because the uniqueness result for the Eq. (7.60) is up to modification, we will show that (7.60) has a suitable modification such that the sample paths are Hölder continuous with a certain degree of Hölder regularity.

Theorem 7.4.1 (Hölder Continuity of Sample Paths) *Let g, h be the continuous versions of functions in $H^2(\mathbb{R}^n)$, and let $f : \mathbb{R} \to \mathbb{R}$ be a Lipschitz continuous function. For each $\alpha \in [0, 1)$ in the case of $n = 1$ and for each $\alpha \in [0, 1/2)$ in the case of $n = 2$, the mild solution to (7.60) has a modification that is (locally) α-Hölder continuous on $\mathbb{R}^+ \times \mathbb{R}^n$ in time and space.*

Remark 7.4.1 There are analogous results for the stochastic heat and wave equations (7.63) and (7.64), but only in one dimension, since existence and uniqueness hold only in one dimension. For the stochastic *heat* equation in $n = 1$ (7.63), there is a modification that is α-Hölder continuous in time and β-Hölder continuous in space for each $\alpha \in [0, 1/4)$ and each $\beta \in [0, 1/2)$. For the stochastic *wave* equation in $n = 1$ (7.64), there is a modification that is α-Hölder continuous in time and space for $\alpha \in [0, 1/2)$. The difference in the degree of Hölder regularity is due to the differences in space-time scaling. We emphasize that our result for the stochastic viscous wave equation (7.60) considers both $n = 1$ and $n = 2$.

The proof of Theorem 7.4.1 follows from the Kolmogorov continuity criterion, stated in Theorem 3.4.2 and Corollary 3.4.1,, which shows that we can estimate increments in space and time to establish the existence of a Hölder continuous modification.

7.4.1 *Proof of Theorem 7.4.1*

We will prove the theorem for $n = 1$ and $n = 2$. Though the specific estimates will be slightly different for each dimension, the general computations are the same for both, and hence we prove the results for $n = 1$ and $n = 2$ simultaneously.

By Corollary 3.4.1, it follows that to prove Theorem 7.4.1, it suffices to show that for all $T > 0$, for all $\delta \in (0, 1)$, and for all $p \geq 2$, there exists a constant $C_{T,p,\delta}$ depending on T, p, and δ such that:

1. The following two estimates hold if $n = 1$:

$$\mathbb{E}(|\eta(t, x) - \eta(t', x)|^p) \leq C_{T,p,\delta}|t - t'|^{\frac{(1+\delta)p}{2}}, \quad \text{for all } t, t' \in [0, T], \text{ and } x \in \mathbb{R}, \tag{7.99}$$

$$\mathbb{E}(|\eta(t, x) - \eta(t, x')|^p) \leq C_{T,p,\delta}|x - x'|^{\frac{(1+\delta)p}{2}}, \quad \text{for all } t \in [0, T], \text{ and } x, x' \in \mathbb{R}. \tag{7.100}$$

2. The following two estimates hold if $n = 2$:

$$\mathbb{E}(|\eta(t, x) - \eta(t', x)|^p) \leq C_{T,p,\delta}|t - t'|^{\frac{\delta p}{2}}, \quad \text{for all } t, t' \in [0, T], \text{ and } x \in \mathbb{R}^2, \tag{7.101}$$

$$\mathbb{E}(|\eta(t, x) - \eta(t, x')|^p) \leq C_{T,p,\delta}|x - x'|^{\frac{\delta p}{2}}, \quad \text{for all } t \in [0, T], \text{ and } x, x' \in \mathbb{R}^2. \tag{7.102}$$

Estimate for the Time Increments To prove estimates (7.99) and (7.101), we consider for $p \geq 2$,

$$\mathbb{E}(|\eta(t,x) - \eta(t',x)|^p), \qquad \text{for } t, t' \in [0,T],$$

for $n = 1, 2$, where $T > 0$ is fixed but arbitrary. Recall from the definition of a mild solution (7.66) that

$$\eta(t,x) = \eta_0(t,x) + \int_0^t \int_{\mathbb{R}^n} K_{t-s}(x-y) f(\eta(s,y)) W(ds,dy),$$

where $\eta_0(t,x)$ is the deterministic function solving the homogeneous deterministic viscous wave equation with initial data g, h. We assume that $0 \leq t' < t \leq T$ and express the time increment as

$$\begin{aligned}\eta(t,x) - \eta(t',x) = {} & \eta_0(t,x) - \eta_0(t',x) \\ & + \int_0^{t'} \int_{\mathbb{R}^n} [K_{t-s}(x-y) - K_{t'-s}(x-y)] f(\eta(s,y)) W(ds,dy) \\ & + \int_{t'}^{t} \int_{\mathbb{R}^n} K_{t-s}(x-y) f(\eta(s,y)) W(ds,dy).\end{aligned}$$

Next, we use the BDG inequality from Theorem 3.7.1, along with $(a+b+c)^p \leq c_p(|a|^p + |b|^p + |c|^p)$ for $a, b, c \geq 0$, to obtain

$$\begin{aligned}\mathbb{E}\left(|\eta(t,x) - \eta(t',x)|^p\right) \leq c_p \Bigg[& |\eta_0(t,x) - \eta_0(t',x)|^p \\ & + \mathbb{E}\left(\int_0^{t'} \int_{\mathbb{R}^n} |K_{t-s}(x-y) - K_{t'-s}(x-y)|^2 |f(\eta(s,y))|^2 dyds \right)^{p/2} \\ & + \mathbb{E}\left(\int_{t'}^{t} \int_{\mathbb{R}^n} |K_{t-s}(x-y)|^2 |f(\eta(s,y))|^2 dyds \right)^{p/2} \Bigg] := c_p(I_1 + I_2 + I_3).\end{aligned} \tag{7.103}$$

By Lemma 7.2.3, there exists C_T such that

$$I_1 = |\eta_0(t,x) - \eta_0(t',x)|^p \leq C_T |t - t'|^p, \qquad \text{in the case of } n = 1, \tag{7.104}$$

and for every $\delta \in (0,1)$, there exists a constant $C_{T,\delta}$ depending only on $T > 0$ and δ such that

$$I_1 = |\eta_0(t,x) - \eta_0(t',x)|^p \leq C_{T,\delta} |t - t'|^{\frac{\delta p}{2}}, \qquad \text{in the case of } n = 2. \tag{7.105}$$

For the integral in I_2 defined in (7.103), we use the same idea as in (7.93) and separate the term involving the kernel into two factors by using Hölder's inequality with $p/2$ and $p/(p-2)$ to obtain

$$I_2 \leq \left(\int_0^{t'} \int_{\mathbb{R}^n} |K_{t-s}(x-y) - K_{t'-s}(x-y)|^2 dyds \right)^{\frac{p}{2}-1}$$
$$\cdot \mathbb{E} \left(\int_0^{t'} \int_{\mathbb{R}^n} |K_{t-s}(x-y) - K_{t'-s}(x-y)|^2 |f(\eta(s,y))|^p dyds \right).$$

In the second factor, we can move the expectation into the integrand, and use the Lipschitz property of f to obtain that for all $s \in [0, t']$, $y \in \mathbb{R}^n$, the following estimate holds:

$$\mathbb{E}\left(|f(\eta(s,y))|^p\right) \leq c_p L^p \mathbb{E}(1 + |\eta(s,y)|^p) = C_{T,p} < \infty, \tag{7.106}$$

where the last inequality follows from the boundedness of pth moments in (7.91). Therefore,

$$I_2 \leq C_{T,p} \left(\int_0^{t'} \int_{\mathbb{R}^n} |K_{t-s}(x-y) - K_{t'-s}(x-y)|^2 dyds \right)^{\frac{p}{2}}.$$

Using Plancherel's theorem in Theorem 2.6.2, recalling the definition of $K_t(x)$ from (7.28), and absorbing constants into $C_{T,p}$ as necessary, we obtain

$$I_2 \leq C_{T,p} \left(\int_0^{t'} \int_{\mathbb{R}^n} \left| e^{-\frac{|\xi|(t-s)}{2}} \frac{\sin\left(\frac{\sqrt{3}}{2}|\xi|(t-s)\right)}{|\xi|} \right.\right.$$
$$\left.\left. - e^{-\frac{|\xi|(t'-s)}{2}} \frac{\sin\left(\frac{\sqrt{3}}{2}|\xi|(t'-s)\right)}{|\xi|} \right|^2 d\xi ds \right)^{\frac{p}{2}}. \tag{7.107}$$

Continuing to absorb constants into $C_{T,p}$ as necessary, we separate this into

$$I_2 \leq C_{T,p}(J_1 + J_2)^{\frac{p}{2}}, \tag{7.108}$$

where

$$J_1 = \int_0^{t'} \int_{\mathbb{R}^n} \frac{\sin^2\left(\frac{\sqrt{3}}{2}|\xi|(t-s)\right)}{|\xi|^2} \left| e^{-\frac{|\xi|(t-s)}{2}} - e^{-\frac{|\xi|(t'-s)}{2}} \right|^2 d\xi ds, \tag{7.109}$$

$$J_2 = \int_0^{t'} \int_{\mathbb{R}^n} e^{-|\xi|(t'-s)} \left(\frac{\sin\left(\frac{\sqrt{3}}{2}|\xi|(t-s)\right)}{|\xi|} - \frac{\sin\left(\frac{\sqrt{3}}{2}|\xi|(t'-s)\right)}{|\xi|} \right)^2 d\xi ds. \tag{7.110}$$

To estimate J_1, we first simplify J_1 to get

$$J_1 = \int_0^{t'} \int_{\mathbb{R}^n} e^{-|\xi|(t'-s)} \frac{\sin^2\left(\frac{\sqrt{3}}{2}|\xi|(t-s)\right)}{|\xi|^2} \left(1 - e^{-\frac{|\xi|(t-t')}{2}}\right)^2 d\xi ds.$$

Next, we use the fact that

$$0 \le 1 - e^{-r} \le \min(1, r), \qquad \text{for all } r \ge 0. \tag{7.111}$$

In addition, there exists a uniform constant C_δ depending only on $\delta \in (0, 1)$ such that

$$r^\delta e^{-r} \le C_\delta, \qquad \text{for all } r > 0, \delta \in (0, 1). \tag{7.112}$$

Thus, for each $\delta \in (0, 1)$, we have:

$$\begin{aligned} J_1 &\le C_\delta \int_0^{t'} \frac{1}{(t'-s)^\delta} \int_{\mathbb{R}^n} \frac{1}{|\xi|^\delta} \cdot \frac{1}{|\xi|^2} \min\left(1, \frac{1}{4}|\xi|^2(t-t')^2\right) d\xi ds \\ &\le C_\delta \int_0^{t'} \frac{1}{(t'-s)^\delta} \int_{\mathbb{R}^n} \frac{1}{|\xi|^{2+\delta}} \min(1, |\xi|^2(t-t')^2) d\xi ds \\ &= C_\delta \left(\int_0^{t'} \frac{1}{(t'-s)^\delta} \left(\int_{|\xi| \le (t-t')^{-1}} \frac{1}{|\xi|^\delta}(t-t')^2 d\xi \right.\right. \\ &\quad \left.\left. + \int_{|\xi| \ge (t-t')^{-1}} \frac{1}{|\xi|^{2+\delta}} d\xi \right) ds \right), \end{aligned} \tag{7.113}$$

where $C_{T,\delta}$ is a constant depending on $\delta \in (0, 1)$, and on the fixed but arbitrary $T > 0$. Note that we have restricted δ to the range of $\delta \in (0, 1)$, so that the appropriate integrals converge in both spatial dimension $n = 1$ and $n = 2$. Computing the integrals in (7.113) gives

$$J_1 \le C_\delta \left(\int_0^{t'} \frac{1}{(t'-s)^\delta} (t-t')^{1+\delta} ds \right) = C_{T,\delta} |t - t'|^{1+\delta}, \qquad \text{for } n = 1, \tag{7.114}$$

$$J_1 \le C_\delta \left(\int_0^{t'} \frac{1}{(t'-s)^\delta} (t-t')^\delta ds \right) = C_{T,\delta} |t - t'|^\delta, \qquad \text{for } n = 2. \tag{7.115}$$

We now consider J_2 as defined in (7.110). By the mean value theorem,

$$\left|\sin\left(\frac{\sqrt{3}}{2}|\xi|(t-s)\right)-\sin\left(\frac{\sqrt{3}}{2}|\xi|(t'-s)\right)\right| \leq \min\left(2, \frac{\sqrt{3}}{2}|\xi|(t-t')\right)$$
$$\leq \min(2, |\xi|(t-t')). \tag{7.116}$$

Combining the estimates (7.112) and (7.116) gives for arbitrary $\delta \in (0, 1)$,

$$J_2 \leq C_\delta \int_0^{t'} \frac{1}{(t'-s)^\delta} \int_{\mathbb{R}^n} \frac{1}{|\xi|^\delta} \cdot \frac{1}{|\xi|^2} \min\left(1, |\xi|^2(t-t')^2\right) d\xi ds.$$

The rest proceeds exactly as for the computation for J_1; see (7.113), and thus we obtain

$$J_2 \leq C_{T,\delta}|t-t'|^{1+\delta} \qquad \text{for } n=1, \qquad\qquad J_2 \leq C_{T,\delta}|t-t'|^{\delta} \qquad \text{for } n=2,$$

for $C_{T,\delta}$ depending only on $\delta \in (0, 1)$ and T. Substituting into (7.108), we have for arbitrary $\delta \in (0, 1)$,

$$I_2 \leq C_{T,p,\delta}|t-t'|^{\frac{(1+\delta)p}{2}} \quad \text{for } n=1, \qquad I_2 \leq C_{T,p,\delta}|t-t'|^{\frac{\delta p}{2}} \quad \text{for } n=2. \tag{7.117}$$

For I_3 as defined in (7.103), we use the idea from (7.93), combined with the Lipschitz property of f, the boundedness of pth moments of $\eta(t, x)$ on finite time intervals, and a calculation similar to (7.74) to obtain for $n = 1, 2$,

$$\begin{aligned}
I_3 &\leq c_p \left(\int_{t'}^t \int_{\mathbb{R}^n} |K_{t-s}(x-y)|^2 dyds\right)^{\frac{p}{2}-1} \left(\int_{t'}^t \int_{\mathbb{R}^n} |K_{t-s}(x-y)|^2 \mathbb{E}(|f(\eta(s,y))|^p) dyds\right)\\
&\leq c_{T,p} \left(\int_{t'}^t \int_{\mathbb{R}^n} |K_{t-s}(x-y)|^2 dyds\right)^{\frac{p}{2}} = c_{T,p} \left(\int_{t'}^t (t-s)^{2-n} ds\right)^{p/2} ||K||^p_{L^2(\mathbb{R}^n)}\\
&= c_{T,p}|t-t'|^{\frac{p}{2}(3-n)}.
\end{aligned} \tag{7.118}$$

The estimates (7.104), (7.105), (7.117), (7.118) for I_1, I_2, and I_3 and (7.103) establish the desired time increment estimates (7.99) for $n = 1$ and (7.101) for $n = 2$.

Estimate for the Spatial Increments We examine the spatial regularity of the stochastic solution $\eta(t, x)$ by establishing (7.100) and (7.102). For $0 \leq t \leq T$ and $x, x' \in \mathbb{R}^n$, we have that

$$\eta(t,x)-\eta(t,x') = (\eta_0(t,x)-\eta_0(t,x'))$$
$$+ \int_0^t \int_{\mathbb{R}^n} (K_{t-s}(x-y)-K_{t-s}(x'-y)) f(\eta(s,y)) W(ds,dy),$$

and hence, for $p \geq 2$,

$$\mathbb{E}(|\eta(t,x)-\eta(t,x')|^p) = c_p\Bigg(|\eta_0(t,x)-\eta_0(t,x')|^p$$

$$+\mathbb{E}\left(\left|\int_0^t\int_{\mathbb{R}^n}(K_{t-s}(x-y)-K_{t-s}(x'-y))f(\eta(s,y))W(ds,dy)\right|^p\right)\Bigg)$$

$$:= c_p(I_4+I_5). \tag{7.119}$$

We bound I_4 by using Lemma 7.2.3 to obtain

$$I_4 = |\eta_0(t,x)-\eta_0(t,x')|^p \leq C_T|x-x'|^p \qquad \text{for } n=1, \tag{7.120}$$

and for arbitrary $\delta \in (0,1)$,

$$I_4 = |\eta_0(t,x)-\eta_0(t,x')|^p \leq C_{T,\delta}|x-x'|^{\frac{\delta p}{2}}, \qquad \text{for } n=2. \tag{7.121}$$

To estimate I_5, we use the BDG inequality stated in Theorem 3.7.1 to obtain

$$I_5 \leq \mathbb{E}\left(\int_0^t\int_{\mathbb{R}^n}|K_{t-s}(x-y)-K_{t-s}(x'-y)|^2|f(\eta(s,y))|^2 dyds\right)^{\frac{p}{2}}.$$

By using the same computation as in (7.93),

$$I_5 \leq \left(\int_0^t\int_{\mathbb{R}^n}|K_{t-s}(x-y)-K_{t-s}(x'-y)|^2 dyds\right)^{\frac{p}{2}-1}$$

$$\cdot\left(\int_0^t\int_{\mathbb{R}^n}|K_{t-s}(x-y)-K_{t-s}(x'-y)|^2\mathbb{E}\left(|f(\eta(s,y))|^p\right)dyds\right).$$

Using the higher moment bound on η as in (7.106), we have

$$I_5 \leq C_{T,p}\left(\int_0^t\int_{\mathbb{R}^n}|K_{t-s}(x-y)-K_{t-s}(x'-y)|^2 dyds\right)^{\frac{p}{2}}$$

$$= C_{T,p}\left(\int_0^t\int_{\mathbb{R}^n}|K_{t-s}(y)-K_{t-s}(y+x'-x)|^2 dyds\right)^{\frac{p}{2}}.$$

Absorbing constants into $C_{T,p}$ as necessary and using Plancherel's formula gives that

$$I_5 \leq C_{T,p} \left(\int_0^t \int_{\mathbb{R}^n} \left| e^{-\frac{|\xi|}{2}(t-s)} \frac{\sin\left(\frac{\sqrt{3}}{2}|\xi|(t-s)\right)}{|\xi|} (1 - e^{i\xi\cdot(x'-x)}) \right|^2 d\xi ds \right)^{\frac{p}{2}}$$

$$= C_{T,p} \left(\int_0^t \int_{\mathbb{R}^n} e^{-|\xi|(t-s)} \frac{1}{|\xi|^2} \left[1 - \cos(\xi \cdot (x'-x))\right] d\xi ds \right)^{\frac{p}{2}}.$$

We use the inequality $1 - \cos(\xi \cdot (x'-x)) \leq \min(2, |\xi|^2|x'-x|^2)$ and (7.112) to obtain for $\delta \in (0, 1)$,

$$I_5 \leq C_{T,p} \left(\int_0^t \int_{\mathbb{R}^n} e^{-|\xi|(t-s)} \frac{1}{|\xi|^2} \left[1 - \cos(\xi \cdot (x'-x))\right] d\xi ds \right)^{\frac{p}{2}}$$

$$\leq C_{T,p,\delta} \left(\int_0^t \frac{1}{(t-s)^\delta} \int_{\mathbb{R}^n} \frac{1}{|\xi|^{2+\delta}} \min(2, |\xi|^2|x'-x|^2) d\xi ds \right)^{\frac{p}{2}}$$

$$\leq C_{T,p,\delta} \left(\int_0^t \frac{1}{(t-s)^\delta} \left(\int_{|\xi| \leq |x-x'|^{-1}} \frac{1}{|\xi|^\delta} |x'-x|^2 d\xi \right.\right.$$

$$\left.\left. + \int_{|\xi| \geq |x-x'|^{-1}} \frac{2}{|\xi|^{2+\delta}} d\xi \right) ds \right)^{\frac{p}{2}}.$$

These integrals converge for $n = 1, 2$ since $\delta \in (0, 1)$. We can compute these integrals to obtain

$$I_5 \leq C_{T,p,\delta} \left(\int_0^t \frac{1}{(t-s)^\delta} |x-x'|^{1+\delta} ds \right)^{\frac{p}{2}} = C_{T,p,\delta} |x-x'|^{\frac{(1+\delta)p}{2}}, \quad \text{for } n = 1, \tag{7.122}$$

$$I_5 \leq C_{T,p,\delta} \left(\int_0^t \frac{1}{(t-s)^\delta} |x-x'|^\delta ds \right)^{\frac{p}{2}} = C_{T,p,\delta} |x-x'|^{\frac{\delta p}{2}}, \quad \text{for } n = 2. \tag{7.123}$$

The estimates (7.119), (7.120), (7.121), (7.122), and (7.123) establish the required spatial increment estimates (7.100) for $n = 1$ and (7.102) for $n = 2$. This completes the proof of Theorem 7.4.1.

7.5 Conclusion

We have shown that a Cauchy problem for the stochastically perturbed viscous wave equation (7.1) has a unique (up to a modification) mild solution in both $n = 1$ and $n = 2$ and that the stochastic mild solution has a modification which is α-Hölder continuous, where the α-Hölder continuity is up to $\alpha = 1$ in $n = 1$, and up to $\alpha = 1/2$ in $n = 2$. This result is significant, especially for $n = 2$, since it indicates that stochastically perturbed fluid-structure interaction problems involving viscous, incompressible fluids at low-to-medium Reynolds numbers, will have Hölder continuous solutions for almost all realizations of sample paths, even in the case when the stochasticity in the forcing (or data) is represented by the very rough space-time white noise. We remark that this would not be the case if the structure itself, modeled by the stochastically perturbed wave equation in $n = 2$, was considered without the fluid, as it is well known that for the space-time white noise perturbed wave and heat equations, stochastic mild solutions do not exist in dimensions $n = 2$ and higher. It is the coupled problem that provides the right space-time scaling and sufficient dissipation that damps high-order frequencies exponentially fast in time, thereby allowing a unique stochastic, Hölder continuous mild solution to exist for almost all realizations.

7.6 Appendix

Proof of Lemma 7.3.1 To prove this lemma, we use induction, presented in several steps below.

Step 1. For the inductive step, suppose that the following properties of η_{k-1} are satisfied:

1. η_{k-1} is adapted to the filtration $\{\mathcal{F}_t\}_{t\geq 0}$,
2. η_{k-1} is jointly measurable,
3. η_{k-1} satisfies for every $T > 0$,

$$\sup_{t\in[0,T]}\sup_{x\in\mathbb{R}^n}\mathbb{E}\left(|\eta_{k-1}(s,y)|^2\right) := C_{k-1,T} < \infty. \tag{7.124}$$

4. η_{k-1} is continuous as a map from $(t,x) \in [0,T] \times \mathbb{R}^n$ to $L^2(\Omega)$, for arbitrary $T > 0$.

Certainly, the base case holds. This is because η_0 is deterministic; hence, it immediately satisfies the adaptedness and joint measurability conditions. For (7.124), we can get rid of the expectation, since η_0 is deterministic. Then, (7.124) follows from the fact that η_0 is bounded by Lemma 7.2.3. The $L^2(\Omega)$ continuity, since η_0 is deterministic, follows from the continuity statement in Lemma 7.2.3.

Step 2. We want to show that with this inductive assumption, the stochastic integral in (7.80) is well-defined. So given arbitrary $t > 0$, we must show that the integrand $K_{t-s}(x-y)f(\eta_{k-1}(y,s))$ for $s \in [0,t)$, $y \in \mathbb{R}^n$, satisfies the conditions in Proposition 3.7.4. Recall from (7.80) that (t,x) is a fixed but arbitrary point in $\mathbb{R}^+ \times \mathbb{R}^n$ and (s,y) here indicates the variables that are integrated against the space-time white noise. Since the kernel $K_{t-s}(x-y)$ is singular at $s = t$ and $x = y$, we first show that the conditions in Proposition 3.7.4 hold for $s \in [0,t)$ and then also in the limit $s \to t$. We start by showing that the conditions in Proposition 3.7.4 hold for $s \in [0,t)$:

- Since $\eta_{k-1}(s, y)$ for $s \in [0, t)$, $y \in \mathbb{R}^n$ is adapted to the filtration $\{\mathcal{F}_t\}_{t\geq 0}$, so is $K_{t-s}(x, y) f(\eta_{k-1}(s, y))$ since f is continuous.
- For each $s \in [0, t)$, $y \in \mathbb{R}^n$, we have

$$\mathbb{E}\left(|K_{t-s}(x-y) f(u_{k-1}(s, y))|^2\right) \leq 2L^2 C_{t-s}\left(1+\mathbb{E}\left(|\eta_{k-1}(s, y)|^2\right)\right) < \infty,$$

by the inductive assumption (7.124). Here, L is the Lipschitz constant for f, and we used the fact that $K_{t-s}(\cdot)$ is bounded by a finite constant C_{t-s} depending on the parameter $t-s$.
- To show that $K_{t-s}(x-y) f(\eta_{k-1}(y, s))$ is $L^2(\Omega)$-continuous for $s \in [0, t)$ and $y \in \mathbb{R}^n$, we fix $s_0 \in [0, t)$ and $y_0 \in \mathbb{R}^n$ and compute

$$\begin{aligned}
&\mathbb{E}(|K_{t-s_1}(x-y_1) f(\eta_{k-1}(s_1, y_1)) - K_{t-s_0}(x-y_0) f(\eta_{k-1}(s_0, y_0))|^2) \\
&\quad \leq 2\mathbb{E}\left(|K_{t-s_1}(x-y_1) f(\eta_{k-1}(s_1, y_1)) - K_{t-s_1}(x-y_1) f(\eta_{k-1}(s_0, y_0))|^2\right) \\
&\quad + 2\mathbb{E}\left(|K_{t-s_1}(x-y_1) f(\eta_{k-1}(s_0, y_0)) - K_{t-s_0}(x-y_0) f(\eta_{k-1}(s_0, y_0))|^2\right).
\end{aligned}$$

Using the Lipschitz condition for f in the first term on the right-hand side, and the fact that f is linearly bounded by $|f(x)| \leq L(1+|x|)$ in the second term on the right-hand side,

$$\begin{aligned}
&\mathbb{E}(|K_{t-s_1}(x-y_1) f(\eta_{k-1}(s_1, y_1)) - K_{t-s_0}(x-y_0) f(\eta_{k-1}(s_0, y_0))|^2) \\
&\quad \leq 2L^2 |K_{t-s_1}(x-y_1)|^2 \mathbb{E}\left(|\eta_{k-1}(s_1, y_1) - \eta_{k-1}(s_0, y_0)|^2\right) \\
&\quad + 4|K_{t-s_1}(x-y_1) - K_{t-s_0}(x-y_0)|^2 \mathbb{E}\left(L^2(1+|u_{k-1}(s_0, y_0)|^2)\right).
\end{aligned}$$

By using the inductive assumption (7.124), in the second term above, we can bound the expectation of $|\eta_{k-1}(s_0, y_0)|^2$ to obtain the following estimate:

$$\begin{aligned}
&\mathbb{E}(|K_{t-s_1}(x-y_1) f(\eta_{k-1}(y_1, s_1)) - K_{t-s_0}(x-y_0) f(\eta_{k-1}(s_0, y_0))|^2) \\
&\quad \leq 2L^2 |K_{t-s_1}(x-y_1)|^2 \mathbb{E}\left(|\eta_{k-1}(s_1, y_1) - \eta_{k-1}(s_0, y_0)|^2\right) \\
&\quad + \tilde{C}_{k-1,t} |K_{t-s_1}(x-y_1) - K_{t-s_0}(x-y_0)|^2, \qquad (7.125)
\end{aligned}$$

for some constant $\tilde{C}_{k-1,t}$ depending only on $k-1$ and t. To show continuity, we want to make the right-hand side of (7.125) arbitrarily small whenever $|(s_1, y_1) - (s_0, y_0)|$ is small. Indeed, in the first term on the right-hand side, $K_{t-s}(x-y)$ is locally bounded for $s \in [0, t)$ and $y \in \mathbb{R}^n$, and η_{k-1} is $L^2(\Omega)$ continuous by the inductive assumption, so the first term on the right-hand side of (7.125) can be made arbitrarily small for $|(s_1, y_1) - (s_0, y_0)| < \delta$, for δ sufficiently small. This is also true for the second term on the right-hand side, because $K_{t-s}(x-y)$ is continuous for $s \in [0,t)$ and $y \in \mathbb{R}^n$. This establishes the claim.

- To check the square integrability condition, we compute

$$\begin{aligned}
&\mathbb{E}\int_0^t \int_{\mathbb{R}^n} |K_{t-s}(x-y)|^2 |f(\eta_{k-1}(s,y))|^2 dyds \\
&\leq 2L^2 \int_0^t \int_{\mathbb{R}^n} |K_{t-s}(x-y)|^2 \left(1+\mathbb{E}\left(|\eta_{k-1}(s,y)|^2\right)\right) dyds \\
&= 2L^2 \left(1 + \sup_{s' \in [0,t], y' \in \mathbb{R}^n} \mathbb{E}\left(|\eta_{k-1}(s',y')|\right)^2\right)\left(\int_0^t (t-s)^{2-n} ds\right) ||K||^2_{L^2(\mathbb{R}^n)} \\
&< \infty,
\end{aligned} \tag{7.126}$$

where we used the identity in (7.74), the fact that $n = 1$ or 2, Lemma 7.2.1, and the inductive assumption (7.124).

To show that the stochastic integral in (7.80) is still well-defined for $s \in [0,t]$, we claim that this stochastic integral can be defined as the $L^2(\Omega)$ limit of stochastic integrals whose integrands are explicitly in the admissible class $\mathcal{P}_W$ of integrands (see Proposition 3.7.4 and the discussion preceding it for the definition of $\mathcal{P}_W$). To see this, choose an increasing sequence $t_i, i = 1, 2, \ldots$ of positive real numbers such that $t_i \to t$ as $i \to \infty$. Note that

$$\int_0^{t_i} \int_{\mathbb{R}^n} K_{t-s}(x-y) f(\eta_{k-1}(s,y)) W(dy,ds) \tag{7.127}$$

is a well-defined stochastic integral by the properties verified above by Proposition 3.7.4. By (7.126),

$$\mathbb{E}\int_{t_i}^t \int_{\mathbb{R}^n} |K_{t-s}(x-y)|^2 |f(\eta_{k-1}(s,y))|^2 dyds \to 0, \text{ as } t_i \to t.$$

Hence, since $\mathcal{P}_W$ is a closed Banach space, the integrand in (7.80) is in $\mathcal{P}_W$ and can be defined rigorously as the limit of the Cauchy sequence (7.127) in $L^2(\Omega)$, by the Itô isometry and the finiteness of the quantity in (7.126).

Step 3. It remains to show that $\eta_k(t,x)$ satisfies the conditions in the inductive assumption in Step 1. Indeed, $\eta_k(t,x)$ is adapted by the construction of the stochastic integral. Joint measurability (up to modification) will follow from the later verification of continuity in $L^2(\Omega)$, as noted in Remark 3.7.2. Thus, properties 1 and 2 in Step 1 are verified.

To verify property 3 in Step 1, we check that for each $T > 0$,

$$\sup_{t \in [0,T]} \sup_{x \in \mathbb{R}^n} \mathbb{E}\left(|\eta_k(t,x)|^2\right) = C_{k,T} < \infty. \tag{7.128}$$

This follows by direct calculation. Fix arbitrary $T > 0$, and consider $t \in [0, T]$, $x \in \mathbb{R}^n$. By (3.25) and (7.80), we get

$$\begin{aligned}\mathbb{E}\left(|\eta_k(t,x)|^2\right) &= 2\mathbb{E}(|\eta_0(t,x)|^2) + 2\mathbb{E}\int_0^t \int_{\mathbb{R}^n} |K_{t-s}(x-y)|^2 \\ &\quad \times |f(\eta_{k-1}(s,y))|^2 dyds \\ &= 2|\eta_0(t,x)|^2 + 2\mathbb{E}\int_0^t \int_{\mathbb{R}^n} |K_{t-s}(x-y)|^2 \\ &\quad \times |f(\eta_{k-1}(s,y))|^2 dyds. \end{aligned} \tag{7.129}$$

Note that by Lemma 7.2.3, $\eta_0(t, x)$ is bounded on $t \in [0, T]$, $x \in \mathbb{R}^n$. So we consider the remaining term. Using the calculation in (7.126) and the bound $|f(x)| \leq L(1 + |x|)$ for some L by the Lipschitz condition,

$$\begin{aligned}&\mathbb{E}\int_0^t \int_{\mathbb{R}^n} |K_{t-s}(x,y)|^2 |f(\eta_{k-1}(s,y))|^2 dyds \\ &\leq 2L^2 \left(1 + \sup_{s' \in [0,t], y' \in \mathbb{R}^n} \mathbb{E}\left(|\eta_{k-1}(s',y')|\right)^2\right) \\ &\quad \times \left(\int_0^t (t-s)^{2-n} ds\right) ||K||^2_{L^2(\mathbb{R}^n)} \leq \tilde{C}_{k,T},\end{aligned}$$

where $\tilde{C}_{k,T}$ is the finite constant, independent of $t \in [0, T]$ and $x \in \mathbb{R}^n$,

$$\begin{aligned}\tilde{C}_{k,T} := 2L^2 &\left(1 + \sup_{s' \in [0,T], y' \in \mathbb{R}^n} \mathbb{E}\left(|\eta_{k-1}(s',y')|\right)^2\right) \\ &\times \left(\int_0^T (T-s)^{2-n} ds\right) ||K||^2_{L^2(\mathbb{R}^n)},\end{aligned}$$

which is finite by the inductive assumption (7.124), Lemma 7.2.1, and the fact that $n = 1$ or 2. This verifies (7.128).

Finally, we show that property 4 in Step 1 holds, namely, that the mapping $(t, x) \mapsto \eta_k(t, x)$ taking values in $L^2(\Omega)$ is continuous on $\mathbb{R}^+ \times \mathbb{R}^n$. We decompose $\eta_k(t, x)$ in (7.80) as

$$\begin{aligned}\eta_k(t,x) &= \eta_0(t,x) + \int_0^t \int_{\mathbb{R}^n} K_{t-s}(x-y) f(\eta_{k-1}(s,y)) W(ds,dy) \\ &:= \eta_0(t,x) + \eta_k^{stoch}(t,x).\end{aligned}$$

Because $\eta_0(t, x)$ is deterministic and continuous by Lemma 7.2.3, it suffices to show that $\eta_k^{stoch}(t, x)$ is continuous in $L^2(\Omega)$. Consider $t_0 > 0$ and $x_0 \in \mathbb{R}^n$. (The argument for $t_0 = 0$ is similar.) Let

$$S_\delta = \{(t, x) \in \mathbb{R}^+ \times \mathbb{R}^n : |t - t_0| < \delta, |x - x_0| < \delta\}. \tag{7.130}$$

Continuity would follow if we can show that given arbitrary $\epsilon > 0$, there exists $\delta > 0$ sufficiently small such that

$$\mathbb{E}\left(|\eta_k^{stoch}(t, x_0) - \eta_k^{stoch}(t_0, x_0)|^2\right) < \epsilon, \qquad \text{for } |t - t_0| < \delta, \tag{7.131}$$

$$\mathbb{E}\left(|\eta_k^{stoch}(t, x_1) - \eta_k^{stoch}(t, x_0)|^2\right) < \epsilon, \qquad \text{for all } (t, x_1), (t, x_0) \in S_\delta. \tag{7.132}$$

Denote

$$T^* = t_0 + 1. \tag{7.133}$$

Let us show the first part of the continuity estimate (7.131). For every $\epsilon > 0$, we need to find a $\delta > 0$ such that (7.131) holds. We begin by first assuming that $\delta > 0$ is such that

$$\delta < \min\left\{1, \frac{t_0}{2}\right\} := \frac{t_0}{2} \wedge 1 \tag{7.134}$$

(the reason for this choice will be clear later), and we denote

$$\tau_m = t \wedge t_0 > 0, \qquad \tau_M = t \vee t_0 > 0, \tag{7.135}$$

where $t \vee t_0 := \max\{t, t_0\}$. By using a change of variables,

$$\begin{aligned}
&|\eta_k^{stoch}(t, x_0) - \eta_k^{stoch}(t_0, x_0)| \\
&= \left|\int_0^{\tau_m} \int_{\mathbb{R}^n} K_{\tau_m - s}(x_0 - y)[f(\eta_{k-1}(s + \tau_M - \tau_m, y)) \right. \\
&\quad \left. - f(\eta_{k-1}(s, y))W(ds, dy)\right| \\
&\quad + \left|\int_0^{\tau_M - \tau_m} \int_{\mathbb{R}^n} K_{\tau_M - s}(x_0 - y) f(\eta_{k-1}(s, y)) W(ds, dy)\right|.
\end{aligned}$$

Using the Lipschitz condition and the growth condition (7.62) on f, together with the Itô isometry (3.25), we can bound the expectation $\mathbb{E}(|\eta_k^{stoch}(t, x_0) - \eta_k^{stoch}(t_0, x_0)|^2)$ by two integrals, J_1 and J_2, one integrated from 0 to τ_m and the other from τ_m to τ_M:

$$\begin{aligned}
&\mathbb{E}(|\eta_k^{stoch}(t, x_0) - \eta_k^{stoch}(t_0, x_0)|^2) \\
&\leq 2L^2 \int_0^{\tau_m} \int_{\mathbb{R}^n} |K_{\tau_m - s}(x_0 - y)|^2 \mathbb{E}\left(|\eta_{k-1}(s + \tau_M - \tau_m, y) - \eta_{k-1}(s, y)|^2\right) dyds \\
&+ 4L^2 \int_0^{\tau_M - \tau_m} \int_{\mathbb{R}^n} |K_{\tau_M - s}(x_0 - y)|^2 \left(1 + \mathbb{E}\left(|\eta_{k-1}(s, y)|^2\right)\right) dyds \\
&:= 2L^2(J_1 + 2J_2).
\end{aligned} \tag{7.136}$$

To handle J_1, as long as the condition (7.134) on δ is satisfied, note that we have $0 < \tau_m < \tau_M \leq T^*$, where T^* is defined in (7.133). Hence, by (7.74),

$$\int_0^{\tau_m} \int_{\mathbb{R}^n} |K_{\tau_m - s}(x_0 - y)|^2 dyds \leq \left(\int_0^{T^*} (T^* - s)^{2-n} ds\right) ||K||^2_{L^2(\mathbb{R}^n)} := C_1. \tag{7.137}$$

We then consider a sufficiently large $M > 0$ that will be chosen later and separate J_1 into two integrals:

$$\begin{aligned}
J_1 &= \int_0^{\tau_m} \int_{|y| \geq M} \cdot + \int_0^{\tau_m} \int_{|y| < M} |K_{\tau_m - s}(x_0 - y)|^2 \mathbb{E}(|\eta_{k-1}(s + \tau_M - \tau_m, y) \\
&\quad - \eta_{k-1}(s, y)|^2) dyds \\
&:= J_{1,1} + J_{1,2}.
\end{aligned}$$

Using the inductive bound (7.124), we obtain

$$J_{1,1} \leq 2C_{k-1,T} \int_0^{\tau_m} \int_{|y| \geq M} |K_{\tau_m - s}(x_0 - y)|^2 dyds,$$

so by (7.137), we can choose M sufficiently large so that

$$J_{1,1} < \frac{\epsilon}{8L^2}. \tag{7.138}$$

Having chosen $M > 0$ sufficiently large so that (7.138) is satisfied, we use continuity to bound $J_{1,2}$ as follows. Since continuous functions are uniformly continuous on compact sets, by using the fact that $\eta_{k-1}(t, x)$ is continuous as an $L^2(\Omega)$-valued function, along with $0 < \tau_m < \tau_M \leq T^*$ and $|\tau_M - \tau_m| < \delta$, we can make

$$J_{1,2} < \frac{\epsilon}{8L^2}, \tag{7.139}$$

by choosing $\delta < \dfrac{t_0}{2} \wedge 1$ sufficiently small so that

$$\mathbb{E}\left(|\eta_{k-1}(t_1, y) - \eta_{k-1}(t_2, y)|^2\right) < C_1^{-1}\frac{\epsilon}{4L^2},$$
$$\text{when } |t_1 - t_2| < \delta,\ t_1, t_2 \in [0, T^*], |y| \le M,$$

where C_1 is the constant from (7.137). To handle J_2, we note that by the inductive assumption (7.128) and a calculation similar to (7.74),

$$\begin{aligned} J_2 &\le (1 + C_{k-1,T^*}) \int_0^{\tau_M - \tau_m} \int_{\mathbb{R}^n} |K_{\tau_M - s}(x_0 - y)|^2 dy ds \\ &= \tilde{C}_{k-1,T^*} \int_0^{\tau_M - \tau_m} (\tau_M - s)^{2-n} ds = \frac{1}{3-n} \tilde{C}_{k-1,T^*}(\tau_M^{3-n} - \tau_m^{3-n}). \end{aligned}$$

Therefore, because $|\tau_M - \tau_m| < \delta$ and $0 \le \tau_m \le \tau_M < T^*$ by (7.133) and (7.134), we can choose δ satisfying condition (7.134) sufficiently small such that

$$J_2 < \frac{\epsilon}{8L^2}. \tag{7.140}$$

So by (7.136), (7.138), (7.139), (7.140), we can choose δ sufficiently small so that (7.131) holds.

Next, we verify (7.132). By the Itô isometry (3.25) and the bound in Lemma 7.3.1,

$$\begin{aligned} &\mathbb{E}\left(|\eta_k^{stoch}(t, x_1) - \eta_k^{stoch}(t, x_0)|^2\right) \\ &= \int_0^t \int_{\mathbb{R}^n} |K_{t-s}(x_1 - y) - K_{t-s}(x_0 - y)|^2 \mathbb{E}\left(|\eta_{k-1}(s, y)|^2\right) dy ds \\ &\le C_{k-1,T^*} \int_0^t \int_{\mathbb{R}^n} |K_{t-s}(x_1 - y) - K_{t-s}(x_0 - y)|^2 dy ds \\ &= C_{k-1,T^*} \int_0^t \int_{\mathbb{R}^n} |K_s(y) - K_s(y + x_0 - x_1)|^2 dy ds. \end{aligned}$$

Recall from (7.28) that the Fourier transform of $K_t(x)$ is $e^{-\frac{|\xi|}{2}t}\frac{\sin\left(\frac{\sqrt{3}}{2}|\xi|t\right)}{\frac{\sqrt{3}}{2}|\xi|}$. Therefore, by Plancherel's formula (Theorem 2.6.2),

$$\begin{aligned} &\mathbb{E}(|\eta_k^{stoch}(t, x_1) - \eta_k^{stoch}(t, x_0)|^2) \\ &\le C_{k-1,T^*} \int_0^t \int_{\mathbb{R}^n} \left| e^{-\frac{|\xi|}{2}s} \frac{\sin\left(\frac{\sqrt{3}}{2}|\xi|s\right)}{\frac{\sqrt{3}}{2}|\xi|} \right|^2 |1 - e^{i(x_0 - x_1)\cdot\xi}|^2 d\xi ds \\ &= 2C_{k-1,T^*} \int_0^t \int_{\mathbb{R}^n} \left| e^{-\frac{|\xi|}{2}s} \frac{\sin\left(\frac{\sqrt{3}}{2}|\xi|s\right)}{\frac{\sqrt{3}}{2}|\xi|} \right|^2 [1 - \cos((x_0 - x_1)\cdot\xi)] d\xi ds \end{aligned}$$

$$
\begin{aligned}
&\leq 2C_{k-1,T^*}\int_0^{T^*}\int_{\mathbb{R}^n}\left|e^{-\frac{|\xi|}{2}s}\frac{\sin\left(\frac{\sqrt{3}}{2}|\xi|s\right)}{\frac{\sqrt{3}}{2}|\xi|}\right|^2[1-\cos((x_0-x_1)\cdot\xi)]d\xi ds\\
&\leq 4C_{k-1,T^*}\int_0^{\tau}\int_{\mathbb{R}^n}\left|e^{-\frac{|\xi|}{2}s}\frac{\sin\left(\frac{\sqrt{3}}{2}|\xi|s\right)}{\frac{\sqrt{3}}{2}|\xi|}\right|^2 d\xi ds\\
&+ 2C_{k-1,T^*}\int_\tau^{T^*}\int_{\mathbb{R}^n}\left|e^{-\frac{|\xi|}{2}s}\frac{\sin\left(\frac{\sqrt{3}}{2}|\xi|s\right)}{\frac{\sqrt{3}}{2}|\xi|}\right|^2[1-\cos((x_0-x_1)\cdot\xi)]d\xi ds\\
&:= 4C_{k-1,T^*}J_3 + 2C_{k-1,T^*}J_4, \qquad (7.141)
\end{aligned}
$$

where $\tau > 0$ will be chosen later. We have repeatedly used the fact that as long as δ is chosen so that it is also less than one (see (7.134)), then $t \in [0, T^*]$ for $t \in S_\delta$. Note that

$$
\int_0^{T^*}\int_{\mathbb{R}^n}\left|e^{-\frac{|\xi|}{2}s}\frac{\sin\left(\frac{\sqrt{3}}{2}|\xi|s\right)}{\frac{\sqrt{3}}{2}|\xi|}\right|^2 d\xi ds = \int_0^{T^*}\int_{\mathbb{R}^n}|K_s(y)|^2 dyds < \infty,
$$

by a calculation similar to (7.74). Therefore, by choosing $\tau \in (0, T^*)$ sufficiently small, we can make

$$
4C_{k-1,T^*}J_3 < \frac{\epsilon}{2}. \qquad (7.142)
$$

Now that we have fixed a choice of τ, we consider J_4. We split it into two integrals, one over the frequencies ξ such that $|\xi| > Ms^{-1}$ and the other over $|\xi| \leq Ms^{-1}$, where $M > 0$ will be chosen later:

$$
\begin{aligned}
J_4 \leq \int_\tau^{T^*}\int_{|\xi|>Ms^{-1}}\cdot + \int_\tau^{T^*}\int_{|\xi|\leq Ms^{-1}}\left|e^{-\frac{|\xi|}{2}s}\frac{\sin\left(\frac{\sqrt{3}}{2}|\xi|s\right)}{\frac{\sqrt{3}}{2}|\xi|}\right|^2\\
\times [1-\cos((x_0-x_1)\cdot\xi)]d\xi ds.
\end{aligned}
$$

By noting that $\frac{\sin\left(\frac{\sqrt{3}}{2}|\xi|s\right)}{\frac{\sqrt{3}}{2}|\xi|} \leq s \leq T^*$ and $0 \leq 1-\cos((x_0-x_1)\cdot\xi) \leq 2$ in the first integral, and $0 \leq 1-\cos(\theta) \leq \frac{1}{2}\theta^2$ in the second integral, we get

$$
J_4 \leq 2(T^*)^2\int_\tau^{T^*}\int_{|\xi|>Ms^{-1}} e^{-|\xi|s}d\xi ds + 2\int_\tau^{T^*}\int_{|\xi|\leq Ms^{-1}} e^{-|\xi|s}|x_0-x_1|^2 d\xi ds
$$

$$
\begin{aligned}
&= 2(T^*)^2 \int_\tau^{T^*} s^{-n} \int_{|\alpha|>M} e^{-|\alpha|} d\alpha ds + 2|x_0 - x_1|^2 \int_\tau^{T^*} s^{-n} \int_{|\alpha|\le M} e^{-|\alpha|} d\alpha ds \\
&\le 2T^* \tau^{-n} \left((T^*)^2 \int_{|\alpha|>M} e^{-|\alpha|} d\alpha + |x_0 - x_1|^2 \int_{\mathbb{R}^n} e^{-|\alpha|} d\alpha \right).
\end{aligned}
$$

By taking M sufficiently large such that

$$
\int_{|\alpha|>M} e^{-|\alpha|} d\alpha < \frac{1}{2C_{k-1,T^*}} \frac{1}{2(T^*)^3 \tau^{-n}} \frac{\epsilon}{4},
$$

and then taking $\delta > 0$ sufficiently small satisfying the condition (7.134), such that

$$
2T^* \tau^{-n} \delta^2 \int_{\mathbb{R}^n} e^{-|\eta|} d\eta < \frac{1}{2C_{k-1,T^*}} \frac{\epsilon}{4},
$$

we have that $2C_{k-1,T^*} J_4 < \frac{\epsilon}{2}$ whenever $(t, x_0), (t, x_1) \in S_\delta$ with $|x_0 - x_1| < \delta$. Using this fact along with (7.142) in (7.141) establishes the desired result (7.132).

□

Chapter 8
Stochastic FSI: A Fully Coupled Model with Linear Coupling

In this chapter, we continue the study of stochastic fluid-structure interaction by considering a more general approach to stochastic FSI. In the previous Chap. 7, we showed well-posedness of a reduced model of stochastic FSI represented by a stochastic viscous wave equation, which is a single self-contained equation that describes the full dynamics of a linearly coupled stochastic fluid-structure system in an unbounded domain. The special geometry in the stochastic reduced model found in Chap. 7, which consists of a fluid residing in the lower half space $(z < 0) \subset \mathbb{R}^3$ interacting with an elastic membrane with reference configuration $(z = 0)$, was essential for reducing the equations down to a single self-contained equation. In addition, the fact that we have a final equation, the stochastic viscous wave equation, posed on $\mathbb{R}^2$ allowed us to appeal to Fourier analysis and fundamental solutions in order to define the solution via a mild formulation, where the solution is defined by convolving the fundamental solution with the stochastic space-time white noise. In fact, the use of Fourier analysis on the space $\mathbb{R}^2$ allowed for a very specific characterization of the fundamental solution for the viscous wave operator, which was essential for our analysis of the reduced stochastic model. However, the unbounded geometry of the model is not feasible for use in real-life applications, which often involve problems posed on finite bounded domains, and attempting to do the same model reduction in the case of even a linearly coupled fluid-structure model posed on a finite domain would be challenging.

Thus, our goal in the current chapter is to develop a more general framework for analyzing stochastic FSI models posed on bounded domains. In this case, we will use the operator splitting method in order to construct random approximate solutions to our linearly coupled fluid-structure problem, where we consider a stochastically perturbed 2D linear Stokes flow describing an incompressible viscous Newtonian fluid (rather than a stationary Stokes flow as in the previous chapter) interacting with a stochastically forced 1D elastic membrane whose transverse displacement is described by a Koiter shell equation. The notion of solution that we will develop here is the notion of a weak solution analytically, where the solution will satisfy

S. Čanić et al., *Deterministic and Stochastic Fluid-Structure Interaction*, Advances in Mathematical Fluid Mechanics, https://doi.org/10.1007/978-3-032-00898-5_8

an appropriate weak formulation almost surely for every admissible deterministic test function, and we will constructively generate this analytically weak random solution through an operator splitting method where we split the problem into two subproblems: a structure and a fluid subproblem, which take into account stochastic effects acting on both the elastodynamics and the fluid dynamics of the system. We will then pass to the limit in these random approximate solutions using probabilistic methods, which differ from the methods used for the corresponding deterministic linearly coupled problem in Chap. 4. In particular, because the uniform boundedness of approximate solutions in finite energy spaces is only in expectation rather than pathwise, we will use compactness arguments to show that the laws of the approximate solutions converge weakly to a limiting law. These compactness arguments are used to establish weak convergence of the laws of the approximate solutions, since one must verify a condition called tightness to obtain a limiting law as a weak limit along a subsequence of laws of the approximate solutions. Since weak convergence of laws is not sufficient to conclude almost sure convergence of the approximate solutions, which is what is needed to pass to the limit in the semidiscrete weak formulations for the approximate solutions, we will need to use probabilistic techniques such as the Skorokhod representation theorem (see Theorem 3.2.5) and the Gyöngy-Krylov lemma (see Theorem 3.2.4) to strengthen the convergence of our approximate solutions to almost sure convergence along a subsequence. This chapter discusses a new general framework for analyzing stochastic fluid-structure systems based on an operator splitting approach and represents the first application of this approach to a stochastic FSI problem, which highlights the versatility of the operator splitting method.

This chapter is based on the results of [217].

8.1 Overview

We introduce a constructive approach to study solutions of stochastic fluid-structure interaction (SFSI) with stochastic noise. We focus on a benchmark problem in which a stochastically forced linearly elastic membrane interacts with the flow of a stochastically perturbed viscous incompressible Newtonian fluid in two spatial dimensions. The membrane is modeled by a Koiter shell equation, while the fluid is modeled by the 2D time-dependent Stokes equations. The problem is forced by "rough" stochastic forcing given by *multiplicative white noise*, which is a time-dependent white noise in time $\dot{W}(t)$, where $\{W(t)\}_{t\geq 0}$ is a given one-dimensional Brownian motion with respect to a complete probability space $(\Omega, \mathcal{F}, \mathbb{P})$ with complete filtration $\{\mathcal{F}_t\}_{t\geq 0}$, scaled by an appropriate nonlinear multiplicative intensity, which depends on the current state of the system. The fluid and the membrane are coupled via a two-way coupling describing continuity of fluid and structure velocities at the fluid-structure interface and continuity of contact forces at the interface. The coupling is calculated at the *linearized, fixed interface*, rendering this problem a *linear stochastic fluid-structure interaction problem*. The goal is to show that despite the rough white noise, the resulting problem is well-posed, which illustrates that the underlying deterministic fluid-structure interaction problem is robust to noise. Indeed, we prove the existence of a weak martingale solution (see Definition 8.3.1 in Sect. 8.3) to this stochastic fluid-structure interaction problem, and in the case of additive noise (in contrast to multiplicative noise), we have the stronger result of existence of a *unique* weak pathwise solution (see Definition 8.3.2 in Sect. 8.3). This means that there exist unique random variables (stochastic processes), describing the fluid velocity $\boldsymbol{u}$, the structure velocity v, and the structure displacement η, such that those stochastic processes are adapted to the filtration $\{\mathcal{F}_t\}_{t\geq 0}$, i.e., they only depend on the past history of the processes up to time t and not on the future, which satisfy the weak formulation of the original problem almost surely. This is the main result of this chapter, stated in Theorem 8.3.1. As a comment about the terminology (*weak martingale solutions* versus *weak pathwise solutions*), we emphasize that the word "weak" refers to the fact that these solutions are *analytically weak* (they satisfy the weak formulation almost surely), whereas "martingale" and "pathwise" are probabilistic terms, which refer to whether the solution is posed on a different probability space (martingale) or on the originally given probability space (pathwise).

We will first prove the existence of a weak martingale solution to the linearly coupled stochastic FSI problem, meaning that we will show existence of a weak (stochastic) solution (satisfying the weak formulation almost surely), where due to the methods used to construct the solution, we may have to transfer the solution to a new probability space $(\tilde{\Omega}, \tilde{\mathcal{F}}, \tilde{\mathbb{P}})$ that may be different from the originally given probability space $(\Omega, \mathcal{F}, \mathbb{P})$. The reason for this is because we use the Skorohod representation theorem in Theorem 3.2.5 to convert convergence in law of the approximate solutions that we construct into almost sure convergence at the price of transferring the problem to a new probability space. The proof of existence of a weak martingale solution will be accomplished via a constructive existence proof, which is based on semi-discretizing the problem in time by dividing the time interval $(0, T)$ into N subintervals of width $\Delta t = T/N$, and using a time-splitting scheme, introduced in [35], to construct approximate solutions. The goal is to show that the approximate solutions converge almost surely with respect to a certain topology, to the unique weak solution as Δt goes to zero. In contrast to the deterministic case, see Chap. 4, where a time discretization via operator splitting approach was used to study existence of weak solutions, our proposed splitting scheme for the current stochastic FSI problem must take into account stochasticity in both the structure and fluid dynamics in each of the structure and fluid subproblems. More precisely, along each time subinterval (t_N^n, t_N^{n+1}), $n = 0, \ldots, N-1$, the following two subproblems are solved to obtain approximate solutions consisting of the fluid and structure velocities, and the structure displacement, $(\boldsymbol{u}, v, \eta)$. First, in **Step 1**, the structure displacement and structure velocity are updated using only the structure displacement and structure velocity from the previous time step, under the dynamics of the homogeneous Koiter shell equation with stochastic forcing and no other external loading. Then, in **Step 2**, the fluid and structure velocities are updated by using the information from the just calculated structure velocity in

Step 1, where we must take into account stochasticity in the fluid dynamics in this weak formulation. We would like to show that the sequence or a subsequence of approximate solutions, which are random functions in an appropriate *phase space*, constructed this way converges in a certain topology to a solution which satisfies the weak formulation of the problem almost surely (a *weak martingale solution*).

Based on this splitting scheme, uniform energy estimates in terms of expectation can be derived, and the novelty in this stochastic splitting method involves obtaining estimates on the total (random) energy of the system, which involves using the BDG inequality (see Theorem 3.6.2) along with numerical dissipation estimates to carefully derive energy estimates for each of the structure/fluid subproblems. In addition to estimating the expectation of the kinetic and elastic energy of the problem, it is important to get a uniform bound on the expectation of the numerical dissipation to show that the numerical dissipation is bounded and that it, in fact, approaches zero as the time step Δt goes to zero, which is crucial in the convergence proof. This is provided in Proposition 8.5.5. We further remark that the numerical dissipation terms in the energy estimate describe dissipation that arises as a result of the time discretization, and they are given explicitly by the sum of the squares of the differences of the approximate solutions on adjacent time steps measured in appropriate energy-level spatial function spaces. Hence, having a uniform bound in expectation of the numerical dissipation terms will be essential for estimates on time shifts of the random approximate solutions (defined as functions of both space and time), see Lemma 8.7.1. Furthermore, another interesting observation is that the energy estimates will have an extra term on the right-hand side which accounts for the energy pumped into the problem by the stochastic noise. This is in addition to the energy/work contributions by the initial and boundary data. These energy estimates define an energy function space for the unknown functions $(\boldsymbol{u}, v, \eta)$. A separable subspace of the energy space, specified in (8.54) in Sect. 8.7.1, is called a *phase space* and is denoted by $\mathcal{X}$.

Uniform estimates are the backbone for weak compactness, giving rise to convergent subsequences whose limits potentially satisfy the original problem in a certain sense. In the deterministic case, for fully linear problems, the uniform energy estimates typically imply existence of weakly- and weakly*-convergent subsequences in the appropriate topologies, which is usually sufficient to pass to the limit in the linear problem and recover the weak solution. However, this approach does not work for the prototypical problem considered in this chapter, due to the nonlinearities in the *multiplicative noise*, where the multiplicative scaling factor of the random noise depends nonlinearly on the fluid/structure velocity and the structure displacement. Thus, we present a general methodology for studying these complex stochastic systems, and we develop a framework for compactness existence using compactness arguments and probabilistic methods from stochastic PDEs that has the further advantage of generalizing to more complex nonlinear stochastic FSI systems. The goal of the compactness arguments that we use is to obtain the existence of weakly convergent subsequences of *probability measures*, or *laws*, describing the *distributions* of the random approximate solutions. Once weak convergence of probability measures is established, one can then pass to the limit in the weak semidiscrete formulation for the approximate solutions to obtain existence of a weak martingale solution (a weak solution satisfying the weak formulation almost surely, potentially on a different probability space). We emphasize that these compactness arguments are very general and can be applied to a wide variety of complex stochastic systems of PDEs, and even in the fully linear case, they provide stronger convergence results almost surely of the (random) approximate solutions than can be deduced just from uniform boundedness results alone.

To establish weak convergence of probability measures, one must show that the probability measures are **tight**. More precisely, one must show that for each $\epsilon > 0$, there exists a **compact set** in the phase space $\mathcal{X}$ of displacements and fluid and structure velocities, such that the probability that our approximate solutions $(\boldsymbol{u}_N, v_N, \eta_N)$ live in that compact set is greater than $1 - \epsilon$. See Definition 3.2.1 for tightness of measures. The proof of tightness of the sequence of probability measures μ_N corresponding to the laws of the approximate solutions $(\boldsymbol{u}_N, v_N, \eta_N)$ will follow from a *deterministic* compactness argument alla Aubin-Lions. The compactness argument will establish the existence of a compact subset of the phase space $\mathcal{X}$ that contains the approximate solutions $(\boldsymbol{u}_N, v_N, \eta_N)$ with probability greater than $1 - \epsilon$, thus verifying the tightness property.

Once we have established the existence of a subsequence of probability measures μ_N that converges weakly to some probability measure μ as $N \to \infty$, or equivalently, as $\Delta t \to 0$, we would like to show that on a further subsequence, the *random variables* $(\boldsymbol{u}_N, v_N, \eta_N)$ will converge almost surely to a random variable with the law μ, with respect to the probability space $(\Omega, \mathcal{F}, \mathbb{P})$. One of the primary goals of this chapter is to present a methodology for using uniform boundedness results to deduce almost sure convergence of approximate solutions. In particular, almost sure convergence is required for passing to the limit in random approximate solutions for the current prototypical stochastic FSI model with multiplicative noise, due to the nonlinear dependence of the multiplicative factor which scales the random noise on the fluid/structure velocity and structure displacement. For our current fully linear benchmark stochastic FSI model, showing this almost sure convergence with respect to the probability space $(\Omega, \mathcal{F}, \mathbb{P})$, will be done in two parts. In the first part, we get a hold of a subsequence of approximate solutions that converge almost surely **but** on another probability space and then use this information in the second part to construct a convergent subsequence of approximate solutions that converge on the *original* probability space. The following is a more detailed albeit succinct description of the two parts.

Part I We use the Skorohod representation theorem to deduce that there exists a sequence of random variables $(\tilde{\boldsymbol{u}}, \tilde{v}, \tilde{\eta})_N$, defined on **a probability space** $(\tilde{\Omega}, \tilde{\mathcal{F}}, \tilde{\mathbb{P}})$, which is *not necessarily the same as* the original probability space $(\Omega, \mathcal{F}, \mathbb{P})$, such that the laws of $(\tilde{\boldsymbol{u}}_N, \tilde{v}_N, \tilde{\eta}_N)$ are μ_N, and $(\tilde{\boldsymbol{u}}_N, \tilde{v}_N, \tilde{\eta}_N)$ converge almost surely to a random variable $(\tilde{\boldsymbol{u}}, \tilde{v}, \tilde{\eta})$ with the law μ, on the "tilde" probability space. On this "tilde" probability space, we also show that the almost sure limit $(\tilde{\boldsymbol{u}}, \tilde{v}, \tilde{\eta})$ satisfies the weak formulation of the original problem almost surely, **but** with respect to the "tilde" probability space. This means that this limit is a **weak solution to**

the original problem in the probabilistically *weak* sense; see Definition 8.3.1. We remark that although transferring all of the probabilistic information to a new probability space $(\tilde{\Omega}, \tilde{\mathcal{F}}, \tilde{\mathbb{P}})$ with equivalence of laws may at first seem to be an abstract construction, we emphasize that this is a mathematically natural construction, as it is well-known that the new probability space can be taken to be one of the simplest probability spaces, in which $\tilde{\Omega} = [0, 1)$, $\mathcal{F}$ is the set of Borel measurable subsets on $[0, 1)$ and $\mathbb{P}$ is the Lebesgue measure; see Theorem 2.4 and its proof in [104].

Part II We would like to be able to prove that our sequence of approximate solutions $(\boldsymbol{u}_N, v_N, \eta_N)$, obtained using our time discretization via operator splitting approach described above, converges almost surely to a random variable $(\boldsymbol{u}, v, \eta)$ on *the original probability space* and satisfies the weak formulation almost surely *on the original probability space*. Namely, we would like to prove that the limit is a **weak solution to the original problem in the probabilistically *strong* sense**. If we could obtain that the sequence $(\boldsymbol{u}_N, v_N, \eta_N)$ converges **in probability** to a random variable on the original probability space $(\Omega, \mathcal{F}, \mathbb{P})$, namely, $(\boldsymbol{u}_N, v_N, \eta_N) \xrightarrow{p} (\boldsymbol{u}, v, \eta)$, then the almost sure convergence along a subsequence will follow immediately. In the case where the nonlinear scaling functions are identically equal to 1 so that the multiplicative noise terms $f(\boldsymbol{u}, v, \eta)dW_1$ and $g(\boldsymbol{u}, v, \eta)dW_2$ are just *additive noise terms* dW_1 and dW_2, we can carry out this procedure and obtain existence of a *unique weak pathwise solution on the original probability space*. To obtain convergence *in probability* of $(\boldsymbol{u}, v, \eta)_N$, we will invoke a standard Gyöngy-Krylov argument [177].

More precisely, to prove that $X_N = (\boldsymbol{u}, v, \eta)_N$ converge *in probability* to some random variable $X^* = (\boldsymbol{u}, v, \eta)$ on $(\Omega, \mathcal{F}, \mathbb{P})$, $X_N \xrightarrow{p} X^*$, based on the Gyöngy-Krylov lemma (Theorem 3.2.4), we need to show that for every two subsequences X_l and X_m, there exists a subsequence $x_k = (X_{l_k}, X_{m_k})$ such that the following two properties hold:

1. The joint laws $\nu_{X_{l_k}, X_{m_k}}$ of the subsequence x_k converge to some probability measure ν as $k \to \infty$;
2. The limiting law is supported on the diagonal: $\nu(\{(X, Y) : X = Y\}) = 1$.

The first property will follow from the tightness of measures μ_l and μ_m, which are the laws associated with the random variables $X_l = (\boldsymbol{u}_l, v_l, \eta_l)$ and $X_m = (\boldsymbol{u}_m, v_m, \eta_m)$. The tightness of the measures μ_l and μ_m implies tightness of the joint measures ν_{X_l, X_m} as well. To show that the second property holds, we will use the result of Part 1 above, combined with a *pathwise uniqueness* argument. Namely, Part 1 gives us the existence of the almost surely convergent subsequences $\tilde{X}_l = (\tilde{\boldsymbol{u}}_l, \tilde{v}_l, \tilde{\eta}_l)$ and $\tilde{X}_m = (\tilde{\boldsymbol{u}}_m, \tilde{v}_m, \tilde{\eta}_m)$ on the "tilde" probability space that have the same laws μ_l and μ_m as $X_l = (\boldsymbol{u}_l, v_l, \eta_l)$ and $X_m = (\boldsymbol{u}_m, v_m, \eta_m)$. Those two "tilde" subsequences of random variables converge to the limits $\tilde{X}^1$ and $\tilde{X}^2$, respectively, each of which has the law μ, and a joint law of $(\tilde{X}^1, \tilde{X}^2)$ equal to ν from Property 1 above. Recall, from Step 1, that both $\tilde{X}^1$ and $\tilde{X}^2$ are weak martingale solutions. To show that this joint law ν is supported on the diagonal, namely, to show Property 2 above, it is sufficient to show that $\tilde{X}^1$ is equal to $\tilde{X}^2$ almost surely, namely, it will be sufficient to show that $\tilde{\mathbb{P}}(\tilde{X}^1 = \tilde{X}^2) = 1$. Indeed, proving the diagonal condition from the Gyöngy-Krylov lemma is associated with proving pathwise uniqueness of weak solutions, which we present in Sect. 8.8.1.

Once the properties from the Gyöngy-Krylov lemma have been verified, we can conclude that there exists a subsequence of $(\boldsymbol{u}_N, v_N, \eta_N)$, which we continue to denote by N, such that $(\boldsymbol{u}_N, v_N, \eta_N) \xrightarrow{P} (\boldsymbol{u}, v, \eta)$, which implies almost sure convergence along a subsequence on the original probability space. This is presented in Sect. 8.8.2.

Finally, the proof that the limiting function $(\boldsymbol{u}, v, \eta)$ recovered above is a weak pathwise solution is presented in Sect. 8.8.3.

In contrast to the results for the stochastic viscous wave equation describing a stochastic reduced model of FSI in Chap. 7, we obtain new results for fully coupled systems in which the coupled dynamics cannot be reduced to a single equation, and we must instead handle a full stochastic *system* of coupled fluid/structure equations. The result shows that our deterministic (linearly coupled) benchmark FSI model (see Chap. 4) is robust to stochastic noise, even in the presence of rough white noise in time. This proof combines stochastic PDE analysis tools with deterministic FSI approaches. Additionally, the constructive proof lays out a framework for the development of a numerical scheme for this class of SFSI problems and provides a framework for the analysis of a broader class of more complex stochastic FSI problems, involving nonlinear dependence on the solution and involving geometric nonlinearities due to the consideration of fluid equations posed on (random) moving fluid domains.

8.2 Description of the Model

The model problem considered here is defined on a fixed fluid domain, which is a rectangle $\mathcal{O}_f = [0, L] \times [0, R]$. The boundary $\partial\mathcal{O}_f$ of the fluid domain consists of four parts: the moving boundary part denoted by Γ (it is the reference configuration of the moving boundary), the bottom of the "channel" denoted by Γ_b, and the inlet and outlet parts of the boundary Γ_{in} and Γ_{out} where the pressure data is prescribed. The flow in the fluid domain $\mathcal{O}_f$ is driven by the inlet and outlet pressure data; see Fig. 8.1. We will use $\boldsymbol{x} = (z, r) \in [0, L] \times [0, R]$ to denote the coordinates of points in the fluid domain.

The Fluid Subproblem The **fluid flow** in $\mathcal{O}_f$ will be modeled by the time-dependent Stokes equations for an incompressible, viscous fluid (see Sect. 2.13.2 for a discussion of the linear time-dependent Stokes equations):

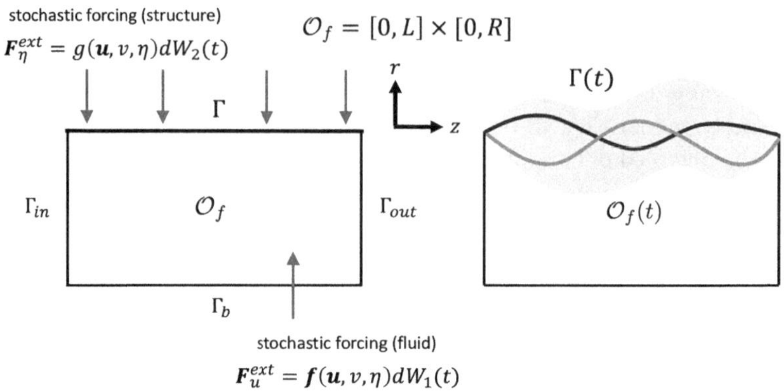

Fig. 8.1 Left: A sketch of the linearly coupled stochastic FSI problem, with $\mathcal{O}_f$ denoting the reference fluid domain, Γ denoting the reference configuration of the structure, and $dW_1(t)$ and $dW_2(t)$ denoting stochastic white noise in time. The stochastic forcing on the fluid and the structure are scaled by nonlinear scaling functions $\boldsymbol{f}(\boldsymbol{u}, v, \eta)$ and $g(\boldsymbol{u}, v, \eta)$, respectively, so that there is multiplicative noise in both the fluid and structure equations. Here v denotes the structure velocity $\partial_t \eta$. Right: Different possible outcomes for the random configuration $\Gamma(t)$ of the structure at some time t. The shaded region represents a depiction of where the structure is likely to be.

$$\left.\begin{aligned} \partial_t \boldsymbol{u} &= \nabla \cdot \boldsymbol{\sigma} + \boldsymbol{F}_u^{ext}, \\ \nabla \cdot \boldsymbol{u} &= 0, \end{aligned}\right\} \quad \text{in } \mathcal{O}_f, \tag{8.1}$$

where $\boldsymbol{u}(t, \boldsymbol{x}) = (u_z(t, \boldsymbol{x}), u_r(t, \boldsymbol{x}))$ is the fluid velocity, $\boldsymbol{\sigma} = -p\boldsymbol{I} + 2\mu \boldsymbol{D}(\boldsymbol{u})$ is the Cauchy stress tensor describing a Newtonian fluid, p is the fluid pressure, and $\boldsymbol{F}_u^{ext}$ is an external source term that will be used to represent the external stochastic forces that will perturb the incompressible fluid dynamics. We will make the definition of $\boldsymbol{F}_u^{ext}$ in terms of stochastic forces explicit later. This gives rise to the following system:

$$\left.\begin{aligned} \partial_t \boldsymbol{u} - \mu \Delta \boldsymbol{u} + \nabla p &= \boldsymbol{F}_u^{ext}, \\ \nabla \cdot \boldsymbol{u} &= 0, \end{aligned}\right\} \quad \text{in } \mathcal{O}_f. \tag{8.2}$$

The Structure Subproblem At the top boundary Γ of the fluid domain, an elastic shell interacts with the fluid flow. We assume that this elastic structure experiences displacement only in the vertical direction from its reference configuration Γ, and we denote the magnitude of this displacement by $\eta(t, z)$. The **elastodynamics of the structure** will be modeled by the following Koiter shell equation:

$$\eta_{tt} - \partial_z^2 \eta + \partial_z^4 \eta = f, \qquad \text{on } \Gamma, \tag{8.3}$$

where f is an external forcing term. We prescribe clamped boundary conditions for the structure displacement on the endpoints of Γ so that $\eta(0) = \eta(L) = \partial_z \eta(0) = \partial_z \eta(L) = 0$.

The Coupling Conditions Between the Fluid and Structure The fluid and structure are coupled via two sets of coupling conditions, the kinematic and dynamic coupling conditions, which are evaluated along the *fixed interface* Γ, rather than the moving interface $\Gamma(t)$. This is known as **linear coupling**. The **kinematic coupling condition** considered in this work describes the continuity of velocities at the fluid-structure interface

$$\boldsymbol{u} = \eta_t \boldsymbol{e}_r, \qquad \text{on } \Gamma, \tag{8.4}$$

which is the *no-slip* condition. The **dynamic coupling condition** describes balance of forces at the interface and identifies the external loading f in (8.3) as a net contribution of the fluid stress and any other external forces on the elastic structure. Namely, it states that the elastodynamics of the thin elastic structure is driven by the jump in the force acting on the structure, coming from the normal component of the normal fluid stress $\boldsymbol{\sigma} \boldsymbol{e_r} \cdot \boldsymbol{e_r}$ on one side, and the external forcing F_η^{ext} on the other:

$$\eta_{tt} - \Delta\eta = -\boldsymbol{\sigma} \boldsymbol{e_r} \cdot \boldsymbol{e_r} + F_\eta^{ext}, \qquad \text{on } \Gamma,$$

where $\boldsymbol{e_r}$ is the unit outer normal to the fixed fluid-structure interface Γ.

The Stochastic (Random) External Forcing In this prototypical model, we consider the external force on the fluid $\boldsymbol{F}_u^{ext}$ and the structure F_η^{ext} to be stochastic (random in time) forces, which represent random perturbations of the fully coupled system. We will consider the stochastic force on the fluid and the structure to be nonlinear rescalings of space-time one-dimensional white noise in time. Consider a probability space with filtration satisfying the usual conditions (see Definition 3.3.2) $(\Omega, \mathcal{F}, \{\mathcal{F}_t\}_{t\geq 0}, \mathbb{P})$. Let W_1 and W_2 be independent one-dimensional $\{\mathcal{F}_t\}_{t\geq 0}$-Wiener processes (see Definition 3.5.2). We assume that

$$\boldsymbol{F}_u^{ext} = \boldsymbol{f}(\boldsymbol{u}, v, \eta) dW_1(t), \qquad F_\eta^{ext} = g(\boldsymbol{u}, v, \eta) dW_2(t), \tag{8.5}$$

where the nonlinear functions

$$\boldsymbol{f} : L^2(\mathcal{O}_f) \times L^2(\Gamma) \times H_0^2(\Gamma) \to L^2(\mathcal{O}_f), \text{ and}$$
$$g : L^2(\mathcal{O}_f) \times L^2(\Gamma) \times H_0^2(\Gamma) \to L^2(\Gamma)$$

satisfy the following boundedness and Lipschitz conditions for a uniform constant C:

$$\|\boldsymbol{f}(\boldsymbol{u}, v, \eta)\|_{L^2(\mathcal{O}_f)} \leq C\left(1 + \|\boldsymbol{u}\|_{L^2(\mathcal{O}_f)} + \|v\|_{L^2(\Gamma)} + \|\eta\|_{H_0^2(\Gamma)}\right), \tag{8.6}$$

$$\|g(\boldsymbol{u}, v, \eta)\|_{L^2(\Gamma)} \leq C\left(1 + \|\boldsymbol{u}\|_{L^2(\mathcal{O}_f)} + \|v\|_{L^2(\Gamma)} + \|\eta\|_{H_0^2(\Gamma)}\right), \tag{8.7}$$

$$\|\boldsymbol{f}(\boldsymbol{u}_1, v_1, \eta_1) - \boldsymbol{f}(\boldsymbol{u}_2, v_2, \eta_2)\|_{L^2(\mathcal{O}_f)} \\ \leq C\left(\|\boldsymbol{u}_1 - \boldsymbol{u}_2\|_{L^2(\mathcal{O}_f)} + \|v_1 - v_2\|_{L^2(\Gamma)} + \|\eta_1 - \eta_2\|_{L^2(\Gamma)}\right), \tag{8.8}$$

$$\|g(\boldsymbol{u}_1, v_1, \eta_1) - g(\boldsymbol{u}_2, v_2, \eta_2)\|_{L^2(\Gamma)} \\ \leq C\left(\|\boldsymbol{u}_1 - \boldsymbol{u}_2\|_{L^2(\mathcal{O}_f)} + \|v_1 - v_2\|_{L^2(\Gamma)} + \|\eta_1 - \eta_2\|_{L^2(\Gamma)}\right). \tag{8.9}$$

Note that the random one-dimensional white noise forcing in time is random yet has no spatial dependence along Γ for the random structure forcing or along $\mathcal{O}_f$ for the random fluid forcing. However, when rescaled by the nonlinear terms $f(\boldsymbol{u}, v, \eta)$ and $g(\boldsymbol{u}, v, \eta)$, the resulting random forcing is no longer spatially constant.

In summary, the prototypical model of linearly coupled stochastic fluid-structure interaction considered in this chapter, supplemented with initial and boundary data, is given by the following: *Find* $(\boldsymbol{u}, \eta)$ *such that*

$$\left.\begin{array}{r}\partial_t \boldsymbol{u} = \nabla\cdot\boldsymbol{\sigma} + \boldsymbol{f}(\boldsymbol{u}, v, \eta)dW_1(t),\\ \nabla\cdot\boldsymbol{u} = 0,\end{array}\right\} \quad \text{in } \mathcal{O}_f, \tag{8.10}$$

$$\left.\begin{array}{r}\eta_{tt} - \partial_z^2\eta + \partial_z^4\eta = -\boldsymbol{\sigma}\boldsymbol{e_r}\cdot\boldsymbol{e_r} + g(\boldsymbol{u}, v, \eta)dW_2(t),\\ \boldsymbol{u} = \eta_t\boldsymbol{e_r},\end{array}\right\} \quad \text{on } \Gamma, \tag{8.11}$$

with boundary data:

$$\left.\begin{array}{r}u_r = 0,\\ p = P_{in/out}(t),\end{array}\right\} \quad \text{on } \Gamma_{in/out}, \qquad u_r = \partial_r u_z = 0, \qquad \text{on } \Gamma_b, \tag{8.12}$$

$$\eta(0) = \eta(L) = \partial_z\eta(0) = \partial_z\eta(L) = 0, \tag{8.13}$$

and the following deterministic initial data:

$$\boldsymbol{u}(0, z, r) = \boldsymbol{u}_0(z, r), \qquad \eta(0, z, R) = \eta_0(z), \qquad \partial_t\eta(0, z, R) = v_0(z), \tag{8.14}$$

where $\boldsymbol{u}_0 \in L^2(\mathcal{O}_f)$, $\eta_0 \in H_0^2(\Gamma)$, *and* $v_0 \in L^2(\Gamma)$, *and* W_1 *and* W_2 *are given independent one-dimensional Brownian motions with respect to the complete probability space* $(\Omega, \mathcal{F}, \mathbb{P})$ *with complete filtration* $\{\mathcal{F}_t\}_{t\geq 0}$.

Thus, the problem is driven by deterministic inlet and outlet pressure data $P_{in/out}(t)$ prescribed on $\Gamma_{in/out}$, with the flow symmetry condition imposed at the bottom boundary Γ_b. To avoid notational ambiguity, we emphasize that throughout this chapter, we will be using Ω to denote the underlying probability space, while $\mathcal{O}_f$ denotes the fluid domain.

8.3 Definition of a Weak Solution and Main Result

To define the space of weak solutions to the above problem, we first introduce the function space for the fluid velocity:

$$\mathcal{V}_f = \{\boldsymbol{u} = (u_z, u_r) \in H^1(\mathcal{O}_f) : \nabla \cdot \boldsymbol{u} = 0,\ u_z = 0 \text{ on } \Gamma,\ u_r = 0 \text{ on } \partial\mathcal{O}_f \backslash \Gamma\}. \tag{8.15}$$

Since the structure subproblem is given by the (fourth order) Koiter shell equation with clamped ends, the natural space of functions for the structure is

$$\mathcal{V}_s = H_0^2(\Gamma). \tag{8.16}$$

Motivated by the energy inequality presented in Sect. 8.4, we introduce the following solution spaces in time for the fluid and structure subproblems:

$$\mathcal{W}_f(0, T) = L^2(\Omega; L^\infty(0, T; L^2(\mathcal{O}_f))) \cap L^2(\Omega; L^2(0, T; \mathcal{V}_f)). \tag{8.17}$$

$$\mathcal{W}_s(0, T) = L^2(\Omega; W^{1,\infty}(0, T; L^2(\Gamma))) \cap L^2(\Omega; L^\infty(0, T; \mathcal{V}_s)). \tag{8.18}$$

We emphasize that $\boldsymbol{u}$ and η are random variables and that the $L^2(\Omega)$ part of the solution spaces reflects the fact that the energy estimate will hold in expectation.

Finally, we introduce the solution space for the *stochastic coupled FSI problem*:

$$\mathcal{W}(0, T) = \{(\boldsymbol{u}, \eta) \in \mathcal{W}_f(0, T) \times \mathcal{W}_s(0, T) : \boldsymbol{u}|_\Gamma = \eta_t \boldsymbol{e_r} \text{ for a.e. } t \in [0, T], \text{ a.s.}\}. \tag{8.19}$$

Notice that in this solution space, the kinematic coupling condition is enforced strongly.

As in the deterministic case, we define weak solutions by integrating in space against an appropriate space of test functions, which we define to be:

$$\mathcal{Q} = \{(\boldsymbol{q}, \psi) \in \mathcal{V}_f \times \mathcal{V}_s : \boldsymbol{q}|_\Gamma = \psi \boldsymbol{e}_r\}, \tag{8.20}$$

where we emphasize that this test space consists of spatial deterministic test functions (which do not vary in time). We can define a fixed test space for all $t \in [0, T]$, due to the fact that we are considering a linearly coupled problem defined on a *fixed fluid domain*, so that the solution space for the fluid velocity can be defined independently of time.

To motivate the definition of a weak solution, we will proceed as in Chap. 4; see also the reference [265]. For the purposes of the derivation of the weak solution, we consider, for the moment, the case of a general deterministic external force on the fluid $\boldsymbol{F}_u^{ext}$ and the structure F_η^{ext} (with both temporal and spatial dependence), so that the structure elastodynamics equation becomes

$$\eta_{tt} - \partial_z^2 \eta + \partial_z^4 \eta = -\boldsymbol{\sigma} \boldsymbol{e_r} \cdot \boldsymbol{e_r} + F_\eta^{ext}, \tag{8.21}$$

and the balance of the momentum equation for the fluid becomes

$$\boldsymbol{u}_t - \nabla \cdot \boldsymbol{\sigma}(\boldsymbol{u}, p) = \boldsymbol{F}_u^{ext}.$$

We will derive the standard deterministic partial differential equation definition of a weak solution, assuming that $\boldsymbol{F}_u^{ext}$ and F_η^{ext} are purely deterministic functions of time and space, and then we generalize this to the stochastic case.

We start by taking a test function $(\boldsymbol{q}, \psi) \in \mathcal{Q}$, and multiplying the linear Stokes equation by $\boldsymbol{q}$ and integrating in space and time. For arbitrary $0 \leq t \leq T$, we obtain:

$$\int_0^t \int_{\mathcal{O}_f} \partial_t \boldsymbol{u} \cdot \boldsymbol{q} d\boldsymbol{x} dt = \int_0^t \int_{\mathcal{O}_f} (\nabla \cdot \boldsymbol{\sigma}) \cdot \boldsymbol{q} d\boldsymbol{x} dt + \int_0^t \int_{\mathcal{O}_f} \boldsymbol{F}_u^{ext} \cdot \boldsymbol{q} d\boldsymbol{x} dt.$$

By integrating the first term in (8.22) in time, we obtain:

$$\int_0^t \int_{\mathcal{O}_f} \partial_t \boldsymbol{u} \cdot \boldsymbol{q} d\boldsymbol{x} ds = \int_{\mathcal{O}_f} \boldsymbol{u}(t) \cdot \boldsymbol{q} d\boldsymbol{x} - \int_{\mathcal{O}_f} \boldsymbol{u}_0 \cdot \boldsymbol{q} d\boldsymbol{x}. \tag{8.22}$$

By integrating the second term in (8.22) by parts in space and using the divergence free condition on $\boldsymbol{q}$, we obtain:

$$\int_{\mathcal{O}_f} (\nabla \cdot \boldsymbol{\sigma}) \cdot \boldsymbol{q} d\boldsymbol{x} = \int_{\partial \mathcal{O}_f} (\boldsymbol{\sigma n}) \cdot \boldsymbol{q} dS - 2\mu \int_{\mathcal{O}_f} \boldsymbol{D}(\boldsymbol{u}) : \boldsymbol{D}(\boldsymbol{q}) d\boldsymbol{x}, \tag{8.23}$$

where $\boldsymbol{D}(\boldsymbol{u})$ and $\boldsymbol{D}(\boldsymbol{q})$ represent the symmetrized gradient. We can compute the contribution of $\int_{\partial \mathcal{O}_f} (\boldsymbol{\sigma n}) \cdot \boldsymbol{q} dS$ along the four parts of the boundary of $\mathcal{O}_f$, namely, Γ, Γ_{in}, Γ_{out}, and Γ_b:

- **Top boundary.** On Γ, we use (8.21) and the fact that the test functions by definition satisfy $\boldsymbol{q} = \psi \boldsymbol{e}_r$ to compute that

$$\begin{aligned}\int_\Gamma (\boldsymbol{\sigma n}) \cdot \boldsymbol{q} dS &= \int_\Gamma (\boldsymbol{\sigma e}_r \cdot \boldsymbol{e}_r) \psi dS = \int_\Gamma (F_\eta^{ext} - \partial_t^2 \eta + \partial_z^2 \eta - \partial_z^4 \eta) \psi dS \\ &= \int_\Gamma F_\eta^{ext} \psi dz - \int_\Gamma (\partial_t v) \psi dz - \int_\Gamma \partial_z \eta \partial_z \psi - \int_\Gamma \partial_z^2 \eta \partial_z^2 \psi.\end{aligned}$$

- **Left and right boundary.** On $\Gamma_{in/out}$, since $q_r = 0$ on $\Gamma_{in/out}$ by the definition of the test space $\mathcal{Q}$, we have that

$$\int_{\Gamma_{in/out}} (\boldsymbol{\sigma n}) \cdot \boldsymbol{q} dS = \mp \int_{\Gamma_{in/out}} (\boldsymbol{\sigma e}_z \cdot \boldsymbol{e}_z) q_z dS.$$

Note that

$$\boldsymbol{D}(\boldsymbol{u})\boldsymbol{e}_z \cdot \boldsymbol{e}_z\Big|_{\Gamma_{in/out}} = \frac{\partial u_z}{\partial z}\Big|_{\Gamma_{in/out}} = 0$$

by the divergence free condition $\nabla \cdot \boldsymbol{u} = 0$ and the fact that $u_r|_{\Gamma_{in/out}} = 0$ and hence $\partial_r u_r|_{\Gamma_{in/out}} = 0$. Hence, using the definition of the Cauchy stress tensor as $\boldsymbol{\sigma} = -p\boldsymbol{I} + 2\mu \boldsymbol{D}(\boldsymbol{u})$, we conclude that:

$$\int_{\Gamma_{in/out}} (\boldsymbol{\sigma}\boldsymbol{n}) \cdot \boldsymbol{q} dS = \pm \int_{\Gamma_{in/out}} p q_z dS.$$

- **Bottom boundary.** On Γ_b, since $q_r = 0$ on Γ_b by the definition of the test space $\mathcal{Q}$, we have that

$$\int_{\Gamma_b} (\boldsymbol{\sigma}\boldsymbol{n}) \cdot \boldsymbol{q} dS = -\int_{\Gamma_b} (\boldsymbol{\sigma}\boldsymbol{e}_r \cdot \boldsymbol{e}_z) q_z dS.$$

We compute that

$$\boldsymbol{D}(\boldsymbol{u})\boldsymbol{e}_r \cdot \boldsymbol{e}_z|_{\Gamma_b} = \frac{1}{2}\left(\frac{\partial u_z}{\partial r} + \frac{\partial u_r}{\partial z}\right)\Big|_{\Gamma_b} = 0,$$

since $u_r = 0$ and $\partial_r u_z = 0$ on Γ_b by the definition of the test space $\mathcal{Q}$. Hence, since $(p\boldsymbol{I})\boldsymbol{e}_r \cdot \boldsymbol{e}_z = 0$, we conclude that $\int_{\Gamma_b} (\boldsymbol{\sigma}\boldsymbol{n}) \cdot \boldsymbol{q} dS = 0$.

Hence, integrating in time in Eq. (8.23) and using the computation above of $\int_{\partial \mathcal{O}_f} (\boldsymbol{\sigma}\boldsymbol{n}) \cdot \boldsymbol{q} dS$, we obtain

$$\int_0^t \int_{\mathcal{O}_f} (\nabla \cdot \boldsymbol{\sigma}) \cdot \boldsymbol{q} d\boldsymbol{x} ds = \int_0^t \int_{\Gamma_{in}} p q_z dr ds - \int_0^t \int_{\Gamma_{out}} p q_z dr ds$$
$$- 2\mu \int_0^t \int_{\mathcal{O}_f} \boldsymbol{D}(\boldsymbol{u}) : \boldsymbol{D}(\boldsymbol{q}) d\boldsymbol{x} ds - \int_0^t \int_{\Gamma} \partial_z \eta \partial_z \psi dz ds$$
$$- \int_0^t \int_{\Gamma} \partial_z^2 \eta \partial_z^2 \psi dz ds - \int_{\Gamma} v(t) \psi dz + \int_{\Gamma} v_0 \psi dz + \int_0^t \int_{\Gamma} F_\eta^{ext} \psi dz ds,$$

where we used (8.21) to evaluate the boundary term on the top boundary Γ, which is the (fixed) linearized fluid-structure interface.

Putting this all together and using the boundary conditions $p = P_{in/out}(t)$ on $\Gamma_{in/out}$, we obtain:

$$\int_{\mathcal{O}_f} \boldsymbol{u}(t) \cdot \boldsymbol{q} d\boldsymbol{x} + \int_{\Gamma} v(t)\psi dz + 2\mu \int_0^t \int_{\mathcal{O}_f} \boldsymbol{D}(\boldsymbol{u}) : \boldsymbol{D}(\boldsymbol{q}) d\boldsymbol{x} ds$$

$$+ \int_0^t \int_{\Gamma} \partial_z \eta \partial_z \psi dz ds + \int_0^t \int_{\Gamma} \partial_z^2 \eta \partial_z^2 \psi dz ds$$

$$= \int_0^t P_{in}(s) \left(\int_{\Gamma_{in}} q_z dr \right) ds - \int_0^t P_{out}(s) \left(\int_{\Gamma_{out}} q_z dr \right) ds$$

$$+ \int_{\mathcal{O}_f} \boldsymbol{u}_0 \cdot \boldsymbol{q}(0) d\boldsymbol{x} + \int_{\Gamma} v_0 \psi(0) dz + \int_0^t \int_{\mathcal{O}_f} \boldsymbol{F}_u^{ext} \cdot \boldsymbol{q} d\boldsymbol{x} ds + \int_0^t \int_{\Gamma} F_\eta^{ext} \psi dz ds.$$

Now, we formally substitute

$$\boldsymbol{F}_u^{ext}(t) = \boldsymbol{f}(\boldsymbol{u}, v, \eta) dW_1(t), \qquad F_\eta^{ext}(t) = g(\boldsymbol{u}, v, \eta) dW_2(t)$$

into the definition of the deterministic weak solution to get that the terms containing $\boldsymbol{F}_u^{ext}$ and F_η^{ext} can be interpreted in the stochastic case as the following stochastic integrals:

$$\int_0^t \int_{\mathcal{O}_f} \boldsymbol{f}(\boldsymbol{u}, v, \eta) \cdot \boldsymbol{q} d\boldsymbol{x} dW_1(s) \quad \text{and} \quad \int_0^t \int_{\Gamma} g(\boldsymbol{u}, v, \eta) \psi dx dW_2(s).$$

Before we give the definition of a weak solution to the stochastic FSI problem above, we recall the definition of a **stochastic basis** (see Definition 3.5.4). A **stochastic basis** $\mathcal{S}$ is an ordered collection

$$\mathcal{S} = (\Omega, \mathcal{F}, \{\mathcal{F}_t\}_{t \geq 0}, \mathbb{P}, W_1, W_2),$$

where $(\Omega, \mathcal{F}, \mathbb{P})$ is a probability space, $\{\mathcal{F}_t\}_{t \geq 0}$ is a complete filtration with respect to this probability space, and $\{W_1\}_{t \geq 0}$ and $\{W_2\}_{t \geq 0}$ are *independent* one-dimensional Brownian motions on the probability space with respect to the filtration $\{\mathcal{F}_t\}_{t \geq 0}$; see Sect. 3.5 for the definition.

We will define two notions of solution: (1) a weak martingale solution and (2) a weak pathwise solution. The second one is stronger than the first, but even to prove existence of a weak pathwise solution, it is customary to first establish existence of a weak martingale solution and then strengthen the result via the Gyöngy-Krylov lemma (Theorem 3.2.4). In this chapter, we will obtain existence of a weak martingale solution to the stochastic FSI problem with multiplicative noise, and furthermore, we will be able to obtain a stronger result of existence and uniqueness of a weak pathwise solution in the case when the noise is purely additive.

Definition 8.3.1 An ordered triple $(\tilde{\mathcal{S}}, \tilde{\boldsymbol{u}}, \tilde{\eta})$ is a *weak martingale solution* if there exists a stochastic basis

$$\tilde{\mathcal{S}} = (\tilde{\Omega}, \tilde{\mathcal{F}}, \{\tilde{\mathcal{F}}_t\}_{t \geq 0}, \tilde{\mathbb{P}}, \tilde{W}_1, \tilde{W}_2)$$

and $(\tilde{\boldsymbol{u}}, \tilde{\eta}) \in \mathcal{W}(0, T)$ with paths almost surely in $C(0, T; \mathcal{Q}')$, which satisfies:

- $(\tilde{\boldsymbol{u}}, \tilde{\eta})$ is adapted to the filtration $\{\tilde{\mathcal{F}}_t\}_{t\geq 0}$,
- $\tilde{\eta}(0) = \eta_0$ almost surely, and
- for all $(\boldsymbol{q}, \psi) \in \mathcal{Q}$ and for all $t \in [0, T]$:

$$\int_{\mathcal{O}_f} \tilde{\boldsymbol{u}}(t) \cdot \boldsymbol{q} d\boldsymbol{x} + \int_\Gamma v(t)\psi dz + 2\mu \int_0^t \int_{\mathcal{O}_f} \boldsymbol{D}(\tilde{\boldsymbol{u}}) : \boldsymbol{D}(\boldsymbol{q}) d\boldsymbol{x} ds$$
$$+ \int_0^t \int_\Gamma \partial_z \tilde{\eta} \partial_z \psi dz ds + \int_0^t \int_\Gamma \partial_z^2 \tilde{\eta} \partial_z^2 \psi dz ds = \int_0^t P_{in}(s) \left(\int_{\Gamma_{in}} q_z dr \right) ds$$
$$- \int_0^t P_{out}(t) \left(\int_{\Gamma_{out}} q_z dr \right) ds + \int_{\mathcal{O}_f} \boldsymbol{u_0} \cdot \boldsymbol{q}(0) d\boldsymbol{x} + \int_\Gamma v_0 \psi(0) dz$$
$$+ \int_0^t \int_{\mathcal{O}_f} \boldsymbol{f}(\tilde{\boldsymbol{u}}, \partial_t \tilde{\eta}, \tilde{\eta}) \cdot \boldsymbol{q} d\boldsymbol{x} d\tilde{W}_1(s) + \int_0^t \int_\Gamma g(\tilde{\boldsymbol{u}}, \partial_t \tilde{\eta}, \tilde{\eta}) \psi dz d\tilde{W}_2(s),$$

$\tilde{\mathbb{P}}$-almost surely.

Definition 8.3.2 Let $\mathcal{S} = (\Omega, \mathcal{F}, \{\mathcal{F}_t\}_{t\geq 0}, \mathbb{P}, W_1, W_2)$ be a stochastic basis with a complete filtration $\{\mathcal{F}_t\}_{t\geq 0}$ and independent one-dimensional Brownian motions $\{W_1(t)\}_{t\geq 0}$ and $\{W_2(t)\}_{t\geq 0}$ on a given probability space $(\Omega, \mathcal{F}, \mathbb{P})$. An ordered pair $(\boldsymbol{u}, \eta)$ is a *weak pathwise solution* if $(\boldsymbol{u}, \eta) \in \mathcal{W}(0, T)$ with paths almost surely in $C(0, T; \mathcal{Q}')$ satisfies:

- $(\boldsymbol{u}, \eta)$ is adapted to the filtration $\{\mathcal{F}_t\}_{t\geq 0}$
- $\eta(0) = \eta_0$ almost surely, and
- for all $(\boldsymbol{q}, \psi) \in \mathcal{Q}$ and $t \in [0, T]$:

$$\int_{\mathcal{O}_f} \boldsymbol{u}(t) \cdot \boldsymbol{q} d\boldsymbol{x} + \int_\Gamma v(t)\psi dz + 2\mu \int_0^t \int_{\mathcal{O}_f} \boldsymbol{D}(\boldsymbol{u}) : \boldsymbol{D}(\boldsymbol{q}) d\boldsymbol{x} ds$$
$$+ \int_0^t \int_\Gamma \partial_z \eta \partial_z \psi dz ds + \int_0^t \int_\Gamma \partial_z^2 \eta \partial_z^2 \psi dz ds = \int_0^t P_{in}(s) \left(\int_{\Gamma_{in}} q_z dr \right) ds$$
$$- \int_0^t P_{out}(t) \left(\int_{\Gamma_{out}} q_z dr \right) ds + \int_{\mathcal{O}_f} \boldsymbol{u_0} \cdot \boldsymbol{q}(0) d\boldsymbol{x} + \int_\Gamma v_0 \psi(0) dz$$
$$+ \int_0^t \int_{\mathcal{O}_f} \boldsymbol{f}(\boldsymbol{u}, \partial_t \eta, \eta) \cdot \boldsymbol{q} d\boldsymbol{x} dW_1(s) + \int_0^t \int_\Gamma g(\boldsymbol{u}, \partial_t \eta, \eta) \psi dz dW_2(s).$$

$\mathbb{P}$-almost surely.

In a weak pathwise solution as in the second definition above, we have a random solution satisfying the initial conditions on the originally given (arbitrary) probability space with the originally given independent one-dimensional Brownian motions $\{W_1(t)\}_{t\geq 0}$, $\{W_2(t)\}_{t\geq 0}$, with respect to the original complete filtration. In a weak martingale solution, we have a weaker requirement that the random solution exists on a *particular (not arbitrary) probability space*, where the initial conditions are satisfied "in law." In general, we will show the existence of a weak martingale solution. To get to that solution, we will introduce a general methodology for passing to the limit almost surely in the (random) approximate solutions that will generalize to more complex stochastic FSI systems, such as in Chap. 9 on stochastic nonlinearly coupled (moving boundary) models. This will involve first showing existence of a convergent subsequence of probability measures corresponding to the laws of the approximate solutions. Then we will construct a weak martingale solution using the Skorohod representation theorem, and we will conclude by using the Gyöngy-Krylov argument [177] to upgrade to a weak pathwise solution, in a special case of additive noise.

The main result of this chapter is stated in the following theorem.

Theorem 8.3.1 (Main Result) *Let* $\boldsymbol{u}_0 \in L^2(\mathcal{O}_f)$, $v_0 \in L^2(\Gamma)$, *and* $\eta_0 \in H_0^2(\Gamma)$. *Let* $P_{in/out} \in L^2_{loc}(0,\infty)$, *and let* $(\Omega, \mathcal{F}, \mathbb{P})$ *be a probability space with two independent one-dimensional Brownian motions* $\{W_1(t)\}_{t\geq 0}$ *and* $\{W_2(t)\}_{t\geq 0}$, *which are Brownian motions with respect to a given complete filtration* $\{\mathcal{F}_t\}_{t\geq 0}$. *Then, for any* $T > 0$, *there exists a weak martingale solution to the given stochastic fluid-structure interaction problem* (8.10)–(8.14). *In the case where the noise is purely additive* ($f = 1$, $g = 1$), *there exists a unique weak pathwise solution.*

8.4 A Priori Energy Estimate

We derive a formal energy estimate by assuming that the solution is pathwise regular enough to justify integration by parts. We use $\|\cdot\|_{L^2(\Gamma)}$ and $(\cdot,\cdot)$ to denote the norm and inner product on $L^2(\Gamma)$, and $\|\cdot\|_{L^2(\mathcal{O}_f)}$ and $\langle\cdot,\cdot\rangle$ to denote the norm and inner product on $L^2(\mathcal{O}_f)$.

We define the total energy at time t by

$$
\begin{aligned}
E(t) &= \frac{1}{2}\int_\Gamma |\partial_z\eta(t,\cdot)|^2 dz + \frac{1}{2}\int_\Gamma |\partial_z^2\eta(t,\cdot)|^2 + \frac{1}{2}\int_\Gamma |v(t,\cdot)|^2 dz \\
&\quad + \frac{1}{2}\int_{\mathcal{O}_f} |\boldsymbol{u}(t,\cdot)|^2 d\boldsymbol{x} \\
&= \frac{1}{2}\Big(\|\partial_z\eta(t,\cdot)\|^2_{L^2(\Gamma)} + \|\partial_z^2\eta(t,\cdot)\|^2_{L^2(\Gamma)} + \|v(t,\cdot)\|^2_{L^2(\Gamma)} + \|\boldsymbol{u}(t,\cdot)\|^2_{L^2(\mathcal{O}_f)}\Big),
\end{aligned}
\tag{8.24}
$$

and the total dissipation by time t by

$$D(t) = \int_0^t \int_{\mathcal{O}_f} |\boldsymbol{D}(\boldsymbol{u})|^2 d\boldsymbol{x} ds. \tag{8.25}$$

To estimate the total energy and dissipation for the stochastic processes $\boldsymbol{u}$, v and η, we rewrite the stochastic fluid-structure interaction problem in the following stochastic differential formulation:

$$\begin{aligned}
d\boldsymbol{u} &= (\nabla \cdot \boldsymbol{\sigma})dt + \boldsymbol{f}(\boldsymbol{u}, v, \eta)dW_1, \\
d\eta &= vdt, \\
dv &= (\partial_z^2 \eta - \partial_z^4 \eta - \boldsymbol{\sigma} \boldsymbol{e_r} \cdot \boldsymbol{e_r})dt + g(\boldsymbol{u}, v, \eta)dW_2.
\end{aligned}$$

Notice that the equation for $d\eta$ implies $d(\partial_z \eta) = (\partial_z v)dt$ and $d(\partial_z^2 \eta) = (\partial_z^2 v)dt$. To obtain an energy estimate, we first apply Itô's formula (see Theorem 3.6.4) to express the differentials of the L^2-norms of the stochastic processes that define the total energy of the problem:

$$\begin{aligned}
&d(\|\boldsymbol{u}\|^2_{L^2(\mathcal{O}_f)}) = \Big(2\langle \nabla \cdot \boldsymbol{\sigma}, \boldsymbol{u}\rangle + \|\boldsymbol{f}(\boldsymbol{u}, v, \eta)\|^2_{L^2(\mathcal{O}_f)}\Big)dt + 2\langle \boldsymbol{f}(\boldsymbol{u}, v, \eta), \boldsymbol{u}\rangle dW_1, \\
&d(\|\partial_z \eta\|^2_{L^2(\Gamma)}) = 2(\partial_z \eta, \partial_z v)dt, \qquad d(\|\partial_z^2 \eta\|^2_{L^2(\Gamma)}) = 2(\partial_z^2 \eta, \partial_z^2 v)dt, \\
&d(\|v\|^2_{L^2(\Gamma)}) = \Big(2(\partial_z^2 \eta, v) - 2(\partial_z^4 \eta, v) - 2(\boldsymbol{\sigma} \boldsymbol{e_r} \cdot \boldsymbol{e_r}, v) + \|g(\boldsymbol{u}, v, \eta)\|^2_{L^2(\Gamma)}\Big)dt \\
&\qquad + 2(g(\boldsymbol{u}, v, \eta), v)dW_2.
\end{aligned}$$

By adding these equations together, we obtain that the differential of the total energy satisfies:

$$\begin{aligned}
&d\Big(\|\partial_z \eta\|^2_{L^2(\Gamma)} + \|\partial_z^2 \eta\|^2_{L^2(\Gamma)} + \|v\|^2_{L^2(\Gamma)} + \|\boldsymbol{u}\|^2_{L^2(\mathcal{O}_f)}\Big) \\
&= \Big(2\langle \nabla \cdot \boldsymbol{\sigma}, \boldsymbol{u}\rangle - 2(\boldsymbol{\sigma} \boldsymbol{e_r} \cdot \boldsymbol{e_r}, v) + \|\boldsymbol{f}(\boldsymbol{u}, v, \eta)\|^2_{L^2(\mathcal{O}_f)} + \|g(\boldsymbol{u}, v, \eta)\|^2_{L^2(\Gamma)}\Big)dt \\
&\qquad + \Big(2\langle f(\boldsymbol{u}, v, \eta), \boldsymbol{u}\rangle\Big)dW_1 + \Big(2(g(\boldsymbol{u}, v, \eta), v)\Big)dW_2,
\end{aligned}$$

where we have used that $(\partial_z^2 \eta, v) = -(\partial_z \eta, \partial_z v)$ and $(\partial_z^4 \eta, v) = (\partial_z^2 \eta, \partial_z^2 v)$, under the assumption that η and v are smooth and the boundary conditions $\eta(0) = \eta(L) = \partial_z \eta(0) = \partial_z \eta(L) = 0$ for η. Recalling the kinematic coupling condition $\boldsymbol{u}|_\Gamma = v\boldsymbol{e}_r$, we obtain that

$$\int_{\mathcal{O}_f} (\nabla \cdot \boldsymbol{\sigma}) \cdot \boldsymbol{u} d\boldsymbol{x} = \int_{\Gamma_{in}} p u_z dr - \int_{\Gamma_{out}} p u_z dr + \int_\Gamma (\boldsymbol{\sigma} \boldsymbol{e_r} \cdot \boldsymbol{e_r}) v dz - 2\mu \int_{\mathcal{O}_f} |\boldsymbol{D}(\boldsymbol{u})|^2 d\boldsymbol{x},$$

which implies

$$
\begin{aligned}
d&\left(\frac{1}{2}\int_\Gamma |\partial_z\eta|^2dz + \frac{1}{2}\int_\Gamma |\partial_z^2\eta|^2 + \frac{1}{2}\int_\Gamma |v|^2dz + \frac{1}{2}\int_{\mathcal{O}_f} |\boldsymbol{u}|^2 d\boldsymbol{x}\right)\\
&= \left(\frac{1}{2}\int_{\mathcal{O}_f} |\boldsymbol{f}(\boldsymbol{u}, v, \eta)|^2 + \frac{1}{2}\int_{\mathcal{O}_f} |g(\boldsymbol{u}, v, \eta)|^2 - 2\mu\int_{\mathcal{O}_f} |\boldsymbol{D}(\boldsymbol{u})|^2 d\boldsymbol{x}\right)dt\\
&\quad + \left(\int_{\Gamma_{in}} pu_z dr - \int_{\Gamma_{out}} pu_z dr\right)dt + \left(\int_\Gamma \boldsymbol{f}(\boldsymbol{u}, v, \eta)\cdot \boldsymbol{u} dz\right) dW_1\\
&\quad + \left(\int_\Gamma g(\boldsymbol{u}, v, \eta) v dz\right) dW_2.
\end{aligned}
$$

Therefore, after integrating in time over the time interval $[0, t]$ for an arbitrary $t \geq 0$, we have

$$
\begin{aligned}
E(t) + 2\mu\int_0^t\int_{\mathcal{O}_f} |\boldsymbol{D}(\boldsymbol{u})|^2 d\boldsymbol{x}dt &= E_0 + \frac{1}{2}\int_0^t\int_{\mathcal{O}_f} |\boldsymbol{f}(\boldsymbol{u}, v, \eta)|^2\\
+ \frac{1}{2}\int_0^t\int_\Gamma |g(\boldsymbol{u}, v, \eta)|^2 &+ \int_0^t\int_{\Gamma_{in}} P_{in}(t) u_z dr dt - \int_0^t\int_{\Gamma_{out}} P_{out}(t) u_z dr dt\\
&+ \int_0^t\int_{\mathcal{O}_f} \boldsymbol{f}(\boldsymbol{u}, v, \eta)\cdot \boldsymbol{u} d\boldsymbol{x} dW_1 + \int_0^t\int_\Gamma g(\boldsymbol{u}, v, \eta) v dz dW_2. \quad (8.26)
\end{aligned}
$$

We estimate the terms on the right-hand side of (8.26) as follows. For the pressure term, we use Hölder's inequality, the trace inequality as stated in Theorem 2.5.1, Poincaré's inequality, and Korn's inequality in Lemma 2.8.4 to get

$$
\begin{aligned}
&\left|\int_0^t \left(\int_{\Gamma_{in}} u_z dr\right) P_{in}(s) ds\right|\\
&\quad \leq C\left|\int_0^t \left(\int_{\Gamma_{in}} |u_z|^2 dr\right)^{1/2} P_{in}(s) ds\right| \leq C\left|\int_0^t \|\nabla \boldsymbol{u}\|_{L^2(\mathcal{O}_f)} P_{in}(s) ds\right|\\
&\quad \leq C\left|\int_0^t \|\boldsymbol{D}(\boldsymbol{u})\|_{L^2(\mathcal{O}_f)} P_{in}(s) ds\right| \leq C(D(t))^{1/2}\|P_{in}(t)\|_{L^2(0,t)}\\
&\quad \leq \epsilon D(t) + C(\epsilon)\|P_{in}(t)\|^2_{L^2(0,t)}, \quad (8.27)
\end{aligned}
$$

where we recall the definition of the dissipation from (8.25). We note that the constant $C(\epsilon)$ depends only on ϵ and the parameters of the problem. The same computation holds for the outlet pressure.

By (8.6) and (8.7), we have that

$$\frac{1}{2}\int_0^t \int_{\mathcal{O}_f} |\boldsymbol{f}(\boldsymbol{u}, v, \eta)|^2 \leq C \int_0^t \left(1 + \|\boldsymbol{u}\|^2_{L^2(\mathcal{O}_f)} + \|v\|^2_{L^2(\Gamma)} + \|\eta\|^2_{H_0^2(\Gamma)}\right)$$
$$\leq C \int_0^t (1 + E(s))ds, \tag{8.28}$$

and

$$\frac{1}{2}\int_0^t \int_{\Gamma} |g(\boldsymbol{u}, v, \eta)|^2 \leq C \int_0^t \left(1 + \|\boldsymbol{u}\|^2_{L^2(\mathcal{O}_f)} + \|v\|^2_{L^2(\Gamma)} + \|\eta\|^2_{H_0^2(\Gamma)}\right)$$
$$\leq C \int_0^t (1 + E(s))ds. \tag{8.29}$$

For the stochastic integrals, we will bound the expectations of the maximum in time of the stochastic integrals

$$\mathbb{E}\left(\max_{0\leq\tau\leq t}\left|\int_0^\tau \int_{\mathcal{O}_f} \boldsymbol{f}(\boldsymbol{u}, v, \eta) \cdot \boldsymbol{q} d\boldsymbol{x} dW_1\right|\right)^p,$$
$$\mathbb{E}\left(\max_{0\leq\tau\leq t}\left|\int_0^\tau \int_{\Gamma} g(\boldsymbol{u}, v, \eta) v dz dW_2\right|\right)^p$$

for $1 \leq p < \infty$, since the final energy estimate will be given in terms of expectation of the total energy and dissipation at time t and will specifically involve the expectation of the maximum in $t \in [0, T]$ of the energy $E(t)$. To bound this quantity, we use the Burkholder-Davis-Gundy (BDG) inequality (Theorem 3.6.2) under the assumption that the process $v = \partial_t \eta$ is a predictable stochastic process with respect to the given filtration $\{\mathcal{F}_t\}_{t\geq 0}$ and assumption (8.6):

$$\mathbb{E}\left[\left(\max_{0\leq\tau\leq t}\left|\int_0^\tau \int_{\mathcal{O}_f} \boldsymbol{f}(\boldsymbol{u}, v, \eta) \cdot \boldsymbol{u} d\boldsymbol{x} dW_1\right|\right)^p\right]$$
$$\leq \mathbb{E}\left(\left|\int_0^t \left(\int_{\mathcal{O}_f} \boldsymbol{f}(\boldsymbol{u}, v, \eta) \cdot \boldsymbol{u} d\boldsymbol{x}\right)^2 ds\right|^{p/2}\right)$$
$$\leq C\mathbb{E}\left(\left|\int_0^t \|\boldsymbol{f}(\boldsymbol{u}, v, \eta)\|^2_{L^2(\mathcal{O}_f)} \cdot \|\boldsymbol{u}\|^2_{L^2(\mathcal{O}_f)} ds\right|^{p/2}\right)$$
$$\leq C\mathbb{E}\left(\max_{0\leq s\leq t} \|\boldsymbol{u}(s)\|^p_{L^2(\mathcal{O}_f)} \left(\int_0^t (1 + E(s))ds\right)^{p/2}\right)$$
$$\leq \epsilon\mathbb{E}\left(\max_{0\leq s\leq t} E(s)\right)^p + C(\epsilon)\mathbb{E}\left(\int_0^t (1 + E(s))ds\right)^p. \tag{8.30}$$

Using the same procedure for the other stochastic integral involving the structure, we obtain a similar estimate:

$$\mathbb{E}\left[\left(\max_{0\le\tau\le t}\left|\int_0^\tau\int_\Gamma g(\boldsymbol{u},v,\eta)v\,dz\,dW_2\right|\right)\right]^p \le \epsilon\mathbb{E}\left(\max_{0\le s\le t}E(s)\right)^p + C(\epsilon)\mathbb{E}\left(\int_0^t(1+E(s))ds\right)^p. \tag{8.31}$$

Now, we use (8.27), (8.28), and (8.29) in (8.26) to obtain

$$E(t)+2\mu D(t)\le E(0)+2\epsilon D(t)+C(\epsilon)\left(\|P_{in}\|^2_{L^2(0,t)}+\|P_{out}\|^2_{L^2(0,t)}\right)$$
$$+\mathbb{E}C\int_0^t(1+E(s))ds+\int_0^t\int_{\mathcal{O}_f}\boldsymbol{f}(\boldsymbol{u},v,\eta)\cdot\boldsymbol{u}\,d\boldsymbol{x}\,dW_1+\int_0^t\int_\Gamma g(\boldsymbol{u},v,\eta)v\,dz\,dW_2.$$

Taking a maximum over times $s\in[0,t]$, raising both sides to the p power for arbitrary $1\le p<\infty$, taking an expectation, and then using the estimate in (8.30) and (8.31), we obtain for all $0\le t\le T$ the following *a priori* energy estimate for the coupled problem (8.10)–(8.14):

$$\mathbb{E}\left(\max_{0\le s\le t}E(s)\right)^p+\mu\mathbb{E}\left(\int_0^t\int_{\mathcal{O}_f}|\boldsymbol{D}(\boldsymbol{u})|^2d\boldsymbol{x}ds\right)^p$$
$$\le C_{p,T}\left[E(0)^p+\|P_{in}\|^{2p}_{L^2(0,t)}+\|P_{out}\|^{2p}_{L^2(0,t)}+\int_0^t\left(1+\mathbb{E}\left(\max_{0\le s\le t}E(s)\right)^p\right)ds\right], \tag{8.32}$$

where $C_{p,T}$ depends only on the parameters of the problem and on $1\le p<\infty$ and the final time T. Then, by Gronwall's inequality (see Theorem 2.12.1), we conclude that there exists a uniform constant $C_{p,T}$, increasing in T, such that

$$\mathbb{E}\left(\max_{0\le t\le T}E(t)\right)^p+\mu\mathbb{E}\left(\int_0^t\int_{\mathcal{O}_f}|\boldsymbol{D}(\boldsymbol{u})|^2d\boldsymbol{x}ds\right)^p\le C_{p,T}.$$

Remark 8.4.1 The right-hand side of the energy estimate (8.32) shows the four sources of energy input into the system: $E(0)$ represents the initial kinetic and potential energy, the next two terms represent the energy input from the inlet and outlet pressure, and the final term represents the energy input from the stochastic forcing on the fluid and the structure.

8.5 The Splitting Scheme

To prove the existence of a weak solution to the given stochastic FSI problem, we adapt a Lie operator splitting scheme that was first designed in the context of (deterministic) nonlinear fluid-structure interaction by Muha and Čanić in [265]; see also [175]. Recall that for *deterministic* prototypical models of FSI with both linear and nonlinear coupling, we have introduced the framework of using Lie operator splitting to construct approximate solutions, in Chaps. 4 and 5. Our goal is to extend this methodology to the study of stochastic FSI by describing a splitting scheme for this stochastic linearly coupled benchmark problem, which shows that the splitting scheme methodology is robust to stochastic noise in the coupled system. This splitting scheme is a two-part splitting scheme that involves a structure subproblem and a fluid subproblem **which gives rise to a stable and convergent scheme**, where the fluid subproblem takes into account the additional influence of stochasticity by using Brownian increments to discretize the stochastic noise in time, acting on both the structure velocity and the fluid velocity.

Given a fixed time $T > 0$, for each positive integer N, let $\Delta t = \frac{T}{N}$ denote the associated time step, and let $t_N^n = n\Delta t$ denote the discrete times for $n = 0, 1, \dots, N-1, N$. At each time step, we update the following vector using a two-step method described below:

$$\boldsymbol{X}_N^{n+\frac{i}{2}} = \begin{pmatrix} \boldsymbol{u}_N^{n+\frac{i}{2}} \\ v_N^{n+\frac{i}{2}} \\ \eta_N^{n+\frac{i}{2}} \end{pmatrix}, \quad n = 0, 1, \dots, N-1, \quad i = 1, 2,$$

where $i = 1$ corresponds to the result after updating the structure subproblem, and $i = 2$ corresponds to the result after updating fluid subproblem. We initialize the scheme with the initial data $\boldsymbol{X}_N^0 = \begin{pmatrix} \boldsymbol{u}_0 \\ v_0 \\ \eta_0 \end{pmatrix}$ for each N.

8.5.1 *The Structure Subproblem*

In this subproblem, we keep the fluid velocity fixed, so that

$$\boldsymbol{u}_N^{n+\frac{1}{2}} = \boldsymbol{u}_N^n,$$

and update the structure displacement and the structure velocity by requiring that $(\eta_N^{n+\frac{1}{2}}, v_N^{n+\frac{1}{2}})$ satisfy the following first-order system in weak variational form:

$$\int_\Gamma \frac{\eta_N^{n+\frac{1}{2}} - \eta_N^n}{\Delta t}\phi dz = \int_\Gamma v_N^{n+\frac{1}{2}}\phi dz, \qquad \text{for all } \phi \in L^2(\Gamma),$$

$$\int_\Gamma \frac{v_N^{n+\frac{1}{2}} - v_N^n}{\Delta t}\psi dz + \int_\Gamma \partial_z \eta_N^{n+\frac{1}{2}} \partial_z \psi dz + \int_\Gamma \partial_z^2 \eta_N^{n+\frac{1}{2}} \partial_z^2 \psi = 0, \text{ for all } \psi \in H_0^2(\Gamma), \tag{8.33}$$

where this system is solved pathwise for each $\omega \in \Omega$ separately. We claim that $(\eta_N^{n+\frac{1}{2}}, v_N^{n+\frac{1}{2}})$ is a random variable taking values in $H_0^2(\Gamma) \times H_0^2(\Gamma)$. To verify this, we must check that it is a measurable function of the probability space.

Proposition 8.5.1 *Suppose that η_N^n and v_N^n are $\mathcal{F}_{t_N^n}$-measurable random variables taking values in $H_0^2(\Gamma)$ and $L^2(\Gamma)$, respectively. Then, the structure problem* (8.33) *has a unique solution $(\eta_N^{n+\frac{1}{2}}, v_N^{n+\frac{1}{2}})$, which is a random variable taking values in $H_0^2(\Gamma) \times H_0^2(\Gamma)$ that is measurable with respect to $\mathcal{F}_{t_N^n}$.*

Proof Let $F_N^n : (\eta_N^n, v_N^n) \to (\eta_N^{n+\frac{1}{2}}, v_N^{n+\frac{1}{2}})$ be the deterministic linear map that sends deterministic data $(\eta_N^n, v_N^n) \in H_0^2(\Gamma) \times L^2(\Gamma)$ to the unique solution $(\eta_N^{n+\frac{1}{2}}, v_N^{n+\frac{1}{2}}) \in H_0^2(\Gamma) \times H_0^2(\Gamma)$ satisfying the weak formulation (8.33) as a deterministic problem. We claim that this deterministic linear map F_N^n : $(\eta_N^n, v_N^n) \to (\eta_N^{n+\frac{1}{2}}, v_N^{n+\frac{1}{2}})$ is a continuous (or equivalently, bounded) linear map from $H_0^2(\Gamma) \times L^2(\Gamma)$ to $H_0^2(\Gamma) \times H_0^2(\Gamma)$.

By plugging the first equation in (8.33) into the second equation, $\eta_N^{n+\frac{1}{2}}$ satisfies the weak formulation:

$$\begin{aligned}\int_\Gamma \eta_N^{n+\frac{1}{2}}\psi dz + (\Delta t)^2 \int_\Gamma \partial_z \eta_N^{n+\frac{1}{2}} \partial_z \psi dz + (\Delta t)^2 \int_\Gamma \partial_z^2 \eta_N^{n+\frac{1}{2}} \partial_z^2 \psi \\ = (\Delta t) \int_\Gamma v_N^n \psi dz + \int_\Gamma \eta_N^n \psi dz, \text{ for all } \psi \in H_0^2(\Gamma).\end{aligned} \tag{8.34}$$

The existence of a unique $\eta_N^{n+\frac{1}{2}} \in H_0^2(\Gamma)$ satisfying the weak formulation (8.34) is given by the Lax-Milgram lemma (Lemma 2.2.1), and then we can set $v_N^{n+\frac{1}{2}} = \dfrac{\eta_N^{n+\frac{1}{2}} - \eta_N^n}{\Delta t} \in H_0^2(\Gamma)$. The linear map $F^n : (\eta_N^n, v_N^n) \to (\eta_N^{n+\frac{1}{2}}, v_N^{n+\frac{1}{2}})$ is a bounded linear map from $H_0^2(\Gamma) \times L^2(\Gamma)$ to $H_0^2(\Gamma) \times H_0^2(\Gamma)$, which can be seen by substituting $\psi = \eta_N^{n+\frac{1}{2}}$ in (8.34). Thus, $F_N^n : (\eta_N^n, v_N^n) \to (\eta_N^{n+\frac{1}{2}}, v_N^{n+\frac{1}{2}})$ is a bounded linear map from $H_0^2(\Gamma) \times L^2(\Gamma)$ to $H_0^2(\Gamma) \times H_0^2(\Gamma)$, and so the result of the structure subproblem, which consists of the random functions $(\eta_N^{n+\frac{1}{2}}, v_N^{n+\frac{1}{2}}) = F_N^n \circ (\eta_N^n, v_N^n)$, is a pair of $\mathcal{F}_{t_N^n}$-measurable random variables, taking values in $H_0^2(\Gamma) \times H_0^2(\Gamma)$. □

To show that the approximate solutions defined by the subproblems converge to the weak solution of the continuous problem as $\Delta t \to 0$, we will need uniform bounds on the approximating sequences, which will follow from the uniform bounds on the discrete energy of the problem. For this purpose, we define the discrete energy at time t_n by

$$E_N^{n+\frac{i}{2}} = \frac{1}{2}\left(\int_{\mathcal{O}_f} |\boldsymbol{u}_N^{n+\frac{i}{2}}|^2 d\boldsymbol{x} + \|v_N^{n+\frac{i}{2}}\|^2_{L^2(\Gamma)} + \|\partial_z \eta_N^{n+\frac{i}{2}}\|^2_{L^2(\Gamma)} + \|\partial_z^2 \eta_N^{n+\frac{i}{2}}\|^2_{L^2(\Gamma)}\right), \tag{8.35}$$

and we define the fluid dissipation at time $t_N^n = n\Delta t$ by

$$D_N^n = (\Delta t)\mu \int_{\mathcal{O}_f} |\boldsymbol{D}(\boldsymbol{u}_N^n)|^2 d\boldsymbol{x}. \tag{8.36}$$

We emphasize that the discrete energy and the discrete fluid dissipation are random variables.

Proposition 8.5.2 *The following discrete energy equality is satisfied pathwise:*

$$E_N^{n+\frac{1}{2}} + \frac{1}{2}\left(\|v_N^{n+\frac{1}{2}} - v_N^n\|^2_{L^2(\Gamma)}\right) + \frac{1}{2}\left(\|\partial_z(\eta_N^{n+\frac{1}{2}} - \eta_N^n)\|^2_{L^2(\Gamma)}\right)$$
$$+ \frac{1}{2}\left(\|\partial_z^2(\eta_N^{n+\frac{1}{2}} - \eta_N^n)\|^2_{L^2(\Gamma)}\right) = E_N^n.$$

Proof Because $v_N^{n+\frac{1}{2}} \in H_0^2(\Gamma)$, we can substitute $\psi = v_N^{n+\frac{1}{2}}$ in the weak formulation to obtain that pathwise,

$$\int_\Gamma (v_N^{n+\frac{1}{2}} - v_N^n) \cdot v_N^{n+\frac{1}{2}} dz + (\Delta t) \int_\Gamma \partial_z \eta_N^{n+\frac{1}{2}} \partial_z v_N^{n+\frac{1}{2}} dz + (\Delta t)$$
$$\times \int_\Gamma \partial_z^2 \eta_N^{n+\frac{1}{2}} \partial_z^2 v_N^{n+\frac{1}{2}} dz = 0.$$

By using the identity $(a-b)\cdot a = \frac{1}{2}(|a|^2 + |a-b|^2 - |b|^2)$, along with the fact that $v_N^{n+\frac{1}{2}} = \dfrac{\eta_N^{n+\frac{1}{2}} - \eta_N^n}{\Delta t}$, we obtain that the following identity holds pathwise:

$$
\begin{aligned}
&\frac{1}{2}\|v_N^{n+\frac{1}{2}}\|^2_{L^2(\Gamma)} + \frac{1}{2}\|\partial_z \eta_N^{n+\frac{1}{2}}\|^2_{L^2(\Gamma)} + \frac{1}{2}\|\partial_z^2 \eta_N^{n+\frac{1}{2}}\|^2_{L^2(\Gamma)} \\
&+\frac{1}{2}\|v_N^{n+\frac{1}{2}} - v_N^n\|^2_{L^2(\Gamma)} + \frac{1}{2}\|\partial_z(\eta_N^{n+\frac{1}{2}} - \eta_N^n)\|^2_{L^2(\Gamma)} + \frac{1}{2}\|\partial_z^2(\eta_N^{n+\frac{1}{2}} - \eta_N^n)\|_{L^2(\Gamma)} \\
&= \frac{1}{2}\|v_N^n\|^2_{L^2(\Gamma)} + \frac{1}{2}\|\partial_z \eta_N^n\|^2_{L^2(\Gamma)} + \frac{1}{2}\|\partial_z^2 \eta_N^n\|^2_{L^2(\Gamma)}.
\end{aligned}
$$

The result follows once we note that $\boldsymbol{u}_N^{n+\frac{1}{2}} = \boldsymbol{u}_N^n$. □

8.5.2 The Fluid Subproblem

In this subproblem, we keep the structure displacement fixed

$$\eta_N^{n+1} = \eta_N^{n+\frac{1}{2}},$$

and update the fluid and structure velocities. To define the problem satisfied by the fluid and structure velocities, we introduce the following notation for the corresponding fixed time function spaces:

$$
\begin{aligned}
\mathcal{V} &= \{(\boldsymbol{u}, v) \in \mathcal{V}_f \times L^2(\Gamma) : \boldsymbol{u}|_\Gamma = v\boldsymbol{e_r}\}, \\
\mathcal{Q} &= \{(\boldsymbol{q}, \psi) \in \mathcal{V}_f \times H_0^2(\Gamma) : \boldsymbol{q}|_\Gamma = \psi \boldsymbol{e}_r\},
\end{aligned}
\tag{8.37}
$$

where $\mathcal{V}_F$ is defined by (8.15). Note that the definition of $\mathcal{V}$ and $\mathcal{Q}$ does not depend on N or n, because we are working with a linearly coupled FSI model, so that the fluid domain is fixed.

Then, the fluid subproblem is to find $(\boldsymbol{u}_N^{n+1}, v_N^{n+1})$ taking values in $\mathcal{V}$ pathwise, such that

$$
\begin{aligned}
&\int_{\mathcal{O}_f} \frac{\boldsymbol{u}_N^{n+1} - \boldsymbol{u}_N^{n+\frac{1}{2}}}{\Delta t} \cdot \boldsymbol{q} d\boldsymbol{x} + 2\mu \int_{\mathcal{O}_f} \boldsymbol{D}(\boldsymbol{u}_N^{n+1}) : \boldsymbol{D}(\boldsymbol{q}) d\boldsymbol{x} + \int_\Gamma \frac{v_N^{n+1} - v_N^{n+\frac{1}{2}}}{\Delta t} \psi dz \\
&\quad = P_{N,in}^n \int_0^R (q_z)|_{z=0} dr - P_{N,out}^n \int_0^R (q_z)|_{z=L} dr \\
&\quad + \frac{1}{\Delta t} \int_{\mathcal{O}_f} \boldsymbol{f}(\boldsymbol{u}_N^{n+\frac{1}{2}}, v_N^{n+\frac{1}{2}}, \eta_N^{n+\frac{1}{2}}) \Delta W_{1,N}^n \cdot \boldsymbol{q} d\boldsymbol{x} \\
&+ \frac{1}{\Delta t} \int_\Gamma g(\boldsymbol{u}_N^{n+\frac{1}{2}}, v_N^{n+\frac{1}{2}}, \eta_N^{n+\frac{1}{2}})(\Delta W_{2,N}^n) \psi dz, \quad \forall (\boldsymbol{q}, \psi) \in \mathcal{Q},
\end{aligned}
\tag{8.38}
$$

pathwise for each outcome $\omega \in \Omega$, where $P^n_{N,in/out} = \frac{1}{\Delta t}\int_{n\Delta t}^{(n+1)\Delta t} P_{in/out}(t)dt$ and $\Delta W^n_{i,N} = W_i((n+1)\Delta t) - W_i(n\Delta t)$ denotes the Brownian increment for each of the Brownian motions $\{W_i(t)\}_{t\geq 0}$ for $i = 1, 2$.

Proposition 8.5.3 *Suppose that $\boldsymbol{u}_N^{n+\frac{1}{2}}$, $\eta_N^{n+\frac{1}{2}}$, and $v_N^{n+\frac{1}{2}}$ are $\mathcal{F}_{t_N^{n+1}}$-measurable random variables taking values in $L^2(\mathcal{O}_f)$, $H_0^2(\Gamma)$, and $L^2(\Gamma)$, respectively. Then, the fluid subproblem* (8.38) *has a unique solution $(\boldsymbol{u}_N^{n+1}, v_N^{n+1})$ that is an $\mathcal{F}_{t_N^{n+1}}$-measurable random variable taking values in $\mathcal{V}$.*

Proof We establish this result again using the Lax Milgram lemma, stated in Lemma 2.2.1. We let $T_N^n : L^2(\mathcal{O}_f) \times L^2(\Gamma) \times \mathbb{R} \times \mathbb{R} \to \mathcal{V}$ denote the deterministic map that sends given data $\left(\boldsymbol{u}_N^{n+\frac{1}{2}}, v_N^{n+\frac{1}{2}}, P^n_{N,in}, P^n_{N,out}, \Delta W^n_{1,N}, \Delta W^n_{2,N}\right) \in L^2(\mathcal{O}_f) \times L^2(\Gamma) \times \mathbb{R} \times \mathbb{R} \times \mathbb{R} \times \mathbb{R}$ to the unique solution $(\boldsymbol{u}_N^{n+1}, v_N^{n+1}) \in \mathcal{V}$ satisfying the weak formulation (8.38). We want to show that the deterministic map $T_N^n : \left(\boldsymbol{u}_N^{n+\frac{1}{2}}, v_N^{n+\frac{1}{2}}, P^n_{N,in}, P^n_{N,out}, \Delta W^n_{1,N}, \Delta W^n_{2,N}\right) \to (\boldsymbol{u}_N^{n+1}, v_N^{n+1})$ is a continuous map, because then when we have that the inputs are $\mathcal{F}_{t_N^{n+1}}$-measurable random variables, the resulting solution $(\boldsymbol{u}_N^{n+1}, v_N^{n+1})$ is also $\mathcal{F}_{t_N^{n+1}}$-measurable by continuity.

We start by showing that the bilinear form $B : \mathcal{V} \times \mathcal{V} \to \mathbb{R}$ given by

$$B((\boldsymbol{u}, v), (\boldsymbol{q}, \psi)) = \int_{\mathcal{O}_f} \boldsymbol{u} \cdot \boldsymbol{q} d\boldsymbol{x} + 2\mu(\Delta t) \int_{\mathcal{O}_f} \boldsymbol{D}(\boldsymbol{u}) : \boldsymbol{D}(\boldsymbol{q}) d\boldsymbol{x} + \int_\Gamma v\psi dz,$$

is coercive and continuous. Coercivity follows from the Korn equality; see, e.g., Lemma 2.8.4, applied to

$$B((\boldsymbol{u}, v), (\boldsymbol{u}, v)) = \int_{\mathcal{O}_f} |\boldsymbol{u}|^2 d\boldsymbol{x} + 2\mu(\Delta t) \int_{\mathcal{O}_f} |\boldsymbol{D}(\boldsymbol{u})|^2 d\boldsymbol{x} + \int_\Gamma v^2 dz,$$

to obtain $\|\nabla \boldsymbol{u}\|^2_{L^2(\mathcal{O}_f)} = 2\|\boldsymbol{D}(\boldsymbol{u})\|^2_{L^2(\mathcal{O}_f)}$. Continuity of the bilinear form B follows from an application of the Cauchy-Schwarz inequality.

Next, one can verify that for fixed but arbitrary $\left(\boldsymbol{u}_N^{n+\frac{1}{2}}, v_N^{n+\frac{1}{2}}, \eta_N^{n+\frac{1}{2}}, P^n_{N,in}, P^n_{N,out}, \Delta W^n_{1,N}, \Delta W^n_{2,N}\right) \in L^2(\mathcal{O}_f) \times L^2(\Gamma) \times H_0^2(\Gamma) \times \mathbb{R} \times \mathbb{R} \times \mathbb{R} \times \mathbb{R}$, the map sending $(\boldsymbol{q}, \psi)$ to

$$\int_{\mathcal{O}_f} \boldsymbol{u}_N^{n+\frac{1}{2}} \cdot \boldsymbol{q} d\boldsymbol{x} + \int_\Gamma v_N^{n+\frac{1}{2}} \psi dz + (\Delta t)\left(P^n_{N,in} \int_0^R (q_z)|_{z=0} dr\right.$$

$$-P^n_{N,out}\int_0^R (q_z)|_{z=L}dr\Big)$$

$$+\int_{\mathcal{O}_f} \boldsymbol{f}(\boldsymbol{u}_N^{n+\frac{1}{2}}, v_N^{n+\frac{1}{2}}, \eta_N^{n+\frac{1}{2}})\Delta W^n_{1,N}\cdot\boldsymbol{q}d\boldsymbol{x}+\int_\Gamma g(\boldsymbol{u}_N^{n+\frac{1}{2}}, v_N^{n+\frac{1}{2}}, \eta_N^{n+\frac{1}{2}})(\Delta W^n_{2,N})\psi dz$$

is a continuous linear functional on $\mathcal{V}$, where we recall the conditions (8.6) and (8.7) on the nonlinear noise coefficients $\boldsymbol{f}$ and g. Thus, the existence of a unique $(\boldsymbol{u}_N^{n+1}, v_N^{n+1}) \in \mathcal{V}$ satisfying (8.38) with the larger space of test functions $(\boldsymbol{q}, \psi) \in \mathcal{V}$ is guaranteed by the Lax-Milgram lemma, Lemma 2.2.1. Note that $\mathcal{V}$ is a larger space than the space $\mathcal{Q}$ required for the test functions in the fluid subproblem (8.38). However, we still have the desired uniqueness of the solution in $\mathcal{V}$ if we restrict the test functions to $\mathcal{Q}$ as in (8.38) because $\mathcal{Q}$ is dense in $\mathcal{V}$.

It only remains to verify that the solution map $T^n : (\boldsymbol{u}_N^{n+\frac{1}{2}}, v_N^{n+\frac{1}{2}}, \eta_N^{n+\frac{1}{2}}, P^n_{N,in}, P^n_{N,out}, \Delta W^n_{1,N}, \Delta W^n_{2,N}) \rightarrow (\boldsymbol{u}_N^{n+1}, v_N^{n+1})$ is continuous. Thus, consider $T^n(\boldsymbol{u}_{N,i}^{n+\frac{1}{2}}, v_{N,i}^{n+\frac{1}{2}}, \eta_{N,i}^{n+\frac{1}{2}}, P^n_{N,in,i}, P^n_{N,out,i}, \Delta W^n_{1,N,i}, \Delta W^n_{2,N,i}) = (\boldsymbol{u}_{N,i}^{n+1}, v_{N,i}^{n+1})$ for $i = 1, 2$. We subtract the weak formulations (8.38) and obtain:

$$\int_{\mathcal{O}_f} (\boldsymbol{u}_{N,1}^{n+1} - \boldsymbol{u}_{N,2}^{n+1}) \cdot \boldsymbol{q}d\boldsymbol{x} + 2\mu(\Delta t)\int_{\mathcal{O}_f} \boldsymbol{D}(\boldsymbol{u}_{N,1}^{n+1} - \boldsymbol{u}_{N,2}^{n+1}) : \boldsymbol{D}(\boldsymbol{q})$$

$$+\int_\Gamma (v_{N,1}^{n+1} - v_{N,2}^{n+1})\psi dz$$

$$= \int_{\mathcal{O}_f} (\boldsymbol{u}_{N,1}^{n+\frac{1}{2}} - \boldsymbol{u}_{N,2}^{n+\frac{1}{2}}) \cdot \boldsymbol{q}d\boldsymbol{x} + \int_\Gamma (v_{N,1}^{n+\frac{1}{2}} - v_{N,2}^{n+\frac{1}{2}})\psi dz$$

$$+ (P^n_{N,in,1} - P^n_{N,in,2})\int_0^R (q_z)|_{z=0}dr - (P^n_{N,out,1} - P^n_{N,out,2})\int_0^R (q_z)|_{z=L}dr$$

$$+\int_{\mathcal{O}_f}\left[\boldsymbol{f}\left(\boldsymbol{u}_{N,1}^{n+\frac{1}{2}}, \eta_{N,1}^{n+\frac{1}{2}}, v_{N,1}^{n+\frac{1}{2}}\right) - \boldsymbol{f}\left(\boldsymbol{u}_{N,2}^{n+\frac{1}{2}}, \eta_{N,2}^{n+\frac{1}{2}}, v_{N,2}^{n+\frac{1}{2}}\right)\right]\Delta W^n_{1,N,1}\cdot\boldsymbol{q}d\boldsymbol{x}$$

$$+\int_\Gamma\left[g\left(\boldsymbol{u}_{N,1}^{n+\frac{1}{2}}, \eta_{N,1}^{n+\frac{1}{2}}, v_{N,1}^{n+\frac{1}{2}}\right) - g\left(\boldsymbol{u}_{N,2}^{n+\frac{1}{2}}, \eta_{N,2}^{n+\frac{1}{2}}, v_{N,2}^{n+\frac{1}{2}}\right)\right](\Delta W^n_{2,N,1})\psi dz$$

$$+\int_{\mathcal{O}_f} \boldsymbol{f}(\boldsymbol{u}_{N,2}^{n+\frac{1}{2}}, \eta_{N,2}^{n+\frac{1}{2}}, v_{N,2}^{n+\frac{1}{2}})(\Delta W^n_{1,N,1} - \Delta W^n_{1,N,2})\cdot\boldsymbol{q}d\boldsymbol{x}$$

$$+\int_\Gamma g(\boldsymbol{u}_{N,2}^{n+\frac{1}{2}}, \eta_{N,2}^{n+\frac{1}{2}}, v_{N,1}^{n+\frac{1}{2}})(\Delta W^n_{2,N,1} - \Delta W^n_{2,N,2})\psi dz,$$

for all $(\boldsymbol{q}, \psi) \in \mathcal{Q}$. Then, substituting $(\boldsymbol{q}, \psi) = (\boldsymbol{u}_{N,1}^{n+1} - \boldsymbol{u}_{N,2}^{n+1}, v_{N,1}^{n+1} - v_{N,2}^{n+1})$, using (8.8) and (8.9), and using the trace inequality and Poincare's inequality, we obtain:

$$\|\boldsymbol{u}_{N,1}^{n+1} - \boldsymbol{u}_{N,2}^{n+1}\|^2_{L^2(\mathcal{O}_f)} + \mu \int_{\mathcal{O}_f} |\boldsymbol{D}(\boldsymbol{u}_{N,1}^{n+1} - \boldsymbol{u}_{N,2}^{n+1})|^2 + \|v_{N,1}^{n+1} - v_{N,2}^{n+1}\|^2_{L^2(\Gamma)}$$
$$\leq \|\boldsymbol{u}_{N,1}^{n+1} - \boldsymbol{u}_{N,2}^{n+1}\|_{L^2(\mathcal{O}_f)} \|\boldsymbol{u}_{N,1}^{n+\frac{1}{2}} - \boldsymbol{u}_{N,2}^{n+\frac{1}{2}}\|_{L^2(\mathcal{O}_f)}$$
$$+ \|v_{N,1}^{n+1} - v_{N,2}^{n+1}\|_{L^2(\Gamma)} \|v_{N,1}^{n+\frac{1}{2}} - v_{N,2}^{n+\frac{1}{2}}\|_{L^2(\Gamma)} + C(|P_{N,in,1}^n - P_{N,in,2}^n|^2$$
$$+ |P_{N,out,1}^n - P_{N,out,2}^n|^2)$$
$$+ C\left(\|\boldsymbol{u}_{N,1}^{n+\frac{1}{2}} - \boldsymbol{u}_{N,2}^{n+\frac{1}{2}}\|_{L^2(\mathcal{O}_f)} + \|v_{N,1}^{n+\frac{1}{2}} - v_{N,2}^{n+\frac{1}{2}}\|_{L^2(\Gamma)} + \|\eta_{N,1}^{n+\frac{1}{2}} - \eta_{N,2}^{n+\frac{1}{2}}\|_{L^2(\Gamma)}\right)$$
$$\cdot \left(\Delta W_{1,N,1}^n \|\boldsymbol{u}_{N,1}^{n+1} - \boldsymbol{u}_{N,2}^{n+1}\|_{L^2(\mathcal{O}_f)} + \Delta W_{2,N,1}^n \|v_{N,1}^{n+1} - v_{N,2}^{n+1}\|_{L^2(\Gamma)}\right)$$
$$+ C\left(1 + \|\boldsymbol{u}_{N,2}^{n+\frac{1}{2}}\|_{L^2(\mathcal{O}_f)} + \|v_{N,2}^{n+\frac{1}{2}}\|_{L^2(\Gamma)} + \|\eta_{N,2}^{n+\frac{1}{2}}\|_{H_0^2(\Gamma)}\right)$$
$$\cdot \left[(\Delta W_{1,N,1}^n - \Delta W_{1,N,2}^n) + (\Delta W_{2,N,1}^n - \Delta W_{2,N,2}^n)\right] \left(\|\boldsymbol{u}_{N,1}^{n+1} - \boldsymbol{u}_{N,2}^{n+1}\|_{L^2(\mathcal{O}_f)}\right.$$
$$\left. + \|v_{N,1}^{n+1} - v_{N,2}^{n+1}\|_{L^2(\Gamma)}\right).$$

Then, using Cauchy's inequality $ab \leq \frac{1}{2}(a^2 + b^2)$, we obtain the estimate:

$$\|\boldsymbol{u}_{N,1}^{n+1} - \boldsymbol{u}_{N,2}^{n+1}\|^2_{L^2(\mathcal{O}_f)} + \|v_{N,1}^{n+1} - v_{N,2}^{n+1}\|^2_{L^2(\Gamma)}$$
$$\leq C(1+(\Delta W_{1,N,1}^n)^2+(\Delta W_{2,N,1}^n)^2) \left(\|\boldsymbol{u}_{N,1}^{n+\frac{1}{2}} - \boldsymbol{u}_{N,2}^{n+\frac{1}{2}}\|^2_{L^2(\mathcal{O}_f)} + \|v_{N,1}^{n+\frac{1}{2}} - v_{N,2}^{n+\frac{1}{2}}\|^2_{L^2(\Gamma)}\right.$$
$$\left. + \|\eta_{N,1}^{n+\frac{1}{2}} - \eta_{N,2}^{n+\frac{1}{2}}\|^2_{L^2(\Gamma)} + |P_{N,in,1}^n - P_{N,in,2}^n|^2 + |P_{N,out,1}^n - P_{N,out,2}^n|^2\right)$$
$$+ C\left(1 + \|\boldsymbol{u}_{N,2}^{n+1}\|_{L^2(\mathcal{O}_f)} + \|v_{N,2}^{n+\frac{1}{2}}\|_{L^2(\Gamma)} + \|\eta_{N,2}^{n+\frac{1}{2}}\|_{H_0^2(\Gamma)}\right)^2 \sum_{i=1}^{2} |\Delta W_{i,N,1}^n - \Delta W_{i,N,2}^n|^2,$$

which shows that $T^n(\boldsymbol{u}_N^{n+\frac{1}{2}}, v_N^{n+\frac{1}{2}}, \eta_N^{n+\frac{1}{2}}, P_{N,in}^n, P_{N,out}^n, \Delta W_{1,N}^n, \Delta W_{2,N}^n) \to (\boldsymbol{u}_N^{n+1}, v_N^{n+1})$ is a continuous map.

Thus, since $\boldsymbol{u}_N^{n+\frac{1}{2}}$, $v_N^{n+\frac{1}{2}}$, $\eta_N^{n+\frac{1}{2}}$, $\Delta W_{1,N}^n$, and $\Delta W_{2,N}^n$ are $\mathcal{F}_{t_N^{n+1}}$-measurable and $P_{N,in}^n$ and $P_{N,out}^n$ are deterministic, the random functions $(\boldsymbol{u}_N^{n+1}, v_N^{n+1}) = T_N^n \circ (\boldsymbol{u}_N^{n+\frac{1}{2}}, v_N^{n+\frac{1}{2}}, \eta_N^{n+\frac{1}{2}}, P_{N,in}^n, P_{N,out}^n, \Delta W_{1,N}^n, \Delta W_{2,N}^n)$, which solve the fluid subproblem, are $\mathcal{F}_{t_N^{n+1}}$ measurable random variables also. □

Proposition 8.5.4 *The following discrete energy identity holds pathwise:*

$$E_N^{n+1} + 2\mu(\Delta t)\int_{\mathcal{O}_f} |\boldsymbol{D}(\boldsymbol{u}_N^{n+1})|^2 d\boldsymbol{x}$$
$$+\frac{1}{2}\left(\|\boldsymbol{u}_N^{n+1} - \boldsymbol{u}_N^{n+\frac{1}{2}}\|^2_{L^2(\mathcal{O}_f)}\right) + \frac{1}{2}\left(\|v_N^{n+1} - v_N^{n+\frac{1}{2}}\|^2_{L^2(\Gamma)}\right)$$
$$= E_N^{n+\frac{1}{2}} + (\Delta t)\left(P^n_{N,in}\int_0^R (\boldsymbol{u}_N^{n+1})_z|_{z=0}dr - P^n_{N,out}\int_0^R (\boldsymbol{u}_N^{n+1})_z|_{z=L}dr\right)$$
$$+\int_{\mathcal{O}_f} \boldsymbol{f}(\boldsymbol{u}_N^{n+\frac{1}{2}}, v_N^{n+\frac{1}{2}}, \eta_N^{n+\frac{1}{2}})\Delta W^n_{1,N}\cdot \boldsymbol{u}_N^{n+1}$$
$$+\int_\Gamma g(\boldsymbol{u}_N^{n+\frac{1}{2}}, v_N^{n+\frac{1}{2}}, \eta_N^{n+\frac{1}{2}})(\Delta W^n_{2,N})v_N^{n+1}dz.$$

Hence, for a constant C that is independent of n and N:

$$E_N^{n+1} + \mu(\Delta t)\int_{\mathcal{O}_f} |\boldsymbol{D}(\boldsymbol{u}_N^{n+1})|^2 d\boldsymbol{x}$$
$$+\frac{1}{4}\left(\|\boldsymbol{u}_N^{n+1} - \boldsymbol{u}_N^{n+\frac{1}{2}}\|^2_{L^2(\mathcal{O}_f)}\right) + \frac{1}{4}\left(\|v_N^{n+1} - v_N^{n+\frac{1}{2}}\|^2_{L^2(\Gamma)}\right)$$
$$\leq E_N^{n+\frac{1}{2}} + C(\Delta t)\Big[(P^n_{N,in})^2 + (P^n_{N,out})^2\Big]$$
$$+\int_{\mathcal{O}_f} |\boldsymbol{f}(\boldsymbol{u}_N^{n+\frac{1}{2}}, v_N^{n+\frac{1}{2}}, \eta_N^{n+\frac{1}{2}})|^2(\Delta W^n_{1,N})^2 + \int_\Gamma |g(\boldsymbol{u}_N^{n+\frac{1}{2}}, v_N^{n+\frac{1}{2}}, \eta_N^{n+\frac{1}{2}})|^2(\Delta W^n_{2,N})^2$$
$$+\left|\int_{\mathcal{O}_f} \boldsymbol{f}(\boldsymbol{u}_N^{n+\frac{1}{2}}, v_N^{n+\frac{1}{2}}, \eta_N^{n+\frac{1}{2}})\Delta W^n_{1,N}\cdot \boldsymbol{u}_N^{n+\frac{1}{2}}\right|$$
$$+\left|\int_\Gamma g(\boldsymbol{u}_N^{n+\frac{1}{2}}, v_N^{n+\frac{1}{2}}, \eta_N^{n+\frac{1}{2}})(\Delta W^n_{2,N})v_N^{n+\frac{1}{2}}\right|.$$

Proof We can substitute $(\boldsymbol{q}, \psi) = (\boldsymbol{u}_N^{n+1}, v_N^{n+1})$ into the weak formulation of the fluid subproblem, since we showed in Proposition 8.5.3 that (8.38) holds more generally for test functions in $\mathcal{V}$. We obtain

$$\int_{\mathcal{O}_f} \frac{\boldsymbol{u}_N^{n+1} - \boldsymbol{u}_N^{n+\frac{1}{2}}}{\Delta t}\cdot \boldsymbol{u}_N^{n+1}d\boldsymbol{x} + 2\mu\int_{\mathcal{O}_f} |\boldsymbol{D}(\boldsymbol{u}_N^{n+1})|^2 d\boldsymbol{x} + \int_\Gamma \frac{v_N^{n+1} - v_N^{n+\frac{1}{2}}}{\Delta t}\cdot v_N^{n+1}dz$$
$$= P^n_{N,in}\int_0^R (\boldsymbol{u}_N^{n+1})_z|_{z=0}dr - P^n_{N,out}\int_0^R (\boldsymbol{u}_N^{n+1})_z|_{z=L}dr$$
$$+\frac{1}{\Delta t}\int_{\mathcal{O}_f} \boldsymbol{f}(\boldsymbol{u}_N^{n+\frac{1}{2}}, v_N^{n+\frac{1}{2}}, \eta_N^{n+\frac{1}{2}})\Delta W^n_{1,N}\cdot \boldsymbol{u}_N^{n+1}d\boldsymbol{x}$$
$$+\frac{1}{\Delta t}\int_\Gamma g(\boldsymbol{u}_N^{n+\frac{1}{2}}, v_N^{n+\frac{1}{2}}, \eta_N^{n+\frac{1}{2}})(\Delta W^n_{2,N})v_N^{n+1}dz.$$

The desired equality follows after multiplication by Δt and by using the identity $(a-b)\cdot a = \frac{1}{2}(|a|^2 + |a-b|^2 - |b|^2)$. The final inequality follows by using the trace, Poincaré, and Korn inequalities to estimate

$$\left|(\Delta t)P^n_{N,in}\int_0^R (\boldsymbol{u}^{n+1}_N)_z|_{z=0}dr\right| \le C(\Delta t)(P^n_{N,in})^2 + (\Delta t)\mu\int_{\mathcal{O}_f}|\boldsymbol{D}(\boldsymbol{u}^{n+1}_N)|^2,$$

and by estimating, for example:

$$\begin{aligned}
&\left|\int_{\mathcal{O}_f}\boldsymbol{f}(\boldsymbol{u}_N^{n+\frac{1}{2}}, v_N^{n+\frac{1}{2}}, \eta_N^{n+\frac{1}{2}})\Delta W^n_{1,N}\cdot \boldsymbol{u}_N^{n+1}\right| \\
&\quad = \left|\int_{\mathcal{O}_f}\boldsymbol{f}(\boldsymbol{u}_N^{n+\frac{1}{2}}, v_N^{n+\frac{1}{2}}, \eta_N^{n+\frac{1}{2}})\Delta W^n_{1,N}\cdot \boldsymbol{u}_N^{n+\frac{1}{2}}\right| \\
&\quad + \left|\int_{\mathcal{O}_f}\boldsymbol{f}(\boldsymbol{u}_N^{n+\frac{1}{2}}, v_N^{n+\frac{1}{2}}, \eta_N^{n+\frac{1}{2}})\Delta W^n_{1,N}\cdot (\boldsymbol{u}_N^{n+1} - \boldsymbol{u}_N^{n+\frac{1}{2}})\right| \\
&\quad \le \left|\int_{\mathcal{O}_f}\boldsymbol{f}(\boldsymbol{u}_N^{n+\frac{1}{2}}, v_N^{n+\frac{1}{2}}, \eta_N^{n+\frac{1}{2}})\Delta W^n_{1,N}\cdot \boldsymbol{u}_N^{n+\frac{1}{2}}\right| \\
&\quad + \int_{\mathcal{O}_f}|\boldsymbol{f}(\boldsymbol{u}_N^{n+\frac{1}{2}}, v_N^{n+\frac{1}{2}}, \eta_N^{n+\frac{1}{2}})|^2(\Delta W^n_{1,N})^2 \\
&\quad + \frac{1}{4}\int_{\mathcal{O}_f}|\boldsymbol{u}_N^{n+1} - \boldsymbol{u}_N^{n+\frac{1}{2}}|^2 d\boldsymbol{x}.
\end{aligned}$$

and absorbing terms into the numerical dissipation $\frac{1}{2}\left(\|\boldsymbol{u}_N^{n+1} - \boldsymbol{u}_N^{n+\frac{1}{2}}\|^2_{L^2(\mathcal{O}_f)}\right) + \frac{1}{2}\left(\|v_N^{n+1} - v_N^{n+\frac{1}{2}}\|_{L^2(\Gamma)}\right)^2$ on the left-hand side. □

8.5.3 The Full Coupled Semidiscrete Problem

Recall the definitions of the spaces $\mathcal{V}$ and $\mathcal{V}_f$ from (8.15) and (8.37). By adding the weak formulation of the fluid subproblem (8.38) and the second equation in the structure subproblem (8.33), we have that the solution to the full semidiscrete problem is $(\boldsymbol{u}_N^{n+1}, v_N^{n+1}) \in \mathcal{V}$, and $(\boldsymbol{u}_N^{n+\frac{1}{2}}, v_N^{n+\frac{1}{2}}, \eta_N^{n+\frac{1}{2}}) \in \mathcal{V}_f \times H^2_0(\Gamma) \times H^2_0(\Gamma)$, satisfying the following equality pathwise:

$$\int_{\mathcal{O}_f} \frac{\boldsymbol{u}_N^{n+1} - \boldsymbol{u}_N^n}{\Delta t} \cdot \boldsymbol{q} d\boldsymbol{x} + 2\mu \int_{\mathcal{O}_f} \boldsymbol{D}(\boldsymbol{u}_N^{n+1}) : \boldsymbol{D}(\boldsymbol{q}) d\boldsymbol{x} + \int_\Gamma \frac{v_N^{n+1} - v_N^n}{\Delta t} \psi dz$$
$$+ \int_\Gamma \partial_z \eta_N^{n+1} \partial_z \psi dz + \int_\Gamma \partial_z^2 \eta_N^{n+1} \partial_z^2 \psi dz$$
$$= \int_{\mathcal{O}_f} \frac{W_1((n+1)\Delta t) - W_1(n\Delta t)}{\Delta t} (\boldsymbol{f}(\boldsymbol{u}_N^{n+\frac{1}{2}}, v_N^{n+\frac{1}{2}}, \eta_N^{n+\frac{1}{2}}) \cdot \boldsymbol{q}) d\boldsymbol{x}$$
$$+ \int_\Gamma \frac{W_2((n+1)\Delta t) - W_2(n\Delta t)}{\Delta t} (g(\boldsymbol{u}_N^{n+\frac{1}{2}}, v_N^{n+\frac{1}{2}}, \eta_N^{n+\frac{1}{2}}) \psi) dz$$
$$+ P_{N,in}^n \int_0^R (q_z)|_{z=0} dr - P_{N,out}^n \int_0^R (q_z)|_{z=L} dr, \quad \text{for all } (\boldsymbol{q}, \psi) \in \mathcal{Q},$$

$$\int_\Gamma \frac{\eta_N^{n+1} - \eta_N^n}{\Delta t} \phi dz = \int_\Gamma v_N^{n+\frac{1}{2}} \phi dz, \quad \forall \phi \in L^2(\Gamma), \tag{8.39}$$

where $P_{N,in/out}^n = \frac{1}{\Delta t} \int_{n\Delta t}^{(n+1)\Delta t} P_{in/out}(t) dt$. Note that $\eta_N^{n+1} = \eta_N^{n+\frac{1}{2}}$ and $\boldsymbol{u}_N^{n+\frac{1}{2}} = \boldsymbol{u}_N^n$ by the way we constructed the splitting scheme.

The following proposition provides uniform estimates on the expectation of the kinetic and elastic energy for the full, semidiscrete coupled problem (uniform in the number of time steps N, or equivalently, uniform in Δt), as well as uniform estimates on the expectation of the numerical dissipation.

Proposition 8.5.5 *Let $N > 0$, and let $\Delta t = \frac{T}{N}$. There exists a constant C independent of N and depending only on the initial data, the parameters of the problem, and $\| P_{in/out} \|_{L^2(0,T)}^2$, such that the following uniform energy estimates hold:*

1. ***Uniform semidiscrete kinetic energy and elastic energy estimates:*** *For all $1 \le p < \infty$, there exists a constant C_p depending only on p such that*

$$\mathbb{E}\left(E_N^{n+\frac{i}{2}}\right)^p \le C_p \qquad \textit{for } n = 0, 1, \dots, N-1 \textit{ and } i = 1, 2.$$

2. ***Uniform semidiscrete viscous fluid dissipation estimate:***

$$\sum_{j=1}^N \mathbb{E}(D_N^j) \le C.$$

3. ***Uniform numerical dissipation estimates:***

$$\sum_{n=0}^{N-1} \mathbb{E}\left(\|v_N^{n+\frac{1}{2}} - v_N^n\|^2_{L^2(\Gamma)}\right) + \sum_{n=0}^{N-1} \mathbb{E}\left(\|\partial_z(\eta_N^{n+\frac{1}{2}} - \eta_N^n)\|^2_{L^2(\Gamma)}\right)$$
$$+ \sum_{n=0}^{N-1} \mathbb{E}\left(\|\partial_z^2(\eta_N^{n+\frac{1}{2}} - \eta_N^n)\|^2_{L^2(\Gamma)}\right) \le C,$$

$$\sum_{n=0}^{N-1} \left(\mathbb{E}\left(\|\boldsymbol{u}_N^{n+1} - \boldsymbol{u}_N^{n+\frac{1}{2}}\|^2_{L^2(\mathcal{O}_f)}\right) + \mathbb{E}\left(\|v_N^{n+1} - v_N^{n+\frac{1}{2}}\|^2_{L^2(\Gamma)}\right)\right) \le C.$$

Proof Recall the definitions of the discrete energy $E_N^{n+\frac{i}{2}}$ and the discrete fluid dissipation D_N^n from (8.35) and (8.36). By summing the structure and fluid discrete energy identities in Propositions 8.5.2 and 8.5.4 and taking expectations, we obtain

$$\mathbb{E}\left(\max_{i=1,2}\left[\max_{n=0,1,\dots,N-1} E_N^{n+\frac{i}{2}}\right]\right)$$
$$+ \sum_{k=0}^{N-1}\left[\mathbb{E}\left(\mu(\Delta t)\int_{\mathcal{O}_f} |\boldsymbol{D}(\boldsymbol{u}_N^{k+1})|^2 d\boldsymbol{x}\right)\right.$$
$$+ \frac{1}{4}\mathbb{E}\left(\|\boldsymbol{u}_N^{k+1} - \boldsymbol{u}_N^{k+\frac{1}{2}}\|^2_{L^2(\mathcal{O}_f)}\right) + \frac{1}{4}\mathbb{E}\left(\|v_N^{k+1} - v_N^{k+\frac{1}{2}}\|^2_{L^2(\Gamma)}\right)$$
$$+ \frac{1}{4}\mathbb{E}\left(\|v_N^{k+\frac{1}{2}} - v_N^k\|^2_{L^2(\Gamma)}\right)$$
$$\left.+ \frac{1}{4}\mathbb{E}\left(\|\partial_z(\eta_N^{k+\frac{1}{2}} - \eta_N^k)\|^2_{L^2(\Gamma)}\right) + \frac{1}{4}\mathbb{E}\left(\|\partial_z^2(\eta_N^{k+\frac{1}{2}} - \eta_N^k)\|^2_{L^2(\Gamma)}\right)\right]$$
$$\le 2E_0 + C(\Delta t)\sum_{k=0}^{N-1}\left[(P_{N,in}^k)^2 + (P_{N,out}^k)^2\right]$$
$$+ 2\mathbb{E}\left[\max_{n=0,1,\dots,N-1}\sum_{k=0}^{n}\left((\Delta W_{1,N}^k)\int_{\mathcal{O}_f} \boldsymbol{u}_N^{k+\frac{1}{2}} \cdot \boldsymbol{f}(\boldsymbol{u}_N^{k+\frac{1}{2}}, v_N^{k+\frac{1}{2}}, \eta_N^{k+\frac{1}{2}}) d\boldsymbol{x}\right.\right.$$
$$\left.\left.+ (\Delta W_{2,N}^k)\int_\Gamma v_N^{k+\frac{1}{2}} g(\boldsymbol{u}_N^{k+\frac{1}{2}}, v_N^{k+\frac{1}{2}}, \eta_N^{k+\frac{1}{2}}) dz\right)\right]$$
$$+ 2\mathbb{E}\left[\sum_{k=0}^{N-1}\left((\Delta W_{1,N}^k)^2\int_{\mathcal{O}_f} |\boldsymbol{f}(\boldsymbol{u}_N^{k+\frac{1}{2}}, v_N^{k+\frac{1}{2}}, \eta_N^{k+\frac{1}{2}})|^2 d\boldsymbol{x} + (\Delta W_{2,N}^k)^2\right.\right.$$
$$\left.\left.\int_\Gamma |g(\boldsymbol{u}_N^{k+\frac{1}{2}}, v_N^{k+\frac{1}{2}}, \eta_N^{k+\frac{1}{2}})|^2 dz\right)\right]. \tag{8.40}$$

Raising both sides to the p power for $1 \le p < \infty$ before taking expectations and using nonnegativity of the fluid/numerical dissipation terms, we obtain:

$$
\begin{aligned}
&\mathbb{E}\left(\max_{i=1,2}\left[\max_{n=0,1,\dots,N-1}\left(E_N^{n+\frac{i}{2}}\right)^p\right]\right)\\
&\le c_p\Bigg[E_0^p + C(\Delta t)^p\left(\sum_{k=0}^{N-1}\left[(P_{N,in}^k)^2 + (P_{N,out}^k)^2\right]\right)^p\\
&+\mathbb{E}\Bigg[\max_{n=0,1,\dots,N-1}\sum_{k=0}^{n}\Bigg((\Delta W_{1,N}^k)\int_{\mathcal{O}_f}\boldsymbol{u}_N^{k+\frac{1}{2}}\cdot\boldsymbol{f}(\boldsymbol{u}_N^{k+\frac{1}{2}}, v_N^{k+\frac{1}{2}}, \eta_N^{k+\frac{1}{2}})d\boldsymbol{x}\\
&+(\Delta W_{2,N}^k)\int_\Gamma v_N^{k+\frac{1}{2}} g(\boldsymbol{u}_N^{k+\frac{1}{2}}, v_N^{k+\frac{1}{2}}, \eta_N^{k+\frac{1}{2}})dz\Bigg)\Bigg]^p\\
&+\mathbb{E}\Bigg[\sum_{k=0}^{N-1}\Bigg((\Delta W_{1,N}^k)^2\int_{\mathcal{O}_f}|\boldsymbol{f}(\boldsymbol{u}_N^{k+\frac{1}{2}}, v_N^{k+\frac{1}{2}}, \eta_N^{k+\frac{1}{2}})|^2 d\boldsymbol{x} + (\Delta W_{2,N}^k)^2\\
&\times\int_\Gamma |g(\boldsymbol{u}_N^{k+\frac{1}{2}}, v_N^{k+\frac{1}{2}}, \eta_N^{k+\frac{1}{2}})|^2 dz\Bigg)\Bigg]^p\Bigg]\\
&=: c_p(E_0^p + I_1 + I_2 + I_3).
\end{aligned}
$$

where

$$
I_1 := C\left((\Delta t)\sum_{k=0}^{N-1}\left[(P_{N,in}^k)^2 + (P_{N,out}^k)^2\right]\right)^p,
$$

$$
\begin{aligned}
I_2 := \mathbb{E}\Bigg[\max_{n=0,1,\dots,N-1}\sum_{k=0}^{n}\Bigg(&(\Delta W_{1,N}^k)\int_{\mathcal{O}_f}\boldsymbol{u}_N^{k+\frac{1}{2}}\cdot\boldsymbol{f}(\boldsymbol{u}_N^{k+\frac{1}{2}}, v_N^{k+\frac{1}{2}}, \eta_N^{k+\frac{1}{2}})d\boldsymbol{x}\\
&+(\Delta W_{2,N}^k)\int_\Gamma v_N^{k+\frac{1}{2}} g(\boldsymbol{u}_N^{k+\frac{1}{2}}, v_N^{k+\frac{1}{2}}, \eta_N^{k+\frac{1}{2}})dz\Bigg)\Bigg]^p,
\end{aligned}
$$

$$
\begin{aligned}
I_3 := \mathbb{E}\Bigg[\sum_{k=0}^{N-1}\Bigg(&(\Delta W_{1,N}^k)^2\int_{\mathcal{O}_f}|\boldsymbol{f}(\boldsymbol{u}_N^{k+\frac{1}{2}}, v_N^{k+\frac{1}{2}}, \eta_N^{k+\frac{1}{2}})|^2 d\boldsymbol{x}\\
&+(\Delta W_{2,N}^k)^2\int_\Gamma |g(\boldsymbol{u}_N^{k+\frac{1}{2}}, v_N^{k+\frac{1}{2}}, \eta_N^{k+\frac{1}{2}})|^2 dz\Bigg)\Bigg]^p.
\end{aligned}
$$

Bound for I_1 Recall that the discretized inlet/outlet pressure is defined by $P^k_{N,in/out} = \frac{1}{\Delta t}\int_{k\Delta t}^{(k+1)\Delta t} P_{in/out}(t)dt$, so that $P^k_{N,in/out}$ is deterministic. Therefore, we have the following bound for I_1:

$$
\begin{aligned}
&C\left((\Delta t)\sum_{k=0}^{N-1}\left[(P^k_{N,in})^2 + (P^k_{N,out})^2\right]\right)^p \\
&\le C\left(\sum_{k=0}^{N-1}\left[\frac{1}{\Delta t}\left(\int_{k\Delta t}^{(k+1)\Delta t} P_{in}(t)dt\right)^2 + \frac{1}{\Delta t}\left(\int_{k\Delta t}^{(k+1)\Delta t} P_{out}(t)dt\right)^2\right]\right)^p \\
&\le C\left(\sum_{k=0}^{N-1}\left[\|P_{in}\|^2_{L^2(k\Delta t,(k+1)\Delta t)} + \|P_{out}\|^2_{L^2(k\Delta t,(k+1)\Delta t)}\right]\right)^p \\
&\le C\left(\|P_{in}\|^{2p}_{L^2(0,T)} + \|P_{out}\|^{2p}_{L^2(0,T)}\right). \qquad (8.41)
\end{aligned}
$$

Bound for I_2 Next, we examine I_2 and start with an estimate involving the absolute values:

$$
\begin{aligned}
I_2 \le c_p\Bigg[&\mathbb{E}\left(\max_{n=0,1,\dots,N-1}\left|\sum_{k=0}^{n}\Delta W^k_{1,N}\left(\int_{\mathcal{O}_f} \boldsymbol{u}_N^{k+\frac{1}{2}}\cdot \boldsymbol{f}(\boldsymbol{u}_N^{k+\frac{1}{2}}, v_N^{k+\frac{1}{2}}, \eta_N^{k+\frac{1}{2}})d\boldsymbol{x}\right)\right|\right)^p \\
&+\mathbb{E}\left(\max_{n=0,1,\dots,N-1}\left|\sum_{k=0}^{n}\Delta W^k_{2,N}\left(\int_0^L v_N^{k+\frac{1}{2}} g(\boldsymbol{u}_N^{k+\frac{1}{2}}, v_N^{k+\frac{1}{2}}, \eta_N^{k+\frac{1}{2}})dz\right)\right|\right)^p\Bigg] \\
:=\; &c_p(I_{2,1}+I_{2,2}). \qquad (8.42)
\end{aligned}
$$

We can estimate the right-hand side of (8.42) by interpreting these terms as stochastic integrals since $\Delta W^k_{i,N} = W_i((k+1)\Delta t) - W_i(k\Delta t)$ for $i = 1, 2$. We can do this, for example, for the first term on the right-hand side $I_{2,1}$ in the following way:

$$
\sum_{k=0}^{n}\left(\int_{\mathcal{O}_f} \boldsymbol{u}_N^{k+\frac{1}{2}}\cdot \boldsymbol{f}(\boldsymbol{u}_N^{k+\frac{1}{2}}, v_N^{k+\frac{1}{2}}, \eta_N^{k+\frac{1}{2}})d\boldsymbol{x}\right)\cdot\Delta W^k_{1,N} = \int_0^{(n+1)\Delta t} f(t)dW_1(t),
$$

where $f(t)$ is the random function on $[0, T]$ defined by:

$$
f(t) = \sum_{k=0}^{N-1}\left(\int_{\mathcal{O}_f} \boldsymbol{u}_N^{k+\frac{1}{2}}\cdot \boldsymbol{f}(\boldsymbol{u}_N^{k+\frac{1}{2}}, v_N^{k+\frac{1}{2}}, \eta_N^{k+\frac{1}{2}})d\boldsymbol{x}\right)\cdot 1_{(k\Delta t,(k+1)\Delta t]}(t). \qquad (8.43)
$$

Because $(\boldsymbol{u}_N^{k+\frac{1}{2}}, v_N^{k+\frac{1}{2}}, \eta_N^{k+\frac{1}{2}})$ is $\mathcal{F}_{t_N^k}$-measurable, see Proposition 8.5.1; this integrand is predictable. Hence, since the stochastic integral is a continuous process in time, we have

$$I_{2,1} \leq \mathbb{E}\left(\max_{0 \leq s \leq T}\left|\int_0^s f(t) dW_1\right|^p\right).$$

and hence, by the BDG inequality (see Theorem 3.6.2), we have that

$$\begin{aligned}
I_{2,1} &\leq \mathbb{E}\left(\int_0^T |f(t)|^2 dt\right)^{p/2} \\
&= \mathbb{E}\left[(\Delta t)^{p/2}\left(\sum_{k=0}^{N-1}\left(\int_{\mathcal{O}_f} \boldsymbol{u}_N^{k+\frac{1}{2}} \cdot \boldsymbol{f}(\boldsymbol{u}_N^{k+\frac{1}{2}}, v_N^{k+\frac{1}{2}}, \eta_N^{k+\frac{1}{2}}) d\boldsymbol{x}\right)^2\right)^{p/2}\right]
\end{aligned}$$

Using the Cauchy-Schwarz inequality and the assumption (8.6), we obtain that

$$\begin{aligned}
I_{2,1} &\leq \mathbb{E}\left[(\Delta t)^{p/2}\left(\sum_{k=0}^{N-1}\left(\|\boldsymbol{u}_N^{k+\frac{1}{2}}\|^2_{L^2(\mathcal{O}_f)}\|\boldsymbol{f}(\boldsymbol{u}_N^{k+\frac{1}{2}}, v_N^{k+\frac{1}{2}}, \eta_N^{k+\frac{1}{2}})\|^2_{L^2(\mathcal{O}_f)}\right)\right)^{p/2}\right] \\
&\leq C\mathbb{E}\left[(\Delta t)^{p/2} \max_{k=0,1,\dots,N-1} \|\boldsymbol{u}_N^{k+\frac{1}{2}}\|^p_{L^2(\mathcal{O}_f)}\left(\sum_{k=0}^{N-1}\left(1 + E_N^{k+\frac{1}{2}}\right)\right)^{p/2}\right] \\
&\leq \epsilon\mathbb{E}\left(\max_{k=0,1,\dots N-1}\left(E_N^{k+\frac{1}{2}}\right)^p\right) + C(\epsilon)\mathbb{E}\left[(\Delta t)^p\left(\sum_{k=0}^{N-1}\left(1 + E_N^{k+\frac{1}{2}}\right)\right)^p\right].
\end{aligned}$$

Using the Cauchy-Schwarz inequality to estimate the finite sum with exponents p and $p/(p-1)$, we obtain:

$$\begin{aligned}
I_{2,1} &\leq \epsilon\mathbb{E}\left(\max_{k=0,1,\dots N-1}\left(E_N^{k+\frac{1}{2}}\right)^p\right) + C(\epsilon) \\
&\quad \times \mathbb{E}\left[(\Delta t)^p N^{p-1}\left(\sum_{k=0}^{N-1}\left(1 + E_N^{k+\frac{1}{2}}\right)^p\right)\right] \\
&\leq \epsilon\mathbb{E}\left(\max_{k=0,1,\dots N-1}\left(E_N^{k+\frac{1}{2}}\right)^p\right) + C(\epsilon) c_p T^{p-1} \\
&\quad \times \left[T + (\Delta t)\sum_{k=0}^{N-1}\mathbb{E}\left(\max_{n=0,1,\dots,k} E_N^{n+\frac{1}{2}}\right)^p\right].
\end{aligned}$$

Using the same procedure as above, we can obtain an analogous estimate for $I_{2,2}$:

$$I_{2,2} \le \epsilon \mathbb{E}\left(\max_{k=0,1,\dots N-1}\left(E_N^{k+\frac{1}{2}}\right)^p\right) + C(\epsilon)c_p T^{p-1} \times \left[T + (\Delta t)\sum_{k=0}^{N-1}\mathbb{E}\left(\max_{n=0,1,\dots,k} E_N^{n+\frac{1}{2}}\right)^p\right].$$

Bound for I_3 For I_3, we separate this into two terms and estimate the first term $I_{3,1}$, since the second term $I_{3,2}$ can be handled analogously:

$$I_3 \le c_p \mathbb{E}\left[\sum_{k=0}^{N-1}\left([W_1((k+1)\Delta t) - W_1(k\Delta t)]^2 \int_{\mathcal{O}_f} |\boldsymbol{f}(\boldsymbol{u}_N^{k+\frac{1}{2}}, v_N^{k+\frac{1}{2}}, \eta_N^{k+\frac{1}{2}})|^2 d\boldsymbol{x}\right)\right]^p$$
$$+ c_p \mathbb{E}\left[\sum_{k=0}^{N-1}\left([W_2((k+1)\Delta t) - W_2(k\Delta t)]^2 \int_{\Gamma} |g(\boldsymbol{u}_N^{k+\frac{1}{2}}, v_N^{k+\frac{1}{2}}, \eta_N^{k+\frac{1}{2}})|^2 dz\right)\right]^p$$
$$:= c_p(I_{3,1} + I_{3,2}).$$

To estimate $I_{3,1}$, we use the Cauchy-Schwarz inequality with the exponents p and $p/(p-1)$ (to estimate the finite sum), the independence properties of Brownian motion, the fact that $(\boldsymbol{u}_N^{k+\frac{1}{2}}, v_N^{k+\frac{1}{2}}, \eta_N^{k+\frac{1}{2}})$ is $\mathcal{F}_{t_N^k}$-measurable (see Proposition 8.5.1), assumption (8.6), and the fact that $\mathbb{E}|W(t) - W(s)|^{2p} = c_p|t-s|^p$ (see Exercise 4.9 in [107]) to obtain:

$$I_{3,1} \le N^{p-1}\mathbb{E}\left[\sum_{k=0}^{N-1}\left(|W_1((k+1)\Delta t) - W_1(k\Delta t)|^{2p} \times \left(\int_{\mathcal{O}_f} |\boldsymbol{f}(\boldsymbol{u}_N^{k+\frac{1}{2}}, v_N^{k+\frac{1}{2}}, \eta_N^{k+\frac{1}{2}})|^2\right)^p\right)\right]$$
$$\le N^{p-1}\sum_{k=0}^{N-1}\left(\mathbb{E}|W_1((k+1)\Delta t) - W_1(k\Delta t)|^{2p} \cdot \mathbb{E}\left(1 + E_N^{k+\frac{1}{2}}\right)^p\right)$$
$$\le c_p N^{p-1}\sum_{k=0}^{N-1}\left[(\Delta t)^p \mathbb{E}\left(1 + E_N^{k+\frac{1}{2}}\right)^p\right]$$
$$\le c_p T^{p-1}(\Delta t)\sum_{k=0}^{N-1}\left[1 + \mathbb{E}\left(\max_{n=0,1,\dots,k} E_N^{n+\frac{1}{2}}\right)^p\right]$$

$$\leq c_p T^{p-1} \left[T + (\Delta t) \sum_{k=0}^{N-1} \mathbb{E} \left(\max_{n=0,1,\dots,k} E_N^{n+\frac{1}{2}} \right)^p \right].$$

Using the same procedure, we can obtain a similar estimate for $I_{3,2}$:

$$I_{3,2} \leq c_p T^{p-1} \left[T + (\Delta t) \sum_{k=0}^{N-1} \mathbb{E} \left(\max_{n=0,1,\dots,k} E_N^{n+\frac{1}{2}} \right)^p \right].$$

Conclusion From the above estimates, we conclude that

$$\mathbb{E} \left(\max_{i=1,2} \left(\max_{n=0,1,\dots,N-1} \left(E_N^{n+\frac{i}{2}} \right)^p \right] \right) \leq c_p \left(E_0^p + \|P_{in}\|^{2p}_{L^2(0,T)} + \|P_{out}\|^{2p}_{L^2(0,T)} \right.$$
$$+ 2\epsilon \mathbb{E} \left(\max_{k=0,1,\dots N-1} \left(E_N^{k+\frac{1}{2}} \right)^p \right) + (2C(\epsilon)c_p + 2c_p) T^{p-1}$$
$$\left. \left[T + (\Delta t) \sum_{k=0}^{N-1} \mathbb{E} \left(\max_{n=0,1,\dots,k} \left(E_N^{n+\frac{1}{2}} \right)^p \right) \right] \right).$$

We then fix $\epsilon > 0$ small enough and move terms involving ϵ from the right-hand side to the left-hand side. We emphasize that this gives a uniform energy estimate, because the choice of ϵ is independent of Δt and hence N, since all constants appearing in the estimate are independent of N, and we obtain:

$$\mathbb{E} \left(\max_{i=1,2} \left[\max_{n=0,1,\dots,N-1} \left(E_N^{n+\frac{i}{2}} \right)^p \right] \right)$$
$$\leq c_{p,T} \left(1 + E_0^p + \|P_{in}\|^{2p}_{L^2(0,T)} + \|P_{out}\|^{2p}_{L^2(0,T)} \right.$$
$$\left. + (\Delta t) \sum_{k=0}^{N-1} \mathbb{E} \left(\max_{n=0,1,\dots,k} \left(E_N^{n+\frac{1}{2}} \right)^p \right) \right),$$

for a constant $c_{p,T}$ depending on p and T, but independent of N, from which the result that $\mathbb{E} \left(\max_{i=1,2} \max_{k=0,1,\dots,N-1} \left(E_N^{k+\frac{i}{2}} \right)^p \right) \leq C_p$ follows from the discrete Gronwall lemma (see Theorem 2.12.2). For the remaining uniform fluid dissipation and numerical dissipation estimates, these follow from the discrete Gronwall estimate in the case where $p = 1$, in which one can check by using the same procedure, using the inequality (8.40) for $p = 1$ which includes the fluid/numerical dissipation terms, that:

$$
\begin{aligned}
&\mathbb{E}\left(\max_{i=1,2}\left[\max_{n=0,1,\dots,N-1} E_N^{n+\frac{i}{2}}\right]\right) \\
&+\sum_{k=0}^{N-1}\Bigg[\mathbb{E}\left(\mu(\Delta t)\int_{\mathcal{O}_f}|\boldsymbol{D}(\boldsymbol{u}_N^{k+1})|^2 d\boldsymbol{x}\right)+\frac{1}{4}\mathbb{E}\left(\|\boldsymbol{u}_N^{k+1}-\boldsymbol{u}_N^{k+\frac{1}{2}}\|^2_{L^2(\mathcal{O}_f)}\right) \\
&\qquad+\frac{1}{4}\mathbb{E}\left(\|v_N^{k+1}-v_N^{k+\frac{1}{2}}\|^2_{L^2(\Gamma)}\right) \\
&+\frac{1}{4}\mathbb{E}\left(\|v_N^{k+\frac{1}{2}}-v_N^k\|^2_{L^2(\Gamma)}\right)+\frac{1}{4}\mathbb{E}\left(\|\partial_z(\eta_N^{k+\frac{1}{2}}-\eta_N^k)\|^2_{L^2(\Gamma)}\right) \\
&\qquad+\frac{1}{4}\mathbb{E}\left(\|\partial_z^2(\eta_N^{k+\frac{1}{2}}-\eta_N^k)\|^2_{L^2(\Gamma)}\right)\Bigg] \\
&\qquad\le C\Bigg(E_0+\|P_{in}\|^2_{L^2(0,T)}+\|P_{out}\|^2_{L^2(0,T)} \\
&+2\epsilon\mathbb{E}\left(\max_{k=0,1,\dots N-1} E_N^{k+\frac{1}{2}}\right)+(2C(\epsilon)+C)\left[T+(\Delta t)\sum_{k=0}^{N-1}\mathbb{E}\left(\max_{n=0,1,\dots,k} E_N^{n+\frac{1}{2}}\right)\right]\Bigg).
\end{aligned}
$$

□

Remark 8.5.1 We remark that the specific form of the splitting scheme enabled us to directly estimate the terms involving the white noise as stochastic integrals, such as the second to last term in estimate (8.43), as a result of measurability considerations. Because $(\boldsymbol{u}_N^{k+\frac{1}{2}}, v_N^{k+\frac{1}{2}}, \eta_N^{k+\frac{1}{2}})$ is $\mathcal{F}_{t_N^k}$-measurable, the stochastic increment $[W_i((k+1)\Delta t)-W_i(k\Delta t)]$ for $i=1,2$ is independent of $(\boldsymbol{u}_N^{k+\frac{1}{2}}, v_N^{k+\frac{1}{2}}, \eta_N^{k+\frac{1}{2}})$, and hence, we were able to rewrite this term as a stochastic integral, see (8.43).

8.6 Approximate Solutions

Currently, we have approximate solutions defined on each time subinterval $[n\Delta t, (n+1)\Delta t]$ separately for each $\Delta t = T/N$, and our next step will be to create (random) approximate solutions on all of $[0, T]$ by gluing together the approximate solutions on the individual subintervals. We use the solutions at fixed times of our semidiscrete scheme, $\boldsymbol{u}_N^{n+\frac{i}{2}}$, $\eta_N^{n+\frac{i}{2}}$, and $v_N^{n+\frac{i}{2}}$ for $i=1,2$. We construct approximate solutions for the given stochastic FSI problem in time on the time interval $[0, T]$ for each N, as piecewise functions of time, so that we can later pass to the limit as $N\to\infty$ in these approximate solutions on $[0, T]$. However, we must be careful in this construction of approximate solutions to make sure that they are adapted to the given filtration $\{\mathcal{F}_t\}_{t\ge 0}$ associated with the given Brownian motions.

8.6.1 Definition of Approximate Solutions

We start with the fluid. We define the approximate random function $\boldsymbol{u}_N$ on $[0, T] \times \mathcal{O}_f$ to be the piecewise constant function

$$\boldsymbol{u}_N(t, \cdot) = \boldsymbol{u}_N^{n-1}, \qquad \text{for } t \in ((n-1)\Delta t, n\Delta t].$$

Note that because $\boldsymbol{u}_N^n$ is $\mathcal{F}_{t_N^n}$-measurable, the choice of $\boldsymbol{u}_N^{n-1}$ instead of $\boldsymbol{u}_N^n$ above is used, so that the resulting process $\boldsymbol{u}_N$ is adapted to the filtration $\{\mathcal{F}_t\}_{t \geq 0}$.

Next, we consider the functions associated with the structure. Note that η_N^{n-1}, $\eta_N^{n-\frac{1}{2}}$, and $v_N^{n-\frac{1}{2}}$ are $\mathcal{F}_{t_N^{n-1}}$-measurable while v_N^n is $\mathcal{F}_{t_N^n}$-measurable. Hence, we define

$$\eta_N(t, \cdot) = \eta_N^{n-1}, \quad v_N(t, \cdot) = v_N^{n-1}, \quad v_N^*(t, \cdot) = v_N^{n-\frac{1}{2}}, \qquad \text{for } t \in ((n-1)\Delta t, n\Delta t], \tag{8.44}$$

and as a result of the measurability properties of the approximate solutions on each time subinterval, these are all adapted to the given filtration $\{\mathcal{F}_t\}_{t \geq 0}$. Note that v_N defined on $[0, T] \times \Gamma$ is pathwise the trace of the fluid velocity $\boldsymbol{u}_N$ defined on $[0, T] \times \mathcal{O}_f$ for all $t \in [0, T]$, but this is not true for v_N^*, since v_N^* is the structure velocity obtained after the structure subproblem in the semidiscrete scheme, which does not update the fluid velocity.

We also introduce a piecewise linear interpolation $\overline{\eta}_N$ of η_N to add additional time regularity to the structure displacement, since we will want the structure displacement to be in $W^{1,\infty}(0, T; L^2(\Gamma))$ almost surely in the limit as $\Delta t \to 0$. Thus, $\overline{\eta}_N$ is piecewise linear such that

$$\overline{\eta}_N(t) = \left(\frac{(n+1)\Delta t - t}{\Delta t}\right) \eta_N^n + \left(\frac{t - n\Delta t}{\Delta t}\right) \eta_N^{n+1}, \qquad \text{for } n\Delta t \leq t \leq (n+1)\Delta t. \tag{8.45}$$

Note that $\overline{\eta}_N$ has Lipschitz continuous paths in time, and furthermore,

$$\partial_t \overline{\eta}_N = v_N^*. \tag{8.46}$$

Because both η_N^n and η_N^{n+1} are $\mathcal{F}_{t_N^n}$-adapted, $\overline{\eta}_N$ is adapted to the filtration $\{\mathcal{F}_t\}_{t \geq 0}$. We will also introduce a piecewise constant function $\eta_N^{\Delta t}$ for the structure displacement, given by

$$\eta_N^{\Delta t}(t, \cdot) = \eta_N^n, \qquad \text{for } t \in ((n-1)\Delta t, n\Delta t]. \tag{8.47}$$

Note that $\eta_N^{\Delta t}$ is adapted to the filtration $\{\mathcal{F}_t\}_{t \geq 0}$ and is a time-shifted version of η_N, which is emphasized in the notation by the superscript of Δt. This time-shifted structure displacement will be useful for passing to the limit in the approximate semidiscrete weak formulations; see Sect. 8.7.3.

We will also consider the corresponding piecewise linear interpolations for the fluid velocity and structure velocity, which satisfy

$$\overline{\boldsymbol{u}}_N(t) = \left(\frac{(n+1)\Delta t - t}{\Delta t}\right)\boldsymbol{u}_N^n + \left(\frac{t - n\Delta t}{\Delta t}\right)\boldsymbol{u}_N^{n+1},$$
$$\overline{v}_N(t) = \left(\frac{(n+1)\Delta t - t}{\Delta t}\right)v_N^n + \left(\frac{t - n\Delta t}{\Delta t}\right)v_N^{n+1}, \qquad \text{for } n\Delta t \le t \le (n+1)\Delta t. \tag{8.48}$$

We will need to consider $\overline{\boldsymbol{u}}_N$ and $\overline{v}_N$, because we will express the discrete time derivatives $\frac{\boldsymbol{u}_N^{n+1} - \boldsymbol{u}_N^n}{\Delta t}$ and $\frac{v_N^{n+1} - v_N^n}{\Delta t}$ in the semidiscrete formulation (8.39) in terms of the time derivatives of $\overline{\boldsymbol{u}}_N$ and $\overline{v}_N$. We will also need to consider piecewise constant time-shifted functions $\boldsymbol{u}_N^{\Delta t}$ and $v_N^{\Delta t}$ for the fluid velocity and the structure velocity, defined by

$$\boldsymbol{u}_N^{\Delta t}(t, \cdot) = \boldsymbol{u}_N^n, \qquad v_N^{\Delta t}(t, \cdot) = v_N^n, \qquad \text{for } t \in ((n-1)\Delta t, n\Delta t]. \tag{8.49}$$

We note that $\boldsymbol{u}_N^{\Delta t}$ and $v_N^{\Delta t}$ are time-shifted versions of $\boldsymbol{u}_N$ and v_N. We will need these time-shifted functions, because the fluid dissipation estimate in Proposition 8.5.5 implies that $\boldsymbol{u}_N^{\Delta t}$, rather than $\boldsymbol{u}_N$, is bounded in $L^2(\Omega; L^2(0, T; H^1(\mathcal{O}_f)))$ independently of N. See Proposition 8.6.2.

We make the following important observation. Unlike $\overline{\eta}_N$, we note that $\overline{\boldsymbol{u}}_N$ and $\overline{v}_N$ are *not* necessarily adapted to the filtration $\{\mathcal{F}_t\}_{t\ge 0}$, even though they can still be considered as random variables taking values in their appropriate path spaces. Similarly, $\boldsymbol{u}_N^{\Delta t}$ and $v_N^{\Delta t}$, unlike $\boldsymbol{u}_N$ and v_N, are *not* necessarily adapted to the filtration $\{\mathcal{F}_t\}_{t\ge 0}$. However, this will not be an issue, because we will see later in Lemma 8.7.1 that $\overline{\boldsymbol{u}}_N$, $\boldsymbol{u}_N^{\Delta t}$, $\overline{v}_N$, and $v_N^{\Delta t}$ are almost surely "close to" the random processes $\boldsymbol{u}_N$ and $\boldsymbol{v}_N$, which are adapted to the filtration $\{\mathcal{F}_t\}_{t\ge 0}$, as $N \to \infty$ along a subsequence.

We summarize some of the previously discussed measure theoretic properties of the stochastic approximate solutions in the following proposition, for future reference.

Proposition 8.6.1 *Recall that W_1 and W_2 are one-dimensional Brownian motions with respect to the probability space with complete filtration, $(\Omega, \mathcal{F}, \{\mathcal{F}_t\}_{t\ge 0}, \mathbb{P})$. For all $N \in \mathbb{N}$, $\boldsymbol{u}_N$, v_N, v_N^*, η_N, and $\overline{\eta}_N$ are adapted to the filtration $\{\mathcal{F}_t\}_{t\ge 0}$ with left continuous paths, with $\overline{\eta}_N$ having continuous paths. In addition, for some fixed $t > 0$ and for each N, define $n_0 = \left\lfloor \frac{t}{\Delta t} \right\rfloor + 1$. Then, $W_1(\tau) - W_1(t)$ and $W_2(\tau) - W_2(t)$ are independent of each other and are independent of each of the random variables in the following collection of random variables for each N and for each $\tau > t$:*

$$\{\boldsymbol{u}_N^{n-1}, v_N^{n-1}, v_N^{n-\frac{1}{2}} : 1 \le n \le n_0\}, \{\eta_N^n : 0 \le n \le n_0\}, \{\overline{\eta}_N(s) : s \in [0, n_0\Delta t]\}.$$

8.6.2 Uniform Boundedness of Approximate Solutions

Using the previous discrete energy estimates in Proposition 8.5.5, we establish uniform boundedness of the approximate solutions in the following proposition. We note that in contrast to the case of deterministic FSI (see Chap. 4), the uniform boundedness of these (random) approximate solutions is only in *expectation*.

Proposition 8.6.2 *The following uniform boundedness results hold for all* $1 \leq p < \infty$*:*

- $(\eta_N)_{N\in\mathbb{N}}$ *is uniformly bounded in* $L^p(\Omega; L^\infty(0, T; H_0^2(\Gamma)))$.
- $(v_N)_{N\in\mathbb{N}}$ *is uniformly bounded in* $L^p(\Omega; L^\infty(0, T; L^2(\Gamma)))$.
- $(v_N^{\Delta t})_{N\in\mathbb{N}}$ *is uniformly bounded in* $L^p(\Omega; L^2(0, T; H^{1/2}(\Gamma)))$.
- $(v_N^*)_{N\in\mathbb{N}}$ *is uniformly bounded in* $L^p(\Omega; L^\infty(0, T; L^2(\Gamma)))$.
- $(\boldsymbol{u}_N)_{N\in\mathbb{N}}$ *is uniformly bounded in* $L^2(\Omega; L^\infty(0, T; L^2(\mathcal{O}_f)))$.
- $(\boldsymbol{u}_N^{\Delta t})_{N\in\mathbb{N}}$ *is uniformly bounded in* $L^p(\Omega; L^2(0, T; H^1(\mathcal{O}_f)))$.

Proof The only part of this result that does not follow directly from Proposition 8.5.5 is to show that $(\boldsymbol{u}_N^{\Delta t})_{N\in\mathbb{N}}$ is uniformly bounded in $L^p(\Omega; L^2(0, T; H^1(\mathcal{O}_f)))$. We compute

$$
\begin{aligned}
&\|\boldsymbol{u}_N^{\Delta t}\|^2_{L^2(\Omega; L^2(0,T; H^1(\mathcal{O}_f)))} \\
&= \mathbb{E}\left(\int_0^T \|\boldsymbol{u}_N^{\Delta t}\|^2_{H^1(\mathcal{O}_f)} dt\right) = (\Delta t)\mathbb{E}\left(\sum_{k=1}^N \|\boldsymbol{u}_N^k\|^2_{H^1(\mathcal{O}_f)}\right) \\
&\leq C(\Delta t)\mathbb{E}\left(\sum_{k=1}^N \|\boldsymbol{u}_N^k\|_{L^2(\mathcal{O}_f)} + \sum_{k=1}^N \|\boldsymbol{D}(\boldsymbol{u}_N^k)\|^2_{L^2(\mathcal{O}_f)}\right),
\end{aligned}
$$

where we used Korn's inequality (see Lemma 2.8.4). The result follows from the uniform boundedness of the sum of the dissipation terms and the uniform boundedness of the kinetic energy of the fluid, stated in Proposition 8.5.5. By taking the trace of the r component of the fluid velocity $\boldsymbol{u}_N^n$, which is in $H^{1/2}(\Gamma)$, we get the corresponding boundedness of $(v_N^{\Delta t})_{N\in\mathbb{N}}$ in $L^p(\Omega; L^2(0, T; H^{1/2}(\Gamma)))$, due to the fact that $\boldsymbol{u}_N^n|_\Gamma = v_N^n \boldsymbol{e}_r$. □

We also state the corresponding uniform boundedness property for the linear interpolations $(\overline{\eta}_N)_{N\in\mathbb{N}}$. Note that in terms of distributional derivatives, $\partial_t \overline{\eta}_N = v_N^*$ holds pathwise for $\omega \in \Omega$. Therefore, we have:

Proposition 8.6.3 *The sequence of linear interpolations of the structure displacements* $(\overline{\eta}_N)_{N\in\mathbb{N}}$ *is uniformly bounded in*
$L^2(\Omega; L^\infty(0, T; H_0^2(\Gamma))) \cap L^2(\Omega; W^{1,\infty}(0, T; L^2(\Gamma)))$.

Remark 8.6.1 To be very precise, one must check that the stochastic approximate solutions are measurable, as random variables taking values in a given path space. The measurability of these stochastic processes is easy to see by using the measurability properties of the functions $\boldsymbol{u}_N^n$, $v_N^{n+\frac{i}{2}}$, and η_N^n. For example, η_N is measurable as a map from the probability space Ω to $L^\infty(0, T; H_0^2(\Gamma))$, because η_N can be considered as the composition of a measurable map F_1 with a continuous map F_2. F_1 is the map from $\omega \in \Omega$ to the space of bounded sequences of length N with values in $H_0^2(\Gamma)$, given by $F_1 : \omega \to (\eta_N^0, \eta_N^1, \ldots, \eta_N^{N-1})$, which is measurable by the measurability properties of each η_N^n. F_2 is the map from the space of bounded sequences of length N with values in $H_0^2(\Gamma)$ to $L^\infty(0, T; H_0^2(\Gamma))$, given by $F_2 : (\eta_N^0, \eta_N^1, \ldots, \eta_N^{N-1}) \to \sum_{k=0}^{N-1} \eta_N^k \cdot 1_{(k\Delta t,(k+1)\Delta t]}(t)$, which is continuous.

Finally, we establish uniform bounds on the approximate (linearly interpolated) fluid velocity and structure velocity $(\overline{\boldsymbol{u}}_N, \overline{v}_N)$ in terms of *equicontinuity*, as these estimates will be useful for compactness results in the limit as $N \to \infty$. For this, we recall the fixed-in-time spatial test space $\mathcal{Q}$, defined in (8.20). We can then denote the *dual space* of the function space $\mathcal{Q}$ by $\mathcal{Q}'$, where we define the dual norm as usual by testing against test functions in $\mathcal{Q}$:

$$\|(\overline{\boldsymbol{u}}_N(t), \overline{v}_N(t))\|_{\mathcal{Q}'} = \sup_{\|(\boldsymbol{q},\psi)\|_{\mathcal{Q}} \le 1} \left| \int_{\mathcal{O}_f} \overline{\boldsymbol{u}}_N(t) \cdot \boldsymbol{q} + \int_\Gamma \overline{v}_N(t) \cdot \psi \right|.$$

We will be able to use the semidiscrete formulation to derive an estimate for the increments of the approximate solutions $(\overline{\boldsymbol{u}}_N, \overline{v}_N)$, which will be useful for obtaining uniform estimates on the fractional time derivatives of these approximate solutions.

Proposition 8.6.4 *For all $1 \le p < \infty$, there exists a constant c_p depending only on p and independent of N, s, and t such that:*

$$\mathbb{E}\left(\|(\overline{\boldsymbol{u}}_N(t) - \overline{\boldsymbol{u}}_N(s), \overline{v}_N(t) - \overline{v}_N(s))\|_{\mathcal{Q}'}^p \right) \le c_p |t - s|^{p/2}, \tag{8.50}$$

for all $0 \le s \le t \le T$.

Proof To establish this estimate on the time increments, we use the weak semidiscrete formulation (8.39) for the approximate solutions.

Step 1. Verification of increment estimate (8.50) **for** $s = n\Delta t$, $t = (n + m)\Delta t$. Let $s = n\Delta t$ and $t = (n + m)\Delta t$ for a positive integer m. Our goal is to verify the increment estimate (8.50) in this case.

Summing the semidiscrete formulation from time step n to time step $n + m - 1$, we obtain:

$$\int_{\mathcal{O}_f} (\boldsymbol{u}_N^{n+m} - \boldsymbol{u}_N^n) \cdot \boldsymbol{q} + \int_\Gamma (v_N^{n+m} - v_N^n)\psi$$

$$= -(\Delta t) \sum_{k=0}^{m-1} \left(2\mu \int_{\mathcal{O}_f} \boldsymbol{D}(\boldsymbol{u}_N^{n+k+1}) : \boldsymbol{D}(\boldsymbol{q}) + \int_\Gamma \partial_z \eta_N^{n+k+1} \partial_z \psi \right.$$
$$\left. + \int_\Gamma \partial_z^2 \eta_N^{n+k+1} \partial_z^2 \psi \right)$$
$$+ \sum_{k=0}^{m-1} \int_{\mathcal{O}_f} \left(\boldsymbol{f}(\boldsymbol{u}_N^{n+k+\frac{1}{2}}, v_N^{n+k+\frac{1}{2}}, \eta_N^{n+k+\frac{1}{2}}) \cdot \boldsymbol{q} \right) \Delta W_{1,N}^{n+k}$$
$$+ \sum_{k=0}^{m-1} \int_\Gamma \left(g(\boldsymbol{u}_N^{n+k+\frac{1}{2}}, v_N^{n+k+\frac{1}{2}}, \eta_N^{n+k+\frac{1}{2}}) \psi \right) \Delta W_{2,N}^{n+k}$$
$$+ (\Delta t) \sum_{k=0}^{m-1} P_{N,in}^{n+k} \left(\int_0^R (q_z)|_{z=0} \right) - (\Delta t) \sum_{k=0}^{m-1} P_{N,out}^{n+k} \left(\int_0^R (q_z)|_{z=L} dr \right),$$

where $\Delta W_{i,N}^{n+k} = W_i((n+k+1)\Delta t) - W_i((n+k)\Delta t)$ for $i = 1, 2$. Because we are considering s and t that are multiples of Δt, we can rewrite the terms involving Brownian increments as stochastic integrals, and in general, we can rewrite the above expression as a continuous time integral:

$$\int_{\mathcal{O}_f} (\boldsymbol{u}_N^{n+m} - \boldsymbol{u}_N^n) \cdot \boldsymbol{q} - \int_\Gamma (v_N^{n+m} - v_N^n)\psi$$

$$= -2\mu \int_{n\Delta t}^{(n+m)\Delta t} \int_{\mathcal{O}_f} \boldsymbol{D}(\boldsymbol{u}_N^{\Delta t}) : \boldsymbol{D}(\boldsymbol{q}) - \int_{n\Delta t}^{(n+m)\Delta t} \int_\Gamma \partial_z \eta_N^{\Delta t} \partial_z \psi$$
$$- \int_{n\Delta t}^{(n+m)\Delta t} \int_\Gamma \partial_z^2 \eta_N^{\Delta t} \partial_z^2 \psi$$
$$+ \int_{n\Delta t}^{(n+m)\Delta t} \left(\int_{\mathcal{O}_f} \boldsymbol{f}(\boldsymbol{u}_N, v_N^*, \eta_N^{\Delta t}) \cdot \boldsymbol{q} \right) dW_1(t)$$
$$+ \int_{n\Delta t}^{(n+m)\Delta t} \left(\int_\Gamma g(\boldsymbol{u}_N, v_N^*, \eta_N^{\Delta t}) \psi \right) dW_2(t)$$
$$+ (\Delta t) \sum_{k=0}^{m-1} P_{N,in}^{n+k} \left(\int_0^R (q_z)|_{z=0} \right) - (\Delta t) \sum_{k=0}^{m-1} P_{N,out}^{n+k} \left(\int_0^R (q_z)|_{z=L} dr \right).$$

Taking the supremum over $(\boldsymbol{q}, \psi) \in \mathcal{Q}$ with $\|(\boldsymbol{q}, \psi)\|_{\mathcal{Q}} \leq 1$ (so that $\|\boldsymbol{q}\|_{H^1(\mathcal{O}_f)} \leq 1$ and $\|\psi\|_{H_0^2(\Gamma)} \leq 1$), we obtain using the noise intensity assumptions (8.6) and (8.7) that:

$$\|(\overline{\boldsymbol{u}}_N((n+m)\Delta t)-\overline{\boldsymbol{u}}_N(n\Delta t),\overline{v}_N((n+m)\Delta t)-\overline{v}_N(n\Delta t))\|_{Q'}$$

$$\leq C\left(\int_{n\Delta t}^{(n+m)\Delta t}\|\boldsymbol{u}_N^{\Delta t}\|_{H^1(\mathcal{O}_f)}+\int_{n\Delta t}^{(n+m)\Delta t}\|\eta_N^{\Delta t}\|_{H_0^2(\Gamma)}\right.$$

$$+\sum_{i=1}^{2}\left|\int_{n\Delta t}^{(n+m)\Delta t}(1+\|\boldsymbol{u}_N\|_{L^2(\mathcal{O}_f)}+\|v_N^*\|_{L^2(\Gamma)}+\|\eta_N^{\Delta t}\|_{H_0^2(\Gamma)})dW_i(t)\right|$$

$$\left.+(\Delta t)\left|\sum_{k=n}^{n+m-1}P_{N,in}^k\right|+(\Delta t)\left|\sum_{k=n}^{n+m-1}P_{N,out}^k\right|\right).$$

We can estimate using the definition $P_{N,in/out}^n=\frac{1}{\Delta t}\int_{n\Delta t}^{(n+1)\Delta t}P_{in/out}(t)dt$ of the discretized pressure that

$$(\Delta t)\left|\sum_{k=n}^{n+m-1}P_{N,in/out}^k\right|=C\left|\int_{n\Delta t}^{(n+m)\Delta t}P_{in/out}(t)dt\right|$$

$$\leq C(m\Delta t)^{1/2}\|P_{in/out}\|_{L^2(0,T)}^2\leq C(m\Delta t)^{1/2},$$

where we absorbed the $L^2(0,T)$ norm of the given inlet/outlet pressure data into the uniform constant C. Hence, we obtain by using the Cauchy-Schwarz inequality:

$$\|(\overline{\boldsymbol{u}}_N((n+m)\Delta t)-\overline{\boldsymbol{u}}_N(n\Delta t),\overline{v}_N((n+m)\Delta t)-\overline{v}_N(n\Delta t)\|_{Q'}$$

$$\leq C(m\Delta t)^{1/2}\left(1+\|\boldsymbol{u}_N^{\Delta t}\|_{L^2(0,T;H^1(\mathcal{O}_f))}+\|\eta_N^{\Delta t}\|_{L^2(0,T;H_0^2(\Gamma))}\right)$$

$$+C\sum_{i=1}^{2}\left|\int_{n\Delta t}^{(n+m)\Delta t}(1+\|\boldsymbol{u}_N\|_{L^2(\mathcal{O}_f)}+\|v_N^*\|_{L^2(\Gamma)}+\|\eta_N^{\Delta t}\|_{H_0^2(\Gamma)})dW_i(t)\right| \tag{8.51}$$

By the BDG inequality (see Theorem 3.6.2), we can estimate for $i=1,2$:

$$\mathbb{E}\left|\int_{n\Delta t}^{(n+m)\Delta t}(1+\|\boldsymbol{u}_N\|_{L^2(\mathcal{O}_f)}+\|v_N^*\|_{L^2(\Gamma)}+\|\eta_N^{\Delta t}\|_{H_0^2(\Gamma)})dW_i(t)\right|^p$$

$$\leq c_p\mathbb{E}\left(\int_{n\Delta t}^{(n+m)\Delta t}(1+\|\boldsymbol{u}_N\|_{L^2(\mathcal{O}_f)}^2+\|v_N^*\|_{L^2(\Gamma)}^2+\|\eta_N^{\Delta t}\|_{H_0^2(\Gamma)}^2)dt\right)^{p/2}$$

$$\leq c_p(m\Delta t)^{p/2}\mathbb{E}\Big(1 + \|\boldsymbol{u}_N\|^2_{L^\infty(0,T;L^2(\mathcal{O}_f))} + \|v^*_N\|^2_{L^\infty(0,T;L^2(\Gamma))}$$
$$+ \|\eta^{\Delta t}_N\|^2_{L^\infty(0,T;H^2_0(\Gamma))}\Big)^{p/2}$$
$$\leq c_p(m\Delta t)^{p/2},$$

where we used the uniform bounds on the approximate solutions in Proposition 8.6.2. So by raising both sides of (8.51) to the p power and taking expectations, we obtain by using the uniform bounds in Proposition 8.6.2 that for $1 \leq p < \infty$:

$$\mathbb{E}\Big(\|(\overline{\boldsymbol{u}}_N((n+m)\Delta t) - \overline{\boldsymbol{u}}_N(n\Delta t), \overline{v}_N((n+m)\Delta t) - \overline{v}_N(n\Delta t)\|^p_{Q'}\Big) \leq c_p(m\Delta t)^{p/2}, \tag{8.52}$$

which establishes (8.50), for the case where the endpoint times s and t are multiples of Δt.

Step 2. Extension of increment estimate (8.50) **to arbitrary** $0 \leq s < t \leq T$. Now consider arbitrary $0 \leq s < t \leq T$, and let $n\Delta t := \lceil s/\Delta t\rceil\Delta t$, $(n+m)\Delta t := \lfloor t/\Delta t\rfloor\Delta t$ so that $0 \leq s \leq n\Delta t \leq (n+m)\Delta t \leq t \leq T$. Then, we estimate

$$\mathbb{E}\Big(\|(\overline{\boldsymbol{u}}_N(t) - \overline{\boldsymbol{u}}_N(s), \overline{v}_N(t) - \overline{v}_N(s)\|^p_{Q'}\Big) \leq c_p(I_1 + I_2 + I_3),$$

for increments I_1, I_2, I_3 on $[n\Delta t, (n+m)\Delta t]$, $[s, n\Delta t]$, and $[(n+m)\Delta t, t]$, respectively. By the previous calculation in Step 1 in (8.52):

$$I_1 := \mathbb{E}\Big(\|(\overline{\boldsymbol{u}}_N(t)((n+m)\Delta t) - \overline{\boldsymbol{u}}_N(t)(n\Delta t), \overline{v}_N((n+m)\Delta t) - \overline{v}(n\Delta t))\|^p_{Q'}\Big)$$
$$\leq c_p(m\Delta t)^{p/2} \leq c_p|t-s|^{p/2}.$$

Then, we can estimate the remaining increment on $[s, n\Delta t]$ by using the fact that $\overline{u}_N$ and $\overline{v}_N$ are defined as linear interpolations in (8.48), along with (8.52) applied to $(n-1)\Delta t$ and $n\Delta t$:

$$I_2 := \mathbb{E}\Big(\|(\overline{\boldsymbol{u}}_N(n\Delta t) - \overline{\boldsymbol{u}}(s), \overline{v}(n\Delta t) - \overline{v}(s))\|^p_{Q'}\Big)$$
$$= \left(\frac{n\Delta t - s}{\Delta t}\right)^p \mathbb{E}\Big(\|(\overline{\boldsymbol{u}}_N(t)(n\Delta t) - \overline{\boldsymbol{u}}((n-1)\Delta t), \overline{v}(n\Delta t) - \overline{v}((n-1)\Delta t))\|^p_{Q'}\Big)$$
$$\leq c_p\left(\frac{n\Delta t - s}{\Delta t}\right)^p (\Delta t)^{p/2} \leq c_p(n\Delta t - s)^{p/2} \leq c_p|t-s|^{p/2}$$

and similarly for I_3:

$$I_3 := \mathbb{E}\Big(\|(\overline{\boldsymbol{u}}_N(t) - \overline{\boldsymbol{u}}_N((n+m)\Delta t), \overline{v}_N(t) - \overline{v}_N((n+m)\Delta t))\|^p_{Q'}\Big) \leq c_p|t-s|^{p/2}.$$

Combining the estimates for I_1, I_2, and I_3 establishes the desired increment estimate (8.50) and hence completes the proof.

□

We then use this increment estimate to establish the following fractional time estimate. We use the framework of fractional Sobolev spaces in time, $W^{\alpha,p}(0,T;\mathcal{Q}')$, where we recall from Definition 2.5.3 that

$$\|(\overline{\boldsymbol{u}}_N,\overline{v}_N)\|_{W^{\alpha,p}(0,T;\mathcal{Q}')}$$
$$=\left(\|(\overline{\boldsymbol{u}}_N,\overline{v}_N)\|^p_{L^p(0,T;\mathcal{Q}')}+\int_0^T\int_0^T\frac{\|(\overline{\boldsymbol{u}}_N(t),\overline{v}_N(t))-(\overline{\boldsymbol{u}}_N(s),\overline{v}_N(s))\|^p_{\mathcal{Q}'}}{|t-s|^{1+\alpha p}}dsdt\right)^{1/p}$$

Proposition 8.6.5 *For any* $0<\alpha<1/2$ *and any* $1\leq p<\infty$, *there exists a uniform constant* $C_{\alpha,p}$ *depending only on* α *and* p, *such that*

$$\mathbb{E}\left(\|(\overline{\boldsymbol{u}}_N,\overline{v}_N)\|^p_{W^{\alpha,p}(0,T;\mathcal{Q}')}\right)\leq C_{\alpha,p}.$$

Proof We first verify that

$$\mathbb{E}\left(\|(\overline{\boldsymbol{u}}_N,\overline{v}_N)\|^p_{L^p(0,T;\mathcal{Q}')}\right)\leq C,$$

uniformly in N. We compute using the fact that $(\overline{\boldsymbol{u}}_N(0),\overline{v}_N(0))=(\boldsymbol{u}_0,v_0)$ for all N almost surely and using the increment estimate (8.50), that

$$\mathbb{E}\left(\|(\overline{\boldsymbol{u}}_N,\overline{v}_N)\|^p_{L^p(0,T;\mathcal{Q}')}\right)=\mathbb{E}\int_0^T\|(\overline{\boldsymbol{u}}_N(t),\overline{v}_N(t))\|^p_{\mathcal{Q}'}dt$$
$$\leq C_p\left(\int_0^T\|(\boldsymbol{u}_0,v_0)\|^p_{\mathcal{Q}'}dt+\int_0^T t^{p/2}dt\right)\leq C_p,$$

where we recall that the initial condition $(\boldsymbol{u}_0,v_0)\in L^2(\mathcal{O}_f)\times L^2(\Gamma)\subset\mathcal{Q}'$ is deterministic. Next, we can estimate using (8.50) that

$$\mathbb{E}\int_0^T\int_0^T\frac{\|(\overline{\boldsymbol{u}}_N(t),\overline{v}_N(t))-(\overline{\boldsymbol{u}}_N(s),\overline{v}(s))\|^p_{\mathcal{Q}'}}{|t-s|^{1+\alpha p}}dsdt$$
$$=\int_0^T\int_0^T\frac{\mathbb{E}\left(\|(\overline{\boldsymbol{u}}_N(t),\overline{v}_N(t))-(\overline{\boldsymbol{u}}_N(s),\overline{v}_N(s))\|^p_{\mathcal{Q}'}\right)}{|t-s|^{1+\alpha p}}dsdt$$
$$\leq C_p\int_0^T\int_0^T\frac{|t-s|^{p/2}}{|t-s|^{1+\alpha p}}dsdt=C_p\int_0^T\int_0^T|t-s|^{-1-\alpha p+\frac{p}{2}}dsdt\leq C_{\alpha,p},$$

due to the assumption that $0 < \alpha < 1/2$ so that $-\alpha p + \dfrac{p}{2} > 0$. This establishes the desired uniform bound. □

8.7 Passage to the Limit

We would like to show that our approximate solutions converge (potentially along a subsequence) in a certain sense to a weak solution of the original problem. While uniform boundedness results give weak convergence of the random approximate solutions in probabilistic function spaces, the equations for the fluid and the structure (8.10) and (8.11) involve multiplicative random noise in the form of the multiplicative functions $\boldsymbol{f}(\boldsymbol{u}, v, \eta)$ and $g(\boldsymbol{u}, v, \eta)$. This requires additional stronger convergences, which we will obtain using compactness arguments, in order to pass to the limit in the semidiscrete formulation.

Since we will have nontrivial multiplicative noise in this problem though, we must use compactness arguments to obtain stronger convergence results. We develop a mathematical framework in the context of stochastic fluid-structure interaction, for strengthening the convergence of the random approximate solutions from weak convergence in probabilistic function spaces to almost sure convergence, which is sufficiently strong to pass to the limit in the semidiscrete formulation (8.39). Starting from the semidiscrete formulation (8.39) and the uniform boundedness (in expectation) of the approximate solutions in Proposition 8.6.2, we will use a compactness argument, which will first imply the existence of a weakly convergent subsequence of the *probability measures* which describe the *laws* or equivalently, the *distributions* of the approximate solutions. From here, we will eventually be able to get to *almost sure* convergence of the stochastic approximate solutions themselves via a Skorokhod representation theorem argument.

We start by designing compactness arguments that will provide weak convergence of the probability measures describing the laws of our random approximate solutions.

8.7.1 Weak Convergence of Measures

We first show that along subsequences, the probability measures, or the laws describing the distributions of our stochastic approximate solutions constructed earlier, converge weakly to a limiting probability measure, as the time step $\Delta t \to 0$, or $N \to \infty$. For this purpose, we recall that we are given a probability space with complete filtration $(\Omega, \mathcal{F}, \{\mathcal{F}_t\}_{t\geq 0}, \mathbb{P})$, with independent one-dimensional Brownian motions $\{W_1(t)\}_{t\geq 0}$, $\{W_2(t)\}_{t\geq 0}$ with respect to the given filtration $\{\mathcal{F}_t\}_{t\geq 0}$. For each N, we define the probability measure (or the law) μ_N:

$$\mu_N = \mu_{\eta_N} \times \mu_{\eta_N^{\Delta t}} \times \mu_{\overline{\eta}_N} \times \mu_{\boldsymbol{u}_N} \times \mu_{\boldsymbol{u}_N^{\Delta t}} \times \mu_{v_N} \times \mu_{v_N^*} \times \mu_{(\overline{\boldsymbol{u}}_N,\overline{v}_N)} \times \mu_{W_1} \times \mu_{W_2}, \tag{8.53}$$

defined on the phase space $\mathcal{X}$:

$$\begin{aligned}\mathcal{X} := {} & [L^2(0,T;L^2(\Gamma)) \cap (L^2(0,T;H_0^2(\Gamma)),w)]^2 \times C(0,T;L^2(\Gamma)) \\ & \times L^2(0,T;L^2(\mathcal{O}_f)) \\ & \times \left(L^2(0,T;L^2(\mathcal{O}_f)) \cap (L^2(0,T;H^1(\mathcal{O}_f)),w)\right) \times [L^2(0,T;L^2(\Gamma))]^2 \\ & \times [C(0,T;\mathcal{Q}') \cap L^2(0,T;L^2(\mathcal{O}_f) \times L^2(\Gamma))] \times [C(0,T;\mathbb{R})]^2. \end{aligned} \tag{8.54}$$

Here, μ_{η_N} denotes the law of η_N on $L^2(0,T;L^2(\Gamma)) \cap (L^2(0,T;H_0^2(\Gamma)),w)$, $\mu_{(\overline{\boldsymbol{u}}_N,\overline{v}_N)}$ denotes the law of $(\overline{\boldsymbol{u}}_N,\overline{v}_N)$ on $C(0,T;\mathcal{Q}') \cap L^2(0,T;L^2(\mathcal{O}_f) \times L^2(\Gamma))$, μ_{W_i} denotes the law of W_i on $C(0,T;\mathbb{R})$, and so on. Therefore, μ_N describes the joint law of the collection of approximate solutions $(\eta_N, \overline{\eta}_N, \boldsymbol{u}_N, \boldsymbol{u}_N^{\Delta t}, v_N, v_N^*, \overline{\boldsymbol{u}}_N, \overline{v}_N, W_1, W_2)$. Furthermore, the notation (B,w) for a Banach space B denotes the Banach space with the weak topology of convergence. As we shall see below, it is easier to consider continuity in time when working with the fluid velocity and the structure velocity in pairs, as we can consider $(\overline{\boldsymbol{u}}_N,\overline{v}_N)$ in $C(0,T;\mathcal{Q}')$, where the dual space $\mathcal{Q}'$ of the test space $\mathcal{Q}$ involves kinematically coupled fluid and structure test functions. This is the reason why in (8.53) above, we consider $\mu_{(\overline{\boldsymbol{u}}_N,\overline{v}_N)}$. The main result of this subsection is the following.

Theorem 8.7.1 *Along a subsequence (which we will continue to denote by N), μ_N converges weakly as probability measures to a probability measure μ on $\mathcal{X}$.*

To show weak convergence of these probability measures along a subsequence, stated in Theorem 8.7.1, we must show that the probability measures are **tight**; see Definition 3.2.1.

To get a hold of the compact subset K_ϵ in Definition 3.2.1, we will need to apply deterministic compactness results for the structure displacements $\{\overline{\eta}_N(\omega)\}$ and for the fluid and structure velocities $\{(\overline{\boldsymbol{u}}_N(\omega), \overline{v}_N(\omega)\}$ and combine these compactness results with uniform estimates from the previous Sect. 8.6.2.

The first lemma, which will be applied to the structure displacements $\{\overline{\eta}_N(\omega)\}$, is the classical Aubin-Lions compactness lemma; see Theorem 2.9.4 which states that the following compact embedding holds:

$$[W^{1,\infty}(0,T;L^2(\Gamma)) \cap L^\infty(0,T;H_0^2(\Gamma))] \subset\subset C(0,T;L^2(\Gamma)).$$

To handle the compactness argument for the structure and fluid velocities, we use the fractional time derivative estimate in Proposition 8.6.5 and the following compact embedding, which is a consequence of Theorem 2.9.9. The following compact embedding holds for $0 < \alpha < 1/2$:

$$L^2(0,T;H^1(\mathcal{O}_f) \times H^{1/2}(\Gamma)) \cap W^{\alpha,2}(0,T;\mathcal{Q}') \subset\subset L^2(0,T;L^2(\mathcal{O}_f) \times L^2(\Gamma)),$$

due to the fact that

$$H^1(\mathcal{O}_f) \times H^{1/2}(\Gamma) \subset\subset L^2(\mathcal{O}_f) \times L^2(\Gamma) \subset \mathcal{Q}'.$$

Because we need the limiting process $(\boldsymbol{u}, v)$ to have continuity in time, which will allow us to deduce desirable measurable properties for the limiting solution such as predictability, we will also require the following compact embedding that will give us continuity in time of the limiting process in the dual of the test space, $\mathcal{Q}'$, which follows from Theorem 2.9.8, for $0 < \alpha < 1$, $p > 1$, and $\alpha p > 1$,

$$C(0, T; L^2(\mathcal{O}_f) \times L^2(\Gamma)) \cap W^{\alpha,p}(0, T; \mathcal{Q}') \subset\subset C(0, T; \mathcal{Q}').$$

Finally, we note that we have obtained compactness results only for the velocity approximate function $\overline{v}_N$, for example, and not v_N^*. In addition, when passing to the limit, we will consider the linear interpolations and time-shifted versions of the fluid velocity and of the structure displacement and velocity, since these appear in the semidiscrete formulation (8.39). We recall that the linear interpolations are piecewise linear functions defined by (8.45), (8.48), and the time-shifted functions are piecewise constant functions defined by (8.47), (8.49). To show all of the different "versions" of the approximate solutions converge to the same limit, we will need the following result, which relies fundamentally on the uniform *numerical dissipation estimates* in Proposition 8.5.5.

Lemma 8.7.1 *We have the following convergence results as $N \to \infty$:*

$$\mathbb{E}\left(\|v_N - v_N^*\|^2_{L^2(0,T;L^2(\Gamma))}\right) \to 0, \qquad \mathbb{E}\left(\|v_N - \overline{v}_N\|^2_{L^2(0,T;L^2(\Gamma))}\right) \to 0,$$

$$\mathbb{E}\left(\|v_N - v_N^{\Delta t}\|^2_{L^2(0,T;L^2(\Gamma))}\right) \to 0, \qquad \mathbb{E}\left(\|\boldsymbol{u}_N - \overline{\boldsymbol{u}}_N\|^2_{L^2(0,T;L^2(\mathcal{O}_f))}\right) \to 0,$$

$$\mathbb{E}\left(\|\boldsymbol{u}_N - \boldsymbol{u}_N^{\Delta t}\|^2_{L^2(0,T;L^2(\mathcal{O}_f))}\right) \to 0, \qquad \mathbb{E}\left(\|\eta_N^{\Delta t} - \overline{\eta}_N\|^2_{L^2(0,T;L^2(\Gamma))}\right) \to 0.$$

Proof To show, for example, the first convergence result, we multiply the uniform numerical dissipation estimate from Proposition 8.5.5 by Δt to obtain

$$\mathbb{E}\left(\Delta t \sum_{n=0}^{N-1} \|v_N^{n+\frac{1}{2}} - v_N^n\|^2_{L^2(\Gamma)}\right) = \mathbb{E}\left(\|v_N - v_N^*\|^2_{L^2(0,T;L^2(\Gamma))}\right) \le C(\Delta t) \to 0, \tag{8.55}$$

as $N \to \infty$. To show the remaining convergence results, we use Proposition 8.5.5 to conclude that there exists a uniform constant C independent of N such that

$$\sum_{n=0}^{N-1} \mathbb{E}\left(\|v_N^{n+1} - v_N^n\|^2_{L^2(\Gamma)}\right) \le C, \quad \sum_{n=0}^{N-1} \mathbb{E}\left(\|\boldsymbol{u}_N^{n+1} - \boldsymbol{u}_N^n\|^2_{L^2(\mathcal{O}_f)}\right) \le C,$$

$$\sum_{n=0}^{N-1}\left[\mathbb{E}\left(\|\partial_z\eta_N^{n+\frac{1}{2}}-\partial_z\eta_N^n\|^2_{L^2(\Gamma)}\right)+\mathbb{E}\left(\|\partial_z^2\eta_N^{n+\frac{1}{2}}-\partial_z^2\eta_N^n\|^2_{L^2(\Gamma)}\right)\right]\le C,$$

where we recall that $\boldsymbol{u}_N^{n+\frac{1}{2}}=\boldsymbol{u}_N^n$ and where we used the triangle inequality to obtain the first estimate. Then, the same argument as above gives the desired result, once we note that as $N\to\infty$,

$$\mathbb{E}\left(\|\overline{v}_N-v_N\|^2_{L^2(0,T;L^2(\Gamma))}\right)\le(\Delta t)\sum_{n=0}^{N-1}\mathbb{E}\left(\|v_N^{n+1}-v_N^n\|^2_{L^2(\Gamma)}\right)\le C(\Delta t)\to 0, \tag{8.56}$$

$$\mathbb{E}\left(\|v_N^{\Delta t}-v_N\|^2_{L^2(0,T;L^2(\Gamma))}\right)=(\Delta t)\sum_{n=0}^{N-1}\mathbb{E}\left(\|v_N^{n+1}-v_N^n\|^2_{L^2(\Gamma)}\right)\le C(\Delta t)\to 0, \tag{8.57}$$

$$\mathbb{E}\left(\|\overline{\boldsymbol{u}}_N-\boldsymbol{u}_N\|^2_{L^2(0,T;L^2(\mathcal{O}_f))}\right)\le(\Delta t)\sum_{n=0}^{N-1}\mathbb{E}\left(\|\boldsymbol{u}_N^{n+1}-\boldsymbol{u}_N^n\|^2_{L^2(\mathcal{O}_f)}\right)\le C(\Delta t)\to 0, \tag{8.58}$$

$$\mathbb{E}\left(\|\boldsymbol{u}_N^{\Delta t}-\boldsymbol{u}_N\|^2_{L^2(0,T;L^2(\mathcal{O}_f))}\right)=(\Delta t)\sum_{n=0}^{N-1}\mathbb{E}\left(\|\boldsymbol{u}_N^{n+1}-\boldsymbol{u}_N^n\|^2_{L^2(\mathcal{O}_f)}\right)\le C(\Delta t)\to 0, \tag{8.59}$$

$$\mathbb{E}\left(\|\eta_N^{\Delta t}-\overline{\eta}_N\|^2_{L^2(0,T;L^2(\Gamma))}\right)\le(\Delta t)\sum_{n=0}^{N-1}\mathbb{E}\left(\|\eta_N^{n+\frac{1}{2}}-\eta_N^n\|^2_{H_0^2(\Gamma)}\right)\le C(\Delta t)\to 0, \tag{8.60}$$

where we used Poincaré's inequality to estimate the $H_0^2(\Gamma)$ norm of $\eta_N^{n+\frac{1}{2}}-\eta_N^n$ in terms of the $L^2(\Gamma)$ norms of its first and second partial derivatives with respect to z. □

Notice that this result follows from the numerical dissipation estimates in Proposition 8.5.5, which imply convergence to zero in expectation, of the numerical dissipation terms, shown in (8.56), (8.57), (8.58), (8.59), and (8.60), from which we were able to deduce the almost sure convergence. This result in particular shows that even though, for example, the approximate structure velocity on $[n\Delta t,(n+1)\Delta t]$ can be defined in several different ways using, for example, v_N^n, $v_N^{n+\frac{1}{2}}$, or v_N^{n+1}, the different versions of the approximate solutions all converge to the same limit as $N\to\infty$ almost surely.

We now have all of the components needed to show the weak convergence of the limiting laws μ_N defined in (8.53) of the approximate solutions in the phase space $\mathcal{X}$ defined in (8.54).

Proof of Theorem 8.7.1 To show weak convergence of probability measures along a subsequence, we must show that the probability measures μ_N are tight, see Definition 3.2.1.

Step 1: Tightness of $\mu_{\overline{\eta}_N}$. We begin by showing the tightness of $\mu_{\overline{\eta}_N}$. We recall the following compact embedding due to the Aubin-Lions theorem (Theorem 2.9.4):

$$W^{1,\infty}(0, T; L^2(\Gamma)) \cap L^\infty(0, T; H_0^2(\Gamma)) \subset\subset C(0, T; L^2(\Gamma)),$$

and we recall from Proposition 8.6.2 the following uniform bounds:

$$\mathbb{E}\left(\|\overline{\eta}_N\|_{W^{1,\infty}(0,T;L^2(\Gamma))}\right) \leq C, \qquad \mathbb{E}\left(\|\overline{\eta}_N\|_{L^\infty(0,T;H_0^2(\Gamma))}\right) \leq C.$$

Hence, if we define the following subset of $C(0, T; L^2(\Gamma))$:

$$A_R = \{\overline{\eta} \in C(0, T; L^2(\Gamma)) : \|\overline{\eta}\|_{W^{1,\infty}(0,T;L^2(\Gamma))} \leq R, \|\overline{\eta}\|_{L^\infty(0,T;H_0^2(\Gamma))} \leq R\},$$

then $\overline{A_R}$ is a compact set in $C(0, T; L^2(\Gamma))$ for each $R > 0$, where the closure is with respect to the norm of $C(0, T; L^2(\Gamma))$. So by Chebychev's inequality and the previous uniform boundedness results, we have that for an arbitrary $\epsilon > 0$,

$$\mu_{\overline{\eta}_N}(\overline{A_R}) > 1 - \epsilon,$$

if R is chosen sufficiently large, which establishes the desired tightness of $\mu_{\overline{\eta}_N}$ as measures on $C(0, T; L^2(\Gamma))$.

Tightness of $\mu_{(\overline{u}_N, \overline{v}_N)}$. To show the tightness of $\mu_{(\overline{u}_N, \overline{v}_N)}$, we recall the following compact embeddings from Theorem 2.9.6 and Theorem 2.9.8 for $0 < \alpha < 1/2$:

$$L^2(0, T; H^1(\mathcal{O}_f) \times H^{1/2}(\Gamma)) \cap W^{\alpha,2}(0, T; \mathcal{Q}') \subset\subset L^2(0, T; L^2(\mathcal{O}_f) \times L^2(\Gamma)),$$

$$C(0, T; L^2(\mathcal{O}_f) \times L^2(\Gamma)) \cap W^{\alpha,p}(0, T; \mathcal{Q}') \subset\subset C(0, T; \mathcal{Q}') \quad \text{for } \alpha p > 1.$$

As a technical remark, since the initial data $\boldsymbol{u}_0$ is only in $L^2(\mathcal{O}_f)$, we do not necessarily have that $(\overline{\boldsymbol{u}}_N, \overline{v}_N) \in L^2(0, T; H^1(\mathcal{O}_f) \times H^{1/2}(\Gamma))$ almost surely, and we can instead consider the time shifts $T_{-\Delta t}(\overline{\boldsymbol{u}}_N, \overline{v}_N)(t) = (\overline{\boldsymbol{u}}_N, \overline{v}_N)(t+\Delta t)$, which are extended to be equal to $(\boldsymbol{u}_N^N, v_N^N)$ on the interval $[(N-1)\Delta t, N\Delta t]$. Then, it is easy to see that by the uniform boundedness results from Proposition 8.6.2 and Proposition 8.6.5:

$$\mathbb{E}\|T_{-\Delta t}(\overline{\boldsymbol{u}}_N, \overline{v}_N)\|_{L^2(0,T;H^1(\mathcal{O}_f)\times H^{1/2}(\Gamma))} \leq C,$$

$$\mathbb{E}\|T_{-\Delta t}(\overline{\boldsymbol{u}}_N, \overline{v}_N)\|_{L^\infty(0,T;L^2(\mathcal{O}_f)\times L^2(\Gamma))} \le C,$$

$$\mathbb{E}\|T_{-\Delta t}(\overline{\boldsymbol{u}}_N, \overline{v}_N)\|_{W^{\alpha,p}(0,T;\mathcal{Q}')} \le C \quad \text{for } 0 < \alpha < 1/2 \text{ and } 1 \le p < \infty.$$

We can then define a subset of $C(0, T; \mathcal{Q}') \cap L^2(0, T; L^2(\mathcal{O}_f) \times L^2(\Gamma))$:

$$\begin{aligned} B_R := \{(\boldsymbol{u}, v) \in C(0, T; \mathcal{Q}') \cap L^2(0, T; L^2(\mathcal{O}_f) \times L^2(\Gamma)) \\ : \|(\boldsymbol{u}, v)\|_{L^2(0,T;H^1(\mathcal{O}_f)\times H^{1/2}(\Gamma))} \le R, \|(\boldsymbol{u}, v)\|_{L^\infty(0,T;L^2(\mathcal{O}_f)\times L^2(\Gamma))} \le R, \\ \|(\boldsymbol{u}, v)\|_{W^{\alpha,2}(0,T;\mathcal{Q}')} \le R, \|(\boldsymbol{u}, v)\|_{W^{\alpha,p}(0,T;\mathcal{Q}')} \le R\}, \end{aligned}$$

for some fixed $\alpha \in (0, 1/2)$ and $1 \le p < \infty$ such that $\alpha p > 1$. We note that $\overline{B_R}$ is a compact subset of $\mathcal{X}_{(\boldsymbol{u},v)} := C(0, T; \mathcal{Q}') \cap L^2(0, T; L^2(\mathcal{O}_f) \times L^2(\Gamma))$, where the closure is with respect to the topology of $\mathcal{X}_{(\boldsymbol{u},v)}$. Then, by Chebychev's inequality, we have that for any $\epsilon > 0$, we can find R sufficiently large such that

$$\mu_{T_{-\Delta t}(\overline{\boldsymbol{u}}_N,\overline{v}_N)}(\overline{B_R}) > 1 - \epsilon,$$

which establishes the desired tightness of the laws of $T_{-\Delta t}(\overline{\boldsymbol{u}}_N, \overline{v}_N)$ in $\mathcal{X}_{(\boldsymbol{u},v)} := C(0, T; \mathcal{Q}') \cap L^2(0, T; L^2(\mathcal{O}_f) \times L^2(\Gamma))$. The tightness of the laws of $(\overline{\boldsymbol{u}}_N, \overline{v}_N)$ (without the time shift) in the same space follows from Proposition 3.2.2 and the fact that $\|T_{-\Delta t}(\overline{\boldsymbol{u}}_N, v_N) - (\overline{\boldsymbol{u}}_N, \overline{v}_N)\|_{\mathcal{X}_{(\boldsymbol{u},v)}} \to 0$ as $N \to \infty$ almost surely, by the numerical dissipation estimates in Lemma 8.7.1 and the continuity estimate in Proposition 8.6.4.

Step 3: Tightness of μ_{η_N}, $\mu_{\eta_N^{\Delta t}}$, $\mu_{(\boldsymbol{u}_N,v_N)}$, $\mu_{(\boldsymbol{u}_N,v_N^*)}$, $\mu_{(\boldsymbol{u}_N^{\Delta t},v_N^{\Delta t})}$. The arguments from Step 1 and Step 2 show that the measures $\mu_{\overline{\eta}_N}$ and $\mu_{(\overline{\boldsymbol{u}}_N,\overline{v}_N)}$ are tight in $L^2(0, T; L^2(\Gamma))$ and $L^2(0, T; L^2(\mathcal{O}_f)) \times L^2(0, T; L^2(\Gamma))$, respectively. Then, recalling the result of Lemma 8.7.1 and recalling that convergence to zero in expectation implies convergence to zero almost surely along a subsequence, we have that, for example,

$$\|\eta_N^{\Delta t} - \overline{\eta}_N\|_{L^2(0,T;L^2(\Gamma))} \to 0, \quad \|(\boldsymbol{u}_N, v_N) - (\overline{\boldsymbol{u}}_N, \overline{v}_N)\|_{L^2(0,T;L^2(\mathcal{O}_f)\times L^2(\Gamma))} \to 0,$$

along a subsequence. So we conclude that the measures $\mu_{\eta_N^{\Delta t}}$ and $\mu_{(\boldsymbol{u}_N,v_N)}$ are tight as probability measures on $L^2(0, T; L^2(\Gamma))$ and $L^2(0, T; L^2(\mathcal{O}_f)) \times L^2(0, T; L^2(\Gamma))$ also, by Proposition 3.2.2. A similar argument will work for showing tightness of μ_{η_N}, $\mu_{(\boldsymbol{u}_N,v_N^*)}$ and $\mu_{(\boldsymbol{u}_N^{\Delta t},v_N^{\Delta t})}$ also.

Step 4: Tightness of full measures μ_N. Note that because the Brownian motions W_1 and W_2 in the approximate solutions are all the same independently of N, the measure $\mu_{W_1} \times \mu_{W_2}$ on $[C(0, T; \mathbb{R})]^2$ is trivially tight. Thus, we have shown in the previous steps that all of the marginals of the approximate laws μ_N are tight in the corresponding function space in the phase space $\mathcal{X}$.

Therefore, the joint laws are tight as probability measures on $\mathcal{X}$, and this hence establishes the desired tightness of the probability measures and completes the proof of Theorem 8.7.1. □

8.7.2 Skorohod Representation Theorem

We now use the Skorohod representation theorem (see Theorem 3.2.5) to translate weak convergence of probability measures to almost sure convergence of random variables, which will allow us to pass to the limit in the semidiscrete weak formulation. However, this will be at the expense of working on a different probability space. Namely, the Skorohod representation theorem provides the *existence* of *a probability space*, on which we will have almost sure convergence of new random variables with the same laws as the original approximate solutions, to a weak solution with the law μ from Theorem 8.7.1. This probability space is not necessarily the same as the original probability space on which our problem is posed. This leads naturally to the concept of a *weak martingale solution*, see Definition 8.3.1, on which we have a stochastic solution to the problem, potentially defined on a new probability space, on which the weak formulation holds almost surely on the potentially new probability space. In certain cases (such as the case of additive noise for our current problem), we can return to the original probability space by using another result, known as the Gyöngy-Krylov lemma in Theorem 3.2.4 (see Sect. 8.8.2) to show that along a subsequence, the original approximate solutions on the original probability space converge almost surely to a limit with the same law μ from Theorem 8.7.1. This gives rise to the stronger concept of a *weak pathwise solution*, one which is defined on the initial probability space and which satisfies the weak formulation on that initial probability space; see Definition 8.3.2.

More precisely, showing convergence of our approximate solutions almost surely to an analytically weak (stochastic) solution consists of two steps:

1. First, we use the **Skorohod representation theorem** to show that *there exists* a probability space, which we denote by "tilde," on which a sequence of random variables that are equal to our approximate solutions *in law* converges almost surely in $\mathcal{X}$ as $N \to \infty$, to a weak solution on the "tilde" probability space, where the law of this weak solution is equal to μ, obtained in Theorem 8.7.1. Thus, in this step, we prove the existence of a weak martingale solution potentially defined on a different probability space; see Definition 8.3.1, to the stochastic FSI problem with multiplicative noise.
2. Then, in step two, in the specific case when the noise is additive (so that the nonlinear functions f and g are identically equal to 1), we continue the analysis and strengthen the existence result by using the **Gyöngy-Krylov lemma** in Theorem 3.2.4. This allows us to transfer the weak martingale solution back to the original probability space, implying that we will have constructed a unique

weak *pathwise* solution; see Definition 8.3.2, of the original continuous problem. As a byproduct of this analysis, we will also be able to obtain *uniqueness* of the weak pathwise solution.

This will complete the existence proof, which is the main result of this chapter.

To achieve these goals, we first obtain almost sure convergence along a subsequence of approximate solutions on a "tilde" probability space using Skorohod's theorem. We note an important feature of the current problem that makes it advantageous for us to use a more general form of the Skorohod representation theorem, which we presented previously in Theorem 3.2.5. In particular, note that the path space $\mathcal{X}$ is not a separable Banach space, due to the presence of $(L^2(0,T;H^1(\mathcal{O}_f)),w)$ endowed with the weak topology. However, this phase space $\mathcal{X}$ is a Jakubowski space (see Sect. 3.2.4), so we must use a version of the Skorohod representation theorem that can handle random variables taking values in Jakubowski spaces.

Before we state the result, we introduce the notation " $=_d$" to denote random variables that are "equal in distribution," i.e., the random variables have the same laws as random variables taking values in the same given phase space $\mathcal{X}$. Namely, we will say that a random variable X is equal in distribution (or equal in law) to the random variable $\tilde{X}$ and denote

$$X =_d \tilde{X} \quad \text{if} \quad \mu_X = \mu_{\tilde{X}},$$

where μ_X, for example, is the probability measure on $\mathcal{X}$ describing the law of the random variable X on $\mathcal{X}$.

Recall again the definition of the laws μ_N corresponding to the approximate solutions (8.53), and the definition of the corresponding phase space $\mathcal{X}$ in (8.54).

Lemma 8.7.2 *Let μ denote the probability measure obtained as a weak limit of the measures μ_N from Theorem 8.7.1. Then, there exists a probability space $(\tilde{\Omega},\tilde{\mathcal{F}},\tilde{\mathbb{P}})$ and $\mathcal{X}$-valued random variables on $(\tilde{\Omega},\tilde{\mathcal{F}},\tilde{\mathbb{P}})$:*

$$\tilde{\mathcal{U}} := (\tilde{\eta}, \tilde{\eta}^{\Delta t}, \tilde{\overline{\eta}}, \tilde{\boldsymbol{u}}, \tilde{v}, \tilde{\boldsymbol{u}}, \tilde{v}^*, \tilde{\boldsymbol{u}}^{\Delta t}, \tilde{v}^{\Delta t}, \tilde{\overline{\boldsymbol{u}}}, \tilde{\overline{v}}, \tilde{W}_1, \tilde{W}_2),$$

and

$$\tilde{\mathcal{U}}_N := (\tilde{\eta}_N, \tilde{\eta}_N^{\Delta t}, \tilde{\overline{\eta}}_N, \tilde{\boldsymbol{u}}_N, \tilde{v}_N, \tilde{\boldsymbol{u}}_N, \tilde{v}_N^*, \tilde{\boldsymbol{u}}_N^{\Delta t}, \tilde{v}_N^{\Delta t}, \tilde{\overline{\boldsymbol{u}}}_N, \tilde{\overline{v}}_N, \tilde{W}_{1,N}, \tilde{W}_{2,N}), \textit{ for each } N,$$

such that

$$\tilde{\mathcal{U}}_N =_d (\eta_N, \eta_N^{\Delta t}, \overline{\eta}_N, \boldsymbol{u}_N, v_N, \boldsymbol{u}_N, v_N^*, \boldsymbol{u}_N^{\Delta t}, v_N^{\Delta t}, \overline{\boldsymbol{u}}_N, \overline{v}_N, W_1, W_2),$$

for all N, and

$$\tilde{\mathcal{U}}_N \to \tilde{\mathcal{U}}, \tag{8.61}$$

$\tilde{\mathbb{P}}$-almost surely in $\mathcal{X}$ (potentially along a subsequence which we continue to denote by N), as $N \to \infty$, where the law of $\tilde{\mathcal{U}}$ is equal to μ. Furthermore, the following properties hold the following:

1. $\tilde{\eta} = \tilde{\eta}^{\Delta t} = \tilde{\tilde{\eta}}$, $\tilde{\boldsymbol{u}} = \tilde{\boldsymbol{u}}^{\Delta t} = \tilde{\tilde{\boldsymbol{u}}}$, *and* $\tilde{v} = \tilde{v}^* = \tilde{v}^{\Delta t} = \tilde{\tilde{v}}$, *$\tilde{\mathbb{P}}$-almost surely.*
2. $\tilde{\eta}(0) = \eta_0$, *$\tilde{\mathbb{P}}$-almost surely.*
3. $\partial_t \tilde{\eta} = \tilde{v}$, *$\tilde{\mathbb{P}}$-almost surely.*
4. $\{\tilde{W}_1(t)\}_{t\geq 0}$ *and* $\{\tilde{W}_2(t)\}_{t\geq 0}$ *are Brownian motions with respect to the filtration* $\tilde{\mathcal{F}}_t$ *defined below. First we let,*

$$\mathcal{F}_t' = \sigma(\tilde{\eta}(s), \tilde{\boldsymbol{u}}(s), \tilde{v}(s), \tilde{W}_1(s), \tilde{W}_2(s) : 0 \leq s \leq t),$$

and then define the filtration $\tilde{\mathcal{F}}_t$ *as follows:*

$$\mathcal{N} := \{\mathcal{A} \in \tilde{\mathcal{F}} \mid \tilde{\mathbb{P}}(\mathcal{A}) = 0\}, \qquad \tilde{\mathcal{F}}_t^0 := \sigma(\mathcal{F}_t' \cup \mathcal{N}), \qquad \tilde{\mathcal{F}}_t := \bigcap_{s\geq t} \tilde{\mathcal{F}}_s^0. \tag{8.62}$$

5. $(\tilde{\boldsymbol{u}}, \tilde{v})$ *and* $\tilde{\eta}$ *are predictable processes with respect to the filtration* $\{\tilde{\mathcal{F}}_t\}_{0\leq t\leq T}$ *as stochastic processes taking values in* $\mathcal{Q}'$.

Proof The existence of the probability space $(\tilde{\Omega}, \tilde{\mathcal{F}}, \tilde{\mathbb{P}})$ and the given random variables follows from the previous result on weak convergence in Theorem 8.7.1 and the Skorohod representation theorem given in Theorem 3.2.5. So it suffices to prove the given properties.

Property 1. By using equivalence of laws, we can transfer the convergences from Lemma 8.7.1 to obtain, for example:

$$\lim_{N\to\infty} \mathbb{E}\left(\|\tilde{\eta}_N - \tilde{\tilde{\eta}}_N\|^2_{L^2(0,T;L^2(\Gamma))}\right) = 0.$$

Because $\tilde{\eta}_N \to \tilde{\eta}$ and $\tilde{\tilde{\eta}}_N \to \tilde{\tilde{\eta}}$ in $L^2(0, T; L^2(\Gamma))$ by the $\tilde{\mathbb{P}}$-almost sure convergence of $\tilde{\mathcal{U}}_N$ to $\tilde{\mathcal{U}}$, we therefore conclude that $\tilde{\eta} = \tilde{\tilde{\eta}}$, $\tilde{\mathbb{P}}$-almost surely, as elements of $L^2(0, T; L^2(\Gamma))$.

Property 2. Note that $\overline{\eta}_N(0) = \eta_0$ for all N on the original probability space so since the phase space $\mathcal{X}_\eta$ for the structure displacement includes $C(0, T; L^2(\Gamma))$, we have by equivalence of laws that $\tilde{\overline{\eta}}_N(0) = \eta_0$ for all N. Then, because $\tilde{\overline{\eta}}_N \to \tilde{\eta}$ in $\mathcal{X}$ and hence in $C(0, T; L^2(\Gamma))$ $\tilde{\mathbb{P}}$-almost surely, we conclude that $\tilde{\eta}(0) = \eta_0$.

Property 3. To prove this property, we recall from the second equation in the semidiscrete formulation (8.39) that

$$\int_\Gamma \frac{\eta_N^{n+1} - \eta_N^n}{\Delta t} \phi dz = \int_\Gamma v_N^{n+\frac{1}{2}} \phi dz,$$

almost surely for all deterministic $\phi \in L^2(\Gamma)$. Integrating in time, we obtain for all N that

$$\int_0^T \int_\Gamma \partial_t \overline{\eta}_N \phi dz dt = \int_0^T \int_\Gamma v_N^* \phi dz dt, \qquad \text{for all } \phi \in C^1([0,T); L^2(\Gamma)),$$

almost surely, where we emphasize that ϕ is deterministic. Because each $\overline{\eta}_N$ is almost surely a piecewise linear continuous function satisfying $\overline{\eta}_N(0) = \eta_0$, we obtain by integration by parts that almost surely, for all $\phi \in C^1([0,T); L^2(\Gamma))$,

$$-\eta_0 \cdot \phi(0) - \int_0^T \int_\Gamma \overline{\eta}_N \partial_t \phi dz dt = \int_0^T \int_\Gamma v_N^* \phi dz dt,$$

and hence, by equivalence of laws,

$$-\eta_0 \cdot \phi(0) - \int_0^T \int_\Gamma \tilde{\overline{\eta}}_N \partial_t \phi dz dt = \int_0^T \int_\Gamma \tilde{v}_N^* \phi dz dt.$$

Passing to the limit and using the convergences $\tilde{\overline{\eta}} \to \tilde{\eta}$ in $C(0,T; L^2(\Gamma))$ and $\tilde{v}_N^* \to \tilde{v}$ in $L^2(0,T; L^2(\Gamma))$, $\tilde{\mathbb{P}}$-almost surely from (8.61), we obtain

$$-\eta_0 \cdot \phi(0) - \int_0^T \int_\Gamma \tilde{\eta} \partial_t \phi dz dt = \int_0^T \int_\Gamma \tilde{v} \phi dz dt,$$

for all deterministic $\phi \in C^1([0,T); L^2(\Gamma))$, $\tilde{\mathbb{P}}$-almost surely. This implies that $\partial_t \tilde{\eta} = \tilde{v}$ holds $\tilde{\mathbb{P}}$-almost surely for the limiting solution, since we showed in Property 2 that $\tilde{\eta}(0) = \eta_0$ almost surely.

Property 4. Let us consider $\tilde{W}_1$, since the same argument will hold for $\tilde{W}_2$. We want to show that $\tilde{W}_1$ is a Brownian motion with respect to the filtration $\tilde{\mathcal{F}}_t$ defined in (8.62). To show this, we note that $\tilde{W}_1(0) = 0$, $\tilde{\mathbb{P}}$-almost surely and $\tilde{W}_1(t) - \tilde{W}_1(s) \sim N(0, t-s)$ for all $0 \le s \le t \le T$, since W_1 and $\tilde{W}_{1,N}$ have the same law in $C(0,T;\mathbb{R})$ for all N. Hence, since W_1 defined on the original probability space is a Brownian motion by assumption, we obtain that $\tilde{W}_{1,N}(0) = 0$, $\tilde{\mathbb{P}}$-almost surely and $\tilde{W}_{1,N}(t) - \tilde{W}_{1,N}(s) \sim N(0, t-s)$ on the new probability space. Thus, we obtain the corresponding properties for $\tilde{W}_1$, since $\tilde{W}_{1,N} \to \tilde{W}_1$ in $C(0,T;\mathbb{R})$, $\tilde{\mathbb{P}}$-almost surely. Furthermore, it is easy to see that $\tilde{W}_1(t)$ is adapted to the filtration $\tilde{\mathcal{F}}_t$ by the way the filtration is defined.

So it suffices to check the independent increments property, namely, that $\tilde{W}_1(t_2) - \tilde{W}_1(t_1)$ is independent of $\tilde{\eta}(s)$, $\tilde{\boldsymbol{u}}(s)$, $\tilde{v}(s)$, $\tilde{W}_1(s)$, and $\tilde{W}_2(s)$, for all $0 \le s \le t_1 \le t_2 \le T$. Since $(W_1, W_2) =_d (\tilde{W}_{1,N}, \tilde{W}_{2,N})$ on $[C(0,T;\mathbb{R})]^2$, it is easy to see that $\tilde{W}_{1,N}(t_2) - \tilde{W}_{1,N}(t_1)$ is independent of $\tilde{W}_{1,N}(s)$ and $\tilde{W}_{2,N}(s)$, since this holds for W_1 and W_2 on the original probability space. We thus obtain the desired independence by using the $\tilde{\mathbb{P}}$-almost sure convergences of $\tilde{W}_{1,N}$ and $\tilde{W}_{2,N}$ to $\tilde{W}_1$ and $\tilde{W}_2$, respectively, in $C(0,T;\mathbb{R})$, along with Lemma 3.2.2.

To verify the independence of $\tilde{W}_1(t_2) - \tilde{W}_1(t_1)$ and the random variable $\tilde{\eta}(s)$, we recall that on the original probability space, W_1 is an $\{\mathcal{F}_t\}_{t\geq 0}$-Brownian motion for an initially given filtration $\{\mathcal{F}_t\}_{t\geq 0}$, and the approximate solution η_N is adapted to this filtration $\{\mathcal{F}_t\}$ for all N, by the measurability properties of approximate solutions constructed by the splitting scheme. Therefore, on the original probability space, $W_1(t_2) - W_1(t_1)$ is independent of $\eta_N(s)$ for all N (see the measurability properties described in Proposition 8.5.1), so by equivalence of laws, $\tilde{W}_{1,N}(t_2) - \tilde{W}_{1,N}(t_1)$ is independent of $\tilde{\eta}_N(s)$ for all N. So since $\tilde{W}_{1,N} \to \tilde{W}_1$ in $C(0, T; \mathbb{R})$ and $\tilde{\eta}_N \to \tilde{\eta}$ in $C(0, T; L^2(\Gamma))$, $\tilde{\mathbb{P}}$-almost surely, we use Lemma 3.2.2 to conclude that $\tilde{W}_1(t_2) - \tilde{W}_1(t_1)$ is independent of $\tilde{\eta}(s)$ for $0 \leq s \leq t_1 \leq t_2 \leq T$, and a similar argument works for $\tilde{\boldsymbol{u}}(s)$ and $\tilde{v}(s)$ also using the convergence $(\tilde{\bar{\boldsymbol{u}}}_N, \tilde{\bar{v}}_N) \to (\tilde{\boldsymbol{u}}, \tilde{v})$ in $C(0, T; \mathcal{Q}')$, $\tilde{\mathbb{P}}$-almost surely. Furthermore, these remarks also imply that $\{\tilde{W}_{1,N}(t)\}_{t\geq 0}$ and $\{\tilde{W}_{2,N}(t)\}_{t\geq 0}$ are Brownian motions with respect to the filtrations $\tilde{\mathcal{F}}^N_t$ defined below. First we let,

$$\mathcal{F}^{N'}_t := \sigma\{\tilde{\eta}_N(s), \tilde{\boldsymbol{u}}_N(s), \tilde{v}_N(s), \tilde{W}_{1,N}(s), \tilde{W}_{2,N}(s) : 0 \leq s \leq t\},$$

and we define

$$\mathcal{N} := \{\mathcal{A} \in \tilde{\mathcal{F}} \mid \tilde{\mathbb{P}}(\mathcal{A}) = 0\}, \qquad \tilde{\mathcal{F}}^0_t := \sigma(\mathcal{F}^{N'}_t \cup \mathcal{N}), \qquad \tilde{\mathcal{F}}^N_t := \bigcap_{s\geq t} \tilde{\mathcal{F}}^0_s. \tag{8.63}$$

also for all N.

Property 5. By the definition of $\{\tilde{\mathcal{F}}_t\}_{t\geq 0}$, the process $(\tilde{\boldsymbol{u}}, \tilde{v}, \tilde{\eta})$ is adapted to $\{\tilde{\mathcal{F}}_t\}_{t\geq 0}$. Note that by the definition of the phase space $\mathcal{X}$ in (8.54), $\tilde{\eta}$ has continuous paths on $[0, T]$, taking values in $L^2(\Gamma)$ and $(\tilde{\boldsymbol{u}}, \tilde{v})$ has continuous paths on $[0, T]$, taking values in $\mathcal{Q}'$, $\tilde{\mathbb{P}}$-almost surely. Since a continuous adapted process is predictable (see Proposition 3.3.3), this establishes the desired property.

This completes the proof of Lemma 8.7.2. □

8.7.3 Passing to the Limit

We now consider the approximate solutions defined as random variables on the probability space $(\tilde{\Omega}, \tilde{\mathcal{F}}, \tilde{\mathbb{P}})$, discussed in Lemma 8.7.2, and show that the almost sure limit obtained in Lemma 8.7.2, satisfies the weak formulation stated in Definition 8.3.1, almost surely on $(\tilde{\Omega}, \tilde{\mathcal{F}}, \tilde{\mathbb{P}})$. For this purpose, we recall the semidiscrete formulation of the problem from (8.39), given by

$$\int_{\mathcal{O}_f} \frac{\boldsymbol{u}_N^{n+1} - \boldsymbol{u}_N^n}{\Delta t} \cdot \boldsymbol{q} d\boldsymbol{x} + 2\mu \int_{\mathcal{O}_f} \boldsymbol{D}(\boldsymbol{u}_N^{n+1}) : \boldsymbol{D}(\boldsymbol{q}) d\boldsymbol{x} + \int_\Gamma \frac{v_N^{n+1} - v_N^n}{\Delta t} \psi dz$$
$$+ \int_\Gamma \partial_z \eta_N^{n+1} \partial_z \psi dz + \int_\Gamma \partial_z^2 \eta_N^{n+1} \partial_z^2 \psi dz$$
$$= \int_{\mathcal{O}_f} \frac{W_1((n+1)\Delta t) - W_1(n\Delta t)}{\Delta t} (\boldsymbol{f}(\boldsymbol{u}_N^{n+\frac{1}{2}}, v_N^{n+\frac{1}{2}}, \eta_N^{n+\frac{1}{2}}) \cdot \boldsymbol{q}) d\boldsymbol{x}$$
$$+ \int_\Gamma \frac{W_2((n+1)\Delta t) - W_2(n\Delta t)}{\Delta t} (g(\boldsymbol{u}_N^{n+\frac{1}{2}}, v_N^{n+\frac{1}{2}}, \eta_N^{n+\frac{1}{2}}) \psi) dz$$
$$+ P_{N,in}^n \int_0^R (q_z)|_{z=0} dr - P_{N,out}^n \int_0^R (q_z)|_{z=L} dr,$$
$$\forall (\boldsymbol{q}, \psi) \in \mathcal{Q},$$

$$\int_\Gamma \frac{\eta_N^{n+1} - \eta_N^n}{\Delta t} \phi dz = \int_\Gamma v_N^{n+\frac{1}{2}} \phi dz, \qquad \forall \phi \in L^2(\Gamma),$$

where $P_{N,in/out}^n = \frac{1}{\Delta t} \int_{n\Delta t}^{(n+1)\Delta t} P_{in/out}(t) dt$. Notice that as stated, this semidiscrete formulation refers to the original variables, defined on the original probability space. Given a general $(\boldsymbol{q}, \psi) \in \mathcal{Q}$, we use the semidiscrete formulation at each fixed time and integrate in time from 0 to T to obtain for all $t \in [0, T]$:

$$\int_{\mathcal{O}_f} \overline{\boldsymbol{u}}_N(t) \cdot \boldsymbol{q} d\boldsymbol{x} + \int_\Gamma \overline{v}_N(t) \psi dz + 2\mu \int_0^t \int_{\mathcal{O}_f} \boldsymbol{D}(\boldsymbol{u}_N^{\Delta t}) : \boldsymbol{D}(\boldsymbol{q}) d\boldsymbol{x} ds$$
$$+ \int_0^t \int_\Gamma \partial_z \eta_N^{\Delta t} \partial_z \psi dz ds + \int_0^t \int_\Gamma \partial_z^2 \eta_N^{\Delta t} \partial_z^2 \psi dz ds = \int_{\mathcal{O}_f} \boldsymbol{u}_0 \cdot \boldsymbol{q} d\boldsymbol{x} + \int_\Gamma v_0 \psi dz$$
$$+ \int_0^t \int_{\mathcal{O}_f} \left(\boldsymbol{f}(\boldsymbol{u}_N, v_N^*, \eta_N^{\Delta t}) \cdot \boldsymbol{q} \right) d\boldsymbol{x} dW_1 + \int_0^t \int_\Gamma \left(g(\boldsymbol{u}_N, v_N^*, \eta_N^{\Delta t}) \psi \right) dz dW_2$$
$$+ \int_0^t P_{in}^n(t) \left(\int_0^R (q_z)|_{z=0} dr \right) ds - \int_0^t P_{out}^n(t) \left(\int_0^R (q_z)|_{z=L} dr \right) ds$$
$$+ (E_N^{stoch,f}(t), \boldsymbol{q})_{L^2(\mathcal{O}_f)} + (E_N^{stoch,g}(t), \psi)_{L^2(\Gamma)}$$
$$+ E_{N,in}^P(t) \left(\int_0^R (q_z)|_{z=0} dr \right) - E_{N,out}^P(t) \left(\int_0^R (q_z)|_{z=L} dr \right),$$

$$\int_0^T \int_\Gamma \partial_t \overline{\eta}_N \phi dz dt = \int_0^T \int_\Gamma v_N^* \phi dz dt, \quad \forall \phi \in L^2(\Gamma),$$

where $\overline{\boldsymbol{u}}_N, \overline{v}_N$ and $\overline{\eta}_N$ are the piecewise linear approximations, given by (8.45) and (8.48), $\boldsymbol{u}_N^{\Delta t}$ and $\eta_N^{\Delta t}$ are the piecewise constant time shifted functions, given by (8.47) and (8.49), and v_N^* is defined by (8.44). Here, the terms $E_N^{stoch}(t)$, $E_{N,in}^P(t)$, and $E_{N,out}^P(t)$ are error terms that arise from considering $t \in [0, T]$ that are in the interior of subintervals (t_n, t_{n+1}), where this error arises from the fact that we are discretizing the Bronwnian increments $W_i((n+1)\Delta t) - W_i(n\Delta t)$ and the inlet/outlet pressure $P_{N,in/out}^n = \frac{1}{\Delta t}\int_{n\Delta t}^{(n+1)\Delta t} P_{in/out}(t)dt$ using the entire subinterval $[t_n, t_{n+1}]$. We can define the error terms as

$$
\begin{aligned}
E_N^{stoch,f} \\
:= \sum_{n=0}^{N-1} \Bigg(& \frac{t - n\Delta t}{\Delta t} (\boldsymbol{f}(\boldsymbol{u}_N^{n+\frac{1}{2}}, v_N^{n+\frac{1}{2}}, \eta_N^{n+\frac{1}{2}}) \Delta W_{1,N}^n \\
& - \int_{n\Delta t}^t \boldsymbol{f}(\boldsymbol{u}_N, v_N^*, \eta_N^{\Delta t}) dW_1 \Bigg) \mathbb{1}_{[n\Delta t, (n+1)\Delta t)}
\end{aligned} \tag{8.64}
$$

$$
\begin{aligned}
E_N^{stoch,g} \\
:= \sum_{n=0}^{N-1} \Bigg(& \frac{t - n\Delta t}{\Delta t} (g(\boldsymbol{u}_N^{n+\frac{1}{2}}, v_N^{n+\frac{1}{2}}, \eta_N^{n+\frac{1}{2}}) \Delta W_{2,N}^n \\
& - \int_{n\Delta t}^t g(\boldsymbol{u}_N, v_N^*, \eta_N^{\Delta t}) dW_2 \Bigg) \mathbb{1}_{[n\Delta t, (n+1)\Delta t)},
\end{aligned} \tag{8.65}
$$

$$
E_{N,in/out}^P := \sum_{n=0}^{N-1} \left((t - n\Delta t) P_{N,in/out}^n - \int_{n\Delta t}^t P_{in/out}(t) dt \right) \mathbb{1}_{[n\Delta t, (n+1)\Delta t)}, \tag{8.66}
$$

and we note that these error terms are identically zero for $t = n\Delta t$. The identity above holds as a result of the following observations:

$$
\sum_{k=0}^{n-1} \int_{\mathcal{O}_f} \left(\boldsymbol{f}(\boldsymbol{u}_N^{k+\frac{1}{2}}, v_N^{k+\frac{1}{2}}, \eta_N^{k+\frac{1}{2}}) \cdot \boldsymbol{q} \right) \Delta W_{1,N}^k = \int_0^{n\Delta t} \boldsymbol{f}(\boldsymbol{u}_N, v_N^*, \eta_N^{\Delta t}) dW_1(t),
$$

$$
\sum_{k=0}^{n-1} \int_{\mathcal{O}_f} \left(g(\boldsymbol{u}_N^{k+\frac{1}{2}}, v_N^{k+\frac{1}{2}}, \eta_N^{k+\frac{1}{2}}) \cdot \boldsymbol{q} \right) \Delta W_{2,N}^k = \int_0^{n\Delta t} g(\boldsymbol{u}_N, v_N^*, \eta_N^{\Delta t}) dW_2(t),
$$

$$
\sum_{k=0}^{n-1} (\Delta t) P_{in/out}^k = \int_0^{n\Delta t} P_{in/out}(t) dt.
$$

Now, we convert to the new probability space by noticing that the same identities hold for the new random variables defined on the "tilde" probability space, since the two sets of random variables have the same law on $\mathcal{X}$. So for all $(\boldsymbol{q}, \psi) \in \mathcal{Q}$, on the new probability space $(\tilde{\Omega}, \tilde{\mathcal{F}}, \tilde{\mathbb{P}})$ with the filtration $\{\tilde{\mathcal{F}}_t\}_{t\geq 0}$ defined in (8.62), we obtain

$$\int_{\mathcal{O}_f} \tilde{\overline{\boldsymbol{u}}}_N(t) \cdot \boldsymbol{q} d\boldsymbol{x} + \int_\Gamma \tilde{\overline{v}}_N(t)\psi dz + 2\mu \int_0^t \int_{\mathcal{O}_f} \boldsymbol{D}(\tilde{\boldsymbol{u}}_N^{\Delta t}) : \boldsymbol{D}(\boldsymbol{q}) d\boldsymbol{x} ds$$

$$+ \int_0^t \int_\Gamma \partial_z \tilde{\eta}_N^{\Delta t} \partial_z \psi dz ds + \int_0^t \int_\Gamma \partial_z^2 \tilde{\eta}_N^{\Delta t} \partial_z^2 \psi dz ds = \int_{\mathcal{O}_f} \boldsymbol{u}_0 \cdot \boldsymbol{q} d\boldsymbol{x} + \int_\Gamma v_0 \psi dz$$

$$+ \int_0^t \int_{\mathcal{O}_f} \left(\boldsymbol{f}(\tilde{\boldsymbol{u}}_N, \tilde{v}_N^*, \tilde{\eta}_N^{\Delta t}) \cdot \boldsymbol{q}\right) d\boldsymbol{x} d\tilde{W}_{1,N} + \int_0^t \int_\Gamma \left(g(\tilde{\boldsymbol{u}}_N, \tilde{v}_N^*, \tilde{\eta}_N^{\Delta t})\psi\right) dz d\tilde{W}_{2,N}$$

$$+ \int_0^t P_{in}^n(t) \left(\int_0^R (q_z)|_{z=0} dr\right) - \int_0^t P_{out}^n(t) \left(\int_0^R (q_z)|_{z=L} dr\right) ds$$

$$+ (\tilde{E}_N^{stoch,f}(t), \boldsymbol{q})_{L^2(\mathcal{O}_f)} + (\tilde{E}_N^{stoch,g}(t), \psi)_{L^2(\Gamma)}$$

$$+ E_{N,in}^P(t) \left(\int_0^R (q_z)|_{z=0} dr\right) - E_{N,out}^P(t) \left(\int_0^R (q_z)|_{z=L} dr\right),$$

$$\int_0^T \int_\Gamma \partial_t \tilde{\overline{\eta}}_N \phi dz dt = \int_0^T \int_\Gamma \tilde{v}_N^* \phi dz dt \quad \forall \phi \in L^2(\Gamma),$$

where $\tilde{E}^{stoch,f}$ and $\tilde{E}^{stoch,g}$ are the corresponding stochastic integral discretization errors on the new probability space; see the definitions of $E^{stoch,f}$ and $E^{stoch,g}$ in (8.64) and (8.65) on the original probability space. We can now pass to the limit in all of the integrals, and we use the almost sure convergence of the "tilde" random variables as follows.

Terms 1 and 2 Note that because $(\tilde{\overline{\boldsymbol{u}}}_N, \tilde{\overline{v}}_N) \to (\tilde{\boldsymbol{u}}, \tilde{v})$ in $C(0, T; \mathcal{Q}')$, $\tilde{\mathbb{P}}$-almost surely, and because $(\boldsymbol{q}, \psi)$ is a deterministic test function in $\mathcal{Q}$, we conclude that $\tilde{\mathbb{P}}$-almost surely:

$$\int_{\mathcal{O}_f} \tilde{\overline{\boldsymbol{u}}}_N \cdot \boldsymbol{q} d\boldsymbol{x} + \int_\Gamma \tilde{\overline{v}}_N(t)\psi dz \to \int_{\mathcal{O}_f} \tilde{\boldsymbol{u}} \cdot \boldsymbol{q} d\boldsymbol{x} + \int_\Gamma \tilde{v}\psi dz.$$

Terms 3–5 Since $\tilde{\boldsymbol{u}}_N^{\Delta t} \rightharpoonup \tilde{\boldsymbol{u}}$ weakly in $L^2(0, T; H^1(\mathcal{O}_f))$ and $\tilde{\eta}_N^{\Delta t} \rightharpoonup \tilde{\eta}$ weakly in $L^2(0, T; H_0^2(\Gamma))$, $\tilde{\mathbb{P}}$-almost surely, we have the following $\tilde{\mathbb{P}}$-almost sure convergences for a deterministic test function $(\boldsymbol{q}, \psi) \in \mathcal{Q}$:

$$2\mu \int_0^t \int_{\mathcal{O}_f} \boldsymbol{D}(\tilde{\boldsymbol{u}}_N^{\Delta t}) : \boldsymbol{D}(\boldsymbol{q}) d\boldsymbol{x} ds \to 2\mu \int_0^t \int_{\mathcal{O}_f} \boldsymbol{D}(\tilde{\boldsymbol{u}}) : \boldsymbol{D}(\boldsymbol{q}) d\boldsymbol{x} ds,$$

$$\int_0^t \int_\Gamma \partial_z \tilde{\eta}_N^{\Delta t} \partial_z \psi dz ds \to \int_0^t \int_\Gamma \partial_z \tilde{\eta} \partial_z \psi,$$

$$\int_0^t \int_\Gamma \partial_z^2 \tilde{\eta}_N^{\Delta t} \partial_z^2 \psi dz ds \to \int_0^t \int_\Gamma \partial_z^2 \tilde{\eta} \partial_z^2 \psi.$$

Passing to the Limit in the Stochastic Integrals We want to pass to the limit in the stochastic integral, and we show that for an arbitrary deterministic test function $(\boldsymbol{q}, \psi) \in \mathcal{Q}$, the following convergences hold $\tilde{\mathbb{P}}$-almost surely as $N \to \infty$:

$$\int_0^t \int_{\mathcal{O}_f} \left(\boldsymbol{f}(\tilde{\boldsymbol{u}}_N, \tilde{v}_N^*, \tilde{\eta}_N^{\Delta t}) \cdot \boldsymbol{q} \right) d\boldsymbol{x} d\tilde{W}_{1,N} \to \int_0^t \int_{\mathcal{O}_f} \left(\boldsymbol{f}(\tilde{\boldsymbol{u}}, \tilde{v}, \tilde{\eta}) \cdot \boldsymbol{q} \right) d\boldsymbol{x} d\tilde{W}_1,$$

$$\int_0^t \int_\Gamma \left(g(\tilde{\boldsymbol{u}}_N, \tilde{v}_N^*, \tilde{\eta}_N^{\Delta t}) \psi \right) dz d\tilde{W}_{2,N} \to \int_0^t \int_\Gamma \left(g(\tilde{\boldsymbol{u}}, \tilde{v}, \tilde{\eta}) \psi \right) dz d\tilde{W}_2.$$

By Lemma 3.6.2, it suffices to show that

$$\int_{\mathcal{O}_f} \left(\boldsymbol{f}(\tilde{\boldsymbol{u}}_N, \tilde{v}_N^*, \tilde{\eta}_N^{\Delta t}) \cdot \boldsymbol{q} \right) d\boldsymbol{x} \to \int_{\mathcal{O}_f} \left(\boldsymbol{f}(\tilde{\boldsymbol{u}}, \tilde{v}, \tilde{\eta}) \cdot \boldsymbol{q} \right) d\boldsymbol{x},$$

$$\int_{\mathcal{O}_f} \left(g(\tilde{\boldsymbol{u}}_N, \tilde{v}_N^*, \tilde{\eta}_N^{\Delta t}) \right) \psi dz \to \int_\Gamma \left(g(\tilde{\boldsymbol{u}}, \tilde{v}, \tilde{\eta}) \right) \psi dz,$$

in $L^2(0, t)$, $\tilde{\mathbb{P}}$-almost surely, since almost sure convergence implies convergence in probability. Let us verify this for the stochastic integral term involving $\boldsymbol{f}$ using assumption (8.8), since the same computation will work to verify the other convergence for g using assumption (8.9). We compute using the assumption (8.8) on the noise coefficient $\boldsymbol{f}$ that

$$\int_0^T \left(\int_{\mathcal{O}_f} \left(\boldsymbol{f}(\tilde{\boldsymbol{u}}, \tilde{v}, \tilde{\eta}) - \boldsymbol{f}(\tilde{\boldsymbol{u}}_N, \tilde{v}_N^*, \tilde{\eta}_N^{\Delta t}) \right) \cdot \boldsymbol{q} d\boldsymbol{x} \right)^2 dt$$

$$\leq C \|\boldsymbol{q}\|_{L^2(\mathcal{O}_f)}^2 \int_0^T \left(\|\tilde{\boldsymbol{u}} - \tilde{\boldsymbol{u}}_N\|_{L^2(\mathcal{O}_f)} + \|\tilde{v} - \tilde{v}_N^*\|_{L^2(\Gamma)} + \|\tilde{\eta} - \tilde{\eta}_N^{\Delta t}\|_{L^2(\Gamma)} \right)^2 dt$$

$$\leq C \|\boldsymbol{q}\|_{L^2(\mathcal{O}_f)}^2 \Big(\|\tilde{\boldsymbol{u}} - \tilde{\boldsymbol{u}}_N\|_{L^2(0,T;L^2(\mathcal{O}_f))}^2 + \|\tilde{v} - \tilde{v}_N^*\|_{L^2(0,T;L^2(\Gamma))}^2 + \|\tilde{\eta} - \tilde{\eta}_N^{\Delta t}\|_{L^2(0,T;L^2(\Gamma))}^2 \Big) \to 0.$$

We have the convergence to zero as $N \to \infty$, $\tilde{\mathbb{P}}$-almost surely, since by (8.61), we have that $\tilde{\boldsymbol{u}}_N \to \tilde{\boldsymbol{u}}$ in $L^2(0, T; L^2(\mathcal{O}_f))$, $\tilde{v}_N^* \to \tilde{v}_N$ in $L^2(0, T; L^2(\Gamma))$, and $\tilde{\eta}_N^{\Delta t} \to \tilde{\eta}$ in $L^2(0, T; L^2(\Gamma))$, $\tilde{\mathbb{P}}$-almost surely. The same argument, using the condition (8.9) on the noise coefficient g, works analogously to show the convergence of the stochastic integral involving g.

Convergence of Stochastic Error Terms We show that on the original probability space $(\Omega, \mathcal{F}, \mathbb{P})$:

$$\mathbb{E}\int_0^T \|E_N^{stoch,f}\|^2_{L^2(\mathcal{O}_f)}dt \to 0, \qquad \mathbb{E}\int_0^T \|E_N^{stoch,g}\|_{L^2(\Gamma)}dt \to 0, \tag{8.67}$$

because this implies by equivalence of laws that $\tilde{E}_N^{stoch,f} \to 0$ in $L^2(\mathcal{O}_f)$ and $\tilde{E}_N^{stoch,g} \to 0$ in $L^2(\Gamma)$ for all $t \in [0,T]$, $\tilde{\mathbb{P}}$-almost surely on the new probability space, which would then imply that $(\tilde{E}_N^{stoch,f}(t), \boldsymbol{q})_{L^2(\mathcal{O}_f)} \to 0$ and $(\tilde{E}_N^{stoch,g}(t), \psi)_{L^2(\Gamma)} \to 0$, $\tilde{\mathbb{P}}$-almost surely for all $t \in [0,T]$, for each fixed but arbitrary deterministic test function $(\boldsymbol{q}, \psi) \in \mathcal{Q}$. Since the argument will be the same for $E_N^{stoch,f}$ and $E_N^{stoch,g}$, we will establish (8.67) for $E_N^{stoch,f}$. Using the definition of $E_N^{stoch,f}$ in (8.64), we separate this calculation into two estimates:

$$\mathbb{E}\int_0^T \|E_N^{f,stoch}(t)\|_{L^2(\mathcal{O}_f)}dt$$
$$\leq 2\left(\mathbb{E}\int_0^T \|E_{1,N}^{f,stoch}\|_{L^2(\mathcal{O}_f)}dt + \int_0^T \|E_{2,N}^{f,stoch}\|^2_{L^2(\mathcal{O}_f)}dt\right), \tag{8.68}$$

where

$$E_{1,N}^{f,stoch} = \sum_{n=0}^{N-1}\left(\frac{t-n\Delta t}{\Delta t}\boldsymbol{f}\left(\boldsymbol{u}_N^{n+\frac{1}{2}}, v_N^{n+\frac{1}{2}}, \eta_N^{n+\frac{1}{2}}\right)\Delta W_{1,N}^n \mathbb{1}_{[n\Delta t,(n+1)\Delta t)}\right)$$

$$E_{2,N}^{f,stoch} = \sum_{n=0}^{N-1}\left(\int_{n\Delta t}^t \boldsymbol{f}(\boldsymbol{u}_N, v_N^*, \eta_N^{\Delta t})dW_1\right)\mathbb{1}_{[n\Delta t,(n+1)\Delta t)}.$$

For $E_{1,N}^{f,stoch}$, we then estimate that

$$\mathbb{E}\int_0^T \|E_{1,N}^{f,stoch}\|^2_{L^2(\mathcal{O}_f)}dt = \mathbb{E}\sum_{n=0}^{N-1}\left(\int_{n\Delta t}^{(n+1)\Delta t}\left(\frac{t-n\Delta t}{\Delta t}\right)^2 dt\right)$$
$$\times\left\|\boldsymbol{f}\left(\boldsymbol{u}_N^{n+\frac{1}{2}}, v_N^{n+\frac{1}{2}}, \eta_N^{n+\frac{1}{2}}\right)\Delta W_{1,N}^n\right\|^2_{L^2(\mathcal{O}_f)}$$
$$\leq \mathbb{E}\sum_{n=0}^{N-1}\frac{\Delta t}{3}\left\|\boldsymbol{f}\left(\boldsymbol{u}_N^{n+\frac{1}{2}}, v_N^{n+\frac{1}{2}}, \eta_N^{n+\frac{1}{2}}\right)\Delta W_{1,N}^n\right\|^2_{L^2(\mathcal{O}_f)}.$$

Using the independence of $f\left(u_N^{n+\frac{1}{2}}, v_N^{n+\frac{1}{2}}, \eta_N^{n+\frac{1}{2}}\right)$ and $\Delta W_{1,N}^n$ (see Proposition 8.5.1), we obtain:

$$\begin{aligned}
&\mathbb{E}\int_0^T \|E_{1,N}^{f,stoch}\|^2_{L^2(\mathcal{O}_f)}dt \\
&\leq \frac{\Delta t}{3}\sum_{n=0}^{N-1}\mathbb{E}\left\|f\left(u_N^{n+\frac{1}{2}}, v_N^{n+\frac{1}{2}}, \eta_N^{n+\frac{1}{2}}\right)\right\|^2_{L^2(\mathcal{O}_f)}\mathbb{E}|\Delta W_{1,N}^n|^2 \\
&\leq C\frac{(\Delta t)^2}{3}\sum_{n=0}^{N-1}\mathbb{E}\left(1+\|u_N^{n+\frac{1}{2}}\|^2_{L^2(\mathcal{O}_f)}+\|v_N^{n+\frac{1}{2}}\|^2_{L^2(\Gamma)}+\|\eta_N^{n+\frac{1}{2}}\|^2_{H_0^2(\Gamma)}\right) \\
&\leq C(\Delta t)\mathbb{E}\Big(1+\|u_N\|^2_{L^\infty(0,T;L^2(\mathcal{O}_f))}+\|v_N^*\|^2_{L^\infty(0,T;L^2(\Gamma))} \\
&\quad +\|\eta_N^{\Delta t}\|^2_{L^\infty(0,T;H_0^2(\Gamma))}\Big)\leq C(\Delta t),
\end{aligned}$$

by (8.6), the fact that $\mathbb{E}|\Delta W_{1,N}^n|^2 = \Delta t$, and the uniform bounds in Proposition 8.6.2.

For $E_{2,N}^{f,stoch}$, we estimate using the Itô isometry (see Proposition 3.6.1) that

$$\begin{aligned}
\mathbb{E}\int_0^T \|E_{2,N}^{f,stoch}\|^2_{L^2(\mathcal{O}_f)}dt &= \mathbb{E}\sum_{n=0}^{N-1}\int_{n\Delta t}^{(n+1)\Delta t}\left\|\int_{n\Delta t}^t f(u_N, v_N^*, \eta_N^{\Delta t})dW\right|^2_{L^2(\mathcal{O}_f)}dt \\
&= \mathbb{E}\sum_{n=0}^{N-1}\int_{n\Delta t}^{(n+1)\Delta t}\int_{n\Delta t}^t \|f(u_N, v_N^*, \eta_N^{\Delta t})\|^2_{L^2(\mathcal{O}_f)}dsdt \\
\leq C\mathbb{E}\sum_{n=0}^{N-1}\int_{n\Delta t}^{(n+1)\Delta t}\int_{n\Delta t}^t &\left(1+\|u_N(s)\|^2_{L^2(\mathcal{O}_f)}+\|v_N^*(s)\|^2_{L^2(\Gamma)}+\|\eta_N^{\Delta t}(s)\|^2_{H_0^2(\Gamma)}\right)dsdt \\
&\leq C(\Delta t)\mathbb{E}\Big(1+\|u_N\|^2_{L^\infty(0,T;L^2(\mathcal{O}_f))}+\|v_N\|^2_{L^\infty(0,T;L^2(\Gamma))} \\
&\qquad +\|\eta_N^{\Delta t}\|^2_{L^\infty(0,T;H_0^2(\Gamma))}\Big)dt\leq C(\Delta t),
\end{aligned}$$

by (8.6) and the uniform estimates in Proposition 8.6.2 combined with equivalence of laws. So by (8.68),

$$\mathbb{E}\int_0^T \|E_N^{f,stoch}\|^2_{L^2(\mathcal{O}_f)} \leq C(\Delta t)\to 0, \qquad \text{as } N\to\infty,$$

for a constant C that is independent of N, which establishes the desired convergence.

Convergence of the Pressure Term Finally, we show that the deterministic pressure error $E^P_{N,in/out}$ satisfies:

$$E^P_{N,in}(t) \to 0, \qquad E^P_{N,out}(t) \to 0 \tag{8.69}$$

as $N \to \infty$, for all $t \in [0, T]$, since this would imply the convergence

$$E^P_{N,in/out}(t)\left(\int_0^R (q_z)|_{z=0} dr\right) \to 0, \qquad \text{as } N \to \infty,$$

for all $t \in [0, T]$ and for each fixed but arbitrary test function $\boldsymbol{q}$. We can compute using the definition (8.66) and the definition of $P^n_{N,in/out} = \frac{1}{\Delta t}\int_{n\Delta t}^{(n+1)\Delta t} P_{in/out}(t)dt$ that

$$|E^P_{N,in/out}(t)| \le \max_{n=0,1,\dots,N-1}\left((\Delta t)|P^n_{N,in/out}| + \int_{n\Delta t}^{(n+1)\Delta t} |P_{in/out}(t)|dt\right)$$

$$\le 2 \max_{n=0,1,\dots,N-1} \int_{n\Delta t}^{(n+1)\Delta t} |P_{in/out}(t)|dt \le 2(\Delta t)^{1/2}\|P_{in/out}\|_{L^2(0,T)} \to 0,$$

as $N \to \infty$. This establishes the desired convergence of the pressure error term.

We have, therefore, established the existence of a weak martingale solution to the stochastic fluid-structure interaction problem, as in Definition 8.3.1. We remark that while the solution that we constructed in Lemma 8.7.2 exists within a phase space $\mathcal{X}$ that does not have the full regularity $L^\infty(0, T; L^2(\mathcal{O}_f))$ for the fluid velocity $\tilde{\boldsymbol{u}}$ and $L^\infty(0, T; H^2_0(\Gamma))$ for the structure displacement $\tilde{\eta}$, for example, since the phase space $\mathcal{X}$ defined in (8.54) only has these functions in $L^2(0, T; L^2(\mathcal{O}_f))$ and $L^2(0, T; H^2_0(\Gamma))$ $\tilde{\mathbb{P}}$-almost surely, respectively, we can still recover the desired finite energy regularity for the solution. To do this, we note that since the solution $(\tilde{\boldsymbol{u}}, \tilde{\eta}, \tilde{v})$ on the new probability space satisfies the desired weak formulation, we can recover the finite energy space regularity for our weak martingale solution on the new probability space by using standard a priori energy estimates via Itô's formula (Theorem 3.6.4).

8.8 Return to the Original Probability Space

We have thus constructed a stochastic process $(\tilde{\boldsymbol{u}}, \tilde{\eta})$, which satisfies the weak formulation of the continuous problem almost surely on the "tilde" probability space determined by the Skorohod representation theorem, which gives us the existence of a weak martingale solution; see Definition 8.3.1, for the linearly coupled stochastic

FSI problem with multiplicative noise. However, we want to bring the solution back to the original probability space if possible, and we also want to show if possible that this solution is *unique*. In particular, we want to see if we can get convergence of the original approximate solutions $(\boldsymbol{u}_N, v_N, \eta_N)$ on the original probability space $(\Omega, \mathcal{F}, \mathbb{P})$ with the originally given complete filtration $\{\mathcal{F}_t\}_{t\geq 0}$ and the original Brownian motions $\{W_1(t)\}_{t\geq 0}$ and $\{W_2(t)\}_{t\geq 0}$. We will be able to carry out this procedure in the case when the nonlinearly noise coefficients f and g are identically equal to one, so that the noise is additive, which will allow us to upgrade the well-posedness result from existence of a weak martingale solution on *some* probability space (see Definition 8.3.1) to existence *and uniqueness* of a weak pathwise solution constructed on the initially given probability space (see Definition 8.3.2).

To do this, we will use a standard *Gyöngy-Krylov argument* based on the Gyöngy-Krylov lemma, discussed in Theorem 3.2.4. As a subtlety, this version of the Gyöngy-Krylov lemma is stated for random variables taking values in separable Banach spaces, but its extension to more general Jakubowski spaces is established for example in Theorem 2.10.3 of [52], which technically is necessary in this case due to the inclusion of the spaces with weak topology in the definition of the phase space (8.54). Recall that the statement of the Gyöngy-Krylov lemma (which involves convergence in probability along a subsequence of a given sequence of random variables $\{X_n\}_{n=1}^{\infty}$) holds if and only if the *diagonal condition* holds: For every two subsequences X_{l_k} and X_{m_k}, there exists a further subsequence such that the joint probability measures associated with $x_k = (X_{l_k}, X_{m_k})$ on $B \times B$, defined by

$$\nu_{x_k} = \nu_{(X_{l_k}, X_{m_k})}(A_1 \times A_2) = \mathbb{P}(X_{l_k} \in A_1, X_{m_k} \in A_2), \quad A_1, A_2 \in \text{Borel}(B),$$

where $\text{Borel}(B)$ is the Borel sigma algebra on B, converge weakly along this *further subsequence* to some probability measure ν, where ν is such that

$$\nu(\{(x, y) \in B \times B : x = y\}) = 1. \tag{8.70}$$

Thus, the limits of any two convergent subsequences have to be "the same" with probability 1.

Once we show convergence in probability of our original sequence using the Gyöngy-Krylov lemma, we will have almost sure convergence along a subsequence of our approximate solutions on the *original probability space*. Then, using the fact that our approximate solutions converge almost surely along a subsequence on the original probability space, we can adapt the arguments in Sect. 8.7.3 in order to show that the limiting weak solution on the original probability space satisfies the weak form of the continuous-in-time problem almost surely, so that the limiting solution is a weak pathwise solution existing on the initially given probability space; see Definition 8.3.2.

Thus, what remains to be shown is that the diagonal condition in the Gyöngy-Krylov lemma holds. Using the Skorohod representation theorem, one can show that the diagonal condition is equivalent to showing *pathwise uniqueness* of solutions. To demonstrate this, we first prove pathwise uniqueness for the problem with additive noise (which is equivalent to deterministic uniqueness) and then use it to show how this implies the diagonal condition.

8.8.1 Uniqueness for the Deterministic FSI Problem

In the specific case where the nonlinear scaling functions $f(\boldsymbol{u}, v, \eta)$ and $g(\boldsymbol{u}, v, \eta)$ are identically one, so that the noise is additive noise instead of multiplicative noise, we show that we can obtain a better existence result: Namely, we can obtain the existence of a *unique weak pathwise solution* (see Definition 8.3.2) to the stochastic FSI problem when $f = 1$ and $g = 1$. This will rely on the following uniqueness result for the deterministic problem, in which there is no noise, as this deterministic uniqueness result will imply a corresponding almost sure uniqueness result for the stochastic problem with additive noise. *More specifically, for additive noise, uniqueness for the deterministic problem implies pathwise uniqueness for the stochastic problem with additive noise.* In particular, we have the following (deterministic) uniqueness result:

Lemma 8.8.1 (Uniqueness for the Deterministic Problem) *Suppose that* $\boldsymbol{u} \in L^\infty(0, T; L^2(\mathcal{O}_f)) \cap L^2(0, T; \mathcal{V}_f)$, $\eta \in W^{1,\infty}(0, T; L^2(\Gamma)) \cap L^\infty(0, T; \mathcal{V}_s)$, *and* $\boldsymbol{u}|_\Gamma = \partial_t \eta \boldsymbol{e}_r$. *Suppose also that* $(\boldsymbol{u}, \partial_t \eta) \in C(0, T; \mathcal{Q}')$, *with* $\eta(0) = 0$. *If for all* $(\boldsymbol{q}, \psi) \in \mathcal{Q}(0, T) := C_c^1([0, T); \mathcal{Q})$,

$$-\int_0^T \int_{\mathcal{O}_f} \boldsymbol{u} \cdot \partial_t \boldsymbol{q} d\boldsymbol{x} dt + 2\mu \int_0^T \int_{\mathcal{O}_f} \boldsymbol{D}(\boldsymbol{u}) : \boldsymbol{D}(\boldsymbol{q}) d\boldsymbol{x} dt - \int_0^T \int_\Gamma \partial_t \eta \partial_t \psi dz dt$$
$$+ \int_0^T \int_\Gamma \partial_z \eta \partial_z \psi dz dt + \int_0^T \int_\Gamma \partial_z^2 \eta \partial_z^2 \psi = 0,$$

then $(\boldsymbol{u}, \eta) = 0$.

Proof Observe first that to get the usual energy equality, we would want to formally substitute in $(\boldsymbol{u}, \partial_t \eta)$ for $(\boldsymbol{q}, \psi)$. However, since $(\boldsymbol{q}, \psi)$ must have $\psi(t) \in H_0^2(\Gamma)$ by the definition of the test space $\mathcal{Q}(0, T)$, we do not have enough regularity to do this. Therefore, we use a different approach of taking an antiderivative, which is an approach used, for example, in establishing uniqueness of weak solutions for general hyperbolic equations (see Section 7.2 in [132]).

Consider an arbitrary s such that $0 \le s \le T$. We use the following test function,

$$(\boldsymbol{q}_0(t), \psi_0(t)) = \begin{cases} \left(\int_t^s \left(\int_0^\tau \boldsymbol{u}(\sigma) d\sigma\right) d\tau, \int_t^s \eta(\tau) d\tau\right) & \text{if } 0 \le t \le s, \\ (0, 0) & \text{if } s \le t \le T. \end{cases}$$

Recall that $\eta(0) = 0$ by assumption. Note that since

$$\int_0^\tau \boldsymbol{u}(\sigma)d\sigma\Big|_\Gamma = \int_0^\tau \partial_t\eta(\sigma)d\sigma = \eta(\tau)$$

for all $\tau \in [0, T]$, the function $(\boldsymbol{q}_0, \psi_0)$ satisfies the necessary kinematic coupling condition for $\mathcal{Q}(0, T)$. While this test function is only piecewise differentiable, it is easy to show by an approximation argument that the weak formulation should still hold with this test function by approximating it with differentiable functions. For notational simplicity, we define

$$\boldsymbol{U}(t) = \int_0^t \boldsymbol{u}(\sigma)d\sigma.$$

Substituting the test function into the weak formulation, we obtain for all $s \in [0, T]$,

$$\int_0^s \int_{\mathcal{O}_f} \boldsymbol{u} \cdot \boldsymbol{U} d\boldsymbol{x}dt + 2\mu \int_0^s \int_{\mathcal{O}_f} \boldsymbol{D}(\boldsymbol{u}) : \boldsymbol{D}(\boldsymbol{q}_0)d\boldsymbol{x}dt + \int_0^s \int_\Gamma \partial_t\eta \cdot \eta dzdt$$
$$+ \int_0^s \int_\Gamma \partial_z\eta\partial_z\psi_0 dzdt + \int_0^s \int_\Gamma \partial_z^2\eta\partial_z^2\psi_0 dzdt = 0,$$

where we note that $\partial_t\boldsymbol{q}_0(t) = -\boldsymbol{U}(t)$ and $\partial_t\psi_0(t) = -\eta(t)$, for $t \in [0, s)$. We handle the four terms on the left-hand side as follows.

- **First term:** We note that $\boldsymbol{u} = \partial_t\boldsymbol{U}$. Hence, using the fact that $\boldsymbol{U}(0) = 0$, we get

$$\int_0^s \int_{\mathcal{O}_f} \boldsymbol{u} \cdot \boldsymbol{U} d\boldsymbol{x}dt = \int_0^s \frac{d}{dt}\left(\frac{1}{2}||\boldsymbol{U}||^2_{L^2(\mathcal{O}_f)}\right)dt$$
$$= \frac{1}{2}||\boldsymbol{U}(s)||^2_{L^2(\mathcal{O}_f)} - \frac{1}{2}||\boldsymbol{U}(0)||^2_{L^2(\mathcal{O}_f)} = \frac{1}{2}||\boldsymbol{U}(s)||^2_{L^2(\mathcal{O}_f)}.$$

- **Second term:** For the second term, we again use that $\boldsymbol{u} = \partial_t\boldsymbol{U}$. Therefore,

$$2\mu \int_0^s \int_{\mathcal{O}_f} \boldsymbol{D}(\boldsymbol{u}) : \boldsymbol{D}(\boldsymbol{q}_0)d\boldsymbol{x}dt = 2\mu \int_0^s \int_{\mathcal{O}_f} \boldsymbol{D}(\partial_t\boldsymbol{U}) : \boldsymbol{D}(\boldsymbol{q}_0)d\boldsymbol{x}dt.$$

  We integrate by parts in time. Note that $\boldsymbol{U}(0) = 0$ and $\boldsymbol{q}_0(s) = 0$, so there are no boundary terms from the integration by parts. Hence, using the fact that $\partial_t\boldsymbol{q}_0 = -\boldsymbol{U}$, we obtain

$$2\mu \int_0^s \int_{\mathcal{O}_f} \boldsymbol{D}(\boldsymbol{u}) : \boldsymbol{D}(\boldsymbol{q}_0)d\boldsymbol{x}dt = -2\mu \int_0^s \int_{\mathcal{O}_f} \boldsymbol{D}(\boldsymbol{U}) : \boldsymbol{D}(\partial_t\boldsymbol{q}_0)d\boldsymbol{x}dt$$
$$= 2\mu \int_0^s \int_{\mathcal{O}_f} |\boldsymbol{D}(\boldsymbol{U})|^2 d\boldsymbol{x}dt.$$

- **Third term:** We immediately have that

$$\int_0^s \int_\Gamma \partial_t \eta \cdot \eta dz dt = \frac{1}{2}||\eta(s)||^2_{L^2(\Gamma)} - \frac{1}{2}||\eta(0)||^2_{L^2(\Gamma)} = \frac{1}{2}||\eta(s)||^2_{L^2(\Gamma)}.$$

- **Fourth and fifth term:** Since $\eta = -\partial_t \psi_0$, we have for $i = 1, 2$ that

$$\int_\Gamma \partial_z^i \eta \partial_z^i \psi_0 dz = -\frac{1}{2}\frac{d}{dt}\left(||\partial_z^i \psi_0||^2_{L^2(\Gamma)}\right),$$

  and hence, using the fact that $\psi_0(s) = 0$, we get that for $i = 1, 2$:

$$\int_0^s \int_\Gamma \partial_z^i \eta \partial_z^i \psi_0 dz dt = \frac{1}{2}||\partial_z^i \psi_0(0)||^2_{L^2(\Gamma)}.$$

Therefore, for all $0 \leq s \leq T$, the entire expression (energy) can now be written as

$$\frac{1}{2}\|\boldsymbol{U}(s)\|^2_{L^2(\mathcal{O}_f)} + 2\mu \int_0^s \int_{\mathcal{O}_f} |\boldsymbol{D}(\boldsymbol{U})|^2 d\boldsymbol{x} dt + \frac{1}{2}\|\eta(s)\|^2_{L^2(\Gamma)}$$
$$+ \frac{1}{2}\|\partial_z \psi_0(0)\|^2_{L^2(\Gamma)} + \frac{1}{2}\|\partial_z^2 \psi_0(0)\|^2_{L^2(\Gamma)} = 0.$$

Thus, we conclude that $\boldsymbol{U}(s) = 0$ and $\eta(s) = 0$ for all $s \in [0, T]$. From the definition of $\boldsymbol{U}$, we conclude that $\boldsymbol{u}(t) = \partial_t \boldsymbol{U}(t) = 0$ for all $t \in [0, T]$ also, which completes the proof. □

8.8.2 *Verifying the Diagonal Condition of the Gyöngy-Krylov Lemma*

Now that we have established a uniqueness result, we can construct a solution on the original probability space $(\Omega, \mathcal{F}, \mathbb{P})$ for the stochastic linearly coupled FSI problem with *additive noise* by invoking a standard argument involving the Gyöngy-Krylov argument (Lemma 3.2.4) to show that the random variables $(\eta_N, \boldsymbol{u}_N, v_N)$ defined on the original probability space converge in probability and hence converge almost surely along a subsequence *in the original topology*.

Because we have already shown pathwise uniqueness in Sect. 8.8.1, it only remains to demonstrate how the Skorohod representation theorem can be used to show that the diagonal condition (8.70) from the Gyöngy-Krylov lemma is equivalent to showing pathwise uniqueness.

For this purpose, denote by $\{X^1_{M_k}\}_{k=1}^\infty$ and $\{X^2_{N_k}\}_{k=1}^\infty$ any two subsequences of our random variables (approximate solutions) defined on the original probability space $(\Omega, \mathcal{F}, \mathbb{P})$:

$$X^1_{M_k} = (\eta_{M_k}, \eta^{\Delta t}_{M_k}, \overline{\eta}_{M_k}, \boldsymbol{u}_{M_k}, \boldsymbol{u}^{\Delta t}_{M_k}, v_{M_k}, v^*_{M_k}, \overline{\boldsymbol{u}}_{M_k}, \overline{v}_{M_k}, W_1, W_2),$$

$$X^2_{N_k} = (\eta_{N_k}, \eta^{\Delta t}_{N_k}, \overline{\eta}_{N_k}, \boldsymbol{u}_{N_k}, \boldsymbol{u}^{\Delta t}_{N_k}, v_{N_k}, v^*_{N_k}, \overline{\boldsymbol{u}}_{N_k}, \overline{v}_{N_k}, W_1, W_2).$$

Recall that the laws corresponding to each of these these two sequences of random variables *individually* converge to the law μ. However, to verify the diagonal condition in the Gyöngy-Krylov lemma, we must examine the *joint laws* of these random variables $(X^1_{M_k}, X^2_{N_k})$.

Hence, we consider the joint probability measures (or joint laws) $\{\nu_{X^1_{M_k}, X^2_{N_k}}\}_{k=1}^{\infty}$ on $\mathcal{X} \times \mathcal{X}$, associated with the subsequence $(X^1_{M_k}, X^2_{N_k})$. By the tightness of the original probability measures μ_N, established in the proof of Theorem 8.7.1, we have that the collection of joint laws $\{\nu_{X^1_{M_k}, X^2_{N_k}}\}_{k=1}^{\infty}$ is also tight and hence converges weakly to a probability measure ν on $\mathcal{X} \times \mathcal{X}$ along a further subsequence, which we will continue to denote by the same indexing for notational simplicity. Then, by the Skorohod representation theorem in Theorem 3.2.5, there exists a probability space $(\tilde{\Omega}, \tilde{\mathcal{F}}, \tilde{\mathbb{P}})$ and random variables

$$\tilde{X}^1_{M_k} = (\tilde{\eta}^1_{M_k}, \tilde{\eta}^{\Delta t,1}_{M_k}, \tilde{\overline{\eta}}^1_{M_k}, \tilde{\boldsymbol{u}}^1_{M_k}, \tilde{\boldsymbol{u}}^{\Delta t,1}_{M_k}, \tilde{v}^1_{M_k}, \tilde{v}^{*,1}_{M_k}, \tilde{\overline{\boldsymbol{u}}}^1_{M_k}, \tilde{\overline{v}}^1_{M_k}, \tilde{W}_{1,M_k}, \tilde{W}_{2,M_k}),$$

$$\tilde{X}^2_{N_k} = (\tilde{\eta}^2_{N_k}, \tilde{\eta}^{\Delta t,2}_{N_k}, \tilde{\overline{\eta}}^2_{N_k}, \tilde{\boldsymbol{u}}^2_{N_k}, \tilde{\boldsymbol{u}}^{\Delta t,2}_{N_k}, \tilde{v}^2_{M_k}, \tilde{v}^{*,2}_{N_k}, \tilde{\overline{\boldsymbol{u}}}^2_{N_k}, \tilde{\overline{v}}^2_{N_k}, \tilde{W}_{1,N_k}, \tilde{W}_{2,N_k}),$$

such that

$$(\tilde{X}^1_{M_k}, \tilde{X}^2_{N_k}) =_d (X^1_{M_k}, X^2_{N_k}), \tag{8.71}$$

and $(\tilde{X}^1_{M_k}, \tilde{X}^2_{N_k}) \to (\tilde{X}^1, \tilde{X}^2)$ in $\mathcal{X} \times \mathcal{X}$ almost surely as $k \to \infty$, where

$$\tilde{X}^1 = (\tilde{\eta}^1, \tilde{\eta}^{\Delta t,1}, \tilde{\overline{\eta}}^1, \tilde{\boldsymbol{u}}^1, \tilde{\boldsymbol{u}}^{\Delta t,1}, \tilde{v}^1, \tilde{v}^{*,1}, \tilde{\overline{\boldsymbol{u}}}^1, \tilde{\overline{v}}^1, \tilde{W}_1, \tilde{W}_2),$$

$$\tilde{X}^2 = (\tilde{\eta}^2, \tilde{\eta}^{\Delta t,2}, \tilde{\overline{\eta}}^2, \tilde{\boldsymbol{u}}^2, \tilde{\boldsymbol{u}}^{\Delta t,2}, \tilde{v}^2, \tilde{v}^{*,2}, \tilde{\overline{\boldsymbol{u}}}^2, \tilde{\overline{v}}^2, \tilde{W}_1, \tilde{W}_2),$$

are random variables on $(\tilde{\Omega}, \tilde{\mathcal{F}}, \tilde{\mathbb{P}})$, and ν is the law of $(\tilde{X}_1, \tilde{X}_2)$.

We want to show that ν is supported on the diagonal. It suffices to show that $\tilde{\mathbb{P}}(\tilde{X}^1 = \tilde{X}^2) = 1$. We do this in two steps.

Step 1. *Show that both $\tilde{X}^1$ and $\tilde{X}^2$ are weak martingale solutions.* First we notice that $\tilde{X}^1$ *is a weak martingale solution with respect to the stochastic basis* $(\tilde{\Omega}, \tilde{\mathcal{F}}, \{\tilde{\mathcal{F}}^1_t\}_{t\geq 0}, \tilde{\mathbb{P}}, \tilde{W}_1)$ in the sense of Definition 8.3.1. This follows from the results of Lemma 8.7.2. Namely, the results of Lemma 8.7.2 imply that $\tilde{\eta}^1 = \tilde{\overline{\eta}}^1 = \tilde{\eta}^{\Delta t,1}$, $\tilde{\boldsymbol{u}}^1 = \tilde{\boldsymbol{u}}^{*,1} = \tilde{\overline{\boldsymbol{u}}}^1 = \tilde{\boldsymbol{u}}^{\Delta t,1}$, $\tilde{v}^1 = \tilde{v}^{*,1} = \tilde{\overline{v}}^1 = \tilde{v}^{\Delta t,1}$, and $\partial_t \tilde{\eta}^1 = \tilde{v}^1$ almost surely. Furthermore, $(\tilde{\boldsymbol{u}}^1, \tilde{\eta}^1) \in \mathcal{W}(0, T)$ and $(\tilde{\boldsymbol{u}}^1, \tilde{v}^1) \in C(0, T; \mathcal{Q}')$, satisfying the initial condition $\tilde{\eta}^1(0) = \eta_0$ almost surely. Furthermore, $\tilde{X}^1$ is a weak martingale solution with respect to the stochastic basis

$(\tilde{\Omega}, \tilde{\mathcal{F}}, \{\tilde{\mathcal{F}}_t^1\}_{t\geq 0}, \tilde{\mathbb{P}}, \tilde{W}_1)$ in the sense of Definition 8.3.1. The same is true for the components of $\tilde{X}^2$, with respect to $(\tilde{\Omega}, \tilde{\mathcal{F}}, \{\tilde{\mathcal{F}}_t^2\}_{t\geq 0}, \tilde{\mathbb{P}}, \tilde{W}_2)$. Here, the filtrations $\{\tilde{\mathcal{F}}_t^1\}_{t\geq 0}$ and $\{\tilde{\mathcal{F}}_t^2\}_{t\geq 0}$ are defined by (8.62) with the appropriate limiting random variables with superscripts "1" and "2," respectively.

Step 2. *Verify the diagonal condition using the uniqueness result of Lemma 8.8.1.* Finally, we use pathwise uniqueness to obtain the diagonal condition. Note that $(\tilde{\boldsymbol{u}}^1, \tilde{\eta}^1)$ and $(\tilde{\boldsymbol{u}}^2, \tilde{\eta}^2)$ both satisfy all of the conditions of Lemma 8.8.1, as they are both analytically weak solutions to the given stochastic FSI problem (with additive noise) with the same initial data. Therefore, by using the uniqueness result in Lemma 8.8.1, we conclude that $\tilde{\boldsymbol{u}}^1 = \tilde{\boldsymbol{u}}^2$ and $\tilde{\eta}_1 = \tilde{\eta}_2$ almost surely. This essentially uses the fact that the noise is assumed to be *additive*, since in this case, the difference of the two solutions $(\tilde{\boldsymbol{u}}_1 - \tilde{\boldsymbol{u}}_2, \tilde{\eta}_1 - \tilde{\eta}_2, \tilde{v}_1 - \tilde{v}_2)$ satisfies the linear *deterministic* FSI problem almost surely, for which we have the uniqueness result stated in Lemma 8.8.1. Therefore, since $\tilde{v}^1 = \partial_t \tilde{\eta}^1$ and $\tilde{v}^2 = \partial_t \tilde{\eta}^2$, we also obtain that $\tilde{v}^1 = \tilde{v}^2$ almost surely. This allows us to conclude that $\tilde{\mathbb{P}}(\tilde{X}^1 = \tilde{X}^2) = 1$, which implies that the limiting joint probability measure (or law) ν is supported on the diagonal.

This completes the verification of the diagonal condition of the Gyöngy-Krylov lemma.

8.8.3 Existence of a Weak Pathwise Solution

The existence of a weak pathwise solution, given by Definition 8.3.2, in the special case of additive noise ($f = 1$ and $g = 1$), now follows from the Gyöngy-Krylov lemma in Theorem 3.2.4. More precisely, by the Gyöngy-Krylov lemma, the original sequence

$$(\eta_N, \eta_N^{\Delta t}, \overline{\eta}_N, \boldsymbol{u}_N, \boldsymbol{u}_N^{\Delta t}, v_N, v_N^*, \overline{\boldsymbol{u}}_N, \overline{v}_N, W_1, W_2)$$

converges in probability to some random variable

$$(\eta, \eta^{\Delta t}, \overline{\eta}, \boldsymbol{u}, \boldsymbol{u}^{\Delta t}, v, v^*, \overline{\boldsymbol{u}}, \overline{v}, W_1, W_2).$$

Since convergence in probability implies almost sure convergence along a subsequence (see Proposition 3.2.1), we conclude that along a subsequence which we continue to denote by N, we have that

$$\begin{aligned}&(\eta_N, \eta_N^{\Delta t}, \overline{\eta}_N, \boldsymbol{u}_N, \boldsymbol{u}_N^{\Delta t}, v_N, v_N^*, W_1, W_2)\\&\quad\to (\eta, \eta^{\Delta t}, \overline{\eta}, \boldsymbol{u}, \boldsymbol{u}_N^{\Delta t}, v, v^*, \overline{\boldsymbol{u}}, \overline{v}, W_1, W_2), \text{ almost surely in } \mathcal{X}. \qquad (8.72)\end{aligned}$$

To show that this limit is a weak solution in the sense of Definition 8.3.2, we use the same arguments as in Lemma 8.7.2. All of the properties from Definition 8.3.2 follow from Lemma 8.7.2, except for uniqueness and showing that $(\boldsymbol{u}, v, \eta)$ is $\mathcal{F}_t$-adapted.

Uniqueness follows from the pathwise uniqueness result of Lemma 8.8.1.

$\mathcal{F}_t$-Adaptedness of $(\boldsymbol{u}, v, \eta)$ Note that this is not provided by Lemma 8.7.2, as we want to show that this solution is adapted to the *original* filtration $\{\mathcal{F}_t\}_{t\geq 0}$, while the filtration defined in (8.62) is not necessarily the same filtration. However, by definition of the phase space $\mathcal{X}$ in (8.54), we have the almost sure convergences of $(\boldsymbol{u}_N, v_N) \to (\boldsymbol{u}, v)$ in $C(0, T; \mathcal{Q}')$ and $\eta_N \to \eta$ in $C(0, T; L^2(\Gamma))$. Since the approximate solution $(\boldsymbol{u}_N, \eta_N, v_N)$ is adapted to the filtration $\{\mathcal{F}_t\}_{t\geq 0}$ for all N, the almost sure limit of $(\boldsymbol{u}_N, v_N)$ and η_N in $C(0, T; \mathcal{Q}')$ and $C(0, T; L^2(\Gamma))$, respectively, which is the limiting solution $(\boldsymbol{u}, \eta, v)$, is also adapted to $\{\mathcal{F}_t\}_{t\geq 0}$, since $\{\mathcal{F}_t\}_{t\geq 0}$ was assumed to be a *complete filtration.*

In conclusion, we have now shown that $(\boldsymbol{u}, v, \eta)$ has all of the required properties needed to be a weak pathwise solution to the given fluid-structure interaction problem with respect to the initially given independent one-dimensional Brownian motions $\{W_1(t)\}_{t\geq 0}$ which are Brownian motions with respect to the initially given complete filtration $\{\mathcal{F}_t\}_{t\geq 0}$, as in Definition 8.3.2. This completes the proof of the main result, stated in Theorem 8.3.1, and restated here:

Theorem 8.8.1 (Main Result) *Let* $\boldsymbol{u}_0 \in L^2(\mathcal{O}_f)$, $v_0 \in L^2(\Gamma)$, *and* $\eta_0 \in H_0^2(\Gamma)$. *Let* $P_{in/out} \in L^2_{loc}(0, \infty)$ *and let* $(\Omega, \mathcal{F}, \mathbb{P})$ *be a probability space with two independent one-dimensional Brownian motions* $\{W_1(t)\}_{t\geq 0}$ *and* $\{W_2(t)\}_{t\geq 0}$ *which are Brownian motions with respect to a given complete filtration* $\{\mathcal{F}_t\}_{t\geq 0}$. *Then, for any* $T > 0$, *there exists a weak martingale solution to the given stochastic fluid-structure interaction problem* (8.10)–(8.14). *In the case where the noise is purely additive* ($f = 1, g = 1$), *there exists a unique weak pathwise solution.*

8.9 Conclusions

In this chapter, we presented a proof of the existence of a weak martingale solution to a benchmark stochastic fluid-structure interaction (SFSI) problem with nonlinear multiplicative noise (8.10)–(8.14), and we show a stronger result of existence of a unique weak pathwise solution, in the case where this noise is additive. An example of such a problem is the flow of blood in coronary arteries that sit on the surface of the heart and contract and expand under the outside forcing due to the heart muscle contraction and expansion. Dynamic patient images show significant stochastic effects in the heart contractions, which can be captured by an SFSI model such as the one studied in this work. Our well-posedness result indicates that stochastic FSI models are robust in the sense that a weak solution in the sense of Definition 8.3.1

(and Definition 8.3.2 in the case of additive noise) will exist even when the problem is stochastically forced by a rough time-dependent white noise, as considered in this work.

In contrast to the deterministic case, the proof based on the operator splitting strategy presented in this work has several new interesting components, which we summarize below.

1. The energy estimates are given in expectation and do not necessarily hold pathwise. Thus, one cannot immediately deduce almost sure convergence of the random approximate solutions along a subsequence immediately from uniform boundedness of the random approximate solutions in the probabilistic finite energy spaces with norms involving expectations. Having such almost sure convergence is essential for passing to the limit in semidiscrete formulation, especially for a more general class of complex stochastic FSI problems involving nonlinearities.
2. The energy estimate has an extra term that accounts for the energy added to the system by the stochastic forcing in expectation.
3. A constructive splitting scheme can be developed for this multiphysical problem that separates the contributions of the elastodynamics and fluid dynamics to the coupled dynamics via two separate subproblems: (1) the structure subproblem and (2) the fluid subproblem. This gives rise to a loosely coupled, stable constructive existence scheme for this stochastic FSI problem that is inherently modular in nature, where each physical component of the problem is treated individually in each subproblem.
4. To establish stronger almost sure convergence of the random approximate solutions, one must first show weak convergence of probability measures, by establishing that the probability measures are **tight**. This requires the use of a *compactness result* alla Aubin-Lions, which form a general framework for constructive existence that generalizes to the well-posedness analysis of more complex nonlinear stochastic FSI systems.
5. Once weak convergence of the probability measures (laws) associated with the approximate solutions is established, probabilistic techniques based on the Skorohod representation theorem (and the Gyöngy-Krylov lemma in the case of additive noise) have to be employed to obtain almost sure convergence along a subsequence to a weak solution, either on a new probability space (for a weak martingale solution) or also on the original probability space (for a weak pathwise solution).

Chapter 9
Stochastic FSI: A Fully Coupled Model with Nonlinear Coupling

There are many applications that suggest and provide evidence to treating moving surfaces as random boundaries in order to deal with numerical, physical, and geometric uncertainties due to insufficient data or measurement errors. In particular, in hemodynamics, the systolic and diastolic rhythm of the heart has a strong stochastic component, giving rise to stochasticity in FSI describing the flow of blood in coronary arteries that sit on the surface of the heart. In fact, in [284], it was observed that the stochastic fluctuations of the single ion channel and the subcellular dynamics in tissue and organ scale get reflected in the macroscopic random cardiac events [284], which should be modeled using stochastic partial differential equations to capture such phenomena. In general, studying stochastic FSI is important because well-posedness of stochastic FSI models provides confidence that the deterministic FSI models are, indeed, robust to stochastic noise that occurs naturally in real-life problems.

In Chap. 8, we studied the existence of a pathwise solution to a *linearly* coupled stochastic FSI model, where the fluid and structure coupling conditions are evaluated along a fixed fluid structure interface, with a stochastically forced membrane and fluid. This is a common assumption when addressing the flow dynamics in smaller vessels positioned several branching levels away from the aorta, as they experience small movement. In this chapter, we will consider the flow dynamics in medium-to-large human arteries, such as the aorta or coronary arteries wherein the motion of fluid domain cannot be neglected.

This chapter presents a method to investigate solutions of a *nonlinearly coupled* stochastic fluid-structure interaction (FSI) problem. The focus is on the benchmark problem considered in Chap. 5, stochastically perturbed, involving a linearly elastic membrane/shell that interacts with a two-dimensional flow of a viscous, incompressible Newtonian fluid across a moving interface, driven by a stochastic noise. The fluid and the structure are coupled across the moving interface via a two-way coupling that ensures continuity of velocities and continuity of contact forces at the fluid-structure interface. The stochastic noise is applied both to the fluid equations as

S. Čanić et al., *Deterministic and Stochastic Fluid-Structure Interaction*, Advances in Mathematical Fluid Mechanics, https://doi.org/10.1007/978-3-032-00898-5_9

a volumetric body force and to the structure as an external forcing to the deformable fluid boundary. The noise is multiplicative depending on the structure displacement, structure velocity, and the fluid velocity, and it is of the form $G(\boldsymbol{u}, v, \eta)dW$, where W is a 1D Brownian motion.

We study the existence of weak martingale solutions (weak in the analytical sense and in the probabilistic sense) to this nonlinear stochastic fluid-structure interaction problem. The proof is based on a Lie-Trotter splitting scheme. However, compared to its deterministic counterpart studied in Chap. 5, the stochasticity leads to further challenges unique to this problem. These difficulties are also unlike the ones faced in Chap. 8, since the fixed geometry in the problem does not lead to the same considerations and issues as the ones in this chapter that arise due to the random and time-dependent motion of the fluid domain.

The results in the context of nonlinearly coupled stochastic FSI problem considered in this chapter were presented in [323].

9.1 Problem Setup

9.1.1 The Deterministic Model and a Weak Formulation

We recall quickly the setup considered in Chap. 5. Consider the flow of an incompressible, viscous fluid in a two-dimensional compliant cylinder with a deformable lateral boundary whose time-dependent motion is described by a one-dimensional membrane/shell equation capturing displacement only in the vertical direction. The left and the right boundary of the cylinder are the inlet and outlet for the time-dependent fluid flow. We assume "axial symmetry" of the data and of the flow, rendering the central horizontal line as the axis of symmetry. This allows us to consider the flow only in the upper half of the domain, with the bottom boundary fixed and equipped with the symmetry boundary conditions (Fig. 9.1).

The time-dependent fluid domain, whose displacement is not known a priori, will be denoted by

$$\mathcal{O}_\eta(t) = \{\mathbf{x} = (z, r) \in \mathbb{R}^2 : z \in (0, L), r \in (0, R + \eta(t, z))\},$$

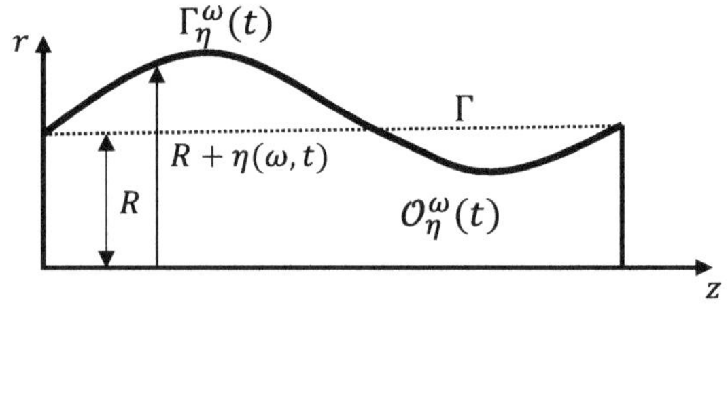

Fig. 9.1 A sketch of a realization of the fluid domain $\mathcal{O}_\eta(t)$ with the elastic lateral boundary $\Gamma_\eta(t)$, the inlet and outlet boundaries Γ_{in} and Γ_{out}, and the bottom (symmetry) boundary Γ_b

where the top lateral boundary is given by

$$\Gamma_\eta(t) = \{(z, r) \in \mathbb{R}^2 : z \in (0, L), r = R + \eta(t, z)\}.$$

The inlet, outlet, and bottom boundaries are $\Gamma_{\text{in}} = \{0\} \times (0, R)$, $\Gamma_{\text{out}} = \{L\} \times (0, R)$, $\Gamma_b = (0, L) \times \{0\}$ respectively.

The Fluid Subproblem The fluid flow is modeled by the incompressible Navier-Stokes equations in the 2D time-dependent domains $\mathcal{O}_\eta(t)$:

$$\begin{aligned} \partial_t \boldsymbol{u} + (\boldsymbol{u} \cdot \nabla)\boldsymbol{u} &= \nabla \cdot \sigma + \boldsymbol{F}_u^{ext} \\ \nabla \cdot \boldsymbol{u} &= 0, \end{aligned} \tag{9.1}$$

where $\boldsymbol{u} = (u_z, u_r)$ is the fluid velocity. The Cauchy stress tensor is $\sigma = -p\boldsymbol{I} + 2\nu\mathbf{D}(\boldsymbol{u})$ where p is the fluid pressure, ν is the kinematic viscosity coefficient, and $\mathbf{D}(\boldsymbol{u}) = \frac{1}{2}(\nabla\boldsymbol{u} + (\nabla\boldsymbol{u})^T)$ is the symmetrized gradient. Here $\boldsymbol{F}_u^{ext}$ represents any external forcing impacting the fluid, which in this chapter will be stochastic. The fluid flow is driven by dynamic pressure data given at the inlet and the outlet boundaries as follows:

$$\begin{aligned} p + \frac{1}{2}|\boldsymbol{u}|^2 &= P_{in/out}(t), \\ u_r = 0 &\quad \text{on } \Gamma_{in/out}. \end{aligned} \tag{9.2}$$

The deterministic inlet and outlet pressure data $P_{in/out} \in L^2(0, T)$ is given. On the bottom boundary Γ_b, we prescribe the symmetry boundary condition:

$$u_r = \partial_r u_z = 0 \quad \text{on } \Gamma_b. \tag{9.3}$$

Note that, in Chap. 8, the time-dependent Stokes equations are used to model the fluid flow, whereas in the present chapter, we consider the Navier-Stokes equations.

The Structure Subproblem The elastodynamics problem is given by the linearly elastic Koiter shell equations that describe the vertical or radial displacement η:

$$\partial_{tt}\eta - \partial_z^2\eta + \partial_z^4\eta = F_\eta \quad \text{in } (0, L), \tag{9.4}$$

where F_η is the vertical component of the total force experienced by the structure. Here, we have set the elasticity coefficients and the density of the structure to 1. As we shall see below, F_η will correspond to the difference between the fluid traction on one side and external random forcing on the other. Equation (9.4) is supplemented with the following boundary conditions:

$$\eta(0) = \eta(L) = \partial_z\eta(0) = \partial_z\eta(L) = 0. \tag{9.5}$$

The Nonlinear Fluid-Structure Coupling The coupling between the fluid and the structure takes place across the current location of the fluid-structure interface, which is simply the current location of the membrane/shell Γ_η. We consider a two-way coupling described by the kinematic and dynamic coupling conditions that describe continuity of velocity and continuity of normal stress at the fluid-structure interface, respectively:

- The kinematic coupling condition is:

$$(0, \partial_t \eta(t, z)) = \boldsymbol{u}(t, z, R + \eta(t, z)). \tag{9.6}$$

- The dynamic coupling condition is:

$$F_\eta = -\mathcal{J}^\eta(t, z)(\sigma \mathbf{n})|_{(t,z,\eta(t,z))} \cdot \mathbf{e}_r + F_\eta^{ext} \cdot \mathbf{e}_r, \tag{9.7}$$

where $\mathbf{n}$ is the unit outward normal to the top boundary, $\mathbf{e}_r$ is the unit vector in the vertical/radial direction, and $\mathcal{J}^\eta(t, z) = \sqrt{1 + (\partial_z \eta(t, z))^2}$ is the Jacobian of the transformation from Eulerian to Lagrangian coordinates. As earlier, F_η^{ext} denotes any external force impacting the structure, which in this work will be a stochastic.

This system is supplemented with the following (deterministic) initial conditions:

$$\boldsymbol{u}(t = 0) = \boldsymbol{u}_0, \eta(t = 0) = \eta_0, \partial_t \eta(t = 0) = v_0. \tag{9.8}$$

Weak Formulation on Moving Domains
Recall that in Chap. 5 Definition 5.2.1, we established the definition of a weak solution to the system (9.1)–(9.8). That is, we look for $(\boldsymbol{u}, \eta) \in \mathcal{V}$ that solves the following weak formulation for all test functions $(\boldsymbol{q}, \psi) \in \mathcal{Q}$:

$$\begin{aligned}
&-\int_0^t \int_{\mathcal{O}_{\eta(s)}} \boldsymbol{u} \cdot \partial_s \boldsymbol{q} d\boldsymbol{x} ds - \int_0^t \int_0^L \partial_s \eta \partial_s \psi dz ds \\
&+ \frac{1}{2} \int_0^t \int_{\mathcal{O}_\eta(s)} ((\boldsymbol{u} \cdot \nabla)\boldsymbol{u} \cdot \mathbf{q} - (\boldsymbol{u} \cdot \nabla)\mathbf{q} \cdot \boldsymbol{u}) d\boldsymbol{x} + 2\nu \int_0^t \int_{\mathcal{O}_\eta(s)} \mathbf{D}(\boldsymbol{u}) \cdot \mathbf{D}(\boldsymbol{q}) d\boldsymbol{x} ds \\
&- \frac{1}{2} \int_0^t \int_0^L (\partial_s \eta)^2 \psi dz ds + \int_0^t \int_0^L (\partial_z \eta \partial_z \psi + \partial_{zz} \eta \partial_{zz} \psi) dz ds \\
&= \int_{\mathcal{O}_{\eta_0}} \boldsymbol{u}_0 \boldsymbol{q}(0) d\boldsymbol{x} + \int_0^L v_0 \psi(0) dz + \int_0^t P_{in}(s) \int_0^1 q_z \Big|_{z=0} dr ds \\
&- \int_0^t P_{out}(s) \int_0^1 q_z \Big|_{z=1} dr ds \\
&+ \int_0^t \int_{\mathcal{O}_\eta(s)} \boldsymbol{q} \cdot \boldsymbol{F}_u^{ext} \, d\boldsymbol{x} ds + \int_0^t \int_0^L \psi F_\eta^{ext} \, dz ds,
\end{aligned} \tag{9.9}$$

where $\boldsymbol{F}_u^{ext}$ is the volumetric external force and F_η^{ext} is the external force applied to the deformable boundary. The spaces $\mathcal{V}$ and $\mathcal{Q}$ are defined in (5.28) and (5.29), respectively.

9.1.2 The Stochastic Framework and Definition of Martingale Solutions

We will incorporate the stochastic effects by considering $\boldsymbol{F}_u^{ext}$ and F_η^{ext} to be multiplicative stochastic forces as in Chap. 8 (see (8.5)). However, for simplicity, these forces will be combined into one multiplicative random force F^{ext} the noise coefficient $G(\boldsymbol{u}, v, \eta)$ of which is defined on a product space and is made precise in Assumption 9.1.1 below, satisfying certain assumptions specified in (9.21). That is, we write the combined stochastic forcing F^{ext} in terms of this G as follows:

$$F^{ext} := G(\boldsymbol{u}, v, \eta)dW, \tag{9.10}$$

where $\boldsymbol{u}$ is the fluid velocity, v is the structure velocity, η is the structure displacement, and W is a standard 1D Brownian motion. More precisely, the stochastic noise term is defined on a filtered probability space $(\Omega, \mathcal{F}, (\mathcal{F}_t)_{t\geq 0}, \mathbb{P})$ that satisfies the usual assumptions (see Definition 3.3.2), i.e., $\mathcal{F}_0$ is complete, and the filtration is right continuous, that is, $\mathcal{F}_t = \cap_{s\geq t}\mathcal{F}_s$ for all $t \geq 0$. We assume that W is a standard 1D Brownian motion with respect to the filtration $(\mathcal{F}_t)_{t\geq 0}$.

We will use this framework to define a martingale solution to our stochastic FSI problem. In order to do this, we will transform the problem defined on moving domains onto a fixed reference domain using a family of Arbitrary Lagrangian-Eulerian (ALE) mappings.

ALE Mappings

To deal with the geometric nonlinearity arising due to the motion of the fluid domain, we will use the Arbitrary Lagrangian-Eulerian (ALE) mappings which are a family of diffeomorphisms, parametrized by time t, from the fixed domain $\mathcal{O} = (0, L) \times (0, 1)$ onto the moving domain $\mathcal{O}_\eta(t)$. The ALE mappings were discussed previously in Sect. 2 and then in Sect. 5.2.1 for the prototypical nonlinearly coupled *deterministic* FSI problem, but we note that an important additional difficulty in the case of the nonlinearly coupled *stochastic* FSI problem considered in this chapter is that the presence of the stochastic forcing implies that the domains $\mathcal{O}_\eta$ are themselves *random*. Hence, to take into account our stochastic setting, we consider for any sample $\omega \in \Omega$, the ALE mappings defined *pathwise*:

$$A_\eta^\omega(t) : \mathcal{O} \to \mathcal{O}_\eta(t, \omega) \text{ given by } A_\eta^\omega(t)(z, r) = (z, (R + \eta(t, z, \omega))r). \tag{9.11}$$

Then, given any $\omega \in \Omega$, for as long as the Jacobian of the ALE mapping

$$|\det \nabla A^{\omega}_{\eta}(t)| = |R + \eta(t, z, \omega)| \tag{9.12}$$

is bounded and bounded away from zero and continuous, the map

$$F_t : L^2(\mathcal{O}_\eta(t)) \to L^2(\mathcal{O}) \text{ given by } F_t(f)(z,r) = f[(A^{\omega}_{\eta}(t)(z,r))] \tag{9.13}$$

is well-defined and

$$\|F_t(f)\|_{L^2(\mathcal{O})} \leq C\|f\|_{L^2(\mathcal{O}_\eta(t))},$$

where the constant $C > 0$ depends on the lower and upper bounds of the Jacobian of the ALE mappings. Hereon, we will suppress the notation ω, and the dependence of the variables on ω will be understood implicitly.

As in Sect. 5.2.1, under the transformation given in (9.11), the pathwise transformed gradient and symmetrized gradient will be given by (cf. (5.32))

$$\nabla^\eta = \left(\partial_z - r\frac{\partial_z \eta}{R+\eta}\partial_r, \frac{1}{R+\eta}\partial_r\right) \text{ and } \mathbf{D}^\eta(\boldsymbol{u}) = \frac{1}{2}(\nabla^\eta \boldsymbol{u} + (\nabla^\eta)^T \boldsymbol{u}).$$

Using $\mathbf{w}^\eta$ to denote the ALE velocity:

$$\mathbf{w}^\eta = (\partial_t \eta) r \mathbf{e}_r,$$

we introduce the advection term as follows:

$$b^\eta(\boldsymbol{u}, \boldsymbol{u}, \boldsymbol{q}) = \frac{1}{2}\int_{\mathcal{O}} (R+\eta)\left(((\boldsymbol{u} - \mathbf{w}^\eta)\cdot\nabla^\eta)\boldsymbol{u}\cdot\boldsymbol{q} - ((\boldsymbol{u} - \mathbf{w}^\eta)\cdot\nabla^\eta)\boldsymbol{q}\cdot\boldsymbol{u}\right). \tag{9.14}$$

We will now define the notion of martingale solutions to our stochastic FSI problem.

Definition of Martingale Solutions

We start by introducing the functional framework for the stochastic problem on the fixed reference domain $\mathcal{O} = (0, L) \times (0, 1)$. We will denote the top boundary by

$$\Gamma = (0, L) \times \{1\}$$

which can be identified with $(0, L)$. The following are the function spaces for the stochastic FSI problem defined on the fixed domain $\mathcal{O}$:

$$V = \{\mathbf{u} = (u_z, u_r) \in \mathbf{H}^1(\mathcal{O}) : u_z = 0 \text{ on } \Gamma, u_r = 0 \text{ on } \partial\mathcal{O} \setminus \Gamma\}, \tag{9.15}$$

$$\mathscr{W}_F = L^2(\Omega; L^\infty(0, T; \mathbf{L}^2(\mathcal{O}))) \cap L^2(\Omega; L^2(0, T; V)), \tag{9.16}$$

$$\mathscr{W}_S = L^2(\Omega; W^{1,\infty}(0, T; L^2(0, L)) \cap L^\infty(0, T; H_0^2(0, L))), \tag{9.17}$$

$$\mathscr{W}(0, T) = \{(\boldsymbol{u}, \eta) \in \mathscr{W}_F \times \mathscr{W}_S : \boldsymbol{u}(t, z, 1) = \partial_t \eta(t, z)\mathbf{e}_r \text{ and } \nabla^\eta \cdot \boldsymbol{u} = 0 \ \mathbb{P} - a.s.\}. \tag{9.18}$$

The test space is defined as follows:

$$\mathscr{D} = \{(\mathbf{q}, \psi) \in V \times H_0^2(0, L) : \mathbf{q}(z, 1) = \psi(z)\mathbf{e}_r\}. \tag{9.19}$$

Very often we will work with the following space for the fluid and structure velocities:

$$\mathscr{U} = \{(\boldsymbol{u}, v) \in V \times L^2(0, L) : \boldsymbol{u}(z, 1) = v(z)\mathbf{e}_r\}. \tag{9.20}$$

We further introduce the following notation:

$$\mathbf{L}^2 := \mathbf{L}^2(\mathcal{O}) \times L^2(0, L).$$

Before we define a martingale solution, we specify the stochastic noise as follows:

Assumption 9.1.1 The noise coefficient G is a function $G : \mathscr{U} \times L^\infty(0, L) \to \mathbf{L}^2$ such that the following Lipschitz continuity assumptions hold:

$$\begin{aligned}
&\|G(\boldsymbol{u}, v, \eta)\|_{\mathbf{L}^2} \leq \|\eta\|_{L^\infty(0,L)}\|\mathbf{u}\|_{\mathbf{L}^2(\mathcal{O})} + \|v\|_{L^2(0,L)},\\
&\|G(\boldsymbol{u}_1, v_1, \eta) - G(\boldsymbol{u}_2, v_2, \eta)\|_{\mathbf{L}^2} \leq \|\eta\|_{L^\infty(0,L)}\|\boldsymbol{u}_1 - \boldsymbol{u}_2\|_{\mathbf{L}^2(\mathcal{O})} + \|v_1 - v_2\|_{L^2(0,L)},\\
&\|G(\boldsymbol{u}, v, \eta_1) - G(\boldsymbol{u}, v, \eta_2)\|_{\mathbf{L}^2} \leq \|\eta_1 - \eta_2\|_{L^\infty(0,L)}\|\boldsymbol{u}\|_{\mathbf{L}^2(\mathcal{O})}.
\end{aligned} \tag{9.21}$$

As example of such a noise coefficient G is a linearly multiplicative noise transformed onto the fixed domain. More precisely, G can take the form

$$(G(\boldsymbol{u}, v, \eta), (\boldsymbol{q}, \psi))_{\mathbf{L}^2} := \alpha\left((R + \eta)\boldsymbol{u}, \boldsymbol{q})_{\mathbf{L}^2(\mathcal{O})} + (v, \psi)_{L^2(0,L)}\right), \tag{9.22}$$

where $\alpha \in \mathbb{R}$.

We remark here that a more general noise, taking values in any Hilbert space, is considered in [323]. We are now in position to define solutions to the system (9.1)–(9.8) in the stochastic setting.

Definition 9.1.1 (Martingale Solution) Let $\boldsymbol{u}_0 \in \mathbf{L}^2(\mathcal{O})$ and $v_0 \in L^2(0, L)$ be deterministic initial data, and let $\eta_0 \in H_0^2(0, L)$ be a compatible initial structure configuration such that for some $\delta > 0$, the initial configuration satisfies

$$\delta < R + \eta_0(z), \quad \forall z \in [0, L], \quad \text{and} \quad \|R + \eta_0\|_{H_0^2(0,L)} < \frac{1}{\delta}. \tag{9.23}$$

We say that $(\mathscr{S}, \boldsymbol{u}, \eta, \tau)$ is a martingale solution to the system (9.1)–(9.8) under the assumptions (9.21) if:

1. $\mathscr{S} = (\Omega, \mathcal{F}, (\mathcal{F}_t)_{t\geq 0}, \mathbb{P}, W)$ is a stochastic basis, that is, $(\Omega, \mathcal{F}, (\mathcal{F}_t)_{t\geq 0}, \mathbb{P})$ is a filtered probability space and W is a 1D Wiener process;
2. $(\boldsymbol{u}, \eta) \in \mathscr{W}(0, T)$;
3. τ is a $\mathbb{P}$-a.s. strictly positive $(\mathcal{F}_t)_{t\geq 0}$-stopping time;
4. $\mathbf{U} = (\boldsymbol{u}, \partial_t \eta)$ and η are $(\mathcal{F}_t)_{t\geq 0}$-progressively measurable;
5. For every $(\mathcal{F}_t)_{t\geq 0}$-adapted, essentially bounded process (see (3.1) with $p = \infty$) $\mathbf{Q} := (\boldsymbol{q}, \psi)$ with C^1 paths in $\mathscr{D}$ such that $\nabla^\eta \cdot \boldsymbol{q} = 0$, the following equation holds $\mathbb{P}$-a.s. for almost every $t \in [0, \tau)$:

$$\begin{aligned}
&\int_{\mathcal{O}} (R + \eta(t))\boldsymbol{u}(t)\boldsymbol{q}(t)d\boldsymbol{x} + \int_0^L \partial_t\eta(t)\psi(t)dz = \int_{\mathcal{O}} (R + \eta_0)\boldsymbol{u}_0\boldsymbol{q}(0)d\boldsymbol{x} \\
&+ \int_0^L v_0\psi(0)dz \\
&+ \int_0^t \int_{\mathcal{O}} (R + \eta(s))\boldsymbol{u}(s) \cdot \partial_s\boldsymbol{q}(s)d\boldsymbol{x}ds + \frac{1}{2}\int_0^t \int_{\mathcal{O}} (\partial_s\eta(s))\boldsymbol{u}(s) \cdot \boldsymbol{q}(s)d\boldsymbol{x}ds \\
&- \int_0^t b^\eta(\boldsymbol{u}(s), \boldsymbol{u}(s), \boldsymbol{q}(s))ds - 2\nu \int_0^t \int_{\mathcal{O}} (R + \eta(s))\mathbf{D}^\eta(\boldsymbol{u}(s)) \cdot \mathbf{D}^\eta(\boldsymbol{q}(s))d\boldsymbol{x}ds \\
&+ \int_0^t \int_0^L \partial_s\eta(s)\partial_s\psi(s)dzds - \int_0^t \int_0^L \partial_z\eta(s)\partial_z\psi(s) + \partial_{zz}\eta(s)\partial_{zz}\psi(s)dzds \\
&+ \int_0^t \left(P_{in} \int_0^1 q_z\Big|_{z=0} dr - P_{out} \int_0^1 q_z\Big|_{z=1} dr\right) ds \\
&+ \int_0^t (\mathbf{Q}(s), G(\mathbf{U}(s), \eta(s)))dW(s).
\end{aligned} \tag{9.24}$$

9.2 Operator Splitting Scheme

In this section, we introduce a Lie operator splitting scheme that defines a sequence of approximate solutions to (9.24) by semi-discretizing the problem in time. We discretize the problem in time by partitioning the time interval $(0, T)$ and construct approximate solutions in the following way. At each time step, the multi-physics problem is split into two subproblems: the structure subproblem and the fluid subproblem. In the first subproblem, the structure displacement and the structure velocity are updated while keeping the fluid velocity the same. In the second subproblem, keeping the structure displacement unchanged, the fluid velocity is updated on an artificially defined domain, while also ensuring that the kinematic coupling condition is satisfied. The final goal is to show that up to a subsequence, as the time-step size goes to 0, approximate solutions converge in a certain sense to a martingale solution of the stochastic FSI problem.

9.2.1 Definition of the Splitting Scheme

We semidiscretize the problem in time and use operator splitting to divide the coupled stochastic problem into two subproblems, a structure and a fluid subproblem. We denote the time step by $\Delta t = \frac{T}{N}$ and use the notation $t_N^n = n\Delta t$ for $n = 0, 1, ..., N$.

Let $(\boldsymbol{u}^0, \eta^0, v^0) = (\boldsymbol{u}_0, \eta_0, v_0)$ be the initial data. At the ith time level, we update the vector $(\boldsymbol{u}_N^{n+\frac{i}{2}}, \eta_N^{n+\frac{i}{2}}, v_N^{n+\frac{i}{2}})$, where $i = 1, 2$ and $n = 0, 1, ..., N-1$, according to the following scheme.

The Structure Subproblem
We update the structure displacement and the structure velocity while keeping the fluid velocity the same. That is, for $n \in \mathbb{N}$ given $(\eta_N^n, v_N^n) \in H_0^2(0, L) \times L^2(0, L)$, we look for a pair $(\eta_N^{n+\frac{1}{2}}, v_N^{n+\frac{1}{2}}) \in H_0^2(0, L) \times H_0^2(0, L)$ that satisfies the following equations pathwise, i.e., for each $\omega \in \Omega$:

$$\begin{aligned}
\boldsymbol{u}_N^{n+\frac{1}{2}} &= \boldsymbol{u}_N^n,\\
\int_0^L (\eta_N^{n+\frac{1}{2}} - \eta_N^n)\phi dz &= (\Delta t)\int_0^L v_N^{n+\frac{1}{2}}\phi dz,\\
\int_0^L \left(v_N^{n+\frac{1}{2}} - v_N^n\right)\psi dz + (\Delta t)\int_0^L &\partial_z \eta_N^{n+\frac{1}{2}}\partial_z\psi + \partial_{zz}\eta_N^{n+\frac{1}{2}}\partial_{zz}\psi dz = 0,
\end{aligned} \tag{9.25}$$

for any $\phi \in L^2(0, L)$ and $\psi \in H_0^2(0, L)$.

Before commenting on the existence of the random variables $\eta_N^{n+\frac{1}{2}}, v_N^{n+\frac{1}{2}}$ and their measurability properties, we introduce the second subproblem.

The Fluid Subproblem
In this step, we update the fluid and the structure velocities while keeping the structure displacement unchanged as done in Chap. 5. In the stochastic case, however, there are two major difficulties that need to be overcome before we can even define the fluid subproblem, which are both due to the fact that the fluid domain is a random variable not known a priori. While similar issues arise in the deterministic case as well, their resolution in the stochastic case is remarkably different.

1. First, as the problem is mapped onto the fixed, reference domain $\mathcal{O}$, see (9.24), the Jacobian of the ALE transformation (9.12) appears is several terms. Unlike in Sect. 5.4.1 of Chap. 5, the fluid domain can possibly degenerate in a random fashion; e.g., the top boundary can touch the bottom of the fluid domain in at a *random* time. That is, Proposition 5.5.1 cannot be applied to obtain a *deterministic* time $T > 0$ until which we can find a uniform lower bound for the Jacobian of the ALE map.

   Handling this no-contact condition in the stochastic case requires a delicate approach, which is done by introducing an appropriate stopped process and a stopping time argument. Hence we introduce an "artificial" structure displacement variable $\eta^n_{*,N}$, which is a discretized version of a stopped process, defined via a cutoff function, of the structure displacement obtained in the structure subproblem (9.25). The cutoff function artificially maintains a minimum distance of $\delta > 0$ between the walls of the tube, preventing them from coming into contact while also providing the required regularity for any realization of the structure. This is crucial as it provides a *deterministic* lower (and upper) bound for the artificial structure displacement. This artificial structure displacement is defined and introduced in the problem in a way that also ensures unconditional stability of our time-marching scheme. At every time step, we then solve the fluid equations with respect to this artificial domain, determined by $\eta^n_{*,N}$, and then show in Lemma 9.4.4 that there exists a time (stopping time) that is a.s. strictly greater than zero, such that the cutoff in the definition of the artificial displacement is, in fact, never used until possibly that stopping time, thereby providing a solution to the original problem without the cutoff until the stopping time.
2. Second, for the problem mapped onto the fixed domain $\mathcal{O}$, the divergence-free condition is a problem when working with the mapped test functions, as we explain below. To avoid working with the divergence-free test functions in the approximate problems, we relax the divergence-free condition by supplementing the weak formulation by a penalty term of the form $\frac{1}{\varepsilon}\int \operatorname{div}^{\eta}\boldsymbol{u}\,\operatorname{div}^{\eta}\boldsymbol{q}$, and work with artificial compressibility until the very end, when we recover the solution to the original incompressible (divergence-free) problem by letting $\varepsilon \to 0$. The test functions in the fluid equations in this subproblem would normally depend on the structure displacement η^n_N found in the previous subproblem—recall the deterministic semi-discrete weak formulation (5.47) and corresponding test space (5.46). This means that we would use test functions $\boldsymbol{q}^{n+1}_N \in \mathscr{U}$ that satisfy $\operatorname{div}^{\eta^n_{*,N}}\boldsymbol{q}^{n+1}_N = 0$. This dependence of the fluid test functions on $\eta^n_{*,N}$, which in our case is a random variable, causes problems in the second part of the existence proof when we apply Theorem 3.2.5 to construct a new probability space and new random variables, with the same laws, that converge almost surely in the topology of the associated phase space. However, it is not clear that all the admissible random test functions on the *new probability space* would necessarily come from the divergence-free test functions built in this subproblem. This is why we introduce artificial compressibility via the penalty term mentioned above

and avoid dealing with the divergence-free condition until the very end, when we recover the divergence-free solution in Sect. 9.4.

Construction of The Cuts =off Displacement For $\delta > 0$, let Θ_δ be the step function such that $\Theta_\delta(x, y) = 1$ if $\delta < x, y < \frac{1}{\delta}$, and $\Theta_\delta(x, y) = 0$ otherwise. Using this function, we define a real-valued function $\theta_\delta(\eta_N^n)$ which tracks all the structure displacements until the time step n and is equal to 1 until the step for which the structure leaves the desired bounds given in terms of δ:

$$\theta_\delta(\eta_N^n) := \min_{k\leq n} \Theta_\delta \left(\inf_{z\in[0,L]} (\eta_N^k(z) + R), \|\eta_N^k + R\|_{H^s(0,L)} \right), \tag{9.26}$$

for a fixed s satisfying $s \in (\frac{3}{2}, 2)$. Now we define the artificial structure displacement random variable as follows:

$$\eta_{*,N}^n(z, \omega) = \eta_N^{\max_{0\leq k\leq n} \theta_\delta(\eta_N^k) k}(z, \omega) \quad \text{for every } \omega \in \Omega. \tag{9.27}$$

Note that the superscript in the definition above indicates the time step and not the power of structure displacement. Compare this definition with Definition 3.3.12.

A Divergence-Free Relaxation via Penalty We introduce a divergence-free penalty term and define the fluid subproblem as follows. Let $\Delta_{n,N} W := W(t_N^{n+1}) - W(t_N^n)$.

Then for $\varepsilon > 0$ and given $\mathbf{U}_N^n = (\boldsymbol{u}_N^n, v_N^n) \in \mathscr{U}$, we look for $(\boldsymbol{u}_N^{n+1}, v_N^{n+1})$ that solves

$$\eta_N^{n+1} := \eta_N^{n+\frac{1}{2}},$$

$$\begin{aligned}
&\int_{\mathcal{O}} (R + \eta_{*,N}^n) \left(\boldsymbol{u}_N^{n+1} - \boldsymbol{u}_N^{n+\frac{1}{2}} \right) \boldsymbol{q} d\boldsymbol{x} + \frac{1}{2} \int_{\mathcal{O}} \left(\eta_{*,N}^{n+1} - \eta_{*,N}^n \right) \boldsymbol{u}_N^{n+1} \cdot \boldsymbol{q} d\boldsymbol{x} \\
&+ \frac{\Delta t}{2} \int_{\mathcal{O}} (R + \eta_{*,N}^n)((\boldsymbol{u}_N^n - v_N^{n+\frac{1}{2}} r \mathbf{e}_r) \cdot \nabla^{\eta_{*,N}^n} \boldsymbol{u}_N^{n+1} \cdot \boldsymbol{q} - (\boldsymbol{u}_N^n - v_N^{n+\frac{1}{2}} r \mathbf{e}_r) \\
&\cdot \nabla^{\eta_{*,N}^n} \boldsymbol{q} \cdot \boldsymbol{u}_N^{n+1}) d\boldsymbol{x} \\
&+ 2\nu(\Delta t) \int_{\mathcal{O}} (R + \eta_{*,N}^n) \mathbf{D}^{\eta_{*,N}^n}(\boldsymbol{u}_N^{n+1}) \cdot \mathbf{D}^{\eta_{*,N}^n}(\boldsymbol{q}) d\boldsymbol{x} \\
&+ \frac{(\Delta t)}{\varepsilon} \int_{\mathcal{O}} \operatorname{div}^{\eta_{*,N}^n} \boldsymbol{u}_N^{n+1} \operatorname{div}^{\eta_{*,N}^n} \boldsymbol{q} d\boldsymbol{x} \\
&+ \int_0^L (v_N^{n+1} - v_N^{n+\frac{1}{2}}) \psi dz \\
&= (\Delta t) \left(P_{N,in}^n \int_0^1 q_z\Big|_{z=0} dr - P_{N,out}^n \int_0^1 q_z\Big|_{z=1} dr \right) + (G(\mathbf{U}_N^n, \eta_{*,N}^n) \Delta_{n,N} W, \mathbf{Q}),
\end{aligned} \tag{9.28}$$

for any $\mathbf{Q} = (\boldsymbol{q}, \psi) \in \mathscr{U}$ with

$$\boldsymbol{u}_N^{n+1}|_\Gamma = v_N^{n+1}\mathbf{e}_r.$$

Here,

$$P_{in/out}^n := \frac{1}{\Delta t}\int_{t_N^n}^{t_N^{n+1}} P_{in/out}\,dt, \quad \mathrm{div}^\eta \boldsymbol{u} = \mathrm{trace}(\nabla^\eta \boldsymbol{u}).$$

Remark 9.2.1 Observe that using the data from the second subproblem, we update η_N^n in the first subproblem and not $\eta_{*,N}^n$. This is crucial for obtaining a discrete version of the energy inequality and a stable time-marching scheme.

Remark 9.2.2 The fluid subproblem defined this way implies that for any $\omega \in \Omega$ and fixed N, after a certain step n, we are possibly producing solutions η_N^n and thus $\boldsymbol{u}_N^n, v_N^n$ that are meaningless. But, as mentioned earlier, that is not an issue, since later we will restrict our attention to only the time spans for which the limiting solutions, corresponding to the approximations constructed in Sect. 9.2.2 using η_N^n's and $\eta_{*,N}^n$'s are equal. This will result in a desired solution until a stopping time.

We are now ready to prove the existence of solutions to the two subproblems. For this purpose, we introduce the following discrete energy and dissipation for $i = 0, 1$:

$$\begin{aligned}
E_N^{n+\frac{i}{2}} &= \frac{1}{2}\Big(\int_{\mathcal{O}} (R + \eta_{*,N}^n)|\boldsymbol{u}_N^{n+\frac{i}{2}}|^2 d\boldsymbol{x} + \|v_N^{n+\frac{i}{2}}\|_{L^2(0,L)}^2 \\
&\quad + \|\partial_z \eta_N^{n+\frac{i}{2}}\|_{L^2(0,L)}^2 + \|\partial_{zz}\eta_N^{n+\frac{i}{2}}\|_{L^2(0,L)}^2\Big), \\
D_N^n &= \Delta t \int_{\mathcal{O}} \Big(2\nu(R + \eta_{N,*}^n)|\mathbf{D}^{\eta_{*,N}^n}(\boldsymbol{u}_N^{n+1})|^2 + \frac{1}{\varepsilon}|\mathrm{div}^{\eta_{*,N}^n}\boldsymbol{u}_N^{n+1}|^2\Big)\,d\boldsymbol{x}.
\end{aligned} \tag{9.29}$$

Lemma 9.2.1 (Existence for the Structure Subproblem) *Assume that η_N^n and v_N^n are $H_0^2(0, L)$ and $L^2(0, L)$-valued $\mathcal{F}_{t_N^n}$-measurable random variables, respectively. Then there exist $H_0^2(0, L)$- valued $\mathcal{F}_{t_N^n}$-measurable random variables $\eta_N^{n+\frac{1}{2}}, v_N^{n+\frac{1}{2}}$ that solve* (9.25)*, and the following semidiscrete energy inequality holds:*

$$E_N^{n+\frac{1}{2}} + C_{1,N}^n = E_N^n, \tag{9.30}$$

where

$$C_{1,N}^n := \frac{1}{2}\|v_N^{n+\frac{1}{2}} - v_N^n\|_{L^2}^2 + \frac{1}{2}\|\partial_z\eta_N^{n+\frac{1}{2}} - \partial_z\eta_N^n\|_{L^2(0,L)}^2 + \frac{1}{2}\|\partial_{zz}\eta_N^{n+\frac{1}{2}} - \partial_{zz}\eta_N^n\|_{L^2(0,L)}^2,$$

corresponds to numerical dissipation.

Proof See Proposition 8.5.1 in Chap. 8. □

Lemma 9.2.2 (Existence for the Fluid Subproblem) *For given* $\delta > 0$, *and given* $\mathcal{F}_{t_N^n}$*-measurable random variables* $(\boldsymbol{u}_N^n, v_N^n)$ *taking values in* $\mathscr{U}$ *and* $v_N^{n+\frac{1}{2}}$ *taking values in* $H_0^2(0, L)$, *there exists an* $\mathcal{F}_{t_N^{n+1}}$*-measurable random variable* $(\boldsymbol{u}_N^{n+1}, v_N^{n+1})$ *taking values in* $\mathscr{U}$ *that solves* (9.28), *and the solution satisfies the following energy estimate for some constant* $C > 0$ *independent of* N:

$$
\begin{aligned}
E_N^{n+1} + D_N^n + C_{2,N}^n \le E_N^{n+\frac{1}{2}} &+ C\Delta t((P_{N,in}^n)^2 + (P_{out}^n)^2) \\
&+ C|\Delta_{n,N} W|^2 \|G(\mathbf{U}_N^n, \eta_{*,N}^n)\|_{\mathbf{L}^2}^2 + |(G(\mathbf{U}_N^n, \eta_{*,N}^n)\Delta_{n,N} W, \mathbf{U}_N^n)| \\
&+ \frac{1}{4}\int_0^L (v_N^{n+\frac{1}{2}} - v_N^n)^2 dz,
\end{aligned}
\tag{9.31}
$$

where

$$
C_{2,N}^n := \frac{1}{4}\int_{\mathcal{O}} (R + \eta_{*,N}^n)\left(|\boldsymbol{u}_N^{n+1} - \boldsymbol{u}_N^n|^2\right) d\boldsymbol{x} + \frac{1}{4}\int_0^L |v_N^{n+1} - v_N^{n+\frac{1}{2}}|^2 dz
$$

is numerical dissipation and $\eta_{*,N}^n$ *is as defined in* (9.27).

Proof Note that $\eta_{*,N}^n + R > \delta$ for every $n \le N$ and for every $N \in \mathbb{N}$. Hence, the existence a unique solution follows from the Lax-Milgram theorem, Lemma 2.2.1, as in Proposition 5.4.3. Measurability of this solution can be established using ideas identical to those in Theorem 8.5.3. We show that the solution satisfies energy estimate (9.31). For this purpose, we will derive a pathwise inequality involving the discrete energies. We take $(\boldsymbol{q}, \psi) = (\boldsymbol{u}^{n+1}, v^{n+1})$ in (9.28) and using the identity $a(a-b) = \frac{1}{2}(|a|^2 - |b|^2 + |a-b|^2)$, we obtain

$$
\begin{aligned}
&\frac{1}{2}\int_{\mathcal{O}} (R + \eta_{*,N}^n)\left(|\boldsymbol{u}_N^{n+1}|^2 - |\boldsymbol{u}_N^{n+\frac{1}{2}}|^2 + |\boldsymbol{u}_N^{n+1} - \boldsymbol{u}_N^{n+\frac{1}{2}}|^2\right) \\
&+ \frac{1}{2}\int_{\mathcal{O}} \left(\eta_{*,N}^{n+1} - \eta_{*,N}^n\right) |\boldsymbol{u}_N^{n+1}|^2 d\boldsymbol{x} \\
&+ 2\nu(\Delta t)\int_{\mathcal{O}} (R + \eta_{*,N}^n)|\mathbf{D}^{\eta_{*,N}^n}(\boldsymbol{u}_N^{n+1})|^2 d\boldsymbol{x} + \frac{(\Delta t)}{\varepsilon}\int_{\mathcal{O}} |\text{div}^{\eta_{*,N}^n} \boldsymbol{u}_N^{n+1}|^2 d\boldsymbol{x} \\
&+ \frac{1}{2}\int_0^L |v_N^{n+1}|^2 - |v_N^{n+\frac{1}{2}}|^2 + |v_N^{n+1} - v_N^{n+\frac{1}{2}}|^2 dz \\
&= (\Delta t)\left(P_{N,in}^n \int_0^1 (u_N^{n+1})_z\Big|_{z=0} dr - P_{N,out}^n \int_0^1 (u_N^{n+1})_z\Big|_{z=1} dr\right) \\
&+ (G(\mathbf{U}_N^n, \eta_{*,N}^n)\Delta_{n,N} W, (\mathbf{U}_N^{n+1} - \mathbf{U}_N^n)) + (G(\mathbf{U}_N^n, \eta_{*,N}^n)\Delta_{n,N} W, \mathbf{U}_N^n).
\end{aligned}
$$

Observe that the discrete stochastic integral is divided into two terms. We estimate the first term by using the Cauchy-Schwarz inequality to obtain that for some $C(\delta) > 0$ independent of n the following holds:

$$
\begin{aligned}
&|(G(\mathbf{U}_N^n, \eta_{*,N}^n)\Delta_{n,N}W, (\mathbf{U}_N^{n+1} - \mathbf{U}_N^n))| \\
&\leq C|\Delta_{n,N}W|^2\|G(\mathbf{U}_N^n, \eta_{*,N}^n)\|_{\mathbf{L}^2}^2 + \frac{1}{4}\int_{\mathcal{O}}(R + \eta_{*,N}^n)\left(\boldsymbol{u}_N^{n+1} - \boldsymbol{u}_N^n\right)^2 d\boldsymbol{x} \\
&\quad + \frac{1}{8}\int_0^L (v_N^{n+1} - v_N^n)^2 dz \\
&\leq C|\Delta_{n,N}W|^2\|G(\mathbf{U}_N^n, \eta_{*,N}^n)\|_{\mathbf{L}^2}^2 + \frac{1}{4}\int_{\mathcal{O}}(R + \eta_{*,N}^n)\left(\boldsymbol{u}_N^{n+1} - \boldsymbol{u}_N^n\right)^2 d\boldsymbol{x} \\
&\quad + \frac{1}{4}\int_0^L (v_N^{n+1} - v_N^{n+\frac{1}{2}})^2 dz + \frac{1}{4}\int_0^L (v_N^{n+\frac{1}{2}} - v_N^n)^2 dz.
\end{aligned}
$$

We treat the terms with $P_{in/out}$ similarly to obtain (9.31). This completes the proof of Lemma 9.2.2. □

Remark 9.2.3 Given any $T > 0$, we note here that we have established the existence of *global-in-time* solutions to the subproblems (9.25) and (9.28). That is, our splitting scheme has a solution $(\boldsymbol{u}_N^n, \eta_N^n)$ for *every* $0 \leq n \leq N$ and *every* $N \in \mathbb{N}$. (Compare with Proposition 5.5.1 in the deterministic setting where $T > 0$ is chosen to be small enough).

Now, we obtain uniform bounds on the expectation of the solutions produced through the operator splitting scheme. These uniform bounds are the analogue of the uniform bounds in expectation obtained in Proposition 8.5.5 in Chap. 8 for the linearly coupled stochastic FSI problem. An important distinction in the current setting of nonlinearly coupled stochastic FSI is that several quantities, such as the kinetic energy of the fluid and the fluid dissipation, are defined via integrals over the *random moving domain* (discretized in time by the operator splitting scheme). The following theorem provides uniform estimates on the expectation of the kinetic and elastic energy for the full, semidiscrete coupled problem (uniform in the number of time steps N and in ε). Uniform estimates on the expectation of the numerical dissipation are derived as well.

Theorem 9.2.1 (Uniform Estimates) *For any $N > 0$, $\Delta t = \frac{T}{N}$, there exists a constant $C > 0$ that depends on the initial data, δ, T, and $P_{in/out}$ and is independent of N and ε such that*

1. $\mathbb{E}\left(\max\limits_{1\leq n\leq N} E_N^n\right) < C, \mathbb{E}\left(\max_{0\leq n\leq N-1} E_N^{n+\frac{1}{2}}\right) < C.$
2. $\mathbb{E}\sum\limits_{n=0}^{N-1} D_N^n < C.$

3. $\mathbb{E}\sum_{n=0}^{N-1}\int_{\mathcal{O}}\left((R+\eta_{*,N}^n)|\boldsymbol{u}_N^{n+1}-\boldsymbol{u}_N^n|^2\right)d\boldsymbol{x}+\int_0^L|v_N^{n+1}-v_N^{n+\frac{1}{2}}|^2dz<C.$

4. $\mathbb{E}\sum_{n=0}^{N-1}\int_0^L\left(|v_N^{n+\frac{1}{2}}-v_N^n|^2+|\partial_z\eta_N^{n+1}-\partial_z\eta_N^n|^2+|\partial_{zz}\eta_N^{n+1}-\partial_{zz}\eta_N^n|^2\right)dz < C,$

where E_N^n and D_N^n are defined in (9.29).

Proof We add the energy estimates for the two subproblems (9.30) and (9.31) to obtain:

$$E_N^{n+1}+D_N^n+C_{1,N}^n+C_{2,N}^n\le E_N^n+C\Delta t((P_{N,in}^n)^2+(P_{N,out}^n)^2)$$
$$+C|\Delta_{n,N}W|^2\|G(\mathbf{U}_N^n,\eta_{*,N}^n)\|_{\mathbf{L}^2}^2+\left|(G(\mathbf{U}_N^n,\eta_{*,N}^n)\Delta_{n,N}W,\mathbf{U}_N^n)\right|. \tag{9.32}$$

Then for any $m\ge 1$, summing $0\le n\le m-1$ gives us

$$E_N^m+\sum_{n=0}^{m-1}D_N^n+\sum_{n=0}^{m-1}C_{1,N}^n+\sum_{n=0}^{m-1}C_{2,N}^n\le E^0+C\,\Delta t\sum_{n=0}^{m-1}\left((P_{N,in}^n)^2+(P_{N,out}^n)^2\right)$$
$$+\sum_{n=0}^{m-1}\left|(G(\mathbf{U}_N^n,\eta_{*,N}^n)\Delta_{n,N}W,\mathbf{U}_N^n)\right|+\sum_{n=0}^{m-1}\|G(\mathbf{U}_N^n,\eta_{*,N}^n)\|_{\mathbf{L}^2}^2|\Delta_{n,N}W|^2. \tag{9.33}$$

Next we take supremum over $1\le m\le N$ and then take expectation on both sides of (9.33). We begin by treating the right-hand side terms. First we have

$$(\Delta t)\sum_{n=0}^{N-1}(P_{N,in/out}^n)^2\le|P_{in/out}|_{L^2(0,T)}^2.$$

Next we apply the Burkholder-Davis-Gundy inequality stated in Theorem 3.6.3 and write $\boldsymbol{u}_N^{n+\frac{1}{2}}=\boldsymbol{u}_N^n$ to obtain for some $C(\delta)>0$ that

$$\mathbb{E}\max_{1\le m\le N}|\sum_{n=0}^{m-1}(G(\mathbf{U}_N^n,\eta_{*,N}^n)\Delta_{n,N}W,\mathbf{U}_N^n)|$$
$$\le C\mathbb{E}\left[\Delta t\sum_{n=0}^{N-1}\|G(\mathbf{U}_N^n,\eta_{*,N}^n)\|_{\mathbf{L}^2}^2\left(\left\|(\sqrt{R+\eta_{*,N}^n})\boldsymbol{u}_N^n\right\|_{\mathbf{L}^2(\mathcal{O})}^2+\|v_N^n\|_{L^2(0,L)}^2\right)\right]^{\frac{1}{2}}$$
$$\le C\mathbb{E}\Bigg[\left(\max_{0\le m\le N}\left\|(\sqrt{R+\eta_{*,N}^m})\boldsymbol{u}_N^m\right\|_{\mathbf{L}^2(\mathcal{O})}^2+\|v_N^m\|_{L^2(0,L)}^2\right).$$

$$\sum_{n=0}^{N-1} \Delta t \left(\| \sqrt{(R + \eta^n_{*,N})} \boldsymbol{u}^n_N \|^2_{\mathbf{L}^2(\mathcal{O})} + \| v^n_N \|^2_{L^2(0,L)} \right) \Bigg]^{\frac{1}{2}}$$

$$\leq \frac{1}{2} \| (\sqrt{R + \eta_0}) \boldsymbol{u}_0 \|^2_{\mathbf{L}^2(\mathcal{O})} + \frac{1}{2} \| v_0 \|^2_{L^2(0,L)}$$

$$+ \frac{1}{2} \mathbb{E} \max_{1 \leq m \leq N} \left[\left\| (\sqrt{R + \eta^m_{*,N}}) \boldsymbol{u}^m_N \right\|^2_{\mathbf{L}^2(\mathcal{O})} + \| v^m_N \|^2_{L^2(0,L)} \right]$$

$$+ C \Delta t \mathbb{E} \left(\sum_{n=0}^{N-1} \| (\sqrt{R + \eta^n_{*,N}}) \boldsymbol{u}^n_N \|^2_{\mathbf{L}^2(\mathcal{O})} + \| v^n_N \|^2_{L^2(0,L)} \right).$$

Moreover, using the tower property and (9.21) for each $n = 0, ..., m - 1$ we write

$$\begin{aligned} \mathbb{E}[\| G(\mathbf{U}^n_N, \eta^n_{*,N}) \|^2_{\mathbf{L}^2} |\Delta_{n,N} W|^2] &= \mathbb{E}[\mathbb{E}[\| G(\mathbf{U}^n_N, \eta^n_{*,N}) \|^2_{\mathbf{L}^2} |\Delta_{n,N} W|^2 | \mathcal{F}_{t^n_N}]] \\ &= \mathbb{E}[\| G(\mathbf{U}^n_N, \eta^n_{*,N}) \|^2_{\mathbf{L}^2} \mathbb{E}[|\Delta_{n,N} W|^2 \| \mathcal{F}_{t^n_N}]] \\ &= (\Delta t) \mathbb{E} \| G(\mathbf{U}^n_N, \eta^n_{*,N}) \|^2_{\mathbf{L}^2} \\ &\leq C \Delta t \mathbb{E}[\| \sqrt{(R + \eta^n_{*,N})} \boldsymbol{u}^n_N \|^2_{\mathbf{L}^2(\mathcal{O})} + \| v^n_N \|^2_{L^2(0,L)}]. \end{aligned} \tag{9.34}$$

Thus we obtain for some $C > 0$ depending on δ that the following holds:

$$\begin{aligned} \mathbb{E} \max_{1 \leq n \leq N} E^n_N + \mathbb{E} \sum_{n=0}^{N-1} D^n_N + \mathbb{E} \sum_{n=0}^{N-1} C^n_{1,N} + \mathbb{E} \sum_{n=0}^{N-1} C^n_{2,N} &\leq C E^0 + C \| P_{in/out} \|^2_{L^2(0,T)} \\ &+ C \Delta t \mathbb{E} \left[\sum_{n=0}^{N-1} \| (\sqrt{R + \eta^n_{*,N}}) \boldsymbol{u}^n_N \|^2_{\mathbf{L}^2(\mathcal{O})} + \| v^n_N \|^2_{L^2(0,L)} \right] \\ &+ \frac{1}{2} \mathbb{E} \max_{1 \leq n \leq N} \left[\left\| (\sqrt{R + \eta^n_{*,N}}) \boldsymbol{u}^n_N \right\|^2_{\mathbf{L}^2(\mathcal{O})} + \| v^n_N \|^2_{L^2(0,L)} \right]. \end{aligned} \tag{9.35}$$

By absorbing the last term on the right-hand side of (9.35), we obtain

$$\begin{aligned} \mathbb{E} \max_{1 \leq n \leq N} \left(\| (\sqrt{R + \eta^n_{*,N}}) \boldsymbol{u}^n_N \|^2_{\mathbf{L}^2(\mathcal{O})} + \| v^n_N \|^2_{L^2(0,L)} \right) &\leq C E^0 + C \| P_{in/out} \|^2_{L^2(0,T)} \\ &+ \sum_{n=1}^{N-1} (\Delta t) \mathbb{E} \max_{1 \leq m \leq n} \left(\| (\sqrt{R + \eta^m_{*,N}}) \boldsymbol{u}^m_N \|^2_{\mathbf{L}^2(\mathcal{O})} + \| v^m_N \|^2_{L^2(0,L)} \right). \end{aligned}$$

Applying the discrete Gronwall inequality (see Theorem 2.12.2) to $\mathbb{E} \max_{1 \leq n \leq N}$ $(\| (\sqrt{R + \eta^n_{*,N}}) \boldsymbol{u}^n_N \|^2_{\mathbf{L}^2(\mathcal{O})} + \| v^n_N \|^2_{L^2(0,L)})$, we obtain

$$\mathbb{E}\max_{1\le n\le N}\left(\|(\sqrt{R+\eta^n_{*,N}})\boldsymbol{u}^n_N\|^2_{\mathbf{L}^2(\mathcal{O})}+\|v^n_N\|^2_{L^2(0,L)}\right)\le Ce^T,$$

where C depends only on the given data and in particular δ.

Hence for E^n_N, D^n_N defined in (9.29), we have obtained the desired energy estimate: For some C independent of N and ε that

$$\mathbb{E}\max_{1\le n\le N}E^n_N+\mathbb{E}\sum_{n=0}^{N-1}D^n_N+\mathbb{E}\sum_{n=0}^{N-1}C^n_{1,N}+\mathbb{E}\sum_{n=0}^{N-1}C^n_{2,N}$$
$$\le C(E^0+\|P_{in}\|^2_{L^2(0,T)}+\|P_{out}\|^2_{L^2(0,T)}+Te^T). \tag{9.36}$$

□

9.2.2 *Approximate Solutions*

In this subsection, we use the solutions $(\boldsymbol{u}_N^{n+\frac{i}{2}},\eta_N^{n+\frac{i}{2}},v_N^{n+\frac{i}{2}})$, $i=0,1$, defined for every $N\in\mathbb{N}\setminus\{0\}$ at discrete times to define approximate solutions on the entire interval $(0,T)$. We start by introducing approximate solutions that are piecewise constant on each subinterval $[n\Delta t,(n+1)\Delta t)$ as

$$\begin{aligned}\boldsymbol{u}_N(t,\cdot)=\boldsymbol{u}^n_N,\quad \eta_N(t,\cdot)=\eta^n_N,\quad \eta^*_N(t,\cdot)=\eta^n_{*,N},\\ v_N(t,\cdot)=v^n_N,\quad v^{\#}_N(t,\cdot)=v_N^{n+\frac{1}{2}}.\end{aligned}\tag{9.37}$$

The benefit and necessity of defining the approximate processes (9.37) are that they are adapted to the given filtration $(\mathcal{F}_t)_{t\ge0}$.

Then as done in (5.54), we also need to define the following time-shifted piecewise constant functions on $t\in[n\Delta t,(n+1)\Delta t)$, which will be useful in obtaining tightness results using compactness methods:

$$\boldsymbol{u}^+_N(t,\cdot)=\boldsymbol{u}^{n+1}_N,\quad \eta^+_N(t,\cdot)=\eta^{n+1}_N,\quad v^+_N(t,\cdot)=v^{n+1}_N.$$

Furthermore, as in (5.55), we define the corresponding piecewise linear interpolations of the structure displacement, velocity, and the fluid velocity: For $t\in[t^n,t^{n+1}]$, we let

$$\begin{aligned}\widetilde{\boldsymbol{u}}_N(t,\cdot)&=\frac{t-t^n_N}{\Delta t}\boldsymbol{u}^{n+1}_N+\frac{t^{n+1}_N-t}{\Delta t}\boldsymbol{u}^n_N,\quad \widetilde{v}_N(t,\cdot)=\frac{t-t^n_N}{\Delta t}v^{n+1}_N+\frac{t^{n+1}_N-t}{\Delta t}v^n_N\\ \widetilde{\eta}_N(t,\cdot)&=\frac{t-t^n_N}{\Delta t}\eta^{n+1}_N+\frac{t^{n+1}_N-t}{\Delta t}\eta^n_N,\quad \widetilde{\eta}^*_N(t,\cdot)=\frac{t-t^n_N}{\Delta t}\eta^{n+1}_{*,N}+\frac{t^{n+1}_N-t}{\Delta t}\eta^n_{*,N}.\end{aligned}\tag{9.38}$$

Observe that

$$\frac{\partial \widetilde{\eta}_N}{\partial t} = v_N^{\#}, \quad \frac{\partial \widetilde{\eta}_N^*}{\partial t} = \sum_{n=0}^{N-1} \theta_\delta(\eta_N^{n+1}) v_N^{\#} \mathbb{1}_{(t_N^n, t_N^{n+1})} := v_N^* \quad a.e. \text{ on } (0, T), \tag{9.39}$$

where $v_N^{\#}$ was introduced in (9.37).

Lemma 9.2.3 *Given $\boldsymbol{u}_0 \in \mathbf{L}^2(\mathcal{O})$, $\eta_0 \in H_0^2(0, L)$, $v_0 \in L^2(0, L)$, for a fixed $\delta > 0$, we have that*

1. *$\{\eta_N\}, \{\eta_N^*\}$ and thus $\{\widetilde{\eta}_N\}, \{\widetilde{\eta}_N^*\}$ are bounded independently of N and ε in $L^2(\Omega; L^\infty(0, T; H_0^2(0, L)))$.*
2. *$\{v_N\}, \{v_N^{\#}\}, \{v_N^*\}, \{v_N^+\}$ are bounded independently of N and ε in $L^2(\Omega; L^\infty(0, T; L^2(0, L)))$.*
3. *$\{\boldsymbol{u}_N\}$ is bounded independently of N and ε in $L^2(\Omega; L^\infty(0, T; \mathbf{L}^2(\mathcal{O})))$.*
4. *$\{\boldsymbol{u}_N^+\}$ is bounded independently of N and ε in $L^2(\Omega; L^2(0, T; V)) \cap L^2(\Omega; L^\infty(0, T; \mathbf{L}^2(\mathcal{O})))$.*
5. *$\{v_N^+\}$ is bounded independently of N and ε in $L^2(\Omega; L^2(0, T; H^{\frac{1}{2}}(0, L)))$.*
6. *$\{\frac{1}{\sqrt{\varepsilon}} \text{div}^{\eta_N^*} \boldsymbol{u}_N^+\}$ is bounded independently of N and ε in $L^2(\Omega; L^2(0, T; L^2(\mathcal{O})))$.*

Proof The proofs of the first three statements and that of (6) follow immediately from Theorem 9.2.1. For the second statement, we write $v_N^*(t, \omega, z) = \theta_N(t, \omega) v_N^{\#}(t, \omega, z)$ where we set

$$\theta_N(t, \omega) = \sum_{n=0}^{N-1} \theta_\delta(\eta_N^{n+1}(\omega)) \mathbb{1}_{[t^n, t^{n+1})}(t).$$

Since $\theta_N(t) \le 1$ for any $t \in [0, T]$, we obtain

$$\mathbb{E}[\|v_N^*\|^2_{L^\infty(0,T;L^2(0,L))}] \le \mathbb{E}[\|v_N^{\#}\|^2_{L^\infty(0,T;L^2(0,L))}] \le C.$$

In order to prove (4), we first observe that Proposition 5.5.2 in the deterministic setting of Chap. 5 cannot be applied due to the penalty term in this problem due to which the fluid variables are not divergence free. Hence, we apply Lemma 2.8.4 instead. Note that the uniform bound $\|R + \eta^n_{*,N}(\omega)\|_{H^s(0,L)} < \frac{1}{\delta}$, for $\frac{3}{2} < s < 2$ and every $\omega \in \Omega$, implies that for some $C(\delta) > 0$

$$\|A^\omega_{\eta^n_{*,N}}\|_{\mathbf{W}^{1,\infty}(\mathcal{O})} < C, \quad \|(A^\omega_{\eta^n_{*,N}})^{-1}\|_{\mathbf{W}^{1,\infty}(\mathcal{O})} < C, \quad n = 1, ..., N, N \in \mathbb{N},$$

for every $\omega \in \Omega$.

Observe also that, for each $\omega \in \Omega$, $\nabla \boldsymbol{u}_N^{n+1} = \nabla^{\eta^n_{*,N}} \boldsymbol{u}_N^{n+1} (\nabla A^\omega_{\eta^n_{*,N}})$. Thus we can write

$$
\begin{aligned}
\delta \mathbb{E} \int_{\mathcal{O}} |\nabla \boldsymbol{u}_N^{n+1}|^2 d\boldsymbol{x} &\leq \mathbb{E} \int_{\mathcal{O}} (R + \eta^n_{*,N}) |\nabla \boldsymbol{u}_N^{n+1}|^2 d\boldsymbol{x} \\
&= \mathbb{E} \int_{\mathcal{O}} (R + \eta^n_{*,N}) |\nabla^{\eta^n_{*,N}} \boldsymbol{u}_N^{n+1} \cdot \nabla A^{\omega}_{\eta^n_{*,N}}|^2 d\boldsymbol{x} \\
&\leq C(\delta) \mathbb{E} \int_{\mathcal{O}} (R + \eta^n_{*,N}) |\nabla^{\eta^n_{*,N}} \boldsymbol{u}_N^{n+1}|^2 d\boldsymbol{x} \\
&\leq K C(\delta) \mathbb{E} \int_{\mathcal{O}} (R + \eta^n_{*,N}) |\mathbf{D}^{\eta^n_{*,N}} (\boldsymbol{u}_N^{n+1})|^2 d\boldsymbol{x},
\end{aligned}
$$

where $K > 0$ is the universal Korn constant that, thanks to Lemma 2.8.4, depends only on the reference domain $\mathcal{O}$. Hence, there exists $C > 0$, independent of N and ε, such that

$$
\mathbb{E} \int_0^T \int_{\mathcal{O}} |\nabla \boldsymbol{u}_N^+|^2 d\boldsymbol{x} ds = \mathbb{E} \sum_{n=0}^{N-1} \Delta t \int_{\mathcal{O}} |\nabla \boldsymbol{u}^{n+1}|^2 d\boldsymbol{x} \leq C(\delta). \tag{9.40}
$$

Note here that the deterministic bounds for the artificial structure displacement play a key role in establishing these bounds. Statement (5) is then a result of statement (4) and the fact that, by construction, v_N^+ is the trace of the vertical component of the time-shifted fluid velocity $\boldsymbol{u}_N^+$ on the top lateral boundary. □

9.3 Passing to the Limit as $N \to \infty$

Our goal is to construct solutions to the coupled FSI problem as limits of subsequences of approximate solutions as $N \to \infty$ and $\varepsilon \to 0$. The δ approximation of the problem via cutoff functions will be dealt with using a stopping time argument at the very end of the chapter. Our approach is to first let $N \to \infty$, obtain approximate solutions for each $\varepsilon > 0$, and then let $\varepsilon \to 0$.

9.3.1 Tightness Results

We first fix $\varepsilon > 0$ and $\delta > 0$ and let $N \to \infty$. For this purpose, we recall the bounds obtained in Lemma 9.2.3 that provide weakly and weakly-* convergent subsequences. In order to upgrade these convergence results to obtain almost sure convergence of the stochastic approximate solutions, which is required to be able to pass $N \to \infty$, we use compactness arguments and establish tightness of the laws of the approximate random variables defined in Sect. 9.2.2 (see Definition 3.2.1). We take a different approach in this chapter to establish tightness of laws of the approximate solutions than the approach taken in Chap. 8 which is necessary since in this chapter the nonlinear coupling results into randomly moving fluid domain. Moreover, due to the penalty term among other things, we cannot apply the ideas used in Theorem 2.10.1 either (note that this theorem requires incompressibility). Hence, we will instead use Theorem 2.9.9 to establish tightness of the laws of the fluid and structure velocities. Note that this result is stronger than Theorem 2.10.1 as this proof, if applied to the deterministic FSI problem, implies that we have uniform bounds for the fractional time derivatives of the approximate and thus limiting solutions, i.e., uniform bounds in an appropriate Nikolskii space in time (see Definition 2.9.3).

Lemma 9.3.1 *For any* $0 \leq \alpha < 1$, *the laws of* $\boldsymbol{u}_N^+$ *and* v_N^+ *are tight in* $L^2(0,T;\mathbf{H}^\alpha(\mathcal{O}))$ *and* $L^2(0,T;L^2(0,L))$, *respectively.*

Proof Recall, from Definition 2.9.3, that T_h is the translation in time by $h \in \mathbb{R}$ of a function f given by $T_h f(t,\cdot) = f(t-h,\cdot)$. For a fixed N and $0 < h < T$, we write

$$h = j\Delta t - s \text{ for some } 1 \leq j \leq N \text{ and } s < \Delta t. \tag{9.41}$$

For simplicity, we will consider the case $s = 0$. The case of an arbitrary $s < \Delta t$ is handled identically to Step 2 of Proposition 8.6.4. The aim of this proof is to apply Theorem 2.9.9 by obtaining appropriate bounds for the following term in terms of powers of h for any $h > 0$, independently of N:

$$\int_h^T \|T_h\boldsymbol{u}_N^+ - \boldsymbol{u}_N^+\|^2_{\mathbf{L}^2(\mathcal{O})} + \|T_h v_N^+ - v_N^+\|^2_{L^2(0,L)}$$
$$= (\Delta t)\sum_{n=j+1}^{N} \|\boldsymbol{u}_N^n - \boldsymbol{u}_N^{n-j}\|^2_{\mathbf{L}^2(\mathcal{O})} + \|v_N^n - v_N^{n-j}\|^2_{L^2(0,L)}, \tag{9.42}$$

For $0 \leq \alpha < 1$ and any $M > 0$, let us introduce the set:

$$\begin{aligned}\mathcal{B}_M :=& \{(\boldsymbol{u}, v) \in L^2(0,T;\mathbf{H}^\alpha(\mathcal{O})) \times L^2(0,T;L^2(0,L)) :\\ &\|\boldsymbol{u}\|^2_{L^2(0,T;\mathbf{H}^1(\mathcal{O}))} + \|v\|^2_{L^2(0,T;H^{\frac{1}{2}}(0,L))}\\ &+ \sup_{0<h<1} h^{-\frac{1}{32}} \int_h^T \left(\|T_h\boldsymbol{u} - \boldsymbol{u}\|^2_{\mathbf{L}^2(\mathcal{O})} + \|T_h v - v\|^2_{L^2(0,L)}\right) \leq M\}.\end{aligned} \tag{9.43}$$

Notice that Theorem 2.9.9 implies that $\mathcal{B}_M$ is compact in $L^2(0,T;\mathbf{H}^\alpha(\mathcal{O})) \times L^2(0,T;L^2(0,L))$ for each $M > 0$. Here in Theorem 2.9.9, we used $B_0 = \mathbf{H}^1(\mathcal{O}) \times H^{\frac{1}{2}}(0,L)$, $B_1 = \mathbf{H}^\alpha(\mathcal{O}) \times L^2(0,L)$, and $B_2 = \mathbf{L}^2(\mathcal{O}) \times L^2(0,L)$. To show tightness of laws of $\boldsymbol{u}_N^+$ and v_N^+, we will then show that there exists $C > 0$ independent of N and ε, such that

$$\mathbb{P}((\boldsymbol{u}_N^+, v_N^+) \notin \mathcal{B}_M) \le \mathbb{P}\left(\|\boldsymbol{u}_N^+\|^2_{L^2(0,T;\mathbf{H}^1(\mathcal{O}))} + \|v_N^+\|^2_{L^2(0,T;H^{\frac{1}{2}}(0,L))} > \frac{M}{2}\right)$$

$$+ \mathbb{P}\left(\sup_{0<h<1} h^{-\frac{1}{32}} \int_h^T \left(\|T_h \boldsymbol{u}_N^+ - \boldsymbol{u}_N^+\|^2_{\mathbf{L}^2(\mathcal{O})} + \|T_h v_N^+ - v_N^+\|^2_{L^2(0,L)}\right) > \frac{M}{2}\right)$$

$$\le \frac{C}{\sqrt{M}}.$$

Notation To keep the presentation clean, for a fixed but arbitrary N, we will ignore the subscript N for the solutions of (9.25) and (9.28) and write $\boldsymbol{u}_N^n, v_N^n, \eta_N^n, \eta_{*,N}^n$ as $\boldsymbol{u}^n, v^n, \eta^n, \eta_*^n$.

To achieve the aforementioned goal, we will construct appropriate test functions for Eqs. (9.25) and (9.28) that will give the term on the right-hand side of (9.42). This has to be done carefully, since the solutions to the fluid subproblem (9.28) are defined on different physical domains. That is, the velocity functions $\boldsymbol{u}^k$, for any $1 \le k \le N$, cannot directly be used as test functions for the equation for $\boldsymbol{u}^n$, for $1 \le n \ne k \le N$. Hence, we must construct the test functions by first transforming $\boldsymbol{u}^k$, such that appropriate boundary conditions are satisfied by these test functions and such that they have "small" divergence, so that the penalty term can be bounded independently of ε. This is achieved by first "squeezing" $\boldsymbol{u}^k$ appropriately (see Definition 5.7.2). However, using the squeezed function directly as a test function would result into testing nth step equation $(9.25)_3$ with v^k, which does not have the required H^2 spatial regularity. Thus, we mollify the squeezed function to get the desired H^2 regularity.

Hence our plan is as follows: assume that, for any $\delta > 0$ used to define our stopped process in (9.27), $\mathcal{O}_\delta$ is the maximal rectangular domain containing all the fluid domains associated with the structure displacement η_N^*, defined in (9.37), for any N and t, ω. We fix an N and $\omega \in \Omega$, and for any $0 \le k, n \le N$, we push $\boldsymbol{u}^k$ onto the physical domain $\mathcal{O}_{\eta_*^{k-1}}$, extend it by its trace outside of $\mathcal{O}_{\eta_*^{k-1}}$ to $\mathcal{O}_\delta$, then "squeeze" it in a way that the L^2-bound on its divergence is preserved, and mollify it. The mollification is done in a way that preserves the zero boundary condition for the radial component of the velocity on the left and right boundaries $\Gamma_{in/out}$, and the zero boundary value for the horizontal velocity component on the fluid-structure interface $\Gamma_{\eta_*^n}$. Then we pull it back to the domain $\mathcal{O}$ via the ALE map $A_{\eta_*^n}^\omega$. This way we have transformed $\boldsymbol{u}^k$, so that it can be used as a test function for Eqs. (9.28) for $\boldsymbol{u}^n$. Details of the construction are presented next.

First, let $\tilde{\boldsymbol{u}}^k = \boldsymbol{u}^k \circ A^{-1}_{\eta_*^{k-1}}$. Here we used "tilde" to denote the function defined on the physical domain. For the purposes of mollification, we define $\tilde{\boldsymbol{u}}^{k,ext}$ to be an extension of $\tilde{\boldsymbol{u}}^k$ to the whole of $\mathbb{R}^2$, as follows:

$$\widetilde{\boldsymbol{u}}^{k,ext} := \begin{cases} (\widetilde{u}_z^k(0,r),0) & \text{for } z \le 0, \\ (\widetilde{u}_z^k(L,r),0) & \text{for } z \le L, \\ (0, v^k) & \text{above } \Gamma_{\eta_*^{k-1}}, \\ 0 & \text{everywhere else.} \end{cases}$$

Then, we introduce the squeezing parameter $\sigma > 1$ and the following squeezing operator:

$$S_\sigma(\widetilde{\boldsymbol{u}}^{k,ext}) := \widetilde{\boldsymbol{u}}_\sigma^k(z,r) := (\sigma \widetilde{u}_z^{k,ext}(z,\sigma r), \widetilde{u}_r^{k,ext}(z,\sigma r)). \tag{9.44}$$

As mentioned earlier, notice that we scaled the r coordinate, i.e., squeezed $\widetilde{\boldsymbol{u}}^{k,ext}$ vertically, so that the resulting function $\widetilde{\boldsymbol{u}}_\sigma^k$ assumes appropriate values on $\Gamma_{\eta_*^n}$ in order for it to be used as a test function for the nth equations. To improve the spatial regularity of its trace on $\Gamma_{\eta_*^n}$, we next introduce $\widetilde{\boldsymbol{u}}_{\sigma,\lambda}^k$, a space mollification of $\widetilde{\boldsymbol{u}}_\sigma^k$, where λ denotes the space mollification parameter. The mollification is obtained using standard 2D mollifiers (2.26) introduced in (2.27) Chap. 2 Sect. 2.11. Based on this space mollification $\widetilde{\boldsymbol{u}}_{\sigma,\lambda}^k$, for any given N and n such that $1 \le n \le N$,, we define

$$\boldsymbol{u}_{\sigma,\lambda}^{k,n} := \widetilde{\boldsymbol{u}}_{\sigma,\lambda}^k \circ A_{\eta_*^n}^\omega, \tag{9.45}$$

where the parameters $\lambda > 0$ and σ are chosen as follows:

1. The random variable σ is defined, so that the squeezed kth fluid velocity function lies entirely within the nth fluid domain; see Fig. 9.2:
$$\sigma = \max_{1 \le n \le N} \max_{n-j \le k \le n} \sup_{z \in [0,L]} \frac{R + \eta_*^k}{R + \eta_*^n - \delta h^{\frac{1}{4}}},$$
where j is as in (9.41).
2. The parameter λ is then chosen to be small enough, so that mollification with parameter λ preserves the kinematic coupling condition; see Fig. 9.2:
$$\lambda = \frac{\delta^2 h^{\frac{1}{4}}}{\sqrt{1+\delta^2}}. \tag{9.46}$$

Due to these choices of σ and λ, we will be able to obtain the desired estimates in terms of the powers of the translation parameter h, as we shall see below. In particular, we observe that for this choice of σ, the vertical distance between $R + \eta_*^n$ and $\frac{R+\eta_*^k}{\sigma}$ is at least $\delta h^{\frac{1}{4}}$. Moreover, thanks to the mean value theorem and the property of the artificial structure variables η_*^n, that $\|\partial_z \eta_*^k\|_{L^\infty(0,L)} \le \frac{1}{\delta}$ for any $k \le N$, we can see that for any $z_0, z \in [0, L]$

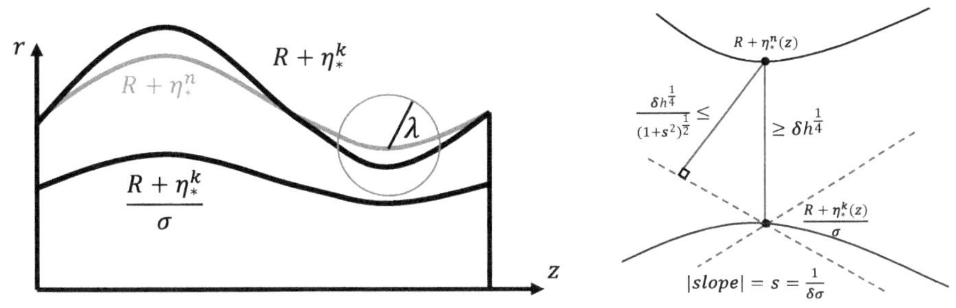

Fig. 9.2 (left) A realization of the vertical squeezing and the mollification operator. (right) Distance between the curves $R + \eta_*^n$ and $\frac{R+\eta_*^k}{\sigma}$

$$\frac{1}{\sigma}(R + \eta_*^k(z)) \leq \frac{1}{\sigma}(R + \eta_*^k(z_0)) + \frac{1}{\delta\sigma}|z - z_0|.$$

Hence, $\frac{1}{\sigma}(R + \eta_*^k(z))$, for any $z \in [0, L]$, remains under the lines of slopes $\pm\frac{1}{\sigma\delta}$ emanating from any point $z_0 \in [0, L]$. Hence, the distance between these two curves $R + \eta_*^n$ and $\frac{R+\eta_*^k}{\sigma}$ is at least $m_{dist} := \frac{\delta h^{\frac{1}{4}}}{\sqrt{1+(\delta\sigma)^{-2}}}$. This motivates our choice of $\lambda < m_{dist}$ (recall that $\sigma > 1$); see Fig. 9.2.

Remark Before we continue, we note here that we additionally need to "squeeze" the function $\tilde{\boldsymbol{u}}^{k,ext}$ horizontally, so that when we mollify the function to obtain the desired regularity, the mollification does not ruin its 0 boundary value for the radial component at the boundaries $\Gamma_{in/out}$ and, for the same reasons, vertically near the bottom boundary Γ_b as well. This can be done using the same argument, i.e., by choosing the horizontal squeezing parameter, say $\sigma_{horizontal}$, and the mollification parameter λ appropriately small. The resulting function is in $\mathbf{H}^s(\mathcal{O}_\delta)$ for any $s < \frac{1}{2}$, and thus the estimates presented below can be obtained the same way. However, we choose to leave this part of the construction out of our discussion for a cleaner presentation.

We now define the structure equation test function v_λ^k to be the space mollification of $v^{k,ext}$ where $v^{k,ext}$ is the extension of v^k to $\mathbb{R}$ by 0. Observe that $\partial_r v_\lambda^k = (\partial_r v^{k,ext})_\lambda = 0$.

Hence, we have that for any $n - j \leq k < n$ (j is defined in (9.41)), the transformed fluid velocity $\boldsymbol{u}_{\sigma,\lambda}^{k,n}$ satisfies:

$$\|\text{div}^{\eta_*^n} \boldsymbol{u}_{\sigma,\lambda}^{k+1,n}\|_{L^2(\mathcal{O})} \leq C(\delta)\sigma \|\text{div}^{\eta_*^k} \boldsymbol{u}^{k+1}\|_{L^2(\mathcal{O})}, \quad \text{and} \quad \boldsymbol{u}_{\sigma,\lambda}^{k,n}|_\Gamma = (0, v_\lambda^k). \tag{9.47}$$

Recall that our goal is to use the modified fluid velocity and structure displacements constructed above as test functions for our coupled problem (9.25) and (9.28), based on which we will obtain the desired estimates of the right-hand side of (9.42) in terms of power of h. For this purpose, we will need the following estimate of σ in terms of powers of h: For any $j \leq n \leq N$, observe that we have

$$
\begin{aligned}
\sigma - 1 &= \max_{1\leq n\leq N}\max_{n-j\leq k\leq n}\sup_{z\in[0,L]} \frac{R+\eta_*^k}{R+\eta_*^n - \delta h^{\frac14}} - 1\\
&= \max_{1\leq n\leq N}\max_{n-j\leq k\leq n}\sup_{z\in[0,L]} \frac{\eta_*^k - \eta_*^n + \delta h^{\frac14}}{R+\eta_*^n - \delta h^{\frac14}}\\
&\leq \max_{1\leq n\leq N}\max_{n-j\leq k\leq n} \frac{\sup_{z\in[0,L]}|\eta_*^n - \eta_*^k| + \delta h^{\frac14}}{\delta}\\
&\leq C(\frac{1}{\delta}\|\widetilde{\eta}_N^*\|_{C^{0,\frac14}(0,T;L^\infty(0,L))} + 1)h^{\frac14},
\end{aligned} \tag{9.48}
$$

where we have used $\|T_h\eta_N^* - \eta_N^*\|_{L^\infty((0,T)\times(0,L))} \leq \|\widetilde{\eta}_N^*\|_{C^{0,\frac14}(0,T;L^\infty(0,L))} h^{\frac14}$ in the last inequality.

In the estimates below, we will also need the estimates on the difference between the transformed and the original fluid velocity functions. For this purpose, we let ρ be the 2D radial function defined in (2.26) in Chap. 2 Sect. 2.11. We denote by $\sigma^*(w) = (w_1, \sigma w_2)$ for $w = (w_1, w_2) \in \mathbb{R}^2$. Then, for any $k \leq N$, $s < \frac12$ we find, for some C depending only on s, δ, L, that

$$
\begin{aligned}
\|\widetilde{\boldsymbol{u}}^k_{\sigma,\lambda} - \widetilde{\boldsymbol{u}}^{k,ext}\|^2_{\mathbf{L}^2(\mathcal{O}_\delta)} &= \int_{\mathcal{O}_\delta}\left|\int_{B(0,1)}\rho(y)(\widetilde{\boldsymbol{u}}^k_\sigma(x+\lambda y) - \widetilde{\boldsymbol{u}}^{k,ext}(x))dy\right|^2 dx\\
\leq \int_{\mathcal{O}_\delta}\int_{B(0,1)} &\frac{|(\widetilde{\boldsymbol{u}}^k_\sigma((x+\lambda y)) - \widetilde{\boldsymbol{u}}^{k,ext}(x))|^2}{|\sigma^*(x+\lambda y) - x|^{2+2s}}|\sigma^*(x+\lambda y) - x|^{2+2s}dydx\\
\leq C((\sigma-1)^{2+2s} + (\sigma\lambda)^{2+2s}) &\int_{\mathcal{O}_\delta}\int_{B(0,1)}\frac{|(\widetilde{\boldsymbol{u}}^k_\sigma(x+\lambda y) - \widetilde{\boldsymbol{u}}^{k,ext}(x))|^2}{|\sigma^*(x+\lambda y) - x|^{2+2s}}dydx.
\end{aligned}
$$

After the change of variables $w = \sigma^*(x+\lambda y)$, we obtain

$$
\begin{aligned}
&\|\widetilde{\boldsymbol{u}}^k_{\sigma,\lambda} - \widetilde{\boldsymbol{u}}^{k,ext}\|^2_{\mathbf{L}^2(\mathcal{O}_\delta)}\\
&\leq C(s,\delta,L)\left((\sigma-1)^{2+2s}\sigma^{2+2s}\right)\sigma^{-1}\lambda^{-2}\\
&\times\int_{\mathcal{O}_\delta}\int_{\mathcal{O}_\delta^{\sigma,\lambda}}\frac{|(\widetilde{\boldsymbol{u}}^{k,ext}(w) - \widetilde{\boldsymbol{u}}^{k,ext}(x))|^2}{|w-x|^{2+2s}}dwdx\\
&\leq C(s,\delta,L)\left(\frac{(\sigma-1)^{2+2s}}{\lambda^2}\sigma^{1+2s}\right)\|\widetilde{\boldsymbol{u}}^{k,ext}\|^2_{\mathbf{H}^s(\mathcal{O}_\delta^{\sigma,\lambda})},
\end{aligned} \tag{9.49}
$$

where $\mathcal{O}_\delta^{\sigma,\lambda} = \sigma^*(\mathcal{O}_\delta) + \lambda B(0,1)$.

Thanks to Theorem 2.8.2, for $s < \frac{1}{2}$, we can estimate the H^s norm of $\widetilde{\boldsymbol{u}}^{k,ext}$ on $\mathcal{O}_\delta^{\sigma,\lambda}$ as follows:

$$\|\widetilde{\boldsymbol{u}}^{k,ext}\|^2_{\mathbf{H}^s(\mathcal{O}_\delta^{\sigma,\lambda})} \leq C(\|\boldsymbol{u}^k\|^2_{\mathbf{H}^1(\mathcal{O})} + \|v^k\|^2_{H^{\frac{1}{2}}(0,L)}),$$

where C depends only on δ and s. Thus, for any $n - j \leq k \leq n$ and $s < \frac{1}{2}$, our choice of σ that satisfies (9.48) gives us

$$\begin{aligned}
\|\boldsymbol{u}^{k,n}_{\sigma,\lambda} - \boldsymbol{u}^k\|_{\mathbf{L}^2(\mathcal{O})} &\leq \|\widetilde{\boldsymbol{u}}^k_{\sigma,\lambda} - \widetilde{\boldsymbol{u}}^{k,ext}\|_{\mathbf{L}^2(\mathcal{O}_{\eta_*^n})} + \|\widetilde{\boldsymbol{u}}^{k,ext} \circ (A_{\eta_*^n} - A_{\eta_*^{k-1}})\|_{\mathbf{L}^2(\mathcal{O})} \\
&\leq \left(\frac{(\sigma-1)^{s+1}}{\lambda}\sigma^{s+\frac{1}{2}} + \lambda^s\right)\|\boldsymbol{u}^k\|_{\mathbf{H}^1(\mathcal{O})} + \|\partial_r \widetilde{\boldsymbol{u}}^{k,ext}\|_{\mathbf{L}^2(\mathcal{O}_\delta)}\|\eta_*^n - \eta_*^{k-1}\|_{L^\infty(0,L)} \\
&\leq \left(\frac{(\sigma-1)^{s+1}}{\lambda}\sigma^{s+\frac{1}{2}} + \lambda^s\right)\|\boldsymbol{u}^k\|_{\mathbf{H}^1(\mathcal{O})} + h^{\frac{1}{4}}\|\widetilde{\eta}_N^*\|_{C^{0,\frac{1}{4}}(0,T;L^\infty(0,L))}\|\boldsymbol{u}^k\|_{\mathbf{H}^1(\mathcal{O})} \\
&\leq h^{\frac{s}{4}}\left(\max\{1, \frac{1}{\delta}\|\widetilde{\eta}_N^*\|^{2s+\frac{3}{2}}_{C^{0,\frac{1}{4}}(0,T;L^\infty(0,L))}\}\right)\|\boldsymbol{u}^k\|_{\mathbf{H}^1(\mathcal{O})} \\
&\quad + h^{\frac{1}{4}}\|\widetilde{\eta}_N^*\|_{C^{0,\frac{1}{4}}(0,T;L^\infty(0,L))}\|\boldsymbol{u}^k\|_{\mathbf{H}^1(\mathcal{O})}.
\end{aligned} \tag{9.50}$$

This estimate of the difference $\|\boldsymbol{u}^{k,n}_{\sigma,\lambda} - \boldsymbol{u}^k\|_{\mathbf{L}^2(\mathcal{O})}$ given in terms of powers of h will be used below in (9.56) to estimate one of the integrals contributing to the final estimate of the time shifts of $\boldsymbol{u}_N$.

Likewise, the following property of mollification along with (9.46) will be used in (9.57):

$$\|v^k_\lambda - v^k\|_{L^2(0,L)} \leq \lambda^s\|v^k\|_{H^s(0,L)} \leq h^{\frac{s}{4}}\|v^k\|_{H^s(0,L)}, \qquad \text{for any } s \leq \frac{1}{2}. \tag{9.51}$$

We are now ready to multiply $(9.25)_3$ and $(9.28)_2$ by the test functions $\boldsymbol{q}_n$ and ψ_n, respectively (see also [81]), where $\boldsymbol{q}_n$ and ψ_n are defined by

$$\mathbf{Q}_n := (\boldsymbol{q}_n, \psi_n) = \left((\Delta t)\sum_{k=n-j+1}^{n}\boldsymbol{u}^{k,n}_{\sigma,\lambda},\ (\Delta t)\sum_{k=n-j+1}^{n} v^k_\lambda\right),$$

for j defined in (9.41).

After adding the resulting equations, we obtain

$$\int_{\mathcal{O}}\left((R+\eta_*^{n+1})\boldsymbol{u}^{n+1} - (R+\eta_*^n)\boldsymbol{u}^n\right)\left(\Delta t\sum_{k=n-j+1}^{n}\boldsymbol{u}^{k,n}_{\sigma,\lambda}\right) \tag{9.52}$$

$$
\begin{aligned}
&+ \int_0^L (v^{n+1} - v^n) \left(\Delta t \sum_{k=n-j+1}^{n} v_\lambda^k \right) \\
&- \frac{1}{2} \int_{\mathcal{O}} \left(\eta_*^{n+1} - \eta_*^n \right) \boldsymbol{u}^{n+1} \cdot \left(\Delta t \sum_{k=n-j+1}^{n} \boldsymbol{u}_{\sigma,\lambda}^{k,n} \right) \\
&+ \frac{(\Delta t)}{\varepsilon} \int_{\mathcal{O}} \operatorname{div}^{\eta_*^n} \boldsymbol{u}^{n+1} \operatorname{div}^{\eta_*^n} \left(\Delta t \sum_{k=n-j+1}^{n} \boldsymbol{u}_{\sigma,\lambda}^{k,n} \right) \\
&+ (\Delta t) b^{\eta_n^*} (\boldsymbol{u}^n, v^{n+\frac{1}{2}} r \mathbf{e}_r, \left(\Delta t \sum_{k=n-j+1}^{n} \boldsymbol{u}_{\sigma,\lambda}^{k,n} \right)) \\
&+ 2\nu (\Delta t) \int_{\mathcal{O}} (R + \eta_*^n) \mathbf{D}^{\eta_*^n} (\boldsymbol{u}^{n+1}) \cdot \mathbf{D}^{\eta_*^n} \left(\Delta t \sum_{k=n-j+1}^{n} \boldsymbol{u}_{\sigma,\lambda}^{k,n} \right) \\
&+ (\Delta t) \int_0^L \partial_z \eta^{n+\frac{1}{2}} \partial_z \left(\Delta t \sum_{k=n-j+1}^{n} v_\lambda^k \right) + \partial_{zz} \eta^{n+\frac{1}{2}} \partial_{zz} \left(\Delta t \sum_{k=n-j+1}^{n} v_\lambda^k \right) \\
&= (\Delta t) \left(P_{in}^n \int_0^1 q_z^n \Big|_{z=0} dr - P_{out}^n \int_0^1 q_z^n \Big|_{z=1} dr \right) + (G(\mathbf{U}^n, \eta_*^n) \Delta_n W, \mathbf{Q}_n).
\end{aligned}
$$

We will apply $\sum_{n=0}^{N}$ to (9.54) and estimate the terms to get the desired estimate for the time shifts. We start by focusing on the first two terms. Applying summation by parts to the first two terms, we obtain:

$$
\begin{aligned}
&- \sum_{n=0}^{N} \int_{\mathcal{O}} \left((R + \eta_*^{n+1}) \boldsymbol{u}^{n+1} - (R + \eta_*^n) \boldsymbol{u}^n \right) \left(\Delta t \sum_{k=n-j+1}^{n} \boldsymbol{u}_{\sigma,\lambda}^{k,n} \right) \\
&- \sum_{n=0}^{N} \int_0^L (v^{n+1} - v^n) \left(\Delta t \sum_{k=n-j+1}^{n} v_\lambda^k \right) \\
&= (\Delta t) \sum_{n=1}^{N} \left(\int_{\mathcal{O}} (R + \eta_*^n) \boldsymbol{u}^n (\boldsymbol{u}_{\sigma,\lambda}^{n,n} - \boldsymbol{u}_{\sigma,\lambda}^{n-j,n}) d\boldsymbol{x} + \int_0^L v^n (v_\lambda^n - v_\lambda^{n-j}) dz \right) \\
&- \int_{\mathcal{O}} (R + \eta_*^{N+1}) \boldsymbol{u}^{N+1} \left(\Delta t \sum_{k=N-j+1}^{N} \boldsymbol{u}_{\sigma,\lambda}^{k,N} \right) - \int_0^L v^N \left(\Delta t \sum_{k=N-j+1}^{N} v_\lambda^k \right).
\end{aligned}
$$

Here the first term on the right side can be written as

$$(\Delta t)\sum_{n=0}^{N}\Big(\int_{\mathcal{O}}(R+\eta_*^n)\boldsymbol{u}^n(\boldsymbol{u}^n-\boldsymbol{u}^{n-j})d\boldsymbol{x} \tag{9.53}$$

$$+\int_{\mathcal{O}}(R+\eta_*^n)\boldsymbol{u}^n(\boldsymbol{u}_{\sigma,\lambda}^{n,n}-\boldsymbol{u}^n-(\boldsymbol{u}_{\sigma,\lambda}^{n-j,n}-\boldsymbol{u}^{n-j}))d\boldsymbol{x}\Big)$$

$$+(\Delta t)\sum_{n=0}^{N}\int_0^L v^n(v^n-v^{n-j})dz+\int_0^L v^n(v_\lambda^n-v^n-(v_\lambda^{n-j}-v^{n-j}))$$

$$=(\Delta t)\sum_{n=0}^{N}\Big(\frac{1}{2}\int_{\mathcal{O}}(R+\eta_*^n)(|\boldsymbol{u}^n|^2-|\boldsymbol{u}^{n-j}|^2+|\boldsymbol{u}^n-\boldsymbol{u}^{n-j}|^2)d\boldsymbol{x}\Big)$$

$$+(\Delta t)\sum_{n=0}^{N}\Big(\int_{\mathcal{O}}(R+\eta_*^n)\boldsymbol{u}^n(\boldsymbol{u}_{\sigma,\lambda}^{n,n}-\boldsymbol{u}^n-(\boldsymbol{u}_{\sigma,\lambda}^{n-j,n}-\boldsymbol{u}^{n-j}))d\boldsymbol{x}\Big)$$

$$+(\Delta t)\sum_{n=0}^{N}\frac{1}{2}\int_0^L(|v^n|^2-|v^{n-j}|^2+|v^n-v^{n-j}|^2)dz$$

$$+(\Delta t)\sum_{n=0}^{N}+\int_0^L v^n(v_\lambda^n-v^n-(v_\lambda^{n-j}-v^{n-j})),$$

where we set $\boldsymbol{u}^n=0$ and $v^n=0$ for $n<0$ and $n>N$. Notice that the terms on the right-hand side are written in terms of time shifts and will be used to obtain the desired estimate. Substituting the above two equations in (9.52), we write

$$\int_h^T \|T_h\boldsymbol{u}_N^+-\boldsymbol{u}_N^+\|_{\mathbf{L}^2(\mathcal{O})}^2+\|T_h v_N-v_N\|_{L^2(0,L)}^2$$

$$\le(\Delta t)\sum_{n=0}^{N}\Big(\frac{1}{2}\int_{\mathcal{O}}(R+\eta_*^n)|\boldsymbol{u}^n-\boldsymbol{u}^{n-j}|^2d\boldsymbol{x}+\int_0^L|v^n-v^{n-j}|^2dz\Big)$$

$$=-(\Delta t)\sum_{n=0}^{N}\Big(\frac{1}{2}\int_{\mathcal{O}}(R+\eta_*^n)(|\boldsymbol{u}^n|^2-|\boldsymbol{u}^{n-j}|^2)d\boldsymbol{x}-\frac{1}{2}\int_0^L(|v^n|^2-|v^{n-j}|^2)dz\Big)$$

$$-(\Delta t)\sum_{n=0}^{N}\Big(\int_{\mathcal{O}}(R+\eta_*^n)\boldsymbol{u}^n(\boldsymbol{u}_{\sigma,\lambda}^{n,n}-\boldsymbol{u}^n-(\boldsymbol{u}_{\sigma,\lambda}^{n-j,n}-\boldsymbol{u}^{n-j}))d\boldsymbol{x}\Big)$$

$$-(\Delta t)\sum_{n=0}^{N}\int_0^L v^n(v_\lambda^n-v^n-(v_\lambda^{n-j}-v^{n-j})),$$

$$+\sum_{n=0}^{N}\int_{\mathcal{O}}(R+\eta_*^{N+1})\boldsymbol{u}^{N+1}\left(\Delta t\sum_{k=N-j+1}^{N}\boldsymbol{u}_{\sigma,\lambda}^{k,N}\right)+\int_0^L v^N\left(\Delta t\sum_{k=N-j+1}^{N}v_\lambda^k\right)$$

$$+\frac{(\Delta t)}{\varepsilon}\sum_{n=0}^{N}\int_{\mathcal{O}}\mathrm{div}^{\eta_*^n}\boldsymbol{u}^{n+1}\mathrm{div}^{\eta_*^n}\left(\Delta t\sum_{k=n-j+1}^{n}\boldsymbol{u}_{\sigma,\lambda}^{k,n}\right)$$

$$-\frac{1}{2}\sum_{n=0}^{N}\int_{\mathcal{O}}\left(\eta_*^{n+1}-\eta_*^n\right)\boldsymbol{u}^{n+1}\cdot\left(\Delta t\sum_{k=n-j+1}^{n}\boldsymbol{u}_{\sigma,\lambda}^{k,n}\right)$$

$$+(\Delta t)\sum_{n=0}^{N}b^{\eta_n^*}(\boldsymbol{u}^n,v^{n+\frac{1}{2}}r\mathbf{e}_r,\left(\Delta t\sum_{k=n-j+1}^{n}\boldsymbol{u}_{\sigma,\lambda}^{k,n}\right))$$

$$+2\nu(\Delta t)\sum_{n=0}^{N}\int_{\mathcal{O}}(R+\eta_*^n)\mathbf{D}^{\eta_*^n}(\boldsymbol{u}^{n+1})\cdot\mathbf{D}^{\eta_*^n}\left(\Delta t\sum_{k=n-j+1}^{n}\boldsymbol{u}_{\sigma,\lambda}^{k,n}\right)$$

$$+(\Delta t)\sum_{n=0}^{N}\int_0^L\partial_z\eta^{n+\frac{1}{2}}\partial_z\left(\Delta t\sum_{k=n-j+1}^{n}v_\lambda^k\right)+\partial_{zz}\eta^{n+\frac{1}{2}}\partial_{zz}\left(\Delta t\sum_{k=n-j+1}^{n}v_\lambda^k\right)$$

$$-(\Delta t)\sum_{n=0}^{N}\left(P_{in}^n\int_0^1 q_z^n\Big|_{z=0}dr-P_{out}^n\int_0^1 q_z^n\Big|_{z=1}dr\right)-\sum_{n=0}^{N}(G(\mathbf{U}^n,\eta_*^n)\Delta_nW,\mathbf{Q}_n)$$

$$=I_1+\ldots+I_{13}. \tag{9.54}$$

To estimate the terms on the right-hand side of (9.53), we observe that the terms containing $|\boldsymbol{u}^n|^2-|\boldsymbol{u}^{n-j}|^2$ can be estimated as follows:

$$I_1:=(\Delta t)\sum_{n=0}^{N}\left(\frac{1}{2}\int_{\mathcal{O}}(R+\eta_*^n)(|\boldsymbol{u}^n|^2-|\boldsymbol{u}^{n-j}|^2)d\boldsymbol{x}\right)$$

$$=\frac{1}{2}(\Delta t)\left(\sum_{n=N-j+1}^{N}\int_{\mathcal{O}}(R+\eta_*^n)|\boldsymbol{u}^n|^2d\boldsymbol{x}+\sum_{n=0}^{N-j}\int_{\mathcal{O}}(\eta_*^n-\eta_*^{n+j})|\boldsymbol{u}^n|^2d\boldsymbol{x}\right)$$

$$=\frac{1}{2}(\Delta t)\left(\sum_{n=N-j+1}^{N}\int_{\mathcal{O}}(R+\eta_*^n)|\boldsymbol{u}^n|^2d\boldsymbol{x}\right.$$

$$\left.+\sum_{n=0}^{N-j}\sum_{k=n}^{n+j-1}\int_{\mathcal{O}}[(\Delta t)\theta_\delta(\eta^{n+1})v^{k+\frac{1}{2}}]|\boldsymbol{u}^n|^2d\boldsymbol{x}\right)$$

$$
\leq \frac{1}{2}(\Delta t)\left(\sum_{n=N-j+1}^{N}\int_{\mathcal{O}}(R+\eta_*^n)|\boldsymbol{u}^n|^2 d\boldsymbol{x}\right.
$$

$$
\left.+(\Delta t)\sum_{n=0}^{N-j}\sum_{k=n}^{n+j-1}\|v^{k+\frac{1}{2}}\|_{L^2(0,L)}\|\boldsymbol{u}^n\|^2_{\mathbf{L}^4(\mathcal{O})}\right)
$$

$$
\leq h\left(\max_{1\leq n\leq N}\int_{\mathcal{O}}(R+\eta_*^n)|\boldsymbol{u}^n|^2 d\boldsymbol{x}+\max_{0\leq k\leq N-1}\|v^{k+\frac{1}{2}}\|_{L^2(0,L)}\sum_{n=0}^{N}(\Delta t)\|\boldsymbol{u}^n\|^2_{\mathbf{H}^1(\mathcal{O})}\right)
$$

We now want to show that the probability $\mathbb{P}(|I_1| > M)$ for any $M > 0$ is bounded by an expression which depends on h^m for $m > 0$ and M^{-1}. For this purpose, we will use the fact that for any two positive random variables A and B, we have that

$$
\{\omega : A + B > M\} \subseteq \{\omega : A > \frac{M}{2}\} \cup \{\omega : B > \frac{M}{2}\}. \tag{9.55}
$$

We will use this property repeatedly throughout the rest of this proof. Property (9.55) implies that

$$
\mathbb{P}(|A| + |B| > M) \leq \mathbb{P}(|A| > \frac{M}{2}) + \mathbb{P}(|B| > \frac{M}{2}).
$$

Similarly we also have

$$
\mathbb{P}(|AB| > M) \leq \mathbb{P}(|A| > \sqrt{M}) + \mathbb{P}(|B| > \sqrt{M}).
$$

Hence,

$$
\mathbb{P}(|I_1| > M) \leq \mathbb{P}\left(h\max_{1\leq n\leq N}\int_{\mathcal{O}}(R+\eta_*^n)|\boldsymbol{u}^n|^2 d\boldsymbol{x} \geq \frac{M}{2}\right)
$$

$$
+\mathbb{P}\left(h\max_{0\leq k\leq N-1}\|v^{k+\frac{1}{2}}\|_{L^2(0,L)}\sum_{n=0}^{N}(\Delta t)\|\boldsymbol{u}^n\|^2_{\mathbf{H}^1(\mathcal{O})} \geq \frac{M}{2}\right)
$$

$$
\leq \mathbb{P}\left(h\max_{1\leq n\leq N}\int_{\mathcal{O}}(R+\eta_*^n)|\boldsymbol{u}^n|^2 d\boldsymbol{x} \geq \frac{M}{2}\right)
$$

$$
+\mathbb{P}\left(\sqrt{h}\max_{0\leq k\leq N-1}\|v^{k+\frac{1}{2}}\|_{L^2(0,L)} \geq \sqrt{\frac{M}{2}}\right)
$$

$$
+\mathbb{P}\left(\sqrt{h}\sum_{n=0}^{N}(\Delta t)\|\boldsymbol{u}^n\|^2_{\mathbf{H}^1(\mathcal{O})} \geq \sqrt{\frac{M}{2}}\right)
$$

$$\leq \frac{\sqrt{h}}{\sqrt{M}}\mathbb{E}\left(\max_{1\leq n\leq N}\|\boldsymbol{u}^n\|^2_{\mathbf{L}^2(\mathcal{O})}+\max_{0\leq k\leq N-1}\|v^{k+\frac{1}{2}}\|_{L^2(0,L)}+\sum_{n=0}^{N}(\Delta t)\|\boldsymbol{u}^n\|^2_{\mathbf{H}^1(\mathcal{O})}\right)$$
$$\leq C\frac{\sqrt{h}}{\sqrt{M}}.$$

The term I_2 is treated identically. Next, we estimate the third term on the right-hand side of (9.53). Thanks to (9.50), we see that for any $s < \frac{1}{2}$ and for some C that depends only on T and δ, we have

$$I_3 := |(\Delta t)\sum_{n=0}^{N}\left(\int_{\mathcal{O}}(R+\eta^n_*)\boldsymbol{u}^n(\boldsymbol{u}^{n,n}_{\sigma,\lambda}-\boldsymbol{u}^n-(\boldsymbol{u}^{n-j,n}_{\sigma,\lambda}-\boldsymbol{u}^{n-j}))d\boldsymbol{x}\right)|$$
$$\leq C(\Delta t)\sum_{n=0}^{N}\|\boldsymbol{u}^n\|_{\mathbf{L}^2(\mathcal{O})}(\|\boldsymbol{u}^{n,n}_{\sigma,\lambda}-\boldsymbol{u}^n\|_{\mathbf{L}^2(\mathcal{O})}+\|\boldsymbol{u}^{n-j,n}_{\sigma,\lambda}-\boldsymbol{u}^{n-j}\|_{\mathbf{L}^2(\mathcal{O})})$$
$$\leq Ch^{\frac{s}{4}}\max\{1,\frac{1}{\delta}\|\widetilde{\eta}^*_N\|^{2s+\frac{3}{2}}_{C^{0,\frac{1}{4}}(0,T;L^\infty(0,L))}\}\max_{1\leq n\leq N}\|\boldsymbol{u}^n\|_{\mathbf{L}^2(\mathcal{O})}\left((\Delta t)\sum_{n=1}^{N}\|\boldsymbol{u}^k\|^2_{\mathbf{H}^1(\mathcal{O})}\right)^{\frac{1}{2}}. \tag{9.56}$$

Hence, we take $s = \frac{1}{4}$ and use Theorem 2.9.11 (1) to observe that $W^{1,\infty}(0,T;L^2(0,L)) \cap L^\infty(0,T;H^2_0(0,L)) \hookrightarrow C^{0,\frac{1}{4}}(0,T;H^{\frac{3}{2}}(0,L)) \hookrightarrow C^{0,\frac{1}{4}}(0,T;L^\infty(0,L))$ and thus obtain

$$\mathbb{P}(|I_3|>M) \leq \frac{h^{\frac{1}{32}}}{M^{\frac{1}{2}}}\mathbb{E}\left(\|\widetilde{\eta}^*_N\|^2_{C^{0,\frac{1}{4}}(0,T;L^\infty(0,L))}+\max_{1\leq n\leq N}\|\boldsymbol{u}^n\|^2_{\mathbf{L}^2(\mathcal{O})}\right.$$
$$\left.+\left((\Delta t)\sum_{n=0}^{N}\|\boldsymbol{u}^n\|^2_{\mathbf{H}^1(\mathcal{O})}\right)\right)$$
$$\leq C\frac{h^{\frac{1}{32}}}{M^{\frac{1}{2}}}.$$

Similarly, the next term on the right-hand side of (9.53) can be estimated using (9.51) to obtain

$$I_4 := |(\Delta t)\sum_{n=0}^{N}\int_0^L v^n(v^n_\lambda - v^n-(v^{n-j}_\lambda - v^{n-j}))|$$
$$\leq C(T)\lambda^{\frac{1}{2}}\max_{0\leq n\leq N}\|v^n\|_{L^2(0,L)}\left((\Delta t)\sum_{n=0}^{N}\|v^n\|^2_{H^{\frac{1}{2}}(0,L)}\right)^{\frac{1}{2}}. \tag{9.57}$$

Hence, by our choice of λ, see (9.46) above, we know that $\lambda \sim h^{1/4}$, and so we get

$$\mathbb{P}(|I_4| > M) \leq \frac{h^{\frac{1}{8}}}{M}\mathbb{E}\left(\max_{0\leq n\leq N} \|v^n\|^2_{L^2(0,L)} + (\Delta t)\sum_{n=0}^{N} \|v^n\|^2_{H^{\frac{1}{2}}(0,L)}\right) \leq C\frac{h^{\frac{1}{8}}}{M}.$$

The term I_5 is treated using similar arguments

Now we go back to (9.54) and estimate the penalty term. Thanks to (9.47) and the bounds (9.48) for σ, we observe that

$$I_6 := \left|\frac{(\Delta t)}{\varepsilon}\sum_{n=0}^{N}\int_{\mathcal{O}} \mathrm{div}^{\eta_*^n}\boldsymbol{u}^{n+1}\left(\Delta t\sum_{k=n-j+1}^{n} \mathrm{div}^{\eta_*^n}(\boldsymbol{u}^{k,n}_{\sigma,\lambda})\right)d\boldsymbol{x}\right|$$

$$\leq \frac{(\Delta t)}{\varepsilon}\sum_{n=0}^{N}\|\mathrm{div}^{\eta_*^n}\boldsymbol{u}^{n+1}\|_{L^2(\mathcal{O})}\left((\Delta t)\sum_{k=n-j+1}^{n}\|\mathrm{div}^{\eta_*^n}(\boldsymbol{u}^{k,n}_{\sigma,\lambda})\|^2_{L^2(\mathcal{O})}\right)^{\frac{1}{2}}\sqrt{h}$$

$$\leq C\sqrt{hT}\,\sigma\left(\frac{(\Delta t)}{\varepsilon}\sum_{n=0}^{N}\|\mathrm{div}^{\eta_*^n}(\boldsymbol{u}^{n+1})\|^2_{L^2(\mathcal{O})}\right)$$

$$\leq C\sqrt{hT}\left(1+\|\widetilde{\eta}^*_N\|_{C^{0,\frac{1}{4}}(0,T;L^\infty(0,L))}h^{\frac{1}{4}}\right)\left(\frac{(\Delta t)}{\varepsilon}\sum_{n=0}^{N}\|\mathrm{div}^{\eta_*^n}(\boldsymbol{u}^{n+1})\|^2_{L^2(\mathcal{O})}\right).$$

Hence,

$$\begin{aligned}\mathbb{P}(I_6 > M) &\leq \frac{\sqrt{h}}{M}\mathbb{E}[1+\|\widetilde{\eta}^*_N\|^2_{C^{0,\frac{1}{4}}(0,T;L^\infty(0,L))}h^{\frac{1}{4}}]\\ &\quad+\frac{h^{\frac{1}{4}}}{M^{\frac{1}{2}}}\mathbb{E}\left(\frac{(\Delta t)}{\varepsilon}\sum_{n=0}^{N}\|\mathrm{div}^{\eta_*^n}(\boldsymbol{u}^{n+1})\|^2_{L^2(\mathcal{O})}\right)\\ &\leq \frac{Ch^{\frac{1}{4}}}{M^{\frac{1}{2}}}.\end{aligned}$$

Notice that due to Theorem 9.2.1 (2), the constant C in the estimate above does not depend on N or ε.

Next, notice that the continuity of the embedding $H^{\frac{1}{2}}(0,L) \hookrightarrow L^p(0,L)$ for any $p < \infty$ implies that

$$\begin{aligned}\|\widetilde{\boldsymbol{u}}^k_\sigma\|_{\mathbf{L}^3(\mathcal{O}_\delta)} &\leq C(\delta)\left(\|\widetilde{\boldsymbol{u}}^{k,ext}\|_{\mathbf{L}^3(\mathcal{O}_{\eta_*^{k-1}})}+\|v^k\|_{L^3(0,L)}\right)\\ &\leq C(\delta)(\|\boldsymbol{u}^k\|_{\mathbf{H}^1(\mathcal{O})}+\|v^k\|_{H^{\frac{1}{2}}(0,L)}).\end{aligned} \tag{9.58}$$

Hence, we estimate

$$I_7 := \left| \sum_{n=0}^{N} \int_{\mathcal{O}} \left(\eta_*^{n+1} - \eta_*^n\right) \boldsymbol{u}^{n+1} \cdot \left(\Delta t \sum_{k=n-j+1}^{n} \boldsymbol{u}_{\sigma,\lambda}^{k,n} \right) d\boldsymbol{x} \right|$$

$$\leq (\Delta t) \sum_{n=0}^{N} \|v^{n+\frac{1}{2}}\|_{L^2(0,L)} \|\boldsymbol{u}^{n+1}\|_{\mathbf{L}^6(\mathcal{O})} \left((\Delta t) \sum_{k=n-j+1}^{n} \|\boldsymbol{u}_{\sigma,\lambda}^{k,n}\|^2_{\mathbf{L}^3(\mathcal{O})} \right)^{\frac{1}{2}} \sqrt{h}$$

$$\leq C\sqrt{hT} \max_{1\leq n\leq N} \|v^{n+\frac{1}{2}}\|_{L^2(0,L)} \left((\Delta t) \sum_{n=0}^{N} \|\boldsymbol{u}^n\|^2_{\mathbf{H}^1(\mathcal{O})} \right).$$

Hence,

$$\mathbb{P}\left(|I_7| \geq M\right) \leq \frac{h^{\frac{1}{4}}}{M^{\frac{1}{2}}} \mathbb{E} \left(\max_{1\leq n\leq N} \|v^{n+\frac{1}{2}}\|_{L^2(0,L)} + (\Delta t) \sum_{n=0}^{N} \|\boldsymbol{u}^{n+1}\|^2_{\mathbf{H}^1(\mathcal{O})} \right) \leq C \frac{h^{\frac{1}{4}}}{M^{\frac{1}{2}}}.$$

Next, since $H^{1+\epsilon}(\mathcal{O}) \hookrightarrow L^\infty(\mathcal{O})$, for any $\epsilon > 0$, we obtain by choosing $\epsilon = \frac{1}{8}$ for some $C > 0$ that depends only on δ that (recall the definition of transformed advection term (9.14)):

$$I_8 := \left| (\Delta t) \sum_{n=0}^{N} b^{\eta_*^n} \left(\boldsymbol{u}^n, v^{n+\frac{1}{2}} r \mathbf{e}_r, \left(\Delta t \sum_{k=n-j+1}^{n} \boldsymbol{u}_{\sigma,\lambda}^{k,n} \right) \right) \right|$$

$$\leq \sqrt{h} (\Delta t) \sum_{n=0}^{N} \|\boldsymbol{u}^n - v^{n+\frac{1}{2}} r \mathbf{e}_r\|_{\mathbf{L}^2(\mathcal{O})} \|\nabla^{\eta_*^n} \boldsymbol{u}^{n+1}\|_{\mathbf{L}^2(\mathcal{O})}$$

$$\times \left(\Delta t \sum_{k=n-j+1}^{n} \|\boldsymbol{u}_{\sigma,\lambda}^{k,n}\|^2_{\mathbf{L}^\infty(\mathcal{O})} \right)^{\frac{1}{2}}$$

$$\leq \sqrt{h} (\Delta t) \sum_{n=0}^{N} \left(\|\boldsymbol{u}^n\|_{\mathbf{L}^2(\mathcal{O})} + \|v^{n+\frac{1}{2}}\|_{L^2(0,L)} \right) \|\nabla^{\eta_*^n} \boldsymbol{u}^{n+1}\|_{\mathbf{L}^2(\mathcal{O})}$$

$$\times \left(\Delta t \sum_{k=n-j+1}^{n} \|\boldsymbol{u}_{\sigma,\lambda}^{k,n}\|^2_{\mathbf{H}^{1+\epsilon}(\mathcal{O})} \right)^{\frac{1}{2}}$$

$$\leq \frac{\sqrt{h}}{\lambda^{1+\epsilon}} (\Delta t) \sum_{n=0}^{N} \left(\|\boldsymbol{u}^n\|_{\mathbf{L}^2(\mathcal{O})} + \|v^{n+\frac{1}{2}}\|_{L^2(0,L)} \right) \|\nabla^{\eta_*^n} \boldsymbol{u}^{n+1}\|_{\mathbf{L}^2(\mathcal{O})}$$

$$\times \left(\Delta t \sum_{k=n-j+1}^{n} \|\boldsymbol{u}_{\sigma}^{k,n}\|_{\mathbf{L}^2(\mathcal{O})}^2 \right)^{\frac{1}{2}}$$

$$\leq C(T,\delta) h^{\frac{1}{8}} \left(\max_{0 \leq n \leq N} \|\boldsymbol{u}^n\|_{\mathbf{L}^2(\mathcal{O})}^2 + \|v^{n+\frac{1}{2}}\|_{L^2(0,L)}^2 \right) \left((\Delta t) \sum_{n=0}^{N} \|\boldsymbol{u}^{n+1}\|_{\mathbf{H}^1(\mathcal{O})}^2 \right)^{\frac{1}{2}}.$$

Here, in the second to last inequality, we also used the property of mollification that $\|\phi_\lambda\|_{H^1} \leq \frac{C}{\lambda} \|\phi\|_{L^2}$ (see Lemma 2.11.1). Hence for any $M > 0$, we have

$$\mathbb{P}(|I_8| \geq M) \leq \frac{h^{\frac{1}{16}}}{M^{\frac{1}{2}}} \mathbb{E} \left(\max_{0 \leq n \leq N} \|\boldsymbol{u}^n\|_{\mathbf{L}^2(\mathcal{O})}^2 + \|v^n\|_{L^2(0,L)}^2 + (\Delta t) \sum_{n=0}^{N} \|\boldsymbol{u}^{n+1}\|_{\mathbf{H}^1(\mathcal{O})}^2 \right)$$

$$\leq C \frac{h^{\frac{1}{16}}}{M^{\frac{1}{2}}}.$$

We next focus of the term with the transformed symmetrized gradients. Similarly as before, since $\|\phi_\lambda\|_{H^1} \leq \frac{C}{\lambda} \|\phi\|_{L^2}$, the term with the transformed symmetrized gradients yields

$$I_9 := |(\Delta t) \sum_{n=0}^{N} \int_{\mathcal{O}} \mathbf{D}^{\eta_*^n} \boldsymbol{u}^{n+1} \left(\Delta t \sum_{k=n-j+1}^{n} \mathbf{D}^{\eta_*^n} (\boldsymbol{u}_{\sigma,\lambda}^{k,n}) \right) d\boldsymbol{x}|$$

$$\leq (\Delta t) \sum_{n=0}^{N} \|\mathbf{D}^{\eta_*^n} \boldsymbol{u}^{n+1}\|_{\mathbf{L}^2(\mathcal{O})} \left((\Delta t) \sum_{k=n-j+1}^{n} \|\mathbf{D}^{\eta_*^n} (\boldsymbol{u}_{\sigma,\lambda}^{k,n})\|_{\mathbf{L}^2(\mathcal{O})} \right)$$

$$\leq \frac{1}{\lambda} (\Delta t) \sum_{n=0}^{N} \|\mathbf{D}^{\eta_*^n} \boldsymbol{u}^{n+1}\|_{\mathbf{L}^2(\mathcal{O})} \left((\Delta t) \sum_{k=n-j+1}^{n} \|\boldsymbol{u}^k\|_{\mathbf{L}^2(\mathcal{O})} + \|v^k\|_{L^2(0,L)} \right)$$

$$\leq C \frac{h}{\lambda} \left(\max_{0 \leq n \leq N} (\|\boldsymbol{u}^n\|_{\mathbf{L}^2(\mathcal{O})} + \|v^k\|_{L^2(0,L)}) (\Delta t) \sum_{n=0}^{N} \|\mathbf{D}^{\eta_*^n} (\boldsymbol{u}^{n+1})\|_{\mathbf{L}^2(\mathcal{O})} \right).$$

Hence, again, since we have $\lambda \sim h^{\frac{1}{4}}$, Young's inequality gives us

$$\mathbb{P}(I_9 > M) \leq \frac{h^{\frac{3}{4}}}{M} \mathbb{E}[\max_{0 \leq n \leq N} (\|\boldsymbol{u}^n\|_{\mathbf{L}^2(\mathcal{O})}^2 + \|v^k\|_{L^2(0,L)}^2)]$$

$$+ T\frac{h^{\frac{3}{4}}}{M}\mathbb{E}\left((\Delta t)\sum_{n=0}^{N}\|\mathbf{D}^{\eta_*^n}(\boldsymbol{u}^{n+1})\|^2_{\mathbf{L}^2(\mathcal{O})}\right)$$
$$\leq C\frac{h^{\frac{3}{4}}}{M}.$$

Next, we focus on the critical integral in (9.54) which contains the second-order derivative of displacement $\partial_{zz}\eta^{n+1}$. Since $\|\phi_\lambda\|_{H^2} \leq \frac{C}{\lambda^2}\|\phi\|_{L^2}$, we obtain

$$I_{11} := |(\Delta t)\sum_{n=0}^{N}\int_0^L \partial_{zz}\eta^{n+1}\partial_{zz}\left(\Delta t\sum_{k=n-j+1}^{n} v_\lambda^k\right)|$$
$$\leq CT\frac{h}{\lambda^2}\max_{0\leq n\leq N}\left(\|\partial_{zz}\eta^n\|_{L^2(0,L)}\|v^n\|_{L^2(0,L)}\right).$$

Now since $\lambda \sim h^{\frac{1}{4}}$, we have

$$\mathbb{P}(I_{11} \geq M) \leq \frac{Ch^{\frac{1}{2}}}{M}\mathbb{E}\left(\max_{0\leq n\leq N}\|\partial_{zz}\eta^n\|^2_{L^2(0,L)} + \max_{0\leq n\leq N}\|v^n\|^2_{L^2(0,L)}\right) \leq C\frac{h^{\frac{1}{2}}}{M}.$$

Finally, we treat the stochastic term using the same argument as above. To bound the expectation, we use Young's inequality and the argument presented in (9.34) to obtain

$$\mathbb{E}\left(\sum_{n=0}^{N}|(G(\mathbf{U}^n,\eta_*^n)\Delta_n W,\mathbf{Q}_n)|\right)$$
$$\leq \mathbb{E}\left(\sum_{n=0}^{N}\|G(\mathbf{U}^n,\eta_*^n)\|_{\mathbf{L}^2}|\Delta_n W|\left((\Delta t)\sum_{k=n-j+1}^{n}\|\boldsymbol{u}^{k,n}_{\sigma,\lambda}\|^2_{\mathbf{L}^2(\mathcal{O})}+\|v_\lambda^k\|^2_{L^2(0,L)}\right)^{\frac{1}{2}}h^{\frac{1}{2}}\right)$$
$$\leq h^{\frac{1}{2}}\mathbb{E}\left(\sum_{n=0}^{N}\|G(\mathbf{U}^n,\eta_*^n)\|^2_{\mathbf{L}^2}|\Delta_n W|^2\right.$$
$$\left.+\left((\Delta t)\sum_{k=n-j+1}^{n}\|\boldsymbol{u}^{k,n}_{\sigma,\lambda}\|^2_{\mathbf{L}^2(\mathcal{O})}+\|v_\lambda^k\|^2_{L^2(0,L)}\right)\right)$$
$$\leq h^{\frac{1}{2}}C(\delta)\mathbb{E}\sum_{n=0}^{N}(\Delta t)[\|\boldsymbol{u}^n\|^2_{\mathbf{L}^2(\mathcal{O})}+\|v^n\|^2_{L^2(0,L)}] \leq Ch^{\frac{1}{2}}.$$

By combining these estimates we are now in a position to show that the laws of the random variables mentioned in the statement of the theorem are tight. Observe that, due to Theorem 2.9.9, $\mathcal{B}_M$, defined in (9.43), is compact in $L^2(0, T; \mathbf{H}^\alpha(\mathcal{O})) \times L^2(0, T; L^2(0, L))$ for each $M > 0$. Now, an application of Chebyshev's inequality (3.2) gives the desired result:

$$
\begin{aligned}
\mathbb{P}((\boldsymbol{u}_N^+, v_N^+) \notin \mathcal{B}_M) &\leq \mathbb{P}\left(\|\boldsymbol{u}_N^+\|^2_{L^2(0,T;\mathbf{H}^1(\mathcal{O}))} + \|v_N^+\|^2_{L^2(0,T;H^{\frac{1}{2}}(0,L))} > \frac{M}{2}\right) \\
&+ \mathbb{P}\left(\sup_{0<h<1} h^{-\frac{1}{32}} \int_h^T \left(\|T_h \boldsymbol{u}_N^+ - \boldsymbol{u}_N^+\|^2_{\mathbf{L}^2(\mathcal{O})} + \|T_h v_N^+ - v_N^+\|^2_{L^2(0,L)}\right) > \frac{M}{2}\right) \\
&\leq \frac{C}{\sqrt{M}},
\end{aligned}
$$

where $C > 0$ depends only on δ and the given data and is independent of N and ε. This completes the proof of Lemma 9.3.1. □

Next we will prove tightness results for the structure displacements. These will be used in Sect. 9.3.2 to obtain almost sure convergence via an application of Theorem 3.2.5.

Lemma 9.3.2 *For fixed $\delta > 0$, the laws of $\{\widetilde{\eta}_N\}_{N\in\mathbb{N}}$ and that of $\{\widetilde{\eta}^*_N\}_{N\in\mathbb{N}}$ are tight in $C([0, T], H^s(0, L))$ for any $s < 2$.*

Proof We observe, thanks to Lemma 9.2.3, that $\widetilde{\eta}_N$ and $\widetilde{\eta}^*_N$ are bounded independently of N and ε in $L^2(\Omega; L^\infty(0, T; H_0^2(0, L)) \cap W^{1,\infty}(0, T; L^2(0, L)))$. Now the Aubin-Lions theorem, Theorem 2.9.4, implies that for $0 < s < 2$

$$
L^\infty(0, T; H_0^2(0, L)) \cap W^{1,\infty}(0, T; L^2(0, L)) \subset\subset C([0, T]; H^s(0, L)).
$$

Hence consider

$$
\begin{aligned}
\mathcal{K}_M := &\{\eta \in L^\infty(0, T; H_0^2(0, L)) \cap W^{1,\infty}(0, T; L^2(0, L)) : \\
&\|\eta\|^2_{L^\infty(0,T;H_0^2(0,L))} + \|\eta\|^2_{W^{1,\infty}(0,T;L^2(0,L))} \leq M\}.
\end{aligned}
$$

Using the Chebyshev inequality (3.2) once again, we obtain for some $C > 0$ independent of N, ε that the following holds:

$$
\begin{aligned}
\mathbb{P}[\widetilde{\eta}_N \notin \mathcal{K}_M] &\leq \mathbb{P}\left[\|\widetilde{\eta}_N\|^2_{L^\infty(0,T;H_0^2(0,L))} \geq \frac{M}{2}\right] + \mathbb{P}\left[\|\widetilde{\eta}_N\|^2_{W^{1,\infty}(0,T;L^2(0,L))} \geq \frac{M}{2}\right] \\
&\leq \frac{4}{M^2}\mathbb{E}\left[\|\widetilde{\eta}_N\|^2_{L^\infty(0,T;H_0^2(0,L))} + \|\widetilde{\eta}_N\|^2_{W^{1,\infty}(0,T;L^2(0,L))}\right] \leq \frac{C}{M^2}.
\end{aligned}
\tag{9.59}
$$

This completes the proof of Lemma 9.3.2. □

Next, we will require analytical weak convergence to handle the dissipation terms in the fluid and the structure equations. For that purpose, we prove the following lemma.

Lemma 9.3.3

1. *The laws of* $\boldsymbol{u}_N^+$ *are tight in* $\left(L^2(0, T; V), w\right)$.
2. *The laws of* η_N, η_N^* *and* η_N^+ *are tight in* $\left(L^2(0, T; H_0^2(0, L)), w\right)$.
3. *Thw laws of* v_N^* *are tight in* $(L^2(0, T; L^2(0, L)), w)$.

Here w *denotes the weak topology.*

Proof The proof of these statements is based on a straightforward application of the Chebyshev inequality (3.2) and the Banach Alaoglu theorem stated in Theorem 2.1.1. Observe that for any $M > 0$, the Chebyshev inequality gives us

$$\mathbb{P}[\|\boldsymbol{u}_N^+\|_{L^2(0,T;V)} > M] \leq \frac{1}{M^2}\mathbb{E}[\|\boldsymbol{u}_N^+\|^2_{L^2(0,T;V)}] \leq \frac{C}{M^2},$$

where the space V is defined in (9.15). Statements (2), (3) follow similarly. □

While formulating an approximate system in terms of the piecewise linear and piecewise constant functions (9.37) and (9.38), respectively, defined in Sect. 9.2.2, we come across typical error terms generated as a result of discretizing the **stochastic term**. To obtain the final convergence result, we need estimates of those terms. We derive those error terms and obtain their estimates next.

Observe that for given N, we can write

$$\widetilde{\eta}_N^* = \eta_N^* + (t - t_N^n)\frac{\partial \widetilde{\eta}_N^*}{\partial t}, \quad t \in (t_N^n, t_N^{n+1}),$$

and that the same equation holds true when η^* is replaced by $\boldsymbol{u}$. This gives for each $\omega \in \Omega$ and $t \in (t_N^n, t_N^{n+1})$ that

$$\begin{aligned}
\frac{\partial((R+\widetilde{\eta}_N^*)\widetilde{\boldsymbol{u}}_N)}{\partial t} &= \frac{\partial \widetilde{\boldsymbol{u}}_N}{\partial t}(R+\widetilde{\eta}_N^*) + \frac{\partial \widetilde{\eta}_N^*}{\partial t}(\widetilde{\boldsymbol{u}}_N) \\
&= \frac{\partial \widetilde{\boldsymbol{u}}_N}{\partial t}(R+\eta_N^* + (t-t_N^n)\frac{\partial \widetilde{\eta}_N^*}{\partial t}) + \frac{\partial \widetilde{\eta}_N^*}{\partial t}(\widetilde{\boldsymbol{u}}_N) \\
&= \frac{\partial \widetilde{\boldsymbol{u}}_N}{\partial t}(R+\eta_N^*) + \frac{\partial \widetilde{\eta}_N^*}{\partial t}(2\widetilde{\boldsymbol{u}}_N - \boldsymbol{u}_N) \\
&= \sum_{n=0}^{N-1} \frac{1}{\Delta t}(\boldsymbol{u}_N^{n+1} - \boldsymbol{u}_N^n)(R+\eta_{*,N}^n)\,\mathbb{1}_{(t_N^n, t_N^{n+1})} + v_N^*(2\widetilde{\boldsymbol{u}}_N - \boldsymbol{u}_N).
\end{aligned}$$

By integrating in time both sides and adding the equation for $\widetilde{v}_N$ from (9.25), we obtain for any $(\boldsymbol{q}, \psi) \in \mathscr{U}_1$:

$$
\begin{aligned}
&((R+\widetilde{\eta}_N^*(t))\widetilde{\boldsymbol{u}}_N(t),\boldsymbol{q})+(\widetilde{v}_N(t),\psi)\\
&=\int_{\mathcal{O}}\boldsymbol{u}_0(R+\eta_0)\cdot\boldsymbol{q}d\boldsymbol{x}+\int_0^L v_0\psi dz-\int_0^t\int_0^L(\partial_z\eta_N^+\partial_z\psi+\partial_{zz}\eta_N^+\partial_{zz}\psi)dzds\\
&+\sum_{n=0}^{N-1}\int_0^t\mathbb{1}_{(t_N^n,t_N^{n+1})}\left(\int_{\mathcal{O}}\frac{(\boldsymbol{u}_N^{n+1}-\boldsymbol{u}_N^n)}{\Delta t}(R+\eta_{*,N}^n)\cdot\boldsymbol{q}\,d\boldsymbol{x}\right.\\
&\left.+\int_0^L\frac{(v_N^{n+\frac{1}{2}}-v_N^n)}{\Delta t}\psi dz\right)ds\\
&+\int_0^t\int_{\mathcal{O}}v_N^*(2\widetilde{\boldsymbol{u}}_N-\boldsymbol{u}_N)\cdot\boldsymbol{q}d\boldsymbol{x}ds\\
&=:(\boldsymbol{u}_0(R+\eta_0),\boldsymbol{q})+(v_0,\psi)+(I_1(t)+I_2(t)+I_3(t),\mathbf{Q}).
\end{aligned}\tag{9.60}
$$

We estimate the right-hand side to be able to pass to the limit as $N \to \infty$. We start by considering the integral I_2. Thanks to (9.28), I_2 on the right-hand side is equal to

$$
\begin{aligned}
&(I_2(t),\mathbf{Q})=-\frac{1}{2}\int_0^t\int_{\mathcal{O}}v_N^*\boldsymbol{u}_N^+\boldsymbol{q}d\boldsymbol{x}ds\\
&-2\nu\int_0^t\int_{\mathcal{O}}(R+\eta_N^*)\mathbf{D}^{\eta_N^*}(\boldsymbol{u}_N^+)\cdot\mathbf{D}^{\eta_N^*}(\boldsymbol{q})d\boldsymbol{x}ds-\frac{1}{\varepsilon}\int_0^t\int_{\mathcal{O}}\mathrm{div}^{\eta_N^*}\boldsymbol{u}_N^+\mathrm{div}^{\eta_N^*}\boldsymbol{q}d\boldsymbol{x}ds\\
&-\frac{1}{2}\int_0^t\int_{\mathcal{O}}(R+\eta_N^*)((\boldsymbol{u}_N-v_N^{\#}r\mathbf{e}_r)\cdot\nabla^{\eta_N^*}\boldsymbol{u}_N\cdot\boldsymbol{q}-(\boldsymbol{u}_N-v_N^{\#}r\mathbf{e}_r)\cdot\nabla^{\eta_N^*}\boldsymbol{q}\\
&\cdot\boldsymbol{u}_N^+)d\boldsymbol{x}ds\\
&+\int_0^t\left(P_{in}\int_0^1 q_z\Big|_{z=0}dr-P_{out}\int_0^1 q_z\Big|_{z=1}dr\right)ds\\
&+\int_0^t(G(\boldsymbol{u}_N,v_N,\eta_N^*),\mathbf{Q})dW+(E_N(t),\mathbf{Q})\\
&=:(I_{2,1}(t)+I_{2,2}(t)+I_{2,3}(t)+I_{2,4}(t)+I_{2,5}(t)+I_{2,6}(t)+E_N(t),\mathbf{Q}),
\end{aligned}\tag{9.61}
$$

where the error term is given by

$$
\begin{aligned}
E_N(t)=&\sum_{m=0}^{N-1}\left(\frac{t-t_N^m}{\Delta t}G(\boldsymbol{u}_N^m,v_N^m,\eta_{*,N}^m)\Delta_m W-\int_{t^m}^t G(\boldsymbol{u}_N^m,v_N^m,\eta_{*,N}^m)dW\right)\mathbb{1}_{[t_N^m,t_N^{m+1})}\\
&+\sum_{m=0}^{N-1}\left((t-t_N^m)P_{N,in/out}^m-\int_{t_N^m}^t P_{N,in/out}(s)ds\right)\mathbb{1}_{[t_N^m,t_N^{m+1})}\\
&=E_N^s+E_N^p.
\end{aligned}
$$

We will study the behavior of each term on the right-hand side of the expression for I_2 as $N \to \infty$. We start by showing that the numerical error term E_N approaches 0 as $N \to \infty$. The remaining integrals will be considered in Lemma 9.3.5 below. Observe that the deterministic numerical error of the pressure discretization E_N^p converges to 0 in $L^2(0, T)$ as $N \to \infty$ (see (8.69) in Sect. 8).

Lemma 9.3.4 *The numerical error E_N^s of the stochastic term has the following property:*

$$\mathbb{E}\int_0^T \|E_N^s(t)\|^2_{\mathbf{L}^2(\mathcal{O})\times L^2(0,L)}dt \to 0 \text{ as } N \to \infty.$$

Proof First, for any N, we have

$$\begin{aligned}E_N^s(t) &= \sum_{m=0}^{N-1}\left(\frac{t-t_N^m}{\Delta t}G(\boldsymbol{u}_N^m, v_N^m, \eta_{*,N}^m)\Delta_{m,N}W - \int_{t_N^m}^t G(\boldsymbol{u}_N^m, v_N^m, \eta_{*,N}^m)dW\right)\mathbb{1}_{[t_N^m, t_N^{m+1})}\\ &=: E_{1,N}^s + E_{2,N}^s.\end{aligned}$$

We estimate $E_{1,N}^s$ and $E_{2,N}^s$. Recall our notation $\mathbf{L}^2 = \mathbf{L}^2(\mathcal{O}) \times L^2(0, L)$. Observe that $E_{1,N}^s$ satisfies (see also (9.34)),

$$\begin{aligned}\mathbb{E}\int_0^T \|E_{1,N}^s\|^2_{\mathbf{L}^2}dt &= \mathbb{E}\sum_{n=0}^{N-1}\|G(\boldsymbol{u}_N^n, v_N^n, \eta_{*,N}^n)\Delta_{n,N}W\|^2_{\mathbf{L}^2}\int_{t_N^n}^{t_N^{n+1}}\left|\frac{t-t_N^m}{\Delta t}\right|^2 dt\\ &= \mathbb{E}\sum_{n=0}^{N-1}\|G(\boldsymbol{u}_N^n, v_N^n, \eta_{*,N}^n)\Delta_{n,N}W\|^2_{\mathbf{L}^2}(\frac{\Delta t}{3})\\ &\le \mathbb{E}\sum_{n=0}^{N-1}\|G(\boldsymbol{u}_N^n, v_N^n, \eta_{*,N}^n)\|^2_{\mathbf{L}^2}|\Delta_{n,N}W|^2\Delta t\\ &= (\Delta t)^2\mathbb{E}\sum_{n=0}^{N-1}\|G(\boldsymbol{u}_N^n, v_N^n, \eta_{*,N}^n)\|^2_{\mathbf{L}^2} \le C\Delta t,\end{aligned}$$

where $C > 0$ does not depend on N or ε, as a consequence of Theorem 9.2.1.

To estimate $E_{2,N}^s$, we use the Itô isometry, Theorem 3.6.1, to obtain

$$\begin{aligned}\mathbb{E}\int_0^T \|E_{2,N}^s(t)\|^2_{\mathbf{L}^2}dt &= \mathbb{E}\sum_{n=0}^{N-1}\int_{t_N^n}^{t_N^{n+1}}\|\int_{t_N^n}^t G(\boldsymbol{u}_N^n, v_N^n, \eta_{*,N}^n)dW\|^2_{\mathbf{L}^2}dt\\ &= \mathbb{E}\sum_{n=0}^{N-1}\int_{t_N^n}^{t_N^{n+1}}\int_{t_N^n}^t \|G(\boldsymbol{u}_N^n, v_N^n, \eta_{*,N}^n)\|^2_{\mathbf{L}^2}dsdt\end{aligned}$$

$$= \frac{1}{2}\mathbb{E}\sum_{n=0}^{N-1} \|G(\boldsymbol{u}_N^n, v_N^n, \eta_{*,N}^n)\|_{\mathbf{L}^2}^2 (\Delta t)^2 \leq C\Delta t.$$

Thus, as $N \to \infty$, we obtain $\mathbb{E}\int_0^T \|E_{2,N}^s(t)\|^2_{\mathbf{L}^2(\mathcal{O})\times L^2(0,L)} dt \to 0$. □

To be able to pass to the limit a $N \to \infty$ in the weak formulation, we consider the following random variable:

$$\widetilde{\mathbf{U}}_N(t) := ((R + \widetilde{\eta}_N^*(t))\widetilde{\boldsymbol{u}}_N(t), \widetilde{v}_N(t)) - E_N(t), \tag{9.62}$$

and study its behavior as $N \to \infty$. We prove the following tightness result, Lemma 9.3.5, for the laws of $\widetilde{\mathbf{U}}_N$ defined in (9.62).
We begin by defining

$$\mathscr{U}_1 := \mathscr{U} \cap (\mathbf{H}^2(\mathcal{O}) \times H_0^2(0, L)). \tag{9.63}$$

where $\mathscr{U}$ is given in (9.20). Next for any $\mathscr{V}_1 \subset\subset \mathscr{U}_1$, we denote by $\mu_N^{u,v}$ the probability measure of $\widetilde{\mathbf{U}}_N$:

$$\mu_N^{u,v} := \mathbb{P}\left(\widetilde{\mathbf{U}}_N \in \cdot\right) \in Pr(C([0, T]; \mathscr{V}_1')),$$

where $Pr(S)$, here and onward, denotes the set of probability measures on a metric space S. Then we have the following tightness result.

Lemma 9.3.5 *For fixed $\varepsilon > 0$ and $\delta > 0$, the laws $\{\mu_N^{u,v}\}_N$ of the random variables $\{\widetilde{\mathbf{U}}_N\}_N$ are tight in $C([0, T]; \mathscr{V}_1')$.*

Proof Recall that the random variables $\{\widetilde{\mathbf{U}}_N\}_N$ consist of two parts, a deterministic part and a stochastic part. We will show that the stochastic terms belong to

$$B_M^S := \{\mathbf{Y} \in C([0, T]; \mathscr{V}_1'); \|\mathbf{Y}\|_{W^{\alpha,q}(0,T;\mathscr{U}_1')} \leq M\}$$

while the deterministic terms belong to

$$B_M^D := \{\mathbf{X} \in C([0, T]; \mathscr{V}_1'); \|\mathbf{X}\|_{H^1(0,T;\mathscr{U}_1')} \leq M\},$$

for some $M > 0$ and α such that $0 < \alpha < \frac{1}{2}$ and $\alpha q > 1$ where $q > 2$. To do this, we first define

$$B_M := B_M^D + B_M^S$$

and note that

$$\{\widetilde{\mathbf{U}}_N \in B_M\} \supseteq \{I_1 + \sum_{i=1}^{5} I_{2,i} + I_3 \in B_M^D\} \cap \{I_{2,6} \in B_M^S\},$$

where the integrals I_1, $I_{2,i}$ and I_3 are defined in (9.60) and (9.61). This implies that for the complement of the set B_M, we have

$$\mu_N^{u,v}\left(B_M^c\right) \leq \mathbb{P}\left(\|I_1 + \sum_{i=1}^{5} I_{2,i} + I_3\|_{H^1(0,T;\mathscr{U}_1')} > M\right) + \mathbb{P}\left(\|I_{2,6}\|_{W^{\alpha,q}(0,T;\mathscr{U}_1')} > M\right).$$

We want to show that for some $C = C(\delta, T, \varepsilon) > 0$ independent of N, we have

$$\mu_N^{u,v}(B_M) \geq 1 - \frac{C}{M}.$$

To do this, we need to bound the integrals I_1, $I_{2,i}$, and I_3. These bounds rely heavily on the uniform bounds obtained in Lemma 9.2.3. To estimate these terms, we introduce the following simplified notation: $\bar{\sup}_{\mathbf{Q}} := \sup_{\mathbf{Q}\in\mathscr{U}_1, |\mathbf{Q}|_{\mathscr{U}_1}=1}$.

We start with I_1. The estimate of I_1 follows by observing that there exists some $C > 0$ independent of N, ε such that

$$\begin{aligned}\mathbb{E}\|\partial_t I_1(t)\|^2_{L^2(0,T;\mathscr{U}_1')} &\leq \mathbb{E}\,\bar{\sup}_{\mathbf{Q}} \int_0^T |\int_0^L (\partial_z \eta_N^+ \partial_z \psi + \partial_{zz}\eta_N^+ \partial_{zz}\psi) dz|^2 dt \\ &\leq \mathbb{E}\int_0^T \|\eta_N^+(t)\|^2_{H_0^2(0,L)} dt \leq C.\end{aligned}$$

Using the continuous Sobolev embedding $\mathbf{H}^2(\mathcal{O}) \hookrightarrow \mathbf{L}^\infty(\mathcal{O})$, we similarly treat $I_{2,1} + I_3$,

$$\begin{aligned}&\mathbb{E}\|\partial_t(I_{2,1} + I_3)\|_{L^2(0,T;\mathscr{U}_1')} \\ &\leq \mathbb{E}\left(\bar{\sup}_{\mathbf{Q}} \int_0^T \|v_N^*\|^2_{L^2(0,L)} \|(2\tilde{\boldsymbol{u}}_N - \boldsymbol{u}_N - \frac{1}{2}\boldsymbol{u}_N^+)\|^2_{\mathbf{L}^2(\mathcal{O})} \|\boldsymbol{q}\|^2_{\mathbf{L}^\infty(\mathcal{O})} dt\right)^{\frac{1}{2}} \\ &\leq \mathbb{E}\left(\|v_N^*\|^2_{L^\infty(0,T;L^2(0,L))}\left(\|\tilde{\boldsymbol{u}}_N\|^2_{L^\infty(0,T;\mathbf{L}^2(\mathcal{O}))} + \|\boldsymbol{u}_N\|^2_{L^\infty(0,T;\mathbf{L}^2(\mathcal{O}))}\right.\right. \\ &\qquad \left.\left. + \|\boldsymbol{u}_N^+\|^2_{L^\infty(0,T;\mathbf{L}^2(\mathcal{O}))}\right)\right)^{\frac{1}{2}} \\ &\leq \left(\mathbb{E}\|v_N^*\|^2_{L^\infty(0,T;L^2(0,L))}\right)^{\frac{1}{2}} \left(\mathbb{E}\|\tilde{\boldsymbol{u}}_N\|^2_{L^\infty(0,T;\mathbf{L}^2(\mathcal{O}))} + \mathbb{E}\|\boldsymbol{u}_N\|^2_{L^\infty(0,T;\mathbf{L}^2(\mathcal{O}))}\right. \\ &\qquad \left. + \mathbb{E}\|\boldsymbol{u}_N^+\|^2_{L^\infty(0,T;\mathbf{L}^2(\mathcal{O}))}\right)^{\frac{1}{2}} \\ &\leq C.\end{aligned}$$

To estimate $I_{2,2}$, we first recall that we can write $\nabla^{\eta_N^*} \boldsymbol{q} = \nabla \boldsymbol{q} (\nabla A^\omega_{\eta_N^*})^{-1}$. Now we note that for $\frac{3}{2} < s < 2$, we have that $\|\eta_N^*\|_{H^s(0,L)} < \frac{1}{\delta}$ for any $\omega \in \Omega$ and $t \in [0, T]$, which implies that $\sup_{t\in[0,T]} \|(\nabla A^\omega_{\eta_N^*})^{-1}\|_{\mathbf{L}^\infty(\mathcal{O})} < C$ and thus,

$$\sup_{t\in[0,T]} \|\nabla^{\eta_N^*} \boldsymbol{q}\|_{\mathbf{L}^2(\mathcal{O})} \leq C \|\nabla \boldsymbol{q}\|_{\mathbf{L}^2(\mathcal{O})} \quad \forall \omega \in \Omega, \tag{9.64}$$

where C depends only on δ. The same argument gives us that

$$\sup_{t\in[0,T]} \|\text{div}^{\eta_N^*} \boldsymbol{q}\|_{L^2(\mathcal{O})} \leq \sup_{t\in[0,T]} \sqrt{2} \|\nabla^{\eta_N^*} \boldsymbol{q}\|_{\mathbf{L}^2(\mathcal{O})} \leq C \|\nabla \boldsymbol{q}\|_{\mathbf{L}^2(\mathcal{O})}.$$

Thus for $I_{2,2}$, we obtain the following bounds:

$$\begin{aligned}
\mathbb{E}\|\partial_t I_{2,2}(t)\|^2_{L^2(0,T;\mathscr{U}_1')} &\leq \mathbb{E}\left(\bar{\sup}_{\mathbf{Q}} \int_0^T \|\mathbf{D}^{\eta_N^*} \boldsymbol{u}_N^+\|^2_{\mathbf{L}^2(\mathcal{O})} \|\mathbf{D}^{\eta_N^*} \boldsymbol{q}\|^2_{\mathbf{L}^2(\mathcal{O})} dt \right) \\
&\leq C(\delta) \mathbb{E}\left(\bar{\sup}_{\mathbf{Q}} \int_0^T \|\mathbf{D}^{\eta_N^*} \boldsymbol{u}_N^+\|^2_{\mathbf{L}^2(\mathcal{O})} \|\nabla \boldsymbol{q}\|^2_{\mathbf{L}^2(\mathcal{O})} dt \right) \leq C(\delta).
\end{aligned}$$

To estimate $I_{2,3}$, we note that for a fixed $\varepsilon > 0$, Lemma 9.2.3(6) implies the existence of a constant C dependent on δ such that:

$$\begin{aligned}
\mathbb{E}\|\partial_t I_{2,3}(t)\|^2_{L^2(0,T;\mathscr{U}_1')} &\leq C(\delta) \frac{1}{\varepsilon^2} \mathbb{E}\left(\bar{\sup}_{\mathbf{Q}} \int_0^T \|\text{div}^{\eta_N^*} \boldsymbol{u}_N^+\|^2_{L^2(\mathcal{O})} \|\text{div}^{\eta_N^*} \boldsymbol{q}\|^2_{L^2(\mathcal{O})} \right) \\
&\leq C(\delta) \frac{1}{\varepsilon^2} \mathbb{E}\left(\int_0^T \|\text{div}^{\eta_N^*} \boldsymbol{u}_N^+\|^2_{L^2(\mathcal{O})} \right) \leq \frac{C(\delta)}{\varepsilon}.
\end{aligned}$$

The integral $I_{2,4}$ can be estimated similarly after the use of the Ladyzenskaya inequality, Theorem 2.5.4 as follows:

$$\begin{aligned}
&\mathbb{E}\|\partial_t I_{2,4}(t)\|_{L^2(0,T;\mathscr{U}_1')} \leq C(\delta) \mathbb{E}\Big(\bar{\sup}_{\mathbf{Q}} \int_0^T \Big((\|\boldsymbol{u}_N\|_{\mathbf{L}^2(\mathcal{O})} + \|v_N^\#\|_{L^2(0,L)}) \times \\
&\qquad (\|\nabla^{\eta_N^*} \boldsymbol{u}_N\|_{\mathbf{L}^2(\mathcal{O})} \|\boldsymbol{q}\|_{\mathbf{L}^\infty(\mathcal{O})} + \|\nabla^{\eta_N^*} \boldsymbol{q}\|_{\mathbf{L}^4(\mathcal{O})} \|\boldsymbol{u}_N^+\|_{\mathbf{L}^4(\mathcal{O})}) \Big)^2 dt \Big)^{\frac{1}{2}} \\
&\leq C(\delta) \mathbb{E}\left(\int_0^T (\|\boldsymbol{u}_N\|_{\mathbf{L}^2(\mathcal{O})} + \|v_N^\#\|_{L^2(0,L)})^2 \|\nabla \boldsymbol{u}_N^+\|^2_{\mathbf{L}^2(\mathcal{O})} dt \right)^{\frac{1}{2}} \\
&\leq C(\delta) \mathbb{E}\left[\left(\|\boldsymbol{u}_N\|_{L^\infty(0,T;\mathbf{L}^2(\mathcal{O}))} + \|v_N^\#\|_{L^\infty(0,T;L^2(0,L))} \right) \left(\int_0^T \|\nabla \boldsymbol{u}_N^+\|^2_{\mathbf{L}^2(\mathcal{O})} dt \right)^{\frac{1}{2}} \right] \\
&\leq C(\delta) \left(\mathbb{E}\left(\|\boldsymbol{u}_N\|^2_{L^\infty(0,T;\mathbf{L}^2(\mathcal{O}))} + \|v_N^\#\|^2_{L^\infty(0,T;L^2(0,L))} \right) \right.
\end{aligned}$$

$$\times\ \mathbb{E}\left(\int_0^T \|\nabla \boldsymbol{u}_N^+\|^2_{\mathbf{L}^2(\mathcal{O})} dt\right)\Bigg)^{\frac{1}{2}} \leq C(\delta).$$

The treatment of the term $I_{2,5}$ is trivial.

Hence, by combining all the bounds derived so far, we obtain the following estimate for the deterministic part of $\widetilde{\mathbf{U}}_N$:

$$\begin{aligned}
&\mathbb{P}\left(\|I_1 + \sum_{i=1}^5 I_{2,i} + I_3\|_{H^1(0,T;\mathscr{U}_1')} > M\right)\\
&\leq \mathbb{P}\left(\|I_1\|_{H^1(0,T;\mathscr{U}_1')} + \|\sum_{i=1}^5 I_{2,i}\|_{H^1(0,T;\mathscr{U}_1')} + \|I_3\|_{H^1(0,T;\mathscr{U}_1')} > M\right)\\
&\leq \frac{1}{M}\mathbb{E}\left(\|I_1\|_{H^1(0,T;\mathscr{U}_1')} + \sum_{i=1}^5 \|I_{2,i}\|_{H^1(0,T;\mathscr{U}_1')} + \|I_3\|_{H^1(0,T;\mathscr{U}_1')}\right)\\
&\leq \frac{C(\delta,\varepsilon)}{M}.
\end{aligned} \tag{9.65}$$

To treat the stochastic part of $\widetilde{\mathbf{U}}_N$, we estimate the stochastic term $I_{2,6}$ as follows. First we use (3.19). To deal with the qth power appearing in the inequality (3.19), we define the following stopping times:

$$\tau_M := \inf_{t\geq 0}\{\sup_{s\in[0,t]} \|(\boldsymbol{u}_N, v_N)\|_{\mathbf{L}^2(\mathcal{O})\times L^2(0,L)} > M\} \wedge T.$$

Observe that, since $(\boldsymbol{u}_N, v_N)$ has right-continuous paths in $\mathbf{L}^2 = \mathbf{L}^2(\mathcal{O}) \times L^2(0, L)$ by definition, and since the filtration $(\mathcal{F}_t)_{t\geq 0}$ is right continuous by assumption, Lemma 3.3.3 implies that τ_M is indeed a stopping time (see Definiton 3.3.12 and Lemma 3.3.3). Using the Chebyshev inequality, we then obtain

$$\begin{aligned}
&\mathbb{P}\left(\|\int_0^\cdot G(\boldsymbol{u}_N, v_N, \eta_N^*)dW\|_{W^{\alpha,q};(0,T;\mathscr{U}')} > M\right)\\
&\leq \mathbb{P}\left(\|\int_0^{\cdot\wedge\tau_M} G(\boldsymbol{u}_N, v_N, \eta_N^*)dW\|_{W^{\alpha,q}(0,T;\mathscr{U}_1')} > M, \tau_M = T\right) + \mathbb{P}(\tau_M < T)\\
&\leq \mathbb{P}\left(\|\int_0^{\cdot\wedge\tau_M} G(\boldsymbol{u}_N, v_N, \eta_N^*)dW\|_{W^{\alpha,q}(0,T;\mathscr{U}_1')} > R, \tau_M = T\right)\\
&\quad + \mathbb{P}\left(\sup_{t\in[0,T]} \|(\boldsymbol{u}_N(t), v_N(t))\|_{\mathbf{L}^2} \geq M\right)
\end{aligned}$$

$$\leq \frac{1}{M^q}\mathbb{E}\left(\| \int_0^{\cdot\wedge\tau_M} G(\boldsymbol{u}_N, v_N, \eta_N^*) dW \|^q_{W^{\alpha,q}(0,T;\mathscr{U}_1')} \right)$$

$$+ \frac{1}{M^2}\mathbb{E} \sup_{t\in[0,T]} \|(\boldsymbol{u}_N(t), v_N(t))\|^2_{\mathbf{L}^2}.$$

Notice that η_N^* and $(\boldsymbol{u}_N, v_N)$ are $(\mathcal{F}_t)_{t\geq 0}$-adapted. Hence using (3.19) and (9.21), we obtain the following upper bounds for the first term on the right-hand side of the above inequality,

$$\leq \frac{C}{M^q}\mathbb{E}\left(\int_0^{T\wedge\tau_M} \|G(\boldsymbol{u}_N, v_N, \eta_N^*)\|^q_{\mathbf{L}^2} \right)$$

$$\leq \frac{C}{M^q}\mathbb{E}\left(\int_0^{T\wedge\tau_M} \|\eta_N^*\|^q_{L^\infty(0,L)} \|\boldsymbol{u}_N\|^q_{\mathbf{L}^2(\mathcal{O})} + \|v_N\|^q_{L^2(0,L)} \right)$$

$$\leq \frac{CTM^{q-2}}{\delta^q M^q}\mathbb{E}[\sup_{s\in[0,T]} \|\boldsymbol{u}_N\|^2_{\mathbf{L}^2(\mathcal{O})} + \|v_N\|^2_{L^2(0,L)}].$$

Hence for any $\delta > 0$, we have

$$\mathbb{P}\left(\| \int_0^{\cdot} G(\boldsymbol{u}_N, v_N, \eta_N^*) dW \|_{W^{\alpha,q};(0,T;\mathscr{U}')} > M \right)$$

$$\leq \frac{CT}{\delta^q M^2}\mathbb{E} \sup_{s\in[0,T]} [\|\boldsymbol{u}_N\|^2_{\mathbf{L}^2(\mathcal{O})} + \|v_N\|^2_{L^2(0,L)}].$$

Finally, the uniform bounds obtained in Lemma 9.2.3 imply

$$\mathbb{P}\left(\|I_{2,6}\|_{W^{\alpha,q}(0,T;\mathscr{U}_1')} > M \right) \leq \frac{CT}{M^2\delta}. \tag{9.66}$$

To sum up, for some $C = C(\delta, T, \varepsilon) > 0$ independent of N, we have obtained that

$$\mu_N^{u,v}(B_M) \geq 1 - \frac{C}{M}.$$

Now, thanks to Theorem 2.9.7, B_M is compactly embedded in $C([0,T]; \mathscr{V}_1')$, and thus we conclude that $\{\mu_N^{u,v}\}_N$ is tight in $C([0,T]; \mathscr{V}_1')$. □

So far we have been successful at obtaining tightness results only for a subset of the random variables defined at the beginning of Sect. 9.2.2. However when passing to the limit, we will also require almost sure convergence of the rest of the random variables for which we will use the following lemma. This result will also help us identify the relation between the limits of the laws of these random variables and, after an application of the Skorohod Representation theorem, the relation between the limits of the new random variables.

Lemma 9.3.6 *For a fixed $\delta > 0$, the following convergence results hold:*

1. $\lim_{N\to\infty} \mathbb{E}\int_0^T \|\boldsymbol{u}_N - \boldsymbol{u}_N^+\|^2_{\mathbf{L}^2(\mathcal{O})}dt = 0, \quad \lim_{N\to\infty} \mathbb{E}\int_0^T \|\boldsymbol{u}_N - \widetilde{\boldsymbol{u}}_N\|^2_{\mathbf{L}^2(\mathcal{O})}dt = 0$
2. $\lim_{N\to\infty} \mathbb{E}\int_0^T \|\eta_N - \widetilde{\eta}_N\|^2_{H_0^2(0,L)}dt = 0, \quad \lim_{N\to\infty} \mathbb{E}\int_0^T \|\eta_N^+ - \widetilde{\eta}_N\|^2_{H_0^2(0,L)}dt = 0$
3. $\lim_{N\to\infty} \mathbb{E}\int_0^T \|\eta_N^* - \widetilde{\eta}_N^*\|^2_{H_0^2(0,L)}dt = 0$
4. $\lim_{N\to\infty} \mathbb{E}\int_0^T \|v_N - \widetilde{v}_N\|^2_{L^2(0,L)}dt = 0, \; \lim_{N\to\infty} \mathbb{E}\int_0^T \|v_N - v_N^{\#}\|^2_{L^2(0,L)}dt = 0.$

Proof See Lemma 8.7.1 in Sect. 8. □

Corollary 9.3.1 *The following statements hold true:*

1. *The laws of $\boldsymbol{u}_N$ are tight in $L^2(0, T; \mathbf{L}^2(\mathcal{O}))$.*
2. *The laws of v_N, $v_N^{\#}$ are tight in $L^2(0, T; L^2(0, L))$.*
3. *The laws of η_N, η_N^+, η_N^* are tight in $L^2(0, T; H^s(0, L))$ for $s \in (\frac{3}{2}, 2)$.*

Proof Due to Lemma 9.3.1, we know that $\boldsymbol{u}_N^+$ is tight in $L^2(0, T; \mathbf{L}^2(\mathcal{O}))$. Recall that due to Prokhorov's theorem stated in Theorem 3.2.2, weak convergence of probability measures up to a subsequence is equivalent to tightness. Thus, thanks to Lemma 9.3.6 and Proposition 3.2.2, we obtain that the laws of $\boldsymbol{u}_N$ and $\widetilde{\boldsymbol{u}}_N$ are also tight in $L^2(0, T; \mathbf{L}^2(\mathcal{O}))$ and converge weakly to the same law as the weak limit of the laws of $\boldsymbol{u}_N^+$. We can apply the same argument to prove statements (2) and (3). See also the proof of Theorem 8.7.1 in Chap. 8. □

9.3.2 Almost Sure Convergence

As in Chap. 8, we let μ_N be the joint law of the random variable $\mathcal{U}_N := (\boldsymbol{u}_N, \boldsymbol{u}_N^+, v_N, v_N^{\#}, v_N^*, \eta_N, \eta_N^*, \eta_N^+, \widetilde{\eta}_N^*, \widetilde{\eta}_N, \widetilde{\mathbf{U}}_N, W)$ taking values in the phase space

$$\Upsilon := L^2(0, T; \mathbf{L}^2(\mathcal{O})) \times \Big(L^2(0, T; \mathbf{L}^2(\mathcal{O})) \cap (L^2(0, T; \mathbf{H}^1(\mathcal{O})), w)\Big)$$
$$\times [L^2(0, T; L^2(0, L))]^2 \times (L^2(0, T); L^2(0, L), w)$$
$$\times [L^2(0, T; H^s(0, L)) \cap (L^2(0, T; H_0^2(0, L)), w)]^3 \times [C([0, T], H^s(0, L))]^2$$
$$\times \Big(C([0, T]; \mathcal{V}_1') \cap L^2(0, T; \mathbf{L}^2(\mathcal{O}) \times L^2(0, L))\Big) \times C([0, T]; \mathbb{R}),$$

for some fixed $\frac{3}{2} < s < 2$.

Observe that the sequence of Borel probability measures, $\mu_N^W(\cdot) := \mathbb{P}(W \in \cdot)$, that are constantly equal to one element, is tight on $C([0, T]; \mathbb{R})$.

Next, recalling Lemmas 9.3.1, 9.3.2, 9.3.3, 9.3.5, and Corollary 9.3.1, and using Tychonoff's theorem, it follows that the sequence of the joint probability measures μ_N of the approximating sequence $\mathcal{U}_N$ is tight on the space Υ. Hence, we will next apply the Skorohod representation theorem stated in Theorem 3.2.5.

Theorem 9.3.1 *There exists a probability space* $(\widetilde{\Omega}, \widetilde{\mathcal{F}}, \widetilde{\mathbb{P}})$[1] *and random variables* $\bar{\mathcal{U}}_N := (\bar{\boldsymbol{u}}_N, \bar{\boldsymbol{u}}_N^+, \bar{v}_N, \bar{v}_N^\#, \bar{v}_N^*, \bar{\eta}_N, \bar{\eta}_N^*, \bar{\eta}_N^+, \bar{\bar{\eta}}_N^*, \bar{\bar{\eta}}_N, \bar{\bar{\mathbf{U}}}_N, \bar{W}_N)$ *and* $\bar{\mathcal{U}} := (\bar{\boldsymbol{u}}, \bar{\boldsymbol{u}}^+, \bar{v}, \bar{v}^\#, \bar{v}^*, \bar{\eta}, \bar{\eta}^*, \bar{\eta}^+, \bar{\bar{\eta}}^*, \bar{\bar{\eta}}, \bar{\bar{\mathbf{U}}}, \bar{W})$ *defined on this new probability space, such that*

1. $\bar{\mathcal{U}}_N =^d \mathcal{U}_N$.
2. $\bar{\mathcal{U}}_N \to \bar{\mathcal{U}}$ $\widetilde{\mathbb{P}}$*-almost surely in the topology of* Υ.
3. $\bar{\boldsymbol{u}} = \bar{\boldsymbol{u}}^+, \quad \bar{\eta} = \bar{\eta}^+ = \bar{\bar{\eta}}, \quad \bar{\eta}^* = \bar{\bar{\eta}}^*, \quad \bar{v} = \bar{v}^\#$ *and* $\bar{\bar{\mathbf{U}}} = ((R + \bar{\eta}^*)\bar{\boldsymbol{u}}, \bar{v})$.
4. $\partial_t \bar{\eta} = \bar{v}$ *and* $\partial_t \bar{\eta}^* = \bar{v}^*$ *in the sense of distributions, almost surely.*

Proof The existence of the probability space and the existence of the new random variables as well as the first two statements of the theorem follow from an application of Theorem 3.2.5. For a proof of statements (3) and (4), see properties 1 and 3 of Lemma 8.7.2 in Chap. 8 . □

Next we will construct a complete, right-continuous filtrations $(\bar{\mathcal{F}}_t)_{t\geq 0}$ and $(\bar{\mathcal{F}}_t^N)_{t\geq 0}$ on the new probability space $(\bar{\Omega}, \bar{\mathcal{F}}, \bar{\mathbb{P}})$ given in Theorem 9.3.1, to which the limiting and approximate noise processes and solutions are adapted, respectively. We denote by $\mathcal{F}_t{}'$ the σ-field generated by the random variables $\bar{\boldsymbol{u}}(s), \bar{v}(s), \bar{\eta}(s), \bar{W}(s)$ for all $s \leq t$. Then we define (as previously done in (8.62) in Chap. 8):

$$\mathcal{N} := \{\mathcal{A} \in \widetilde{\mathcal{F}} \mid \widetilde{\mathbb{P}}(\mathcal{A}) = 0\}, \qquad \bar{\mathcal{F}}_t^0 := \sigma(\mathcal{F}_t' \cup \mathcal{N}), \qquad \bar{\mathcal{F}}_t := \bigcap_{s\geq t} \bar{\mathcal{F}}_s^0. \tag{9.67}$$

We note here that (9.70) gives us stochastic processes $(R + \bar{\eta}^*)\bar{\boldsymbol{u}}$ and $\bar{v}$ that are $(\bar{\mathcal{F}}_t)_{t\geq 0}$-progressively measurable and thus helps in identifying the limit of the stochastic integral as we pass to the limit as $N \to \infty$ (see also Lemma 9.4.3 where we pass to the limit as $\varepsilon \to 0$). The filtrations $(\bar{\mathcal{F}}_t^N)_{t\geq 0}$ are constructed similarly with $(\bar{\boldsymbol{u}}, \bar{v}, \bar{\eta}, \bar{W})$ replaced by $(\bar{\boldsymbol{u}}_N, \bar{v}_N, \bar{\eta}_N, \bar{W}_N)$ in the definition above.

Next, thanks to Proposition 3.6.2, relative to the new stochastic basis $(\widetilde{\Omega}, \widetilde{\mathcal{F}}, (\bar{\mathcal{F}}_t^N)_{t\geq 0}, \widetilde{\mathbb{P}}, \bar{W}_N)$, we can see that for each N, the following equation holds $\widetilde{\mathbb{P}}$-a.s. for every $t \in [0, T]$ and any $\mathbf{Q} \in \mathscr{D}$:

[1] Recall that in Theorem 3.2.5, $\widetilde{\Omega} = [0, 1)$, $\widetilde{\mathcal{F}}$ is the Borel algebra on $\widetilde{\Omega}$ and $\widetilde{\mathbb{P}}$ is the Lebesgue measure on $\widetilde{\Omega}$ and thus $(\widetilde{\Omega}, \widetilde{\mathcal{F}}, \widetilde{\mathbb{P}})$ is independent of the approximate parameters N and ε.

$$
\begin{aligned}
(\bar{\tilde{\mathbf{U}}}_N, \mathbf{Q}) &= (\boldsymbol{u}_0(R+\eta_0), \boldsymbol{q}) + (v_0, \psi) - \int_0^t \int_0^L (\partial_z \bar{\eta}_N^+ \partial_z \psi + \partial_{zz} \bar{\eta}_N^+ \partial_{zz} \psi) \\
&- \frac{1}{2} \int_0^t \int_{\mathcal{O}} \bar{v}_N^* \bar{\boldsymbol{u}}_N^+ \cdot \boldsymbol{q} + \int_0^t \int_{\mathcal{O}} \bar{v}_N^* (2\bar{\tilde{\boldsymbol{u}}}_N - \bar{\boldsymbol{u}}_N) \cdot \boldsymbol{q} \\
&- \frac{1}{2} \int_0^t \int_{\mathcal{O}} (R+\bar{\eta}_N^*)((\bar{\boldsymbol{u}}_N - \bar{v}_N^\# r \mathbf{e}_r) \cdot \nabla^{\bar{\eta}_N^*} \bar{\boldsymbol{u}}_N \cdot \boldsymbol{q} - (\bar{\boldsymbol{u}}_N - \bar{v}_N^\# r \mathbf{e}_r) \cdot \nabla^{\bar{\eta}_N^*} \boldsymbol{q} \cdot \bar{\boldsymbol{u}}_N^+) \\
&- 2\nu \int_0^t \int_{\mathcal{O}} (R+\bar{\eta}_N^*) \mathbf{D}^{\bar{\eta}_N^*}(\bar{\boldsymbol{u}}_N^+) \cdot \mathbf{D}^{\bar{\eta}_N^*}(\boldsymbol{q}) dx ds - \frac{1}{\varepsilon} \int_0^t \int_{\mathcal{O}} \operatorname{div}^{\bar{\eta}_N^*} \bar{\boldsymbol{u}}_N^+ \operatorname{div}^{\bar{\eta}_N^*} \boldsymbol{q} \\
&+ \int_0^t \left(P_{in} \int_0^1 q_z \Big|_{z=0} - P_{out} \int_0^1 q_z \Big|_{z=1} \right) + \int_0^t (G(\bar{\boldsymbol{u}}_N, \bar{v}_N, \bar{\eta}_N^*) d\bar{W}_N, \mathbf{Q}).
\end{aligned}
\tag{9.68}
$$

Observe also that the bounds obtained in Lemma 9.2.3 hold for the new random variables $\bar{\mathcal{U}}_N$ as well. As a result, weak convergence results up to a subsequence (due to the bounds in Lemma 9.2.3) give us the following: for any $\alpha \in (0, \frac{1}{2})$

$$
\begin{aligned}
&\bar{\boldsymbol{u}} \in L^2(\tilde{\Omega}; L^\infty(0,T; \mathbf{L}^2(\mathcal{O}))) \cap L^2(\tilde{\Omega}; L^2(0,T; V)), \\
&\bar{v}, \bar{v}^* \in L^2(\tilde{\Omega}; L^\infty(0,T; L^2(0,L))), \\
&\bar{\eta}, \bar{\eta}^* \in L^2(\tilde{\Omega}; L^\infty(0,T; H_0^2(0,L))) \cap L^2(\tilde{\Omega}; W^{1,\infty}(0,T; L^2(0,L))),
\end{aligned}
\tag{9.69}
$$

and

$$((R+\bar{\eta}^*)\bar{\boldsymbol{u}}, \bar{v}) \in C([0,T]; \mathcal{V}_1') \quad \tilde{\mathbb{P}}-a.s. \tag{9.70}$$

where $\bar{\boldsymbol{u}}$, $\bar{v}$, $\bar{v}^*$, $\bar{\eta}$, and $\bar{\eta}^*$ are the limits defined in Theorem 9.3.1. We also have the following upgraded convergence results for the displacements. Namely, notice that Theorem 9.3.1 statement (2) and Lemma 5.7.1 imply that for given $\frac{3}{2} < s < 2$,

$$\bar{\eta}_N \to \bar{\eta} \text{ and } \bar{\eta}_N^* \to \bar{\eta}^* \quad \text{in } L^\infty(0,T; H^s(0,L))\, a.s. \tag{9.71}$$

and, since for any $s \in (\frac{3}{2}, 2)$, $H^s(0,L) \hookrightarrow C^1([0,L])$ (see Theorem 2.5.2), the following uniform convergence result holds

$$\bar{\eta}_N \to \bar{\eta} \text{ and } \bar{\eta}_N^* \to \bar{\eta}^* \quad \text{in } L^\infty(0,T; C^1[0,L])\, a.s. \tag{9.72}$$

Combining these regularity and convergence results, we can now pass to the limit as $N \to \infty$ in the weak formulations of the approximate problems (9.68) stated on the fixed reference domain $\mathcal{O}$, defined on the probability space $(\tilde{\Omega}, \tilde{\mathcal{F}}, \tilde{\mathbb{P}})$ while keeping $\varepsilon > 0$ fixed. We consider the random variables $\bar{\mathcal{U}}_N$ defined on the probability space $(\tilde{\Omega}, \tilde{\mathcal{F}}, \tilde{\mathbb{P}})$ and pass to the limit $N \to \infty$ in (9.68) and show that the limit satisfies the weak formulation (9.28) with divergence-free penalty. Except for the divergence-free penalty term, this is the same weak formulation as stated in the definition of martingale solutions Definition 9.1.1. More precisely, the following theorem holds true.

Theorem 9.3.2 (Existence for the Problem with Penalized Compressibility) *For the stochastic basis $(\tilde{\Omega}, \tilde{\mathcal{F}}, (\bar{\mathcal{F}}_t)_{t\geq 0}, \tilde{\mathbb{P}}, \bar{W})$ as found in Theorem 9.3.1, given any fixed $\varepsilon, \delta > 0$, the processes $(\bar{\boldsymbol{u}}, \bar{\eta}, \bar{\eta}^*)$ obtained in Theorem 9.3.1 (see (9.69)) are such that $\bar{\eta}^*$ and $((R+\bar{\eta}^*)\bar{\boldsymbol{u}}, \partial_t \bar{\eta})$ are $(\bar{\mathcal{F}}_t)_{t\geq 0}$-progressively measurable with $\tilde{\mathbb{P}}$-a.s. continuous paths in $H^s(0, L)$, $\frac{3}{2} < s < 2$ and $\mathscr{V}_1'$, respectively, and such that the following weak formulation holds $\tilde{\mathbb{P}}$-a.s. for every $t \in [0, T]$ and for every $\mathbf{Q} \in \mathscr{D}$:*

$$
\begin{aligned}
&((R + \bar{\eta}^*(t))\bar{\boldsymbol{u}}(t), \boldsymbol{q}) + (\partial_t \bar{\eta}(t), \psi) = (\boldsymbol{u}_0(R + \eta_0), \boldsymbol{q}) + (v_0, \psi) \\
&- \int_0^t \int_0^L \partial_z \bar{\eta} \partial_z \psi + \partial_{zz} \bar{\eta} \partial_{zz} \psi + \frac{1}{2} \int_0^t \int_{\mathcal{O}} \partial_s \bar{\eta}^* \bar{\boldsymbol{u}} \cdot \boldsymbol{q} \\
&- \frac{1}{2} \int_0^t \int_{\mathcal{O}} (R + \bar{\eta}^*)((\bar{\boldsymbol{u}} - \partial_t \bar{\eta} r \mathbf{e}_r) \cdot \nabla^{\bar{\eta}^*} \bar{\boldsymbol{u}} \cdot \boldsymbol{q} - (\bar{\boldsymbol{u}} - \partial_t \bar{\eta} r \mathbf{e}_r) \cdot \nabla^{\bar{\eta}^*} \boldsymbol{q} \cdot \bar{\boldsymbol{u}}) \\
&- 2\nu \int_0^t \int_{\mathcal{O}} (R + \bar{\eta}^*) \mathbf{D}^{\bar{\eta}^*}(\bar{\boldsymbol{u}}) \cdot \mathbf{D}^{\bar{\eta}^*}(\boldsymbol{q}) + \frac{1}{\varepsilon} \int_0^t \int_{\mathcal{O}} div^{\bar{\eta}^*} \bar{\boldsymbol{u}} \, div^{\bar{\eta}^*} \boldsymbol{q} \\
&+ \int_0^t \left(P_{in} \int_0^1 q_z \Big|_{z=0} - P_{out} \int_0^1 q_z \Big|_{z=1} \right) + \int_0^t (G(\bar{\boldsymbol{u}}, \bar{\boldsymbol{v}}, \bar{\eta}^*) d\bar{W}, \mathbf{Q}).
\end{aligned}
\tag{9.73}
$$

The proof of the above result is similar to and slightly simpler than the proof of Theorem 9.4.2, and thus we refer the reader to the proof of Theorem 9.4.2 for the details.

Before we proceed and consider the limit as the divergence-free penalty parameter $\varepsilon \to 0$, we argue that

$$
\bar{\eta}^*(t) = \bar{\eta}(t) \quad \text{for any } t < \tau, \quad \tilde{\mathbb{P}} - a.s. \tag{9.74}
$$

where for the fixed $\frac{3}{2} < s < 2$ τ is an $(\bar{\mathcal{F}}_t)_{t\in[0,T]}$-stopping time (see Definition 3.3.12) given by,

$$
\tau := T \wedge \inf\{t > 0 : \inf_{z\in[0,L]} (R + \bar{\eta}(t, z)) \leq \delta \text{ or } \|R + \bar{\eta}(t)\|_{H^s(0,L)} \geq \frac{1}{\delta}\}.
$$

Indeed, to show that this is true, let us introduce the following $(\bar{\mathcal{F}}_t)_{t\in[0,T]}$-stopping times. For $\frac{3}{2} < s < 2$, we define

$$
\tau_N := T \wedge \inf\{t > 0 : \inf_{z\in[0,L]} (R + \bar{\eta}_N(t, z)) \leq \delta \text{ or } \|R + \bar{\eta}_N(t)\|_{H^s(0,L)} \geq \frac{1}{\delta}\}.
$$

Note that, thanks to Proposition 3.3.4 and equivalence of laws,

$$\bar{\eta}_N^*(t) = \bar{\eta}_N(t) \quad \text{for every } t < \tau_N, \quad \widetilde{\mathbb{P}} - a.s.$$

Observe that for almost any $\omega \in \widetilde{\Omega}$ and $t < \tau$, and for any $\epsilon > 0$, there exists an N such that

$$\begin{aligned} \|\bar{\eta}(t) - \bar{\eta}^*(t)\|_{H^s(0,L)} &\leq \|\bar{\eta}(t) - \bar{\eta}_N(t)\|_{H^s(0,L)} + \|\bar{\eta}_N^*(t) - \bar{\eta}_N(t)\|_{H^s(0,L)} \\ &\quad + \|\bar{\eta}^*(t) - \bar{\eta}_N^*(t)\|_{H^s(0,L)} \\ &< \epsilon. \end{aligned} \tag{9.75}$$

This is true because for any $\epsilon > 0$, there exists an $N_1 \in \mathbb{N}$ such that the first and the last terms on the right side of the above inequality are each bounded by $\frac{\epsilon}{2}$ for any $N \geq N_1$, thanks to the uniform convergence (9.72).

Furthermore, for any $t < \tau$, it is true that there exists $N_0 \in \mathbb{N}$ such that $t < \tau_N$, for all $N > N_0$. Indeed, for a fixed $t < \tau$, because of the uniform convergence result (9.72), we know that for $\epsilon_0 = \frac{1}{\delta} - \|R + \bar{\eta}(t)\|_{H^s(0,L)} > 0$, there exists N_0 such that for all $N > N_0$,

$$\|R + \bar{\eta}_N(t)\|_{H^s(0,L)} < \epsilon_0 + \|R + \bar{\eta}(t)\|_{H^s(0,L)} = \frac{1}{\delta}.$$

Hence for all $N > N_0$ we have that $t < \tau_N$ (which also implies that $\tau \leq \liminf \tau_N$) and thus $\bar{\eta}_N^*(t) = \bar{\eta}_N(t)$, rendering the second term, in the right-hand side of (9.75), 0.

9.4 Passing to the Limit as $\varepsilon \to 0$

In this section, to emphasize the dependence on the parameter $\varepsilon > 0$, we will use the notation $(\bar{\boldsymbol{u}}_\varepsilon, \bar{v}_\varepsilon, \bar{v}_\varepsilon^*, \bar{\eta}_\varepsilon, \bar{\eta}_\varepsilon^*, \bar{W}_\varepsilon)$ to denote the solution on the filtered probability space $(\widetilde{\Omega}, \widetilde{\mathcal{F}}, (\widetilde{\mathcal{F}}_t^\varepsilon)_{t\geq 0}, \widetilde{\mathbb{P}})$ found in the previous section. In what follows, we will pass $\varepsilon \to 0$ in (9.73) with appropriate test functions. Most of the results in the first half of this section can be proven as in the previous section. Hence we only summarize the important theorems without proof here.

First observe that, thanks to the weak lower-semicontinuity of norm, the uniform estimates obtained in the previous section still hold.

Lemma 9.4.1 (Uniform Boundedness) *For a fixed $\delta > 0$ and $0 < \alpha < \frac{1}{2}$, we have for some $C > 0$* ***independent of*** *ε that*

1. $\widetilde{\mathbb{E}}\|\bar{\boldsymbol{u}}_\varepsilon\|^2_{L^\infty(0,T;\mathbf{L}^2(\mathcal{O}))\cap L^2(0,T;V)\cap H^\alpha(0,T;\mathbf{L}^2(\mathcal{O}))} < C.$
2. $\widetilde{\mathbb{E}}\|\bar{v}_\varepsilon\|^2_{L^\infty(0,T;L^2(0,L))\cap L^2(0,T;H^{\frac{1}{2}}(0,L))\cap H^\alpha(0,T;L^2(0,L))} < C.$
3. $\widetilde{\mathbb{E}}\|\bar{\eta}_\varepsilon\|^2_{L^\infty(0,T;H_0^2(0,L)\cap W^{1,\infty}(0,T;L^2(0,L)))} < C.$

4. $\widetilde{\mathbb{E}}\|\bar{\eta}_\varepsilon^*\|^2_{L^\infty(0,T;H_0^2(0,L))\cap W^{1,\infty}(0,T;L^2(0,L)))} < C.$
5. $\widetilde{\mathbb{E}}\|div^{\bar{\eta}_\varepsilon^*}\bar{\boldsymbol{u}}_\varepsilon\|^2_{L^2(0,T;L^2(\mathcal{O}))} < C\varepsilon.$

These bounds together with Lemma 2.9.9 and Theorem 2.9.4 immediately give us the following result.

Lemma 9.4.2 (Tightness of the Laws)

1. *The laws of $\bar{\boldsymbol{u}}_\varepsilon$ are tight in $L^2(0,T;\mathbf{H}^\gamma(\mathcal{O})) \cap (L^2(0,T;\mathbf{H}^1(\mathcal{O})), w)$ for any $\gamma < 1$.*
2. *The laws of $\bar{v}_\varepsilon$ and that of $\bar{v}_\varepsilon^*$ are tight in $L^2(0,T;H^\gamma(0,L))$ for any $\gamma < 1$ and in $(L^2(0,T;L^2(0,L)), w)$, respectively.*
3. *The laws of $\bar{\eta}_\varepsilon$ and that of $\bar{\eta}_\varepsilon^*$ are tight in $C([0,T];H^s(0,L))$ for $\frac{3}{2} < s < 2$.*

Now for an infinite denumerable set of indices Λ, we let μ_ε be the law of the random variable $\bar{\mathcal{U}}_\varepsilon := (\bar{\boldsymbol{u}}_\varepsilon, \bar{\eta}_\varepsilon, \bar{\eta}_\varepsilon^*, \bar{W}_\varepsilon)$ taking values in the phase space

$$\begin{aligned}\mathcal{S} &:= L^2(0,T;\mathbf{H}^{\frac{3}{4}}(\mathcal{O})) \cap (L^2(0,T;\mathbf{H}^1(\mathcal{O})), w)\\ &\times [C([0,T],H^s(0,L)) \cap H^1(0,T;L^2(0,L))]\\ &\times [C([0,T],H^s(0,L)) \cap (H^1(0,T;L^2(0,L)), w) \cap (L^2(0,T;H_0^2(\Gamma)), w)]\\ &\times C([0,T];\mathbb{R}),\end{aligned}$$

for $\frac{3}{2} < s < 2$.

Then tightness of μ_ε on $\mathcal{S}$ and an application of Theorem 3.2.5 gives us the following almost sure representation and convergence.

Theorem 9.4.1 (Almost Sure Convergence in ε) *There exists a probability space $(\widetilde{\Omega}, \widetilde{\mathcal{F}}, \widetilde{\mathbb{P}})$ and random variables $\widehat{\mathcal{U}}_\varepsilon = (\widehat{\boldsymbol{u}}_\varepsilon, \widehat{\eta}_\varepsilon, \widehat{\eta}_\varepsilon^*, \widehat{W}_\varepsilon)$ and $\widehat{\mathcal{U}} := (\widehat{\boldsymbol{u}}, \widehat{\eta}, \widehat{\eta}^*, \widehat{W})$ such that*

1. *$\widehat{\mathcal{U}}_\varepsilon =^d \bar{\mathcal{U}}_\varepsilon$ for every $\varepsilon \in \Lambda$.*
2. *$\widehat{\mathcal{U}}_\varepsilon \to \widehat{\mathcal{U}}$ $\widetilde{\mathbb{P}}$-a.s. in the topology of $\mathcal{S}$ as $\varepsilon \to 0$.*

We let

$$\widehat{v}_\varepsilon := \partial_t \widehat{\eta}_\varepsilon, \qquad \widehat{v}_\varepsilon^* := \partial_t \widehat{\eta}_\varepsilon^*.$$

The filtration $(\widehat{\mathcal{F}}_t^\varepsilon)_{t\in[0,T]}$ is constructed in the same way as (9.67). We denote by $\mathcal{F}_t^{\varepsilon\prime}$ the σ-field generated by the random variables $\widehat{\boldsymbol{u}}_\varepsilon(s), \widehat{v}_\varepsilon(s), \widehat{\eta}_\varepsilon(s), \widehat{W}_\varepsilon(s)$ for all $s \le t$. Then we define

$$\mathcal{N} := \{\mathcal{A} \in \widetilde{\mathcal{F}} \mid \widetilde{\mathbb{P}}(\mathcal{A}) = 0\}, \qquad \mathcal{F}_t^\varepsilon := \sigma(\mathcal{F}_t^{\varepsilon\prime} \cup \mathcal{N}), \qquad \widehat{\mathcal{F}}_t^\varepsilon := \bigcap_{s\ge t} \mathcal{F}_s^\varepsilon. \tag{9.76}$$

Observe also that, since $\mathcal{O}$ is a Lipschitz domain, thanks to Theorem 2.5.2, the embedding $\mathbf{H}^{\gamma}(\mathcal{O}) \hookrightarrow \mathbf{L}^{q}(\mathcal{O})$ is continuous for any $q \in [1, \frac{2}{1-\gamma}]$ and $\gamma < 1$, and thus statement (2) implies that

$$\hat{\boldsymbol{u}}_{\varepsilon} \to \hat{\boldsymbol{u}} \quad \text{in } L^2(0,T;\mathbf{L}^4(\mathcal{O})), \quad \widetilde{\mathbb{P}}-a.s. \tag{9.77}$$

Next, we discuss additional regularity and convergence properties for the random variables discussed in Theorem 9.4.1, which we will use to pass to the limit as $\varepsilon \to 0$.

First, we notice that equivalence of laws implies that for some $C > 0$ independent of ε,

$$\begin{aligned}
&\widetilde{\mathbb{E}}\|\hat{\boldsymbol{u}}_{\varepsilon}\|^2_{L^\infty(0,T;\mathbf{L}^2(\mathcal{O}))\cap L^2(0,T;V)} < C.\\
&\widetilde{\mathbb{E}}\|\hat{v}_{\varepsilon}\|^2_{L^\infty(0,T;L^2(0,L))} < C, \quad \widetilde{\mathbb{E}}\|\hat{v}^*_{\varepsilon}\|^2_{L^\infty(0,T;L^2(0,L))} < C.\\
&\widetilde{\mathbb{E}}\|\hat{\eta}_{\varepsilon}\|^2_{L^\infty(0,T;H^2_0(0,L))\cap W^{1,\infty}(0,T;L^2(0,L))} < C,\\
&\widetilde{\mathbb{E}}\|\hat{\eta}^*_{\varepsilon}\|^2_{L^\infty(0,T;H^2_0(0,L))\cap W^{1,\infty}(0,T;L^2(0,L))} < C.\\
&\widetilde{\mathbb{E}}\|\mathrm{div}^{\hat{\eta}^*_{\varepsilon}}\hat{\boldsymbol{u}}_{\varepsilon}\|^2_{L^2(0,T;L^2(\mathcal{O}))} < C\varepsilon.
\end{aligned} \tag{9.78}$$

Furthermore, for the limiting functions, we have

$$\begin{aligned}
&\hat{\boldsymbol{u}} \in L^2(\widetilde{\Omega};L^\infty(0,T;\mathbf{L}^2(\mathcal{O}))) \cap L^2(\widetilde{\Omega};L^2(0,T;V)),\\
&\hat{v}, \bar{v}^* \in L^2(\widetilde{\Omega};L^\infty(0,T;L^2(0,L))),\\
&\hat{\eta}, \hat{\eta}^* \in L^2(\widetilde{\Omega};L^\infty(0,T;H^2_0(0,L))) \cap L^2(\widetilde{\Omega};W^{1,\infty}(0,T;L^2(0,L))).
\end{aligned} \tag{9.79}$$

Now, notice that $(9.78)_5$ implies that, up to a subsequence, we have

$$\mathrm{div}^{\hat{\eta}^*_{\varepsilon}}\hat{\boldsymbol{u}}_{\varepsilon} \to 0 \quad \text{in } L^2(0,T;L^2(\mathcal{O})), \quad \widetilde{\mathbb{P}}-a.s. \tag{9.80}$$

Furthermore, as shown in the previous section, we also have that

$$\hat{\eta}_{\varepsilon} \to \hat{\eta} \text{ and } \hat{\eta}^*_{\varepsilon} \to \hat{\eta}^* \quad \text{in } L^\infty(0,T;H^s(0,L)) \text{ and } L^\infty(0,T;C^1([0,L]))\ \widetilde{\mathbb{P}}-a.s. \tag{9.81}$$

and, for given $\frac{3}{2} < s < 2$ we have that

$$\inf_{t\in[0,T],z\in[0,L]}(R+\hat{\eta}^*(t,z)) \geq \delta \quad \text{and} \quad \|R+\hat{\eta}^*\|_{L^\infty(0,T;H^s(0,L))} \leq \frac{1}{\delta}, \quad \widetilde{\mathbb{P}}-a.s. \tag{9.82}$$

Finally, we note that $(\hat{\boldsymbol{u}}_\varepsilon, \hat{v}_\varepsilon, \hat{v}^*_\varepsilon, \hat{\eta}_\varepsilon, \hat{\eta}^*_\varepsilon, \hat{W}_\varepsilon)$ solves the following weak formulation:

$$\begin{aligned}
&((R + \hat{\eta}^*_\varepsilon(t))\hat{\boldsymbol{u}}_\varepsilon(t), \boldsymbol{q}) + (\partial_t \hat{\eta}_\varepsilon(t), \psi) = (\boldsymbol{u}_0(R + \eta_0), \boldsymbol{q}) + (v_0, \psi) \\
&- \int_0^t \int_0^L \partial_z \hat{\eta}_\varepsilon \partial_z \psi + \partial_{zz} \hat{\eta}_\varepsilon \partial_{zz} \psi + \frac{1}{2} \int_0^t \int_{\mathcal{O}} \partial_s \hat{\eta}^*_\varepsilon \hat{\boldsymbol{u}}_\varepsilon \cdot \boldsymbol{q} \\
&- \frac{1}{2} \int_0^t \int_{\mathcal{O}} (R + \hat{\eta}^*_\varepsilon)((\hat{\boldsymbol{u}}_\varepsilon - \partial_t \hat{\eta}_\varepsilon r \mathbf{e}_r) \cdot \nabla^{\hat{\eta}^*_\varepsilon} \hat{\boldsymbol{u}}_\varepsilon \cdot \boldsymbol{q} - (\hat{\boldsymbol{u}}_\varepsilon - \partial_t \hat{\eta}_\varepsilon r \mathbf{e}_r) \cdot \nabla^{\hat{\eta}^*_\varepsilon} \boldsymbol{q} \cdot \hat{\boldsymbol{u}}_\varepsilon) \\
&- 2\nu \int_0^t \int_{\mathcal{O}} (R + \hat{\eta}^*_\varepsilon) \mathbf{D}^{\hat{\eta}^*_\varepsilon}(\hat{\boldsymbol{u}}_\varepsilon) \cdot \mathbf{D}^{\hat{\eta}^*_\varepsilon}(\boldsymbol{q}) + \frac{1}{\varepsilon} \int_0^t \int_{\mathcal{O}} \operatorname{div}^{\hat{\eta}^*_\varepsilon} \hat{\boldsymbol{u}}_\varepsilon \operatorname{div}^{\hat{\eta}^*_\varepsilon} \boldsymbol{q} \\
&+ \int_0^t \left(P_{in} \int_0^1 q_z \Big|_{z=0} - P_{out} \int_0^1 q_z \Big|_{z=1} \right) + \int_0^t (G(\hat{\boldsymbol{u}}_\varepsilon, \hat{v}_\varepsilon, \hat{\eta}^*_\varepsilon) d\hat{W}_\varepsilon, \mathbf{Q}),
\end{aligned} \tag{9.83}$$

$\tilde{\mathbb{P}}$-a.s. for every $t \in [0, T]$ and for every $\mathbf{Q} = (\boldsymbol{q}, \psi) \in \mathscr{D}$.

Observe that the candidate solution for fluid and structure velocities, $(\hat{\boldsymbol{u}}, \hat{v})$, is not regular enough to be a stochastic process in the classical sense (see Sect. 3.3.1) as it only belongs to the space $L^2(0, T; \mathbf{L}^2(\mathcal{O})) \times L^2(0, T; L^2(0, L))$. Hence we recall the definition of random distributions, Definition 3.3.9 and its history Definition 3.3.10 . Let $\hat{\mathcal{F}}'_t$ be the σ- field generated by the random variables $\hat{\eta}(s), \hat{W}(s)$ for all $0 \le s \le t$ (see (3.5)). Then we define

$$\hat{\mathcal{F}}^0_t := \bigcap_{s \ge t} \sigma(\hat{\mathcal{F}}'_s \cup \mathcal{N}), \qquad \hat{\mathcal{F}}_t := \sigma(\sigma_t((\hat{\boldsymbol{u}}, \hat{v})) \cup \hat{\mathcal{F}}^0_t), \tag{9.84}$$

where $\mathcal{N} = \{\mathcal{A} \in \tilde{\mathcal{F}} \mid \tilde{\mathbb{P}}(\mathcal{A}) = 0\}$ and $\sigma_t(\mathbf{U})$ is the history of $\mathbf{U}$ in the sense of Definition 3.3.10. This gives a complete, right-continuous filtration $(\hat{\mathcal{F}}_t)_{t \ge 0}$, on the probability space $(\tilde{\Omega}, \tilde{\mathcal{F}}, \tilde{\mathbb{P}})$, to which the noise processes, and the candidate solutions are adapted. Moreover, for any $t > s$, $\sigma(\hat{W}(t) - \hat{W}(s))$ is independent of $\hat{\mathcal{F}}_t$, and we can show that $\hat{W}$ is an $(\hat{\mathcal{F}}_t)_{t \ge 0}$-Wiener process. Now, Lemma 3.3.1 says that there exists a stochastic process in $L^2(0, T; \mathbf{L}^2(\mathcal{O}) \times L^2(0, L))$ a.s. which is a $(\hat{\mathcal{F}}_t)_{t \ge 0}$-progressively measurable representative of $(\hat{\boldsymbol{u}}, \hat{v})$.

Remark 9.4.1 Thanks to this consequence of Lemma 3.3.1 and continuity of the coefficient G, we can deduce that $G(\hat{\boldsymbol{u}}(s), \hat{v}(s), \hat{\eta}^*(s))$ is $\{\hat{\mathcal{F}}_t\}_{t \ge 0}$-progressively measurable. This measurability property along with the growth assumptions (9.21) implies that the stochastic integral $\int_0^t (G(\hat{\boldsymbol{u}}(s), \hat{v}(s), \hat{\eta}^*(s)) d\hat{W}(s), \mathbf{Q}(s))$ is well-defined for any $(\hat{\mathcal{F}}_t)_{t \ge 0}$-adapted continuous $\mathscr{D}$-valued process $\mathbf{Q}$, where $\mathscr{D}$ is defined in (9.19).

We are now ready to state the main result of this section, which shows that the limiting stochastic processes $(\hat{\boldsymbol{u}}_\varepsilon, \hat{\eta}_\varepsilon, \hat{W}_\varepsilon)$ for each fixed $\delta > 0$, converge to a martingale solution of our stochastic FSI problem, as $\varepsilon \to 0$.

The following theorem is a key intermediate step in establishing the final existence result as it shows that the processes $(\hat{\boldsymbol{u}}, \hat{\eta}, \hat{\eta}^*)$, obtained in Theorem 9.4.1, solve the weak formulation (9.24) for almost every time $t \in [0, T]$, except with $\widehat{\eta}$ replaced by the artificial structure random variable $\widehat{\eta}^*$ in the appropriate terms. This implies that we obtain the desired martingale solution in the sense of Definition 9.1.1 for as long as $\widehat{\eta}$ and $\widehat{\eta}^*$ are equal.

Theorem 9.4.2 (Main Convergence Result as $\varepsilon \to 0$) *For any fixed $\delta > 0$, the stochastic processes $(\widehat{\boldsymbol{u}}, \widehat{\eta}, \widehat{\eta}^*)$ constructed in Theorem 9.4.1 satisfy*

$$
\begin{aligned}
&((R+\widehat{\eta}^*(t))\widehat{\boldsymbol{u}}(t), \boldsymbol{q}(t)) + (\partial_t\widehat{\eta}(t), \psi(t)) = (\boldsymbol{u}_0(R+\eta_0), \boldsymbol{q}(0)) + (v_0, \psi(0)) \\
&+ \int_0^t \int_{\mathcal{O}} (R+\widehat{\eta}^*)\widehat{\boldsymbol{u}} \cdot \partial_t \boldsymbol{q} + \int_0^t \int_0^L \partial_t\widehat{\eta}\, \partial_t\psi \\
&- \int_0^t \int_0^L \partial_z\widehat{\eta}\partial_z\psi + \partial_{zz}\widehat{\eta}\partial_{zz}\psi + \frac{1}{2}\int_0^t \int_{\mathcal{O}} \partial_t\widehat{\eta}^*\widehat{\boldsymbol{u}} \cdot \boldsymbol{q} \\
&- \frac{1}{2}\int_0^t \int_{\mathcal{O}} (R+\widehat{\eta}^*)((\widehat{\boldsymbol{u}} - \partial_t\widehat{\eta} r\mathbf{e}_r) \cdot \nabla^{\widehat{\eta}^*}\widehat{\boldsymbol{u}} \cdot \boldsymbol{q} - (\widehat{\boldsymbol{u}} - \partial_t\widehat{\eta} r\mathbf{e}_r) \cdot \nabla^{\widehat{\eta}^*}\boldsymbol{q} \cdot \widehat{\boldsymbol{u}}) \\
&- 2\nu \int_0^t \int_{\mathcal{O}} (R+\widehat{\eta}^*)\mathbf{D}^{\widehat{\eta}^*}(\widehat{\boldsymbol{u}}) \cdot \mathbf{D}^{\widehat{\eta}^*}(\boldsymbol{q}) \\
&+ \int_0^t \left(P_{in}\int_0^1 q_z\Big|_{z=0} dr - P_{out}\int_0^1 q_z\Big|_{z=1} dr \right) ds + \int_0^t (G(\widehat{\boldsymbol{u}}, \widehat{v}, \widehat{\eta}^*)d\widehat{W}, \mathbf{Q}),
\end{aligned}
\tag{9.85}
$$

$\widetilde{\mathbb{P}}$-a.s. for almost every $t \in [0, T]$ and for any $(\widehat{\mathcal{F}}_t)_{t\geq 0}$-adapted process $\mathbf{Q} = (\boldsymbol{q}, \psi)$ with continuous paths in $\mathscr{D}$ such that $\text{div}^{\widehat{\eta}^*}\boldsymbol{q} = 0$ for every $(t, \omega) \in [0, T] \times \Omega$.

9.4.1 *Proof of Main Result*

In this section, we will give a proof of Theorem 9.4.2 by passing to the limit as $\varepsilon \to 0$ in (9.83). This will be done in the following four steps. In Step 1, we will construct appropriate test functions $\mathbf{Q}_\varepsilon$ for the weak formulation (9.83) that will approximate any admissible test function for the limiting weak formulation. In Step 2, we discuss the limit $\varepsilon \to 0$ of the stochastic integral and of the remaining terms, of which the nonlinear advection term is the only one which requires discussion. In this, we will also show that the limit $\widehat{\boldsymbol{u}}$ satisfies the transformed incompressibily condition. In Step 3, we will discuss the stopping time argument. We present these steps next.

Step 1: Construction of Test Functions

We begin by taking any pair of $(\widehat{\mathcal{F}}_t)_{t\in[0,T]}$-adapted processes $(\boldsymbol{q}, \psi)$ with C^1 paths in $\mathscr{D}$ such that $\boldsymbol{q}(t, z, 1) = \psi(t, z)\mathbf{e}_r$ and $\text{div}^{\widehat{\eta}^*}\boldsymbol{q} = 0$ for every $\omega \in \widetilde{\Omega}$. This pair serves as an admissible test function for the limiting weak formulation (9.85) satisfied by $(\widehat{\boldsymbol{u}}, \widehat{\eta}, \widehat{\eta}^*)$. The aim of Step 1 is to approximate this process $(\boldsymbol{q}, \psi)$ by $(\widehat{\mathcal{F}}_t^\varepsilon)_{t\in[0,T]}$-adapted processes $(\boldsymbol{q}_\varepsilon, \psi_\varepsilon)$ taking values in $\mathscr{D}$ (see (9.19)) such that $\text{div}^{\widehat{\eta}_\varepsilon^*}\boldsymbol{q}_\varepsilon = 0$ which will serve as test functions for the approximate weak formulation (9.83). Notice that since $\boldsymbol{q}_\varepsilon$ satisfies the transformed divergence-free condition, the penalty term in (9.83) drops out.

Before we construct these test functions, we will prove that composing with the ALE map (and its inverse) preserves adaptability, as we shall use this fact multiple times in the rest of this step. Consider an $(\widehat{\mathcal{F}}_t)_{t\in[0,T]}$-adapted $\mathbf{r} = (r_1, r_2)$ on $\bar{\mathcal{O}}_{\widehat{\eta}^*}$ such that $\nabla \cdot \mathbf{r} = 0$ and such that $\mathbf{r}$ satisfies the required boundary conditions $r_2 = 0$ on $z = 0, L, r = 0$ and $\partial_r r_1 = 0$ on Γ_b. Assume also that on the top lateral boundary of the moving domain associated with $\widehat{\eta}^*$, the function $\mathbf{r}$ satisfies $\mathbf{r}(t, z, R + \widehat{\eta}^*(t, z)) = \psi(t, z)\mathbf{e}_r$. Next, we define

$$\boldsymbol{q}(t, z, r, \omega) = \mathbf{r}(t, \omega) \circ A^\omega_{\widehat{\eta}^*}(t)(z, r).$$

Now, we will show that $(\boldsymbol{q}, \psi)$ is $(\widehat{\mathcal{F}}_t)_{t\in[0,T]}$-adapted and thus is an appropriate test function as it appears in the statement of Theorem 9.4.2.

For that purpose, we fix some arbitrary $t \in [0, T]$ and for the process $\mathbf{r}$ mentioned above, we let $\mathcal{C}_\mathbf{r} : \widetilde{\Omega} \times C([0, L]) \to \mathbf{C}^1(\bar{\mathcal{O}})$ be defined as

$$\mathcal{C}_\mathbf{r}(\omega, \eta) = F_\eta(\mathbf{r}(t, \omega)),$$

where $F_\eta(\mathbf{f})(z, r) := \mathbf{f}(z, (R + \eta(z))r)$ is a well-defined map from $\mathbf{C}(\bar{\mathcal{O}}_\eta)$ to $\mathbf{C}(\bar{\mathcal{O}})$ for any $\eta \in C([0, L])$. We observe the following two properties:

- Thanks to the continuity of the composition operator F_η and the assumption that $\mathbf{r}(t)$ is $\widehat{\mathcal{F}}_t$-measurable, we obtain that, for any η, the $\mathbf{C}^1(\bar{\mathcal{O}})$-valued map $\omega \mapsto \mathcal{C}_\mathbf{r}(\omega, \eta)$ is measurable as a map from $(\Omega, \widehat{\mathcal{F}}_t)$ to $\mathbf{C}^1(\bar{\mathcal{O}})$ endowed with Borel σ-algebra.
- For any fixed ω, the map $\eta \mapsto \mathcal{C}_\mathbf{r}(\omega, \eta)$ is continuous.

Hence we deduce, for any $t \in [0, T]$, that

$$\mathcal{C}_\mathbf{r} : (\widetilde{\Omega} \times C([0, L]), \mathcal{F}_t \otimes \text{Borel}(C([0, L]))) \to (\mathbf{C}^1(\bar{\mathcal{O}}), \text{Borel}(\mathbf{C}^1(\bar{\mathcal{O}})))$$
$$\text{is a Carathéodory function (see Definition 3.3.11).}$$

Now by the construction of the filtration $(\widehat{\mathcal{F}}_t)_{t\in[0,T]}$ in (9.84), we know that $\widehat{\eta}^*$ is $(\widehat{\mathcal{F}}_t)_{t\in[0,T]}$-adapted. Therefore, thanks to Lemma 3.3.2, we conclude that the $\mathbf{C}^1(\bar{\mathcal{O}})$-valued process $\boldsymbol{q}$, which by definition is

$$\boldsymbol{q}(t, \omega) = \mathcal{C}_\mathbf{r}(\omega, \widehat{\eta}^*(t, \omega)),$$

is $(\widehat{\mathcal{F}}_t)_{t\geq 0}$-adapted as well.

Now we will construct the appropriate test functions $(\boldsymbol{q}_\varepsilon, \psi_\varepsilon)$ for the approximate system (9.83).

Main Idea To construct $(\boldsymbol{q}_\varepsilon, \psi_\varepsilon)$ the idea is to first transform $(\boldsymbol{q}, \psi)$ into an $(\widehat{\mathcal{F}}_t)_{t\in[0,T]}$-adapted process $(\widetilde{\boldsymbol{q}}, \psi)$ taking values in $\mathscr{D}$ such that $\text{div}\widetilde{\boldsymbol{q}} = 0$ on $\mathcal{O}$ (henceforth denoted by $\text{div}_{\mathcal{O}}$ to avoid any confusion). This will be done by using the cofactor matrix used in the definition of the Piola transform (2.11). Then $(\widetilde{\boldsymbol{q}}, \psi)$ will be approximated by $(\widehat{\mathcal{F}}_t^\varepsilon)_{t\in[0,T]}$-adapted processes $(\widetilde{\boldsymbol{q}}_\varepsilon, \psi_\varepsilon)$ with values in $C^1(0,T;\mathscr{D})$ satisfying $\text{div}_{\mathcal{O}}\widetilde{\boldsymbol{q}}_\varepsilon = 0$. We will again use the inverse of the Piola transform (2.11) to transform $\widetilde{\boldsymbol{q}}_\varepsilon$ into $\boldsymbol{q}_\varepsilon$ such that $(\boldsymbol{q}_\varepsilon, \psi_\varepsilon)$ is an $(\widehat{\mathcal{F}}_t^\varepsilon)_{t\in[0,T]}$-adapted process satisfying the kinematic coupling condition and $\text{div}^{\widehat{\eta}_\varepsilon^*}\boldsymbol{q}_\varepsilon = 0$.

However, while doing so, we must be careful since applying the Piola transform will not preserve the time regularity of the test function, since the artificial structure velocity $\widehat{v}^*$ has space regularity of $H^{\frac{1}{2}}(0,L)$ and not the desired $H^1(0,L)$. Hence, we will first find a regularized lower bound for the structure displacements, denoted by $\widehat{\eta}_\kappa^*$, and then modify $(\boldsymbol{q}, \psi)$ as follows: First we extend $\boldsymbol{q}$ outside of $\mathcal{O}$ by $\psi\mathbf{e}_r$ when $r \geq 1$. Next, for any squeezing parameter $\kappa > 1$, we choose $\widehat{\eta}_\kappa^*$ to be an $(\widehat{\mathcal{F}}_t)_{t\in[0,T]}$-adapted smooth structure displacement such that

$$\frac{R+\widehat{\eta}^*}{\kappa} < R + \widehat{\eta}_\kappa^* < R + \widehat{\eta}^*, \qquad \text{almost everywhere in } [0,T]\times[0,L]\times\widetilde{\Omega}.$$

Such a function is constructed in Appendix 9.5. The fact that it is $(\widehat{\mathcal{F}}_t)_{t\in[0,T]}$-adapted follows by the same arguments as above. Then we define

$$\boldsymbol{q}_\kappa = S_\kappa \boldsymbol{q} = (\kappa q_1(t,z,\kappa r), q_2(t,z,\kappa r)), \tag{9.86}$$

where the squeezing operator S_κ is defined in (9.44) (and previously in Defintion 5.5) although on physical domains. Note that, for any $\kappa > 1$, the squeezing operator preserves the coupling and divergence conditions on the fixed domain:

$$\boldsymbol{q}_\kappa(t,z,1) = \psi(t,z)\mathbf{e}_r \text{ and } \text{div}^{\widehat{\eta}^*}\boldsymbol{q}_\kappa = 0.$$

Next, we let

$$\bar{\boldsymbol{q}}_\kappa = (\boldsymbol{q}_\kappa \circ A_{\widehat{\eta}^*}^{-1}) \circ A_{\widehat{\eta}_\kappa^*},$$

More explicitly,

$$\bar{\boldsymbol{q}}_\kappa(t,z,r) = \left(\kappa q_1\left(t,z,\kappa\frac{R+\widehat{\eta}_\kappa^*(t,z)}{R+\widehat{\eta}^*(t,z)}r\right), q_2\left(t,z,\kappa\frac{R+\widehat{\eta}_\kappa^*(t,z)}{R+\widehat{\eta}^*(t,z)}r\right)\right).$$

Since $\boldsymbol{q}$ takes values in $C^1(0,T;\mathbf{H}^1(\mathcal{O}))$, we can see that $\bar{\boldsymbol{q}}_\kappa$ takes values in $C(0,T;\mathbf{H}^1(\mathcal{O})) \cap W^{1,\infty}(0,T;\mathbf{L}^2(\mathcal{O}))$. Moreover,

$$\bar{\boldsymbol{q}}_\kappa(t,z,1) = \psi(t,z)\mathbf{e}_r \text{ and } \operatorname{div}^{\widehat{\eta}^*_\kappa}\bar{\boldsymbol{q}}_\kappa = 0.$$

Next, we let

$$\widetilde{\boldsymbol{q}}_\kappa = (R+\widehat{\eta}^*_\kappa)\nabla A^{-1}_{\widehat{\eta}^*_\kappa} \cdot \bar{\boldsymbol{q}}_\kappa. \tag{9.87}$$

The idea here is to apply the Piola transform (2.11) to the pull back of $\bar{\boldsymbol{q}}_\kappa$ on $\mathcal{O}_{\widehat{\eta}^*_\kappa}$, i.e., $\bar{\boldsymbol{q}}_\kappa \circ A^{-1}_{\widehat{\eta}^*_\kappa}$.

Note the following properties:

1. Observe that

$$\widetilde{\boldsymbol{q}}_\kappa(t,z,1) = \begin{pmatrix} R+\widehat{\eta}^*_\kappa(t,z) & 0 \\ -\partial_z\widehat{\eta}^*_\kappa(t,z) & 1 \end{pmatrix} \begin{pmatrix} 0 \\ \psi(t,z) \end{pmatrix} = \psi(t,z)\mathbf{e}_r.$$

   Hence, $(\widetilde{\boldsymbol{q}}_\kappa, \psi)$ takes values in $C(0,T;\mathscr{D}) \cap W^{1,\infty}(0,T;\mathbf{L}^2(\mathcal{O}) \times L^2(0,L))$.
2. Observe that (see also Lemma 2.8.1) we have

$$\operatorname{div}_{\mathcal{O}}\widetilde{\boldsymbol{q}}_\kappa = \operatorname{div}^{\widehat{\eta}^*_\kappa}\boldsymbol{q}_\kappa = 0. \tag{9.88}$$

3. $\widetilde{\boldsymbol{q}}_\kappa$ is adapted to the filtration $(\widehat{\mathcal{F}}_t)_{t\in[0,T]}$ since $\widehat{\eta}^*_\kappa$ and $\boldsymbol{q}_\kappa$ are both adapted to $(\widehat{\mathcal{F}}_t)_{t\in[0,T]}$ (see above for an explanation).

Now consider the following Hilbert space:

$$\mathscr{D}_\sigma = \{(\boldsymbol{q},\psi) \in \mathscr{D} : \operatorname{div}_{\mathcal{O}}\boldsymbol{q} = 0\}, \tag{9.89}$$

where $\mathscr{D}$ is defined in (9.19).

Since $(\widetilde{\boldsymbol{q}}_\kappa, \psi)$ is an $(\widehat{\mathcal{F}}_t)_{t\in[0,T]}$-adapted process taking values in $C(0,T;\mathscr{D}_\sigma) \cap W^{1,\infty}(0,T;\mathbf{L}^2(\mathcal{O}) \times L^2(0,L))$, we can apply Proposition 3.3.2 (see also Remark 3.3.1), to obtain the existence of processes $(\widetilde{\boldsymbol{q}}_{\kappa,\varepsilon}, \psi_{\kappa,\varepsilon})$ such that

1. $(\widetilde{\boldsymbol{q}}_{\kappa,\varepsilon}, \psi_{\kappa,\varepsilon}) \in C(0,T;\mathscr{D}_\sigma) \cap W^{1,\infty}(0,T;\mathbf{L}^2(\mathcal{O}) \times L^2(0,L))$ is an $(\widehat{\mathcal{F}}^\varepsilon_t)_{t\in[0,T]}$-adapted process.
2. $(\widetilde{\boldsymbol{q}}_{\kappa,\varepsilon}, \psi_{\kappa,\varepsilon}) \to (\widetilde{\boldsymbol{q}}_\kappa, \psi)$ in $C(0,T;\mathscr{D}_\sigma) \cap W^{1,\infty}(0,T;\mathbf{L}^2(\mathcal{O}) \times L^2(0,L))$ almost surely.

Next, we use the Piola transform again and define the test process $\bar{\boldsymbol{q}}_{\kappa,\varepsilon}$ as follows:

$$\bar{\boldsymbol{q}}_{\kappa,\varepsilon} = \frac{1}{(R+\widehat{\eta}^*_{\kappa,\varepsilon})}\nabla A_{\widehat{\eta}^*_{\kappa,\varepsilon}} \cdot \widetilde{\boldsymbol{q}}_{\kappa,\varepsilon}. \tag{9.90}$$

Here $\widehat{\eta}^*_{\kappa,\varepsilon}$ are smooth bounding functions adapted to $(\widehat{\mathcal{F}}^\varepsilon_t)_{t\in[0,T]}$ as constructed in Appendix 9.5 satifying, for every $\varepsilon > 0$, the following condition

$$\frac{R+\widehat{\eta}^*_\varepsilon}{\kappa} < R+\widehat{\eta}^*_{\kappa,\varepsilon} < R+\widehat{\eta}^*_\varepsilon, \qquad \text{almost everywhere in } [0,T]\times[0,L]\times\widetilde{\Omega}.$$

Observe that $\text{div}^{\widehat{\eta}^*_{\kappa,\varepsilon}}\bar{\boldsymbol{q}}_{\kappa,\varepsilon} = 0$. Hence to obtain the incompressibility condition with respect to the correct structure displacement $\widehat{\eta}^*_\varepsilon$, we first extend $\bar{\boldsymbol{q}}_{\kappa,\varepsilon}$ outside of $\mathcal{O}$ by $\psi_{\kappa,\varepsilon}\mathbf{e}_r$ when $r \geq 1$ and then define

$$\boldsymbol{q}_{\kappa,\varepsilon} = \left(\bar{\boldsymbol{q}}_{\kappa,\varepsilon} \circ A^{-1}_{\widehat{\eta}^*_{\kappa,\varepsilon}}\right) \circ A_{\widehat{\eta}^*_\varepsilon}. \tag{9.91}$$

We will now list the properties of the test process $(\boldsymbol{q}_{\kappa,\varepsilon}, \psi_{\kappa,\varepsilon})$.

1. $(\boldsymbol{q}_{\kappa,\varepsilon}, \psi_{\kappa,\varepsilon})$ is adapted to $(\widehat{\mathcal{F}}^\varepsilon_t)_{t\in[0,T]}$.
2. The kinematic coupling condition is satisfied, i.e.,

$$\boldsymbol{q}_{\kappa,\varepsilon}(t,z,1) = \psi_{\kappa,\varepsilon}(t,z)\mathbf{e}_r.$$

   Furthermore, due to (2.8.1), we have that the following transformed incompressibility condition is satisfied almost surely and for almost every $t \in [0,T]$:

$$\text{div}^{\widehat{\eta}^*_\varepsilon}\boldsymbol{q}_{\kappa,\varepsilon} = 0.$$

3. Due to (9.81), for any fixed $\kappa > 1$ as $\varepsilon \to 0$, we have that $\widehat{\eta}^*_{\kappa,\varepsilon} \to \widehat{\eta}^*_\kappa$ in $L^\infty(0,T;C^a[0,L])$ almost surely for any $a \geq 1$. Hence, as $\varepsilon \to 0$,

$$\boldsymbol{q}_{\kappa,\varepsilon} \to \boldsymbol{q}_\kappa \qquad \text{almost surely in } L^\infty(0,T;\mathbf{H}^1(\mathcal{O})). \tag{9.92}$$

   Moreover, since $\widehat{v}^*_\varepsilon \rightharpoonup \widehat{v}^*$ weakly in $L^2(0,T;L^2(0,L))$ a.s. we have for any $p < 2$ that

$$\partial_t\boldsymbol{q}_{\kappa,\varepsilon} \rightharpoonup \partial_t\boldsymbol{q}_\kappa, \qquad \text{almost surely weakly in } L^2(0,T;\mathbf{L}^p(\mathcal{O})). \tag{9.93}$$

Hence, $(\boldsymbol{q}_{\kappa,\varepsilon}, \psi_{\kappa,\varepsilon})$ is an appropriate test function for (9.83).

We are now in a position to take the limit as $\varepsilon \to 0$ in the weak formulation (9.83). Namely, we consider (9.83) and use $\mathbf{Q}_{\kappa,\varepsilon} = (\boldsymbol{q}_{\kappa,\varepsilon}, \psi_{\kappa,\varepsilon})$ as the test functions. This requires a special version of the Itô product rule which can be proven by using a regularization argument as outlined in Lemma 5.1 in [63]. We obtain that

$$
\begin{aligned}
&((R+\widehat{\eta}_\varepsilon^*(t))\widehat{\boldsymbol{u}}_\varepsilon(t), \boldsymbol{q}_{\kappa,\varepsilon}(t))+(\widehat{v}_\varepsilon(t), \psi_{\kappa,\varepsilon}(t))=(\boldsymbol{u}_0(R+\eta_0), \boldsymbol{q}_{\kappa,\varepsilon}(0))+(v_0, \psi_{\kappa,\varepsilon}(0))\\
&+\int_0^t\int_{\mathcal{O}}(R+\widehat{\eta}_\varepsilon^*)\widehat{\boldsymbol{u}}_\varepsilon\partial_t\boldsymbol{q}_{\kappa,\varepsilon}+\int_0^t\int_0^L \widehat{v}_\varepsilon\partial_t\psi_{\kappa,\varepsilon}\\
&-\int_0^t\int_0^L(\partial_z\widehat{\eta}_\varepsilon\partial_z\psi+\partial_{zz}\widehat{\eta}_\varepsilon\partial_{zz}\psi_{\kappa,\varepsilon})+\frac{1}{2}\int_0^t\int_{\mathcal{O}}\widehat{v}_\varepsilon^*\widehat{\boldsymbol{u}}_\varepsilon\cdot\boldsymbol{q}_{\kappa,\varepsilon}\\
&-\frac{1}{2}\int_0^t\int_{\mathcal{O}}(R+\widehat{\eta}_\varepsilon^*)((\widehat{\boldsymbol{u}}_\varepsilon-\widehat{v}_\varepsilon r\mathbf{e}_r)\cdot\nabla^{\widehat{\eta}_\varepsilon^*}\widehat{\boldsymbol{u}}_\varepsilon\cdot\boldsymbol{q}_{\kappa,\varepsilon}-(\widehat{\boldsymbol{u}}_\varepsilon-\widehat{v}_\varepsilon r\mathbf{e}_r)\cdot\nabla^{\widehat{\eta}_\varepsilon^*}\boldsymbol{q}_{\kappa,\varepsilon}\cdot\widehat{\boldsymbol{u}}_\varepsilon)\\
&-2\nu\int_0^t\int_{\mathcal{O}}(R+\widehat{\eta}_\varepsilon^*)\mathbf{D}^{\widehat{\eta}_\varepsilon^*}(\widehat{\boldsymbol{u}}_\varepsilon)\cdot\mathbf{D}^{\widehat{\eta}_\varepsilon^*}(\boldsymbol{q}_{\kappa,\varepsilon})\\
&+\int_0^t\left(P_{in}\int_0^1(q_{\kappa,\varepsilon})_z\Big|_{z=0}dr-P_{out}\int_0^1(q_{\kappa,\varepsilon})_z\Big|_{z=1}dr\right)ds\\
&+\int_0^t(G(\widehat{\boldsymbol{u}}_\varepsilon,\widehat{v}_\varepsilon,\widehat{\eta}_\varepsilon^*)d\widehat{W}_\varepsilon,\mathbf{Q}_{\kappa,\varepsilon}),
\end{aligned}
\tag{9.94}
$$

holds $\widetilde{\mathbb{P}}$-a.s. for every $t \in [0, T]$.

Now we begin passing $\varepsilon \to 0$ in (9.94).

Step 2: Passing $\varepsilon \to 0$ in the Stochastic Integral and Other Deterministic Terms

We start with the stochastic integral. Recall Remark 9.4.1 and the fact that, by construction, $\mathbf{Q}_\kappa = (\boldsymbol{q}_\kappa, \psi)$ is $(\widehat{\mathcal{F}}_t)_{t\in[0,T]}$-adapted.

Lemma 9.4.3 *The processes* $\left(\int_0^t(G(\widehat{\boldsymbol{u}}_\varepsilon(s),\widehat{v}_\varepsilon(s),\widehat{\eta}_\varepsilon^*(s))d\widehat{W}_\varepsilon(s),\mathbf{Q}_{\kappa,\varepsilon}(s))\right)_{t\in[0,T]}$ *converge to* $\left(\int_0^t(G(\widehat{\boldsymbol{u}}(s),\widehat{v}(s),\widehat{\eta}^*(s))d\widehat{W}(s),\mathbf{Q}_\kappa(s))\right)_{t\in[0,T]}$ *in* $L^1(\widetilde{\Omega}; L^1(0,T;\mathbb{R}))$ *as* $\varepsilon \to 0$.

Proof First, by using $(9.21)_{2,3}$, we observe that

$$
\begin{aligned}
&\int_0^T |(G(\widehat{\boldsymbol{u}}_\varepsilon,\widehat{v}_\varepsilon,\widehat{\eta}_\varepsilon^*),\mathbf{Q}_{\kappa,\varepsilon})-(G(\widehat{\boldsymbol{u}},\widehat{v},\widehat{\eta}^*),\mathbf{Q}_\kappa)|^2 ds\\
&=\int_0^T |(G(\widehat{\boldsymbol{u}}_\varepsilon,\widehat{v}_\varepsilon,\widehat{\eta}_\varepsilon^*)-G(\widehat{\boldsymbol{u}},\widehat{v},\widehat{\eta}_\varepsilon^*),\mathbf{Q}_{\kappa,\varepsilon})|^2+\int_0^T |(G(\widehat{\boldsymbol{u}},\widehat{v},\widehat{\eta}_\varepsilon^*),\mathbf{Q}_{\kappa,\varepsilon}-\mathbf{Q}_\kappa)|^2\\
&\quad+\int_0^T |(G(\widehat{\boldsymbol{u}},\widehat{v},\widehat{\eta}_\varepsilon^*)-G(\widehat{\boldsymbol{u}},\widehat{v},\widehat{\eta}^*),\mathbf{Q}_\kappa)|^2\\
&\le\int_0^T\left(\|\widehat{\eta}_\varepsilon^*\|^2_{L^\infty(0,L)}\|\widehat{\boldsymbol{u}}_\varepsilon-\widehat{\boldsymbol{u}}\|^2_{\mathbf{L}^2(\mathcal{O})}+\|\widehat{v}_\varepsilon-\widehat{v}\|^2_{L^2(0,L)}\right)\|\mathbf{Q}_{\kappa,\varepsilon}\|^2_{\mathbf{L}^2}
\end{aligned}
$$

$$
\begin{aligned}
&+\int_0^T \left(\|\widehat{\eta}_\varepsilon^*\|_{L^\infty(0,L)}^2\|\widehat{\boldsymbol{u}}\|_{\mathbf{L}^2(\mathcal{O})}^2+\|\widehat{v}\|_{L^2(0,L)}^2\right)\|\mathbf{Q}_{\kappa,\varepsilon}-\mathbf{Q}_\kappa\|_{\mathbf{L}^2}^2\\
&+\int_0^T \|\widehat{\eta}_\varepsilon^*-\widehat{\eta}^*\|_{L^\infty(0,L)}^2\|\widehat{\boldsymbol{u}}\|_{\mathbf{L}^2(\mathcal{O})}^2\|\mathbf{Q}_\kappa\|_{\mathbf{L}^2}^2\\
\le{}& \frac{1}{\delta}\|\mathbf{Q}_{\kappa,\varepsilon}\|_{L^2(0,T;\mathbf{L}^2)}^2\left(\|\widehat{\boldsymbol{u}}_\varepsilon-\widehat{\boldsymbol{u}}\|_{L^2(0,T;\mathbf{L}^2(\mathcal{O}))}^2+\|\widehat{v}_\varepsilon-\widehat{v}\|_{L^2(0,T;L^2(0,L))}^2\right)\\
&+\frac{1}{\delta}\|\mathbf{Q}_{\kappa,\varepsilon}-\mathbf{Q}_\kappa\|_{L^2(0,T;\mathbf{L}^2)}^2\left(\|\widehat{\boldsymbol{u}}\|_{L^2(0,T;\mathbf{L}^2(\mathcal{O}))}^2+\|\widehat{v}\|_{L^2(0,T;L^2(0,L))}^2\right)\\
&+\|\widehat{\eta}_\varepsilon^*-\widehat{\eta}^*\|_{L^\infty((0,T)\times(0,L))}\|\widehat{\boldsymbol{u}}\|_{L^2(0,T;\mathbf{L}^2(\mathcal{O}))}^2\|\mathbf{Q}_\kappa\|_{L^2(0,T;\mathbf{L}^2)}^2.
\end{aligned}
$$

Thanks to Theorem 9.4.1, the right-hand side of the inequality above converges to 0, $\widetilde{\mathbb{P}}$-a.s. as $\varepsilon \to 0$. Hence, we have proven that

$$
(G(\widehat{\boldsymbol{u}}_\varepsilon,\widehat{v}_\varepsilon,\widehat{\eta}_\varepsilon^*),\mathbf{Q}_{\kappa,\varepsilon}) \to (G(\widehat{\boldsymbol{u}},\widehat{v},\widehat{\eta}^*),\mathbf{Q}_\kappa), \qquad \widetilde{\mathbb{P}}-\text{a.s} \quad \text{in } L^2(0,T;\mathbb{R}). \tag{9.95}
$$

Now we apply Lemma 3.6.2 to obtain

$$
\int_0^t (G(\widehat{\boldsymbol{u}}_\varepsilon,\widehat{v}_\varepsilon,\widehat{\eta}_\varepsilon^*)d\widehat{W}_\varepsilon,\mathbf{Q}_{\kappa,\varepsilon}) \to \int_0^t (G(\widehat{\boldsymbol{u}},\widehat{v},\widehat{\eta}^*)d\widehat{W},\mathbf{Q}_\kappa) \tag{9.96}
$$

in probability in $L^2(0,T;\mathbb{R})$.

Furthermore, observe that for some $C > 0$ independent of ε, we have the following bounds which follow from the Itô isometry:

$$
\begin{aligned}
&\widetilde{\mathbb{E}}\int_0^T \left|\int_0^t (G(\widehat{\boldsymbol{u}}_\varepsilon,\widehat{v}_\varepsilon,\widehat{\eta}_\varepsilon^*)d\widehat{W}_\varepsilon,\mathbf{Q}_{\kappa,\varepsilon})\right|^2 dt\\
&\quad=\int_0^T \widetilde{\mathbb{E}}\int_0^t |(G(\widehat{\boldsymbol{u}}_\varepsilon,\widehat{v}_\varepsilon,\widehat{\eta}_\varepsilon^*),\mathbf{Q}_{\kappa,\varepsilon})|^2 ds dt\\
&\quad\le T\widetilde{\mathbb{E}}\left(\int_0^T \left(\|\widehat{\eta}_\varepsilon^*\|_{L^\infty(0,L)}^2\|\widehat{\boldsymbol{u}}_\varepsilon\|_{\mathbf{L}^2(\mathcal{O})}^2+\|\widehat{v}_\varepsilon\|_{L^2(0,L)}^2\right)ds\right)\\
&\quad\le C(\delta).
\end{aligned}
\tag{9.97}
$$

Combining (9.96), (9.97) and using the Vitali convergence theorem, we thus conclude the proof of Lemma 9.4.3. □

We now show that the limit $\widehat{\boldsymbol{u}}$ solves the transformed incompressibily condition. To do that, observe that $(9.78)_4$ implies that for any measurable set $A \subset \widetilde{\Omega} \times [0,T]$, we have $\widetilde{\mathbb{E}}\int_0^T \mathbb{1}_A\left(\text{div}^{\widehat{\eta}_\varepsilon^*}\widehat{\boldsymbol{u}}_\varepsilon,\varphi\right) \to 0$ for any $\varphi \in L^2(\mathcal{O})$. Observe also that, thanks to (9.81) and (9.82), we can infer that $\frac{\partial_z\widehat{\eta}_\varepsilon^*}{R+\widehat{\eta}_\varepsilon^*} \to \frac{\partial_z\widehat{\eta}^*}{R+\widehat{\eta}^*}$ uniformly on $[0,T]\times\mathcal{O}$ a.s.

and also that $\widetilde{\mathbb{E}}\|\frac{\partial_z \widehat{\eta}^*_\varepsilon}{R+\widehat{\eta}^*_\varepsilon}\|^p_{L^\infty(0,T;L^\infty(\mathcal{O}))} < C$ for any $p > 2$. An application of the Vitali convergence theorem thus gives us that

$$\frac{\partial_z \widehat{\eta}^*_\varepsilon}{R+\widehat{\eta}^*_\varepsilon} \to \frac{\partial_z \widehat{\eta}^*}{R+\widehat{\eta}^*} \quad \text{in } L^2(\widetilde{\Omega}; L^\infty(0,T; L^\infty(\mathcal{O}))).$$

Combining this with the weak convergence (up to a subsequence) result that we obtain as a consequence of (9.78)$_4$, we deduce that $\widetilde{\mathbb{E}} \int_0^T \mathbb{1}_A \left(\text{div}^{\widehat{\eta}^*} \widehat{\boldsymbol{u}}, \varphi\right) = 0$ for any $\varphi \in L^2(\mathcal{O})$ and measurable set A. Thus, $\text{div}^{\widehat{\eta}^*(t)} \widehat{\boldsymbol{u}}(t) = 0$ for almost every $t \in [0,T]$ almost surely.

We now pass to the limit $\varepsilon \to 0$ in the remaining terms. For the first term, we write

$$\begin{aligned}|\int_0^t \int_{\mathcal{O}} \widehat{\boldsymbol{u}}_\varepsilon \partial_t \boldsymbol{q}_{\kappa,\varepsilon} - \widehat{\boldsymbol{u}} \partial_t \boldsymbol{q}_\kappa| &\le |\int_0^t \int_{\mathcal{O}} (\widehat{\boldsymbol{u}}_\varepsilon - \widehat{\boldsymbol{u}}) \partial_t \boldsymbol{q}_{\kappa,\varepsilon}| \\ &\quad + |\int_0^t \int_{\mathcal{O}} \widehat{\boldsymbol{u}} (\partial_t \boldsymbol{q}_{\kappa,\varepsilon} - \partial_t \boldsymbol{q}_\kappa)| \to 0,\end{aligned}$$

$\widetilde{\mathbb{P}}$-a.s. in $L^\infty(0,T)$. The first term on the right side converges to 0 in $L^\infty(0,T)$ almost surely, thanks to (9.77), whereas the second term converges to 0, thanks to the weak convergence (9.93) in $L^2(0,T;\mathbf{L}^{\frac{4}{3}}(\mathcal{O}))$. We now focus on a part of the nonlinear advection term on the right side of (9.94) and leave the rest of the proof of convergence to the reader, which is carried out in the same way using the results Theorem 9.4.1, (9.77), (9.80), (9.81), (9.92), and (9.93).

Observe that by integrating by parts, we can write

$$\begin{aligned}&\int_0^t \int_{\mathcal{O}} (R+\widehat{\eta}^*_\varepsilon)((\widehat{\boldsymbol{u}}_\varepsilon \cdot \nabla^{\widehat{\eta}^*_\varepsilon}) \widehat{\boldsymbol{u}}_\varepsilon \cdot \boldsymbol{q}_{\kappa,\varepsilon} - (\widehat{\boldsymbol{u}}_\varepsilon \cdot \nabla^{\widehat{\eta}^*_\varepsilon}) \boldsymbol{q}_{\kappa,\varepsilon} \cdot \widehat{\boldsymbol{u}}_\varepsilon) \\ &= -2 \int_0^t \int_{\mathcal{O}} (R+\widehat{\eta}^*_\varepsilon)(\widehat{\boldsymbol{u}}_\varepsilon \cdot \nabla^{\widehat{\eta}^*_\varepsilon}) \boldsymbol{q}_{\kappa,\varepsilon} \cdot \widehat{\boldsymbol{u}}_\varepsilon - \int_0^t \int_{\mathcal{O}} (R+\widehat{\eta}^*_\varepsilon) \text{div}^{\widehat{\eta}^*_\varepsilon} \widehat{\boldsymbol{u}}_\varepsilon \, \widehat{\boldsymbol{u}}_\varepsilon \cdot \boldsymbol{q}_{\kappa,\varepsilon} \\ &\quad + \int_0^t \int_0^L (\widehat{v}_\varepsilon)^2 \psi.\end{aligned}$$

It is easy to see that the last term on the right-hand side converges to $\int_0^t \int_0^L \widehat{v}^2 \psi dz$ in $L^\infty(0,T)$ a.s. Now, for the other terms, writing $\widehat{\boldsymbol{u}}_\varepsilon = (\widehat{u}^1_\varepsilon, \widehat{u}^2_\varepsilon)$ (likewise $\widehat{\boldsymbol{u}}$) and $\nabla^{\widehat{\eta}^*_\varepsilon} = (\partial_1^{\widehat{\eta}^*_\varepsilon}, \partial_2^{\widehat{\eta}^*_\varepsilon})$, we observe that

$$|\int_0^t \int_{\mathcal{O}} (R+\widehat{\eta}^*_\varepsilon)(\widehat{\boldsymbol{u}}_\varepsilon \cdot \nabla^{\widehat{\eta}^*_\varepsilon}) \boldsymbol{q}_{\kappa,\varepsilon} \cdot \widehat{\boldsymbol{u}}_\varepsilon - \int_0^t \int_{\mathcal{O}} (R+\widehat{\eta}^*)(\widehat{\boldsymbol{u}} \cdot \nabla^{\widehat{\eta}^*}) \boldsymbol{q}_\kappa \cdot \widehat{\boldsymbol{u}}|$$

$$= |\int_0^t\int_{\mathcal{O}}(R+\widehat{\eta}_\varepsilon^*)\sum_{i,j=1}^2(\widehat{u}_\varepsilon^i\widehat{u}_\varepsilon^j-\widehat{u}^i\widehat{u}^j)\partial_i^{\widehat{\eta}_\varepsilon^*}q_{\kappa,\varepsilon}^j+\int_0^t\int_{\mathcal{O}}(\widehat{\eta}_\varepsilon^*-\widehat{\eta}^*)(\widehat{\boldsymbol{u}}\cdot\nabla^{\widehat{\eta}^*})\boldsymbol{q}_\kappa\cdot\widehat{\boldsymbol{u}}$$

$$+\int_0^t\int_{\mathcal{O}}(R+\widehat{\eta}_\varepsilon^*)(\widehat{\boldsymbol{u}}\cdot(\nabla^{\widehat{\eta}_\varepsilon^*}\boldsymbol{q}_{\kappa,\varepsilon}-\nabla^{\widehat{\eta}^*}\boldsymbol{q}_\kappa))\cdot\widehat{\boldsymbol{u}}|.$$

The first two terms on the right side converge to 0 a.s., thanks to Theorem 9.4.1 and (9.92). Similarly for the third term, we use (9.92) and obtain that

$$|\int_0^t\int_{\mathcal{O}}(R+\widehat{\eta}_\varepsilon^*)(\widehat{\boldsymbol{u}}\cdot(\nabla^{\widehat{\eta}_\varepsilon^*}\boldsymbol{q}_{\kappa,\varepsilon}-\nabla^{\widehat{\eta}^*}\boldsymbol{q}_\kappa))\cdot\widehat{\boldsymbol{u}}|$$
$$\le C\|\widehat{\boldsymbol{u}}\|^2_{L^2(0,T;\mathbf{L}^4(\mathcal{O}))}\|\nabla^{\widehat{\eta}_\varepsilon^*}\boldsymbol{q}_{\kappa,\varepsilon}-\nabla^{\widehat{\eta}^*}\boldsymbol{q}_\kappa\|_{L^\infty(0;T;\mathbf{L}^2(\mathcal{O}))}\to 0,\qquad \widetilde{\mathbb{P}}-a.s.$$

Almost sure convergence of the third integral to 0 in $L^\infty(0,T)$ is immediate. Finally, we have

$$\sup_{t\in[0,T]}|\int_0^t\int_{\mathcal{O}}(R+\widehat{\eta}_\varepsilon^*)\mathrm{div}^{\widehat{\eta}_\varepsilon^*}\widehat{\boldsymbol{u}}_\varepsilon\widehat{\boldsymbol{u}}_\varepsilon\cdot\boldsymbol{q}_{\kappa,\varepsilon}|\le\int_0^T\int_{\mathcal{O}}|(R+\widehat{\eta}_\varepsilon^*)\mathrm{div}^{\widehat{\eta}_\varepsilon^*}\widehat{\boldsymbol{u}}_\varepsilon\widehat{\boldsymbol{u}}_\varepsilon\cdot\boldsymbol{q}_{\kappa,\varepsilon}|$$
$$\le C(\delta)\|\mathrm{div}^{\widehat{\eta}_\varepsilon^*}\widehat{\boldsymbol{u}}_\varepsilon\|_{L^2(0,T;L^2(\mathcal{O}))}\|\widehat{\boldsymbol{u}}_\varepsilon\|_{L^2(0,T;\mathbf{L}^4(\mathcal{O}))}\|\boldsymbol{q}_{\kappa,\varepsilon}\|_{L^\infty(0,T;\mathbf{L}^4(\mathcal{O}))}$$
$$\to 0,\qquad \widetilde{\mathbb{P}}-a.s.$$

Step 3: Passing $\kappa\to 1$

We have now proved that, for any fixed $\delta>0$, the stochastic processes $(\widehat{\boldsymbol{u}},\widehat{\eta},\widehat{\eta}^*)$ constructed in Theorem 9.4.1 and $\mathbf{Q}_\kappa=(\boldsymbol{q}_\kappa,\psi)$, where $\boldsymbol{q}_\kappa$ is defined in (9.86), satisfy the following equation $\widetilde{\mathbb{P}}$-a.s.,

$$\begin{aligned}&((R+\widehat{\eta}^*(t))\widehat{\boldsymbol{u}}(t),\boldsymbol{q}_\kappa(t))+(\partial_t\widehat{\eta}(t),\psi(t))=(\boldsymbol{u}_0(R+\eta_0),\boldsymbol{q}_\kappa(0))+(v_0,\psi(0))\\&+\int_0^t\int_{\mathcal{O}}(R+\widehat{\eta}^*)\widehat{\boldsymbol{u}}\cdot\partial_t\boldsymbol{q}_\kappa+\int_0^t\int_0^L\partial_t\widehat{\eta}\,\partial_t\psi\\&-\int_0^t\int_0^L\partial_z\widehat{\eta}\partial_z\psi+\partial_{zz}\widehat{\eta}\partial_{zz}\psi+\frac{1}{2}\int_0^t\int_{\mathcal{O}}\partial_t\widehat{\eta}^*\widehat{\boldsymbol{u}}\cdot\boldsymbol{q}_\kappa\\&-\frac{1}{2}\int_0^t\int_{\mathcal{O}}(R+\widehat{\eta}^*)((\widehat{\boldsymbol{u}}-\partial_t\widehat{\eta}r\mathbf{e}_r)\cdot\nabla^{\widehat{\eta}^*}\widehat{\boldsymbol{u}}\cdot\boldsymbol{q}_\kappa-(\widehat{\boldsymbol{u}}-\partial_t\widehat{\eta}r\mathbf{e}_r)\cdot\nabla^{\widehat{\eta}^*}\boldsymbol{q}_\kappa\cdot\widehat{\boldsymbol{u}})\\&-2\nu\int_0^t\int_{\mathcal{O}}(R+\widehat{\eta}^*)\mathbf{D}^{\widehat{\eta}^*}(\widehat{\boldsymbol{u}})\cdot\mathbf{D}^{\widehat{\eta}^*}(\boldsymbol{q}_\kappa)\\&+\int_0^t\left(P_{in}\int_0^1 q_{\kappa z}\Big|_{z=0}dr-P_{out}\int_0^1(q_\kappa)_z\Big|_{z=1}dr\right)ds+\int_0^t(G(\widehat{\boldsymbol{u}},\widehat{v},\widehat{\eta}^*)d\widehat{W},\mathbf{Q}_\kappa).\end{aligned}\tag{9.98}$$

We pass $\kappa \to 1$ in (9.98) by observing that, for $\boldsymbol{q}_\kappa$ defined (9.86), we have

$$\boldsymbol{q}_\kappa \to \boldsymbol{q}, \quad \text{in } C^1(0,T;\mathbf{H}^1(\mathcal{O})) \quad \widetilde{\mathbb{P}} - \text{almost surely}.$$

This concludes the proof of Theorem 9.4.2.

Step 4: Stopping Time Argument: Agreement of the Structure and the Artificial Structure Displacements

Notice that in the statement of Theorem 9.4.2, we still have the function $\widehat{\eta}^*(t)$ in the weak formulation, which keeps the displacement uniformly bounded. We now show that in fact $\widehat{\eta}^*(t)$ can be replaced by the limiting stochastic process $\widehat{\eta}(t)$ to obtain the desired weak formulation and martingale solution until some stopping time τ, which we show is strictly greater than zero almost surely.

Lemma 9.4.4 (Stopping Time) *Let the deterministic initial data η_0 satisfy the assumptions* (9.23). *Then, for any $\delta > 0$ and for a given $\frac{3}{2} < s < 2$, there exists an almost surely positive stopping time τ, given by*

$$\tau := T \wedge \inf\{t > 0 : \inf_{z\in[0,L]} (R+\widehat{\eta}(t,z)) \le \delta\} \wedge \inf\{t > 0 : \|R+\widehat{\eta}(t)\|_{H^s(0,L)} \ge \frac{1}{\delta}\}, \tag{9.99}$$

such that

$$\widehat{\eta}^*(t) = \widehat{\eta}(t) \quad \textit{for } t < \tau. \tag{9.100}$$

Proof In **Step 1** below we first show that the stopping time (9.99) is strictly positive a.s., and then in **Step 2**, we show that $\widehat{\eta}^*(t) = \widehat{\eta}(t) \quad \text{for } t < \tau$.

Step 1. We start by showing that under the assumptions (9.23) on the deterministic initial data η_0, the stopping time τ is almost surely strictly positive for any $\delta > 0$. For this purpose, let us write the stopping time as

$$\tau = T \wedge \tau^{(1)} \wedge \tau^{(2)},$$

where $\tau^{(1)}$ and $\tau^{(2)}$ are defined by:

$$\begin{aligned}\tau = &T \wedge \inf\{t > 0 : \inf_{z\in[0,L]} (R+\widehat{\eta}(t,z)) \le \delta\} \wedge \inf\{t > 0 : \|R + \widehat{\eta}(t)\|_{H^s(0,L)} \ge \frac{1}{\delta}\} \\ &=: T \wedge \tau^{(1)} \wedge \tau^{(2)}.\end{aligned}$$

We start with $\tau^{(2)}$. Observe that using the triangle inequality and an interpolation inequality, for any $\delta_0 > \delta$, we obtain

$$\widetilde{\mathbb{P}}[\tau^{(2)}=0, \|R+\eta_0\|_{H^2(0,L)}<\frac{1}{\delta_0}]=\lim_{\epsilon\to 0}\widetilde{\mathbb{P}}[\tau^{(2)}<\epsilon, \|R+\eta_0\|_{H^2(0,L)}<\frac{1}{\delta_0}]$$

$$\le \limsup_{\epsilon\to 0^+}\widetilde{\mathbb{P}}[\sup_{t\in[0,\epsilon)}\|R+\widehat{\eta}(t)\|_{H^s(0,L)}>\frac{1}{\delta}, \|R+\eta_0\|_{H^2(0,L)}<\frac{1}{\delta_0}]$$

$$\le \limsup_{\epsilon\to 0^+}\widetilde{\mathbb{P}}[\sup_{t\in[0,\epsilon)}\|\widehat{\eta}(t)-\eta_0\|_{H^s(0,L)}>\frac{1}{\delta}-\frac{1}{\delta_0}]$$

$$\le \frac{1}{(\frac{1}{\delta}-\frac{1}{\delta_0})}\limsup_{\epsilon\to 0}\widetilde{\mathbb{E}}[\sup_{t\in[0,\epsilon)}\|\widehat{\eta}(t)-\eta_0\|_{H^s(0,L)}]$$

$$\le \frac{1}{(\frac{1}{\delta}-\frac{1}{\delta_0})}\limsup_{\epsilon\to 0}\widetilde{\mathbb{E}}[\sup_{t\in[0,\epsilon)}\|\widehat{\eta}(t)-\eta_0\|^{1-\frac{s}{2}}_{L^2(0,L)}\|\widehat{\eta}(t)-\eta_0\|^{\frac{s}{2}}_{H^2(0,L)}]$$

$$\le \frac{1}{(\frac{1}{\delta}-\frac{1}{\delta_0})}\limsup_{\epsilon\to 0}\widetilde{\mathbb{E}}[\sup_{t\in[0,\epsilon)}\epsilon\|\widehat{v}(t)\|^{1-\frac{s}{2}}_{L^2(0,L)}\|\widehat{\eta}(t)-\eta_0\|^{\frac{s}{2}}_{H^2(0,L)}]$$

$$\le \limsup_{\epsilon\to 0}\frac{\epsilon}{(\frac{1}{\delta}-\frac{1}{\delta_0})}\left(\widetilde{\mathbb{E}}[\sup_{t\in[0,\epsilon)}\|\widehat{v}(t)\|^2_{L^2(0,L)}]\right)^{\frac{2-s}{4}}\left(\widetilde{\mathbb{E}}[\sup_{t\in(0,\epsilon)}\|\widehat{\eta}(t)-\eta_0\|^2_{H^2(0,L)}]\right)^{\frac{s}{4}}$$

$$=0.$$

Hence, by continuity from below, we deduce that for any $\delta>0$,

$$\widetilde{\mathbb{P}}[\tau^{(2)}=0, \|R+\eta_0\|_{H^2(0,L)}<\frac{1}{\delta}]=0. \tag{9.101}$$

To estimate $\tau^{(1)}$ we observe that, similarly, since for any $t\in[0,T]$, we have that $\inf_{z\in(0,L)}(R+\widehat{\eta}(t,z))\ge\inf_{z\in(0,L)}(R+\eta_0(z))-\|\widehat{\eta}(t)-\eta_0\|_{L^\infty(0,L)}$, we write for any $\delta_0>\delta$

$$\widetilde{\mathbb{P}}[\tau^{(1)}=0, \inf_{z\in(0,L)}(R+\eta_0)>\delta_0]$$

$$\le \limsup_{\epsilon\to 0^+}\widetilde{\mathbb{P}}[\inf_{t\in[0,\epsilon)}\inf_{z\in(0,L)}(R+\widehat{\eta}(t))<\delta, \inf_{z\in(0,L)}(R+\eta_0)>\delta_0]$$

$$\le \limsup_{\epsilon\to 0^+}\widetilde{\mathbb{P}}[\sup_{t\in[0,\epsilon)}\|\widehat{\eta}(t)-\eta_0\|_{L^\infty(0,L)}>\delta_0-\delta]$$

$$\le \frac{1}{(\delta_0-\delta)^2}\limsup_{\epsilon\to 0}\widetilde{\mathbb{E}}[\sup_{t\in[0,\epsilon)}\|\widehat{\eta}(t)-\eta_0\|^2_{H^1(0,L)}]$$

$$=0.$$

Hence for given $\delta>0$,

$$\widetilde{\mathbb{P}}[\tau^{(1)} = 0, \inf_{z \in (0,L)} (R + \eta_0) > \delta] = 0. \tag{9.102}$$

That is, we have

$$\widetilde{\mathbb{P}}[\tau = 0, \inf_{z \in (0,L)} (R + \eta_0) > \delta, \|R + \eta_0\|_{H^2(0,L)} < \frac{1}{\delta}] = 0. \tag{9.103}$$

Step 2. The statement $\widehat{\eta}^*(t) = \widehat{\eta}(t)$ if $t < \tau$, can be obtained the same way as in (9.74).

□

Thus, by combining Theorem 9.4.2, Lemma 9.4.4, and (9.100), we summarize our findings in the following result.

Theorem 9.4.3 *For any given $\delta > 0$, if the deterministic initial data η_0 satisfies* (9.23)*, then the stochastic processes $(\widehat{\boldsymbol{u}}, \widehat{\eta}, \tau)$ obtained in the limit specified in Theorem 9.4.1, along with the stochastic basis constructed in Theorem 9.4.1, determine a martingale solution in the sense of Definition 9.1.1 of the stochastic FSI problem* (9.1)–(9.8)*, with the stochastic forcing given by* (9.10).

9.5 Appendix

As part of the test function construction for the nonlinearly coupled stochastic FSI problem, given a limiting (random) structure displacement $\widehat{\eta}^*_\kappa$, we must construct a corresponding smooth (regularized) $\widehat{\eta}^*_\kappa$ for each squeezing parameter $\kappa > 1$ such that

$$\frac{R + \widehat{\eta}^*}{\kappa} < R + \widehat{\eta}^*_\kappa < R + \widehat{\eta}^*, \tag{9.104}$$

for almost every $(t, z, \omega) \in [0, T] \times [0, L] \times \widetilde{\Omega}$. In this appendix, we show that this is possible.

A Smooth Lower Bound Function for Deterministic Structure Displacements For each possible (deterministic, fixed-in-time) structure displacement $\eta \in C(0, T; H^s(\Gamma))$ for any fixed $s \in (3/2, 2)$, satisfying

$$\inf_{t \in [0,T], z \in [0,L]} R + \eta(t, z) \geq \delta, \qquad \|R + \eta\|_{C(0,T;H^s(\Gamma))} \leq \frac{1}{\delta}, \tag{9.105}$$

we will associate a corresponding smooth function $\eta^\flat_\alpha : [0, T] \times \Gamma \to \mathbb{R}$ for a parameter $\alpha > 0$ such that

$$\eta^\flat_\alpha < \eta, \qquad \text{and} \qquad \|\eta - \eta^\flat_\alpha\|_{C(\Gamma)} \to 0 \text{ as } \alpha \to 0.$$

Note that the uniform bound in (9.105) implies that η is uniformly Lipschitz in the spatial variable (uniformly in time), so that there exists a constant C_s depending only on $s \in (3/2, 2)$ such that

$$|\eta(z_1) - \eta(z_2)| \le C_s |z_1 - z_2|, \qquad \text{for } z_1, z_2 \in [0, L], \tag{9.106}$$

due to one-dimensional Sobolev embeddings. We then define a downward translation of η given a parameter $\alpha > 0$, by

$$\eta_\alpha^\flat := R + \eta - 2C_s\alpha, \tag{9.107}$$

where C_s is the Lipschitz constant from estimate (9.106). Let $\zeta : \mathbb{R} \to \mathbb{R}$ be a standard convolution kernel in one dimension (a smooth, compactly supported, even function, with total integral one), and define $\zeta_\alpha(w) = \alpha^{-1}\zeta(w/\alpha)$. We then define

$$a_\alpha^\eta := \eta_\alpha^\flat * \zeta_\alpha, \tag{9.108}$$

where $*$ denotes a spatial convolution. To handle the spatial convolution near the endpoints $z = 0$ and $z = L$, we can define $\eta_\alpha^\flat$ to be constant outside of $[0, L]$, where $\eta_\alpha^\flat(z) = \eta_\alpha^\flat(0)$ for $z \le 0$ and $\eta_\alpha^\flat(z) = \eta_\alpha^\flat(L)$ for $z \ge L$. Note that a_α^η is a spatially smooth function, as a result of the properties of spatial convolution, since the convolution kernel is smooth.

We have the following important property of the lower bounding function a_α^η, which shows how the parameter $\alpha > 0$ affects the distance between η and a_α^η.

Proposition 9.5.1 *Given $\eta \in C(0, T; H^s(\Gamma))$ for $s \in (3/2, 2)$ satisfying the bounds (9.105), we have that for any parameter $\alpha > 0$, a_α^η is a spatially smooth function that is continuous in time, satisfying*

$$C_s\alpha \le \eta(t, z) - a_\alpha^\eta(t, z) \le 3C_s\alpha, \qquad \textit{for all } t \in [0, T] \textit{ and } z \in [0, L].$$

In particular,

$$\|\eta - a_\alpha^\eta\|_{C(0,T;C(\Gamma))} \to 0, \qquad \textit{as } \alpha \to 0.$$

Proof Note that $\eta_\alpha^\flat$ defined in (9.107) also satisfies the Lipschitz estimate (9.106), since it is a translation of η, so

$$|\eta_\alpha^\flat(t, z_1) - \eta_\alpha^\flat(t, z_2)| \le C_s|z_1 - z_2|, \qquad \text{for all } t \in [0, T] \text{ and } z \in [0, L].$$

Therefore, we observe that

$$|a_\alpha^\eta(t, z) - \eta_\alpha^\flat(t, z)| \le C_s\alpha, \qquad \text{for all } t \in [0, T] \text{ and } z \in [0, L].$$

This is because

$$|\eta^{\flat}_{\alpha}(z+h) - \eta^{\flat}_{\alpha}(z)| \le C_s h, \qquad \text{for all } |h| \le \alpha,$$

so therefore, by the definition of a^{η}_{α} as a convolution,

$$\begin{aligned}|a^{\eta}_{\alpha}(t,z) - \eta^{\flat}_{\alpha}(t,z)| &= \left|\int_{-\alpha}^{\alpha} \left(\eta^{\flat}_{\alpha}(t,z+w) - \eta^{\flat}_{\alpha}(z)\right)\zeta_{\alpha}(w)dw\right| \\ &\le \int_{-\alpha}^{\alpha} |\eta^{\flat}_{\alpha}(t,z+w) - \eta^{\flat}_{\alpha}(z)|\zeta_{\alpha}(w)dw \le C_s\alpha,\end{aligned}$$

where we used the fact that the convolution kernel $\zeta_{\alpha}(w) := \alpha^{-1}\zeta(w/\alpha)$ has unit integral, is compactly supported on $[-\alpha, \alpha]$ and is nonnegative. So since

$$|a^{\eta}_{\alpha}(t,z) - \eta^{\flat}_{\alpha}(t,z)| \le C_s\alpha,$$

combining this with the fact that $\eta - \eta^{\flat}_{\alpha} = 2C_s\alpha$, by the definition of $\eta^{\flat}_{\alpha}$ in (9.107), we obtain the desired result.

□

Next, we discuss regularity and convergence properties of a^{η}_{α} for fixed $\alpha > 0$.

Proposition 9.5.2 *Given $\eta \in C(0,T; H^s(\Gamma))$ for $s \in (3/2, 2)$ satisfying the bounds in* (9.105), *$a^{\eta}_{\alpha} \in C(0,T; C^k(\Gamma))$ for all positive integers k, with*

$$\|a^{\eta}_{\alpha}\|_{C(0,T;C^k(\Gamma))} \le C(k, \alpha, \|R+\eta\|_{C(0,T;H^s(\Gamma))}).$$

If $\eta_n \to \eta$ in $C(0,T; C(\Gamma))$ as $n \to \infty$, where $\{\eta_n\}_{n=1}^{\infty}$ and η all satisfy the uniform bounds (9.105), *then $a^{\eta_n}_{\alpha} \to a^{\eta}_{\alpha}$ in $C(0,T; C^k(\Gamma))$ for all positive integers k.*

Proof Note that by properties of convolution and the definition of a^{η}_{α} in (9.108) and the definition of $\eta^{\flat}_{\alpha}$ in (9.107),

$$\begin{aligned}\|a^{\eta}_{\alpha}\|_{C(0,T;C^k(\Gamma))} &\le C_{k,\alpha}\|\eta^{\flat}_{\alpha}\|_{C(0,T;C(\Gamma))} \\ &\le C_k\|R+\eta\|_{C(0,T;C(\Gamma))} \le C_k\|R+\eta\|_{C(0,T;H^s(\Gamma))},\end{aligned}$$

for a constant C_k depending on k. Similarly, for the convergence result, note that by (9.107), $(\eta_n)^{\flat}_{\alpha} \to \eta^{\flat}_{\alpha}$ as $n \to \infty$, in $C(0,T; C(\Gamma))$ for fixed $\alpha > 0$, and hence, by properties of convolution, we obtain the desired convergence

$$a^{\eta_n}_{\alpha} = (\eta_n)^{\flat}_{\alpha} * \zeta_{\alpha} \to a^{\eta}_{\alpha} = \eta^{\flat}_{\alpha} * \zeta_{\alpha}, \qquad \text{in } C(0,T; C^k(\Gamma)), \text{ as } n \to \infty,$$

for every positive integer k. □

Application to the Construction of Test Functions Given a (random) structure displacement $\widehat{\eta}^*$, which satisfies the uniform bounds in (9.105) $\widetilde{\mathbb{P}}$-almost surely and for all $t \in [0, T]$, we note that there is a positive (deterministic) constant $c_\delta > 0$ depending only on δ for which for any fixed $\kappa > 1$,

$$\min_{z \in [0,L]} \left(R + \widehat{\eta}^* - \frac{R + \widehat{\eta}^*}{\kappa} \right) \geq c_\delta > 0,$$

$\widetilde{\mathbb{P}}$-almost surely. By choosing α so that $3C_s\alpha := c_\delta/2$, and defining $\widehat{\eta}^*_\kappa$ so that

$$R + \widehat{\eta}^*_\kappa := a_\alpha^{\widehat{\eta}^*},$$

we obtain the desired chain of inequalities in (9.104) from Proposition 9.5.1. Here, we note that $a_\alpha^{\widehat{\eta}^*}$ is now a random (spatially smooth) lower bound function, defined by mapping pathwise for each outcome $\widetilde{\omega} \in \widetilde{\Omega}$ in the probability space, $\widehat{\eta}^*(\widetilde{\omega}) \to a_\alpha^{\widehat{\eta}^*(\widetilde{\omega})}$, using the map defined for deterministic functions η satisfying (9.105).

9.6 Bibliographical Notes

Stochastic Navier-Stokes Equations There is a vast literature on the existence theory for the stochastic Navier-Stokes equations and related models (on fixed domains) dating back to the 1970s seminal paper by Temam and Bensoussan [36]. The first proof of existence of weak martingale solutions to the 2D Navier-Stokes equations driven by a white-in-time multiplicative noise with a special noise coefficient was given by Capinski and Cutland in [75] and was then generalized by Flandoli and Gatarek in [139] (see also [34]) and by Mikulevicius and Rozovskii in [258]. The existence of a weak pathwise solution in a 2D bounded domain was then given in [51] (see also [255]) and that of local in time, strong pathwise solution in 3D was given in [164]. More recently, the authors of [226, 227] proved the existence of a strong pathwise solution on a 3D torus with L^p data. For Navier-Stokes equations driven by space-time white noise, see [103].

A fundamental work in the mathematical analysis of stochastic *compressible* fluid flows is [53], which studies the stochastic isentropic compressible Navier-Stokes equations with adiabatic constant $\gamma > 3/2$ and with multiplicative noise (see also [52, 309]). There is also a well-posedness result on stochastic inviscid compressible fluid flow in [37], where solutions known as "martingale weak entropy solutions" are proven to exist for the stochastically forced compressible Euler equations for an isentropic fluid in one spatial dimension, on a one-dimensional spatial torus.

Stochastic Fluid-Structure Interaction Work on stochastic incompressible fluid-structure interaction began first with a fully coupled model [217] involving the coupled interaction between the linear Stokes equations and a wave equation with external additive stochastic structure forcing, which forms the basis for results discussed in Chap. 8. These results were then extended to the more challenging case of stochastic incompressible fluid-structure interaction involving the Navier-Stokes equations interacting with an elastic shell, with multiplicative stochastic forcing acting on both the fluid/structure in [338], where the existence of weak martingale solutions is established. This model is discussed in Chap. 9. The well-posedness of an analogous moving-boundary stochastic model was also established in the different context of noise of transport-type perturbing the structure elastodynamics [55]. This noise provides pathwise control on the estimates for the fluid-structure coupled system, and thus the issues faced in this paper are different than the ones discussed in Chap. 9. Later, existence of weak martingale solutions to stochastic coupled moving boundary systems of incompressible Navier-Stokes equations interacting with elastic shells was established for more general models, including models with unrestricted structure displacements [322], that is, when the structure displacement is a vector function η with components in all $\mathbb{R}^2$, directions. Then, in [321], the Navier-slip boundary condition (6.1) is imposed at the fluid-structure interface (instead of (9.6)), and the existence of a martingale solution is proved. More recently in [215], the existence of a martingale solution to an FSI system consisting of an isentropic *compressible* fluid and a viscoelastic plate is established.

Part IV
Deterministic Fluid-Poroelastic Structure Interaction (FPSI)

Introduction to FPSI with Linear and Nonlinear Coupling

The two chapters that follow are devoted to the study of coupled fluid-poroelastic structure interaction (FPSI), which describes the coupled dynamical interaction between poroelastic structures and fluids. Fluid-poroelastic structure interaction problems arise in various fields where the interaction between fluid flow and porous elastic materials is critical. Examples of such fields include biomedical engineering, geomechanics and petroleum engineering, civil engineering, environmental engineering, aerospace and marine engineering, and energy systems. More specific example of bioengineering applications include blood flow in soft tissues or cerebrospinal fluid interacting with brain tissue. In geomechanics and petroleum engineering, FPSI models are essential in studying how fluids move through porous rock formations during oil or gas extraction, where the displacement of the porous matrix and fluid pressure affect the overall behavior of the reservoir. In environmental engineering, FPSI is involved in modeling contaminant transport describing the migration of contaminants through groundwater systems, where the flow of fluids through porous media (like soil or aquifers) interacts with the elastic properties of the subsurface.

Mathematically, poroelastic materials are modeled as porous materials with elastic properties that are coupled to the flow of fluid through their pores. These materials are described by a set of (macro-scale) equations known as the **Biot equations**. The Biot equations are composed of two coupled equations; one is based on the second Newton's law of motion describing the elastodynamics of the poroelastic matrix (structure) containing the fluid, and the other is based on the conservation of mass principle describing the fluid flow in the poroelastic matrix. The first equation is given in terms of the structure displacement while the second equation in terms of the fluid pressure. To recover the fluid velocity from the solution of the Biot system, one uses the so-called Darcy law, relating the fluid velocity to the pressure gradient via a permeability tensor. See (10.5).

In this chapter, we are interested in studying the interaction of the poroelastic medium flow and displacement, with the flow of free (unrestricted) fluid. The free fluid flow will be described by either the time-dependent Stokes equations in the case of linear coupling or the Navier-Stokes equations in the case of nonlinear coupling. As mentioned earlier in Chap. 4, linear coupling refers to a FSI problem in which the fluid and structure are coupled via a fixed interface, representing a first, linearized approximation of the true interface motion. This results in the fluid (and structure equations) being defined on fixed domains. By contrast, nonlinear coupling refers to a FSI problem in which the fluid and structure are coupled via the current location of the interface, resulting in the fluid (and possibly structure) equations being defined on the current domain, which is time dependent and not known a priori. As discussed earlier, this gives rise to additional geometric nonlinearities in the problem that are exceedingly difficult to deal with from the analysis point of view. We will see in this chapter that in addition to the "classical" difficulties associated with this geometric nonlinearity in FSI problems, poroelastic FSI presents an additional set of difficulties associated with the geometric nonlinearities introduced by nonlinear coupling, which arise due to the fact that the elastic structure itself now contains fluid and its motion now suffers from certain regularity issues, not observed before.

In the next two chapters, we present the state-of-the-art results on the well-posedness analysis of FPSI with linear (Chap. 10) and nonlinear coupling (Chap. 11), in the context of finite energy solutions. We will consider poroelastic structures that are multi-layered consisting of a thick layer modeled by the Biot equations and a thin layer located at the interface between the Biot medium and the flow of the free fluid. The thin layer will be modeled by the poroelastic plate/reticular plate models. These models were motivated by biomedical applications where a thin poroelastic interface sits between the flow of a biofluid such as blood and a thick poroelastic medium, such as a bioartificial pancreas [69,348]. Mathematically, the presence of a thin interface with mass/inertia and elastic energy allows us to define traces of fluid and Biot structure velocities at the interface and provides certain regularizing effects on the coupled FPSI solution.

Chapter 10
FPSI with Linear Coupling

In this chapter, we will consider a linearly coupled FPSI system consisting of an incompressible, viscous fluid modeled by the time-dependent Stokes equations interacting with a multilayered poroelastic structure, consisting of a thin poroelastic plate and a thick Biot poroelastic material modeled by the Biot equations. This FPSI system will be linearly coupled so that both the fluid domain and the Biot domain where the equations are defined are fixed and the coupling conditions are evaluated at a fixed interface. The interface displacement, as well as the Biot medium displacement, will be calculated along the fixed domains, thereby approximating the true physical motion in case the displacement and deformation gradient are small. The fluid flow and the elastodynamics of the multilayered poroelastic structure will be fully coupled across a fixed interface through physical coupling conditions (including the Beavers-Joseph-Saffman condition), which present mathematical challenges related to the regularity of associated velocity traces. We will show the proof of the existence of weak, finite energy solutions to this linearly coupled FPSI problem in two case scenarios. First, for the thick poroelastic medium, we will consider a linear, dynamic Biot model. Then, for the thick poroelastic medium, we will consider a nonlinear quasi-static Biot model, where one of the coefficients in the model corresponding to the so-called permeability of the poroelastic medium is a nonlinear function of a particular combination of the unknowns in the problem called fluid content (as motivated by biological applications). The content of this chapter is adapted from the paper [43], with added insights and expanded discussions.

In both cases, the proofs will be based on constructing approximate solutions through Rothe's method, as already described in the previous chapters of this book, and on using energy methods, combined with a version of the Aubin-Lions compactness lemma (in the nonlinear case) to recover the weak solution as the limit of approximate subsequences. We will also provide uniqueness criteria and

S. Čanić et al., *Deterministic and Stochastic Fluid-Structure Interaction*, Advances in Mathematical Fluid Mechanics, https://doi.org/10.1007/978-3-032-00898-5_10

show that the constructed weak solutions are indeed strong solutions to the coupled problem if one assumes additional regularity.

We start by presenting the model and the spatial domain on which the model equations are defined.

10.1 Mathematical Formulation

10.1.1 Spatial Domain

We consider the simplest domain $\Omega \subset \mathbb{R}^3$ that can be used to describe the coupled problem between the flow of an incompressible, viscous fluid, and a thick poroelastic medium, where the coupling occurs through a thin poroelastic plate, serving as a fluid-structure interface with mass. Namely, our domain is a rectangular prism: $\Omega = (0, 1) \times (0, 1) \times (-1, 1)$, with three distinct sub-regions (see Fig. 10.1):

- $\Omega_f = (0, 1)^2 \times (-1, 0)$, which is occupied by a viscous, incompressible fluid
- $\Omega_b = (0, 1)^2 \times (0, 1)$ corresponding to a (thick) poroelastic medium modeled by the Biot model
- Ω_p corresponding to a thin poroelastic plate of thickness h, whose middle surface, denoted by ω_p, separates the regions Ω_f and Ω_b Here $\omega_p = (0, 1)^2 \times \{0\}$, $\Omega_p = \omega_p \times (-h/2, h/2) = (0, 1)^2 \times (-h/2, h/2)$.

In Cartesian coordinates (x_1, x_2, x_3), the physical domain Ω is the union

$$\Omega = \Omega_f \cup \omega_p \cup \Omega_b.$$

The variable x_3 will be referred to as the "transverse" variable $x_3 \in [-1, 1]$ and $x_1 \in [0, 1]$ and $x_2 \in [0, 1]$ will be called the "longitudinal" and "lateral" variables, respectively.

To study filtration through the thin, poroelastic plate of thickness h, we will consider a reduced "2.5-dimensional model," derived and rigorously justified by Marciniak-Cochra and Mikelić in [250], obtained via asymptotic reduction from the full Biot model. In the poroelastic plate model, the elastodynamics of the thin plate is described in terms of the displacement of the middle surface of the plate from its reference (flat) configuration. The middle surface is described in terms of x_1 and x_2 variables.

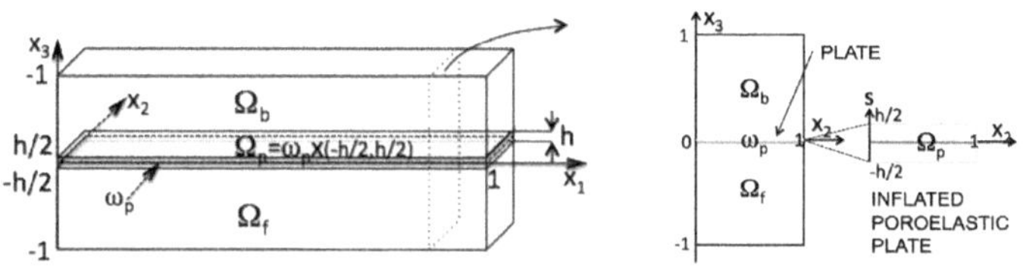

Fig. 10.1 Left: 3D domain Ω. Right: 2D vertical cross-section through domain Ω

Filtration, however, needs to be described in terms of an additional variable $s \in [-h/2, h/2]$, since it was shown in [250] that filtration through a thin poroelastic isotropic structure is dominant in the transverse direction. Therefore, we denote by ω_p the location of the middle surface of the elastic plate and by $\Omega_p = \omega_p \times (-h/2, h/2)$ the "inflated" mathematical domain on which the filtration velocity and pore fluid pressure will be defined. In Sect. 10.1.2 we present the model and provide more details about the approach.

The fluid contained in region Ω_f will be modeled by the time-dependent Stokes equations, with periodic boundary conditions at $x_1 = 0, x_1 = 1, x_2 = 0$ and $x_2 = 1$ and with the no-slip boundary condition at the bottom boundary $x_3 = -1$, which is considered fixed. The boundary conditions at the top fluid domain boundary, where the bulk fluid flow in Ω_f is coupled to the filtration flow through the poroelastic membrane Ω_p, will be described in Sect. 10.1.2.

Finally, Ω_b represents a thick poroelastic region relative to the thin poroelastic plate, with filtration flow and structure displacement satisfying Biot's equations. The thick structure displacement in Ω_b, as well as the filtrate flow, will satisfy periodic boundary conditions at $x_1 = 0, x_1 = 1, x_2 = 0$ and $x_2 = 1$. At the top boundary of Ω_b, with the reference configuration described by $x_3 = 1$, we prescribe zero elastic stress and no filtration through the wall. At the bottom boundary of Ω_b, both the elastodynamics of the thick poroelastic medium and filtration flow in Ω_b are fully coupled to the elastodynamics and filtration flow in the thin poroelastic plate Ω_p, which is, in turn, fully coupled to the fluid flow in Ω_f. The filtration flow in Ω_b is driven by the pressure gradient between the bottom boundary $x_3 = 0$ and the top boundary $x_3 = 1$ and by the time-dependent change in structure displacement.

In summary, the boundary of Ω consists of the periodic boundary at $x_1 = 0, x_1 = 1, x_2 = 0$ and $x_2 = 1$, a fixed bottom boundary $x_3 = -1$, and the moving top boundary with the reference configuration described by $x_3 = 1$. Detailed boundary conditions, and the coupling between different subdomains of Ω, will be described in the next section.

10.1.2 Mathematical Models

In this section, we describe the mathematical models, holding in each of the sub-domains specified above, and specify the coupling conditions. We will use $\mathbf{x}$ to denote points in Cartesian coordinates $\mathbf{x} = (x_1, x_2, x_3)$, the subscript b to denote quantities in the Biot domain Ω_b, and subscript p to denote the quantities corresponding to the poroelastic plate.

Biot Model in Ω_b

The Biot equations are a set of two coupled PDEs for the displacement $\boldsymbol{\eta} : \Omega_b \to \mathbb{R}^n$ of a poroelastic material from its reference configuration Ω_b and its pore pressure $p_b : \Omega_b \to \mathbb{R}$, given by

$$\rho_b \partial_{tt} \boldsymbol{\eta} - \nabla \cdot \boldsymbol{\sigma}_b(\nabla \boldsymbol{\eta}, p_b) = F_{\mathbf{b}}, \quad \text{in } \Omega_b, \tag{10.1}$$

$$c_b \partial_t p_b + \alpha_b (\nabla \cdot \boldsymbol{\eta})_t - \nabla \cdot (\kappa \nabla p_b) = S, \quad \text{in } \Omega_b, \tag{10.2}$$

where the poroelastic Cauchy stress tensor is given by

$$\boldsymbol{\sigma}_b(\nabla \boldsymbol{\eta}, p_b) = \boldsymbol{\sigma}^E(\nabla \boldsymbol{\eta}, p_b) - \alpha p_b \boldsymbol{I}. \tag{10.3}$$

Here, $\boldsymbol{\sigma}^E(\nabla \boldsymbol{\eta}, p_b)$ is the elastic part describing the elastic properties of the poroelastic matrix, specified by a constitutive law. On the right-hand side, $\mathbf{F}_b = \mathbf{F}_b(\mathbf{x}, t)$ denotes the body force per unit volume, and $S = S(\mathbf{x}, t)$ is the net volumetric fluid production rate.

The first equation, (10.1), describes the elastodynamics of the poroelastic material, captured by Newton's second law of motion. Here, the influence of the fluid pressure on the elastodynamics appears in the term $-\alpha_b \boldsymbol{I}$ in $\boldsymbol{\sigma}_b(\nabla \boldsymbol{\eta}, p_b)$. The elastic properties of the poroelastic matrix, captured by $\boldsymbol{\sigma}^E(\nabla \boldsymbol{\eta}, p_b)$, will be modeled in this book by the constitutive law associated with the Saint Venant-Kirchhoff material:

$$\boldsymbol{\sigma}^E = 2\mu_b \boldsymbol{D}(\boldsymbol{\eta}) + \lambda_b \mathrm{tr} \boldsymbol{D}(\boldsymbol{\eta}) \boldsymbol{I} = 2\mu_b \boldsymbol{D}(\boldsymbol{\eta}) + \lambda_b \nabla \cdot \boldsymbol{\eta} \boldsymbol{I}, \tag{10.4}$$

where λ_b, μ_b denote the Lamé constants. This constitutive law, describing linear elasticity, assumes small deformations and displacements.

The second equation, (10.2), describes the conservation of mass principle of the fluid component. It states that the rate of change of the quantity $c_b p_b + \alpha_b(\nabla \cdot \boldsymbol{\eta})$, known as the fluid content, is equal to the divergence of $\kappa \nabla p_b$, which is related to the velocity of the fluid flow passing through the pores of the poroelastic material, via **Darcy's law**:

$$\mathbf{u}_b = -\boldsymbol{K} \nabla p_b, \quad \text{with} \quad \boldsymbol{K} = k_b \boldsymbol{I}, \tag{10.5}$$

where $\boldsymbol{K}$ is the permeability tensor. Thus, the second equation states that the rate of change of fluid content equals negative divergence of the fluid velocity, which is a classical conservation of mass principle.

The parameter coefficients in this model are as follows:

Coefficients	Name	
ρ_b	Density of poroelastic matrix;	
λ_b, μ_b	Lamé constants;	
α_b	Biot-Willis coefficient [303];	(10.6)
c_b	Constrained storage coefficient;	
k_b	Permeability of the poroelastic matrix.	

Generally, one can take $\alpha_b > 0$, $k_b > 0$, and $c_b \geq 0$, though in many works it is assumed that c_b is a strictly positive constant. The constant ρ_b is nonnegative, and if $\rho_b = 0$, then the resulting equations are known as the **quasistatic Biot equations**, which describe the situation when the inertial effects of the structure are negligible.

In this chapter, we will allow the permeability k_b of the poroelastic matrix to be a nonlinear function of one variable, which we specify below in Assumption (1). We will also consider the case when the permeability k_b is a given function of $\mathbf{x}$ and t.

By taking into account the elastic stress tensor (10.4), system (10.1) and (10.2) can be written as follows:

$$\begin{cases} \rho_b \boldsymbol{\eta}_{tt} - \mu_b \Delta \boldsymbol{\eta} - (\mu_b + \lambda_b)\nabla(\nabla \cdot \boldsymbol{\eta}) + \alpha_b \nabla p_b = \mathbf{F}_b, & \text{in } \Omega_b \times (0, T), \\ [c_b p_b + \alpha_b \nabla \cdot \boldsymbol{\eta}]_t - \nabla \cdot (k_b \nabla p_b) = S, & \text{in } \Omega_b \times (0, T). \end{cases} \tag{10.7}$$

Notice that system (10.7) is obtained under the assumption of small deformations and displacements.

The quantity $c_b p_b + \alpha_b \nabla \cdot \boldsymbol{\eta}$, appearing in the pressure equation above, is called the **fluid content**, and it will be denoted by ζ. More precisely, ζ can be considered as a constitutive law $\zeta = \zeta(p_b, \boldsymbol{\eta})$, defining the fluid content in terms of the fluid pore pressure, and the "compressibility of the solid matrix" $\nabla \cdot \boldsymbol{\eta}$, i.e., the change in the pore volume:

$$\zeta = c_b p_b + \alpha_b \nabla \cdot \boldsymbol{\eta}, \tag{10.8}$$

where $c_b \geq 0$ is the constrained storage coefficient and $\alpha_b > 0$ is the Biot-Willis coefficient [303]. Note that the case $c_b = 0$, corresponding to *incompressible constituents* in the poroelastic material, will be permitted below.

System (10.1), (10.2), and, therefore, system (10.7), can be obtained from the following system describing the balance of linear momentum for the fluid-solid mixture and conservation of mass for the fluid content ζ:

$$\begin{cases} \rho_b \boldsymbol{\eta}_{tt} - \nabla \cdot \boldsymbol{\sigma}_b = \mathbf{F}_b(\mathbf{x}, t), & \text{in } \Omega_b \times (0, T), \\ \zeta_t + \nabla \cdot \mathbf{u}_b = S(\mathbf{x}, t), & \text{in } \Omega_b \times (0, T), \end{cases} \tag{10.9}$$

where $\mathbf{u}_b$ is the discharge/filtration fluid velocity associated with the pore pressure p_b and $\boldsymbol{\sigma}_b$ is the poroelastic stress given in (10.3).

For completeness, we recall here the definitions of *porosity* ϕ and *fluid content* ζ given in terms of porosity. Porosity ϕ is defined as the volumetric fraction

$$\phi = \frac{V_f(\mathbf{x}, t)}{V(\mathbf{x}, t)} \tag{10.10}$$

of the volume $V_f(\mathbf{x}, t)$ occupied by the fluid within the poroelastic medium, over the representative elementary volume of the poroelastic medium $V(\mathbf{x}, t)$, centered at $\mathbf{x} \in \Omega_b$ at time t.

Similarly, if we denote by $V_s(\mathbf{x}, t)$ the volume occupied by the solid, then under the assumption of fully saturated mixture, the volumetric fraction

$$V_s(\mathbf{x}, t)/V(\mathbf{x}, t) = 1 - \phi(\mathbf{x}, t).$$

Fluid content ζ can then be defined as the increment in the volumetric fraction of the fluid component ϕ with respect to its baseline value ϕ_0:

$$\zeta(\mathbf{x}, t) = \phi(\mathbf{x}, t) - \phi_0(\mathbf{x}). \tag{10.11}$$

Notice that the constitutive law for the fluid content (10.8) implies that the porosity is given by

$$\phi = \phi(p_b, \boldsymbol{\eta}) = \phi_0 + c_b p_b + \alpha_b \nabla \cdot \boldsymbol{\eta},$$

where c_b is the constrained storage coefficient and α_b is the Biot-Willis coefficient.

As mentioned earlier, in this book, we will be considering two cases of permeability tensors $\boldsymbol{K}$: a linear case when $\boldsymbol{K} = \boldsymbol{K}(\boldsymbol{x}, t)$ is a function of $\boldsymbol{x}$ and t and a nonlinear case.

The *nonlinear* case is made more precise in the following assumption.

Assumption 1 When considering the case of nonlinear permeability, we assume that the permeability tensor $\boldsymbol{K}$ is a *nonlinear function* of the *porosity* ϕ:

$$\boldsymbol{K} = k_b \boldsymbol{I}, \text{ with } k_b = k_{\text{ref}} f_k(\phi), \tag{10.12}$$

where k_{ref} is a reference value for the permeability of the mixture. The particular form of the relationship between the permeability k_b and the porosity ϕ is represented by the function $f_k(\phi)$, and it depends on the geometrical architecture of the pores inside the matrix and the physical properties of the fluid [79, 186].

Thus, under the nonlinearity assumption, specified in Assumption 1, we have:

$$k_b = k_b(\phi) = k_b(\phi(p_b, \nabla \cdot \boldsymbol{\eta})) = k_b(c_b p_b + \alpha_b \nabla \cdot \boldsymbol{\eta}). \tag{10.13}$$

For analysis purposes, we will be assuming the following properties of the permeability function k_b:

Assumption 2 We assume that the permeability function $k_b : \mathbb{R} \to \mathbb{R}$ is continuous and that there exist constants $k_{\min} > 0$ and $k_{\max} > 0$ satisfying

$$0 < k_{\min} \le k_b(\zeta) \le k_{\max}, \quad \forall \zeta \in \mathbb{R}.$$

In the nonlinear considerations, we will denote by $k_b(\Psi(x))$ the Nemytskii operator associated with k_b. Recall that a Nemytskii operator is defined as follows.

Definition 10.1.1 Let X and Y be function spaces, and let $f : \mathbb{R} \to \mathbb{R}$ be a continuous function. The *Nemytskii operator* T_f associated with f acts on a function $u \in X$ as follows:

$$T_f(u)(x) = f(u(x)), \ \forall x \in \Omega$$

where Ω is the domain of the function u.

Namely, the Nemytskii operator, also known as the *superposition operator*, takes a function u and creates a new function by applying the scalar function f pointwise to u's values.

In our case, the assumptions on the function k_b, and the theory of superposition operators [289, 331], will yield that the operator k_b is bounded and continuous from $L^2(\Omega \times (0,T))$ into $L^2(\Omega \times (0,T))$. In our considerations of uniqueness for the nonlinear problem below, we will assume that k_b is a globally Lipschitz function on $\mathbb{R}$.

Remark 10.1.1 As in [42], we could also consider a poroviscoelastic material, for which the stress tensor $\boldsymbol{\sigma}_b$ would include a viscoelastic component such as:

$$\boldsymbol{\sigma}_b = \boldsymbol{\sigma}^E + \boldsymbol{\sigma}^V - \alpha_b p_b \mathbf{I}, \qquad \boldsymbol{\sigma}^V = 2\mu_V \boldsymbol{D}(\boldsymbol{\eta}_t) + \lambda_V \mathrm{tr} \boldsymbol{D}(\boldsymbol{\eta}_t) \boldsymbol{I}, \tag{10.14}$$

where λ_V and μ_V are independent Lamé parameters for viscoelasticity. All the results in this book would hold for the viscoporoelastic case as well.

We next give a description of the poroelastic plate model located at the interface between the free fluid flow of a viscous incompressible fluid modeled by the time-dependent Stokes equations and the poroelastic medium modeled by the Biot equations.

Poroelastic Plate Model in Ω_p

Domain Ω_p denotes a thin, poroelastic plate, separating the fluid flow region Ω_f, from the thick, poroelastic structure in Ω_b. A plate is a thin, 3D region, bounded by two surfaces of small curvature, whose distance defines the thickness of the plate. In our case, the plate will be assumed to be flat and of uniform thickness h, with the two surfaces located at $h/2$ distance on either side of the *middle surface* of the plate, denoted by ω_p, so that

$$\Omega_p = \omega_p \times (-h/2, h/2).$$

The reference configuration of the middle surface ω_p is $x_3 = 0$.

We will be assuming that the elastic plate is porous, isotropic, and saturated by a viscous fluid. To capture the elastic deformation of the poroelastic skeleton, as well as the fluid flow through the skeleton pores, we will employ a dynamic version of the quasi-static Biot poroelastic plate model studied by Biot in [41] and rigorously justified by Marciniak-Czochra and Mikelić in [250]. This model was obtained under the following two hypotheses:

1. The Kirchhoff hypothesis: Every straight line in the plate that was originally perpendicular to the plate's middle surface remains straight and perpendicular to the deflected middle surface after deformation.
2. The fluid velocity derivatives in the longitudinal and lateral directions are small compared to the transverse one.

Therefore, the dominant flow is in the transverse direction.

The resulting equations are given in terms of:

- The displacement $\boldsymbol{w} = \boldsymbol{w}(x_1, x_2)$ of the plate's middle surface from its reference configuration ω_p, where $(x_1, x_2) \in \omega_p$
- The fluid pore pressure p_p, where p_p is a function of three variables: two corresponding to the coordinates along the middle surface ω_p and the third one, denoted by $s \in (-h/2, h/2)$, corresponding to the *local transverse* coordinate, so that $p_p = p_p(x_1, x_2, s)$, where $(x_1, x_2, s) \in \Omega_p = \omega_p \times (-h/2, h/2)$.

According to [250], for small deflections and pressure gradients in the "transverse" (normal) s direction, the in-plane and transverse plate dynamics fully decouple. Since in our problem the stress in the transverse direction is larger than the stress in the longitudinal and lateral directions, we consider only the *transverse plate displacement*, which we denote by $w(x_1, x_2, t)$, where $(x_1, x_2) \in \omega_p$. For each s slice in $(-h/2, h/2)$, the w dynamics is constant. However, the fluid pore pressure $p_p(x_1, x_2, s)$ can vary for $s \in (-h/2, h/2)$. The resulting reduced Biot model for poroelastic plate is then given by two equations, one describing poroelastodynamics, which will be defined in ω_p, and one describing conservation of mass of the fluid phase, given in terms of the evolution of p_p, defined in $\Omega_p = \omega_p \times (-h/2, h/2)$ [41, 101, 142, 250]:

$$\begin{cases} \rho_p w_{tt} + D\Delta^2_{\omega_p} w + \gamma w + \alpha_p \Delta_{\omega_p} \displaystyle\int_{-h/2}^{h/2} s p_p \, ds = F_p(x_1, x_2, t) \quad \text{in} \quad \omega_p \times (0, T), \\ \partial_t [c_p p_p - \alpha_p s \Delta_{\omega_p} w] - \partial_s (k_p \partial_s p_p) = 0 \quad \text{in} \quad \Omega_p \times (0, T). \end{cases} \tag{10.15}$$

All the Laplacians above refer to the in-plane Laplacian, i.e., $\Delta_{\omega_p} = \partial^2_{x_1} + \partial^2_{x_2}$. The constant $D > 0$ is the elastic stiffness coefficient for the plate, and $\gamma > 0$ is an [elastic] coefficient that we added for technical reasons to provide coercivity, since coercivity is not automatically satisfied in the periodic framework. The constants ρ_p, α_p, c_p and k_p are the same as those in (10.6), taken for the poroelastic plate. One distinction for the poroelastic plate is that our results below will require $c_p > 0$.

The source term F_p corresponds to the loading of the poroelastic plate, which will come from the jump in the normal components of the normal stress (traction) between the fluid on one side and at the thick Biot poroelastic structure on the other: $F_p(x_1, x_2, t) = \boldsymbol{\sigma}_b \mathbf{e}_3 \cdot \mathbf{e}_3\big|_{\omega_p} - \boldsymbol{\sigma}_f \mathbf{e}_3 \cdot \mathbf{e}_3|_{\omega_p}$. This will be specified below in the coupling conditions (10.23). As before, Darcy's law, relating the filtration velocity to the pressure gradient, holds here:

$$u_p = -k_p \partial_s p_p, \quad \text{in} \quad \Omega_p \times (0, T), \tag{10.16}$$

where u_p denotes the transverse (dominant) component of filtration velocity, relative to the motion of the poroelastic matrix. Notice that the gradient of pressure from Darcy's law (10.5) is now given in terms of the derivative ∂_s with respect to the transverse (normal) direction to the plate.

Permeability $k_p > 0$ is considered to be a constant.

Time-Dependent Stokes Model in Ω_f

In the lower half space, denoted by Ω_f, we consider the flow of a viscous, incompressible fluid modeled by the time-dependent Stokes equations :

$$\left.\begin{aligned} &\rho_f \frac{\partial \mathbf{u}}{\partial t} = \nabla \cdot \boldsymbol{\sigma}_f + \boldsymbol{f} \\ &\nabla \cdot \mathbf{u} = 0 \end{aligned}\right\} \text{ in } \Omega_f \times (0, T), \tag{10.17}$$

where $\mathbf{u}$ is the fluid velocity, $\boldsymbol{\sigma}_f = \boldsymbol{\sigma}_f(\mathbf{u}, p_f)$ is the Cauchy stress tensor, p_f is the fluid pressure, ρ_f is the fluid density, and μ_f is the fluid viscosity with $\boldsymbol{D}(\mathbf{u})$, denoting the fluid strain-rate tensor (symmetrized gradient of velocity). In this book, we will be assuming that the fluid is Newtonian, which specifies the following constitutive law for the Cauchy stress:

$$\boldsymbol{\sigma}_f = 2\mu_f \boldsymbol{D}(\mathbf{u}) - p_f \boldsymbol{I}.$$

The fluid interacts with the multilayered poroelastic structure by exerting stress onto the poroelastic structure that causes deformation of the poroelastic skeleton while at the same time generating the pressure difference, or more generally, the normal stress difference across the poroelastic plate, causing filtration flow through the poroelastic plate and on to the Biot poroelastic medium. The poroelastic plate serves as a fluid-structure interface with mass, which will provide a regularizing mechanism in the analysis of this fluid-structure interaction problem.

Summary of the Three Model Equations

The following are the three models holding in the three different subdomains of Ω:

$$\begin{cases} \rho_b \boldsymbol{\eta}_{tt} - \mu_b \Delta \boldsymbol{\eta} - (\mu_b + \lambda_b)\nabla(\nabla \cdot \boldsymbol{\eta}) + \alpha_b \nabla p_b = \mathbf{F}_b, & (x_1, x_2, x_3) \in \Omega_b, \\ [c_b p_b + \alpha_b \nabla \cdot \boldsymbol{\eta}]_t - \nabla \cdot (k_b \nabla p_b) = S, & (x_1, x_2, x_3) \in \Omega_b, \end{cases} \tag{10.18}$$

$$\begin{cases} \rho_p w_{tt} + D\Delta^2_{\omega_p} w + \gamma w + \alpha_p \Delta_{\omega_p} \displaystyle\int_{-h/2}^{h/2} s p_p \, ds = F_p, & (x_1, x_2) \in \omega_p, \\ [c_p p_p - \alpha_p s \Delta_{\omega_p} w]_t - \partial_s (k_p \partial_s p_p) = 0, & (x_1, x_2, s) \in \Omega_p, \end{cases} \tag{10.19}$$

$$\begin{cases} \rho_f \mathbf{u}_t - \mu_f \Delta \boldsymbol{u} + \nabla p_f = \mathbf{f}, & (x_1, x_2, x_3) \in \Omega_f, \\ \nabla \cdot \boldsymbol{u} = 0, & (x_1, x_2, x_3) \in \Omega_f. \end{cases} \tag{10.20}$$

Remark 10.1.2 We use the notation k_b here (and below) to denote both the linear and nonlinear case of permeability. It will be clear from the context which of the two cases we are working with, as we shall see below.

Coupling Conditions

The coupling between the fluid and multilayered poroelastic structure is assumed at the poroelastic plate, which serves as a fluid-structure interface with mass and poroelastic energy. Similar to the coupling conditions in the fluid-structure interaction problems involving elastic structures, discussed in Chaps. 4 and 5, we will impose two sets of coupling conditions: the kinematic coupling conditions describing the relationship between the kinematic quantities such as velocities, and the dynamic coupling conditions describing balance of forces. The concrete coupling conditions specified below are motivated by the physics of the problem and by the weak formulation of the problem, which, for the conditions specified below, will give rise to the total energy of the problem, which will be bounded only by the data given in the problem.

Describing coupling conditions for the fluid flow quantities such as the fluid velocity and pressure requires special attention. This is because the plate fluid flow occurs in the vertical/transverse/normal direction, and so we need to consider the "inflated" plate domain Ω_p, where the filtration velocity and pore pressure are defined. The inflated domain Ω_p has two interfaces: the top interface across which the plate flow communicates with the Biot medium flow, denoted by ω_p^+, and the bottom interface across which the plate flow communicates with free fluid flow, denoted by ω_p^-. Thus, to describe the pressure coupling across the plate interface, the plate's pore fluid pressure p_p (and Darcy velocity u_p) will be evaluated at the two "inflated" surfaces:

$$\omega_p^+ = \omega_p \times \{h/2\}, \quad \text{and} \quad \omega_p^- = \omega_p \times \{-h/2\}. \tag{10.21}$$

The pressure p_p and the Darcy velocity u_p will be coupled to the corresponding quantities coming from the Biot model defined on Ω_b on one side and Stokes fluid defined on Ω_F on the other. Since Ω_b and Ω_f are defined all the way up to ω_p, the traces of the pressure and fluid velocities coming from these two domains will be evaluated at ω_p in the coupling conditions. More precisely, we have the following coupling conditions between the three models summarized in (10.18)–(10.20).

1. *Coupling determining elastodynamics of the plate's middle surface* ω_p: The coupling across the middle surface of the plate captures the main features of the coupling determining the plate's elastodynamics. As mentioned above, we have two sets of coupling conditions, the kinematic and dynamic. For the kinematic coupling condition, we assume continuity of the plate and Biot displacement velocities, which implies continuity of the plate and Biot displacements evaluated at the interface ω_p. For the dynamic coupling, we state that the elastodynamics of the poroelastic plate is driven by the jump in the normal components of normal stress between the fluid and the Biot poroelastic medium, where the jump is evaluated at the two-dimensional fluid-structure interface ω_p:

$$\langle 0, 0, w\rangle = \boldsymbol{\eta}\big|_{\omega_p}, \tag{10.22}$$

$$F_p = \boldsymbol{\sigma}_b \mathbf{e}_3 \cdot \mathbf{e}_3\big|_{\omega_p} - \boldsymbol{\sigma}_f \mathbf{e}_3 \cdot \mathbf{e}_3|_{\omega_p}. \tag{10.23}$$

   Here F_p is the forcing term on the right-hand side of the plate equation (10.19). Notice that we have accounted only for the normal component of the normal stress here since only the normal component of the plate displacement is captured by the plate equation (10.19).
2. *The Biot-plate coupling of the fluid quantities:* This coupling captures the fluid flow coupling between the Biot model and the plate model. For this purpose, we evaluate the plate quantities at the inflated surface ω_p^+, defined in (10.21), and the Biot quantities at the interface portion of the boundary of the Biot domain Ω_b, namely, on ω_p. In particular, we assume continuity of filtration velocities, which are defined relative to the motion of the poroelastic matrix, $\mathbf{u}_b \cdot \mathbf{e}_3 = u_p$ (kinematic coupling), and continuity of pore pressures (dynamic coupling). Using Darcy's laws (10.5) and (10.16) and the fact that the "inflation" variable s is aligned with $\mathbf{e}_3$, the two coupling conditions read:

$$k_p \partial_s p_p\big|_{\omega_p^+} = k_b \partial_s p_b\big|_{\omega_p} \tag{10.24}$$

$$p_p\big|_{\omega_p^+} = p_b\big|_{\omega_p}. \tag{10.25}$$

3. *The Stokes-plate coupling of the fluid quantities:* Similarly to the coupling presented above, here we specify the coupling of the fluid quantities defined on the "inflated" plate domain Ω_p with the Stokes flow. Again, we evaluate the plate quantities at the inflated surface ω_p^- adjacent to the fluid domain Ω_f and the Stokes quantities at the interface boundary of the fluid domain Ω_f, namely, on ω_p. In particular, the kinematic coupling conditions here are (1) continuity of the *normal* components of the fluid velocities relative to the interface velocity w_t, $w_t - (\boldsymbol{u} \cdot \mathbf{e}_3)\big|_{\omega_p} = u_p$, and (2) slip in the *tangential* component of the fluid velocity at ω_p coming from the Stokes flow, known as the Beavers-Joseph-

Saffman condition. The dynamic coupling condition describes continuity of the normal components of the normal fluid stress. Using Darcy's law (10.16), the three coupling conditions read:

$$\begin{cases} w_t - (\boldsymbol{u} \cdot \mathbf{e}_3)\big|_{\omega_p} = k_p \partial_s p_p\big|_{\omega_p^-}, \\ \qquad \beta \mathbf{u}\big|_{\omega_p} \cdot \mathbf{e}_i = -[\boldsymbol{\sigma}_f\big|_{\omega_p} \mathbf{e}_3] \cdot \mathbf{e}_i, \quad i = 1, 2, \end{cases} \tag{10.26}$$

$$-\boldsymbol{\sigma}_f \mathbf{e}_3 \cdot \mathbf{e}_3\big|_{\omega_p} = p_p\big|_{\omega_p^-}. \tag{10.27}$$

Remark 10.1.3 By combining the coupling conditions for the fluid pressure given in (10.25) and (10.27), we see that across the FPSI interface, the pressure jumps from $-\boldsymbol{\sigma}_f \mathbf{e}_3 \cdot \mathbf{e}_3\big|_{\omega_p}$ on the fluid side to $p_b\big|_{\omega_p}$ on the Biot side. In between, a pressure gradient-driven vertical/transverse flow through the poroelastic plate takes place from $p_p\big|_{\omega_p^-}$ to $p_p\big|_{\omega_p^+}$. Thus, the jump in the pressure at the interface ω_p given by $[p_b - \boldsymbol{\sigma}_f \mathbf{e}_3 \cdot \mathbf{e}_3]$ on ω_p drives the filtration flow through the interface, i.e., through the poroelastic plate.

Boundary and Initial Conditions

Boundary Conditions on $\partial\Omega$ We impose periodic boundary conditions for all the quantities of interest in the longitudinal and lateral directions and identify $x_i = 0$ with $x_i = 1$, for $i = 1, 2$.

For $x_3 = -1$, we consider the standard *no-slip* boundary conditions for the Stokes fluid, i.e., $\mathbf{u} \equiv \mathbf{0}\big|_{x_3=-1}$. For $x_3 = 1$, we take Neumann-type (homogeneous) boundary conditions for the Biot system.

We summarize the boundary conditions in the following table:

Region	Functions	Boundary Condition	
			(10.28)
$x_i = 0,\ x_i = 1, i = 1, 2$	$\boldsymbol{\eta}, p_b, w, p_p, \mathbf{u}$	periodic	(10.29)
$x_3 = -1$	$\mathbf{u}$	$\mathbf{u}\big|_{x_3=-1} = \mathbf{0}$	(10.30)
$x_3 = 1$	$\boldsymbol{\eta},\ p_b$	$\boldsymbol{\sigma}(\boldsymbol{\eta}, p_b)\mathbf{e}_3 = \mathbf{0},\ \partial_{x_3} p_b = 0$	(10.31)

Initial Conditions at $t = 0$

$$\text{Fluid}: \ \boldsymbol{u}(.,0) = \boldsymbol{u}^0, \tag{10.32}$$

$$\text{Poroelastic plate}: \; w(.,0) = w^0, \;\; \partial_t w(.,0) = \dot{w}^0, \;\; p_p(.,0) = p_p^0, \tag{10.33}$$

$$\text{Biot equations}: \; \boldsymbol{\eta}(.,0) = \boldsymbol{\eta}^0, \;\; \partial_t \boldsymbol{\eta}(,.0) = \dot{\boldsymbol{\eta}}^0, \;\; p_b(.,0) = p_b^0. \tag{10.34}$$

Here, we used $\dot{w}^0$ and $\dot{\boldsymbol{\eta}}^0$ to denote the initial data for the time derivative of w and $\boldsymbol{\eta}$, respectively.

10.1.3 Quasi-static Problem

The *quasi-static* version of problem (10.18)–(10.34) refers to the problem in which the inertial effects in the Biot model (10.7) are neglected, as described by the following assumption.

Assumption 3 The quasi-static FPSI coupled model refers to the coupled problem (10.18)–(10.34) with $\rho_b = 0$ in (10.18) and $\rho_p \geq 0$ in (10.19).

This quasi-static assumption arises naturally in the classical Biot model of consolidation for a linearly elastic and a porous solid, which is saturated by a slightly compressible viscous fluid. See, e.g., [303]. The quasi-static assumption requires modifying the initial conditions but leaves the coupling conditions the same. As we shall see in Sect. 10.5, we can extend our linear existence result to the nonlinear case *when* Assumption 3 is taken. On the other hand, it has been noted by many authors that the elliptic-parabolic nature of the quasi-static Biot problem presents its own challenges [44, 303].

The new *initial conditions*, owing to loss of acceleration terms, are given by:

- If $\rho_b = 0$, then the initial conditions (10.34) are replaced with the initial condition for the fluid content:

$$\zeta_b(.,0) = (c_b p_b + \alpha_b \nabla \cdot \boldsymbol{\eta})(.,0) = \zeta_b^0. \tag{10.35}$$

- If $\rho_p = 0$, then the initial conditions (10.33) are replaced with the initial condition for the poroelastic plate fluid content:

$$\zeta_p(.,0) = (c_p p_p - \alpha_p s \Delta_{\omega_p} w)(.,0) = \zeta_p^0. \tag{10.36}$$

In the analysis of such quasi-static Biot models (see [42, 44, 74, 278, 303, 345]), the structure of the initial condition is a non-trivial issue. In this book, we only consider ζ_b^0 for which there exist $\boldsymbol{\eta}^0 = \boldsymbol{\eta}(0)$ and $p_b^0 = p_b(0)$ satisfying $c_b p_b^0 + \alpha_b \nabla \cdot \boldsymbol{\eta}^0 = \zeta_b^0$. Such a hypothesis is common; see, e.g., [42, 74, 345].

Remark 10.1.4 For a very weak notion of solution in the linear case, the results in [302, 303] only require initial conditions as specified above, which is, to say, for the respective fluid contents alone. However, in nearly every other construction, additional regularity of the initial condition is required. This is a by-product of using a priori estimates to construct solutions. See [345, Remark 2] and [303, p.327] where this is discussed in depth.

Details of the existence result are given in Sect. 10.5.

10.2 Energy, Function Spaces, and Weak Formulation

In this section, we derive a formal energy estimate, present functions spaces for the solution, and define a weak solution. For this purpose, we recall the principal solution variables in each sub-region:

- $(\boldsymbol{u}, p_f)$—stokes fluid velocity and pressure defined on Ω_f
- (w, w_t, p_p)—transverse displacement and velocity of the poroelastic plate and poroelastic plate pressure, defined on ω_p and Ω_p, respectively
- $(\boldsymbol{\eta}, \boldsymbol{\eta}_t, p_b)$—poroelastic displacement, velocity, and pressure defined on Ω_b (thick Biot region). The quantity $\boldsymbol{u}_b = -\boldsymbol{K}\nabla p_b = -k_b(c_b p_b + \alpha_b \nabla \cdot \boldsymbol{\eta})\nabla p_b$ is the discharge velocity in thick poroelastic material.

10.2.1 A Formal Energy Inequality

By testing Equations (10.18)–(10.20) by the solution variables, specified above, and by using coupling conditions (10.22)–(10.27), and periodic boundary conditions, we obtain the following formal energy identity:

$$
\begin{aligned}
&\frac{1}{2}\frac{d}{dt}\left(\rho_f \|\boldsymbol{u}\|^2_{L^2(\Omega_f)} + \rho_p \|w_t\|^2_{L^2(\omega_p)} + \rho_b \|\partial_t \boldsymbol{\eta}\|^2_{L^2(\Omega_b)}\right) \\
&+ \frac{1}{2}\frac{d}{dt}\left(c_p \|p_p\|^2_{L^2(\Omega_p)} + \|\boldsymbol{\eta}\|^2_E + c_b \|p_b\|^2_{L^2(\Omega_b)} + D\|\Delta w\|^2_{L^2(\omega_p)} + \gamma \|w\|^2_{L^2(\omega_p)}\right) \\
&+ 2\mu_f \|\boldsymbol{D}(\boldsymbol{u})\|^2_{L^2(\Omega_f)} + \beta \|\boldsymbol{u}\cdot \boldsymbol{t}\|^2_{L^2(\omega_p)} \\
&+ \|k_p^{1/2} \partial_s p_p\|^2_{L^2(\Omega_p)} + \|k_b^{1/2} \nabla p_b\|^2_{L^2(\Omega_b)}\| \le C_{data},
\end{aligned}
\tag{10.37}
$$

where

$$
\|\boldsymbol{\eta}\|^2_E = \int_{\Omega_b} [\boldsymbol{\sigma}^E : \nabla \boldsymbol{\eta}] d\mathbf{x} = 2\mu_E(\boldsymbol{D}(\boldsymbol{\eta}), \boldsymbol{D}(\boldsymbol{\eta}))_{L^2(\Omega_b)} + \lambda_E(\nabla\cdot\boldsymbol{\eta}, \nabla\cdot\boldsymbol{\eta})_{L^2(\Omega_b)},
\tag{10.38}
$$

and

$$\|\boldsymbol{u} \cdot \boldsymbol{t}\|^2_{L^2(\omega_p)} = \sum_{i=1,2} \|\boldsymbol{u} \cdot \boldsymbol{t}_i\|^2,$$

where $\boldsymbol{t}_i$ are the tangent vectors, given by $\mathbf{e}_1$, $\mathbf{e}_2$ on ω_p in this case. Note that the formal energy identity for the quasi-static dynamics can be obtained by letting $\rho_b = 0$, and this is valid for $\rho_p = c_b = 0$ as well. Our results will include all these cases.

Remark 10.2.1 We shall see later that this energy inequality will hold for constructed solutions but need not hold in the general case for weak solutions, even in the linear dynamics.

10.2.2 Function Spaces for Weak Solutions

For any function space X, we define $X_\#$ as the space of all functions from space X that are 1-periodic in directions x_1 and x_2:

$$X_\# := \{f \in X : f\big|_{x_1=0} = f\big|_{x_1=1},\ f\big|_{x_2=0} = f\big|_{x_2=1}\}.$$

We now define the underlying spaces from which we will build our weak solution space. We start with standard spaces for the Stokes equation:

$$\begin{aligned} H &= \{\boldsymbol{u} \in \mathbf{L}^2(\Omega_f) : \nabla \cdot \boldsymbol{u} = 0,\ (\boldsymbol{u} \cdot \mathbf{e}_3)\big|_{x_3=-1} = 0\}, \\ V &= \{\boldsymbol{u} \in \mathbf{H}^1(\Omega_f) : \nabla \cdot \boldsymbol{u} = 0,\ \boldsymbol{u}\big|_{x_3=-1} = 0\}, \\ \mathcal{V}_f &= L^\infty(0,T; H_\#) \cap L^2(0,T; V_\#). \end{aligned}$$

The function spaces for the poroelastic plate are:

$$\begin{aligned} H^{0,0,1}_\# &= \{p \in L^2_\#(\Omega_p) : \partial_s p \in L^2(\Omega_p)\} \\ \mathcal{V}_p &= \{w \in L^\infty(0,T; H^2_\#(\omega_p)) : \rho_p w \in W^{1,\infty}(0,T; L^2_\#(\omega_p))\}, \\ \mathcal{Q}_p &= \big\{p \in L^2(0,T; H^{0,0,1}_\#) : c_p p \in L^\infty(0,T; L^2_\#(\Omega_p))\big\}. \end{aligned}$$

The functions spaces for the Biot dynamics are:

$$\begin{aligned} \mathcal{V}_b &= \{\boldsymbol{\eta} \in L^\infty(0,T; \mathbf{H}^1_\#(\Omega_b)) : \rho_b \boldsymbol{\eta} \in W^{1,\infty}(0,T; \mathbf{L}^2_\#(\Omega_b))\}, \\ \mathcal{Q}_b &= \{p \in L^2(0,T; H^1_\#(\Omega_b)) : c_b p \in L^\infty(0,T; L^2_\#(\Omega_b))\}. \end{aligned}$$

Using the spaces introduced above, we define the (weak) *solution space* by the following:

$$\mathcal{V}_{\mathrm{sol}} = \big\{(\boldsymbol{u}, w, p_p, \boldsymbol{\eta}, p_b) \in \mathcal{V}_f \times \mathcal{V}_p \times \mathcal{Q}_p \times \mathcal{V}_b \times \mathcal{Q}_b \;:\; p_p\big|_{s=h/2} = p_b\big|_{x_3=0},\ \boldsymbol{\eta}\big|_{x_3=0} = w\mathbf{e}_3\big\}. \tag{10.39}$$

Remark 10.2.2 We included the "inertial" and "compressibility" constants in the definition of the solution spaces in order to unify the exposition and, at the same time, point to the different regularity of solution for the different scenarios.

In conjunction with $\mathcal{V}_{\mathrm{sol}}$, we define the *test space* as follows:

$$\mathcal{V}_{\mathrm{test}} = \big\{(\boldsymbol{v}, z, q_p, \boldsymbol{\psi}, q_b) \in C_c^1\big([0,T); V_\# \times H^2_\#(\omega_p) \times H^{0,0,1}_\# \times \mathbf{H}^1_\#(\Omega_b) \times H^1_\#(\Omega_b)\big) : q_p\big|_{s=h/2} = q_b\big|_{x_3=0},\ \boldsymbol{\psi}\big|_{x_3=0} = z\mathbf{e}_3\big\}. \tag{10.40}$$

10.2.3 General Weak Formulation

We aim to construct a function $(\boldsymbol{u}, w, p_p, \boldsymbol{\eta}, p_b) \in \mathcal{V}_{\mathrm{sol}}$, call a *weak solution*, such that for every smooth test function $(\boldsymbol{v}, z, q_p, \boldsymbol{\psi}, q_b) \in \mathcal{V}_{\mathrm{test}}$, the *weak formulation* specified below in Definition 10.2.1 holds. To formulate the weak formulation, we introduce the following notation:

- $((\cdot,\cdot))_{\mathcal{O}}$ denotes an inner product on $L^2(0,T,L^2(\mathcal{O}))$, for a domain $\mathcal{O}$.
- $\langle\cdot,\cdot\rangle$ denotes a particular "duality" pairing whose precise interpretation will be specified below.

To obtain the weak formulation for the coupled system, we proceed by formally multiplying the equations in (10.18)–(10.20) by test functions and integrating by parts as described below.

We begin by using the test function $\boldsymbol{\psi} \in C_c^1\big([0,T); \mathbf{H}^1_\#(\Omega_b)\big)$, with $\boldsymbol{\psi}\big|_{x_3=0} = z\mathbf{e}_3$ for the balance of linear momentum equation in the Biot model (10.18), to obtain

$$-\rho_b((\boldsymbol{\eta}_t, \boldsymbol{\psi}_t))_{\Omega_b} + ((\boldsymbol{\sigma}_b(\boldsymbol{\eta}, p_b), \nabla\boldsymbol{\psi}))_{\Omega_b} - \int_0^T \int_{\partial\Omega_b} \boldsymbol{\sigma}_b(\boldsymbol{\eta}, p_b)\mathbf{n}\cdot\boldsymbol{\psi} = \rho_b(\boldsymbol{\eta}_t, \boldsymbol{\psi})\big|_{t=0} + ((\mathbf{F}_b, \boldsymbol{\psi}))_{\Omega_b}. \tag{10.41}$$

For the balance of mass equation, we use the test function $q_b \in C_c^1\big([0,T); H^1_\#(\Omega_b)\big)$ and obtain

$$-((c_b p_b + \alpha_b \nabla \cdot \boldsymbol{\eta}, \partial_t q_b))_{\Omega_b} + ((k_b \nabla p_b, \nabla q_b))_{\Omega_b} - \int_0^T \int_{\partial\Omega_b} k_b \nabla p_b \cdot \mathbf{n}\, q_b$$
$$= (c_b p_b + \alpha_b \nabla \cdot \boldsymbol{\eta}, q_b)\big|_{t=0} + ((S, q_b))_{\Omega_b}. \quad (10.42)$$

For the equation describing the poroelastic plate dynamics (10.19), we use test function $z \in C_c^1\big([0,T); H^2_\#(\omega_p)\big)$ to obtain

$$-\rho_p((w_t, \partial_t z))_{\omega_p} + D((\Delta w, \Delta z))_{\omega_p} + \alpha_p\Big(\Big(\int_{-h/2}^{h/2} (s p_p) ds, \Delta z\Big)\Big)_{\omega_p} + \gamma((w, z))_{\omega_p}$$
$$= \rho_p(w_t, z)\big|_{t=0} + \int_0^T \int_{\omega_p} F_p z. \quad (10.43)$$

In the conservation of mass of the fluid phase for the poroelastic plate, we use the test function $q_p \in C_c^1\big([0,T); H_\#^{0,0,1}\big)$ to obtain

$$-((c_p p_p - \alpha_p s \Delta w, \partial_t q_p))_{\Omega_p} + ((k_p \partial_s p_p, \partial_s q_p))_{\Omega_p} - \int_0^T \int_{\partial\Omega_p} k_p \partial_s p_p n_3 q_p \quad (10.44)$$
$$= (c_p p_p - \alpha_p s \Delta w, q_p)_{\Omega_p}\big|_{t=0}.$$

Finally, we use test function $\boldsymbol{v} \in C_c^1\big([0,T); V_\#\big)$ in the Stokes equation (10.20), which provides

$$-\rho_f((\boldsymbol{u}, \boldsymbol{v}_t))_{\Omega_f} + \mu_f((\boldsymbol{D}(\boldsymbol{u}), \boldsymbol{D}(\boldsymbol{v})))_{\Omega_f} + \int_0^T \int_{\partial\Omega_f} \boldsymbol{\sigma}_f \mathbf{n} \cdot \boldsymbol{v}$$
$$= \rho_f(\boldsymbol{u}, \boldsymbol{v})\big|_{t=0} + ((\mathbf{f}, \boldsymbol{v}))_{\Omega_f}. \quad (10.45)$$

Note that when we combine (10.41)–(10.45), the collection of boundary terms is represented by

$$I_{\text{bdry}} = -\int_0^T \int_{\partial\Omega_b} \boldsymbol{\sigma}_b(\boldsymbol{\eta}, p_b)\mathbf{n} \cdot \boldsymbol{\psi} - \int_0^T \int_{\omega_p} \boldsymbol{\sigma}_b \mathbf{e}_3 \cdot \mathbf{e}_3\big|_{\omega_p} z \quad (10.46)$$
$$-\int_0^T \int_{\partial\Omega_b} k_b \nabla p_b \cdot \mathbf{n}\, q_b - \int_0^T \int_{\partial\Omega_p} k_p \partial_s p_p n_3 q_p$$
$$+\int_0^T \int_{\omega_p} \boldsymbol{\sigma}_f \mathbf{e}_3 \cdot \mathbf{e}_3\big|_{\omega_p} z + \int_0^T \int_{\partial\Omega_f} \boldsymbol{\sigma}_f \mathbf{n} \cdot \boldsymbol{v}.$$

After using the boundary condition $\boldsymbol{\sigma}(\boldsymbol{\eta}, p_b)\mathbf{e}_3 = \mathbf{0}$ and the fact that $\boldsymbol{\psi}\big|_{x_3=0} = z\mathbf{e}_3$, the first term in (10.46) becomes

$$-\int_0^T\int_{\partial\Omega_b}\boldsymbol{\sigma}_b(\boldsymbol{\eta},p_b)\mathbf{n}\cdot\boldsymbol{\psi}=-\int_{x_3=1}\boldsymbol{\sigma}_b\mathbf{e}_3\cdot\boldsymbol{\psi}+\int_{\omega_p}\boldsymbol{\sigma}_b\mathbf{e}_3\cdot\boldsymbol{\psi}=\int_0^T\int_{\omega_p}\boldsymbol{\sigma}_b\mathbf{e}_3\cdot\mathbf{e}_3\big|_{\omega_p}z, \tag{10.47}$$

and therefore the first two terms cancel each other. The next two terms in (10.46) can be rewritten as follows:

$$\begin{aligned}&-\int_0^T\int_{\partial\Omega_b}k_b\nabla p_b\cdot\mathbf{n}\,q_b-\int_0^T\int_{\partial\Omega_p}k_p\partial_s p_p n_3 q_p \qquad (10.48)\\ &=-\int_{x_3=1}k_b\partial_{x_3}p_b q_b+\int_{\omega_p}k_b\partial_{x_3}p_b q_b+\int_{s=-h/2}k_p\partial_s p_p q_p-\int_{s=h/2}k_p\partial_s p_p q_p.\end{aligned}$$

By using the boundary condition $\partial_{x_3}p_b=0$ at $x_3=1$, the coupling across the Biot-plate interface $k_p\partial_s p_p\big|_{\omega_p^+}=k_b\partial_s p_b\big|_{\omega_p}$, and the coupling across the Stokes-plate interface $w_t-(\boldsymbol{u}\cdot\mathbf{e}_3)\big|_{\omega_p}=k_p\partial_s p_p\big|_{\omega_p^-}$, Eq. (10.48) simplifies to

$$-\int_0^T\int_{\partial\Omega_b}k_b\nabla p_b\cdot\mathbf{n}\,q_b-\int_0^T\int_{\partial\Omega_p}k_p\partial_s p_p n_3 q_p=\int_{s=-h/2}(w_t-(\boldsymbol{u}\cdot\mathbf{e}_3)\big|_{\omega_p})q_p. \tag{10.49}$$

Using the second coupling at the Stokes-plate interface $-\boldsymbol{\sigma}_f\mathbf{e}_3\cdot\mathbf{e}_3\big|_{\omega_p}=p_p\big|_{\omega_p^-}$, we get

$$\int_0^T\int_{\omega_p}\boldsymbol{\sigma}_f\mathbf{e}_3\cdot\mathbf{e}_3\big|_{\omega_p}z=-\int_0^T\int_{\omega_p}p_p\big|_{\omega_p^-}z. \tag{10.50}$$

Lastly, using again the coupling at the Stokes-plate interface $-\boldsymbol{\sigma}_f\mathbf{e}_3\cdot\mathbf{e}_3\big|_{\omega_p}=p_p\big|_{\omega_p^-}$, the boundary condition $\boldsymbol{v}=0$ on $\{x_3=-1\}$, and the coupling across the plate's middle surface ω_p, $\beta\mathbf{u}\big|_{\omega_p}\cdot\mathbf{e}_i=-[\boldsymbol{\sigma}_f\big|_{\omega_p}\mathbf{e}_3]\cdot\mathbf{e}_i$, $i=1,2$, we can simplify the boundary term associated with the fluid as follows:

$$\begin{aligned}\int_0^T\int_{\partial\Omega_f}\boldsymbol{\sigma}_f\mathbf{n}\cdot\boldsymbol{v}&=\int_{x_3=-1}\boldsymbol{\sigma}_f\mathbf{e}_3\cdot\boldsymbol{v}-\int_{\omega_p}\boldsymbol{\sigma}_f\mathbf{e}_3\cdot\boldsymbol{v}\\ &=-\int_{\omega_p}[\boldsymbol{\sigma}_f\mathbf{e}_3\cdot\boldsymbol{e}_1(\boldsymbol{v}\cdot\boldsymbol{e}_1)+\boldsymbol{\sigma}_f\mathbf{e}_3\cdot\boldsymbol{e}_2(\boldsymbol{v}\cdot\boldsymbol{e}_2)\\ &\quad+\boldsymbol{\sigma}_f\mathbf{e}_3\cdot\mathbf{e}_3(\boldsymbol{v}\cdot\mathbf{e}_3)]\\ &=\int_{\omega_p}\beta(\boldsymbol{u}\cdot\boldsymbol{t})(\boldsymbol{v}\cdot\boldsymbol{t})+\int_{\omega_p}p_p\big|_{\omega_p^-}(\boldsymbol{v}\cdot\mathbf{e}_3).\end{aligned} \tag{10.51}$$

Combining (10.46) with (10.47), (10.49), (10.50), and (10.51), we obtain that the sum of the boundary terms in the variational form simplifies to

$$I_{\text{bdry}} = -((p_p\big|_{s=-h/2}, z - \boldsymbol{v}\cdot\mathbf{e}_3))_{\omega_p} + ((\langle w_t - \boldsymbol{u}\cdot\mathbf{e}_3, q_p\big|_{s=-h/2}\rangle))_{\omega_p} + \beta((\boldsymbol{u}\cdot\boldsymbol{t}, \boldsymbol{v}\cdot\boldsymbol{t}))_{\omega_p}.$$

We can now introduce the definition of weak solution for the fully coupled system. The following weak form applies to all cases of admissible values of ρ_b, ρ_p, c_b, c_p. Namely, the solution and test spaces are built to accommodate subsets of these parameters vanishing.

Definition 10.2.1 We say that $(\boldsymbol{u}, w, p_p, \boldsymbol{\eta}, p_b) \in \mathcal{V}_{\text{sol}}$ is a *weak solution* to (10.18)–(10.34), if for every test function $(\boldsymbol{v}, z, q_p, \boldsymbol{\psi}, q_b) \in \mathcal{V}_{\text{test}}$ the following identity holds:

$$\begin{aligned}
&- \rho_b((\boldsymbol{\eta}_t, \boldsymbol{\psi}_t))_{\Omega_b} + ((\boldsymbol{\sigma}_b(\boldsymbol{\eta}, p_b), \nabla\boldsymbol{\psi}))_{\Omega_b} - ((c_b p_b + \alpha_b \nabla\cdot\boldsymbol{\eta}, \partial_t q_b))_{\Omega_b} \\
&+ ((k_b \nabla p_b, \nabla q_b))_{\Omega_b} \\
&- ((c_p p_p - \alpha_p s \Delta w, \partial_t q_p))_{\Omega_p} + ((k_p \partial_s p_p, \partial_s q_p))_{\Omega_p} \\
&- \rho_p((w_t, \partial_t z))_{\omega_p} + D((\Delta w, \Delta z))_{\omega_p} + \alpha_p((\int_{-h/2}^{h/2}[s p_p] ds, \Delta z))_{\omega_p} + \gamma((w, z))_{\omega_p} \\
&- ((p_p\big|_{s=-h/2}, z - \boldsymbol{v}\cdot\mathbf{e}_3))_{\omega_p} + ((\langle w_t - \boldsymbol{u}\cdot\mathbf{e}_3, q_p\big|_{s=-h/2}\rangle))_{\omega_p}
\end{aligned} \tag{10.52}$$

$$\begin{aligned}
&- \rho_f((\boldsymbol{u}, \boldsymbol{v}_t))_{\Omega_f} + \mu_f((\boldsymbol{D}(\boldsymbol{u}), \boldsymbol{D}(\boldsymbol{v})))_{\Omega_f} + \beta((\boldsymbol{u}\cdot\boldsymbol{t}, \boldsymbol{v}\cdot\boldsymbol{t}))_{\omega_p} \\
&= \rho_b(\boldsymbol{\eta}_t, \boldsymbol{\psi})\big|_{t=0} + (c_b p_b + \alpha_b \nabla\cdot\boldsymbol{\eta}, q_b)\big|_{t=0} + (c_p p_p - \alpha_p s \Delta w, q_p)_{\Omega_p}\big|_{t=0} \\
&+ \rho_p(w_t, z)\big|_{t=0} + \rho_f(\boldsymbol{u}, \boldsymbol{v})\big|_{t=0} + ((\mathbf{F}_b, \boldsymbol{\psi}))_{\Omega_b} + ((S, q_b))_{\Omega_b} + ((\mathbf{f}, \boldsymbol{v}))_{\Omega_f},
\end{aligned}$$

where

$$\rho_b \boldsymbol{\eta}(.,0) = \rho_b \boldsymbol{\eta}^0,\ \rho_b \boldsymbol{\eta}_t(.,0) = \rho_b \boldsymbol{\eta}_t^0;\ \rho_p w(.,0) = \rho_p w^0,\ \rho_p w_t(.,0) = \rho_p w_t^0,$$

$$c_b p_b(.,0) + \alpha_b \nabla\cdot\boldsymbol{\eta}(.,0) = \zeta_b^0;\ c_p p_p(.,0) - \alpha_p \Delta w(.,0) = \zeta_p^0.$$

The interpretation of the dual pairing in term $\langle w_t - \boldsymbol{u}\cdot\mathbf{e}_3, q_p\big|_{s=-h/2}\rangle_{\omega_p}$ is the following:

- If $\rho_p > 0$, the dual pairing is just $L^2(0,T; L^2(\omega_p))$ scalar product, which is well defined since $w_t \in L^2(0,T; L^2(\omega_p))$ (see definition of the solution space).
- If $\rho_p = 0$, the term is to be interpreted in the following way:

$$\begin{aligned}
((\langle w_t - \boldsymbol{u}\cdot\mathbf{e}_3, q_p\big|_{s=-h/2}\rangle))_{\omega_p} &= -((w, \partial_t q_p\big|_{s=-h/2}))_{\omega_p} - ((\boldsymbol{u}\cdot\mathbf{e}_3, q_p\big|_{s=-h/2}))_{\omega_p} \\
&+ (w, q_p\big|_{s=-h/2})\big|_{t=0}.
\end{aligned} \tag{10.53}$$

Remark 10.2.3 We work in the periodic settings to avoid unnecessary technical difficulties. However, there are some coercivity issues related to this setting. Namely, to ensure coercivity, we require $c_p > 0$. This is because in the degenerate case $c_p = c_b = 0$, the L^2 norm of p_p is not bounded by the energy inequality. More precisely, in the degenerate case, the energy estimate provides the bounds for $\|\partial_s p_p\|_{L^2(\Omega_p)}$ and $\|p_p|_{s=-h/2}\|_{H^{-1}((0,T)\times\omega_p\times\{-h/2\})}$, where the latter bound is a consequence of the coupling condition (10.27) and can be proved by taking $(\boldsymbol{v}, 0, 0, 0, 0)$ as a test function in the weak formulation (10.52). However, this is not enough to control the high oscillations in the x_1 and x_2 direction. For example, we can take $p_p = \sin(\pi n x_1)$ to see that its L^2 norm is not controlled by the energy.

10.2.4 Differential (Strong) Formulation from Weak Form

We note that strong solution is recovered from weak solution with sufficient regularity. Indeed, if a weak solution (as in Definition 10.2.1) $(\boldsymbol{u}, w, p_p, \boldsymbol{\eta}, p_b) \in \mathcal{V}_{\text{sol}}$ is sufficiently smooth in space and time, then $(\boldsymbol{u}, w, p_p, \boldsymbol{\eta}, p_b) \in \mathcal{V}_{\text{sol}}$ satisfies the PDEs (10.18)–(10.20) in a pointwise sense, as well as the coupling conditions (10.22)–(10.27). Reconstruction of the differential formulation proceeds in the following way:

1. Take $\boldsymbol{\psi}$, q_b, z, q_p identically zero on their respective domains in the variational form of Definition 1. Invoke the formal integration by parts with C_c^∞- test function $\boldsymbol{v}$ to observe that the Stokes equation is satisfied point-wise a.e. in Ω_f. Then using "standard" test functions $\boldsymbol{v}$ from the test space $\mathcal{V}_{\text{test}}$, the following integral over ω_p remains:

$$\int_{\omega_p} \Big[\boldsymbol{\sigma}_f(\boldsymbol{u}, p_f)\mathbf{e}_3 \cdot \boldsymbol{v} + \beta(\boldsymbol{u}\cdot\boldsymbol{t})(\boldsymbol{v}\cdot\boldsymbol{t}) + p_p\big|_{s=-h/2}\boldsymbol{v}\cdot\mathbf{e}_3\Big] = 0.$$

   Since $\boldsymbol{v}$ is arbitrary, we obtain:

   (a) The Beavers-Joseph-Saffman condition $\boldsymbol{\sigma}_f\mathbf{n}\cdot\boldsymbol{t}+\beta\boldsymbol{u}\cdot\boldsymbol{t} = 0$ on ω_p, by taking $\boldsymbol{v}$ to be purely tangential
   (b) The pressure condition $p_p|_{s=-h/2} = -\boldsymbol{\sigma}_f\mathbf{e}_3\cdot\mathbf{e}_3$ on ω_p, by taking $\boldsymbol{v} = \mathbf{e}_3$

2. Now take all test function components zero, except z and $\boldsymbol{\psi}$. This gives the coupled dynamic equations for the poroelastic plate displacement w and the Biot displacement $\boldsymbol{\eta}$:

$$\begin{aligned}
&-\rho_b((\boldsymbol{\eta}_t, \boldsymbol{\psi}_t))_{\Omega_b} + ((\boldsymbol{\sigma}_b(\boldsymbol{\eta}, p_b), \nabla\boldsymbol{\psi}))_{\Omega_b} - \rho_p((w_t, \partial_t z))_{\omega_p} + D((\Delta w, \Delta z))_{\omega_p}\\
&= \alpha_p((\int_{-h/2}^{h/2}[s p_p]ds, \Delta z))_{\omega_p} + ((p_p\big|_{s=-h/2}, z))_{\omega_p}\\
&+ \rho_b(\boldsymbol{\eta}_t, \boldsymbol{\psi})\big|_{t=0} + \rho_p(w_t, z)\big|_{t=0} + ((\mathbf{F}_b, \boldsymbol{\psi}))_{\Omega_b}.
\end{aligned}$$

Similarly as above, to obtain the interior differential equations holding almost everywhere, we use the C_c^∞ test functions and integrate back by parts. Then, by using the standard test functions z and $\boldsymbol{\psi}$ from $\mathcal{V}_{\text{test}}$, after integrating back by parts in the $\boldsymbol{\eta}$ equation, we obtain the boundary terms on the right-hand side given by

$$p_p\Big|_{s=-h/2} + \boldsymbol{\sigma}_b(\boldsymbol{\eta}, p_b)\mathbf{e}_3 \cdot \mathbf{e}_3.$$

Next, we return to the full weak formulation, including the Stokes equation, and take only $z,\ p_p,\ \boldsymbol{v}$ to be different from zero. Undoing the integration by parts in the $\boldsymbol{u}$ equation gives the trace terms

$$-\boldsymbol{\sigma}_f(\boldsymbol{u}, p_f)\mathbf{e}_3 \cdot \mathbf{e}_3 + \boldsymbol{\sigma}_b(\boldsymbol{\eta}, p_b)\mathbf{e}_3 \cdot \mathbf{e}_3.$$

Finally, with the interior equations satisfied a.e., we obtain the following identity holding for the interface terms, since the appropriate integral equality holds true for all test functions:

$$p_p\Big|_{s=-h/2} + \boldsymbol{\sigma}_b(\boldsymbol{\eta}, p_b)\mathbf{e}_3 \cdot \mathbf{e}_3 = -\boldsymbol{\sigma}_f(\boldsymbol{u}, p_f)\mathbf{e}_3 \cdot \mathbf{e}_3 + \boldsymbol{\sigma}_b(\boldsymbol{\eta}, p_b)\mathbf{e}_3 \cdot \mathbf{e}_3.$$

This is exactly the dynamic coupling condition, i.e., the balance of forces acting across the poroelastic plate.

3. Lastly, let us take $q_p \neq 0$ such that $q_p|_{s=h/2} = 0$, and $q_b = 0$. Again, after integrating by parts and taking into account that the interior equations are satisfied point-wise, the following boundary terms remain:

$$-k_p \int_{\omega_p} \partial_{x_3} p_p|_{s=-h/2} q_p|_{s=-h/2} + \int_{\omega_p} [w_t - \boldsymbol{u} \cdot \mathbf{e}_3] q_p|_{s=-h/2} = 0.$$

   Therefore, we obtain $\partial_t w = u_3 + k_p \partial_3 p_p = u_3 - (\boldsymbol{u}_p)_3$, where $(\boldsymbol{u}_p)_3$ is the third component of the discharge velocity $\boldsymbol{u}_p = -k_p \partial_s p_p$ and u_3 is the third component of the Stokes field $\boldsymbol{u} = (u_1, u_2, u_3)$. This is exactly the kinematic coupling condition at the fluid-poroplate interface.

Note that the remaining "essential" type boundary coupling conditions are built in the compatibility requirements on $\mathcal{V}_{\text{sol}}$ and $\mathcal{V}_{\text{test}}$.

10.3 Statement of Main Results and Proof Strategy

In this section, we informally state the following two main results of this chapter:

1. The existence of a weak solution to the linear evolution problem: in particular, the Biot coefficient (diagonal tensor) k_b is a given function of t and $\mathbf{x}$. In this case,

the following parameters are allowed to be non-negative: $\rho_b, \rho_p, c_b, c_\rho \geq 0$, with only the poroelastic plate's compressibility parameter $c_p > 0$.
2. The existence of a weak solution to the nonlinear quasi-static problem: the coefficient k_b depends on the solution, in particular on the fluid content $\zeta_b = c_b p + \alpha_b \nabla \cdot \eta$. In this case, we consider the quasi-static Biot model, namely, $\rho_b = 0$, with the coefficients $\rho_p, c_b \geq 0$ and $c_p > 0$.

For the main results, we adopt the proof based on Rothe's method , as discussed in the previous chapters (see, e.g., Chap. 2). As mentioned earlier in this book, in the context of FSI problems, Rothe's method was developed and used to study moving boundary problems involving linear and nonlinear elastic structures [265, 267]. In the present chapter, we show how to extend this approach to deal with FPSI problems, showing that this approach is quite robust and applicable to a large-class FSI problems. Nevertheless, we emphasize that in the present chapter, the presence of the thin poroelastic plate is of particular importance since it serves as a regularizing mechanism [14, 264] in the existence proof while also allowing perfusion/filtration through the interface.

The proof relies on the following main steps:

1. Construction of approximate solutions.
   (a) Derive a semi-discrete in time and linearized version of the weak formulation, with the time discretization parameter $\Delta t = \frac{T}{N}$. In the semi-discretized formulation, the linearization is performed by evaluating the nonlinear coefficient functions at the previous time step. This defines a sequence of linear elliptic problems for solutions at time $t^n = n\Delta t$, with data given at $t^{n-1} = (n-1)\Delta t$.
   (b) Prove existence of a unique solution to the elliptic problem defined for every $n = 1, \ldots, N$ by using the Lax-Milgram lemma.
   (c) Define approximate *solutions*

$$(\boldsymbol{u}^{[N]}, w^{[N]}, p_p^{[N]}, \boldsymbol{\eta}^{[N]}, p_b^{[N]})$$

   as functions that are piece-wise constant in time, extrapolating the values obtained from the time discretization and satisfying the approximate weak formulation.

2. Derive uniform estimates in Δt through appropriate discretized test functions.
3. Obtain weak solution by letting $N \to \infty$, i.e., $\Delta t \to 0$, utilizing compactness criteria to push limits on the nonlinear term involving k_b.

Since in Steps 1 and 2 we use the linearized problem, these steps are the same for both the linear evolution problem and non-linear quasi-static problem. However, Step 3 deviates, because in the non-linear case, we need to prove additional strong convergence properties. In Step 3, we will utilize piecewise constant approximations and an elegant compactness result for piecewise constant approximate solutions

[126], which greatly simplifies the construction from past work on nonlinear Biot problems [42, 74, 345].

We begin with construction of the approximate solutions that covers both cases. Namely, for the linear problem, simply take $k_b \equiv k_b(\mathbf{x}, t)$ as a given function.

Finally, we address uniqueness criteria for solutions that exhibit more regularity than our weak solutions. In particular, the result is of weak-strong uniqueness type, wherein weak solutions from a certain class (see Assumption 4) are unique if there is a single solution in that class exhibiting additional regularity (see Proposition 10.5.1).

10.4 Main Result I: Linear Existence

In this section, we consider $k_b = k_b(\mathbf{x}, t) \in L^\infty(0, T; L^\infty(\mathbb{R}))$ as a given function of space and time.

The proof of the existence of a weak solution is based on temporal semi-discretization along with formal energy estimates. It is a direct application of Rothe's method to the coupled linear Stokes-Biot-poroelastic plate problem. We begin by the construction of approximate solutions, given a discretization of the time interval based on a time step Δt.

10.4.1 Construction of the Approximate Solutions

Let $\Delta t > 0 = T/N$, $N \in \mathbb{N}$. For every fixed N, we inductively define a sequence of approximate solutions $(\boldsymbol{u}_N^n, w_N^n, (p_p)_N^n, \boldsymbol{\eta}_N^n, (p_b)_N^n)$, $n = 0, \ldots, N$. For this purpose, we use implicit Euler method to discretize the time derivatives, making use of the following notation:

$$\dot{\boldsymbol{\eta}}^{n+1} = \frac{\boldsymbol{\eta}^{n+1} - \boldsymbol{\eta}^n}{\Delta t}, \quad \dot{w}^{n+1} = \frac{w^{n+1} - w^n}{\Delta t}. \tag{10.54}$$

Thus, at every fixed time t^n, we have to "solve" a semi-discretized problem (specified below in (10.59)), with the solution (depending only on the spatial variables) belonging to the following *weak solution space*:

$$\mathcal{V}_{\text{sd}} = \Big\{(\boldsymbol{u}, w, p_p, \boldsymbol{\eta}, p_b) \in V_\# \times H^2_\#(\omega_p) \times H^{0,0,1}_\# \times H^1_\#(\Omega_b) \times H^1_\#(\Omega_b) \ : \tag{10.55}$$

$$p_p\big|_{s=h/2} = p_b\big|_{x_3=0}, \ \boldsymbol{\eta}\big|_{x_3=0} = w\mathbf{e}_3\Big\},$$

equipped with the following norm:

$$\|(\boldsymbol{u}, w, p_p, \boldsymbol{\eta}, p_b)\|_{sd}^2 := \|\boldsymbol{u}\|_{H^1(\Omega_f)}^2 + \|\Delta w\|_{L^2(\omega_p)}^2 + \|w\|_{L^2(\omega_p)}^2 + \|p_p\|_{L^2(\Omega_p)}^2 + \|\partial_s p_p\|_{L^2(\Omega_p)}^2 + \|\boldsymbol{\eta}\|_E^2 + \|\nabla p_b\|_{L^2(\Omega_b)}^2, \tag{10.56}$$

where we recall that the norm $\|\cdot\|_E$ was defined in (10.38).

Lemma 10.4.1 $(\mathcal{V}_{\rm sd}, \|.\|_{sd})$ *is a Hilbert space. In particular, the norm* $\|\cdot\|_{\mathcal{V}_{\rm sd}}$ *defined in* (10.56) *is equivalent to the usual norm in the space* $V_{\#} \times H^2_{\#}(\omega_p) \times H^{0,0,1}_{\#} \times H^1_{\#}(\Omega_b) \times H^1_{\#}(\Omega_b)$.

Proof We first notice that $\|\Delta w\|_{L^2(\omega_p)} + \|w\|_{L^2(\omega_p)}$ is a norm equivalent to the H^2 norm on $H^2_{\#}(\omega_p)$. This follows by applying elliptic regularity to the case of periodic boundary conditions associated with the operator $-\Delta w + \gamma w \in L^2(\Omega)$, giving full control of the $H^2(\omega_p)$ norm.

Next, we observe that the L^2 norms of $\boldsymbol{\eta}$ and p_b (Biot components) are not included in the definition of $\|.\|_{sd}$, and so we need to show the following Poincaré-type inequalities:

$$\|\boldsymbol{\eta}\|_{L^2(\Omega_b)}^2 \le C\|(\boldsymbol{u}, w, p_p, \boldsymbol{\eta}, p_b)\|_{sd}^2, \quad \boldsymbol{\eta} \in H^1_{\#}(\Omega_b), \tag{10.57}$$

$$\|p_b\|_{L^2(\Omega_b)}^2 \le C\|(\boldsymbol{u}, w, p_p, \boldsymbol{\eta}, p_b)\|_{sd}^2, \quad p_b \in H^1_{\#}(\Omega_b). \tag{10.58}$$

In contrast with the homogeneous Dirichlet case, the proof of the Poincaré inequalities will follow by employing the interface conditions incorporated in the definition of $\mathcal{V}_{sd}$. We will prove the inequalities for smooth functions since general case follows by density argument in the standard way. To prove the first inequality in (10.57), we notice

$$|\boldsymbol{\eta}(x_1, x_2, x_3, t)|^2 = |\boldsymbol{\eta}(x_1, x_2, 0, t) + \int_0^{x_3} \partial_{x_3}\boldsymbol{\eta}(x_1, x_2, z, t)dz|^2 \le C\big(|\boldsymbol{\eta}(x_1, x_2, 0, t|^2 + \int_0^{x_3} |\partial_{x_3}\boldsymbol{\eta}(x_1, x_2, z, t)|^2 dz\big).$$

Integrate the above inequality over Ω_b and invoke the interface coupling $\boldsymbol{\eta}|_{\omega_p} = \langle 0, 0, w\rangle$ to obtain:

$$\|\boldsymbol{\eta}\|_{L^2(\Omega_b)}^2 \le C\big(\|\boldsymbol{\eta}\|_{L^2(w_p)}^2 + \|\partial_{x_3}\boldsymbol{\eta}\|_{L^2(\Omega_b)}^2\big) \le C\big(\|w\|_{L^2(w_p)}^2 + \|\boldsymbol{\eta}\|_E^2\big) \le C\|(\boldsymbol{u}, w, p_p, \boldsymbol{\eta}, p_b)\|_{sd}^2.$$

This proves the first inequality.

To prove the second inequality, we first need to estimate the trace of p_b on $x_3 = 0$:

$$p_b^2(x_1, x_2, 0, t) = p_p^2(x_1, x_2, h/2, t) = \Big(\frac{2}{h}\int_0^{h/2} \partial_s(sp_p(x_1, x_2, z, t))dz\Big)^2.$$

We continue, as before, by integrating the above inequality over ω_p to obtain:

$$\|p_b\|^2_{L^2(w_p)} \leq C\big(\|p_p\|^2_{L^2(\Omega_p)} + \|\partial_s p_p\|^2_{L^2(\Omega_p)}\big).$$

Thus, we get:

$$\|p_b\|^2_{L^2(\Omega_b)} \leq C\big(\|p_b\|^2_{L^2(w_p)} + \|\nabla p_b\|^2_{L^2(\Omega_b)}\big) \leq C\|(\boldsymbol{u}, w, p_p, \boldsymbol{\eta}, p_b)\|^2_{sd}.$$

□

Remark 10.4.1 If there were homogeneous Dirichlet boundary conditions on any sub-portion of the boundary for $\boldsymbol{\eta}$ and p_b, the classical Poincaré and Korn inequalities would yield norm equivalences for those variables. In the present configuration, the L^2 control in (10.57) for $\boldsymbol{\eta}$ and p_b is obtained through interface conditions, which gives equivalence of the full norms, as stated in Lemma 10.4.1.

We now define weak formulation for the semi-discrete linear problem at time $t^n = n\Delta t$ for each $n = 1, \dots, N$. The weak formulation defines semi-discretized approximations of the solution.

Definition 10.4.1 (Weak Formulation for the Semi-Discrete Problem) The following is the semi-discretized weak formulation of the linear version of problem (10.18)–(10.34) where $k_b^n = k_b(\mathbf{x}, t^n)$:

Find $(\boldsymbol{u}^n, w^n, (p_p)^n, \boldsymbol{\eta}^n, (p_b)^n) \in \mathcal{V}_{\text{sd}}$ such that for every $(\boldsymbol{v}, z, q_p, \boldsymbol{\psi}, q_b) \in \mathcal{V}_{\text{sd}}$, the following holds:

$$\begin{aligned}
&\rho_b(\dot{\boldsymbol{\eta}}^{n+1} - \dot{\boldsymbol{\eta}}^n, \boldsymbol{\psi})_{\Omega_b} + \Delta t(\boldsymbol{\sigma}_b(\boldsymbol{\eta}^{n+1}, p_b^{n+1}), \nabla\boldsymbol{\psi})_{\Omega_b} + \Delta t(c_b\dot{p}_b^{n+1} \\
&+ \alpha_b\nabla\cdot\dot{\boldsymbol{\eta}}^{n+1}, q_b)_{\Omega_b} \\
&+ \Delta t(k_b^n\nabla p_b^{n+1}, \nabla q_b)_{\Omega_b} + \Delta t(c_p\dot{p}_p^{n+1} - \alpha_p s\Delta\dot{w}^{n+1}, q_p)_{\Omega_p} \\
&+ \Delta t(k_p\partial_s p_p^{n+1}, \partial_s q_p)_{\Omega_p} \\
&+ \rho_p(\dot{w}^{n+1} - \dot{w}^n, z)_{\omega_p} + \Delta t D(\Delta w^{n+1}, \Delta z)_{\omega_p} + \Delta t\alpha_p\Big(\int_{-h/2}^{h/2}[sp_p^{n+1}]ds, \Delta z\Big)_{\omega_p} \\
&+ \gamma[\Delta t](w^{n+1}, z)_{\omega_p} - \Delta t\big(p_p^{n+1}\big|_{s=-h/2}, z - \boldsymbol{v}\cdot\mathbf{e}_3\big)_{\omega_p} + \Delta t(\dot{w}^{n+1} \\
&- \boldsymbol{u}^{n+1}\cdot\mathbf{e}_3, q_p\big|_{s=-h/2}) \\
&+ \rho_f(\boldsymbol{u}^{n+1} - \boldsymbol{u}^n, \boldsymbol{v})_{\Omega_f} + \Delta t\mu_f(\boldsymbol{D}(\boldsymbol{u}^{n+1}), \boldsymbol{D}(\boldsymbol{v}))_{\Omega_f} + \Delta t\beta(\boldsymbol{u}^{n+1}\cdot\boldsymbol{t}, \boldsymbol{v}\cdot\boldsymbol{t})_{\omega_p} \\
&= \Delta t(\mathbf{F}_b^{n+1}, \boldsymbol{\psi})_{\Omega_b} + \Delta t(S^{n+1}, q_b)_{\Omega_b} + \Delta t(\mathbf{f}^{n+1}, \boldsymbol{v})_{\Omega_f},
\end{aligned} \tag{10.59}$$

with the initial data specified in (10.34) and the discretized time derivative $\dot{\boldsymbol{\eta}}^n$ defined in (10.54).

Notice that in the linear case, k_b^n is the value of $k_b(x, t)$ at the previous time step t^n. In the nonlinear case, we will take $k_b^n = (c_b p_b^n + \alpha_b \nabla \cdot \boldsymbol{\eta}^n)$.

To write this weak formulation entirely in terms of the unknown functions $(\boldsymbol{u}^n, w^n, (p_p)^n, \boldsymbol{\eta}^n, (p_b)^n)$ belonging to the function space $\mathcal{V}_{\text{sd}}$, we express the terms $\dot{\boldsymbol{\eta}}^n$ and $\dot{\boldsymbol{\eta}}^{n+1}$ in terms of $\boldsymbol{\eta}$ using (10.54) to obtain:

$$
\begin{aligned}
&\rho_b(\boldsymbol{\eta}^{n+1}, \boldsymbol{\psi})_{\Omega_b} + \Delta t^2(\boldsymbol{\sigma}_b(\boldsymbol{\eta}^{n+1}, p_b^{n+1}), \nabla \boldsymbol{\psi})_{\Omega_b} + \Delta t(c_b p_b^{n+1}, q_b)_{\Omega_b} \\
&+ \Delta t(\alpha_b \nabla \boldsymbol{\eta}^{n+1}, q_b)_{\Omega_b} \\
&+ \Delta t^2(k_b^n \nabla p_b^{n+1}, \nabla q_b)_{\Omega_b} + \Delta t(c_p p_p^{n+1} - \alpha_p s \Delta w^{n+1}, q_p)_{\Omega_p} \\
&+ \Delta t^2(k_p \partial_s p_p^{n+1}, \partial_s q_p)_{\Omega_p} \\
&+ \rho_p(w^{n+1}, z)_{\omega_p} + \Delta t^2 D(\Delta w^{n+1}, \Delta z)_{\omega_p} + \Delta t^2 \alpha_p\Big(\int_{-h/2}^{h/2} [s p_p^{n+1}] ds, \Delta z\Big)_{\omega_p} \\
&+ \gamma \Delta t^2 (w^{n+1}, z)_{\omega_p} \\
&- \Delta t^2\Big(p_p^{n+1}\Big|_{s=-h/2}, z\Big)_{\omega_p} + \Delta t^2\Big(p_p^{n+1}\Big|_{s=-h/2}, \boldsymbol{v} \cdot \mathbf{e}_3\Big)_{\omega_p} + \Delta t\Big(w^{n+1}, q_p\Big|_{s=-h/2}\Big) \\
&- \Delta t^2\Big(\boldsymbol{u}^{n+1} \cdot \mathbf{e}_3, q_p\Big|_{s=-h/2}\Big) + \Delta t \rho_f(\boldsymbol{u}^{n+1}, \boldsymbol{v})_{\Omega_f} + \Delta t^2 \mu_f(\boldsymbol{D}(\boldsymbol{u}^{n+1}), \boldsymbol{D}(\boldsymbol{v}))_{\Omega_f} \\
&+ \Delta t^2 \beta(\boldsymbol{u}^{n+1} \cdot \boldsymbol{t}, \boldsymbol{v} \cdot \boldsymbol{t})_{\omega_p} \\
&= \Delta t^2(\mathbf{F}_b^{n+1}, \boldsymbol{\psi})_{\Omega_b} + \Delta t^2(S^{n+1}, q_b)_{\Omega_b} + \Delta t^2(\mathbf{f}^{n+1}, \boldsymbol{v})_{\Omega_f} \\
&+ \Delta t(\dot{\boldsymbol{\eta}}^n, \boldsymbol{\psi})_{\Omega_b} + \Delta t(\dot{w}^n, z)_{\omega_p} + \Delta t \rho_f(\boldsymbol{u}^n, \boldsymbol{v})_{\Omega_f} + \rho_b(\boldsymbol{\eta}^n, \boldsymbol{\psi})_{\Omega_b} \\
&+ \Delta t(c_b p_b^n + \alpha_b \nabla \boldsymbol{\eta}^n, q_b)_{\Omega_b} \\
&+ \Delta t(c_p p_p^n - \alpha_p s \Delta w^n, q_p)_{\Omega_p} + \rho_p(w^n, z)_{\omega_p} + \Delta t\Big(w^n, q_p\Big|_{s=-h/2}\Big)
\end{aligned}
\tag{10.60}
$$

We note that the above problem, although equivalent to the discrete weak form in (10.59), does not have a coercive structure. This is a consequence of the mixed parabolic-hyperbolic nature of the problem, the particular coupling in our multilayer problem, and the use of mixed-velocity/solution test functions in obtaining the energy estimates. To circumvent this issue, we modify the test functions (i.e., the variational problem we are trying to solve) in the following way:

$$
q_b \mapsto \Delta t q_b; \quad q_p \mapsto \Delta t q_p; \quad \boldsymbol{v} \mapsto \Delta t \boldsymbol{v}. \tag{10.61}
$$

This re-scaling can be undone, of course. This is because the re-scaling of test functions does not affect admissability in the space $\mathcal{V}_{\text{sd}}$, and after the use of the Lax-

Milgram lemma to obtain existence, as described below, the variational problem in (10.60) is recovered through the "for all" quantifier in the definition of a weak solution. As a consequence, the solution to (10.60) recovers the solution to the original semi-discretized problem (10.59).

With the re-scaled test functions (10.61), the weak formulation of the semi-discretized linear problem is given by the following:

Definition 10.4.2 (Weak Formulation with Rescaled Test Functions) We say that a function $(\boldsymbol{u}^n, w^n, (p_p)^n, \boldsymbol{\eta}^n, (p_b)^n) \in \mathcal{V}_{\rm sd}$ is a weak solution to the semi-discretized linear problem (10.18)–(10.34) if for every test function $(\Delta t\boldsymbol{v}, z, \Delta t q_p, \boldsymbol{\psi}, \Delta t q_b) \in \mathcal{V}_{\rm sd}$ the following holds:

$$
\begin{aligned}
&\rho_b(\boldsymbol{\eta}^{n+1}, \boldsymbol{\psi})_{\Omega_b} + \Delta t^2(\boldsymbol{\sigma}_b(\boldsymbol{\eta}^{n+1}, p_b^{n+1}), \nabla\boldsymbol{\psi})_{\Omega_b} + \Delta t(c_b p_b^{n+1}, \Delta t q_b)_{\Omega_b}\\
&+ \Delta t(\alpha_b \nabla\boldsymbol{\eta}^{n+1}, \Delta t q_b)_{\Omega_b}\\
&+ \Delta t^2(k_b^n \nabla p_b^{n+1}, \Delta t \nabla q_b)_{\Omega_b} + \Delta t(c_p p_p^{n+1} - \alpha_p s\Delta w^{n+1}, \Delta t q_p)_{\Omega_p}\\
&+ \Delta t^2(k_p \partial_s p_p^{n+1}, \Delta t \partial_s q_p)_{\Omega_p}\\
&+ \rho_p(w^{n+1}, z)_{\omega_p} + \Delta t^2 D(\Delta w^{n+1}, \Delta z)_{\omega_p} + \Delta t^2 \alpha_p\Big(\int_{-h/2}^{h/2} [s p_p^{n+1}] ds, \Delta z\Big)_{\omega_p}\\
&+ \gamma \Delta t^2 (w^{n+1}, z)_{\omega_p}\\
&- \Delta t^2 \Big(p_p^{n+1}\Big|_{s=-h/2}, z\Big)_{\omega_p} + \Delta t^2\Big(p_p^{n+1}\Big|_{s=-h/2}, \Delta t\boldsymbol{v}\cdot \mathbf{e}_3\Big)_{\omega_p}\\
&+ \Delta t\Big(w^{n+1}, \Delta t q_p\Big|_{s=-h/2}\Big)\\
&- \Delta t^2\Big(\boldsymbol{u}^{n+1}\cdot \mathbf{e}_3, \Delta t q_p\Big|_{s=-h/2}\Big)\\
&+ \Delta t\Big[\Delta t \rho_f(\boldsymbol{u}^{n+1}, \boldsymbol{v})_{\Omega_f} + \Delta t^2 \mu_f(\boldsymbol{D}(\boldsymbol{u}^{n+1}), \boldsymbol{D}(\boldsymbol{v}))_{\Omega_f} + \Delta t^2 \beta(\boldsymbol{u}^{n+1}\cdot \boldsymbol{t}, \boldsymbol{v}\cdot \boldsymbol{t})_{\omega_p}\Big]
\end{aligned} \tag{10.62}
$$

$$
\begin{aligned}
&= \Delta t^2(\mathbf{F}_b^{n+1}, \boldsymbol{\psi})_{\Omega_b} + \Delta t^2(S^{n+1}, \Delta t q_b)_{\Omega_b} + \Delta t^2(\mathbf{f}^{n+1}, \Delta t\boldsymbol{v})_{\Omega_f}\\
&+ \Delta t(\dot{\boldsymbol{\eta}}^n, \boldsymbol{\psi})_{\Omega_b} + \Delta t(\dot{w}^n, z)_{\omega_p} + \Delta t\rho_f(\boldsymbol{u}^n, \Delta t\boldsymbol{v})_{\Omega_f} + \rho_b(\boldsymbol{\eta}^n, \boldsymbol{\psi})_{\Omega_b} + \rho_p(w^n, z)_{\omega_p}\\
&+ \Delta t(c_b p_b^n + \alpha_b \nabla\boldsymbol{\eta}^n, \Delta t q_b)_{\Omega_b} + \Delta t(c_p p_p^n - \alpha_p s\Delta w^n, \Delta t q_p)_{\Omega_p}\\
&+ \Delta t\Big(w^n, \Delta t q_p\Big|_{s=-h/2}\Big),
\end{aligned}
$$

with the initial data specified in (10.34).

We aim at using the Lax-Milgram lemma to prove existence of a weak solution specified in Definition 10.4.2. More precisely, we will prove the following result.

Theorem 10.4.1 *There exists a unique weak solution to* (10.62).

To prove this result using a Lax-Milgram approach, we introduce a bilinear form on $\mathcal{V}_{\text{sd}}$, associated with the left hand-side of (10.62), defining our elliptic operator, and a functional on $\mathcal{V}_{\text{sd}}$, associated with the right hand-side of (10.62) given in terms of $\boldsymbol{\eta}^n$, $\dot{\boldsymbol{\eta}}^n$, w^n, $\dot{w}^n$, $\boldsymbol{u}^n$, p_b^n, p_p^n and the source terms $\mathbf{F}_b^{n+1}$, S^{n+1}, $\mathbf{f}^{n+1}$, treated as given data.

More precisely, given k_b^n, we introduce a bilinear form $a_n(\cdot,\cdot)$ on $\mathcal{V}_{\text{sd}} \times \mathcal{V}_{\text{sd}}$ as:

$$
\begin{aligned}
&a_n\left((\boldsymbol{\eta}, p_b, w, p_p, \boldsymbol{u}), (\boldsymbol{\psi}, q_b, z, q_p, \boldsymbol{v})\right) := \rho_b(\boldsymbol{\eta}^{n+1}, \boldsymbol{\psi})_{\Omega_b} \\
&+ \Delta t^2(\boldsymbol{\sigma}_b(\boldsymbol{\eta}^{n+1}, p_b^{n+1}), \nabla\boldsymbol{\psi})_{\Omega_b} \\
&+ \Delta t^2(c_b p_b^{n+1}, q_b)_{\Omega_b} + \Delta t^2(\alpha_b \nabla\boldsymbol{\eta}^{n+1}, q_b)_{\Omega_b} + \Delta t^3(k_b^n \nabla p_b^{n+1}, \nabla q_b)_{\Omega_b} \\
&+ \Delta t^2(c_p p_p^{n+1} - \alpha_p s \Delta w^{n+1}, q_p)_{\Omega_p} + \Delta t^3(k_p \partial_s p_p^{n+1}, \partial_s q_p)_{\Omega_p} + \rho_p(w^{n+1}, z)_{\omega_p} \\
&+ \Delta t^2 D(\Delta w^{n+1}, \Delta z)_{\omega_p} + \gamma \Delta t^2(w^{n+1}, z)_{\omega_p}, + \Delta t^2 \alpha_p (\int_{-h/2}^{h/2} [s p_p^{n+1}] ds, \Delta z)_{\omega_p} \\
&- \Delta t^2 (p_p^{n+1}\Big|_{s=-h/2}, z)_{\omega_p} + \Delta t^3 (p_p^{n+1}\Big|_{s=-h/2}, \boldsymbol{v}\cdot\mathbf{e}_3)_{\omega_p} + \Delta t^2 (w^{n+1}, q_p\Big|_{s=-h/2}) \\
&- \Delta t^3 (\boldsymbol{u}^{n+1}\cdot\mathbf{e}_3, q_p\Big|_{s=-h/2}) + \Delta t^2 \rho_f(\boldsymbol{u}^{n+1}, \boldsymbol{v})_{\Omega_f} + \Delta t^3 \mu_f(\boldsymbol{D}(\boldsymbol{u}^{n+1}), \boldsymbol{D}(\boldsymbol{v}))_{\Omega_f} \\
&+ \Delta t^3 \beta(\boldsymbol{u}^{n+1}\cdot\boldsymbol{t}, \boldsymbol{v}\cdot\boldsymbol{t})_{\omega_p}
\end{aligned} \tag{10.63}
$$

Notice again that the dependence of the bilinear form $a_n(\cdot,\cdot)$ on n comes specifically through the linear coefficient k_b^n in the case when this function exhibits any temporal dependence.

Similarly, we introduce the functional $\mathscr{F}_n$ by:

$$
\begin{aligned}
\mathscr{F}_n\left(\boldsymbol{\psi}, q_b, z, q_p, \boldsymbol{v}\right) := {} & \Delta t^2(\mathbf{F}_b^{n+1}, \boldsymbol{\psi})_{\Omega_b} + \Delta t^3(S^{n+1}, q_b)_{\Omega_b} + \Delta t^3(\mathbf{f}^{n+1}, \boldsymbol{v})_{\Omega_f} \\
& + \Delta t(\dot{\boldsymbol{\eta}}^n, \boldsymbol{\psi})_{\Omega_b} + \Delta t(\dot{w}^n, z)_{\omega_p} + \Delta t^2 \rho_f(\boldsymbol{u}^n, \boldsymbol{v})_{\Omega_f} + \rho_b(\boldsymbol{\eta}^n, \boldsymbol{\psi})_{\Omega_b} \\
& + \Delta t^2(c_b p_b^n + \alpha_b \nabla\boldsymbol{\eta}^n, q_b)_{\Omega_b} \\
& + \Delta t^2(c_p p_p^n - \alpha_p s \Delta w^n, q_p)_{\Omega_p} + \rho_p(w^n, z)_{\omega_p} + \Delta t^2(w^n, q_p\Big|_{s=-h/2}),
\end{aligned} \tag{10.64}
$$

which depends on Δt, the source terms $\mathbf{F}_b$, S, $\mathbf{f}$ at t^{n+1}, and solution evaluated at the previous time step t^n.

We have the following lemma:

Lemma 10.4.2 *Suppose that $c_p > 0$ and $\rho_b, \rho_p, c_b, c_p \geq 0$. Consider $a_n : \mathcal{V}_{\text{sd}} \times \mathcal{V}_{\text{sd}} \to \mathbb{R}$ and $\mathscr{F}_n \in \mathscr{L}(\mathcal{V}_{\text{sd}}, \mathbb{R})$, as defined in* (10.63) *and* (10.64)*. Then, as a bilinear form, $a_n(\cdot,\cdot)$ is continuous and coercive, and $\mathscr{F}_n \in [\mathcal{V}_{\text{sd}}]'$.*

Notice that in using Lax-Milgram with our particular rescaling, the setup is robust in that it allows c_b, ρ_b, ρ_p to vanish.

Proof First we demonstrate coercivity of $a_n(\cdot, \cdot)$ on $\mathcal{V}_{\rm sd} \times \mathcal{V}_{\rm sd}$. We choose $\boldsymbol{\psi} = \boldsymbol{\eta}^{n+1}$, $q_b = p_b^{n+1}$, $q_p = p_p^{n+1}$, and $\boldsymbol{v} = \boldsymbol{u}^{n+1}$ and calculate:

$$
\begin{aligned}
& a_n(\boldsymbol{u}^{n+1}, w^{n+1}, p_p^{n+1}, \boldsymbol{\eta}^{n+1}, p_b^{n+1}), (\boldsymbol{u}^{n+1}, w^{n+1}, p_p^{n+1}, \boldsymbol{\eta}^{n+1}, p_b^{n+1})) \\
&= \rho_b \|\boldsymbol{\eta}^{n+1}\|^2_{\Omega_b} + \Delta t^2 \|\boldsymbol{\eta}^{n+1}\|^2_E \\
&- \Delta t^2 \alpha_b (p_b^{n+1}, \nabla \boldsymbol{\eta}^{n+1})_{\Omega_b} + \Delta t^2 (\alpha_b \nabla \boldsymbol{\eta}^{n+1}, p_b^{n+1})_{\Omega_b} + \Delta t^2 c_b \|p_b^{n+1}\|^2_{\Omega_b} \\
&+ \Delta t^3 \|(k_b^n)^{1/2} \nabla p_b^{n+1}\|^2_{\Omega_b} + \Delta t^2 c_p \|p_p^{n+1}\|^2_{\Omega_p} - \Delta t^2 \alpha_p (s \Delta w^{n+1}, p_p^{n+1})_{\Omega_p} \\
&+ \Delta t^3 \|k_p^{1/2} \|\partial_s p_p^{n+1}\|^2_{\Omega_p} \\
&+ \rho_p \|w^{n+1}\|^2_{\omega_p} + \Delta t^2 D \|\Delta w^{n+1}\|^2_{\omega_p} + \gamma \Delta t^2 \|w^{n+1}\|^2_{\omega_p} \\
&+ \Delta t^2 \alpha_p \left(\int_{-h/2}^{h/2} [s p^{n+1}] ds, \Delta w^{n+1} \right)_{\omega_p} \\
&- \Delta t^2 (p_p^{n+1}\Big|_{s=-h/2}, w^{n+1})_{\omega_p} + \Delta t^3 (p_p^{n+1}\Big|_{s=-h/2}, \boldsymbol{u}^{n+1} \cdot \mathbf{e}_3)_{\omega_p} \\
&+ \Delta t^2 (w^{n+1}, p_p^{n+1}\Big|_{s=-h/2}) - \Delta t^3 (\boldsymbol{u}^{n+1} \cdot \mathbf{e}_3, p_p^{n+1}\Big|_{s=-h/2}) \\
&+ \Delta t^2 \rho_f (\boldsymbol{u}^{n+1}, \boldsymbol{u}^{n+1})_{\Omega_f} + \Delta t^3 \mu_f (\boldsymbol{D}(\boldsymbol{u}^{n+1}), \boldsymbol{D}(\boldsymbol{u}^{n+1}))_{\Omega_f} \\
&+ \Delta t^3 \beta (\boldsymbol{u}^{n+1} \cdot \boldsymbol{t}, \boldsymbol{u}^{n+1} \cdot \boldsymbol{t})_{\omega_p}.
\end{aligned}
$$

Now, because of the correct scaling in terms of Δt, we can cancel out both Biot type mixed terms and boundary coupling terms, to obtain:

$$
\begin{aligned}
& a_n(\boldsymbol{u}^{n+1}, w^{n+1}, p_p^{n+1}, \boldsymbol{\eta}^{n+1}, p_b^{n+1}), (\boldsymbol{u}^{n+1}, w^{n+1}, p_p^{n+1}, \boldsymbol{\eta}^{n+1}, p_b^{n+1})) \\
&= \rho_b \|\boldsymbol{\eta}^{n+1}\|^2_{\Omega_b} + \Delta t^2 \|\boldsymbol{\eta}^{n+1}\|^2_E \\
&+ \Delta t^2 c_b \|p_b^{n+1}\|^2_{\Omega_b} + \Delta t^3 \|[k_b^n]^{1/2} \nabla p_b^{n+1}\|^2_{\Omega_b} + \Delta t^2 c_p \|p_p^{n+1}\|^2_{\Omega_p} \\
&+ \Delta t^3 \|[k_p]^{1/2} \partial_s p_p^{n+1}\|^2_{\Omega_p} \\
&+ \rho_p \|w^{n+1}\|^2_{\omega_p} + \Delta t^2 D \|\Delta w^{n+1}\|^2_{\omega_p} + \gamma \Delta t^2 \|w^{n+1}\|^2 \\
&+ \Delta t^2 \|\boldsymbol{u}^{n+1}\|^2_{\Omega_f} + \Delta t^3 \mu_f \|\boldsymbol{D}(\boldsymbol{u}^{n+1})\|^2_{\Omega_f} \\
&+ \Delta t^3 \beta \|\boldsymbol{u}^{n+1} \cdot \boldsymbol{t}\|^2_{\omega_p}.
\end{aligned}
$$

We can now estimate the right-hand side to obtain coercivity on $\mathcal{V}_{\rm sd}$:

$$
\begin{aligned}
& a_n(\boldsymbol{u}^{n+1}, w^{n+1}, p_p^{n+1}, \boldsymbol{\eta}^{n+1}, p_b^{n+1}), (\boldsymbol{u}^{n+1}, w^{n+1}, p_p^{n+1}, \boldsymbol{\eta}^{n+1}, p_b^{n+1})) \\
& \gtrsim \Delta t^2 \|\boldsymbol{\eta}^{n+1}\|^2_{\mathbf{H}^1(\Omega_b)} + \Delta t^3 \|p_b^{n+1}\|^2_{H^1(\Omega_b)} + \Delta t^3 \|p_p^{n+1}\|^2_{H^1(\Omega_p)} \\
& \quad + \Delta t^2 \|w^{n+1}\|^2_{H^2(\omega_p)} + \Delta t^3 \|\boldsymbol{u}\|^2_{\mathbf{H}^1(\Omega_f)} \\
& \gtrsim (\Delta t^2 + \Delta t^3) \|[\boldsymbol{u}^{n+1}, w^{n+1}, p_p^{n+1}, \boldsymbol{\eta}^{n+1}, p_b^{n+1}]\|^2_{\mathcal{V}_{\text{sd}}}.
\end{aligned}
$$

Here, we have implicitly used the lower-bound $k_b(\mathbf{x}, t) \geq k_{\min} > 0$ and the notation $\gtrsim$, which represents a bound by a generic constant depending on $\rho_f, \rho_b, c_b, \alpha_b, c_p, \alpha_p, \rho_p$, as well as the Korn's and Poincaré constants, and possibly Δt.

This estimate shows coercivity for the variable $(\boldsymbol{u}^{n+1}, w^{n+1}, p_p^{n+1}, \boldsymbol{\eta}^{n+1}, p_b^{n+1})$ in the $\mathcal{V}_{\text{sd}}$ norm. The quantities $\dot{\boldsymbol{\eta}}^{n+1}$ and $\dot{w}^{n+1}$ are recovered from the formulae in (10.54), recalling that $\boldsymbol{\eta}^n \in \mathbf{L}^2_\#(\Omega_b)$ and $w^n \in L^2_\#(\omega_p)$ are given as data.

To see that the functional $\mathscr{F}_n$ belongs to $[\mathcal{V}_{\text{sd}}]'$, we derive a straightforward estimate in terms of L^2 norms:

$$
\begin{aligned}
\|\mathscr{F}\|_{[\mathcal{V}_{sd}]'} \lesssim\ & \Delta t^2 \big[\|\mathbf{F}_b^{n+1}\|_{\Omega_b} + \Delta t \|S^{n+1}\|_{\Omega_b} + \Delta t \|\mathbf{f}^{n+1}\|_{\Omega_f} \big] \\
& + \Delta t \Big[\|\dot{\boldsymbol{\eta}}^n\|_{\Omega_b} + \|\dot{w}^n\|_{\omega_p} + \Delta t \|\boldsymbol{u}^n\|_{\Omega_f} \Big] \\
& + \Delta t^2 \big[\|p_b^n\|_{\Omega_b} + \|\boldsymbol{\eta}^n\|_E + \|p_p^n\|_{\Omega_p} + \|\Delta w^n\|_{\omega_p} \big] \\
& + \|\boldsymbol{\eta}^n\|_{\Omega_b} + \|w^n\|_{\omega_p}.
\end{aligned}
$$

□

The proof of Theorem 10.4.1 now follows by applying the Lax-Milgram lemma to the problem

$$
\begin{aligned}
& a_n \Big((\boldsymbol{u}^{n+1}, w^{n+1}, p_p^{n+1}, \boldsymbol{\eta}^{n+1}, p_b^{n+1}), (\boldsymbol{\psi}, q_b, z, q_p, \boldsymbol{v}) \Big) \\
& = \mathscr{F}_n \left(\boldsymbol{\psi}, q_b, z, q_p, \boldsymbol{v} \right), \quad \forall (\boldsymbol{\psi}, q_b, z, q_p, \boldsymbol{v}) \in \mathcal{V}_{sd},
\end{aligned}
$$

defined on the Hilbert space $\mathcal{V}_{sd}$.

We emphasize that the approximate solution lies in the test space, as is necessary for obtaining estimates later on.

Remark 10.4.2 We note that in the above construction, the continuity and coercivity constants may depend on Δt in a singular way. However, we utilize the Lax-Milgram lemma only for existence of an approximate solution to the re-scaled problem (10.62). After obtaining the solution at the time t^{n+1} from the data at t^n, we recover the solution to the original problem (10.59) Once we have the solution to (10.59), we obtain *a priori* estimates for problem (10.59). We note that the approximate solution lies in the test space, which will be important for the choice of appropriate multiplies in the derivation of *a priori* estimates. The resulting bounds will be uniform in Δt. This feature is present in the related considerations [42, 44, 345].

Theorem 10.4.1 provides a sequence of approximate functions, which are functions of the spatial variable only, defined for every t^n, $n = 0, \ldots, N$. To define *approximate solutions*, we extend these functions to the entire time interval $[0, T]$ by extrapolation as piecewise constant functions in time, to obtain the following *approximate solutions*:

$$(\boldsymbol{u}^{[N]}, w^{[N]}, p_p^{[N]}, \boldsymbol{\eta}^{[N]}, p_b^{[N]})(t) = (\boldsymbol{u}_N^n, w_N^n, (p_p)_N^n, \boldsymbol{\eta}_N^n, (p_b)_N^n),$$
$$t \in (t^{n-1}, t^n], \; n = 1, \ldots, N. \tag{10.65}$$

Since we will have to work with derivatives of approximate functions, we introduce the following definition.

Definition 10.4.3 For a piecewise constant function $f^{[N]}$, the time derivative of $f^{[N]}$ is a function denoted by $\dot{f}^{[N]}(t)$, defined via backward difference quotients:

$$\dot{f}^{[N]}(t) = \frac{f_N^n - f_N^{n-1}}{\Delta t}, \quad t \in (t^{n-1}, t^n]. \tag{10.66}$$

With this notion of the time derivative, one can easily see that the approximate solutions satisfy the following weak formulation:

Proposition 10.4.1 *The approximate solutions* $(\boldsymbol{u}^{[N]}, w^{[N]}, p_p^{[N]}, \boldsymbol{\eta}^{[N]}, p_b^{[N]})$, *defined by* (10.65), *satisfy the following weak formulation for all test functions* $(\boldsymbol{v}, z, q_p, \boldsymbol{\psi}, q_b) \in \mathcal{V}_{test}$*:*

$$\begin{aligned}
&\rho_b((\ddot{\boldsymbol{\eta}}^{[N]}, \boldsymbol{\psi}))_{\Omega_b} + ((\boldsymbol{\sigma}_b(\boldsymbol{\eta}^{[N]}, p_b^{[N]}), \nabla\boldsymbol{\psi}))_{\Omega_b} + ((c_b\dot{p}_b^{[N]} + \alpha_b\nabla\cdot\dot{\boldsymbol{\eta}}^{[N]}, q_b))_{\Omega_b}\\
&+ ((k_b(\tau_{\Delta t}(c_b p_b^{[N]} + \alpha_b\nabla\cdot\boldsymbol{\eta}^{[N]})\nabla p_b^{[N]}, \nabla q_b))_{\Omega_b}\\
&+ ((c_p\dot{p}_p^{[N]} - \alpha_p s\Delta\dot{w}^{[N]}, q_p))_{\Omega_p} + ((k_p\partial_s p_p^{[N]}, \partial_s q_p))_{\Omega_p}\\
&+ \rho_p((\ddot{w}^{[N]}, z))_{\omega_p} + D((\Delta w^{[N]}, \Delta z))_{\omega_p} + \alpha_p((\int_{-h/2}^{h/2}[s p_p^{[N]}]ds, \Delta z))_{\omega_p}\\
&- ((p_p^{[N]}\Big|_{s=-h/2}, z - \boldsymbol{v}\cdot\mathbf{e}_3))_{\omega_p} + ((\dot{w}^{[N]} - \boldsymbol{u}^{[N]}\cdot\mathbf{e}_3, q_p\Big|_{s=-h/2}))\\
&+ \rho_f((\dot{\boldsymbol{u}}^{[N]}, \boldsymbol{v}))_{\Omega_f} + \mu_f((\boldsymbol{D}(\boldsymbol{u}^{[N]}), \boldsymbol{D}(\boldsymbol{v})))_{\Omega_f} + \beta(\boldsymbol{u}^{[N]}\cdot\boldsymbol{t}, \boldsymbol{v}\cdot\boldsymbol{t}))_{\omega_p}\\
&= ((\mathbf{F}_b^{[N]}, \boldsymbol{\psi}))_{\Omega_b} + ((S^{[N]}, q_b))_{\Omega_b} + ((\mathbf{f}^{[N]}, \boldsymbol{v}))_{\Omega_f},
\end{aligned} \tag{10.67}$$

where $\tau_{\Delta t}$ *is the translation operator defined by* $\tau_{\Delta t} f = f(t - \Delta t)$.

To see that this is true, we first focus on the time interval (t^n, t^{n+1}) and the weak formulation (10.59). Let $(\boldsymbol{v}, z, q_p, \boldsymbol{\psi}, q_b) \in \mathcal{V}_{test}$, where $\mathcal{V}_{test}$ is defined by (10.40). Notice that $(\boldsymbol{v}, z, q_p, \boldsymbol{\psi}, q_b)(t)$ for $t \in (t^n, t^{n+1})$, is an admissible test function for the weak formulation (10.59). Thus, take it as a test function in (10.59), integrate the resulting formulation in time from t^n to t^{n+1}, and then sum with respect to $n = 1, \dots, N$, to obtain that the approximate solutions satisfy the approximate weak form (10.67).

10.4.2 *Uniform* a Priori *Estimates for the Approximate Solutions*

We show here that the following estimates, uniform in Δt, or equivalently, independent of N, hold for approximate solutions.

Lemma 10.4.3 *There exists a constant $C > 0$, independent of N, such that approximate solutions $\boldsymbol{u}^{[N]}$, $w^{[N]}$, $p_p^{[N]}$, $\boldsymbol{\eta}^{[N]}$, and $p_b^{[N]}$ satisfy the following estimates:*

$$\rho_f \|\boldsymbol{u}^{[N]}\|^2_{L^\infty_t L^2_x} + \rho_p \|\dot{w}^{[N]}\|^2_{L^\infty_t L^2_x} + \rho_b \|\dot{\boldsymbol{\eta}}^{[N]}\|^2_{L^\infty_t L^2_x} \leq C,$$

$$c_p \|p_p^{[N]}\|^2_{L^\infty_t L^2_x} + \|\boldsymbol{\eta}^{[N]}\|_{L^\infty_t E} + c_b \|p_b^{[N]}\|^2_{L^\infty_t L^2_x} + D \|\Delta w^{[N]}\|^2_{L^\infty_t L^2_x} \leq C,$$

$$\mu_f \|\boldsymbol{D}(\boldsymbol{u}^{[N]})\|^2_{L^2_t L^2_x} + \beta \|\boldsymbol{u}^{[N]} \cdot \boldsymbol{t}\|^2_{L^2_t L^2_x(\omega_p)} + \|\partial_{x_3} p_p^{[N]}\|^2_{L^2_t L^2_x} + \|\nabla p_b^{[N]}\|^2_{L^2_t L^2_x} \leq C.$$

The shorthand notation $L^2_t L^2_x$, for instance, indicates the space $L^2(0, T; L^2(\Omega))$.

Proof To simplify notation, in this proof, we will be assuming that all the physical constants are equal to 1, except for the constants c_p, c_b, ρ_p, ρ_b, which can vanish (in the quasistatic case).

We start by obtaining an *a priori* estimate on the functions $\boldsymbol{u}^n$, w^n, $(p_p)^n$, $\boldsymbol{\eta}^n$, $(p_b)^n$, satisfying the semi-discretized problem (10.59). For this purpose, we use the following test functions in (10.59): $\boldsymbol{\psi} = \dot{\eta}^{n+1}$, $q_b = p_b^{n+1}$, $q_p = p_p^{n+1}$, $z = \dot{w}^{n+1}$, and $\boldsymbol{v} = \boldsymbol{u}^{n+1}$. Using Young's inequality and absorbing all constants that do not depend on Δt, we get:

$$\begin{aligned}
&\rho_b \|\dot{\boldsymbol{\eta}}^{n+1}\|^2_{\Omega_b} + \Delta t (\sigma^E(\boldsymbol{\eta}^{n+1}), \nabla \dot{\boldsymbol{\eta}}^{n+1})_{\Omega_b} + c_b \Delta t (\dot{p}_b^{n+1}, p_b^{n+1})_{\Omega_b} \\
&+ \Delta t \|(k_b^n)^{1/2} \nabla p_b^{n+1}\|^2_{\Omega_b} \\
&+ c_p \Delta t (\dot{p}_p^{n+1}, p_p^{n+1})_{\Omega_p} + \Delta t \|k_p^{1/2} \partial_s p_p^{n+1}\|^2_{\Omega_p} + \rho_p \|\dot{w}^{n+1}\|^2_{\omega_p} \\
&+ \Delta t (\Delta w^{n+1}, \Delta \dot{w}^{n+1})_{\omega_p} \\
&+ \|\boldsymbol{u}^{n+1}\|^2_{\Omega_f} + \Delta t \|\boldsymbol{D}(\boldsymbol{u}^{n+1})\|^2_{\Omega_f} + \Delta t \|\boldsymbol{u}^{n+1} \cdot \boldsymbol{t}\|^2_{\omega_p}
\end{aligned}$$

$$\lesssim \rho_b \|\dot{\boldsymbol{\eta}}^n\|^2_{\Omega_b} + \rho_p \|\dot{w}^n\|^2_{\omega_p} + \|\boldsymbol{u}^n\|^2_{\Omega_f} + \Delta t \|\mathbf{F}_b^{n+1}\|^2_{\Omega_b} + \Delta t \|S^{n+1}\|^2_{\Omega_b}$$
$$+ \Delta t \|\mathbf{f}^{n+1}\|^2_{\Omega_f}.$$

By invoking the definition of difference quotients, as well as Young's inequality, we obtain the following a priori estimate:

$$\begin{aligned}
&\rho_b \|\dot{\boldsymbol{\eta}}^{n+1}\|^2_{\Omega_b} + \|\boldsymbol{\eta}^{n+1}\|^2_E + c_b \|p_b^{n+1}\|_{\Omega_b} + \Delta t \|(k_b^n)^{1/2} \nabla p_b^{n+1}\|^2_{\Omega_b} \\
&+ c_p \|p_p^{n+1}\|^2_{\Omega_p} + \Delta t \|k_p^{1/2} \partial_s p_p^{n+1}\|^2_{\Omega_p} \\
&+ \rho_p \|\dot{w}^{n+1}\|^2_{\omega_p} + \|\Delta w^{n+1}\|^2_{\omega_p} + \|\boldsymbol{u}^{n+1}\|^2_{\Omega_f} + \Delta t \|\boldsymbol{D}(\boldsymbol{u}^{n+1})\|^2_{\Omega_f} \\
&+ \Delta t \|\boldsymbol{u}^{n+1} \cdot \boldsymbol{t}\|^2_{\omega_p} \\
\lesssim\ &\rho_b \|\dot{\boldsymbol{\eta}}^n\|^2_{\Omega_b} + \|\boldsymbol{\eta}^n\|^2_E + c_b \|p_b^n\|^2_{\Omega_b} + c_p \|p_p^n\|^2_{\Omega_p} + \rho_p \|\dot{w}^n\|^2_{\omega_p} + \|\Delta w^n\|^2_{\omega_p} \\
&+ \Delta t \|\mathbf{F}_b^{n+1}\|^2_{\Omega_b} + \Delta t \|S^{n+1}\|^2_{\Omega_b} + \Delta t \|\mathbf{f}^{n+1}\|^2_{\Omega_f}.
\end{aligned}$$

This estimate can be rearranged and expressed as follows:

$$\begin{aligned}
&\rho_b \|\dot{\boldsymbol{\eta}}^{n+1}\|^2_{\Omega_b} + \|\boldsymbol{\eta}^{n+1}\|^2_E + c_b \|p_b^{n+1}\|_{\Omega_b} + c_p \|p_p^{n+1}\|^2_{\Omega_p} + \rho_p \|\dot{w}^{n+1}\|^2_{\omega_p} \\
&+ \|\Delta w^{n+1}\|^2_{\omega_p} + \|\boldsymbol{u}^{n+1}\|^2_{\Omega_f} \\
\lesssim\ &\rho_b \|\dot{\boldsymbol{\eta}}^n\|^2_{\Omega_b} + \|\boldsymbol{\eta}^n\|^2_E + c_b \|p_b^n\|^2_{\Omega_b} + c_p \|p_p^n\|^2_{\Omega_p} + \rho_p \|\dot{w}^n\|^2_{\omega_p} + \|\Delta w^n\|^2_{\omega_p} \\
&+ \Delta t \|\mathbf{F}_b^{n+1}\|^2_{\Omega_b} + \Delta t \|S^{n+1}\|^2_{\Omega_b} + \Delta t \|\mathbf{f}^{n+1}\|^2_{\Omega_f} \\
&- \Delta t \|k_b^{1/2} \nabla p_b^{n+1}\|^2_{\Omega_b} - \Delta t \|k_p^{1/2} \partial_s p_p^{n+1}\|^2_{\Omega_p} - \Delta t \|\boldsymbol{D}(\boldsymbol{u}^{n+1})\|^2_{\Omega_f} \\
&- \Delta t \|\boldsymbol{u}^{n+1} \cdot \boldsymbol{t}\|^2_{\omega_p}.
\end{aligned}$$

Inductively, after rearrangement, we obtain:

$$\begin{aligned}
&\|\dot{\boldsymbol{\eta}}^{N+1}\|^2_{\Omega_b} + \|\boldsymbol{\eta}^{N+1}\|^2_E + c_b \|p_b^{N+1}\|_{\Omega_b} + c_p \|p_p^{N+1}\|^2_{\Omega_p} + \rho_p \|\dot{w}^{N+1}\|^2_{\omega_p} \\
&+ \|\Delta w^{N+1}\|^2_{\omega_p} + \|\boldsymbol{u}^{N+1}\|^2_{\Omega_f} \\
&+ \sum_{n=0}^{N} \Big[\|k_b^{1/2} \nabla p_b^{n+1}\|^2_{\Omega_b} + \|k_p^{1/2} \partial_s p_p^{n+1}\|^2_{\Omega_p} + \|\boldsymbol{D}(\boldsymbol{u}^{n+1})\|^2_{\Omega_f} + \|\boldsymbol{u}^{n+1} \cdot \boldsymbol{t}\|^2_{\omega_p} \Big] \Delta t \\
\lesssim\ &\|\dot{\boldsymbol{\eta}}^0\|^2_{\Omega_b} + \|\boldsymbol{\eta}^0\|^2_E + c_b \|p_b^0\|^2_{\Omega_b} + c_p \|p_p^0\|^2_{\Omega_p} + \|\dot{w}^0\|^2_{\omega_p} + \rho_p \|\Delta w^0\|^2_{\omega_p} \\
&+ \sum_{n=0}^{N} \Big[\|\mathbf{F}_b^{n+1}\|^2_{\Omega_b} + \|S^{n+1}\|^2_{\Omega_b} + \|\mathbf{f}^{n+1}\|^2_{\Omega_f} \Big] \Delta t.
\end{aligned} \tag{10.68}$$

Noting the lower bound on $k_b \geq k_{\min} > 0$ again, we obtain the uniform estimates for the approximate solutions.

□

We will need two more estimates to be able to pass to the limit as $N \to \infty$ (or, equivalently, as $\Delta t \to 0$) in the evolution case when we have to deal with the terms $\dot{\boldsymbol{\eta}}^{[N]}$ and $\dot{w}^{[N]}$. In this case, we require bounds on $\ddot{\boldsymbol{\eta}}^{[N]}$, $\ddot{w}^{[N]}$ in the associated dual, H^{-1}-like spaces. Additionally, we will need an estimate for the time derivative of the fluid content in the associated dual space.

Having established the solution, such estimates will follow immediately from (10.59). More precisely, since some test functions are coupled, we need to consider duals of product spaces to establish these estimates. In particular, consider first taking all test functions in (10.59) zero except for the "inertial test functions" $(\boldsymbol{\psi}, z) \in V_\# \times H^2_\#(\omega_p)$ such that $\boldsymbol{\psi}|_{x_3=0} = z\mathbf{e}_3$. Then, dividing (10.59) by Δt, we obtain the following equality for the solution at any t^{n+1}:

$$\rho_b(\ddot{\boldsymbol{\eta}}^{n+1}, \boldsymbol{\psi})_{\Omega_b} + \rho_p(\ddot{w}^{n+1}, z)_{\omega_p} = (\boldsymbol{\sigma}_b(\boldsymbol{\eta}^{n+1}, p_b^{n+1}), \nabla\boldsymbol{\psi}))_{\Omega_b} + (\mathbf{F}_b^{n+1}, \boldsymbol{\psi})_{\Omega_b}$$
$$- D(\Delta w^{n+1}, \Delta z)_{\omega_p} - \alpha_p(\int_{-h/2}^{h/2} [s p_p^{n+1}] ds, \Delta z)_{\omega_p} - (p_p^{n+1}\Big|_{s=-h/2}, z)_{\omega_p}.$$

Similar calculation can be done for the fluid content by taking all test functions zero except for the "fluid content test functions" $(q_p, q_b) \in H^1_\#(\Omega_b) \times H^{0,0,1}_\#$ such that $q_b|_{s=h/2} = q_p$. After integrating the obtained equalities with respect to time from $t^n = n\Delta t$ to $t^{n+1} = (n+1)\Delta t$, and taking the sum over n, for $n = 0, \ldots, N-1$, the estimates in Lemma 10.4.3 give the following result:

Lemma 10.4.4 *Let* $\mathcal{V}_{\text{in}} = \{(\boldsymbol{\psi}, z) \in V_\# \times H^2_\#(\omega_p) : \boldsymbol{\psi}|_{x_3=0} = z\mathbf{e}_3\}$ *and* $\mathcal{V}_{\text{fc}} = \{(q_p, q_b) \in H^1_\#(\Omega_b) \times H^{0,0,1}_\# : q_b|_{s=h/2} = q_p\}$. *Then there exists constant C independent on N such that:*

$$\left\| \left(\rho_b \ddot{\boldsymbol{\eta}}^{[N]}, \rho_p \ddot{w}^{[N]}\right) \right\|_{\mathcal{V}'_{\text{in}}} \leq C,$$

$$\left\| \left(c_b \dot{p}_b^{[N]} + \alpha_b \nabla \cdot \dot{\boldsymbol{\eta}}^{[N]}, c_p \dot{p}_p^{[N]} - \alpha_p \Delta \dot{w}^{[N]}\right) \right\|_{\mathcal{V}'_{\text{fc}}} \leq C.$$

10.4.3 Limit N → ∞ and Recovery of Linear Weak Solution

Existence

To prove the existence of a weak solution to the linear Stokes-Biot-poroplate fluid-structure interaction problem (10.18)–(10.34), we would like to show that there exist subsequences of approximate solutions $\{(\boldsymbol{u}^{[N]}, w^{[N]}, p_p^{[N]}, \boldsymbol{\eta}^{[N]}, p_b^{[N]})\}_{N\in\mathbb{N}}$, which converge, as $N \to \infty$ (or equivalently, as $\Delta t \to 0$), to a function that satisfies the weak formulation of the continuous problem specified in Definition 10.2.1. Since the problem is linear, weak convergence of subsequences of approximate solutions $\{(\boldsymbol{u}^{[N]}, w^{[N]}, p_p^{[N]}, \boldsymbol{\eta}^{[N]}, p_b^{[N]})\}_{N\in\mathbb{N}}$ is sufficient to pass to the limit in the weak formulations (10.67) satisfied by the approximate functions and recover the weak formulation of the continuous problem, satisfied by the limiting function. Indeed, the uniform bounds presented in Lemmas 10.4.3 and 10.4.4 imply existence of subsequences of $\{(\boldsymbol{u}^{[N]}, w^{[N]}, p_p^{[N]}, \boldsymbol{\eta}^{[N]}, p_b^{[N]})\}_{N\in\mathbb{N}}$, denoted again with $\{(\boldsymbol{u}^{[N]}, w^{[N]}, p_p^{[N]}, \boldsymbol{\eta}^{[N]}, p_b^{[N]})\}_{N\in\mathbb{N}}$ (with a slight abuse of notation), such that the following weak convergence is true for those subsequences:

$$(\boldsymbol{u}^{[N]}, w^{[N]}, p_p^{[N]}, \boldsymbol{\eta}^{[N]}, p_b^{[N]}) \rightharpoonup (\boldsymbol{u}, w, p_p, \boldsymbol{\eta}, p_b) \quad \text{weakly in } \mathcal{V}_{sol}, \tag{10.69}$$

and

$$(\rho_b \ddot{\boldsymbol{\eta}}^{[N]}, \rho_p \ddot{w}^{[N]}) \overset{*}{\rightharpoonup} (\rho_b \ddot{\boldsymbol{\eta}}, \rho_p \ddot{w}) \quad \text{weakly} * \text{ in } \mathcal{V}_{\text{in}}$$

$$(c_b \dot{p}_b^{[N]} + \alpha_b \nabla \cdot \dot{\boldsymbol{\eta}}^{[N]}, c_p \dot{p}_p^{[N]} - \alpha_p \Delta \dot{w}^{[N]}), \overset{*}{\rightharpoonup} (c_b \dot{p}_b^{+} \alpha_b \nabla \cdot \dot{\boldsymbol{\eta}}, c_p \dot{p}_p - \alpha_p \Delta \dot{w})$$
$$\text{weakly} * \text{ in } \mathcal{V}_{\text{fc}}.$$

This is sufficient to pass to the limit in (10.67) and obtain that the limiting function satisfies the continuous weak formulation (10.52). This calculation is straightforward. Therefore, the uniform estimates from Lemmas 10.4.3 and 10.4.4 imply the following existence result:

Theorem 10.4.2 *Let $c_p > 0$ and $\rho_b, \rho_p, c_b, \rho_f \geq 0$, $\alpha_n, \alpha_p, \mu_E, \lambda_E, D, \mu, k_p > 0$, and $T > 0$. Moreover, let $k_b \in L^\infty(0, T; L^\infty(\Omega_b))$ such that $0 < k_{\min} \leq k_b \leq k_{\max}$. Then there exists a weak solution on $(0, T)$ in the sense of Definition 10.2.1.*

Uniqueness

We conclude this section with a brief discussion of the issue of (linear) uniqueness. First, we note that even in the linear case, uniqueness of solutions is a non-trivial issue. This primarily stems from the fact that the energy estimates obtained for the constructed solution above need not apply to a general weak solution. Hence, given two weak solutions, an *a priori* estimate on the difference is not immediately obtainable via formal estimation as was done on approximate solutions in the construction of our particular weak solution. This issue is two-pronged: first, the hyperbolic nature of the inertial problem (when ρ_b and/or ρ_p non-zero) has challenges associated with justifying velocity multipliers (test functions) $\boldsymbol{\eta}_t$ and w_t, owing to the higher-order terms appearing from the Biot structure of the solid dynamics. Second, Biot problems with time-dependent coefficients (even in the quasi-static case) have issues associated with multiplier analysis to obtain a priori estimates. The effect of this can be noted, for instance, in [302, pp.110–117], where abstract implicit evolutions are discussed and "much non-uniqueness" is possible, since "...the stability estimate...was shown to hold for *some* solution, not every solution...."

We provide a lemma that yields uniqueness, under the hypothesis that formal test functions $\boldsymbol{\eta}_t$ and w_t are valid. Let us first formulate the hypothesis:

Assumption 4 Assume that the function $(\boldsymbol{u}, w, p_p, \boldsymbol{\eta}, p_b) \in \mathcal{V}_{\text{sol}}$ has the additional property that

$$(\rho_b w_{tt} + D\Delta^2 w, w_t)_{\omega_p} = \frac{1}{2}\frac{d}{dt}\Big\{\rho_p \|w_t\|^2_{L^2(\omega_p)} + D\|\Delta w\|^2_{L^2(\omega_p)}\Big\} \tag{10.70}$$

$$(\rho_p \boldsymbol{\eta}_{tt} + \boldsymbol{\sigma}^E, \boldsymbol{\eta}_t)_{\Omega_p} = \frac{1}{2}\frac{d}{dt}\Big\{\rho_b \|\boldsymbol{\eta}_t\|^2_{L^2(\Omega_b)} + \|\boldsymbol{\eta}\|^2_E\Big\}. \tag{10.71}$$

Then, the following uniqueness result holds.

Lemma 10.4.5 (Uniqueness of Weak Solutions) *Let $c_p > 0$ and $\rho_b, \rho_p, c_b, \rho_f \geq 0$, $\alpha_b, \alpha_p, \mu_E, \lambda_E, D, \mu, k_p > 0$, and $T > 0$. Moreover, let $k_b \in L^\infty(\Omega_b \times (0, T))$ such that $0 < k_1 \leq k_b \leq k_2$. Suppose further that a weak solution $(\boldsymbol{u}, w, p_p, \boldsymbol{\eta}, p_b) \in \mathcal{V}_{\text{sol}}$ has the additional property specified in Assumption 4. Then, for the given data, such a solution is unique.*

Proof The above lemma follows immediately from the formal energy estimates presented in the preceding section applied to the difference of two solutions to the linear problem. □

Remark 10.4.3 Note that Assumption 4 is not obvious when, for instance, ρ_p or ρ_b is zero. Furthermore, the above lemma would hold if in the assumption above the L^2 inner products in (10.70)–(10.71) are replaced with the appropriate duality pairings.

10.5 Main Result II: Quasistatic Nonlinear Problem

In this section, we consider the nonlinear case where:

- The permeability function k_b depends explicitly on $\zeta_b = c_b p_b + \alpha_b \nabla \cdot \boldsymbol{\eta}$, as specified in (10.13)
- The Biot model (10.18) is quasi-static, as specified in Assumption 3, namely, $\rho_b = 0$ in (10.18)
- We allow $\rho_p \geq 0$ in the poroelastic plate equations (10.19).

The main nonlinear existence result is the following:

Theorem 10.5.1 (Nonlinear Existence of Weak Solutions) *Let $\rho_b = 0$, $c_p > 0$, and $\rho_p, c_b, \rho_f \geq 0$, $\alpha_n, \alpha_p, \mu_E, \lambda_E, D, \mu, k_p > 0$, and $T > 0$. Moreover,* *let $k_b(\cdot) \in L^\infty(\mathbb{R})$ be such that $0 < k_{\min} \le k_b \le k_{\max}$. Then, there exists a weak solution to the nonlinear quasi-static problem on $(0, T)$ in the sense of Definition 10.2.1, with $\rho_b = 0$.*

To prove this result, we first notice that all the steps in the existence proof performed for the linear case still hold, except the last step involving passage to the limit. To pass to the limit in the nonlinear problem, we will use a version of Aubin-Lions lemma, which is particularly suitable for piece-wise constant approximations in time of the approximate solutions constructed using Rothe's method, described in Theorem 2.9.5.

We will apply the compactness result in Theorem 2.9.5 to the sequence $\left[c_b p_b^{[N]} + \alpha_b \nabla \cdot \boldsymbol{\eta}^{[N]}\right]$, with the corresponding Hilbert spaces $W = H^{-1}(\Omega_b)$, $V = H^\epsilon(\Omega_b)$, and $H = L^2(\Omega_b)$. We start by showing the following uniform estimate on $\left[c_b \dot{p}_b{}^{[N]} + \alpha_b \nabla \cdot \dot{\boldsymbol{\eta}}^{[N]}\right]$:

Lemma 10.5.1 *There exists a constant C independent of N such that*

$$\|c_b \dot{p}_b{}^{[N]} + \alpha_b \nabla \cdot \dot{\boldsymbol{\eta}}^{[N]}\|_{L^2(0,T;H^{-1}(\Omega_b))} \le C.$$

Proof Let $q_b \in C_c^1((0, T); H_0^1(\Omega_b))$. Then $(0, 0, 0, 0, q_b)$ belongs to the space of test functions V_{test}. Therefore, we can use this test function in the weak formulation (10.67) for approximate solutions to obtain (due to the second equation in (10.18)):

$$((c_b \dot{p}_b^{[N]} + \alpha_b \nabla \cdot \dot{\boldsymbol{\eta}}^{[N]}, q_b))_{\Omega_b} = -((k_b(\tau_{\Delta t}(c_b p_b^{[N]} + \alpha_b \nabla \cdot \boldsymbol{\eta}^{[N]}))\nabla p_b^{[N]}, \nabla q_b))_{\Omega_b} + ((S^{[N]}, q_b))_{\Omega_b}.$$

By using uniform estimates presented in Lemma 10.4.3, we obtain the following estimate:

$$|((c_b \dot{p}_b^{[N]} + \alpha_b \nabla \cdot \dot{\boldsymbol{\eta}}^{[N]}, q_b))_{\Omega_b}| \le k_{max} \|\nabla p_b^{[N]}\|_{L_t^2 L_x^2} \|q_b\|_{L_t^2 H_x^1} + C\|S^{[N]}\|_{L_t^2 L_x^2} \|q_b\|_{L_t^2 H_x^1} \le C\|q_b\|_{L_t^2 H_x^1}.$$

Now the conclusion of the lemma follows by density arguments and from the converse of Hölder's inequality. □

In order to prove the compactness result, we need the regularity result for the Biot displacement.

Theorem 10.5.2 *In the quasistatic case ($\rho_b = 0$), the Biot displacement satisfies the following additional regularity property: the sequence $\boldsymbol{\eta}^{[N]}$ is uniformly bounded in $L^2(0, T; \mathbf{H}^2(\Omega_b))$.*

Proof We use the arguments and notation similar to [42, Section 3.3]. We denote by $\mathcal{E} : \mathcal{D}(\mathcal{E}) \subset L^2(\Omega_b) \to \mathbf{L}^2(\Omega_b)$ the Lamé operator:

$$\mathcal{E}(\boldsymbol{\eta}) := -\nabla \cdot (2\mu_b \boldsymbol{D}(\boldsymbol{\eta}) + \lambda_b \nabla \cdot \boldsymbol{\eta} \boldsymbol{I})$$

defined on the domain

$$\mathcal{D}(\mathcal{E}) = \{\boldsymbol{\eta} \in \mathbf{H}^1_\#(\Omega_b) \mid \nabla \cdot (2\mu_b \boldsymbol{D}(\boldsymbol{\eta}) + \lambda_b \nabla \cdot \boldsymbol{\eta} \boldsymbol{I}) \in \mathbf{L}^2(\Omega_b)\}.$$

Then $\boldsymbol{\eta}^{[N]}$ satisfies the following equation:

$$\mathcal{E}\boldsymbol{\eta}^{[N]} = -\nabla p^{[N]} \quad \text{in} \quad (0, T) \times \Omega_b,$$

with the following mixed boundary conditions:

1. Periodic at $x_i = 0, x_i = 1,$ for $i = 1, 2$ (see (10.29));
2. Zero traction condition at $x_3 = 1$ (see (10.31));
3. Dirichlet condition on $x_3 = 0$, i.e. on ω_p : $\boldsymbol{\eta}^{[N]} = w^{[N]}\mathbf{e}_3$ (see (10.22)).

From Lemma 10.4.3, we see that:

- The right hand-side ∇p is uniformly bounded in $L^2(0, T; L^2(\Omega_b))$.
- The sequence $w^{[N]}$ in the Dirichlet boundary condition is uniformly bounded in $L^\infty(0, T; H^2(\omega_p))$.

Now the conclusion of the Lemma follows directly from standard elliptic regularity on a rectangular prism, as in [42, Lemma 2]. □

Remark 10.5.1 First, it is important to note here that due to the way how the problem is set up, our domain is smooth without corners. Second, the full elliptic regularity recovered in Theorem 10.5.2 is more than is needed in the construction of solutions; indeed, if $\boldsymbol{\eta}^{[N]}$ is uniformly bounded in $L^2(0, T; \mathbf{H}^{1+\epsilon}(\Omega_b))$, $\epsilon > 0$, then the convergence properties stated below follow as in [42].

Corollary 10.5.1 *The following convergence properties hold:*

$$c_b {p_b}^{[N]} + \alpha_b \nabla \cdot \boldsymbol{\eta}^{[N]} \to c_b p_b + \alpha_b \nabla \cdot \boldsymbol{\eta} \text{ strongly in } L^2(0, T; L^2(\Omega_b))$$

$$k_b\big(\tau_{\Delta t}(c_b p_b^{[N]} + \alpha_b \nabla \cdot \boldsymbol{\eta}^{[N]})\big) \to k_b\big(c_b p_b + \alpha_b \nabla \cdot \boldsymbol{\eta}\big) \text{ strongly in } L^2(0, T; L^2(\Omega_b))$$

Proof Uniform energy estimates from Lemma 10.4.3 imply that sequence $p_b^{[N]}$ is uniformly bounded in $L^2(0, T; H^1(\Omega_b))$. Therefore, by Theorem 10.5.2, sequence $c_b {p_b}^{[N]} + \alpha_b \nabla \cdot \boldsymbol{\eta}^{[N]}$ is uniformly bounded in $L^2(0, T; H^\epsilon(\Omega_b))$. Moreover, by Lemma 10.5.1, sequence $\left(c_b {p_b}^{[N]} + \alpha_b \nabla \cdot \boldsymbol{\eta}^{[N]}\right)^{\dot{}}$ is uniformly bounded in $L^2(0, T; H^{-1}(\Omega_b))$.

The strong convergence of $\left[c_b {p_b}^{[N]} + \alpha_b \nabla \cdot \boldsymbol{\eta}^{[N]}\right]$ then follows by direct application of Theorem 2.9.5.

The second convergence result follows by noticing that $k_b(\cdot) : L^2(\Omega_b) \to L^2(\Omega_b)$ represents a Nemytskii operator (see Definition 10.1.1 and monograph [302]). This follows from the bounds $0 < k_{\min} \leq k_b(s) \leq \kappa_{\max}$ for all $s \in \mathbb{R}$, as well as the continuity of $k_b(\cdot)$. Thus

$$k_b\big(\tau_{\Delta t}(c_b p_b^{[N]} + \alpha_b \nabla \cdot \boldsymbol{\eta}^{[N]})\big) \to k_b\big(c_b p_b + \alpha_b \nabla \cdot \boldsymbol{\eta}\big),$$

as desired. □

To complete the proof of the existence of weak solutions, what remains to show is how to pass to the limit in the nonlinear term in (10.67). For this purpose, we recall the following convergence properties of the sequences appearing in the nonlinear term:

$$k_b\big(\tau_{\Delta t}(c_b p_b^{[N]} + \alpha_b \nabla \cdot \boldsymbol{\eta}^{[N]})\big) \to k_b\big(c_b p_b + \alpha_b \nabla \cdot \boldsymbol{\eta}\big) \text{ in } L^2(0, T; L^2(\Omega_b)),$$

$$\nabla p_b^{[N]} \rightharpoonup \nabla p_b \quad \text{weakly in } L^2(0, T; L^2(\Omega_b)).$$

Now, by the convergence of weak-strong products, we get:

$$k_b\big(\tau_{\Delta t}(c_b p_b^{[N]}+\alpha_b \nabla\cdot\boldsymbol{\eta}^{[N]})\big)\nabla p_b^{[N]} \rightharpoonup k_b\big(c_b p_b+\alpha_b \nabla\cdot\boldsymbol{\eta}\big)\nabla p_b \text{ in } L^1(0, T; L^1(\Omega_b)).$$

To see that this sequence converges weakly in $L^2(0, T; L^2(\Omega_b))$, we recall that $k_b\big(\tau_{\Delta t}(c_b p_b^{[N]} + \alpha_b \nabla \cdot \boldsymbol{\eta}^{[N]})\big)\nabla p_b^{[N]}$ is a bounded sequence in $L^2(0, T; L^2(\Omega_b))$, and therefore it has a subsequence which converges weakly in $L^2(0, T; L^2(\Omega_b))$. Because of the uniqueness of the limit in $\mathcal{D}'$, its limit is exactly $k_b\big(c_b p_b + \alpha_b \nabla \cdot \boldsymbol{\eta}\big)\nabla p_b$. □

10.5.1 Uniqueness

Lastly, we consider the issue of uniqueness for nonlinear solutions. At present, even for the nonlinear Biot system alone, a full uniqueness result does exist in the literature. In [74], a *central* hypothesis for the entire well-posedness analysis is that $c_b > 0$; for uniqueness, the permeability function k_b is assumed to be Lipschitz continuous, *and* a smallness hypothesis is also imposed on the Biot pressure.

Along these lines, we present here a strong-weak uniqueness result for mild solutions. More precisely, in addition to assuming that our weak solutions satisfy an energy-type equality, specified in Assumption 4, which is related to allowing $\boldsymbol{\eta}_t$ to serve as a test function (multiplier), we also require some additional regularity on the Biot pressure, as specified below in Lemma 10.5.1.

More precisely, we introduce the space

$$\mathcal{W} \subset \mathcal{V}_{\text{sol}} \text{ such that Assumption 4 is satisfied,}$$

and consider solutions in $\mathcal{W}$ with the additional regularity for p_b as follows:

Proposition 10.5.1 (Uniqueness of Weak Solutions) *Let $c_p > 0$ with $\rho_b = 0$, and $\rho_p, c_b, \rho_f \geq 0$, $\alpha_n, \alpha_p, \mu_E, \lambda_E, D, \mu, k_p > 0$. Moreover, let $k_b \in L^\infty(\mathbb{R}) \cap Lip(\mathbb{R})$ such that $0 < k_{\min} \leq k_b \leq k_{\max}$. Suppose there is a weak solution $(\boldsymbol{u}, w, p_p, \boldsymbol{\eta}, p_b) \in \mathcal{W}$ and a time $T^* > 0$ so that the following holds:*

$$\nabla p_b \in L^2(0, T^*; \mathbf{L}^\infty(\Omega_b)).$$

Then, all weak solutions in $\mathcal{W}$ are equal to $(\boldsymbol{u}, w, p_p, \boldsymbol{\eta}, p_b)$ on any interval $t \in [0, T]$ with $T < T^*$.

Proof Assume that there are two solutions $\mathbf{s}^i = (\boldsymbol{u}^i, w^i, p_p^i, \boldsymbol{\eta}^i, p_b^i) \in \mathcal{W}$ for $i = 1, 2$ with $\mathbf{s}^1 = (\boldsymbol{u}^1, w^1, p_p^1, \boldsymbol{\eta}^1, p_b^1)$. Under the hypotheses that solutions belong to $\mathcal{W}$, each solution satisfies the formal energy inequality (10.37). Consider the difference $\overline{\mathbf{s}} = \mathbf{s}^1 - \mathbf{s}^2$ (with the "bar" notation for each coordinate as well) as a weak solution to the *difference* equation with the same data $\mathbf{F}_b, S, F_p, \mathbf{f}$, utilizing (in the appropriate sense) $\overline{\mathbf{s}}$ as a test function. Then, for any $T < T^*$, $\overline{\mathbf{s}}$ satisfies the inequality:

$$\begin{aligned}\mathcal{E}(\overline{\mathbf{s}}(T)) + \int_0^T \Big[2\mu_f\|\boldsymbol{D}(\overline{\boldsymbol{u}}(t))\|^2_{L^2(\Omega_f)} + \beta\|\overline{\boldsymbol{u}}(t)\cdot\boldsymbol{t}\|^2_{L^2(\omega_p)} + \|k_p^{1/2}\partial_s\overline{p}_p(t)\|^2_{L^2(\Omega_p)}\Big]dt \\ \leq \mathcal{E}(\overline{\mathbf{s}}(0)) - \int_0^T \Big([k_b^1(t)\nabla p_b^1(t) - k_b^2(t)\nabla p_b^2(t)], \nabla[p_b^1(t) - p_b^2(t)]\Big)_{\Omega_b} dt\end{aligned} \tag{10.72}$$

where we denoted

$$\begin{aligned}\mathcal{E}(\mathbf{s}^i) \equiv \frac{1}{2}\Big(&\rho_f\|\boldsymbol{u}^i\|^2_{L^2(\Omega_f)} + \rho_p\|w_t^i\|^2_{L^2(\omega_p)} + c_p\|p_p^i\|^2_{L^2(\Omega_p)} \\ &+ \|\boldsymbol{\eta}^i\|^2_E + c_b\|p_b^i\|^2_{L^2(\Omega_b)} + D\|\Delta w^i\|^2_{L^2(\omega_p)} + \gamma\|w^i\|^2_{L^2(\omega_p)}\Big)\end{aligned}$$

and

$$k_b^i(t) \equiv k_b\big(c_b p^i(t) + \alpha_b\nabla\cdot\boldsymbol{\eta}^i(t)\big).$$

We focus on the nonlinear term with the aim to obtain an estimate on $\overline{\mathbf{s}}$ through the functional $\mathcal{E}$. We can rewrite the nonlinear term as

$$\begin{aligned}&\int_0^T \Big([k_b^1(t)\nabla p_b^1(t) - k_b^2(t)\nabla p_b^2(t)], \nabla\overline{p}_b\Big)_{\Omega_b} dt \\ &= \int_0^T \big([k_b^1(t) - k_b^2(t)]\nabla p_b^1(t), \nabla\overline{p}_b(t)\big)_{L^2(\Omega_b)} dt \\ &+ \int_0^T \big(k_b^2(t)\nabla\overline{p}_b(t), \nabla\overline{p}_b(t)\big)_{L^2(\Omega_b)} dt,\end{aligned}$$

where we used $\overline{p} = p_b^1 - p_b^2$. Using the lower bound on the permeability function $0 < k_{\min} \le k_b$ and discarding the additional dissipation terms, we obtain the inequality

$$\mathcal{E}(\overline{\mathbf{s}}(T)) + k_{\min} \int_0^T \|\nabla \overline{p}_b\|_{L^2(\Omega_b)}^2 dt \le \mathcal{E}(\overline{\mathbf{s}}(0)) - \int_0^T \big([k_b^1(t) - k_b^2(t)]\nabla p_b^1(t), \nabla \overline{p}_b(t)\big) dt.$$

The nonlinear term on the right-hand side can be estimated in two ways, yielding the two cases in this theorem. In both cases, we will use the Lipschitz hypothesis on the function k_b. Namely, let $L_k > 0$ be the global Lipschitz constant so that

$$\begin{aligned} |k_b^1(t) - k_b^2(t)| &= \big|k_b\big(c_b p_b^1(t) + \alpha_b \nabla \cdot \boldsymbol{\eta}^1(t)\big) - k_b\big(c_b p_b^2(t) + \alpha_b \nabla \cdot \boldsymbol{\eta}^2(t)\big)\big| \\ &\le L_k |c_b \overline{p}_b(t) + \alpha_b \nabla \cdot \overline{\boldsymbol{\eta}}(t)|. \end{aligned}$$

By the Cauchy-Schwarz inequality, we get

$$\int_0^T \big([k_b^1(t) - k_b^2(t)]\nabla p_b^1(t), \nabla \overline{p}_b(t)\big) dt \le L_k \int_0^T \|\nabla p_b^1\|_{L^\infty(\Omega_b)} \|c_b \overline{p}_b(t) + \alpha_b \nabla \cdot \overline{\boldsymbol{\eta}}(t)\|_{L^2(\Omega_b)} \|\nabla \overline{p}_b\|_{L^2(\Omega_b)} dt.$$

Now we estimate the right-hand side by using regularity hypothesis, i.e., $\nabla p_b^1 \in L^2(0, T^*; \mathbb{L}^\infty(\Omega_b))$. Namely, from Young's and triangle inequalities, we have the following:

$$\begin{aligned} &\int_0^T \big([k_b^1(t) - k_b^2(t)]\nabla p_b^1(t), \nabla \overline{p}_b(t)\big) dt \le \int_0^T L_k \|\nabla p_b^1\|_{L^\infty(\Omega_b)} \|c_b \overline{p}_b \\ &\quad + \alpha_b \nabla \cdot \overline{\boldsymbol{\eta}}(t)\|_{L^2(\Omega_b)} \|\nabla \overline{p}_b\|_{L^2(\Omega_b)} dt \\ &\le \frac{L_k^2}{2\epsilon} \|\nabla p_b^1\|_{L^\infty(0,T;L^\infty(\Omega_b))}^2 \left\{ \int_0^T \|c_b \overline{p}_b\|_{L^2(\Omega_b)}^2 dt + \int_0^T \|\alpha_b \nabla \cdot \overline{\boldsymbol{\eta}}(t)\|_{L^2(\Omega_b)}^2 dt \right\} \\ &\quad + \frac{\epsilon}{2} \int_0^T \|\nabla \overline{p}_b\|_{L^2(\Omega_b)}^2 dt. \end{aligned}$$

Choosing $\epsilon = k_{\min}$ to absorb the pressure term into the left-hand side, and by bounding the elasticity term from above by the full $\|\cdot\|_E$ norm, we have the following estimate:

$$\begin{aligned} &\int_0^T \big([k_b^1(t) - k_b^2(t)]\nabla p_b^1(t), \nabla \overline{p}_b(t)\big) dt \le \int_0^T L_k \|\nabla p_b^1\|_{L^\infty(\Omega_b)} \|c_b \overline{p}_b \\ &\quad + \alpha_b \nabla \cdot \overline{\boldsymbol{\eta}}(t)\|_{L^2(\Omega_b)} \|\nabla \overline{p}_b\|_{L^2(\Omega_b)} dt \end{aligned}$$

$$\leq \int_0^T \frac{L_k[c_b + \alpha_b^2]}{2k_{\min}} \|\nabla p_b^1(t)\|^2_{L^\infty(\Omega_b)} \left\{ c_b \|\overline{p}_b(t)\|^2_{L^2(\Omega_b)} + \|\overline{\boldsymbol{\eta}}(t)\|^2_E \right\} dt$$
$$+ \frac{k_{\min}}{2} \int_0^T \|\nabla \overline{p}_b\|^2_{L^2(\Omega_b)} dt.$$

After discarding the dissipation term, we get another Grönwall-type estimate:

$$\mathcal{E}(\overline{\mathbf{s}}(T)) \leq C\mathcal{E}(\overline{\mathbf{s}}(0)) + C \frac{L_k^2[c_b + \alpha_b^2]}{2k_{\min}} \int_0^T \|\nabla p_b^1(t)\|^2_{L^\infty(\Omega_b)} \mathcal{E}(\overline{\mathbf{s}}(t)) dt \tag{10.73}$$

Uniqueness follows by applying the L^2-version of Grönwall's inequality [125].

□

We conclude this section by mentioning that in 3D, hypothesis of Proposition 10.5.1 would be satisfied, for instance, if $p_b^1 \in L^2(0, T; H^{2.5+\epsilon}(\Omega_b))$.

This concludes the chapter on linearly coupled fluid-poroelastic structure interaction.

Next, we discuss a nonlinearly coupled problem.

Chapter 11
FPSI with Nonlinear Coupling

In this chapter, we will consider a nonlinearly coupled FPSI system consisting of an incompressible fluid modeled by the Navier-Stokes equations interacting with a multilayered poroelastic structure, consisting of a thin reticular plate and a thick Biot poroelastic material modeled by the Biot equations. This FPSI system will be nonlinearly coupled, so that both the fluid domain and the Biot domain are moving domains, where the time-dependent configuration of the fluid domain is determined by the thin plate displacement and the time-dependent configuration of the Biot domain is determined by the displacement of the Biot material from its reference configuration. The nonlinearly coupled nature of this problem and the additional geometric nonlinearities arising from considering the Biot equations and the Navier-Stokes equations on moving domains make the analysis of this problem complicated. In this chapter, our goal is to develop a well-posedness theory for this problem, involving weak solutions to a regularized FPSI problem. We will then show that our well-posedness theory is compatible with real-life dynamics, by establishing a weak-classical consistency result that states that weak solutions to the regularized FPSI problem coincide with smooth solutions to the original problem when smooth solutions exist, as the regularization parameter goes to zero. This chapter is based on the work presented in [219].

Although the Biot equations have already been discussed in the previous chapter, for completeness, we begin this chapter by describing the Biot equations again, defined on a fixed domain, and then we move onto the specifics of the nonlinearly coupled FPSI problems and the modified equations that will be used in the analysis of the nonlinearly coupled problem. We precisely state the specific nonlinearly coupled FPSI model that we will consider and then carry out a priori estimates and define the concept of a weak solution to a regularized form of the problem. We use a splitting scheme to construct approximate solutions and then use compactness arguments for functions defined on moving domains, where there are additional subtleties that make this constructive existence proof distinct from

S. Čanić et al., *Deterministic and Stochastic Fluid-Structure Interaction*, Advances in Mathematical Fluid Mechanics, https://doi.org/10.1007/978-3-032-00898-5_11

the corresponding existence proof for the prototypical model of nonlinearly coupled FSI. We end this chapter by stating and establishing the weak-classical consistency result for the regularized FPSI problem.

11.1 Introduction

We study a time-dependent nonlinearly coupled fluid-structure interaction problem between the flow of an incompressible, viscous fluid, modeled by the Navier-Stokes equations, and bulk poroviscoelasticity modeled by the Biot equations. Bulk poroviscoelasticity means that the dimensions of the "free fluid flow" domain and that of the poroviscoelastic medium domain are the same. In particular, in this chapter, we consider the 2D case (see Fig. 11.1), which captures the main mathematical difficulties of such coupling. The free fluid flow and the Biot poro(visco)elastic medium are coupled across the current location of the interface, which has inertia and elastic energy, modeled by the reticular plate equation. A reticular plate is a lattice-type structure characterized by two properties: periodicity and small thickness, where periodicity refers to periodic cells (holes) distributed in all directions [94]. The reticular plate interface is transparent to fluid flow. We are interested in the existence of finite energy weak solutions (of the Leray-Hopf type; see Theorem 2.13.1 in the Preliminaries Sect. 2).

The problem we study here arises in many applications. In particular, we mention encapsulation of bioartificial organs [348] and blood flow in arteries that are modeled as poro(visco)elastic media to study drug transport through the vascular walls [22, 71, 341]. The reticular plate can be used to capture the elastodynamics behavior of the intima/elastic laminae layer of arterial walls, which is in direct contact with the blood flow on one side and a poroelastic medium consisting of the arterial media/adventitia complex on the other side.

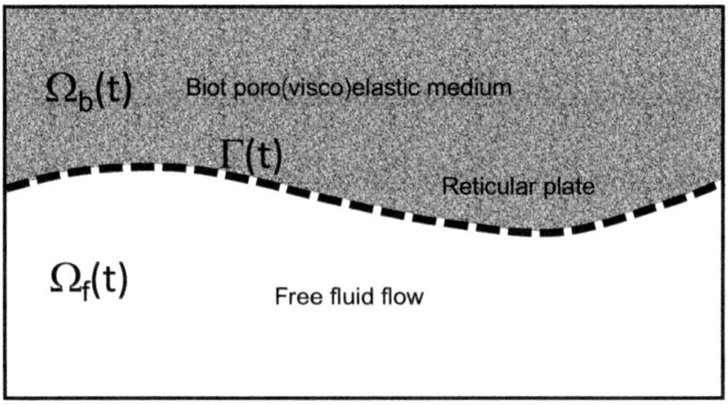

Fig. 11.1 A sketch of the fluid-poroelastic structure interaction domain

From a mathematical point of view, the primary difficulties in studying the 2D or 3D Navier-Stokes equations nonlinearly coupled to the 2D or 3D bulk poro(visco)elasticity arise from the fact that the finite energy solutions do not possess sufficient regularity to (1) define the moving domain (boundary) and the corresponding traces and (2) guarantee that all the integrals in the weak formulation of the problem are well defined. The first issue is related to the difficulties associated with 3D-3D fluid-structure coupling. The second issue is a consequence of the geometric nonlinearities associated with moving domain problems.

To get around these difficulties, we take the following approaches. First, the reticular plate at the interface associates mass and elastic energy to the interface and regularizes the boundary of the fluid domain. This is, however, not sufficient to deal with the fact that in the nonlinearly coupled problem, the Biot domain is also moving, and as a result certain integrals in the weak formulation over the moving Biot domain are not well defined. This is why we introduce a "consistent **regularized** weak formulation" of the coupled problem by introducing a suitably constructed convolution in spatial variables and regularizing only the problematic terms in the weak formulation of the coupled problem. We then prove the existence of a weak solution to the regularized problem and show that as the regularization parameter tends to zero, this solution converges to the solution of the original nonregularized problem in the case when the original problem has a classical solution and the Biot poroelastic matrix is viscoelastic. Here, by a classical solution, we mean a solution that is smooth both temporally and spatially that hence satisfies the original system of PDEs for the original FPSI problem pointwise. We call this type of result a **weak-classical consistency** result. Namely, we prove that if a classical solution for the fluid-poroviscoelastic structure interaction (FPSI) problem without regularization exists on time-interval $[0, T]$, then a sequence of weak solutions to the regularized FPSI problem constructed here converges to the classical solution on $[0, T]$ as the regularization parameter converges to 0.

In this chapter, *we prove the existence of a weak solution to a nonlinearly coupled fluid-structure interaction problem between the flow of an incompressible, viscous fluid modeled by the Navier-Stokes equations and a structure consisting of two solids—a thick poroviscoelastic medium modeled by the Biot equations and a thin interface with mass modeled by a reticular plate equation.* We mention that no viscoelasticity is needed for the proof of existence of a weak solution to the regularized problem. A weak solution to the regularized problem exists in the purely poroelastic case (and in the viscoelstic case). Viscoelasticity of the Biot poroelastic matrix is needed only in the proof of weak-classical consistency.

One of the interesting features of the approach presented here is that the proof of the existence of a weak solution is constructive. The main steps of the proof can be used to construct a numerical scheme to capture the physical solution to the problem [296]. The main idea of the proof is based on semidiscretizing the regularized FPSI problem in time by subdividing the time interval into N subintervals of width Δt. At each time step, we split the reticular plate subproblem from the regularized fluid-Biot subproblem using a Lie operator splitting strategy [165] (see Sect. 2.15.2 in Chap. 2). To deal with the moving domains, we use the Lagrangian map for the Biot domain and an Arbitrary Lagrangian-Eulerian mapping for the fluid domain, which maps a fixed, reference domain onto the current, physical domain (see Sect. 2.8.1 in Chap. 2). We switch between the reference domain formulation and moving domain formulation in the proof as needed. For each Δt, approximate solutions are constructed by "solving" the sequence of semidiscretized (linearized) problems defined on the current (approximate) moving domain for each $t_n = n\Delta t, n = 1, \ldots, N$. For each fixed Δt, we obtain energy estimates uniform in Δt, which allow us to deduce the existence of weakly convergent subsequences. Since the problem is highly nonlinear, these are not sufficient to pass to the limit in the weak formulations of the approximate problems. Strong convergence of approximate sequences is then obtained by using several compactness results: (1) For the Biot displacement, we will use the classical Aubin-Lions compactness lemma (see Theorem 2.9.4 in Chap. 2); (2) for the plate displacement, we will use Arzela-Ascoli Theorem (see Theorem 2.9.3 in Chap. 2); (3) for the Biot and plate velocity and pore pressure, we will use a Dreher and Jüngel's compactness result [126] (see Theorem 2.9.5 in Chap. 2); and (4) for the most involved part, which is the free fluid velocity defined on different time-dependent fluid domains, we will use the generalized Aubin-Lions-Simon compactness result (see Theorem 2.10.1 in Chap. 2). Once the strongly convergent subsequences are obtained from the compactness results, one would like to pass to the limit in the weak formulation to show that the limits of the subsequences are weak solutions to the regularized fluid-poroelastic structure interaction problem. However, this cannot be done yet, since the fluid velocity test functions are also defined on moving domains and we need to construct "appropriate" test functions, which can be compared for different domains and for which we can show convergence to a test function of the limiting, continuous problem. Luckily, in contrast with the classical fluid-elastic structure interaction problems, in our case, the fluid test functions decouple from the structure problem, and so it is slightly easier to construct appropriate test functions for which one can show uniform convergence to a test function for the continuous problem. With this final step, we can pass to the limit in the weak formulations of approximate problems and show that the limits of the approximate subsequences satisfy the continuous weak formulation of the regularized problem. This existence result is local in time because we can guarantee the nondegeneracy of the fluid domains, both for the free fluid flow and the filtrating flow through the poroelastic medium, only locally in time. However, using the approaches presented in [81, Section 5], the time of existence can be extended to the maximal time until either (1) the moving fluid domain or Biot domain degenerates (e.g., the interface touches the bottom of the fluid domain or the top of the Biot domain), (2) the pores in the poroelastic matrix degenerate in the sense that the Lagrangian mapping stops being injective, or (3) $T = \infty$.

We will finish this chapter by addressing the weak-classical consistency of the regularized problem, namely, we prove, using a Gronwall-type estimate, that the energy of the difference between the weak solution of the regularized problem and the classical (temporally and spatially smooth) solution to the original, non-regularized problem with viscoelastic Biot poroelastic matrix, converges to zero as the regularization parameter tends to zero. While the main idea is simple, the estimates are quite nontrivial due to the fact that we need to work with the integrals over regularized Biot domains and compare them with the integrals over the nonregularized moving domains. Details are presented in Sect. 11.9.

We emphasize that the mathematical results presented in this chapter are of significant, practical importance, and they have direct applications to real-life engineering and science problems. There has been a need in applications to understand not just poroelastic materials by themselves but the interaction between poroelastic materials and fluids. Recent progress in the design of bioartificial organs (see, e.g., [348]) sparked the need to study FPSI problems in which the fluid-structure interface has mass and elastic or poroelastic properties itself. Namely, in the recent work on the design of a bioartificial pancreas [348], the bioartifical pancreas consists of an *encapsulated* poroelastic agarose gel containing transplanted pancreatic cells, where the capsule containing the poroelastic medium is itself poroelastic, and it is designed to protect the transplanted cells within the poroelastic agarose gel from the patient's own immune cells while allowing the passage of oxygen and nutrients to the cells for long-time viability. This capsule is a thin poroelastic membrane/plate, which sits at the interface between the poroelastic gel containing the transplanted cells and the flow of blood carrying oxygen and nutrients to the bioartificial organ. The resulting mathematical problem in [348] is a fluid-poroelastic structure interaction problem in which the structure consists of two layers: a thin poroelastic plate located at the interface between the free fluid flow and a thick poroelastic medium modeled by the linear and nonlinear Biot equations, coupled over a fixed interface (linearized coupling). The well-posedness of this problem was studied in Chap. 10 for both the linear and nonlinear Biot equations, where the nonlinearity arises from the dependence of the permeability tensor on the fluid content. In Chap. 10, the fluid-structure interface with mass acts as a regularizing mechanism, providing sufficient information about the regularity of the interface and the free fluid domain to establish the existence of a finite-energy weak solution.

The model presented in this chapter is particularly important for real-life modeling and simulation because, to date, no existing work on weak solutions to fluid-structure interaction (FSI) problems between the flow of an incompressible, viscous fluid and a poroelastic solid has accounted for nonlinear coupling across a moving interface. In these nonlinearly coupled FSI problems, the complex motion of the moving boundaries can have a significant impact on the coupled dynamics. Such problems are especially relevant to applications such as the flow of blood through coronary arteries located on the surface of the heart, which contract and relax with each heartbeat [70, 341]. In the remainder of this chapter, we will present a well-posedness theory for a nonlinearly coupled (moving boundary) fluid-poroelastic structure interaction problem by constructing new tools for dealing with the equations of poroelasticity defined on a priori unknown and time-dependent domains.

11.2 Description of the Main Problem

We study fluid-poroelastic structure interaction between the flow of an incompressible, viscous fluid and a multilayered poro(visco)elastic structure consisting of two layers: a thick poro(visco)elastic layer modeled by the Biot equations and a thin elastic layer modeled by the reticular plate equation. The problem is set on a two-dimensional domain, which embodies all the main mathematical difficulties associated with the analysis of this problem. The entire two-dimensional domain $\hat{\Omega}$ is a union of the reference domain for the fluid subproblem $\hat{\Omega}_f$, the reference domain for the Biot poroviscoelastic material $\hat{\Omega}_b$, and the reference domain $\hat{\Gamma}$ of the elastic reticular plate, which serves as the interface separating the free fluid flow and the Biot medium:

$$\hat{\Omega} = \hat{\Omega}_b \cup \hat{\Omega}_f \cup \hat{\Gamma}, \text{ where } \hat{\Omega}_b = (0, L) \times (0, R), \ \hat{\Gamma} = (0, L) \times \{0\},$$
$$\hat{\Omega}_f = (0, L) \times (-R, 0).$$

These domains will evolve in time, giving rise to the time-dependent $\Omega(t) = \Omega_b(t) \cup \Omega_f(t) \cup \Gamma(t)$. We will be using the hat notation to denote objects associated with the reference domain. On each subdomain we will consider the following mathematical models.

11.2.1 The Biot Equations on a Moving Domain

We start by recalling the Biot equations defined on a fixed domain, which were introduced in the previous chapter (see (10.1) and (10.2)). We recall that these equations were first formulated in the context of geoscience by Maurice A. Biot in the seminal works [40] and [41], where these equations for poroelasticity arose from Biot's interest in modeling soil consolidation. Our nonlinearly coupled model of FPSI will involve the Biot equations posed on a *moving (time-dependent) domain*. Before we define the moving-domain version of the Biot equation, we first discuss the Biot equations defined on a fixed domain Ω_b, where Ω_b is a bounded domain in $\mathbb{R}^n$.

The Biot equations are a set of two coupled PDEs for the displacement $\boldsymbol{\eta} : \Omega_b \to \mathbb{R}^n$ of a poroelastic material from its reference configuration Ω_b and its pore pressure $p : \Omega_b \to \mathbb{R}$, given by

$$\rho_b \partial_{tt} \boldsymbol{\eta} - \nabla \cdot \boldsymbol{\sigma}(\nabla \boldsymbol{\eta}, p) = 0, \quad \text{in } \Omega_b, \tag{11.1}$$

$$c_0 \partial_t p + \alpha(\nabla \cdot \boldsymbol{\eta})_t - \nabla \cdot (\kappa \nabla p) = 0, \quad \text{in } \Omega_b, \tag{11.2}$$

where the Cauchy stress tensor is given by

$$\boldsymbol{\sigma}(\nabla \boldsymbol{\eta}, p) = \boldsymbol{\sigma}^E(\nabla \boldsymbol{\eta}, p) - \alpha p \boldsymbol{I}.$$

Here, $\boldsymbol{\sigma}^E(\nabla \boldsymbol{\eta}, p)$ is the elastic part of the full stress tensor, which can be given, for example, by the usual Piola-Kirchhoff stress tensor in linear elasticity.

The first equation, (11.1), describes the elastodynamics of the poroelastic material, arising from balance of momentum, where the influence of the pore pressure on the elastodynamics appears in the term $\alpha p \boldsymbol{I}$ in the stress tensor $\boldsymbol{\sigma}(\nabla \boldsymbol{\eta}, p)$ for the poroelastic material. The second equation, (11.2), describes conservation of mass, where the quantity $\kappa \nabla p$ is related to the filtration velocity $\boldsymbol{q}$ or the velocity of fluid flow through the pores of the poroelastic material, via **Darcy's law**:

$$\boldsymbol{q} = -\kappa \nabla p, \tag{11.3}$$

where $\kappa > 0$ is a positive constant. If we use Darcy's law to rewrite the term $-\nabla \cdot (\kappa \nabla p)$ appearing in the second equation (11.2) as $\nabla \cdot \boldsymbol{q}$, we see that the second equation (11.2) relates the local change in the volume occupied by the fluid in the pores of the material to the local expansion/contraction of the poroelastic material and the change in the pore pressure.

The **Biot equations on a moving domain** will be defined as follows. The elastodynamics equation (11.1) will be defined on the Lagrangian domain $\hat{\Omega}_b$ and the fluid equation (11.2) on the Eulerian domain $\Omega_b(t)$.

Let $\hat{\boldsymbol{\eta}} : [0, T] \times \hat{\Omega}_b \to \mathbb{R}^2$ denote the displacement of the Biot poroviscoelastic matrix from its reference configuration, and let $\hat{p} : \hat{\Omega}_b \to \mathbb{R}$ denote the fluid pore pressure. To specify the fluid equation given in terms of the fluid pore pressure in Eulerian formulation, we introduce the **Lagrangian map** by

$$\hat{\boldsymbol{\Phi}}_b^{\eta}(t, \cdot) = \mathrm{Id} + \hat{\boldsymbol{\eta}}(t, \cdot) : \hat{\Omega}_b \to \Omega_b(t), \tag{11.4}$$

with $(\boldsymbol{\Phi}_b^{\eta})^{-1}(t, \cdot) : \Omega_b(t) \to \hat{\Omega}_b$ denoting its inverse. The Biot equations are then given by:

$$\rho_b \partial_{tt} \hat{\boldsymbol{\eta}} = \hat{\nabla} \cdot \hat{S}_b(\hat{\nabla} \hat{\boldsymbol{\eta}}, \hat{p}) \qquad \text{in } \hat{\Omega}_b, \tag{11.5}$$

$$\frac{c_0}{[\det(\hat{\nabla} \hat{\boldsymbol{\Phi}}_b^{\eta})] \circ (\boldsymbol{\Phi}_b^{\eta})^{-1}} \frac{D}{Dt} p + \alpha \nabla \cdot \frac{D}{Dt} \boldsymbol{\eta} = \nabla \cdot (\kappa \nabla p) \qquad \text{in } \Omega_b(t), \tag{11.6}$$

where $\frac{D}{Dt} = \frac{d}{dt} + \big((\partial_t \boldsymbol{\eta}(t, \cdot) \circ (\boldsymbol{\Phi}_b^{\eta})^{-1}(t, \cdot)) \cdot \nabla \big)$ is the material derivative. See, e.g., [300, 351]. To recover the filtration fluid velocity $\boldsymbol{q}$, Darcy's law is used:

$$\boldsymbol{q} = -\kappa \nabla p \qquad \text{on } \Omega_b(t), \tag{11.7}$$

where κ is a positive permeability constant.

In this book we will consider both the viscoelastic and the purely elastic constitutive models for the Biot poroelastic matrix with the Piola-Kirchhoff stress tensor for the viscoelastic case given by

$$\hat{S}_b(\nabla\boldsymbol{\eta}, p) = 2\mu_e \hat{\boldsymbol{D}}(\hat{\boldsymbol{\eta}}) + \lambda_e(\hat{\nabla}\cdot\hat{\boldsymbol{\eta}})\boldsymbol{I} + 2\mu_v \hat{\boldsymbol{D}}(\hat{\boldsymbol{\eta}}_t) + \lambda_v(\hat{\nabla}\cdot\hat{\boldsymbol{\eta}}_t)\boldsymbol{I} - \alpha \det(\hat{\nabla}\hat{\boldsymbol{\Phi}}_b^{\eta})\hat{p}(\hat{\nabla}\hat{\boldsymbol{\Phi}}_b^{\eta})^{-t}, \tag{11.8}$$

where superscript t denotes matrix transposition and $A^{-t} = (A^{-1})^t$. The purely elastic case has the coefficients λ_v and μ_v equal to zero. Here, $\boldsymbol{D}$ denotes the symmetrized gradient, μ_e and λ_e are the Lamé parameters related to the elastic stress, μ_v and λ_v are the corresponding parameters related to the viscoelastic stress, and $\hat{\boldsymbol{\Phi}}_b^{\eta}$ is the Lagrangian map defined above. In Eq. (11.6), the Biot material displacement $\boldsymbol{\eta}$ and the pore pressure p are defined on the physical domain $\Omega_b(t)$ as

$$\boldsymbol{\eta}(t,\cdot) = \hat{\boldsymbol{\eta}}(t, (\boldsymbol{\Phi}_b^{\eta})^{-1}(t,\cdot)), \quad p(t,\cdot) = \hat{p}(t, (\boldsymbol{\Phi}_b^{\eta})^{-1}(t,\cdot)), \text{ where } \Omega_b(t) = \hat{\boldsymbol{\Phi}}_b^{\eta}(t, \hat{\Omega}_b).$$

We remark that in the last term of the Piola-Kirchhoff stress tensor (11.8), we have used the Piola transform (see (2.11) in Chap. 2), which is a transformation that maps tensors in Lagrangian coordinates to corresponding tensors in Eulerian coordinates in such a way that divergence-free tensors in Lagrangian coordinates remain divergence-free in Eulerian coordinates (see Lemma 2.8.1 in Chap. 2).

We note that *a priori* the notion of $\Omega_b(t)$ is not entirely clear, unless $\hat{\boldsymbol{\eta}}$ is sufficiently regular, and furthermore, the formulation of this problem makes sense only if the map $\hat{\boldsymbol{\Phi}}_b^{\eta} = \mathrm{Id} + \hat{\boldsymbol{\eta}}$ is an injective map from $\hat{\Omega}_b$ to $\Omega_b(t)$. We address these important issues later.

11.2.2 The Reticular Plate Equation

A reticular plate is a lattice-type structure characterized by two properties, periodicity and small thickness, where periodicity refers to periodic cells (holes) distributed in all directions [94]. Reticular plates, shells, or membranes are models for reticular tissue, which is a connective tissue made up of a network of supportive fibers that provide a framework for soft organs. The elastodynamics of reticular plates, studied in [94] using homogenization, is governed by a plate-type equation, defined on the equilibrium middle surface $\hat{\Gamma}$ of the homogenized plate or shell. The homogenized equation is given in terms of transverse displacement $\hat{\boldsymbol{\omega}} = \hat{\omega}\boldsymbol{e}_y$ from the reference configuration:

$$\rho_p \partial_{tt}\hat{\omega} + \hat{\Delta}^2\hat{\omega} = \hat{F}_p, \qquad \text{on } \hat{\Gamma}, \tag{11.9}$$

where ρ_p is the plate density coefficient and $\hat{F}_p$ is the external forcing on the plate in y direction, to be specified later in the coupling conditions. The constant ρ_p is the "average" plate density, which depends on the periodic structure. The in-plane bi-Laplacian $\hat{\Delta}^2$ (Laplace-Beltrami operator for curved $\hat{\Gamma}$'s) is associated with the elastic energy of the plate. Typically, there is a coefficient $\tilde{D}$ in front of the bi-Laplacian, which contains information about the periodicity of the structure and its stiffness properties [94]. Without loss of generality, we will assume that it is equal to 1. The source term $\hat{F}_p$ corresponds to the loading of the poroelastic plate, which will come from the jump in the normal stress (traction) between the free fluid on one side and the thick Biot poroelastic structure on the other (see (7)).

In our problem, the reticular plate separates the regions of free fluid flow and the Biot poroviscoelastic medium and is transparent to the flow between the two. This means, in particular, that there is no resistance to the fluid flow passing through the reticular place. However, due to the inertia and elastic energy of the plate, the analysis of the problem will be simplified due to the regularizing effects of the plate inertia and elastic energy, as we shall see below (see, e.g., Remark 11.4.1).

The time-dependent configuration of the plate

$$\Gamma(t) = \{(x, y) : 0 < x < L,\ y = \hat{\omega}(t, x)\},$$

forms the bottom boundary of the moving Biot domain $\Omega_b(t)$, and the remaining left, top, and right boundaries of the moving Biot domain $\Omega_b(t)$ are fixed in time. Hence, we impose $\boldsymbol{\eta} = 0$ on the left, top, and right boundaries of $\Omega_b(t)$. See Fig. 11.1. Hence, we can describe the moving domain $\Omega_b(t)$ as

$$\Omega_b(t) = \{(x, y) : 0 < x < L,\ \hat{\omega}(t, x) < y < R\}.$$

11.2.3 The Navier-Stokes Equations on a Moving Domain

The free flow of an incompressible, viscous fluid will be modeled by the Navier-Stokes equations

$$\left.\begin{aligned} \partial_t \boldsymbol{u} + (\boldsymbol{u} \cdot \nabla)\boldsymbol{u} &= \nabla \cdot \boldsymbol{\sigma}_f(\nabla \boldsymbol{u}, \pi) \\ \nabla \cdot \boldsymbol{u} &= 0 \end{aligned}\right\} \quad \text{in } \Omega_f(t), \tag{11.10}$$

where $\boldsymbol{u}$ is the fluid velocity and π is the fluid pressure. The Cauchy stress tensor is given by

$$\boldsymbol{\sigma}_f(\nabla \boldsymbol{u}, \pi) = 2\nu \boldsymbol{D}(\boldsymbol{u}) - \pi \boldsymbol{I},$$

where π is the fluid pressure and ν is kinematic viscosity coefficient. Notice that the fluid problem is defined on a moving domain, which is not known *a priori*. The moving fluid domain $\Omega_f(t)$ is a function of time, and it is determined by the plate displacement $\hat{\omega}$ as follows:

$$\Omega_f(t) = \{(x, y) : 0 < x < L, -R < y < \hat{\omega}(t, x)\}.$$

The fact that the free fluid domain depends on one of the unknowns in the problem presents a geometric nonlinearity that is difficult to deal with. We will be using the following **Arbitrary Lagrangian Eulerian (ALE) mapping** $\hat{\mathbf{\Phi}}_f^{\omega} : \hat{\Omega}_f \to \Omega_f(t)$ to map the fixed reference domain $\hat{\Omega}_f$ onto the current, physical domain $\Omega_f(t)$:

$$\hat{\mathbf{\Phi}}_f^{\omega}(\hat{x}, \hat{y}) = \left(\hat{x}, \hat{y} + \left(1 + \frac{\hat{y}}{R}\right)\hat{\omega}\right), \qquad (\hat{x}, \hat{y}) \in \hat{\Omega}_f. \tag{11.11}$$

In our analysis, we will use this ALE mapping to switch between the fixed and moving boundary formulations of the coupled problem as needed.

Remark 11.2.1 In numerical computations, it is typical to employ harmonic extension to construct the arbitrary Lagrangian-Eulerian (ALE) mapping. However, given the simplicity of our geometry, we chose to utilize the explicit formula for extension to simplify the calculations related to the change of variables. Since our methodology is not contingent on the particular selection of the ALE map, in scenarios involving more complex geometries where an explicit formula is not viable, alternatives such as harmonic extension can also be utilized (see Remark 2.8.1 in Chap. 2).

11.2.4 The Coupling Conditions

The Navier-Stokes equations (11.10), the Biot equations (11.5), (11.6), and the reticular plate equation (11.9) are coupled across the moving reticular plate interface $\Gamma(t)$ via two sets of coupling conditions: the kinematic and dynamic coupling conditions. To state these conditions, we introduce the following notation:

- The Biot Cauchy stress tensor defined on the physical domain $S_b(\nabla\boldsymbol{\eta}, p)$ is obtained by applying the Piola transform to the Biot Cauchy stress tensor $\hat{S}_b(\nabla\boldsymbol{\eta}, p)$ defined on the reference domain, to obtain:

$$\begin{aligned} S_b(\nabla\boldsymbol{\eta}, p) &= [\det(\hat{\nabla}\hat{\mathbf{\Phi}}_b^{\eta})^{-1}\hat{S}_b(\hat{\nabla}\hat{\boldsymbol{\eta}}, \hat{p})(\hat{\nabla}\hat{\mathbf{\Phi}}_b^{\eta})^t] \circ (\mathbf{\Phi}_b^{\eta})^{-1} \\ &= \left(\frac{1}{\det(\hat{\nabla}\hat{\mathbf{\Phi}}_b^{\eta})}\Big[2\mu_e\hat{\boldsymbol{D}}(\hat{\boldsymbol{\eta}}) + \lambda_e(\hat{\nabla}\cdot\hat{\boldsymbol{\eta}}) + 2\mu_v\hat{\boldsymbol{D}}(\hat{\boldsymbol{\eta}}_t)\right. \\ &\quad \left. +\lambda_v(\hat{\nabla}\cdot\hat{\boldsymbol{\eta}}_t)\Big](\hat{\nabla}\hat{\mathbf{\Phi}}_b^{\eta})^t\right) \circ (\mathbf{\Phi}_b^{\eta})^{-1} - \alpha p\boldsymbol{I}. \end{aligned} \tag{11.12}$$

- The Eulerian structure velocity of the Biot poroviscoelastic matrix is given at each point of the physical domain $\Omega_b(t)$ by

$$\boldsymbol{\xi}(t,\cdot) = \partial_t\hat{\boldsymbol{\eta}}\left(t, (\mathbf{\Phi}_b^{\eta})^{-1}(t,\cdot)\right). \tag{11.13}$$

- The normal unit vector to the moving interface $\Gamma(t)$ will be denoted by $\boldsymbol{n}(t)$, and the normal unit vector to the reference configuration of the interface $\hat{\Gamma}$ will be denoted by $\hat{\boldsymbol{n}}$. Note that $\hat{\boldsymbol{n}} = \boldsymbol{e}_y$. The vectors $\boldsymbol{n}(t)$ and $\hat{\boldsymbol{n}}$ point outward from $\Omega_f(t)$ and Ω_f and inward toward $\Omega_b(t)$ and Ω_b.

The following two sets of coupling conditions give rise to a well-defined bounded energy of the coupled problem:

(I) Kinematic coupling conditions:

- Continuity of normal components of velocity (conservation of mass of the fluid):

$$\boldsymbol{u} \cdot \boldsymbol{n}(t) = (\boldsymbol{q} + \boldsymbol{\xi}) \cdot \boldsymbol{n}(t), \qquad \text{on } (0, T) \times \Gamma(t). \tag{11.14}$$

- Slip in the tangential component of free fluid velocity, known as the Beavers-Joseph-Saffman condition [195, 196]:

$$\beta(\boldsymbol{\xi} - \boldsymbol{u}) \cdot \boldsymbol{\tau}(t) = \boldsymbol{\sigma}_f \boldsymbol{n}(t) \cdot \boldsymbol{\tau}(t), \qquad \text{on } (0, T) \times \Gamma(t), \tag{11.15}$$

  where $\beta \geq 0$ is a constant and $\boldsymbol{\tau}(t)$ is the rightward pointing unit tangent vector to $\Gamma(t)$.
- Continuity of displacements:

$$\hat{\boldsymbol{\eta}} = \hat{\omega} \boldsymbol{e}_y, \qquad \text{on } (0, T) \times \hat{\Gamma}. \tag{11.16}$$

(II) Dynamic coupling conditions:

- Balance of forces describing the body forcing on the plate as the difference between the normal components of normal stress coming from the Biot medium on one side and free fluid flow on the other:

$$\begin{aligned} \hat{F}_p = & - \det(\nabla \hat{\boldsymbol{\Phi}}^{\omega}_f)[\boldsymbol{\sigma}_f(\nabla \boldsymbol{u}, \pi) \circ \hat{\boldsymbol{\Phi}}^{\omega}_f](\nabla \hat{\boldsymbol{\Phi}}^{\omega}_f)^{-t} \hat{\boldsymbol{n}} \cdot \hat{\boldsymbol{n}} \\ & + \hat{S}_b(\hat{\nabla} \hat{\boldsymbol{\eta}}, \hat{p}) \hat{\boldsymbol{n}} \cdot \hat{\boldsymbol{n}}|_{\hat{\Gamma}}, \quad \text{on } (0, T) \times \hat{\Gamma}, \end{aligned} \tag{11.17}$$

  where $\hat{\boldsymbol{\Phi}}^{\omega}_f$ is the Arbitrary Lagrangian-Eulerian (ALE) mapping defined in (11.11).
- Balance of pressure at the interface:

$$-\boldsymbol{\sigma}_f(\nabla \boldsymbol{u}, \pi) \boldsymbol{n}(t) \cdot \boldsymbol{n}(t) + \frac{1}{2}|\boldsymbol{u}|^2 = p, \qquad \text{on } (0, T) \times \Gamma(t). \tag{11.18}$$

  See [97, 122, 265, 333] for the use of the dynamic (total) pressure $\pi + \frac{1}{2}|\boldsymbol{u}|^2$ on the left-hand side of (11.18).

11.2.5 The Initial and Boundary Conditions

For the fluid, we will assume rigid walls on $\partial\Omega_f(t) \setminus \Gamma(t)$ and impose a no-slip condition

$$\boldsymbol{u} = 0, \qquad \text{on } \partial\Omega_f(t) \setminus \Gamma(t).$$

Similarly, we will assume that the boundaries of the Biot poroviscoelastic medium, excluding the interface $\Gamma(t)$, are rigid and impose

$$\hat{\boldsymbol{\eta}} = 0 \;\text{ and }\; \hat{p} = 0, \qquad \text{on } \partial\hat{\Omega}_b \setminus \hat{\Gamma}.$$

Finally, we prescribe the following initial conditions:

$$\begin{aligned} \boldsymbol{u}(0) = \boldsymbol{u}_0 \quad &\text{in } \Omega_f(0), \\ \hat{\boldsymbol{\eta}}(0) = \hat{\boldsymbol{\eta}}_0,\; \partial_t\hat{\boldsymbol{\eta}}(0) = \hat{\boldsymbol{\xi}}_0 \quad &\text{in } \hat{\Omega}_b, \\ \hat{\omega}(0) = \hat{\omega}_0,\; \partial_t\hat{\omega}(0) = \hat{\zeta}_0 \quad &\text{in } \hat{\Gamma}, \\ \hat{p}(0) = \hat{p}_0 \quad &\text{in } \hat{\Omega}_b. \end{aligned}$$

11.2.6 Preview of the Main Results

Our first main result is the existence of a weak solution to a regularized FPSI problem, with a regularization parameter $\delta > 0$. The regularization will involve spatially regularizing the Biot displacement by extending the displacement $\hat{\boldsymbol{\eta}}$ on $\hat{\Omega}_b$ to a larger domain and using spatial convolution by a smooth compactly supported kernel, scaled by δ. This regularized FPSI problem will be introduced in Sect. 11.4. The existence result for the regularized FPSI problem holds for both elastic and viscoelastic Biot material. Here we state the theorem informally and refer the reader to Theorem 11.4.1 for the precise statement.

Theorem 11.2.1 (Existence of a Weak Solution to the Regularized Problem) *Let $\rho_b, \mu_e, \lambda_e, \alpha, \rho_p, \nu > 0$ and $\mu_v, \lambda_v \geq 0$. Moreover, assume that initial data are in the finite energy class and that, initially, the interface does not touch the bottom boundary of the fluid domain and the top boundary of the Biot domain and assume that certain compatibility conditions are satisfied. Then for every regularization parameter $\delta > 0$, there exists $T > 0$ (potentially depending on $\delta > 0$) such that there is a weak solution on $[0, T]$ to the regularized problem with regularization parameter δ. Furthermore, the weak solution to the regularized problem exists on a* *maximal time interval* $[0, T]$, *where either (1)* $T = \infty$ *or (2)* T *is finite and is the time at which either:*

- *The fluid or Biot domain degenerates so that the moving interface collides with the bottom boundary of* $\hat{\Omega}_f$ *or the top boundary of* $\hat{\Omega}_b$
- *The (regularized) Lagrangian mapping* $\hat{\mathbf{\Phi}}_b^{\eta^\delta}$ *for the Biot domain is no longer injective*

The second main result of this chapter is a weak-classical consistency result. Namely, in order to justify our regularization procedure and the corresponding definition of weak solutions to the regularized problem, we prove that weak solutions to the regularized problem indeed converge to the solution of the original (non-regularized) FPSI problem when a classical solution to the original FPSI problem exists. Namely, we prove the following result, made precise in Theorem 11.9.1.

Theorem 11.2.2 (Weak-Classical Consistency) *Assume that a classical (smooth) solution to the FPSI problem with a Biot poroviscoelastic medium exists on time-interval [0,T] for the case for which the viscoelasticity parameters* $\mu_v, \lambda_v > 0$. *Then every sequence of weak solutions to the regularized problem with regularization parameter* $\delta > 0$ *converges to the classical solution on* $[0, T]$ *as the regularization parameter* δ *converges to* 0. *In particular, the time interval of existence for the weak solutions to the regularized problem is uniform in regularization parameter and solutions to the regularized problem exist on the same time interval where the classical solution exists.*

Remark 11.2.2 An alternative formulation for Theorems 11.2.1 and 11.2.2 is that there exists a weak solution to an approximate problem of the original FPSI problem. Specifically, Theorem 11.2.2 asserts that the strong solution can be approximated by solutions to the regularized problem, the existence of which is guaranteed by Theorem 11.2.1.

The heart of the proof of this theorem is a bootstrap argument presented in Sect. 11.9.4. Namely, the main issue is that geometric quantities, such as the determinant of the displacement, cannot be estimated by the energy, and thus are not uniformly bounded in the regularization parameter δ. We derive appropriate bounds by using a bootstrap argument in combination with optimal convergence rate estimates for the convolution regularization. The main technical issue in comparing the classical solution with weak solutions to the regularized problem is the fact that they are defined on different domains. Therefore, we use a change of variables that transfers fluid velocities as vector fields and preserves the divergence-free condition. This transformation was introduced in [194] and was used in proving weak-strong type of results in the context of FSI in [83, 271, 298]. The corresponding estimates are carried out in section "Proof of Gronwall's Estimate".

11.3 Definition of a Weak Solution

Because the problem under consideration is nonlinearly coupled, the fluid domain $\Omega_f(t)$ and the Biot poroviscoelastic domain $\Omega_b(t)$ in physical space are time-dependent and not known a priori. To handle the moving domains, it is useful to introduce the mappings that map the reference domains $\hat{\Omega}_b$, $\hat{\Gamma}$, and $\hat{\Omega}_f$ onto the moving domains that depend on time and on the solution itself.

11.3.1 Mappings Between Reference and Physical Domains

Let

$$\hat{\mathbf{\Phi}}_b^\eta(t,\cdot):\hat{\Omega}_b\to\Omega_b(t),\qquad \hat{\mathbf{\Phi}}_\Gamma^\omega(t,\cdot):\hat{\Gamma}\to\Gamma(t),\qquad \hat{\mathbf{\Phi}}_f^\omega(t,\cdot):\hat{\Omega}_f\to\Omega_f(t),$$

be such that

$$\begin{aligned}
&\hat{\mathbf{\Phi}}_b^\eta=\mathrm{Id}+\hat{\boldsymbol{\eta}}(\hat{x},\hat{y}), && (\hat{x},\hat{y})\in\hat{\Omega}_b\\
&\hat{\mathbf{\Phi}}_\Gamma^\omega(\hat{x},0)=(\hat{x},\hat{\omega}(\hat{x})), && \hat{x}\in\hat{\Gamma}\\
&\hat{\mathbf{\Phi}}_f^\omega(\hat{x},\hat{y})=\left(\hat{x},\hat{y}+\left(1+\tfrac{\hat{y}}{R}\right)\hat{\omega}(\hat{x})\right), && (\hat{x},\hat{y})\in\hat{\Omega}_f,
\end{aligned}\tag{11.19}$$

with the inverse

$$(\mathbf{\Phi}_f^\omega)^{-1}(x,y)=\left(x,-R+\frac{R}{R+\hat{\omega}}(R+y)\right).\tag{11.20}$$

We are using $(\hat{x},\hat{y})$ to denote the coordinates on the reference domain and (x,y) the coordinates on the physical domain. Note that these mappings are *time dependent*, even though in the rest of this chapter we will not explicitly notate this time dependence for ease of notation.

The **Jacobians of the transformations** are given by:

$$\hat{\mathcal{J}}_f^\omega=1+\tfrac{\hat{\omega}}{R},\qquad \hat{\mathcal{J}}_b^\eta=\det(\boldsymbol{I}+\hat{\nabla}\hat{\boldsymbol{\eta}}),\qquad \hat{\mathcal{J}}_\Gamma^\omega=\sqrt{1+|\partial_{\hat{x}}\hat{\omega}|^2},\tag{11.21}$$

where $\hat{\mathcal{J}}_\Gamma^\omega$ measures the arc length difference between the reference and deformed configuration of the plate. Notice that in the Jacobian $\hat{\mathcal{J}}_f^\omega$, we dropped the absolute value sign since our results will hold up until the time of domain degeneracy when $|\hat{\omega}|\geq R$.

Under these mappings, the functions are transformed as follows.

Tranformations Under $\mathbf{\Phi}_f^\omega$ The fluid velocity $\boldsymbol{u}$ defined on $\Omega_f(t)$ is transferred to the fixed reference domain $\hat{\Omega}_f$ by

$$\hat{\boldsymbol{u}}(t, \hat{x}, \hat{y}) = \boldsymbol{u} \circ \hat{\boldsymbol{\Phi}}_f, \qquad \text{for } (\hat{x}, \hat{y}) \in \hat{\Omega}_f.$$

Recall that on the moving domain $\Omega_f(t)$, the fluid velocity $\boldsymbol{u}$ is divergence-free, i.e., $\nabla \cdot \boldsymbol{u} = 0$. However, when we pull the fluid velocity back to the reference domain, $\hat{\boldsymbol{u}}$ is not necessarily divergence-free on $\hat{\Omega}_f$. Hence, we want to reformulate the divergence-free condition on the fixed reference domain.

The Divergence Free Condition Let g be an arbitrary differentiable function defined on $\Omega_f(t)$. Then

$$\nabla g = \nabla \left(\hat{g} \circ (\boldsymbol{\Phi}_f^{\omega})^{-1} \right) = (\hat{\nabla}_f^{\omega} \hat{g}) \circ (\boldsymbol{\Phi}_f^{\omega})^{-1},$$

where $\hat{\nabla}_f^{\omega}$ is the *transformed gradient operator*:

$$\hat{\nabla}_f^{\omega} = \begin{pmatrix} \partial_{\hat{x}} - (R + y)\partial_{\hat{x}}\hat{\omega}\frac{R}{(R+\hat{\omega})^2}\partial_{\hat{y}} \\ \frac{R}{R+\hat{\omega}}\partial_{\hat{y}} \end{pmatrix} \quad \text{where} \quad y = \hat{y} + \left(1 + \frac{\hat{y}}{R}\right)\hat{\omega}. \tag{11.22}$$

Therefore, the divergence-free condition and the symmetrized gradient on the fixed reference domain $\hat{\Omega}_f$ are:

$$\hat{\nabla}_f^{\omega} \cdot \hat{\boldsymbol{u}} = 0, \qquad \hat{\boldsymbol{D}}_f^{\omega}(\hat{\boldsymbol{u}}) = \frac{1}{2}\left(\hat{\nabla}_f^{\omega}\hat{\boldsymbol{u}} + (\hat{\nabla}_f^{\omega}\hat{\boldsymbol{u}})^t\right).$$

Time Derivatives The time derivative transforms under the map $\hat{\boldsymbol{\Phi}}_f^{\omega}$ as follows:

$$\partial_t \boldsymbol{u} = \partial_t \hat{\boldsymbol{u}} - (\hat{\boldsymbol{w}} \cdot \hat{\nabla}_f^{\omega})\hat{\boldsymbol{u}} \quad \text{where} \quad \hat{\boldsymbol{w}} = \frac{R + \hat{y}}{R}\partial_t \hat{\omega} \boldsymbol{e}_y. \tag{11.23}$$

Transformations Under $\boldsymbol{\Phi}_b^{\omega}$ Given a scalar function g defined on $\Omega_b(t)$, the pullback of g to the reference domain $\hat{\Omega}_b$ is given by

$$\hat{g} = g \circ \hat{\boldsymbol{\Phi}}_b^{\eta}.$$

We claim that for some differential operator $\hat{\nabla}_b^{\eta}$, which we will determine below,

$$\nabla g = \nabla \left(\hat{g} \circ (\boldsymbol{\Phi}_b^{\eta})^{-1} \right) = (\hat{\nabla}_b^{\eta} \hat{g}) \circ (\boldsymbol{\Phi}_b^{\eta})^{-1},$$

where ∇ is a gradient on the physical domain, $\hat{\nabla}$ is a gradient on the reference domain, and $\hat{\nabla}_b^{\eta}$ is a differential operator (different from $\hat{\nabla}$) on the reference domain. For any function g defined on the physical domain, we have that

$$\hat{\nabla}\left(g \circ \hat{\boldsymbol{\Phi}}_b^{\eta}\right) = [(\nabla g) \circ \hat{\boldsymbol{\Phi}}_b^{\eta}] \cdot (\boldsymbol{I} + \hat{\nabla}\hat{\boldsymbol{\eta}}).$$

Hence, for

$$\hat{\nabla}_b^\eta \hat{g} = (\nabla g) \circ \hat{\boldsymbol{\Phi}}_b^\eta,$$

we get the following explicit formula for the *transformed gradient operator* $\hat{\nabla}_b^\eta$ on $\hat{\Omega}_b$:

$$\hat{\nabla}_b^\eta \hat{g} = \left(\frac{\partial \hat{g}}{\partial \hat{x}}, \frac{\partial \hat{g}}{\partial \hat{y}}\right) \cdot (\boldsymbol{I} + \hat{\nabla}\hat{\boldsymbol{\eta}})^{-1}. \tag{11.24}$$

Notice that the invertibility of the matrix $\boldsymbol{I} + \hat{\nabla}\hat{\boldsymbol{\eta}}$ will be related to whether the map $(\hat{x}, \hat{y}) \to (\hat{x}, \hat{y}) + \hat{\boldsymbol{\eta}}(\hat{x}, \hat{y})$ is a bijection between $\hat{\Omega}_b$ and $\Omega_b(t)$.

11.3.2 Weak Solution

We now derive the definition of a weak solution to the given FPSI problem, by means of the following formal calculation. We start with the fluid equations and multiply by a test function $\boldsymbol{v}$. Recall the definition of the Eulerian structure velocity $\boldsymbol{\xi}$ from (11.13). For the inertia term of the Navier-Stokes equations, using Reynold's transport theorem (see Theorem 2.8.1 in Chap. 2) and integration by parts, we obtain:

$$\begin{aligned}
\int_{\Omega_f(t)} (\partial_t \boldsymbol{u} + (\boldsymbol{u} \cdot \nabla)\boldsymbol{u})) \cdot \boldsymbol{v} &= \frac{d}{dt}\int_{\Omega_f(t)} \boldsymbol{u} \cdot \boldsymbol{v} - \int_{\Omega_f(t)} \boldsymbol{u} \cdot \partial_t \boldsymbol{v} - \int_{\Gamma(t)} (\boldsymbol{\xi} \cdot \boldsymbol{n})\boldsymbol{u} \cdot \boldsymbol{v} \\
&\quad + \frac{1}{2}\int_{\Omega_f(t)} [((\boldsymbol{u} \cdot \nabla)\boldsymbol{u}) \cdot \boldsymbol{v} - (\boldsymbol{u} \cdot \nabla)\boldsymbol{v}) \cdot \boldsymbol{u}] + \frac{1}{2}\int_{\Gamma(t)} (\boldsymbol{u} \cdot \boldsymbol{n})\boldsymbol{u} \cdot \boldsymbol{v} \\
&= \frac{d}{dt}\int_{\Omega_f(t)} \boldsymbol{u} \cdot \boldsymbol{v} - \int_{\Omega_f(t)} \boldsymbol{u} \cdot \partial_t \boldsymbol{v} + \frac{1}{2}\int_{\Omega_f(t)} [((\boldsymbol{u} \cdot \nabla)\boldsymbol{u}) \cdot \boldsymbol{v} - ((\boldsymbol{u} \cdot \nabla)\boldsymbol{v}) \cdot \boldsymbol{u}] \\
&\quad + \frac{1}{2}\int_{\Gamma(t)} (\boldsymbol{u} \cdot \boldsymbol{n} - 2\boldsymbol{\xi} \cdot \boldsymbol{n})\boldsymbol{u} \cdot \boldsymbol{v}.
\end{aligned}$$

For the diffusive term of the Navier Stokes equations, we integrate by parts to obtain

$$-\int_{\Omega_f(t)} (\nabla \cdot \boldsymbol{\sigma}_f(\nabla \boldsymbol{u}, \pi)) \cdot \boldsymbol{v} = 2\nu \int_{\Omega_f(t)} \boldsymbol{D}(\boldsymbol{u}) : \boldsymbol{D}(\boldsymbol{v}) - \int_{\Gamma(t)} \boldsymbol{\sigma}_f(\nabla \boldsymbol{u}, \pi)\boldsymbol{n} \cdot \boldsymbol{v},$$

where we used the fact that the test function $\boldsymbol{v}$ is divergence-free to eliminate the pressure, and we use the fact that the test function satisfies $\boldsymbol{v} = 0$ on $\partial\Omega_f(t) \setminus \Gamma(t)$ due to the boundary conditions for $\boldsymbol{u}$.

Next, we multiply the structure equation (11.5) by a test function $\hat{\boldsymbol{\psi}}$ to obtain

$$\int_{\hat{\Omega}_b} (\rho_b \partial_{tt}\hat{\boldsymbol{\eta}} - \hat{\nabla} \cdot \hat{S}_b(\hat{\nabla}\hat{\boldsymbol{\eta}}, \hat{p})) \cdot \hat{\boldsymbol{\psi}} = \rho_b \left(\frac{d}{dt} \int_{\hat{\Omega}_b} \partial_t \hat{\boldsymbol{\eta}} \cdot \hat{\boldsymbol{\psi}} - \int_{\hat{\Omega}_b} \partial_t \hat{\boldsymbol{\eta}} \cdot \partial_t \hat{\boldsymbol{\psi}} \right)$$

$$+ \int_{\hat{\Omega}_b} \hat{S}_b(\hat{\nabla}\hat{\boldsymbol{\eta}}, \hat{p}) : \hat{\nabla}\hat{\boldsymbol{\psi}} + \int_{\hat{\Gamma}} \hat{S}_b(\hat{\nabla}\hat{\boldsymbol{\eta}}, \hat{p}) \boldsymbol{e}_y \cdot \hat{\boldsymbol{\psi}}$$

$$= \rho_b \left(\frac{d}{dt} \int_{\hat{\Omega}_b} \partial_t \hat{\boldsymbol{\eta}} \cdot \hat{\boldsymbol{\psi}} - \int_{\hat{\Omega}_b} \partial_t \hat{\boldsymbol{\eta}} \cdot \partial_t \hat{\boldsymbol{\psi}} \right)$$

$$+ \int_{\hat{\Omega}_b} (2\mu_e \hat{\boldsymbol{D}}(\hat{\boldsymbol{\eta}}) : \hat{\boldsymbol{D}}(\hat{\boldsymbol{\psi}}) + \lambda_e (\hat{\nabla} \cdot \hat{\boldsymbol{\eta}})(\hat{\nabla} \cdot \hat{\boldsymbol{\psi}}) + 2\mu_v \hat{\boldsymbol{D}}(\partial_t \hat{\boldsymbol{\eta}}) : \hat{\boldsymbol{D}}(\hat{\boldsymbol{\psi}})$$

$$+ \lambda_v (\hat{\nabla} \cdot \partial_t \hat{\boldsymbol{\eta}})(\hat{\nabla} \cdot \hat{\boldsymbol{\psi}}))$$

$$- \alpha \int_{\Omega_b(t)} p(\nabla \cdot \boldsymbol{\psi}) + \int_{\hat{\Gamma}} \hat{S}_b(\nabla\hat{\boldsymbol{\eta}}, \hat{p}) \boldsymbol{e}_r \cdot \hat{\boldsymbol{\psi}}.$$

Except on $\hat{\Gamma}$, there are no boundary terms, because $\hat{\boldsymbol{\eta}} = 0$ on the left, top, and right boundaries of $\hat{\Omega}_b$, and hence the same condition holds for the corresponding test function $\hat{\boldsymbol{\psi}}$. Note that in the integral over $\Omega_b(t)$, $\boldsymbol{\psi} := \hat{\boldsymbol{\psi}} \circ (\boldsymbol{\Phi}_b^{\eta})^{-1}$.

Finally, we test the second equation (11.6) corresponding to the evolution of the pore pressure for the Biot poroviscoelastic medium with a test function r and recall the definition of the Darcy velocity $\boldsymbol{q}$ from (11.7), keeping in mind that $\boldsymbol{n}$ is the *inward* normal vector to $\Omega_b(t)$:

$$\int_{\Omega_b(t)} \left(\frac{c_0}{[\det(\hat{\nabla}\hat{\boldsymbol{\Phi}}_b^{\eta})] \circ (\boldsymbol{\Phi}_b^{\eta})^{-1}} \frac{D}{Dt} p + \alpha \nabla \cdot \frac{D}{Dt} \boldsymbol{\eta} - \nabla \cdot (\kappa \nabla p) \right) r$$

$$= \int_{\hat{\Omega}_b} c_0 \partial_t \hat{p} \cdot \hat{r} + \int_{\Omega_b(t)} \alpha \left(\nabla \cdot \frac{D}{Dt} \boldsymbol{\eta} \right) r + \int_{\Omega_b(t)} \kappa \nabla p \cdot \nabla r - \int_{\Gamma(t)} (\boldsymbol{q} \cdot \boldsymbol{n}) r$$

$$= \frac{d}{dt} \int_{\hat{\Omega}_b} c_0 \hat{p} \cdot \hat{r} - \int_{\hat{\Omega}_b} c_0 \hat{p} \cdot \partial_t \hat{r} - \int_{\Omega_b(t)} \alpha \frac{D}{Dt} \boldsymbol{\eta} \cdot \nabla r - \alpha \int_{\Gamma(t)} (\boldsymbol{\xi} \cdot \boldsymbol{n}) r$$

$$+ \int_{\Omega_b(t)} \kappa \nabla p \cdot \nabla r - \int_{\Gamma(t)} (\boldsymbol{q} \cdot \boldsymbol{n}) r.$$

There are no boundary terms except on $\Gamma(t)$ from the application of integration by parts in the integral involving α and in the integral involving κ because of the Dirichlet boundary condition $r = 0$ (since $p = 0$) on the left, top, and right boundaries of $\hat{\Omega}_b$.

After adding the two stress terms and recalling the definition of $\hat{\boldsymbol{\Phi}}_{\Gamma}^{\omega}$ in (11.19) and $\hat{\mathcal{J}}_{\Gamma}^{\omega}$ in (11.21), we obtain:

$$- \int_{\Gamma(t)} \boldsymbol{\sigma}_f(\nabla \boldsymbol{u}, \pi)\boldsymbol{n} \cdot \boldsymbol{v} + \int_{\hat{\Gamma}} \hat{S}_b(\hat{\nabla}\hat{\boldsymbol{\eta}}, \hat{p})\boldsymbol{e}_y \cdot \hat{\boldsymbol{\psi}}$$

$$= \int_{\Gamma(t)} \boldsymbol{\sigma}_f(\nabla \boldsymbol{u}, \pi)\boldsymbol{n}\cdot(\boldsymbol{\psi} - \boldsymbol{v}) + \int_{\hat{\Gamma}} (\hat{S}_b(\hat{\nabla}\hat{\boldsymbol{\eta}}, \hat{p})\boldsymbol{e}_y - \hat{\mathcal{J}}_{\Gamma}^{\omega} \cdot (\boldsymbol{\sigma}_f(\nabla \boldsymbol{u}, \pi)\boldsymbol{n}|_{\Gamma(t)} \circ \boldsymbol{\Phi}_{\Gamma}^{\omega}) \cdot \hat{\boldsymbol{\psi}}.$$

Since the displacement of the plate is only in the y direction, so that $\hat{\boldsymbol{\eta}} = \hat{\omega}\boldsymbol{e}_y$ on $\hat{\Gamma}$, the test function $\hat{\boldsymbol{\psi}}$ points in the y direction on $\hat{\Gamma}$ as well. We will denote by $\hat{\varphi}$ the magnitude of $\hat{\boldsymbol{\psi}}|_{\hat{\Gamma}}$ so that $\hat{\boldsymbol{\psi}} = \hat{\varphi}\boldsymbol{e}_y$ on $\hat{\Gamma}$. By the dynamic coupling condition (11.17), we have that the previous expression is equal to

$$= \int_{\Gamma(t)} \boldsymbol{\sigma}_f(\nabla \boldsymbol{u}, \pi)\boldsymbol{n} \cdot (\boldsymbol{\psi} - \boldsymbol{v}) + \int_{\hat{\Gamma}} \hat{F}_p \cdot \hat{\varphi} = \int_{\Gamma(t)} \boldsymbol{\sigma}_f(\nabla \boldsymbol{u}, \pi)\boldsymbol{n} \cdot (\boldsymbol{\psi} - \boldsymbol{v})$$
$$+ \int_{\hat{\Gamma}} (\rho_p \partial_{tt}\hat{\omega} + \hat{\Delta}^2\hat{\omega})\hat{\varphi}$$
$$= \int_{\Gamma(t)} \boldsymbol{\sigma}_f(\nabla \boldsymbol{u}, \pi)\boldsymbol{n} \cdot \boldsymbol{n}(\psi_n - v_n) + \int_{\Gamma(t)} \boldsymbol{\sigma}_f(\nabla \boldsymbol{u}, \pi)\boldsymbol{n} \cdot \boldsymbol{\tau}(\psi_\tau - v_\tau)$$
$$+ \int_{\hat{\Gamma}} (\rho_p \partial_{tt}\hat{\omega} + \hat{\Delta}^2\hat{\omega})\hat{\varphi}$$
$$= \int_{\Gamma(t)} \boldsymbol{\sigma}_f(\nabla \boldsymbol{u}, \pi)\boldsymbol{n} \cdot \boldsymbol{n}(\psi_n - v_n) + \int_{\Gamma(t)} \beta(\boldsymbol{\xi} - \boldsymbol{u}) \cdot \boldsymbol{\tau}(\psi_\tau - v_\tau)$$
$$+ \int_{\hat{\Gamma}} (\rho_p \partial_{tt}\hat{\omega} + \hat{\Delta}^2\hat{\omega})\hat{\varphi}$$
$$= \int_{\Gamma(t)} \left(\frac{1}{2}|\boldsymbol{u}|^2 - p\right)(\psi_n - v_n) + \int_{\Gamma(t)} \beta(\boldsymbol{\xi} - \boldsymbol{u}) \cdot \boldsymbol{\tau}(\psi_\tau - v_\tau)$$
$$+ \frac{d}{dt}\left(\int_{\hat{\Gamma}} \rho_p \partial_t\hat{\omega} \cdot \hat{\varphi}\right) - \int_{\hat{\Gamma}} \rho_p \partial_t\hat{\omega} \cdot \partial_t\hat{\varphi} + \int_{\hat{\Gamma}} \hat{\Delta}\hat{\omega} \cdot \hat{\Delta}\hat{\varphi},$$

where we used the coupling conditions (11.15) and (11.18) in the last step. For clarity, we note that in the preceding calculation and in the remainder of the chapter, $\boldsymbol{n}$ denotes the unit normal vector along $\Gamma(t)$ that points upward toward $\Omega_b(t)$, and $\boldsymbol{\tau}$ denotes the unit tangential vector along $\Gamma(t)$ that points to the right.

The weak formulation then follows by summing everything together. Before we state the definition of a weak solution to our FPSI problem, we introduce the following notation. Let $\hat{\zeta}$ denote the **transverse velocity** of the plate so that

$$\partial_t\hat{\omega} = \hat{\zeta}, \tag{11.25}$$

and let $\zeta = \hat{\zeta} \circ (\boldsymbol{\Phi}_{\Gamma}^{\omega})^{-1}$.

Definition 11.3.1 The ordered four-tuple $(\boldsymbol{u}, \hat{\omega}, \hat{\boldsymbol{\eta}}, p)$ satisfies the weak formulation to the nonlinearly coupled FPSI problem if for every test function $(\boldsymbol{v}, \hat{\varphi}, \hat{\boldsymbol{\psi}}, r)$

that is C_c^1 in time on $[0, T)$ taking values in the test space, satisfying $\hat{\boldsymbol{\psi}} = \hat{\varphi}\boldsymbol{e}_y$ on $\hat{\Gamma}$, we have that

$$\begin{aligned}
&-\int_0^T\int_{\Omega_f(t)} \boldsymbol{u}\cdot\partial_t\boldsymbol{v} + \frac{1}{2}\int_0^T\int_{\Omega_f(t)} [((\boldsymbol{u}\cdot\nabla)\boldsymbol{u})\cdot\boldsymbol{v} - ((\boldsymbol{u}\cdot\nabla)\boldsymbol{v})\cdot\boldsymbol{u}] \\
&\qquad + \frac{1}{2}\int_0^T\int_{\Gamma(t)} (\boldsymbol{u}\cdot\boldsymbol{n} - 2\zeta\boldsymbol{e}_y\cdot\boldsymbol{n})\boldsymbol{u}\cdot\boldsymbol{v} \\
&\quad + 2\nu\int_0^T\int_{\Omega_f(t)} \boldsymbol{D}(\boldsymbol{u}) : \boldsymbol{D}(\boldsymbol{v}) + \int_0^T\int_{\Gamma(t)}\left(\frac{1}{2}|\boldsymbol{u}|^2 - p\right)(\psi_n - v_n) \\
&\qquad + \beta\int_0^T\int_{\Gamma(t)} (\zeta\boldsymbol{e}_y - \boldsymbol{u})\cdot\boldsymbol{\tau}(\boldsymbol{\psi} - \boldsymbol{v})\cdot\boldsymbol{\tau} \\
&\quad - \rho_p\int_0^T\int_{\hat{\Gamma}} \partial_t\hat{\omega}\cdot\partial_t\hat{\varphi} + \int_0^T\int_{\hat{\Gamma}} \hat{\Delta}\hat{\omega}\cdot\hat{\Delta}\hat{\varphi} - \rho_b\int_0^T\int_{\hat{\Omega}_b} \partial_t\hat{\boldsymbol{\eta}}\cdot\partial_t\hat{\boldsymbol{\psi}} \\
&\qquad + 2\mu_e\int_0^T\int_{\hat{\Omega}_b} \hat{\boldsymbol{D}}(\hat{\boldsymbol{\eta}}) : \hat{\boldsymbol{D}}(\hat{\boldsymbol{\psi}}) \\
&+\lambda_e\int_0^T\int_{\hat{\Omega}_b}(\hat{\nabla}\cdot\hat{\boldsymbol{\eta}})(\hat{\nabla}\cdot\hat{\boldsymbol{\psi}}) + 2\mu_v\int_0^T\int_{\hat{\Omega}_b}\hat{\boldsymbol{D}}(\partial_t\hat{\boldsymbol{\eta}}) : \hat{\boldsymbol{D}}(\hat{\boldsymbol{\psi}}) + \lambda_v\int_0^T\int_{\hat{\Omega}_b}(\hat{\nabla}\cdot\partial_t\hat{\boldsymbol{\eta}})(\hat{\nabla}\cdot\hat{\boldsymbol{\psi}}) \\
&-\alpha\int_0^T\int_{\Omega_b(t)} p\nabla\cdot\boldsymbol{\psi} - c_0\int_0^T\int_{\hat{\Omega}_b}\hat{p}\cdot\partial_t\hat{r} - \alpha\int_0^T\int_{\Omega_b(t)}\frac{D}{Dt}\boldsymbol{\eta}\cdot\nabla r - \alpha\int_0^T\int_{\Gamma(t)}(\zeta\boldsymbol{e}_y\cdot\boldsymbol{n})r \\
&\qquad + \kappa\int_0^T\int_{\Omega_b(t)}\nabla p\cdot\nabla r - \int_0^T\int_{\Gamma(t)}((\boldsymbol{u} - \zeta\boldsymbol{e}_y)\cdot\boldsymbol{n})r \\
&= \int_{\Omega_f(0)}\boldsymbol{u}(0)\cdot\boldsymbol{v}(0) + \rho_p\int_{\hat{\Gamma}}\partial_t\hat{\omega}(0)\cdot\hat{\varphi}(0) + \rho_b\int_{\hat{\Omega}_b}\partial_t\hat{\boldsymbol{\eta}}(0)\cdot\hat{\boldsymbol{\psi}}(0) + c_0\int_{\hat{\Omega}_b}\hat{p}(0)\cdot\hat{r}(0).
\end{aligned} \tag{11.26}$$

Remark 11.3.1 It is immediate to see that a classical (temporally and spatially smooth) solution to the FPSI problem satisfies the weak formulation stated above. However, when considering less regular solutions (in particular, weak solutions in the class of finite-energy solutions), the above weak formulation *is inadequate for the regularity of finite-energy solutions* for the following reason. By the energy estimates (see Sect. 11.4.2), the regularity of the structure displacement $\hat{\boldsymbol{\eta}}$ on $\hat{\Omega}_b$ is $L^\infty(0, T, H^1(\hat{\Omega}_b))$, which is not enough regularity to make sense of the term

$$\alpha\int_{\Omega_b(t)} p\nabla\cdot\boldsymbol{\psi},$$

since the test function has regularity $\hat{\boldsymbol{\psi}} \in H^1(\hat{\Omega}_b)$ on the *fixed reference domain*, due to the corresponding finite energy regularity of $\hat{\boldsymbol{\eta}}$. Hence, after changing variables, which adds an extra factor of $\det(\boldsymbol{I} + \hat{\nabla}\hat{\boldsymbol{\eta}})$ arising from the Jacobian, which is only in $L^\infty(0, T; L^1(\hat{\Omega}_b))$ in two dimensions, there is not enough regularity to guarantee that this integral is finite. Therefore, we cannot interpret the above notion of weak solution properly in the space of finite energy solutions, as the finite energy space does not have enough regularity to make sense of certain integrals in the weak formulation, involving the deformed domain $\Omega_b(t)$.

This is why we introduce a regularized problem, which is consistent with the original problem in the sense that weak solutions to the regularized problem converge, as the regularization parameter tends to zero, to a smooth solution of the original, nonregularized problem, when a smooth solution exists. This weak-classical consistency result will be shown in Sect. 11.9.

11.4 Regularized Weak Solution and Statement of Existence Result

Since all the mathematical challenges related to the inability to properly interpret all of the terms in the weak solution arise fundamentally from the *lack of regularity* of $\hat{\boldsymbol{\eta}}$ on $\hat{\Omega}_b$, we will regularize $\hat{\boldsymbol{\eta}}$ via a convolution with a smooth, compactly supported kernel and introduce an appropriate *regularized weak formulation* of the original FPSI problem. Because we are working on a bounded domain $\hat{\Omega}_b$, we must be careful to introduce the convolution in a way that preserves the Dirichlet condition on the left, top, and right boundaries of $\hat{\Omega}_b = (0, L) \times (0, R)$.

This is why we define an *extended domain* $\tilde{\Omega}_b$:

$$\tilde{\Omega}_b = [-L, 2L] \times [-R, 2R],$$

so that for $\delta < \min(L, R)$ the convolution of a function on $\tilde{\Omega}_b$ with a smooth function of compact support in the closed ball of radius δ gives a function defined on $\hat{\Omega}_b$. We then introduce an odd extension along the lines $\hat{x} = 0$, $\hat{x} = L$, $\hat{y} = 0$, and $\hat{y} = R$ as follows.

Definition 11.4.1 Given $\hat{\boldsymbol{\eta}}$ defined on $\hat{\Omega}_b$ satisfying $\hat{\boldsymbol{\eta}} = 0$ on $\hat{x} = 0$, $\hat{x} = L$, and $\hat{y} = R$, and $\hat{\boldsymbol{\eta}} = \hat{\omega}\boldsymbol{e}_y$ on $\hat{y} = 0$, define the **odd extension of $\hat{\boldsymbol{\eta}}$ to $\tilde{\Omega}_b$** by keeping $\hat{\boldsymbol{\eta}}$ the same on $\hat{\Omega}_b = [0, L] \times [0, R]$ and defining $\hat{\boldsymbol{\eta}}$ outside of the closure of $\hat{\Omega}_b$ as follows:

1. On $[0, L] \times [-R, 0]$, set $\hat{\boldsymbol{\eta}}(\hat{x}, \hat{y}) = \hat{\omega}(\hat{x})\boldsymbol{e}_y + (\hat{\omega}(\hat{x})\boldsymbol{e}_y - \hat{\boldsymbol{\eta}}(\hat{x}, -\hat{y}))$.
2. On $[0, L] \times [R, 2R]$, set $\hat{\boldsymbol{\eta}}(\hat{x}, \hat{y}) = -\hat{\boldsymbol{\eta}}(\hat{x}, 2R - \hat{y})$.
3. On $[-L, 0] \times [-R, 2R]$, set $\hat{\boldsymbol{\eta}}(\hat{x}, \hat{y}) = -\hat{\boldsymbol{\eta}}(-\hat{x}, \hat{y})$.
4. On $[L, 2L] \times [-R, 2R]$, set $\hat{\boldsymbol{\eta}}(\hat{x}, \hat{y}) = -\hat{\boldsymbol{\eta}}(2L - \hat{x}, \hat{y})$.

Let σ be a radially symmetric function on $\mathbb{R}^2$ with compact support in the closed ball of radius one such that $\int_{\mathbb{R}^2} \sigma = 1$, and define

$$\sigma_\delta = \delta^{-2}\sigma(\delta^{-1}\boldsymbol{x}), \qquad \text{on } \mathbb{R}^2.$$

Definition 11.4.2 We define the following **regularized functions**, which are spatially smooth on $\hat{\Omega}_b$:

- The regularized Biot displacement, which is obtained by extending $\hat{\boldsymbol{\eta}}$ to $\tilde{\Omega}_b$ by odd extension and defining:

$$\hat{\boldsymbol{\eta}}^\delta(t, \hat{x}, \hat{y}) = (\hat{\boldsymbol{\eta}} * \sigma_\delta)(t, \hat{x}, \hat{y}) := \int_{\mathbb{R}^{2d}} \hat{\boldsymbol{\eta}}(t, \hat{x}-z, \hat{y}-w)\sigma_\delta(z, w)dzdw, \qquad \text{on } \hat{\Omega}_b, \tag{11.27}$$

- The regularized Lagrangian mapping:

$$\hat{\boldsymbol{\Phi}}_b^{\eta^\delta}(t, \cdot) = \text{Id} + \hat{\boldsymbol{\eta}}^\delta(t, \cdot), \tag{11.28}$$

- The regularized moving Biot domain:

$$\Omega_b^\delta(t) = \hat{\boldsymbol{\Phi}}_b^{\eta^\delta}(t, \hat{\Omega}_b). \tag{11.29}$$

Note that even though the kinematic coupling condition holds for $\hat{\boldsymbol{\eta}}$ in the sense that $\hat{\boldsymbol{\eta}}|_{\hat{\Gamma}} = \hat{\omega}\boldsymbol{e}_y$, it is *not necessarily true* that $\hat{\boldsymbol{\eta}}^\delta|_{\hat{\Gamma}} = \hat{\omega}\boldsymbol{e}_y$. Therefore, we will also define:

- The regularized moving interface:

$$\Gamma^\delta(t) = \hat{\boldsymbol{\Phi}}_b^{\eta^\delta}(t, \hat{\Gamma}).$$

Alternatively, $\hat{\Gamma}^\delta$ is the plate interface if it were displaced from the reference configuration $\hat{\Gamma}$ in the direction $\hat{\boldsymbol{\eta}}^\delta|_{\hat{\Gamma}}$, which is a purely transverse y displacement, as one can verify.

Note that by the way we extended $\hat{\boldsymbol{\eta}}$ to the larger domain $\tilde{\Omega}_b$, we have that

$$\hat{\boldsymbol{\eta}}^\delta = 0 \quad \text{on} \quad \partial\hat{\Omega}_b \setminus \hat{\Gamma}.$$

With these regularized versions of the Biot structure displacement and velocity, we can now define the notion of a *weak solution to the regularized weak FPSI problem with the regularization parameter* δ. We start by defining the solution and test space, which are motivated by the energy estimates in Sect. 11.4.2, and then we state the regularized weak formulation in the moving domain framework and in the fixed reference domain framework.

Remark 11.4.1 We have regularized the physical Biot domain using the regularized Biot displacement, which results in the regularized moving Biot domain $\Omega_b^\delta(t)$ as stated in (11.29). We emphasize that the main reason for this regularization is because the structure displacement $\boldsymbol{\eta}$, which is in the finite energy space $H^1(\Omega_b)$ (without regularization), does not possess sufficient regularity to make sense of the definition of $\Omega_b^\delta(t)$. However, while the moving Biot domain requires regularization, we emphasize that there is no need to regularize the fluid domain $\Omega_f(t)$ because the fluid domain $\Omega_f(t)$ can be defined without explicit reference to the Biot displacement $\boldsymbol{\eta}$, and hence it is not affected by the regularity issues associated with $\boldsymbol{\eta}$. In particular, $\Omega_f(t)$ can already be well-defined by using just the plate displacement ω. Even though ω is the trace of $\boldsymbol{\eta}$ along Γ, the plate displacement ω has additional regularity in the finite energy space, i.e., $\omega \in H_0^2(\Gamma)$, due to the fact that ω itself satisfies the plate equation. This makes ω a continuous function on Γ, which allows us to define $\Omega_f(t)$ unambiguously.

11.4.1 Functional Spaces and Definition of Weak Solutions

Definition 11.4.3 (Solution and Test Spaces for the Regularized Problem)

- *Fluid function space (moving domain/Eulerian formulation).*

$$V_f(t) = \{\boldsymbol{u} = (u_x, u_y) \in H^1(\Omega_f(t)) : \nabla \cdot \boldsymbol{u} = 0, \text{ and } \boldsymbol{u} = 0 \text{ when } x = 0, x = L, y = -R\}, \tag{11.30}$$

$$\mathcal{V}_f = L^\infty(0, T; L^2(\Omega_f(t))) \cap L^2(0, T; V_f(t)). \tag{11.31}$$

- *Fluid function space (fixed domain/Lagrangian formulation).*

$$V_f^\omega = \{\hat{\boldsymbol{u}} = (\hat{u}_x, \hat{u}_y) \in H^1(\hat{\Omega}_f) : \hat{\nabla}_f^\omega \cdot \hat{\boldsymbol{u}} = 0, \text{ and } \hat{\boldsymbol{u}} = 0 \text{ when } \hat{x} = 0, \hat{x} = L, \hat{y} = -R\}, \tag{11.32}$$

$$\mathcal{V}_f^\omega = L^\infty(0, T; L^2(\hat{\Omega}_f)) \cap L^2(0, T; V_f^\omega). \tag{11.33}$$

- *Plate function space.*

$$\mathcal{V}_\omega = W^{1,\infty}(0, T; L^2(\hat{\Gamma})) \cap L^\infty(0, T; H_0^2(\hat{\Gamma})). \tag{11.34}$$

- *Biot displacement function space.*

$$V_d = \{\hat{\boldsymbol{\eta}} = (\hat{\eta}_x, \hat{\eta}_y) \in H^1(\hat{\Omega}_b) : \hat{\boldsymbol{\eta}} = 0 \text{ for } \hat{x} = 0, \hat{x} = L, \hat{y} = R, \text{ and } \hat{\eta}_x = 0 \text{ on } \hat{\Gamma}\}, \tag{11.35}$$

$$\mathcal{V}_b = W^{1,\infty}(0, T; L^2(\hat{\Omega}_b)) \cap L^\infty(0, T; V_d) \cap H^1(0, T; V_d). \tag{11.36}$$

- *Biot pore pressure function space.*

$$V_p = \{\hat{p} \in H^1(\hat{\Omega}_b) : \hat{p} = 0 \text{ for } \hat{x} = 0, \hat{x} = L, \hat{y} = R\}, \tag{11.37}$$

$$\mathcal{Q}_b = L^\infty(0, T; L^2(\hat{\Omega}_b)) \cap L^2(0, T; V_p). \tag{11.38}$$

- *Weak solution space (moving domain).*

$$\mathcal{V}_{\text{sol}} = \{(\boldsymbol{u}, \hat{\omega}, \hat{\boldsymbol{\eta}}, \hat{p}) \in \mathcal{V}_f \times \mathcal{V}_\omega \times \mathcal{V}_b \times \mathcal{Q}_b : \hat{\boldsymbol{\eta}} = \hat{\omega}\boldsymbol{e}_y \text{ on } \hat{\Gamma}\}. \tag{11.39}$$

- *Weak solution space (fixed domain).*

$$\mathcal{V}^\omega_{\text{sol}} = \{(\hat{\boldsymbol{u}}, \hat{\omega}, \hat{\boldsymbol{\eta}}, \hat{p}) \in \mathcal{V}^\omega_f \times \mathcal{V}_\omega \times \mathcal{V}_b \times \mathcal{Q}_b : \hat{\boldsymbol{\eta}} = \hat{\omega}\boldsymbol{e}_y \text{ on } \hat{\Gamma}\}. \tag{11.40}$$

- *Test space (moving domain).*

$$\mathcal{V}_{\text{test}} = \{(\boldsymbol{v}, \hat{\varphi}, \hat{\boldsymbol{\psi}}, \hat{r}) \in C^1_c([0, T); V_f(t) \times H^2_0(\hat{\Gamma}) \times V_d \times V_p) : \hat{\boldsymbol{\psi}} = \hat{\varphi}\boldsymbol{e}_y \text{ on } \hat{\Gamma}\}. \tag{11.41}$$

- *Test space (fixed domain).*

$$\mathcal{V}^\omega_{\text{test}} = \{(\hat{\boldsymbol{v}}, \hat{\varphi}, \hat{\boldsymbol{\psi}}, \hat{r}) \in C^1_c([0, T); V^\omega_f \times H^2_0(\hat{\Gamma}) \times V_d \times V_p) : \hat{\boldsymbol{\psi}} = \hat{\varphi}\boldsymbol{e}_y \text{ on } \hat{\Gamma}\}. \tag{11.42}$$

Remark 11.4.2 Because $\hat{\Gamma}$ is one dimensional, for plate displacements $\hat{\omega} \in \mathcal{V}_\omega$, we have that $\hat{\omega} \in C(0, T; C^1(\hat{\Gamma}))$, and hence, there is a one-to-one correspondence between functions in $\mathcal{V}_{\text{sol}}$ and $\mathcal{V}^\omega_{\text{sol}}$ and functions in $\mathcal{V}_{\text{test}}$ and $\mathcal{V}^\omega_{\text{test}}$, given by composition with the ALE mapping (11.19).

Next, we state the weak formulation to the regularized problem as follows.

Definition 11.4.4 (Weak Solution to the Regularized Problem, Moving Fluid Domain Formulation) An ordered four-tuple $(\boldsymbol{u}, \hat{\omega}, \hat{\boldsymbol{\eta}}, p) \in \mathcal{V}_{\text{sol}}$ is a *weak solution to the regularized nonlinearly coupled FPSI problem with regularization parameter* δ if for every test function $(\boldsymbol{v}, \hat{\varphi}, \hat{\boldsymbol{\psi}}, \hat{r}) \in \mathcal{V}_{\text{test}}$,

$$-\int_0^T \int_{\Omega_f(t)} \boldsymbol{u} \cdot \partial_t \boldsymbol{v} + \frac{1}{2}\int_0^T \int_{\Omega_f(t)} [((\boldsymbol{u} \cdot \nabla)\boldsymbol{u}) \cdot \boldsymbol{v} - ((\boldsymbol{u} \cdot \nabla)\boldsymbol{v}) \cdot \boldsymbol{u}]$$

$$+ \frac{1}{2}\int_0^T \int_{\Gamma(t)} (\boldsymbol{u} \cdot \boldsymbol{n} - 2\zeta \boldsymbol{e}_y \cdot \boldsymbol{n})\boldsymbol{u} \cdot \boldsymbol{v}$$

$$+ 2\nu \int_0^T \int_{\Omega_f(t)} \boldsymbol{D}(\boldsymbol{u}) : \boldsymbol{D}(\boldsymbol{v}) + \int_0^T \int_{\Gamma(t)} \left(\frac{1}{2}|\boldsymbol{u}|^2 - p\right)(\psi_n - v_n)$$

$$
\begin{aligned}
&+\beta \int_0^T \int_{\Gamma(t)} (\zeta \boldsymbol{e}_y - \boldsymbol{u}) \cdot \boldsymbol{\tau} (\boldsymbol{\psi} - \boldsymbol{v}) \cdot \boldsymbol{\tau} \\
&- \rho_p \int_0^T \int_{\hat{\Gamma}} \partial_t \hat{\omega} \cdot \partial_t \hat{\varphi} + \int_0^T \int_{\hat{\Gamma}} \hat{\Delta} \hat{\omega} \cdot \hat{\Delta} \hat{\varphi} - \rho_b \int_0^T \int_{\hat{\Omega}_b} \partial_t \hat{\boldsymbol{\eta}} \cdot \partial_t \hat{\boldsymbol{\psi}} \\
&+ 2\mu_e \int_0^T \int_{\hat{\Omega}_b} \hat{\boldsymbol{D}}(\hat{\boldsymbol{\eta}}) : \hat{\boldsymbol{D}}(\hat{\boldsymbol{\psi}}) \\
&+\lambda_e \int_0^T \int_{\hat{\Omega}_b} (\hat{\nabla} \cdot \hat{\boldsymbol{\eta}})(\hat{\nabla} \cdot \hat{\boldsymbol{\psi}}) + 2\mu_v \int_0^T \int_{\hat{\Omega}_b} \hat{\boldsymbol{D}}(\partial_t \hat{\boldsymbol{\eta}}) : \hat{\boldsymbol{D}}(\hat{\boldsymbol{\psi}}) + \lambda_v \int_0^T \int_{\hat{\Omega}_b} (\hat{\nabla} \cdot \partial_t \hat{\boldsymbol{\eta}})(\hat{\nabla} \cdot \hat{\boldsymbol{\psi}}) \\
&- \alpha \int_0^T \int_{\Omega_b^\delta(t)} p \nabla \cdot \boldsymbol{\psi} - c_0 \int_0^T \int_{\hat{\Omega}_b} \hat{p} \cdot \partial_t \hat{r} - \alpha \int_0^T \int_{\Omega_b^\delta(t)} \frac{D^\delta}{Dt} \boldsymbol{\eta} \cdot \nabla r \\
&- \alpha \int_0^T \int_{\Gamma^\delta(t)} (\zeta \boldsymbol{e}_y \cdot \boldsymbol{n}^\delta) r \\
&+ \kappa \int_0^T \int_{\Omega_b^\delta(t)} \nabla p \cdot \nabla r - \int_0^T \int_{\Gamma(t)} ((\boldsymbol{u} - \zeta \boldsymbol{e}_y) \cdot \boldsymbol{n}) r \\
&= \int_{\Omega_f(0)} \boldsymbol{u}(0) \cdot \boldsymbol{v}(0) + \rho_p \int_{\hat{\Gamma}} \partial_t \hat{\omega}(0) \cdot \hat{\varphi}(0) + \rho_b \int_{\hat{\Omega}_b} \partial_t \hat{\boldsymbol{\eta}}(0) \cdot \hat{\boldsymbol{\psi}}(0) + c_0 \int_{\hat{\Omega}_b} \hat{p}(0) \cdot \hat{r}(0),
\end{aligned}
\tag{11.43}
$$

where $\frac{D^\delta}{Dt} = \frac{d}{dt} + (\boldsymbol{\xi}^\delta \cdot \nabla)$ with $\boldsymbol{\xi}^\delta(t, \cdot) = \partial_t \hat{\boldsymbol{\eta}}^\delta(t, (\boldsymbol{\Phi}_b^{\eta^\delta})^{-1}(t, \cdot))$ is the material derivative with respect to the regularized displacement, $\boldsymbol{n}$ denotes the upward pointing normal vector to $\Gamma(t)$, and $\boldsymbol{n}^\delta$ denotes the upward pointing normal vector to $\Gamma^\delta(t)$.

Notice that only four terms contain regularization via convolution with parameter δ. While there are many different ways to write the regularized weak formulation, the regularization presented above is a regularization that deviates from the original, nonregularized problem, in the smallest possible number of terms, and is still consistent with the original, nonregularized problem, as we show later.

Remark 11.4.3 While the solution to the regularized problem above depends on the regularization parameter δ implicitly, to simplify notation, we will drop the δ notation whenever it is clear from the context that we are working with the solution to the regularized problem.

Remark 11.4.4 We simplify notation by omitting the explicit compositions with the maps $\hat{\boldsymbol{\Phi}}_f^\omega$, $\hat{\boldsymbol{\Phi}}_\Gamma^\omega$, $\hat{\boldsymbol{\Phi}}_b^\eta$, and $\hat{\boldsymbol{\Phi}}_b^{\eta^\delta}$, and their inverses. The necessary compositions with such mappings will be clear from the context. For example,

$$-\alpha \int_0^T \int_{\Omega_b^\delta(t)} p\nabla\cdot\boldsymbol{\psi} \quad \text{means} \quad -\alpha \int_0^T \int_{\Omega_b^\delta(t)} \left(\hat{p}\circ(\boldsymbol{\Phi}_b^{\eta^\delta})^{-1}\right)\nabla\cdot\left(\hat{\boldsymbol{\psi}}\circ(\boldsymbol{\Phi}_b^{\eta^\delta})^{-1}\right),$$

and

$$-\int_0^T \int_{\Gamma(t)} ((\boldsymbol{u}-\zeta\boldsymbol{e}_y)\cdot\boldsymbol{n})r \quad \text{means}$$
$$-\int_0^T \int_{\Gamma(t)} \left(\left(\boldsymbol{u}-(\zeta\circ(\boldsymbol{\Phi}_\Gamma^\omega)^{-1})\boldsymbol{e}_y\right)\cdot\boldsymbol{n}\right)\left(\hat{r}\circ(\boldsymbol{\Phi}_b^\eta)^{-1}\right).$$

Notice that here we tacitly assume that $\boldsymbol{\Phi}_b^{\eta^\delta}$ is invertible. We will later justify this assumption by proving that it holds on some time interval $[0, T_\delta]$, where the time T_δ may depend on the regularization parameter δ. Next, we reformulate the definition of a regularized weak solution on the **fixed reference domain**. Recall that the Jacobians $\hat{\mathcal{J}}_f^\omega$, $\hat{\mathcal{J}}_b^\eta$, and $\hat{\mathcal{J}}_\Gamma^\omega$ in (11.21) will appear upon using a change of variables to map the problem onto the reference domain. To transform the first term in the weak formulation (11.43) above, we use (11.23) to transform the time derivatives and assume that $|\hat{\omega}| < R$ so that there is no domain degeneracy. After using (11.22) and (11.23), we get

$$\begin{aligned}
\int_{\Omega_f(t)} \boldsymbol{u}\cdot\partial_t\boldsymbol{v} &= \int_{\hat{\Omega}_f}\left(1+\frac{\hat{\omega}}{R}\right)\hat{\boldsymbol{u}}\cdot\partial_t\hat{\boldsymbol{v}} - \int_{\hat{\Omega}_f}\left(1+\frac{\hat{\omega}}{R}\right)\hat{\boldsymbol{u}}\cdot[(\hat{\boldsymbol{w}}\cdot\hat{\nabla}_f^\omega)\hat{\boldsymbol{v}}]\\
&= \int_{\hat{\Omega}_f}\left(1+\frac{\hat{\omega}}{R}\right)\hat{\boldsymbol{u}}\cdot\partial_t\hat{\boldsymbol{v}} - \frac{1}{R}\int_{\hat{\Omega}_f}\hat{\boldsymbol{u}}\cdot[(R+\hat{y})\partial_t\hat{\omega}\partial_{\hat{y}}\hat{\boldsymbol{v}}]\\
&= \int_{\hat{\Omega}_f}\left(1+\frac{\hat{\omega}}{R}\right)\hat{\boldsymbol{u}}\cdot\partial_t\hat{\boldsymbol{v}} - \frac{1}{2R}\int_{\hat{\Omega}_f}\hat{\boldsymbol{u}}\cdot[(R+\hat{y})\partial_t\hat{\omega}\partial_{\hat{y}}\hat{\boldsymbol{v}}] + \frac{1}{2R}\int_{\hat{\Omega}_f}(\partial_t\hat{\omega})\hat{\boldsymbol{u}}\cdot\hat{\boldsymbol{v}}\\
&\quad+\frac{1}{2R}\int_{\hat{\Omega}_f}[(R+\hat{y})\partial_t\hat{\omega}\partial_{\hat{y}}\hat{\boldsymbol{u}}]\cdot\hat{\boldsymbol{v}} - \frac{1}{2}\int_{\hat{\Gamma}}(\hat{\boldsymbol{u}}\cdot\hat{\boldsymbol{v}})\partial_t\hat{\omega}\\
&= \int_{\hat{\Omega}_f}\left(1+\frac{\hat{\omega}}{R}\right)\hat{\boldsymbol{u}}\cdot\partial_t\hat{\boldsymbol{v}} - \frac{1}{2}\int_{\hat{\Omega}_f}\left(1+\frac{\hat{\omega}}{R}\right)[((\hat{\boldsymbol{w}}\cdot\hat{\nabla}_f^\omega)\hat{\boldsymbol{v}})\cdot\hat{\boldsymbol{u}} - ((\hat{\boldsymbol{w}}\cdot\hat{\nabla}_f^\omega)\hat{\boldsymbol{u}})\cdot\hat{\boldsymbol{v}}]\\
&\quad+\frac{1}{2R}\int_{\hat{\Omega}_f}(\partial_t\hat{\omega})\hat{\boldsymbol{u}}\cdot\hat{\boldsymbol{v}} - \frac{1}{2}\int_{\hat{\Gamma}}(\hat{\boldsymbol{u}}\cdot\hat{\boldsymbol{v}})\partial_t\hat{\omega},
\end{aligned} \tag{11.44}$$

where we integrated by parts in the $\hat{y}$ direction. Note that the final term in (11.44) will combine with the following term in (11.43):

$$\int_0^T\int_{\Gamma(t)}(\zeta\boldsymbol{e}_y\cdot\boldsymbol{n})\boldsymbol{u}\cdot\boldsymbol{v} = \int_0^T\int_{\hat{\Gamma}}(\hat{\boldsymbol{u}}\cdot\hat{\boldsymbol{v}})\partial_t\hat{\omega}, \tag{11.45}$$

where we used $\boldsymbol{n} = (-\partial_{\hat{x}}\hat{\omega}, 1)/\hat{\mathcal{J}}_\Gamma^\omega$ for the normal vector to the interface and $\zeta \boldsymbol{e}_y|_{\Gamma(t)} = \partial_t \hat{\omega} \boldsymbol{e}_y$. Because the transformation from $\Gamma(t)$ to $\hat{\Gamma}$ cancels out the factor of $\hat{\mathcal{J}}_\Gamma^\omega$ in the unit normal vector, it is useful to define the following renormalized normal and tangent vectors:

$$\hat{\boldsymbol{n}}^\omega = (-\partial_{\hat{x}}\hat{\omega}, 1), \qquad \hat{\boldsymbol{\tau}}^\omega = (1, \partial_{\hat{x}}\hat{\omega}). \tag{11.46}$$

We similarly define

$$\hat{\boldsymbol{n}}^{\omega^\delta} = (-\partial_{\hat{x}}(\hat{\boldsymbol{\eta}}^\delta|_{\hat{\Gamma}}), 1). \tag{11.47}$$

We are now ready to state the definition of a weak solution to the regularized problem on the fixed reference domain.

Definition 11.4.5 (Weak Solution to the Regularized Problem, Fixed Fluid Domain Formulation) An ordered four-tuple $(\hat{\boldsymbol{u}}, \hat{\omega}, \hat{\boldsymbol{\eta}}, \hat{p}) \in \mathcal{V}_{\text{sol}}^\omega$ is a *weak solution to the regularized nonlinearly coupled FPSI problem with regularization parameter* δ if for all test functions $(\hat{\boldsymbol{v}}, \hat{\varphi}, \hat{\boldsymbol{\psi}}, \hat{r}) \in \mathcal{V}_{\text{test}}^\omega$, the following equality holds:

$$
\begin{aligned}
&-\int_0^T \int_{\hat{\Omega}_f} \left(1+\frac{\hat{\omega}}{R}\right) \hat{\boldsymbol{u}} \cdot \partial_t \hat{\boldsymbol{v}} + \frac{1}{2}\int_0^T \int_{\hat{\Omega}_f} \left(1+\frac{\hat{\omega}}{R}\right) [((\hat{\boldsymbol{u}} - \hat{\boldsymbol{w}}) \cdot \hat{\nabla}_f^\omega \hat{\boldsymbol{u}}) \cdot \hat{\boldsymbol{v}} - ((\hat{\boldsymbol{u}} \\
&\qquad - \hat{\boldsymbol{w}}) \cdot \hat{\nabla}_f^\omega \hat{\boldsymbol{v}}) \cdot \hat{\boldsymbol{u}}] \\
&\qquad - \frac{1}{2R}\int_0^T \int_{\hat{\Omega}_f} (\partial_t \hat{\omega}) \hat{\boldsymbol{u}} \cdot \hat{\boldsymbol{v}} + \frac{1}{2}\int_0^T \int_{\hat{\Gamma}} (\hat{\boldsymbol{u}} \cdot \hat{\boldsymbol{n}}^\omega - \hat{\zeta} \boldsymbol{e}_y \cdot \hat{\boldsymbol{n}}^\omega) \hat{\boldsymbol{u}} \cdot \hat{\boldsymbol{v}} \\
&\qquad + 2\nu \int_0^T \int_{\hat{\Omega}_f} \left(1+\frac{\hat{\omega}}{R}\right) \hat{\boldsymbol{D}}(\hat{\boldsymbol{u}}) : \hat{\boldsymbol{D}}(\hat{\boldsymbol{v}}) \\
&+ \int_0^T \int_{\hat{\Gamma}} \left(\frac{1}{2}|\hat{\boldsymbol{u}}|^2 - \hat{p}\right) (\hat{\boldsymbol{\psi}} - \hat{\boldsymbol{v}}) \cdot \hat{\boldsymbol{n}}^\omega + \beta \int_0^T \int_{\hat{\Gamma}} (\hat{\mathcal{J}}_\Gamma^\omega)^{-1} (\hat{\zeta} \boldsymbol{e}_y - \hat{\boldsymbol{u}}) \cdot \hat{\boldsymbol{\tau}}^\omega (\hat{\boldsymbol{\psi}} - \hat{\boldsymbol{v}}) \cdot \hat{\boldsymbol{\tau}}^\omega \\
&\qquad - \rho_p \int_0^T \int_{\hat{\Gamma}} \partial_t \hat{\omega} \cdot \partial_t \hat{\varphi} + \int_0^T \int_{\hat{\Gamma}} \hat{\Delta}\hat{\omega} \cdot \hat{\Delta}\hat{\varphi} - \rho_b \int_0^T \int_{\hat{\Omega}_b} \partial_t \hat{\boldsymbol{\eta}} \cdot \partial_t \hat{\boldsymbol{\psi}} \\
&\qquad + 2\mu_e \int_0^T \int_{\hat{\Omega}_b} \hat{\boldsymbol{D}}(\hat{\boldsymbol{\eta}}) : \hat{\boldsymbol{D}}(\hat{\boldsymbol{\psi}}) \\
&+ \lambda_e \int_0^T \int_{\hat{\Omega}_b} (\hat{\nabla} \cdot \hat{\boldsymbol{\eta}})(\hat{\nabla} \cdot \hat{\boldsymbol{\psi}}) + 2\mu_v \int_0^T \int_{\hat{\Omega}_b} \hat{\boldsymbol{D}}(\partial_t \hat{\boldsymbol{\eta}}) : \hat{\boldsymbol{D}}(\hat{\boldsymbol{\psi}}) + \lambda_v \int_0^T \int_{\hat{\Omega}_b} (\hat{\nabla} \cdot \partial_t \hat{\boldsymbol{\eta}})(\hat{\nabla} \cdot \hat{\boldsymbol{\psi}}) \\
&\qquad - \alpha \int_0^T \int_{\hat{\Omega}_b} \hat{\mathcal{J}}_b^{\eta^\delta} \hat{p} \hat{\nabla}_b^{\eta^\delta} \cdot \hat{\boldsymbol{\psi}} - c_0 \int_0^T \int_{\hat{\Omega}_b} \hat{p} \cdot \partial_t \hat{r} - \alpha \int_0^T \int_{\hat{\Omega}_b} \hat{\mathcal{J}}_b^{\eta^\delta} \partial_t \hat{\boldsymbol{\eta}} \cdot \hat{\nabla}_b^{\eta^\delta} \hat{r}
\end{aligned}
$$

$$-\alpha\int_0^T\int_{\hat\Gamma}(\hat\zeta \boldsymbol{e}_y\cdot\hat{\boldsymbol{n}}^{\omega^\delta})\hat r+\kappa\int_0^T\int_{\hat\Omega_b}\hat{\mathcal{J}}_b^{\eta^\delta}\hat\nabla_b^{\eta^\delta}\hat p\cdot\hat\nabla_b^{\eta^\delta}\hat r-\int_0^T\int_{\hat\Gamma}((\hat{\boldsymbol{u}}-\hat\zeta\boldsymbol{e}_y)\cdot\hat{\boldsymbol{n}}^{\omega})\hat r$$

$$=\int_{\Omega_f(0)}\boldsymbol{u}(0)\cdot\boldsymbol{v}(0)+\rho_p\int_{\hat\Gamma}\partial_t\hat\omega(0)\cdot\hat\varphi(0)+\rho_b\int_{\hat\Omega_b}\partial_t\hat{\boldsymbol{\eta}}(0)\cdot\hat{\boldsymbol{\psi}}(0)+c_0\int_{\hat\Omega_b}\hat p(0)\cdot\hat r(0), \tag{11.48}$$

where $\hat{\mathcal{J}}_b^{\eta^\delta}$ and $\hat{\mathcal{J}}_\Gamma^{\omega}$ are defined in (11.21), $\hat{\boldsymbol{w}}$ is defined in (11.23), $\hat\nabla_f^{\omega}$ in (11.22), $\hat\nabla_b^{\eta^\delta}\hat g$ in (11.24), and $\hat\zeta$ in (11.25).

11.4.2 Formal Energy Inequality

Here we show that the regularized problem is defined in a way that still gives rise to an energy equality, which in fact is the same energy inequality that one would formally obtain for the original problem, except with integrals over the moving Biot domain $\Omega_b(t)$ becoming integrals over the regularized moving Biot domain $\Omega_b^\delta(t)$. More precisely, we formally prove that a weak solution to the regularized problem satisfies the following energy equality.

Lemma 11.4.1 *Assuming that a weak solution exists, the following energy equality holds:*

$$E^K(T)+E^E(T)+\int_0^T\left(D_f^V(t)+D_b^V(t)+D_{fb}^V(t)+D_\beta^V(t)\right)dt=E^K(0)+E^E(0) \tag{11.49}$$

where

$$E^K(t)=\frac{1}{2}\int_{\Omega_f(t)}|\boldsymbol{u}(t)|^2+\frac{1}{2}\rho_b\int_{\hat\Omega_b}|\partial_t\hat{\boldsymbol{\eta}}(t)|^2+\frac{1}{2}c_0\int_{\hat\Omega_b}|\hat p(t)|^2+\frac{1}{2}\rho_p\int_{\hat\Gamma}|\partial_t\hat\omega(t)|^2$$

is the sum of the kinetic energy of the fluid, the kinetic energy of the Biot poroviscoelastic matrix motion, the kinetic energy of the filtrating fluid flow in the Biot medium, and the kinetic energy of the plate motion, $E^E(t)$ is defined by

$$E^E(t)=2\mu_e\int_{\hat\Omega_b}|\hat{\boldsymbol{D}}(\hat{\boldsymbol{\eta}})(t)|^2+2\lambda_e\int_{\hat\Omega_b}|\hat\nabla\cdot\hat{\boldsymbol{\eta}}(t)|^2+\int_{\hat\Gamma}|\hat\Delta\hat\omega(t)|^2,$$

which corresponds to the elastic energy of the Biot poroviscoelastic matrix and the elastic energy of the plate, and

$$D_f^V(t)=2\nu\int_{\Omega_f(t)}|\boldsymbol{D}(\boldsymbol{u})|^2,\quad D_b^V(t)=2\mu_v\int_{\hat\Omega_b}|\hat{\boldsymbol{D}}(\partial_t\hat{\boldsymbol{\eta}})|^2+\lambda_v\int_{\hat\Omega_b}|\hat\nabla\cdot\partial_t\hat{\boldsymbol{\eta}}|^2,$$

$$D^V_{f_b}(t) = \kappa \int_{\Omega_b^\delta(t)} |\nabla p|^2, \quad D^V_\beta(t) = \beta \int_{\Gamma(t)} |(\boldsymbol{\xi} - \boldsymbol{u}) \cdot \boldsymbol{\tau}|^2$$

correspond to dissipation due to fluid viscosity, viscosity of the Biot poroviscoelastic matrix, dissipation due to permeability effects, and dissipation due to friction in the Beavers-Joseph-Saffman slip condition.

Proof To derive this energy equality, we start by substituting $(\hat{\boldsymbol{v}}, \hat{\varphi}, \hat{\boldsymbol{\psi}}, \hat{r}) = (\hat{\boldsymbol{u}}, \hat{\zeta}, \partial_t \hat{\boldsymbol{\eta}}, \hat{p})$ into the regularized weak formulation (11.48) defined on the fixed reference domain and calculate

$$\frac{1}{2}\int_{\hat{\Gamma}} (\hat{\boldsymbol{u}} - \hat{\zeta}\boldsymbol{e}_y) \cdot \hat{\boldsymbol{n}}^\omega |\hat{\boldsymbol{u}}|^2 + \int_{\hat{\Gamma}} \left(\frac{1}{2}|\hat{\boldsymbol{u}}|^2 - \hat{p}\right)(\hat{\zeta}\boldsymbol{e}_y - \hat{\boldsymbol{u}}) \cdot \hat{\boldsymbol{n}}^\omega - \int_{\hat{\Gamma}} ((\hat{\boldsymbol{u}} - \hat{\zeta}\boldsymbol{e}_y) \cdot \hat{\boldsymbol{n}}^\omega)\hat{p} = 0.$$

Furthermore, using integration by parts, one obtains

$$\begin{aligned}
&\alpha \left(\int_{\hat{\Omega}_b} \hat{\mathcal{J}}_b^{\eta^\delta} \hat{p} \hat{\nabla}_b^{\eta^\delta} \cdot \partial_t \hat{\boldsymbol{\eta}} + \int_{\hat{\Omega}_b} \hat{\mathcal{J}}_b^{\eta^\delta} \partial_t \hat{\boldsymbol{\eta}} \cdot \hat{\nabla}_b^{\eta^\delta} \hat{p} + \int_{\hat{\Gamma}} (\hat{\zeta}\boldsymbol{e}_y \cdot \hat{\boldsymbol{n}}^{\omega^\delta})\hat{p} \right) \\
&= \alpha \left(\int_{\Omega_b^\delta(t)} p \nabla \cdot \boldsymbol{\xi} + \int_{\Omega_b^\delta(t)} \boldsymbol{\xi} \cdot \nabla p + \int_{\Gamma^\delta(t)} (\zeta \boldsymbol{e}_y \cdot \boldsymbol{n}^\delta) p \right) = 0,
\end{aligned}$$

where $\boldsymbol{n}^\delta$ is the upward pointing unit normal vector to $\Gamma^\delta(t)$. Finally, by Reynold's transport theorem

$$\int_0^T \int_{\Omega_f(t)} \boldsymbol{u} \cdot \partial_t \boldsymbol{u} + \frac{1}{2}\int_0^T \int_{\Gamma(t)} (\zeta \boldsymbol{e}_y \cdot \boldsymbol{n})|\boldsymbol{u}|^2 = \frac{1}{2}\int_{\Omega_f(T)} |\boldsymbol{u}|^2 - \frac{1}{2}\int_{\Omega_f(0)} |\boldsymbol{u}|^2.$$

By combining these calculations, one obtains the final energy estimate:

$$\begin{aligned}
&\frac{1}{2}\int_{\Omega_f(T)} |\boldsymbol{u}(T)|^2 + 2\nu \int_0^T \int_{\Omega_f(t)} |\boldsymbol{D}(\boldsymbol{u})|^2 + \beta \int_0^T \int_{\Gamma(t)} |(\boldsymbol{\xi} - \boldsymbol{u}) \cdot \boldsymbol{\tau}|^2 \\
&\quad + \frac{1}{2}\rho_p \int_{\hat{\Gamma}} |\partial_t \hat{\omega}(T)|^2 + \int_{\hat{\Gamma}} |\hat{\Delta}\hat{\omega}(T)|^2 \\
&+ \frac{1}{2}\rho_b \int_{\hat{\Omega}_b} |\partial_t \hat{\boldsymbol{\eta}}(T)|^2 + 2\mu_e \int_{\hat{\Omega}_b} |\hat{\boldsymbol{D}}(\hat{\boldsymbol{\eta}})(T)|^2 + 2\lambda_e \int_{\hat{\Omega}_b} |\hat{\nabla} \cdot \hat{\boldsymbol{\eta}}(T)|^2 \\
&\quad + 2\mu_v \int_0^T \int_{\hat{\Omega}_b} |\hat{\boldsymbol{D}}(\partial_t \hat{\boldsymbol{\eta}})|^2 \\
&+ \lambda_v \int_0^T \int_{\hat{\Omega}_b} |\hat{\nabla} \cdot \partial_t \hat{\boldsymbol{\eta}}|^2 + \frac{1}{2}c_0 \int_{\hat{\Omega}_b} |\hat{p}(T)|^2 + \kappa \int_0^T \int_{\Omega_b^\delta(t)} |\nabla p|^2
\end{aligned}$$

$$= \frac{1}{2}\int_{\Omega_f(0)} |\boldsymbol{u}(0)|^2 + \frac{1}{2}\rho_p \int_{\hat{\Gamma}} |\partial_t \hat{\omega}(0)|^2$$

$$+ \int_{\hat{\Gamma}} |\hat{\Delta}\hat{\omega}(0)|^2 + \frac{1}{2}\rho_b \int_{\hat{\Omega}_b} |\partial_t \hat{\boldsymbol{\eta}}(0)|^2 + 2\mu_e \int_{\hat{\Omega}_b} |\hat{\boldsymbol{D}}(\hat{\boldsymbol{\eta}})(0)|^2$$

$$+ 2\lambda_e \int_{\hat{\Omega}_b} |\hat{\nabla} \cdot \hat{\boldsymbol{\eta}}(0)|^2 + \frac{1}{2}c_0 \int_{\hat{\Omega}_b} |\hat{p}(0)|^2.$$

□

11.4.3 Statement of the Main Existence Result for the Regularized Problem

We now state the main result on the existence of a weak solution to the regularized problem.

Theorem 11.4.1 *Let $\rho_b, \mu_e, \lambda_e, \alpha, \rho_p, \nu > 0$ and $\mu_v, \lambda_v \geq 0$. Consider initial data for the plate displacement $\hat{\omega}_0 \in H_0^2(\hat{\Gamma})$, plate velocity $\hat{\zeta}_0 \in L^2(\hat{\Gamma})$, Biot displacement $\hat{\boldsymbol{\eta}}_0 \in H^1(\hat{\Omega}_b)$, Biot velocity $\hat{\boldsymbol{\xi}}_0 \in H^1(\hat{\Omega}_b)$ in the case of a viscoelastic Biot medium $\mu_v, \lambda_v > 0$ and $\hat{\boldsymbol{\xi}}_0 \in L^2(\hat{\Omega}_b)$ otherwise for the case of a purely elastic Biot medium, Biot pore pressure $\hat{p}_0 \in L^2(\hat{\Omega}_b)$, and fluid velocity $\boldsymbol{u}_0 \in H^1(\Omega_f(0))$, which is divergence-free. Suppose further that $|\hat{\omega}_0| \leq R_0 < R$ for some R_0, $\hat{\boldsymbol{\eta}}_0|_\Gamma = \hat{\omega}_0 \boldsymbol{e}_y$, and $\hat{\boldsymbol{\xi}}_0|_\Gamma = \hat{\zeta}_0 \boldsymbol{e}_y$, and for some arbitrary but fixed regularization parameter $\delta > 0$, suppose that $Id + \hat{\boldsymbol{\eta}}_0^\delta$ is an invertible map with $\det(\boldsymbol{I} + \nabla \hat{\boldsymbol{\eta}}_0^\delta) > 0$. Then, there exists a weak solution $(\boldsymbol{u}, \hat{\omega}, \hat{\boldsymbol{\eta}}, \hat{p})$ to the regularized FPSI problem with regularization parameter δ on some time interval $[0, T]$, for some $T > 0$.*

While T in general depends on δ, we will show that if there exists a smooth solution to the nonregularized FPSI problem, then this time T for the regularized problem is independent of δ. This will allow us to pass to the limit as $\delta \to 0$ and show that weak solutions to the regularized FPSI problems constructed in this book converge to a smooth solution of the original, nonregularized problem, when a smooth solution to the nonregularized problem exists. On the other hand, without the additional assumption of the existence of a strong solution, one cannot draw any conclusions about the limit as $\delta \to 0$. This assumption plays a crucial role in demonstrating that T remains independent of δ. Furthermore, as elucidated in Remark 11.3.1, energy estimates alone are insufficient to take the limit in certain terms in the weak formulation. Therefore, in order to pass to the limit as $\delta \to 0$, one would need to prove additional regularity estimates (beyond energy estimates) for weak solutions, which appears to be beyond the current state-of-the-art techniques.

Remark 11.4.5 The result above is a local result, since it holds up to some time $T > 0$, which needs to be sufficiently small. However, it is easy to show that this $T > 0$ can be made maximal, in the sense that it holds until the time for which $\mathrm{Id} + \hat{\boldsymbol{\eta}}^\delta$ fails to be invertible or $|\hat{\omega}(\cdot, x)| = R$ for some $x \in \hat{\Gamma}$ when the reticular plate collides with the boundary. This can be shown using a standard method (see, e.g., pg. 397–398 of [81], or the proof of Theorem 7.1 in [265]).

An Important Notational Convention For notational simplicity, in the remainder of the chapter, we will omit the "hat" notation to distinguish between functions and domains in the physical and reference configurations: for example, we will denote both the pore pressure p on $\Omega_b(t)$ and $\hat{p}$ on $\hat{\Omega}_b$ by p, as the distinction between these two will be clear from context. In addition, we will remove the "hat" convention from the reference domains, and for example, we will denote the reference domain $\hat{\Omega}_b$ for the Biot medium by Ω_b. We will follow this notational convention for the rest of the chapter.

The proof of Theorem 11.4.1 is constructive and based on an operator splitting scheme. This is an approach that has been used in constructive existence proofs of weak solutions for a variety of FSI problems (see, e.g., Chap. 5).

11.5 The Splitting Scheme

We use an adaptation of splitting scheme described in Sects. 2.15.3 and 2.15.2 in Chap. 2, which is defined as follows. First, semidiscretize the problem in time by introducing the time step $\Delta t = T/N$, and subdivide the time interval $[0, T]$ into N subintervals, each of width Δt. On each subinterval, we will run two subproblems: (1) a plate subproblem that takes into account the elastodynamics of the reticular plate and updates the plate displacement and the plate velocity and (2) a fluid/Biot subproblem that updates the fluid velocity, the Biot displacement, the Biot pore pressure, and the plate velocity. Hence, each subinterval $[n\Delta t, (n+1)\Delta t]$ involves an iteration of the plate subproblem and then an iteration of the fluid/Biot subproblem, and the solution from the previous subproblem is used as data for the subsequent subproblem. The approximations of the fluid velocity, plate displacement and velocity, and Biot poroviscoelastic material displacement and pressure will be denoted by

$$(\boldsymbol{u}_N^{n+\frac{i}{2}}, \omega_N^{n+\frac{i}{2}}, \zeta_N^{n+\frac{i}{2}}, \boldsymbol{\eta}_N^{n+\frac{i}{2}}, p_N^{n+\frac{i}{2}}), \qquad \text{for } n = 0, 1, \ldots, N \text{ and } i = 0, 1,$$

where they are all defined on the given time subinterval $[n\Delta t, (n+1)\Delta t]$. Here, the quantities with the superscript $n + 1/2$ denote the resulting approximate solutions obtained after the plate subproblem is solved, and the quantities with the superscript $n + 1$ denote the resulting approximate solutions obtained after the fluid/Biot subproblem is solved. For the splitting scheme, we will work on the fixed reference domain, and hence, we will semidiscretize the regularized weak formulation (11.48) on the fixed reference domain. Backward Euler discretization will be used to approximate time derivatives, with the following shorthand notation:

$$\dot{f}_N^{n+\frac{i}{2}} = \frac{f_N^{n+\frac{i}{2}} - f_N^{n+\frac{i}{2}-1}}{\Delta t}. \tag{11.50}$$

As a technical comment, in the description of the subproblems below, the backward Euler discretization (11.50) can potentially give rise to negative subscripts, so when relevant, we will explicitly define f_{-1} and $f_{-1/2}$ depending on the context.

11.5.1 The Plate Subproblem

Only the plate displacement and velocity $\omega_N^{n+\frac{1}{2}}$ and $\zeta_N^{n+\frac{1}{2}}$ are updated in this subproblem, leaving the remaining variables unchanged:

$$\boldsymbol{u}_N^{n+\frac{1}{2}} = \boldsymbol{u}_N^n, \qquad \boldsymbol{\eta}_N^{n+\frac{1}{2}} = \boldsymbol{\eta}_N^n, \qquad p_N^{n+\frac{1}{2}} = p_N^n.$$

The new plate displacement and velocity are calculated from the following weak formulation of the plate subproblems: find $\omega_N^{n+\frac{1}{2}} \in H_0^2(\Gamma)$ and $\zeta_N^{n+\frac{1}{2}} \in H_0^2(\Gamma)$, such that

$$\int_\Gamma \left(\frac{\omega_N^{n+\frac{1}{2}} - \omega_N^{n-\frac{1}{2}}}{\Delta t} \right) \cdot \phi = \int_\Gamma \zeta_N^{n+\frac{1}{2}} \cdot \phi, \qquad \text{for all } \phi \in L^2(\Gamma), \tag{11.51}$$

$$\rho_p \int_\Gamma \left(\frac{\zeta_N^{n+\frac{1}{2}} - \zeta_N^n}{\Delta t} \right) \cdot \varphi + \int_\Gamma \Delta \omega_N^{n+\frac{1}{2}} \cdot \Delta \varphi = 0, \qquad \text{for all } \varphi \in H_0^2(\Gamma). \tag{11.52}$$

When $n = 0$, we set $\omega_N^{-\frac{1}{2}} = \omega(0)$ and $\zeta_N^0 = \zeta(0)$. In particular, $\omega(0)\boldsymbol{e}_y = \boldsymbol{\eta}(0)|_\Gamma$ and $\zeta(0)\boldsymbol{e}_y = \boldsymbol{\xi}(0)$.

Lemma 11.5.1 *Problem* (11.51), (11.52) *has a unique solution that satisfies the following energy equality:*

$$\begin{aligned} &\frac{1}{2}\rho_p \int_\Gamma |\zeta_N^{n+\frac{1}{2}}|^2 + \frac{1}{2}\rho_p \int_\Gamma |\zeta_N^{n+\frac{1}{2}} - \zeta_N^n|^2 + \frac{1}{2}\int_\Gamma |\Delta \omega_N^{n+\frac{1}{2}}|^2 \\ &+ \frac{1}{2}\int_\Gamma |\Delta(\omega_N^{n+\frac{1}{2}} - \omega_N^{n-\frac{1}{2}})|^2 \\ &= \frac{1}{2}\rho_p \int_\Gamma |\zeta_N^{n-\frac{1}{2}}|^2 + \frac{1}{2}\int_\Gamma |\Delta \omega_N^{n-\frac{1}{2}}|^2. \end{aligned} \tag{11.53}$$

Proof To prove this, we first notice that (11.51) immediately implies that

$$\zeta_N^{n+\frac{1}{2}} = \frac{\omega_N^{n+\frac{1}{2}} - \omega_N^{n-\frac{1}{2}}}{\Delta t} \tag{11.54}$$

so that, by substituting into (11.52), it suffices to find $\omega_N^{n+\frac{1}{2}} \in H_0^2(\Gamma)$, which satisfies:

$$\rho_p \int_\Gamma \omega_N^{n+\frac{1}{2}} \cdot \varphi + (\Delta t)^2 \int_\Gamma \Delta \omega_N^{n+\frac{1}{2}} \cdot \Delta \varphi = \rho_p \int_\Gamma (\omega_N^{n-\frac{1}{2}} + (\Delta t) \zeta_N^n) \cdot \varphi, \quad \text{for all } \varphi \in H_0^2(\Gamma).$$

The bilinear form

$$B[\omega, \varphi] = \rho_p \int_\Gamma \omega \cdot \varphi + (\Delta t)^2 \int_\Gamma \Delta \omega \cdot \Delta \varphi$$

is coercive on $H_0^2(\Gamma)$, and

$$\varphi \to \rho_p \int_\Gamma \left(\omega_N^{n-\frac{1}{2}} + (\Delta t) \zeta_N^n \right) \cdot \varphi$$

is a continuous linear functional on $H_0^2(\Gamma)$, since we will have $\omega_N^{n-\frac{1}{2}} \in H_0^2(\Gamma)$ and $\zeta_N^n \in L^2(\Gamma)$ by the way our splitting scheme is defined. Thus, by the Lax-Milgram lemma, there exists a unique solution $\omega_N^{n+\frac{1}{2}} \in H_0^2(\Gamma)$, from which we also recover $\zeta_N^{n+\frac{1}{2}} \in H_0^2(\Gamma)$ using (11.54) above.

The energy equality above follows by substituting $\varphi = \zeta_N^{n+\frac{1}{2}} = \frac{\omega_N^{n+\frac{1}{2}} - \omega_N^{n-\frac{1}{2}}}{\Delta t} \in H_0^2(\Gamma)$ into the weak formulation and using the identity

$$(a - b) \cdot a = \frac{1}{2}(|a|^2 + |a - b|^2 - |b|^2).$$

□

11.5.2 The Fluid and Biot Subproblem

For the fluid and Biot subproblem, we update the quantities related to the fluid and the Biot medium. Due to the kinematic coupling between the Biot medium displacement and the plate displacement, we must also update the plate velocity, as the dynamics of the Biot medium affect the kinematics of the plate. In this step, only the plate displacement remains unchanged:

$$\omega_N^{n+1} = \omega_N^{n+\frac{1}{2}}.$$

To state the weak formulation of the fluid and Biot subproblem, we define the solution and test spaces, respectively:

$$\mathcal{V}_N^{n+1} = \{(\boldsymbol{u}, \zeta, \boldsymbol{\eta}, p) \in \mathcal{V}_f^{\omega_N^n} \times H_0^2(\Gamma) \times V_d \times V_p\}, \tag{11.55}$$

$$\mathcal{Q}_N^{n+1} = \{(\boldsymbol{v}, \varphi, \boldsymbol{\psi}, r) \in V_f^{\omega_N^n} \times H_0^2(\Gamma) \times V_d \times V_p : \boldsymbol{\psi} = \varphi \boldsymbol{e}_y \text{ on } \Gamma\}, \tag{11.56}$$

where V_f^{ω}, V_d, and V_p are defined in (11.32), (11.35), and (11.37).

The weak formulation now reads: find $(\boldsymbol{u}_N^{n+1}, \zeta_N^{n+1}, \boldsymbol{\eta}_N^{n+1}, p_N^{n+1}) \in \mathcal{V}_N^{n+1}$ defined on the reference domain, such that for all test functions $(\boldsymbol{v}, \varphi, \boldsymbol{\psi}, r) \in \mathcal{Q}_N^{n+1}$ defined on the reference domain, the following holds:

$$\begin{aligned}
&\int_{\Omega_f} \left(1 + \frac{\omega_N^n}{R}\right) \dot{\boldsymbol{u}}_N^{n+1} \cdot \boldsymbol{v} + 2\nu \int_{\Omega_f} \left(1 + \frac{\omega_N^n}{R}\right) \boldsymbol{D}_f^{\omega_N^n}(\boldsymbol{u}_N^{n+1}) : \boldsymbol{D}_f^{\omega_N^n}(\boldsymbol{v}) \\
&+ \int_\Gamma \left(\frac{1}{2}\boldsymbol{u}_N^{n+1} \cdot \boldsymbol{u}_N^n - p_N^{n+1}\right)(\boldsymbol{\psi} - \boldsymbol{v}) \cdot \boldsymbol{n}^{\omega_N^n} \\
&+ \frac{1}{2} \int_{\Omega_f} \left(1 + \frac{\omega_N^n}{R}\right) \Bigg[\left(\left(\boldsymbol{u}_N^n - \zeta_N^{n+\frac{1}{2}} \frac{R+y}{R}\boldsymbol{e}_y\right) \cdot \nabla_f^{\omega_N^n} \boldsymbol{u}_N^{n+1}\right) \cdot \boldsymbol{v} \\
&- \left(\left(\boldsymbol{u}_N^n - \zeta_N^{n+\frac{1}{2}} \frac{R+y}{R}\boldsymbol{e}_y\right) \cdot \nabla_f^{\omega_N^n} \boldsymbol{v}\right) \cdot \boldsymbol{u}_N^{n+1}\Bigg] \\
&+ \frac{1}{2R} \int_{\Omega_f} \zeta_N^{n+\frac{1}{2}} \boldsymbol{u}_N^{n+1} \cdot \boldsymbol{v} + \frac{1}{2} \int_\Gamma (\boldsymbol{u}_N^{n+1} - \dot{\boldsymbol{\eta}}_N^{n+1}) \cdot \boldsymbol{n}^{\omega_N^n} (\boldsymbol{u}_N^n \cdot \boldsymbol{v}) \\
&+ \beta \int_\Gamma (\mathcal{J}_\Gamma^{\omega_N^n})^{-1} (\dot{\boldsymbol{\eta}}_N^{n+1} - \boldsymbol{u}_N^{n+1}) \cdot \boldsymbol{\tau}^{\omega_N^n} (\boldsymbol{\psi} - \boldsymbol{v}) \cdot \boldsymbol{\tau}^{\omega_N^n} + \rho_b \int_{\Omega_b} \left(\frac{\dot{\boldsymbol{\eta}}_N^{n+1} - \dot{\boldsymbol{\eta}}_N^n}{\Delta t}\right) \cdot \boldsymbol{\psi} \\
&+ \rho_p \int_\Gamma \left(\frac{\zeta_N^{n+1} - \zeta_N^{n+\frac{1}{2}}}{\Delta t}\right) \varphi + 2\mu_e \int_{\Omega_b} \boldsymbol{D}(\boldsymbol{\eta}_N^{n+1}) : \boldsymbol{D}(\boldsymbol{\psi}) \\
&+ \lambda_e \int_{\Omega_b} (\nabla \cdot \boldsymbol{\eta}_N^{n+1})(\nabla \cdot \boldsymbol{\psi}) \\
&+ 2\mu_v \int_{\Omega_b} \boldsymbol{D}(\dot{\boldsymbol{\eta}}_N^{n+1}) : \boldsymbol{D}(\boldsymbol{\psi}) + \lambda_v \int_{\Omega_b} (\nabla \cdot \dot{\boldsymbol{\eta}}_N^{n+1})(\nabla \cdot \boldsymbol{\psi}) \\
&- \alpha \int_{\Omega_b} \mathcal{J}_b^{(\eta_N^n)^\delta} p_N^{n+1} \nabla_b^{(\eta_N^n)^\delta} \cdot \boldsymbol{\psi} \\
&+ c_0 \int_{\Omega_b} \frac{p_N^{n+1} - p_N^n}{\Delta t} r - \alpha \int_{\Omega_b} \mathcal{J}_b^{(\eta_N^n)^\delta} \dot{\boldsymbol{\eta}}_N^{n+1} \cdot \nabla_b^{(\eta_N^n)^\delta} r - \alpha \int_\Gamma (\dot{\boldsymbol{\eta}}_N^{n+1} \cdot \boldsymbol{n}^{(\omega_N^n)^\delta}) r \\
&+ \kappa \int_{\Omega_b} \mathcal{J}_b^{(\eta_N^n)^\delta} \nabla_b^{(\eta_N^n)^\delta} p_N^{n+1} \cdot \nabla_b^{(\eta_N^n)^\delta} r - \int_\Gamma [(\boldsymbol{u}_N^{n+1} - \dot{\boldsymbol{\eta}}_N^{n+1}) \cdot \boldsymbol{n}^{\omega_N^n}] r = 0,
\end{aligned} \tag{11.57}$$

and

$$\int_{\Gamma} \left(\frac{\boldsymbol{\eta}_N^{n+1} - \boldsymbol{\eta}_N^n}{\Delta t} \right) \cdot \boldsymbol{\phi} = \int_{\Gamma} \zeta_N^{n+1} \boldsymbol{e}_y \cdot \boldsymbol{\phi}, \qquad \text{for all } \boldsymbol{\phi} \in L^2(\Gamma). \tag{11.58}$$

We remark that when $n = 0$, the backward Euler discretization (11.50) involves a negative subscript in the definition of $\dot{\boldsymbol{\eta}}_N^0$, so in this case, we will instead set $\dot{\boldsymbol{\eta}}_N^0$ to be the initial plate velocity ζ_0.

Lemma 11.5.2 *Problem* (11.57), (11.58) *has a unique solution, provided that the following assumptions hold:*

1. ASSUMPTION 1A*: Boundedness of the plate displacement away from R. There exists a positive constant* R_{max} *such that*

$$|\omega_N^{k+\frac{i}{2}}| \leq R_{max} < R, \qquad \textit{for all } k = 0, 1, \ldots, n \textit{ and } i = 0, 1. \tag{11.59}$$

2. ASSUMPTION 2A: *Invertibility of the map from fixed to moving Biot domain. The map*

$$Id + (\boldsymbol{\eta}_N^n)^\delta : \Omega_b \to (\Omega_b)_N^{n,\delta} \qquad \textit{is invertible,} \tag{11.60}$$

where we define $(\Omega_b)_N^{n,\delta}$ *to be the image of* Ω_b *under the map* $Id + (\boldsymbol{\eta}_N^n)^\delta$.

Additionally, the weak solution satisfies the following energy equality:

$$\begin{aligned}
&\frac{1}{2} \int_{\Omega_f} \left(1 + \frac{\omega_N^{n+1}}{R} \right) |\boldsymbol{u}_N^{n+1}|^2 + \frac{1}{2}\rho_b \int_{\Omega_b} |\dot{\boldsymbol{\eta}}_N^{n+1}|^2 + \frac{1}{2} c_0 \int_{\Omega_b} |p_N^{n+1}|^2 \\
&+ \mu_e \int_{\Omega_b} |\boldsymbol{D}(\boldsymbol{\eta}_N^{n+1})|^2 + \frac{1}{2}\lambda_e \int_{\Omega_b} |\nabla \cdot \boldsymbol{\eta}_N^{n+1}|^2 \\
&+ \frac{1}{2}\rho_p \int_{\Gamma} |\zeta_N^{n+1}|^2 + 2\mu_v(\Delta t) \int_{\Omega_b} |\boldsymbol{D}(\dot{\boldsymbol{\eta}}_N^{n+1})|^2 + \lambda_v(\Delta t) \int_{\Omega_b} |\nabla \cdot \dot{\boldsymbol{\eta}}_N^{n+1}|^2 \\
&+ \kappa(\Delta t) \int_{\Omega_b} \mathcal{J}_b^{(\eta_N^n)^\delta} |\nabla_b^{(\eta_N^n)^\delta} p_N^{n+1}|^2 \\
&+ \beta(\Delta t) \int_{\Gamma} (\mathcal{J}_\Gamma^{\omega_N^n})^{-1} |(\dot{\boldsymbol{\eta}}_N^{n+1} - \boldsymbol{u}_N^{n+1}) \cdot \boldsymbol{\tau}^{\omega_N^n}|^2 \\
&+ \frac{1}{2}\rho_b \int_{\Omega_b} |\dot{\boldsymbol{\eta}}_N^{n+1} - \dot{\boldsymbol{\eta}}_N^n|^2 + \frac{1}{2} c_0 \int_{\Omega_b} |p_N^{n+1} - p_N^n|^2 \\
&+ \mu_e \int_{\Omega_b} |\boldsymbol{D}(\boldsymbol{\eta}_N^{n+1} - \boldsymbol{\eta}_N^n)|^2
\end{aligned}$$

$$+\frac{1}{2}\lambda_e \int_{\Omega_b} |\nabla \cdot (\boldsymbol{\eta}_N^{n+1} - \boldsymbol{\eta}_N^n)|^2 = \frac{1}{2}\int_{\Omega_f} \left(1 + \frac{\omega_N^n}{R}\right) |\boldsymbol{u}_N^n|^2 + \frac{1}{2}\rho_b \int_{\Omega_b} |\dot{\boldsymbol{\eta}}_N^n|^2$$
$$+\frac{1}{2}c_0 \int_{\Omega_b} |p_N^n|^2 + \mu_e \int_{\Omega_b} |\boldsymbol{D}(\boldsymbol{\eta}_N^n)|^2$$
$$+\frac{1}{2}\lambda_e \int_{\Omega_b} |\nabla \cdot \boldsymbol{\eta}_N^n|^2 + \frac{1}{2}\rho_p \int_{\Gamma} |\zeta_N^{n+\frac{1}{2}}|^2.$$

The proof is based on using the Lax-Milgram Lemma. However, in this case, the proof is more involved for two reasons. First, the bilinear form associated with problem (11.57) and (11.58) is not coercive on the Hilbert space $\mathcal{V}_f^{\omega_N^n} \times V_d \times V_p$, because of a mismatch between the hyperbolic and parabolic scaling in the problem. The second reason is that it is not *a priori* clear that Korn's inequality, which is needed in the proof of the existence, holds for the Biot domain. To deal with the first difficulty and recover the coercive structure of the problem, the test functions can be rescaled by the factor Δt so that

$$\boldsymbol{v} \to (\Delta t)\boldsymbol{v}, \qquad r \to (\Delta t) r. \tag{11.61}$$

This scaling of the test functions is valid because if $(\boldsymbol{v}, \varphi, \boldsymbol{\psi}, v) \in \mathcal{Q}_N^{n+1}$, then the rescaled test function satisfies $((\Delta t)^{-1}\boldsymbol{v}, \varphi, \boldsymbol{\psi}, (\Delta t)^{-1} r) \in \mathcal{Q}_N^{n+1}$ also. To deal with the second difficulty, one can show by explicit calculation that the following Korn's inequality holds for this problem. We refer the reader to Section 11.2 and Corollary 11.2.22 in [59] for a more general proof of the Korn inequality.

Proposition 11.5.1 (Korn's Inequality for the Biot Poroviscoelastic Domain) *For all $\boldsymbol{\eta} \in V_d$,*

$$\int_{\Omega_b} |\boldsymbol{D}(\boldsymbol{\eta})|^2 \geq \frac{1}{2}\int_{\Omega_b} |\nabla \boldsymbol{\eta}|^2.$$

Proof By a standard approximation argument, it suffices to assume that $\boldsymbol{\eta}$ is smooth. Because $\eta_x = 0$ on Γ and because $\boldsymbol{\eta} = 0$ on the left, top, and right boundaries of Ω_b, we have from integration by parts that

$$\int_{\Omega_b} \frac{\partial \eta_x}{\partial y}\frac{\partial \eta_y}{\partial x} = -\int_{\Omega_b} \eta_x \frac{\partial^2 \eta_y}{\partial x \partial y} = \int_{\Omega_b} \frac{\partial \eta_x}{\partial x}\frac{\partial \eta_y}{\partial y}.$$

Therefore, by using the inequality $a^2 + 2ab + b^2 \geq 0$, we obtain

$$\int_{\Omega_b} |\boldsymbol{D}(\boldsymbol{\eta})|^2 = \int_{\Omega_b} \left(\frac{\partial \eta_x}{\partial x}\right)^2 + \left(\frac{\partial \eta_y}{\partial y}\right)^2 + 2\left[\frac{1}{2}\left(\frac{\partial \eta_x}{\partial y} + \frac{\partial \eta_y}{\partial x}\right)\right]^2$$

$$
\begin{aligned}
&= \int_{\Omega_b} \left(\frac{\partial \eta_x}{\partial x}\right)^2 + \left(\frac{\partial \eta_y}{\partial y}\right)^2 + \frac{1}{2}\left(\frac{\partial \eta_x}{\partial y} + \frac{\partial \eta_y}{\partial x}\right)^2 \\
&= \int_{\Omega_b} \left(\frac{\partial \eta_x}{\partial x}\right)^2 + \left(\frac{\partial \eta_y}{\partial y}\right)^2 + \frac{1}{2}\left[\left(\frac{\partial \eta_x}{\partial y}\right)^2 + \left(\frac{\partial \eta_y}{\partial x}\right)^2\right] + \frac{\partial \eta_x}{\partial y}\frac{\partial \eta_y}{\partial x} \\
&= \int_{\Omega_b} \left(\frac{\partial \eta_x}{\partial x}\right)^2 + \frac{\partial \eta_x}{\partial x}\frac{\partial \eta_y}{\partial y} + \left(\frac{\partial \eta_y}{\partial y}\right)^2 + \frac{1}{2}\left[\left(\frac{\partial \eta_x}{\partial y}\right)^2 + \left(\frac{\partial \eta_y}{\partial x}\right)^2\right] \\
&\geq \frac{1}{2}\int_{\Omega_b} |\nabla \boldsymbol{\eta}|^2.
\end{aligned}
$$

□

Proof ***(Proof of Lemma 11.5.2)*** Rewrite the weak formulation (11.57) and (11.58) so that all of the functions at the $(n+1)$st time step are on the left-hand side, while all other quantities are on the right-hand side. In addition, we rewrite ζ_N^{n+1} in terms of $\boldsymbol{\eta}_N^n$ and $\boldsymbol{\eta}_N^{n+1}$ by using (11.58):

$$
\zeta_N^{n+1}\boldsymbol{e}_y = \left.\frac{\boldsymbol{\eta}_N^{n+1} - \boldsymbol{\eta}_N^n}{\Delta t}\right|_\Gamma.
$$

After using the rescaling (11.61) of the test functions, the weak formulation involves the following coercive and continuous bilinear form $B : H \times H \to \mathbb{R}$, where H is the Hilbert space $\mathcal{V}_f^{\omega_N^n} \times V_d \times V_p$:

$$
\begin{aligned}
&B([\boldsymbol{u}, \boldsymbol{\eta}, p), (\boldsymbol{v}, \boldsymbol{\psi}, r)] := (\Delta t)^2 \int_{\Omega_f} \left(1 + \frac{\omega_N^n}{R}\right) \boldsymbol{u}\cdot\boldsymbol{v} \\
&+ \frac{1}{2}(\Delta t)^3 \int_{\Omega_f} \left(1 + \frac{\omega_N^n}{R}\right)\left[\left(\left(\boldsymbol{u}_N^n - \zeta_N^{n+\frac{1}{2}}\frac{R+y}{R}\boldsymbol{e}_y\right)\cdot\nabla^{\omega_N^n}\boldsymbol{u}\right)\cdot\boldsymbol{v}\right. \\
&\left.- \left(\left(\boldsymbol{u}_N^n - \zeta_N^{n+\frac{1}{2}}\frac{R+y}{R}\boldsymbol{e}_y\right)\cdot\nabla^{\omega_N^n}\boldsymbol{v}\right)\cdot\boldsymbol{u}\right] \\
&+ (\Delta t)^3 \cdot \frac{1}{2R}\int_{\Omega_f} \zeta_N^{n+\frac{1}{2}}\boldsymbol{u}\cdot\boldsymbol{v} + \frac{1}{2}(\Delta t)^3 \int_\Gamma (\boldsymbol{u} - (\Delta t)^{-1}\boldsymbol{\eta})\cdot\boldsymbol{n}^{\omega_N^n}(\boldsymbol{u}_N^n\cdot\boldsymbol{v}) \\
&+ 2\nu(\Delta t)^3 \int_{\Omega_f}\left(1 + \frac{\omega_N^n}{R}\right)\boldsymbol{D}_f^{\omega_N^n}(\boldsymbol{u}) : \boldsymbol{D}_f^{\omega_N^n}(\boldsymbol{v}) \\
&+ (\Delta t)^2 \int_\Gamma \left(\frac{1}{2}\boldsymbol{u}\cdot\boldsymbol{u}_N^n - p\right)(\boldsymbol{\psi} - (\Delta t)\boldsymbol{v})\cdot\boldsymbol{n}^{\omega_N^n} \\
&+ \beta(\Delta t)^2 \int_\Gamma (\mathcal{J}_\Gamma^{\omega_N^n})^{-1}[(\Delta t)^{-1}\boldsymbol{\eta} - \boldsymbol{u}]\cdot\boldsymbol{\tau}^{\omega_N^n}(\boldsymbol{\psi} - (\Delta t)\boldsymbol{v})\cdot\boldsymbol{\tau}^{\omega_N^n}
\end{aligned}
$$

$$+ \rho_b \int_{\Omega_b} \boldsymbol{\eta} \cdot \boldsymbol{\psi} + \rho_p \int_\Gamma \boldsymbol{\eta} \cdot \boldsymbol{\psi}$$

$$+ (2\mu_e(\Delta t)^2 + 2\mu_v(\Delta t)) \int_{\Omega_b} \boldsymbol{D}(\boldsymbol{\eta}) : \boldsymbol{D}(\boldsymbol{\psi}) + (\lambda_e(\Delta t)^2$$

$$+ \lambda_v(\Delta t)) \int_{\Omega_b} (\nabla \cdot \boldsymbol{\eta})(\nabla \cdot \boldsymbol{\psi})$$

$$- \alpha(\Delta t)^2 \int_{\Omega_b} \mathcal{J}_b^{(\eta_N^n)^\delta} p \nabla_b^{(\eta_N^n)^\delta} \cdot \boldsymbol{\psi} + c_0(\Delta t)^2 \int_{\Omega_b} pr$$

$$- \alpha(\Delta t)^2 \int_{\Omega_b} \mathcal{J}_b^{(\eta_N^n)^\delta} \boldsymbol{\eta} \cdot \nabla_b^{(\eta_N^n)^\delta} r$$

$$- \alpha(\Delta t)^2 \int_\Gamma (\boldsymbol{\eta} \cdot \boldsymbol{n}^{(\omega_N^n)^\delta}) r + \kappa(\Delta t)^3 \int_{\Omega_b} \mathcal{J}_b^{(\eta_N^n)^\delta} \nabla_b^{(\eta_N^n)^\delta} p \cdot \nabla_b^{(\eta_N^n)^\delta} r$$

$$- (\Delta t)^3 \int_\Gamma [(\boldsymbol{u} - (\Delta t)^{-1}\boldsymbol{\eta}) \cdot \boldsymbol{n}^{\omega_N^n}] r.$$

With this notation, the weak formulation reads: find $(\boldsymbol{u}_N^{n+1}, \boldsymbol{\eta}_N^{n+1}, p_N^{n+1}) \in \mathcal{V}_f^{\omega_N^n} \times V_d \times V_p$ such that for all test functions $(\boldsymbol{v}, \boldsymbol{\psi}, r) \in \mathcal{V}_f^{\omega_N^n} \times V_d \times V_p$,

$$\begin{aligned} B[(\boldsymbol{u}_N^{n+1}, \boldsymbol{\eta}_N^{n+1}, p_N^{n+1}), (\boldsymbol{v}, \boldsymbol{\psi}, r)] &= (\Delta t)^2 \int_{\Omega_f} \left(1 + \frac{\omega_N^n}{R}\right) \boldsymbol{u}_N^n \cdot \boldsymbol{v} \\ &- \frac{1}{2}(\Delta t)^2 \int_\Gamma \boldsymbol{\eta}_N^n \cdot \boldsymbol{n}^{\omega_N^n} (\boldsymbol{u}_N^n \cdot \boldsymbol{v}) \\ + \beta(\Delta t) \int_\Gamma (\mathcal{J}_\Gamma^{\omega_N^n})^{-1} \boldsymbol{\eta}_N^n \cdot \boldsymbol{\tau}^{\omega_N^n} (\boldsymbol{\psi} - (\Delta t)\boldsymbol{v}) \cdot \boldsymbol{\tau}^{\omega_N^n} &+ \rho_b \int_{\Omega_b} (2\boldsymbol{\eta}_N^n - \boldsymbol{\eta}_N^{n-1}) \cdot \boldsymbol{\psi} \\ &+ \rho_p \int_\Gamma (\boldsymbol{\eta}_N^n + (\Delta t)\zeta_N^{n+\frac{1}{2}} \boldsymbol{e}_y) \cdot \boldsymbol{\psi} \\ + 2\mu_v(\Delta t) \int_{\Omega_b} \boldsymbol{D}(\boldsymbol{\eta}_N^n) : \boldsymbol{D}(\boldsymbol{\psi}) &+ \lambda_v(\Delta t) \int_{\Omega_b} (\nabla \cdot \boldsymbol{\eta}_N^n)(\nabla \cdot \boldsymbol{\psi}) \\ + c_0(\Delta t)^2 \int_{\Omega_b} p_N^n r - \alpha(\Delta t)^2 \int_{\Omega_b} \mathcal{J}_b^{(\eta_N^n)^\delta} \boldsymbol{\eta}_N^n \cdot \nabla_b^{(\eta_N^n)^\delta} r &- \alpha(\Delta t)^2 \int_\Gamma (\boldsymbol{\eta}_N^n \cdot \boldsymbol{n}^{(\omega_N^n)^\delta}) r \\ &+ (\Delta t)^2 \int_\Gamma (\boldsymbol{\eta}_N^n \cdot \boldsymbol{n}^{\omega_N^n}) r. \end{aligned} \tag{11.62}$$

We now show that the bilinear form $B[(\boldsymbol{u}, \boldsymbol{\eta}, p), (\boldsymbol{v}, \boldsymbol{\psi}, r)]$ is coercive and continuous on the Hilbert space $\mathcal{V}_f^{\omega_N^n} \times V_d \times V_p$, with the inner product given by

$$\langle(\boldsymbol{u},\boldsymbol{\eta},p),(\boldsymbol{v},\boldsymbol{\psi},r)\rangle = \int_{\Omega_f}(\boldsymbol{u}\cdot\boldsymbol{v}+\nabla\boldsymbol{u}:\nabla\boldsymbol{v}) + \int_{\Omega_b}(\boldsymbol{\eta}\cdot\boldsymbol{\psi}+\nabla\boldsymbol{\eta}:\nabla\boldsymbol{\psi})$$
$$+\int_{\Omega_b}(p\cdot r+\nabla p\cdot\nabla r).$$

We focus on establishing coercivity, since continuity follows by standard arguments. To show coercivity, we calculate $B[(\boldsymbol{u},\boldsymbol{v},p),(\boldsymbol{u},\boldsymbol{\eta},p)]$. In this calculation, we note that after integrating by parts, the sum of the following terms becomes zero:

$$-\alpha(\Delta t)^2\int_{\Omega_b}\mathcal{J}_b^{(\eta_N^n)^\delta}p\nabla_b^{(\eta_N^n)^\delta}\cdot\boldsymbol{\eta}-\alpha(\Delta t)^2\int_{\Omega_b}\mathcal{J}_b^{(\eta_N^n)^\delta}\boldsymbol{\eta}\cdot\nabla_b^{(\eta_N^n)^\delta}p$$
$$-\alpha(\Delta t)^2\int_\Gamma\left(\boldsymbol{\eta}\cdot\boldsymbol{n}^{(\omega_N^n)^\delta}\right)p=0.$$

Indeed, to see this, we bring the integrals back to the time-dependent physical domain, **which we can do as long as $(\eta_N^n)^\delta$ is a bijection from Ω_b to $(\Omega_b)_N^{n,\delta}$**, which is provided by Assumption 2A (11.60), and perform the following computation:

$$-\alpha(\Delta t)^2\int_{\Omega_b}\mathcal{J}_b^{(\eta_N^n)^\delta}p\nabla_b^{(\eta_N^n)^\delta}\cdot\boldsymbol{\eta}-\alpha(\Delta t)^2\int_{\Omega_b}\mathcal{J}_b^{(\eta_N^n)^\delta}\boldsymbol{\eta}\cdot\nabla_b^{(\eta_N^n)^\delta}p$$
$$-\alpha(\Delta t)^2\int_\Gamma\left(\boldsymbol{\eta}\cdot\boldsymbol{n}^{(\omega_N^n)^\delta}\right)p$$
$$=-\alpha(\Delta t)^2\left(\int_{(\Omega_b)_N^{n,\delta}}p\nabla\cdot\boldsymbol{\eta}+\int_{(\Omega_b)_N^{n,\delta}}\boldsymbol{\eta}\cdot\nabla p+\int_{\Gamma_N^{n,\delta}}(\boldsymbol{\eta}\cdot\boldsymbol{n})p\right)=0,$$

where we used integration by parts, the fact that $\boldsymbol{n}$ points outward from Ω_f and hence inward toward Ω_b and also use that $\boldsymbol{\eta}=0$ on the left, right, and top boundaries of Ω_b. Combining this with the fact that $(\Delta t)\zeta_N^{n+\frac{1}{2}}=\omega_N^{n+\frac{1}{2}}-\omega_N^{n-\frac{1}{2}}=\omega_N^{n+1}-\omega_N^n$, we obtain

$$B[(\boldsymbol{u},\boldsymbol{\eta},p),(\boldsymbol{u},\boldsymbol{\eta},p)]:=(\Delta t)^2\int_{\Omega_f}\left(1+\frac{\omega_N^n+\omega_N^{n+1}}{2R}\right)|\boldsymbol{u}|^2$$
$$+2\nu(\Delta t)^3\int_{\Omega_f}\left(1+\frac{\omega_N^n}{R}\right)\left|\boldsymbol{D}_f^{\omega_N^n}(\boldsymbol{u})\right|^2$$
$$+\beta(\Delta t)\int_\Gamma(\mathcal{J}_\Gamma^{\omega_N^n})^{-1}\left|(\boldsymbol{\eta}-(\Delta t)\boldsymbol{u})\cdot\boldsymbol{\tau}^{\omega_N^n}\right|^2+\rho_b\int_{\Omega_b}|\boldsymbol{\eta}|^2+\rho_p\int_\Gamma|\boldsymbol{\eta}|^2$$
$$+(2\mu_e(\Delta t)^2+2\mu_v(\Delta t))\int_{\Omega_b}|\boldsymbol{D}(\boldsymbol{\eta})|^2$$

$$+(\lambda_e(\Delta t)^2+\lambda_v(\Delta t))\int_{\Omega_b}|\nabla\cdot\boldsymbol{\eta}|^2+c_0(\Delta t)^2\int_{\Omega_b}|p|^2+\kappa(\Delta t)^3\int_{\Omega_b}\mathcal{J}_b^{(\eta_N^n)^\delta}|\nabla_b^{(\eta_N^n)^\delta}p|^2.$$

Coercivity of this form follows from the fact that $|\omega_N^{k+\frac{i}{2}}| < R$ (see Assumption 1A in (11.59)) and Korn inequality (see Proposition 11.5.1) once we handle the last term and show that

$$\kappa(\Delta t)^3\int_{\Omega_b}\mathcal{J}_b^{(\eta_N^n)^\delta}|\nabla_b^{(\eta_N^n)^\delta}p|^2 \geq c\int_{\Omega_b}|\nabla p|^2,$$

for some positive constant $c > 0$. To show this, we first recall the definitions

$$\mathcal{J}_b^{(\eta_N^n)^\delta} = \det(\boldsymbol{I}+\nabla(\boldsymbol{\eta}_N^n)^\delta), \qquad \nabla_b^{(\eta_N^n)^\delta}p = \nabla p\cdot(\boldsymbol{I}+\nabla(\boldsymbol{\eta}_N^n)^\delta)^{-1}.$$

Then, letting $|\cdot|$ denote the matrix norm, we have

$$\kappa(\Delta t)^3\int_{\Omega_b}\mathcal{J}_b^{(\eta_N^n)^\delta}|\nabla_b^{(\eta_N^n)^\delta}p|^2 \geq \kappa(\Delta t)^3\int_{\Omega_b}\mathcal{J}_b^{(\eta_N^n)^\delta}|\boldsymbol{I}+\nabla(\boldsymbol{\eta}_N^n)^\delta|^{-2}|\nabla p|^2. \tag{11.63}$$

Assumption 2A (11.60) implies that $\boldsymbol{I}+(\boldsymbol{\eta}_N^n)^\delta$ is an invertible map from Ω_b to $(\Omega_b)_N^{n,\delta}$, and we further note that $|\boldsymbol{I}+\nabla(\boldsymbol{\eta}_N^n)^\delta|$ is continuous on $\overline{\Omega_b}$ and hence is bounded from above. Thus, $|\boldsymbol{I}+\nabla(\boldsymbol{\eta}_N^n)^\delta|^{-2} \geq c_0 > 0$ for some positive constant c_0. The assumption that $\boldsymbol{I}+(\boldsymbol{\eta}_N^n)^\delta$ is invertible implies that $\det(\boldsymbol{I}+\nabla(\boldsymbol{\eta}_N^n)^\delta) > 0$. However, since this determinant is a continuous function on the compact set $\overline{\Omega_b}$, we conclude that there exists a positive constant $c_1 > 0$ such that $\det(\boldsymbol{I}+\nabla(\boldsymbol{\eta}_N^n)^\delta) \geq c_1 > 0$. This establishes coercivity.

Existence of a unique weak solution $(\boldsymbol{u}_N^{n+1}, \boldsymbol{\eta}_N^{n+1}, p_N^{n+1}) \in \mathcal{V}_f^{\omega_N^n} \times V_d \times V_p$ now follows from the Lax-Milgram lemma. From here, we recover ζ_N^{n+1}, by using $\zeta_N^{n+1}\boldsymbol{e}_y = \left.\frac{\boldsymbol{\eta}_N^{n+1}-\boldsymbol{\eta}_N^n}{\Delta t}\right|_\Gamma$. Note that $\left.\frac{\boldsymbol{\eta}_N^{n+1}-\boldsymbol{\eta}_N^n}{\Delta t}\right|_\Gamma$ points in the y direction because the trace of any function $\boldsymbol{\eta} \in V_d$ on Γ points in the y direction by definition (see (11.35)).

Energy Equality We substitute $\boldsymbol{v} = \boldsymbol{u}_N^{n+1}$, $\varphi = \zeta_N^{n+1}$, $\boldsymbol{\psi} = \dot{\boldsymbol{\eta}}_N^{n+1}$, and $r = p_N^{n+1}$ into (11.57) and use the identity

$$(a-b)\cdot a = \frac{1}{2}(|a|^2+|a-b|^2-|b|^2).$$

Since $\omega_N^{n+1} = \omega_N^{n+\frac{1}{2}}$ and $(\Delta t)\zeta_N^{n+\frac{1}{2}} = \omega_N^{n+\frac{1}{2}} - \omega_N^n$, we obtain the following energy equality:

$$\frac{1}{2}\int_{\Omega_f}\left(1+\frac{\omega_N^{n+1}}{R}\right)|\boldsymbol{u}_N^{n+1}|^2+\frac{1}{2}\rho_b\int_{\Omega_b}|\dot{\boldsymbol{\eta}}_N^{n+1}|^2+\frac{1}{2}c_0\int_{\Omega_b}|p_N^{n+1}|^2$$
$$+\mu_e\int_{\Omega_b}|\boldsymbol{D}(\boldsymbol{\eta}_N^{n+1})|^2+\frac{1}{2}\lambda_e\int_{\Omega_b}|\nabla\cdot\boldsymbol{\eta}_N^{n+1}|^2$$
$$+\frac{1}{2}\rho_p\int_\Gamma|\zeta_N^{n+1}|^2+2\mu_v(\Delta t)\int_{\Omega_b}|\boldsymbol{D}(\dot{\boldsymbol{\eta}}_N^{n+1})|^2+\lambda_v(\Delta t)\int_{\Omega_b}|\nabla\cdot\dot{\boldsymbol{\eta}}_N^{n+1}|^2$$
$$+\kappa(\Delta t)\int_{\Omega_b}\mathcal{J}_b^{(\eta_N^n)^\delta}|\nabla_b^{(\eta_N^n)^\delta}p_N^{n+1}|^2$$
$$+\beta(\Delta t)\int_\Gamma(\mathcal{J}_\Gamma^{\omega_N^n})^{-1}|(\dot{\boldsymbol{\eta}}_N^{n+1}-\boldsymbol{u}_N^{n+1})\cdot\boldsymbol{\tau}^{\omega_N^n}|^2+\frac{1}{2}\rho_b\int_{\Omega_b}|\dot{\boldsymbol{\eta}}_N^{n+1}-\dot{\boldsymbol{\eta}}_N^n|^2$$
$$+\frac{1}{2}c_0\int_{\Omega_b}|p_N^{n+1}-p_N^n|^2+\mu_e\int_{\Omega_b}|\boldsymbol{D}(\boldsymbol{\eta}_N^{n+1}-\boldsymbol{\eta}_N^n)|^2$$
$$+\frac{1}{2}\lambda_e\int_{\Omega_b}|\nabla\cdot(\boldsymbol{\eta}_N^{n+1}-\boldsymbol{\eta}_N^n)|^2=\frac{1}{2}\int_{\Omega_f}\left(1+\frac{\omega_N^n}{R}\right)|\boldsymbol{u}_N^n|^2$$
$$+\frac{1}{2}\rho_b\int_{\Omega_b}|\dot{\boldsymbol{\eta}}_N^n|^2+\frac{1}{2}c_0\int_{\Omega_b}|p_N^n|^2+\mu_e\int_{\Omega_b}|\boldsymbol{D}(\boldsymbol{\eta}_N^n)|^2$$
$$+\frac{1}{2}\lambda_e\int_{\Omega_b}|\nabla\cdot\boldsymbol{\eta}_N^n|^2+\frac{1}{2}\rho_p\int_\Gamma|\zeta_N^{n+\frac{1}{2}}|^2,$$

where the terms containing parameter α cancel out after bringing the integrals back to the time-dependent domain, integrating by parts, and recalling that the normal vector points inward toward the Biot domain:

$$-\alpha\int_{\Omega_b}\mathcal{J}_b^{(\eta_N^n)^\delta}p_N^{n+1}\nabla_b^{(\eta_N^n)^\delta}\cdot\dot{\boldsymbol{\eta}}_N^{n+1}-\alpha\int_{\Omega_b}\mathcal{J}_b^{(\eta_N^n)^\delta}\dot{\boldsymbol{\eta}}_N^{n+1}\cdot\nabla_b^{(\eta_N^n)^\delta}p_N^{n+1}$$
$$-\alpha\int_\Gamma\left(\dot{\boldsymbol{\eta}}_N^{n+1}\cdot\boldsymbol{n}^{(\omega_N^n)^\delta}\right)p_N^{n+1}$$
$$=-\alpha\int_{(\Omega_b)_N^{n,\delta}}p_N^{n+1}(\nabla\cdot\dot{\boldsymbol{\eta}}_N^{n+1})-\alpha\int_{(\Omega_b)_N^{n,\delta}}\dot{\boldsymbol{\eta}}_N^{n+1}\cdot\nabla p_N^{n+1}-\alpha\int_{\Gamma_N^{n,\delta}}(\dot{\boldsymbol{\eta}}_N^{n+1}\cdot\boldsymbol{n})p_N^{n+1}=0.$$

This completes the proof of the Lemma. □

11.5.3 The Coupled Semi-discrete Problem: Weak Formulation and Energy

To obtain uniform energy estimates for approximate solutions of our semidiscretized scheme, it is useful to present the scheme in monolithic form:

$$
\begin{aligned}
&\int_{\Omega_f}\left(1+\frac{\omega_N^n}{R}\right)\dot{\boldsymbol{u}}_N^{n+1}\cdot\boldsymbol{v}+2\nu\int_{\Omega_f}\left(1+\frac{\omega_N^n}{R}\right)\boldsymbol{D}_f^{\omega_N^n}(\boldsymbol{u}_N^{n+1}):\boldsymbol{D}_f^{\omega_N^n}(\boldsymbol{v})\\
&+\int_\Gamma\left(\frac{1}{2}\boldsymbol{u}_N^{n+1}\cdot\boldsymbol{u}_N^n-p_N^{n+1}\right)(\boldsymbol{\psi}-\boldsymbol{v})\cdot\boldsymbol{n}^{\omega_N^n}\\
&+\frac{1}{2}\int_{\Omega_f}\left(1+\frac{\omega_N^n}{R}\right)\left[\left(\left(\boldsymbol{u}_N^n-\zeta_N^{n+\frac{1}{2}}\frac{R+y}{R}\boldsymbol{e}_y\right)\cdot\nabla_f^{\omega_N^n}\boldsymbol{u}_N^{n+1}\right)\cdot\boldsymbol{v}\right.\\
&\left.-\left(\left(\boldsymbol{u}_N^n-\zeta_N^{n+\frac{1}{2}}\frac{R+y}{R}\boldsymbol{e}_y\right)\cdot\nabla_f^{\omega_N^n}\boldsymbol{v}\right)\cdot\boldsymbol{u}_N^{n+1}\right]\\
&+\frac{1}{2R}\int_{\Omega_f}\zeta_N^{n+\frac{1}{2}}\boldsymbol{u}_N^{n+1}\cdot\boldsymbol{v}+\frac{1}{2}\int_\Gamma(\boldsymbol{u}_N^{n+1}-\dot{\boldsymbol{\eta}}_N^{n+1})\cdot\boldsymbol{n}^{\omega_N^n}(\boldsymbol{u}_N^n\cdot\boldsymbol{v})\\
&+\beta\int_\Gamma(\mathcal{J}_\Gamma^{\omega_N^n})^{-1}(\dot{\boldsymbol{\eta}}_N^{n+1}-\boldsymbol{u}_N^{n+1})\cdot\boldsymbol{\tau}^{\omega_N^n}(\boldsymbol{\psi}-\boldsymbol{v})\cdot\boldsymbol{\tau}^{\omega_N^n}\\
&+\rho_b\int_{\Omega_b}\left(\frac{\dot{\boldsymbol{\eta}}_N^{n+1}-\dot{\boldsymbol{\eta}}_N^n}{\Delta t}\right)\cdot\boldsymbol{\psi}+\rho_p\int_\Gamma\left(\frac{\zeta_N^{n+1}-\zeta_N^n}{\Delta t}\right)\varphi+2\mu_e\int_{\Omega_b}\boldsymbol{D}(\boldsymbol{\eta}_N^{n+1}):\boldsymbol{D}(\boldsymbol{\psi})\\
&+\lambda_e\int_{\Omega_b}(\nabla\cdot\boldsymbol{\eta}_N^{n+1})(\nabla\cdot\boldsymbol{\psi})\\
&+2\mu_v\int_{\Omega_b}\boldsymbol{D}(\dot{\boldsymbol{\eta}}_N^{n+1}):\boldsymbol{D}(\boldsymbol{\psi})+\lambda_v\int_{\Omega_b}(\nabla\cdot\dot{\boldsymbol{\eta}}_N^{n+1})(\nabla\cdot\boldsymbol{\psi})\\
&-\alpha\int_{\Omega_b}\mathcal{J}_b^{(\eta_N^n)^\delta}p_N^{n+1}\nabla_b^{(\eta_N^n)^\delta}\cdot\boldsymbol{\psi}+c_0\int_{\Omega_b}\frac{p_N^{n+1}-p_N^n}{\Delta t}r\\
&-\alpha\int_{\Omega_b}\mathcal{J}_b^{(\eta_N^n)^\delta}\dot{\boldsymbol{\eta}}_N^{n+1}\cdot\nabla_b^{(\eta_N^n)^\delta}r-\alpha\int_\Gamma(\dot{\boldsymbol{\eta}}_N^{n+1}\cdot\boldsymbol{n}^{(\omega_N^n)^\delta})r\\
&+\kappa\int_{\Omega_b}\mathcal{J}_b^{(\eta_N^n)^\delta}\nabla_b^{(\eta_N^n)^\delta}p_N^{n+1}\cdot\nabla_b^{(\eta_N^n)^\delta}r\\
&-\int_\Gamma[(\boldsymbol{u}_N^{n+1}-\dot{\boldsymbol{\eta}}_N^{n+1})\cdot\boldsymbol{n}^{\omega_N^n}]r+\int_\Gamma\Delta\omega_N^{n+\frac{1}{2}}\cdot\Delta\varphi=0,\ \forall(\boldsymbol{v},\varphi,\boldsymbol{\psi},r)\in\mathcal{Q}_N^{n+1},
\end{aligned}
\tag{11.64}
$$

$$\int_\Gamma \left(\frac{\omega_N^{n+\frac{1}{2}} - \omega_N^{n-\frac{1}{2}}}{\Delta t} \right) \phi = \int_\Gamma \zeta_N^{n+\frac{1}{2}} \phi,$$

$$\int_\Gamma \left(\frac{\boldsymbol{\eta}_N^{n+1} - \boldsymbol{\eta}_N^n}{\Delta t} \right) \cdot \boldsymbol{\phi} = \int_\Gamma \zeta_N^{n+1} \boldsymbol{e}_y \cdot \boldsymbol{\phi}, \ \forall \phi, \boldsymbol{\phi} \in L^2(\Gamma). \tag{11.65}$$

Next, we will obtain uniform energy estimates for the approximate solutions generated from the splitting scheme. To do this, we define the discrete energy $E_N^{n+\frac{i}{2}}$ and discrete dissipation D_N^{n+1} as follows:

$$\begin{aligned} E_N^{n+\frac{i}{2}} &= \frac{1}{2} \int_{\Omega_f} \left(1 + \frac{\omega_N^n}{R}\right) |\boldsymbol{u}_N^{n+\frac{i}{2}}|^2 + \frac{1}{2}\rho_b \int_{\Omega_b} |\dot{\boldsymbol{\eta}}_N^{n+\frac{i}{2}}|^2 + \frac{1}{2} c_0 \int_{\Omega_b} |p_N^{n+\frac{i}{2}}|^2 \\ &+ \mu_e \int_{\Omega_b} |\boldsymbol{D}(\boldsymbol{\eta}_N^{n+\frac{i}{2}})|^2 \\ &+ \frac{1}{2}\lambda_e \int_{\Omega_b} |\nabla \cdot \boldsymbol{\eta}_N^{n+\frac{i}{2}}|^2 + \frac{1}{2}\rho_p \int_\Gamma |\zeta_N^{n+\frac{i}{2}}|^2 + \frac{1}{2}\int_\Gamma |\Delta \omega_N^{n+\frac{i}{2}}|^2, \ i = 0, 1. \\ D_N^{n+1} &= 2\nu(\Delta t) \int_{\Omega_f} \left(1 + \frac{\omega_N^n}{R}\right) \left| \boldsymbol{D}_f^{\omega_N^n}(\boldsymbol{u}_N^{n+1}) \right|^2 + 2\mu_v(\Delta t) \int_{\Omega_b} |\boldsymbol{D}(\dot{\boldsymbol{\eta}}_N^{n+1})|^2 \\ &+ \lambda_v(\Delta t) \int_{\Omega_b} |\nabla \cdot \dot{\boldsymbol{\eta}}_N^{n+1}|^2 \\ &+ \kappa(\Delta t) \int_{\Omega_b} \mathcal{J}_b^{(\eta_N^n)^\delta} |\nabla_b^{(\eta_N^n)^\delta} p_N^{n+1}|^2 + \frac{\beta(\Delta t)}{\mathcal{J}_\Gamma^{\omega_N^n}} \int_\Gamma \left| (\dot{\boldsymbol{\eta}}_N^{n+1} - \boldsymbol{u}_N^{n+1}) \cdot \boldsymbol{\tau}^{\omega_N^n} \right|^2. \end{aligned} \tag{11.66}$$

Then, the semidiscrete weak formulations (11.64) and (11.65) imply the following uniform estimates on the discretized energy and dissipation.

Lemma 11.5.3 *The following* ***discrete energy equalities*** *hold for the semidiscretized formulations* (11.64) *and* (11.65)*:*

$$E_N^{n+\frac{1}{2}} + \frac{1}{2}\rho_p \int_\Gamma \left| \zeta_N^{n+\frac{1}{2}} - \zeta_N^n \right|^2 + \frac{1}{2} \int_\Gamma \left| \Delta(\omega_N^{n+\frac{1}{2}} - \omega_N^{n-\frac{1}{2}}) \right|^2 = E_N^n \tag{11.67}$$

$$\begin{aligned} E_N^{n+1} + D_N^{n+1} &+ \frac{1}{2} \int_{\Omega_f} \left(1 + \frac{\omega_N^n}{R}\right) \left| \boldsymbol{u}_N^{n+1} - \boldsymbol{u}_N^n \right|^2 + \frac{1}{2}\rho_b \int_{\Omega_b} \left| \dot{\boldsymbol{\eta}}_N^{n+1} - \dot{\boldsymbol{\eta}}_N^n \right|^2 \\ &+ \frac{1}{2} c_0 \int_{\Omega_b} \left| p_N^{n+1} - p_N^n \right|^2 \end{aligned}$$

$$+ \mu_e \int_{\Omega_b} \left| \boldsymbol{D}(\boldsymbol{\eta}_N^{n+1} - \boldsymbol{\eta}_N^n) \right|^2 + \frac{1}{2}\lambda_e \int_{\Omega_b} \left| \nabla \cdot (\boldsymbol{\eta}_N^{n+1} - \boldsymbol{\eta}_N^n) \right|^2$$

$$+ \frac{1}{2}\rho_p \int_\Gamma |\zeta_N^{n+1} - \zeta_N^{n+\frac{1}{2}}|^2 = E_N^{n+\frac{1}{2}}, \qquad (11.68)$$

where the discrete energy $E_N^{n+\frac{i}{2}}$ *and the discrete dissipation* D_N^{n+1} *are defined in* (11.66).

We remark that the terms not included in the definition of $E_N^{n+\frac{i}{2}}$ and D_N^{n+1}, appearing in (11.67) and (11.68), are **numerical dissipation terms**.

These energy identities immediately imply that $E_N^{n+\frac{i}{2}}$ and $\sum_{n=1}^N D_N^n$ are **uniformly bounded** by a constant C independent of n and N.

The semidiscretized splitting scheme defines semidiscretized approximations of the solution to the regularized problem at discrete time points. To work with approximate functions and show that they converge to the solution of the continuous problem, we need to extend the semidiscrete approximations to the entire time interval and investigate uniform boundedness of those approximate solution functions. This is done next.

11.6 Approximate Solutions

We now use the method described in Sect. 2.15 of Chap. 2 to define the approximate solutions, for which we will derive uniform estimates based on our previous energy estimates.

- Piecewise constant approximate solutions, for $(n-1)\Delta t < t \leq n\Delta t$:

$$\boldsymbol{u}_N(t) = \boldsymbol{u}_N^n, \quad \boldsymbol{\eta}_N(t) = \boldsymbol{\eta}_N^n, \quad p_N(t) = p_N^n, \quad \omega_N(t) = \omega_N^{n-\frac{1}{2}},$$

$$\zeta_N(t) = \zeta_N^{n-\frac{1}{2}}, \quad \zeta_N^*(t) = \zeta_N^n; \qquad (11.69)$$

- Linear interpolations:

$$\overline{\boldsymbol{\eta}}_N(n\Delta t) = \boldsymbol{\eta}_N^n, \qquad \overline{p}_N(n\Delta t) = p_N^n, \qquad \overline{\omega}_N(n\Delta t) = \omega_N^{n-\frac{1}{2}},$$

$$\text{for } n = 0, 1, \ldots, N, \qquad (11.70)$$

where we formally set $\omega_N^{-\frac{1}{2}} = \omega_0$.

Note that by construction, we have that

$$\partial_t \overline{\omega}_N = \zeta_N, \qquad \partial_t \overline{\boldsymbol{\eta}}_N|_\Gamma = \zeta_N^* \boldsymbol{e}_y. \tag{11.71}$$

From the preceding energy estimates, we have the following lemma on uniform boundedness.

Lemma 11.6.1 (Uniform Boundedness of Approximate Solutions) *Assume:*

1. ASSUMPTION 1B: UNIFORM BOUNDEDNESS OF PLATE DISPLACEMENTS. *There exists a positive constant R_{max} such that for all N,*

$$|\omega_N^{n-\frac{1}{2}}| \le R_{max} < R, \qquad \textit{for all } n = 0, 1, \ldots, N, \tag{11.72}$$

$$\left|(\boldsymbol{\eta}_N^n)^\delta|_\Gamma\right| \le R_{max} < R, \qquad \textit{for all } n = 0, 1, \ldots, N. \tag{11.73}$$

2. ASSUMPTION 2B: UNIFORM INVERTIBILITY OF THE LAGRANGIAN MAP (JACOBIAN). *There exists a positive constant c_0 such that for all N,*

$$\det(\boldsymbol{I} + \nabla(\boldsymbol{\eta}_N^n)^\delta) \ge c_0 > 0, \qquad \textit{for all } n = 0, 1, \ldots, N. \tag{11.74}$$

3. ASSUMPTION 2C: UNIFORM BOUNDEDNESS OF THE LAGRANGIAN MAP (MATRIX NORM). *There exists positive constants c_1 and c_2 such that for all N,*

$$|(\boldsymbol{I} + \nabla(\boldsymbol{\eta}_N^n)^\delta)^{-1}| \le c_1, \qquad |\boldsymbol{I} + \nabla(\boldsymbol{\eta}_N^n)^\delta| \le c_2, \qquad \textit{for all } n = 0, 1, \ldots, N. \tag{11.75}$$

Then for all N:

- $\boldsymbol{u}_N$ *is uniformly bounded in $L^\infty(0, T; L^2(\Omega_f))$ and $L^2(0, T; H^1(\Omega_f))$.*
- $\boldsymbol{\eta}_N$ *is uniformly bounded in $L^\infty(0, T; H^1(\Omega_b))$.*
- p_N *is uniformly bounded in $L^\infty(0, T; L^2(\Omega_b))$ and $L^2(0, T; H^1(\Omega_b))$.*
- ω_N *is uniformly bounded in $L^\infty(0, T; H_0^2(\Gamma))$.*

In addition, we have the following estimates on the linear interpolations.

- $\overline{\boldsymbol{\eta}}_N$ *is uniformly bounded in $W^{1,\infty}(0, T; L^2(\Omega_b))$.*
- $\overline{\boldsymbol{\omega}}_N$ *is uniformly bounded in $W^{1,\infty}(0, T; L^2(\Gamma))$.*

Remark 11.6.1 (A Crucial Remark About Invertibility) At first, it would appear that to show the uniform boundedness result above, we would also need to have a fourth assumption, which is Assumption 2A (11.60) from before, that the map $\mathrm{Id} + (\boldsymbol{\eta}_N^n)^\delta : \Omega_b \to \mathbb{R}^2$ is injective (and is hence a bijection onto its image), for each $n = 0, 1, \ldots, N$ and for all N. However, this is implied by an injectivity theorem (see Theorem 2.14.3 in Chap. 2). Note also that Assumption 1A (11.59) from before is automatically satisfied once we verify Assumption 1B (11.72), (11.73). In particular, this injectivity theorem is as follows. Since $\det(\boldsymbol{I} + \nabla(\boldsymbol{\eta}_N^n)^\delta) > 0$ by Assumption 2B (11.74), it suffices to show that $\mathrm{Id} + (\boldsymbol{\eta}_N^n)^\delta = \boldsymbol{\varphi}_0$ on $\partial\Omega_b$, for some injective mapping $\varphi_0 : \overline{\Omega_b} \to \mathbb{R}^2$, for example, a standard ALE mapping $\varphi_0(x, y) = \left(x, y + \left(1 - \frac{y}{R}\right)\omega\right)$ can be used. This implies the very useful fact that $(\mathrm{Id} + (\eta_N^n)^\delta)(\overline{\Omega_b}) = \varphi_0(\overline{\Omega_b})$, which means that the *deformed configuration is fully determined by the behavior on the boundary*.

Proof The uniform boundedness of approximate solutions follows from the uniform energy estimates. More precisely, the uniform boundedness of $\boldsymbol{u}_N$ in $L^\infty(0, T; L^2(\Omega_f))$ follows from Assumption 1B (11.72). The uniform boundedness of $\boldsymbol{u}_N$ in $L^2(0, T; H^1(\Omega_f))$ follows from Korn's inequality on the fluid domain. The uniform boundedness of $\boldsymbol{\eta}_N$ in $L^\infty(0, T; H^1(\Omega_b))$ follows from combining the uniform energy estimates with Korn's inequality, stated in Proposition 11.5.1. To establish the uniform boundedness of p_N in $L^2(0, T; H^1(\Omega_b))$, we note that the discrete energy estimates in Lemma 11.5.3 imply the following uniform discrete dissipation bound:

$$\sum_{n=0}^{N-1} \kappa(\Delta t) \int_{\Omega_b} \mathcal{J}_b^{(\eta_N^n)^\delta} |\nabla_b^{(\eta_N^n)^\delta} p_N^{n+1}|^2 \le C,$$

for some constant C uniform in N, where $\mathcal{J}_b^{(\eta_N^n)^\delta} = \det(\boldsymbol{I} + \nabla(\boldsymbol{\eta}_N^n)^\delta)$, and $\nabla_b^{(\eta_N^n)^\delta} r = \nabla r \cdot (\boldsymbol{I} + \nabla(\boldsymbol{\eta}_N^n)^\delta)^{-1}$ on Ω_b. By Assumption 2B (11.74), we conclude that

$$(\Delta t) \sum_{n=0}^{N-1} \int_{\Omega_b} |\nabla_b^{(\eta_N^n)^\delta} p_N^{n+1}|^2 \le C.$$

Since on Ω_b, we have that $\nabla p_N^{n+1} = \nabla_b^{(\eta_N^n)^\delta} p_N^{n+1} \cdot (\boldsymbol{I} + \nabla(\boldsymbol{\eta}_N^n)^\delta)$, we use Assumption 2C (11.75), which implies $|\boldsymbol{I} + \nabla(\boldsymbol{\eta}_N^n)^\delta| \le c_2$, and obtain the estimate

$$(\Delta t) \sum_{n=0}^{N-1} \int_{\Omega_b} |\nabla p_N^{n+1}|^2 \le |\boldsymbol{I} + \nabla(\boldsymbol{\eta}_N^n)^\delta|^2 \cdot (\Delta t) \sum_{n=0}^{N-1} \int_{\Omega_b} |\nabla_b^{(\eta_N^n)^\delta} p_N^{n+1}|^2 \le C,$$

for a constant C independent of N. Thus, p_N is uniformly bounded in $L^2(0, T; H^1(\Omega_b))$, since by the definition of the piecewise constant approximate solution p_N in (11.69), we have that

$$\|p_N\|^2_{L^2(0,T;H^1(\Omega_b))} = (\Delta t) \sum_{n=0}^{N-1} \int_{\Omega_b} |\nabla p_N^{n+1}|^2.$$

□

The above uniform boundedness result implies the following weak convergence results.

Proposition 11.6.1 *Assume that the three assumptions listed in Lemma 11.6.1 hold. Then, there exists a subsequence such that the following weak convergence results hold:*

- $\boldsymbol{u}_N \rightharpoonup \boldsymbol{u}$ *weakly* in* $L^\infty(0, T; L^2(\Omega_f))$,
  $\boldsymbol{u}_N \rightharpoonup \boldsymbol{u}$ *weakly in* $L^2(0, T; H^1(\Omega_f))$.
- $\boldsymbol{\eta}_N \rightharpoonup \boldsymbol{\eta}$ *weakly* in* $L^\infty(0, T; H^1(\Omega_b))$,
  $\overline{\boldsymbol{\eta}}_N \rightharpoonup \overline{\boldsymbol{\eta}}$ *weakly* in* $W^{1,\infty}(0, T; L^2(\Omega_b))$.
- $p_N \rightharpoonup p$ *weakly* in* $L^\infty(0, T; L^2(\Omega_b))$,
  $p_N \rightharpoonup p$ *weakly in* $L^2(0, T; H^1(\Omega_b))$.
- $\omega_N \rightharpoonup \omega$ *weakly* in* $L^\infty(0, T; H_0^2(\Gamma))$,
  $\overline{\omega}_N \rightharpoonup \overline{\omega}$ *weakly* in* $W^{1,\infty}(0, T; L^2(\Gamma))$.

Furthermore, $\boldsymbol{\eta} = \overline{\boldsymbol{\eta}}$ *and* $\omega = \overline{\omega}$.

To use these results and to be able to construct approximate solutions, it is essential to show that the assumptions from Lemma 11.6.1 hold. This is given by the following lemma.

Lemma 11.6.2 *Suppose that the initial data satisfies* $|\omega_0| \leq R_0 < R$ *for some* R_0*, and suppose that* $\boldsymbol{\eta}_0$ *has the property that* $Id + (\boldsymbol{\eta}_0)^\delta$ *is invertible with* $\det(\boldsymbol{I} + \nabla(\boldsymbol{\eta}_0)^\delta) \geq c_0 > 0$ *on* Ω_b *for some positive constant* c_0*. Then,* ***there exists a sufficiently small time*** $T > 0$ *such that for all* N*, all three assumptions in Lemma 11.6.1 hold, and the splitting scheme is well defined until time* T*.*

Proof First, notice that the assumptions on the initial data immediately imply that the three assumptions from Lemma 11.6.1 hold for the initial data, i.e., for $n = 0$. In particular, there exist constants α_0, α_1, and α_2 such that

$$\det(\boldsymbol{I} + \nabla(\boldsymbol{\eta}_0)^\delta) \geq \alpha_0 > 0, \tag{11.76}$$

$$|\boldsymbol{I} + \nabla(\boldsymbol{\eta}_0)^\delta| \geq \alpha_1 > 0, \qquad |(\boldsymbol{I} + \nabla(\boldsymbol{\eta}_0)^\delta)^{-1}| \geq \alpha_2 > 0. \tag{11.77}$$

This is because $\det(\boldsymbol{I} + \nabla(\boldsymbol{\eta}_0)^\delta)$, $|\boldsymbol{I} + \nabla(\boldsymbol{\eta}_0)^\delta|$, and $|(\boldsymbol{I} + \nabla(\boldsymbol{\eta}_0)^\delta)^{-1}|$ are positive continuous functions on the compact set $\overline{\Omega_b}$.

Next, we want to define an appropriate time $T > 0$ such that the three assumptions hold uniformly for all N and $n\Delta t$ up to time T. To do this, we use the energy estimates. Define the initial energy determined by the initial data by E_0. Then, by the uniform energy estimates, we have that

$$E_N^{k+\frac{1}{2}} \leq E_0, \qquad E_N^{k+1} \leq E_0, \qquad \text{for all } k = 0, 1, \ldots, N-1.$$

Therefore, after completing both subproblems of the scheme on the time step $[k\Delta t, (k+1)\Delta t]$, we obtain that

$$\|\dot{\boldsymbol{\eta}}_N^n\|_{L^2(\Omega_b)} \leq C, \qquad \text{for } n = 0, 1, \ldots, k+1, \tag{11.78}$$

$$\|\omega_N^{n+\frac{1}{2}}\|_{H_0^2(\Gamma)} \leq C, \qquad \text{for } n = 0, 1, \ldots, k, \tag{11.79}$$

$$\|\zeta_N^{n+\frac{i}{2}}\|_{L^2(\Gamma)} \leq C, \qquad \text{for } 0 \leq n + \frac{i}{2} \leq k+1 \qquad \text{and } i = 0, 1, \tag{11.80}$$

for a constant C depending only on the initial energy E_0.

Step 1 (Uniform Bound on the Plate Displacements $\omega_N^{n-\frac{1}{2}}$) We first find a condition on T such that Assumption 1B (11.72) is satisfied. Suppose that the linear interpolation $\overline{\omega}_N$ is defined up to time $(k+1)\Delta t$, where we recall that the linear interpolation is defined via (11.70). Then, by (11.79) and (11.80) and the fact that $\partial_t \overline{\omega}_N = \zeta_N$ from (11.71), we have

$$\|\overline{\omega}_N\|_{W^{1,\infty}(0,(k+1)\Delta t; L^2(\Gamma))} \leq C, \tag{11.81}$$

$$\|\overline{\omega}_N\|_{L^\infty(0,(k+1)\Delta t; H_0^2(\Gamma))} \leq C, \tag{11.82}$$

where C depends only on E_0 and is independent of N. Thus, following the method in [265] (see in particular equation (73) in [265]), we obtain by an interpolation inequality that for all $t, t+\tau \in [0, (k+1)\Delta t]$ with $\tau > 0$,

$$\|\overline{\omega}_N(t+\tau)-\overline{\omega}_N(t)\|_{H^1(\Gamma)} \leq C\|\overline{\omega}_N(t+\tau)-\overline{\omega}_N(t)\|_{L^2(\Gamma)}^{1/2}\|\overline{\omega}_N(t+\tau)-\overline{\omega}_N(t)\|_{H^2(\Gamma)}^{1/2}. \tag{11.83}$$

Here, we used a Sobolev interpolation inequality (see, e.g., Theorem 4.17 (pg. 79) of [1]). By the Lipschitz continuity of $\overline{\omega}_N$ taking values in $L^2(\Gamma)$ given by (11.81) and by the boundedness of $\overline{\omega}_N$ in $H_0^2(\Gamma)$ given by (11.82),

$$\|\overline{\omega}_N(t+\tau) - \overline{\omega}_N(t)\|_{H^1(\Gamma)} \leq C \cdot \tau^{1/2} \tag{11.84}$$

for a constant C depending only on E_0 (and in particular, not depending on k or N). Therefore, setting $t = 0$ and $\tau = (k+1)\Delta t$ and using the continuous embedding of $H^1(\Gamma)$ into $C(\Gamma)$ (see, e.g., [1, Chapter V] for Sobolev embeddings),

$$\|\omega_N^{k+1} - \omega_0\|_{C(\Gamma)} \leq C \cdot [(k+1)\Delta t]^{1/2} \leq C \cdot T^{1/2}, \tag{11.85}$$

where C depends only E_0. Because $|\omega_0| < R$, we can choose $T > 0$ sufficiently small so that

$$C \cdot T^{1/2} < R - \|\omega_0\|_{C(\Gamma)}. \tag{11.86}$$

This will give the first part of Assumption 1B, which is (11.72).

Step 2 (Bound on the Trace of the Biot Displacements η_N^n and the Lagrangian map) Next, we find a condition on T so that the remaining assumptions (11.73), (11.74), and (11.75) are satisfied. We do this by controlling the behavior of the structure displacement $\boldsymbol{\eta}$. First note that

$$\|\boldsymbol{\eta}_N^{k+1} - \boldsymbol{\eta}_0\|_{L^2(\Omega_b)} \leq (\Delta t) \sum_{n=1}^{k+1} \|\dot{\boldsymbol{\eta}}_N^n\|_{L^2(\Omega_b)} \leq C(k+1)(\Delta t) \leq CT,$$

for C depending only on E_0, where the first inequality follows from the triangle inequality and the definition of $\dot{\boldsymbol{\eta}}_N^n = \dfrac{\eta_N^n - \eta_N^{n-1}}{\Delta t}$ and the second inequality follows from (11.78). By the odd extension defined in Definition 11.4.1,

$$\|\boldsymbol{\eta}_N^{k+1} - \boldsymbol{\eta}_0\|_{L^2(\tilde{\Omega}_b)} \leq C\left(\|\boldsymbol{\eta}_N^{k+1} - \boldsymbol{\eta}_0\|_{L^2(\Omega_b)} + \|\omega_N^{k+1} - \omega_0\|_{L^2(\Gamma)}\right) \leq CT,$$

for a constant C depending only on E_0, where the estimate $\|\omega_N^{k+1} - \omega_0\|_{L^2(\Gamma)} \leq CT$ follows from the Lipschitz estimate (11.81). By regularization, we then have that for a constant depending only on δ and E_0,

$$\|(\boldsymbol{\eta}_N^{k+1})^\delta - (\boldsymbol{\eta}_0)^\delta\|_{H^3(\Omega_b)} \leq C(\delta, E_0) \cdot T.$$

By using the trace theorem and the continuous embedding of $H^2(\Gamma)$ into $C(\Gamma)$, we thus conclude that

$$\|(\boldsymbol{\eta}_N^{k+1})^\delta|_\Gamma - (\boldsymbol{\eta}_0)^\delta|_\Gamma\|_{C(\Gamma)} \leq C(\delta, E_0) \cdot T. \tag{11.87}$$

Since $H^2(\Omega_b)$ embeds continuously into $C(\Omega_b)$, we also have that

$$\|\nabla(\boldsymbol{\eta}_N^{k+1})^\delta - \nabla(\boldsymbol{\eta}_0)^\delta\|_{C(\Omega_b)} \leq C(\delta, E_0) \cdot T. \tag{11.88}$$

Note that $\det(\boldsymbol{I} + \boldsymbol{A})$ is a continuous function of the entries of $\boldsymbol{A}$. Also note that the matrix norms $|\boldsymbol{I} + \boldsymbol{A}|$ and $|(\boldsymbol{I} + \boldsymbol{A})^{-1}|$ are continuous functions of the matrix $\boldsymbol{A}$. Furthermore, we emphasize that the constant $C(\delta, E_0)$ depends only on δ and E_0 and hence is independent of k and N. This dependence on δ is allowable, since for this existence proof, δ is an arbitrary but fixed regularization parameter.

Thus, there exists T sufficiently small so that by (11.87) and (11.88), the remaining assumptions (11.73), (11.74), and (11.75) are satisfied, since these assumptions are all satisfied for the initial displacement $\boldsymbol{\eta}_0$. Furthermore, we can choose the constants c_0, c_1, c_2, and R_{max} (defined in the statement of those assumptions) independently of N and $n = 0, 1, \ldots, N$, because of the fact that the constant $C(\delta, E_0)$ in our estimates does not depend on k (satisfying $(k+1)\Delta t \leq T$) or N. □

11.7 Compactness Arguments

We next want to pass to the limit in the semidiscrete formulation for the approximate solutions, stated in (11.64) and (11.65). Because this is a nonlinear problem with geometric nonlinearities, we must obtain stronger convergence than just weak and weak* convergence in Proposition 11.6.1, in order to pass to the limit. To do this, we will use compactness arguments of two types: the classical Aubin-Lions compactness theorem for functions defined on fixed domains and *generalized Aubin-Lions compactness arguments* (see Theorem 2.10.1 in Chap. 2). We will first deal with compactness arguments for the plate displacement and the Biot domain displacement. Then, we will deal with compactness arguments for the fluid velocity defined on moving domains.

11.7.1 Compactness for Biot Poroelastic Medium Displacement

We show strong convergence of the Biot structure displacements $\overline{\eta}_N$ by using a standard Aubin-Lions compactness argument (see Theorem 2.9.4). In particular, we have the following strong convergence result for the Biot medium displacement:

Lemma 11.7.1 *The following compact embedding holds true* $W^{1,\infty}(0, T; L^2(\Omega_b)) \cap L^\infty(0, T; H^1(\Omega_b)) \subset\subset C(0, T; L^2(\Omega_b))$, *which implies the existence of a subsequence such that*

$$\overline{\eta}_N \to \eta \text{ strongly in } C(0, T; L^2(\Omega_b)).$$

Proof The compact embedding above is a direct consequence of the standard Aubin-Lions compactness lemma in the case of $p = \infty$, which gives a stronger compact embedding into $C(0, T; L^2(\Omega_b))$ rather than just $L^\infty(0, T; L^2(\Omega_b))$. The fact that we can find a strongly convergent subsequence follows from this compact embedding, once we recall that $\{\overline{\eta}_N\}_{N=1}^\infty$ are uniformly bounded in the Banach space $W^{1,\infty}(0, T; L^2(\Omega_b)) \cap L^\infty(0, T; H^1(\Omega_b))$ by the uniform energy estimates. □

11.7.2 Compactness for the Plate Displacement

The uniform boundedness of the linear interpolation of the plate displacement $\overline{\omega}_N$ in $W^{1,\infty}(0, T; L^2(\Gamma))$ and $L^\infty(0, T; H_0^2(\Gamma))$ implies strong convergence of $\overline{\omega}_N$ in $C(0, T; H^s(\Gamma))$. Even though the plate displacements are uniformly bounded in $L^\infty(0, T; H_0^2(\Gamma))$, we only get convergence in $C(0, T; H^s(\Gamma))$ for $0 < s < 2$. This is because we will be using Arzela-Ascoli theorem (Theorem 2.9.3), and hence we will lose regularity due to the compact embedding of $H^2(\Gamma)$ into $H^s(\Gamma)$ for

$0 < s < 2$. The precise statement of the compactness results for the approximate plate displacements is as follows:

Proposition 11.7.1 *Given arbitrary* $0 < s < 2$*, there exists a subsequence such that the following strong convergences hold:*

$$\overline{\omega}_N \to \omega, \qquad \textit{in } C(0, T; H^s(\Gamma)),$$

$$\omega_N \to \omega, \qquad \textit{in } L^\infty(0, T; H^s(\Gamma)).$$

Proof Using the same argument as in Step 1 of the proof of Lemma 11.6.2, one can show the following uniform estimate for the linear interpolations $\overline{\omega}_N$ and $\tau > 0$, $t, t+\tau \in [0, T]$:

$$\begin{aligned} \|\overline{\omega}_N(t+\tau) - \overline{\omega}_N(t)\|_{H^{2\alpha}(\Gamma)} &\le \|\overline{\omega}_N(t+\tau) - \overline{\omega}_N(t)\|^{1-\alpha}_{L^2(\Gamma)} \\ \|\overline{\omega}_N(t+\tau) - \overline{\omega}_N(t)\|^{\alpha}_{H^2(\Gamma)} &\le C\tau^{1-\alpha}, \qquad \text{for } 0 < \alpha < 1, \end{aligned} \tag{11.89}$$

where the constant C is independent of N but can depend on the choice of α. The first inequality in (11.89) follows from an interpolation estimate for Sobolev spaces (see Theorem 4.17, pg. 79 of [1]), and as in the proof of estimate (11.83) from Lemma 11.6.2, the second inequality follows from the uniform Lipschitz estimate (11.81) and the uniform boundedness estimate (11.82). Because the constant C in (11.89) is independent of N, this estimate implies that for a given arbitrary $\alpha \in (0, 1)$, the functions $\overline{\omega}_N$ are uniformly bounded as functions in $C^{0,1-\alpha}(0, T; H^{2\alpha}(\Gamma))$. Hence, the strong convergence of $\overline{\omega}_N$ follows directly from the Arzela-Ascoli theorem and the fact that $H^{2\alpha}$ embeds compactly into any $H^{2\alpha-\epsilon}$ for $\epsilon > 0$, once we choose $\alpha \in (0, 1)$ and $\epsilon > 0$ appropriately so that $2\alpha - \epsilon = s$ for a given arbitrary $0 < s < 2$. Hence, we obtain the desired strong convergence, as the equicontinuity condition for the Arzela-Ascoli theorem follows from the above estimate.

To show a similar strong convergence result for ω_N, we must show that

$$\|\omega_N(t) - \overline{\omega}_N(t)\|_{L^\infty(0,T;H^s(\Gamma))} \to 0,$$

for arbitrary $0 < s < 2$. Once we observe that $\overline{\omega}_N(n\Delta t) = \omega_N(t)$ for $n\Delta t \le t < (n+1)\Delta t$, this follows immediately from the above Hölder continuity estimate (11.89), as

$$\|\omega_N(t) - \overline{\omega}_N(t)\|_{L^\infty(0,T;H^s(\Gamma))} \le C(\Delta t)^{1-\frac{s}{2}} \to 0, \qquad \text{as } N \to \infty.$$

Thus, ω_N and $\overline{\omega}_N$ have the same limit in $L^\infty(0, T; H^s(\Gamma))$ for $0 < s < 2$.

□

Next, we will obtain compactness for the Biot velocity, plate velocity, pore pressure, and fluid velocity. Because the test space (11.56) has the pore pressure and fluid velocity decoupled from the Biot/plate velocity, we can handle the compactness argument for each of these quantities separately. In particular, we recall the definition of the discrete test space from (11.56) and note that we can decouple this test space into three smaller test spaces, one for the Biot/plate displacement/velocity, one for the pore pressure, and one for the fluid velocity. In the next section, we show compactness results for the Biot velocity and plate velocity, which must be treated together since they are coupled by a kinematic coupling condition at the plate interface Γ.

11.7.3 Compactness for the Biot Velocity and Plate Velocity

Here, we will state and prove a compactness result for the Biot and plate velocities $(\boldsymbol{\xi}_N, \zeta_N)$, by showing the existence of convergent subsequences that converge in $L^2(0, T; H^{-s}(\Omega_b) \times H^{-s}(\Gamma))$ for $-1/2 < s < 0$. We remark that we must consider negative spatial Sobolev spaces for the Biot/plate velocities for the following two reasons:

- First, our existence result in Theorem 11.4.1 includes the purely elastic case in which the viscoelasticity coefficients μ_v, λ_v are allowed to be zero. Hence, we can only expect the Biot velocities in the finite-energy spaces to have spatial regularity of at most $L^2(\Omega_b)$.
- Second, for the plate velocities ζ_N, we must consider negative spatial Sobolev spaces on Γ since by the coupling conditions (11.14) and (11.15), it is **not true** that the plate velocities ζ_N are equal to the traces of the fluid velocities $\boldsymbol{u}_N \in H^1(\Omega_f)$, which is typically the case in FSI with purely elastic structures and no-slip condition. Therefore, we do not get any higher regularity of the plate velocities than what we get from the finite energy spaces, which implies that the plate velocities ζ_N are only at most $L^2(\Gamma)$.

The main compactness result for the Biot/plate velocities is as follows:

Theorem 11.7.1 *For* $-1/2 < s < 0$, *there exists a subsequence such that*

$$(\boldsymbol{\xi}_N, \zeta_N) \to (\boldsymbol{\xi}, \zeta) \text{ strongly in } L^2(0, T; H^{-s}(\Omega_b) \times H^{-s}(\Gamma)).$$

Proof We will establish this result by using a compactness criterion for piecewise constant functions, given in Theorem 2.9.5. To simplify arguments, we define a slightly more regular Biot/plate velocity test space:

$$\mathcal{Q}_v = \{(\boldsymbol{\psi}, \varphi) \in (V_d \cap H^2(\Omega_b)) \times H_0^2(\Gamma) : \boldsymbol{\psi} = \varphi \boldsymbol{e}_y \text{ on } \Gamma\}. \tag{11.90}$$

We will use the following chain of embeddings

$$L^2(\Omega_b) \times L^2(\Gamma) \subset\subset H^{-s}(\Omega_b) \times H^{-s}(\Gamma) \subset \mathcal{Q}_v',$$

where the first embedding is compact.

Let $\tau_{\Delta t}$ denote the time shift $\tau_{\Delta t} f(t, \cdot) = f(t - \Delta t, \cdot)$ for a function f defined on $[0, T]$. As required by the Dreher-Jüngel compactness criterion (Theorem 2.9.5), to obtain compactness, we must verify that the following inequality is satisfied for a uniform constant C and for all $\Delta t = T/N$:

$$\left\| \frac{\tau_{\Delta t}(\boldsymbol{\xi}_N, \zeta_N) - (\boldsymbol{\xi}_N, \zeta_N)}{\Delta t} \right\|_{L^1(\Delta t, T; \mathcal{Q}_v')} + \|(\boldsymbol{\xi}_N, \zeta_N)\|_{L^\infty(0,T;L^2(\Omega_b)\times L^2(\Gamma))} \leq C. \tag{11.91}$$

The second term in this inequality is uniformly bounded by Lemma 11.6.1, which gives exactly the uniform boundednenss of $(\boldsymbol{\xi}_N, \zeta_N)$ in $L^\infty(0, T; L^2(\Omega_b) \times L^2(\Gamma))$.

To deal with the first term in (11.91), we use the coupled semidiscrete formulations (11.64) and (11.65) and set the test functions $\boldsymbol{v}$ and r for the fluid velocity and Biot pore pressure to be zero because we are considering only the Biot and plate velocities. We obtain that for all test functions $(\boldsymbol{\psi}, \varphi) \in \mathcal{Q}_v$, where $\mathcal{Q}_v$ is defined in (11.90), the following holds:

$$\begin{aligned}
&\rho_b \int_{\Omega_b} \left(\frac{\boldsymbol{\xi}_N^{n+1} - \boldsymbol{\xi}_N^n}{\Delta t} \right) \cdot \boldsymbol{\psi} + \rho_p \int_\Gamma \left(\frac{\zeta_N^{n+1} - \zeta_N^n}{\Delta t} \right) \cdot \varphi \\
&= - \int_\Gamma \left(\frac{1}{2} \boldsymbol{u}_N^{n+1} \cdot \boldsymbol{u}_N^n - p_N^{n+1} \right) (\boldsymbol{\psi} \cdot \boldsymbol{n}^{\omega_N^n}) \\
&- \int_\Gamma \frac{\beta}{\mathcal{J}_\Gamma^{\omega_N^n}} (\zeta_N^{n+1} \boldsymbol{e}_y - \boldsymbol{u}_N^{n+1}) \cdot \boldsymbol{\tau}^{\omega_N^n} (\boldsymbol{\psi} \cdot \boldsymbol{\tau}^{\omega_N^n}) \\
&- 2\mu_e \int_{\Omega_b} \boldsymbol{D}(\boldsymbol{\eta}_N^{n+1}) : \boldsymbol{D}(\boldsymbol{\psi}) - \lambda_e \int_{\Omega_b} (\nabla \cdot \boldsymbol{\eta}_N^{n+1})(\nabla \cdot \boldsymbol{\psi}) \\
&- 2\mu_v \int_{\Omega_b} \boldsymbol{D}(\boldsymbol{\xi}_N^{n+1}) : \boldsymbol{D}(\boldsymbol{\psi}) - \lambda_v \int_{\Omega_b} (\nabla \cdot \boldsymbol{\xi}_N^{n+1})(\nabla \cdot \boldsymbol{\psi}) \\
&+ \alpha \int_{\Omega_b} \mathcal{J}_b^{(\eta_N^n)^\delta} p_N^{n+1} \nabla_b^{(\eta_N^n)^\delta} \cdot \boldsymbol{\psi} - \int_\Gamma \Delta \omega_N^{n+\frac{1}{2}} \cdot \Delta \varphi.
\end{aligned}$$

The estimate for the first term in (11.91) will follow if we can estimate the right-hand side in terms of the $\mathcal{Q}_v'$ norm. For this purpose, consider an arbitrary $\|(\boldsymbol{\psi}, \varphi)\|_{\mathcal{Q}_v} \leq 1$, so that $\|\boldsymbol{\psi}\|_{H^2(\Omega_b)} \leq 1$ and $\|\varphi\|_{H_0^2(\Gamma)} \leq 1$. By the uniform estimates in Lemma 11.6.1 and the regularity of the test functions in (11.90), it is clear that the terms on the right-hand side are all uniformly bounded by a constant C, independent of $\|(\boldsymbol{\psi}, \varphi)\|_{\mathcal{Q}_v} \leq 1$, except possibly the term

$$\alpha \int_{\Omega_b} \mathcal{J}_b^{(\eta_N^n)^\delta} p_N^{n+1} \nabla_b^{(\eta_N^n)^\delta} \cdot \boldsymbol{\psi}.$$

To estimate this term, we recall the definitions

$$\mathcal{J}_b^{(\eta_N^n)^\delta} = \det(\boldsymbol{I} + \nabla(\boldsymbol{\eta}_N^n)^\delta), \qquad \nabla^{(\eta_N^n)^\delta} \cdot \boldsymbol{\psi} = \text{tr}\left[\nabla\boldsymbol{\psi} \cdot (I + \nabla(\boldsymbol{\eta}_N^n)^\delta)^{-1}\right].$$

By assumption 2C (11.75) and the fact that $\|\boldsymbol{\psi}\|_{H^1(\Omega_b)} \leq 1$, we have that $\|\nabla^{(\eta_N^n)^\delta} \cdot \boldsymbol{\psi}\|_{L^2(\Omega_b)}$ is uniformly bounded, while by the boundedness of $\boldsymbol{\eta}_N^n$ in $H^1(\Omega_b)$, we have that $|\mathcal{J}_b^{(\eta_N^n)^\delta}| \leq C$. Therefore, using the fact that p_N is uniformly bounded in $L^\infty(0, T; L^2(\Omega_b))$, we obtain the desired estimate

$$\left|\alpha \int_{\Omega_b} \mathcal{J}_b^{(\eta_N^n)^\delta} p_N^{n+1} \nabla_b^{(\eta_N^n)^\delta} \cdot \boldsymbol{\psi}\right| \leq C.$$

Finally, we conclude that

$$\left\|\frac{(\boldsymbol{\xi}_N^{n+1}, \zeta_N^{n+1}) - (\boldsymbol{\xi}_N^n, \zeta_N^n)}{\Delta t}\right\|_{\mathcal{Q}_v'} \leq C, \text{ for a constant } C \text{that is independent of } n \text{ and} N,$$

and since

$$\sum_{n=1}^{N-1} (\Delta t) \left\|\frac{(\boldsymbol{\xi}_N^{n+1}, \zeta_N^{n+1}) - (\boldsymbol{\xi}_N^n, \zeta_N^n)}{\Delta t}\right\|_{\mathcal{Q}_v'} \leq (\Delta t) \sum_{n=1}^{N-1} C \leq CT,$$

we conclude that (11.91) holds for a uniform constant C. This establishes the desired result. □

11.7.4 Compactness for the Pore Pressure

Theorem 11.7.2 *There exists a subsequence such that*

$$p_N \to p \text{ strongly in } L^2(0, T; L^2(\Omega_b)).$$

Proof The proof is based on a similar application of the Dreher-Jüngel compactness criterion for piecewise constant functions in Theorem 2.9.5, as in the previous compactness result for the approximate Biot/plate velocities. We first observe that we have the following chain of embeddings $H^1(\Omega_b) \subset\subset L^2(\Omega_b) \subset (V_p \cap$

$H^2(\Omega_b))'$, and so by Theorem 2.9.5, it suffices to show that the following inequality holds for a constant C independent of N:

$$\left\| \frac{\tau_{\Delta t} p_N - p_N}{\Delta t} \right\|_{L^1(\Delta t, T; (V_p \cap H^2(\Omega_b))')} + \|p_N\|_{L^2(0,T;H^1(\Omega_b))} \leq C. \tag{11.92}$$

To obtain this estimate, we observe that the approximate solutions for the pore pressure satisfy the following weak formulation for all test functions $r \in V_p$, where V_p is defined by (11.37):

$$c_0 \int_{\Omega_b} \left(\frac{p_N^{n+1} - p_N^n}{\Delta t} \right) \cdot r - \alpha \int_{\Omega_b} \mathcal{J}_b^{(\eta_N^n)^\delta} \dot{\boldsymbol{\eta}}_N^{n+1} \cdot \nabla_b^{(\eta_N^n)^\delta} r - \alpha \int_\Gamma (\dot{\boldsymbol{\eta}}_N^{n+1} \cdot \boldsymbol{n}^{(\omega_N^n)^\delta}) r$$
$$+ \kappa \int_{\Omega_b} \mathcal{J}_b^{(\eta_N^n)^\delta} \nabla_b^{(\eta_N^n)^\delta} p_N^{n+1} \cdot \nabla_b^{(\eta_N^n)^\delta} r - \int_\Gamma [(\boldsymbol{u}_N^{n+1} - \dot{\boldsymbol{\eta}}_N^{n+1}) \cdot \boldsymbol{n}^{\omega_N^n}] r = 0.$$

We use more regularity for the test space $V_p \cap H^2(\Omega_b)$ to make the following estimates simpler. We compute that for any $r \in V_p \cap H^2(\Omega_b)$, we have

$$c_0 \int_{\Omega_b} \left(\frac{p_N^{n+1} - p_N^n}{\Delta t} \right) \cdot r = \alpha \int_{\Omega_b} \mathcal{J}_b^{(\eta_N^n)^\delta} \boldsymbol{\xi}_N^{n+1} \cdot \nabla_b^{(\eta_N^n)^\delta} r + \alpha \int_\Gamma (\zeta_N^{n+1} \boldsymbol{e}_y \cdot \boldsymbol{n}^{(\omega_N^n)^\delta}) r$$
$$- \kappa \int_{\Omega_b} \mathcal{J}_b^{(\eta_N^n)^\delta} \nabla_b^{(\eta_N^n)^\delta} p_N^{n+1} \cdot \nabla_b^{(\eta_N^n)^\delta} r + \int_\Gamma [(\boldsymbol{u}_N^{n+1} - \zeta_N^{n+1} \boldsymbol{e}_y) \cdot \boldsymbol{n}^{\omega_N^n}] r.$$

We estimate the right-hand side for $\|r\|_{V_p \cap H^2(\Omega_b)} \leq 1$. Recall that $\mathcal{J}_b^{(\eta_N^n)^\delta} = \det(\boldsymbol{I} + \nabla(\boldsymbol{\eta}_N^n)^\delta)$,

$$\nabla_b^{(\eta_N^n)^\delta} r = \left(\frac{\partial r}{\partial \tilde{x}}, \frac{\partial r}{\partial \tilde{y}} \right) \cdot (\boldsymbol{I} + \nabla(\boldsymbol{\eta}_N^n)^\delta)^{-1}, \quad \text{and}$$
$$\nabla_b^{(\eta_N^n)^\delta} p_N^{n+1} = \left(\frac{\partial p_N^{n+1}}{\partial \tilde{x}}, \frac{\partial p_N^{n+1}}{\partial \tilde{y}} \right) \cdot (\boldsymbol{I} + \nabla(\boldsymbol{\eta}_N^n)^\delta)^{-1}.$$

We have by Assumption 2C (11.75) that $|(\boldsymbol{I} + \nabla(\boldsymbol{\eta}_N^n)^\delta)^{-1}|$ is uniformly bounded, and furthermore, $\mathcal{J}_b^{(\eta_N^n)^\delta}$ is positive and bounded above. By combining these facts with standard estimates, we obtain that

$$\left\| \frac{p_N^{n+1} - p_N^n}{\Delta t} \right\|_{(V_p \cap H^2(\Omega_b))'} \leq C \quad \text{for a constant } C \text{ that is independent of } n \text{ and } N.$$

Combining this with the fact that p_N is uniformly bounded in $L^2(0, T; H^1(\Omega_b))$ gives the desired estimate (11.92). □

11.7.5 Compactness for the Fluid Velocity

We will obtain convergence of the fluid velocity along a subsequence by using a *generalized Aubin-Lions compactness theorem* for functions defined on moving domains (see Theorem 2.10.1 in Chap. 2). The reason we must use the generalized Aubin-Lions compactness theorem is that the approximate fluid velocities are defined on different time-dependent fluid domains. To prepare for an application of the generalized Aubin-Lions compactness argument, we will *map our approximate fluid problem back onto the physical domain*

$$\Omega^n_{f,N} = \{(x, y) \in \mathbb{R}^2 : 0 \leq x \leq L, -R \leq y \leq \omega^n_N(x)\},$$

where we redefine the fluid velocity solution and test spaces as follows:

$$V^{n+1}_N = \{\boldsymbol{u} \in H^1(\Omega^n_{f,N}) : \nabla \cdot \boldsymbol{u} = 0 \text{ on } \Omega^n_{f,N}, \boldsymbol{u} = 0 \text{ on } \partial\Omega^n_{f,N} \setminus \Gamma^n_N\},$$
$$Q^n_N = V^{n+1}_N \cap H^3(\Omega^n_{f,N}). \tag{11.93}$$

The approximate fluid velocity $\boldsymbol{u}^{n+1}_N \in V^{n+1}_N$ on the physical domain satisfies the following semidiscrete formulation:

$$\begin{aligned}
&\int_{\Omega^n_{f,N}} \frac{\boldsymbol{u}^{n+1}_N - \tilde{\boldsymbol{u}}^n_N}{\Delta t} \cdot \boldsymbol{v} + 2\nu \int_{\Omega^n_{f,N}} \boldsymbol{D}(\boldsymbol{u}^{n+1}_N) : \boldsymbol{D}(\boldsymbol{v}) \\
&+ \frac{1}{2} \int_{\Omega^n_{f,N}} \left[\left(\left(\tilde{\boldsymbol{u}}^n_N - \zeta^{n+\frac{1}{2}}_N \frac{R+y}{R+\omega^n_N} \boldsymbol{e}_y \right) \cdot \nabla \boldsymbol{u}^{n+1}_N \right) \cdot \boldsymbol{v} \right. \\
&- \left. \left(\left(\tilde{\boldsymbol{u}}^n_N - \zeta^{n+\frac{1}{2}}_N \frac{R+y}{R+\omega^n_N} \boldsymbol{e}_y \right) \cdot \nabla \boldsymbol{v} \right) \cdot \boldsymbol{u}^{n+1}_N \right] \\
&+ \frac{1}{2R} \int_{\Omega^n_{f,N}} \frac{R}{R+\omega^n_N} \zeta^{n+\frac{1}{2}}_N \boldsymbol{u}^{n+1}_N \cdot \boldsymbol{v} + \frac{1}{2} \int_{\Gamma^n_N} (\boldsymbol{u}^{n+1}_N - \dot{\boldsymbol{\eta}}^{n+1}_N) \cdot \boldsymbol{n} (\tilde{\boldsymbol{u}}^n_N \cdot \boldsymbol{v}) \\
&- \int_{\Gamma^n_N} \left(\frac{1}{2} \boldsymbol{u}^{n+1}_N \cdot \tilde{\boldsymbol{u}}^n_N - p^{n+1}_N \right) (\boldsymbol{v} \cdot \boldsymbol{n}) \\
&- \beta \int_{\Gamma^n_N} (\dot{\boldsymbol{\eta}}^{n+1}_N - \boldsymbol{u}^{n+1}_N) \cdot \boldsymbol{\tau} (\boldsymbol{v} \cdot \boldsymbol{\tau}) = 0, \ \forall \boldsymbol{v} \in Q^n_N
\end{aligned} \tag{11.94}$$

where

$$\tilde{\boldsymbol{u}}_N^n = \boldsymbol{u}_N^n \circ \boldsymbol{\Phi}_f^{\omega_N^{n-1}} \circ (\boldsymbol{\Phi}_f^{\omega_N^n})^{-1}, \tag{11.95}$$

$\boldsymbol{u}_N^n$ is originally defined on $\Omega_{f,N}^{n-1}$, and the ALE map $\boldsymbol{\Phi}_f^{\omega_N^n} : \Omega_f \to \Omega_{f,N}^n$ is defined by (11.19).

To be able to compare functions on different physical domains we introduce a maximal domain Ω_f^M which contains all the physical domains. The existence of such a domain, and the extensions of the velocity functions onto the maximal domain are discussed next.

Extension to Maximal Domain

We consider the following *maximal fluid domain*, which contains all the physical fluid domains:

$$\Omega_f^M = \{(x, y) \in \mathbb{R}^2 : 0 \leq x \leq L, -R \leq y \leq M(x)\},$$

where the function $M(x)$ is obtained from the following proposition, established in Lemma 2.5 in [333] and Lemma 4.5 in [270] in the context of fluid-structure interaction between an incompressible viscous Newtonian fluid and an elastic Koiter shell:

Proposition 11.7.2 *There exists smooth functions $m(x)$ and $M(x)$ defined on $\Gamma = [0, L]$, satisfying $m(0) = m(L) = M(0) = M(L) = 0$, such that*

$$m(x) \leq \omega_N^n(x) \leq M(x), \qquad \textit{for all } x \in [0, L], N, \textit{ and } n = 0, 1, \ldots, N.$$

Furthermore, there exist smooth functions $m_N^{n,l}(x)$ and $M_N^{n,l}(x)$ defined for positive integers N, $n = 0, 1, \ldots, N-1$ and $l = 0, 1, \ldots, N-n$, such that

1. $m_N^{n,l}(x) \leq \omega_N^{n+i}(x) \leq M_N^{n,l}(x)$, *for all $x \in [0, L]$ and $i = 0, 1, \ldots, l$.*
2. $M_N^{n,l}(x) - m_N^{n,l}(x) \leq C\sqrt{l\Delta t}$, *for all $x \in [0, L]$.*
3. $\|M_N^{n,l}(x) - m_N^{n,l}(x)\|_{L^2(\Gamma)} \leq C(l\Delta t)$,

where C is independent of n, l, and N. Finally, the functions $M_N^{n,l}(x)$ and $m_N^{n,l}(x)$ for all n, l, and N are Lipschitz continuous with a Lipschitz constant that is uniformly bounded above by some constant $L > 0$ independent of n, l, and N.

Once the maximal fluid domain is defined, we can extend the fluid velocities $\boldsymbol{u}_N^n$ from $\Omega_{f,N}^n$ to this common maximal domain Ω_f^M, using extensions by zero in $\Omega_f^M \cap (\Omega_{f,N}^n)^c$. Notice that since $\omega_N^n(x)$ are all uniformly Lipschitz, the extensions by zero of the H^1 functions $\boldsymbol{u}_N^n$ defined on Lipschitz domains to Ω_f^M are uniformly bounded in $H^s(\Omega_f^M)$ for all s such that $0 < s < 1/2$. Indeed, we have the following lemma, which follows from Theorem 2.8.2.

Lemma 11.7.2 *The approximate fluid velocities $\{\boldsymbol{u}_N\}_{N=1}^{\infty}$ defined on the maximal fluid domain Ω_f^M by extension by zero are uniformly bounded in $L^2(0, T; H^s(\Omega_f^M))$ for $s \in (0, 1/2)$.*

Velocity Convergence via a Generalized Aubin-Lions Compactness Argument

We now show strong convergence as $N \to \infty$ along a subsequence of the approximate fluid velocities $\boldsymbol{u}_N$, which are now functions in time defined on the fixed maximal domain Ω_f^M.

Proposition 11.7.3 *The sequence $\boldsymbol{u}_N$ is relatively compact in $L^2(0, T; L^2(\Omega_f^M))$.*

Proof The proof is based on using the generalized Aubin-Lions compactness theorem for problems on moving domains (see Theorem 2.10.1 in Chap. 2). For this purpose, we define the Hilbert spaces V and H from the statement of the theorem to be

$$H = L^2(\Omega_f^M), \qquad V = H^s(\Omega_f^M), \qquad \text{for } 0 < s < 1/2,$$

where we note that indeed $V \subset\subset H$ as required by Theorem 2.10.1. Additionally, the spaces $(V_{\Delta t}^n, Q_{\Delta t}^n)$ from the statement of the theorem correspond to our spaces (V_N^n, Q_N^n) defined by (11.93). Notice that $V_N^n \times Q_N^n$ embeds continuously into $V \times V$ as required by the statement of Theorem 2.10.1, where the embedding can be achieved by the extension by zero operator to the maximal domain Ω_f^M, uniformly in n and N.

To obtain compactness of the sequence $\boldsymbol{u}_N$ in $L^2(0, T, H)$, by Theorem 2.10.1, seven properties need to be satisfied by the sequence $\boldsymbol{u}_N$ and the spaces V_N^n and Q_N^n. They are called Properties A1–3, B, and C1–3.

The proof that approximate solutions $\boldsymbol{u}_N$ satisfy Properties A1–3 and C1–3 is analogous to the corresponding proof of Theorem 5.7.1 in Chap. 5. The main difficulty is to verify Property B, which is a condition on equicontinuity of $\boldsymbol{u}_N$, stated as follows:

Property B, [270] There exists a constant $C > 0$ independent of N such that

$$\left\| P_N^n \frac{\boldsymbol{u}_N^{n+1} - \boldsymbol{u}_N^n}{\Delta t} \right\|_{(Q_N^n)'} \le C\left(1 + \|\boldsymbol{u}_N^{n+1}\|_{V_N^{n+1}}\right), \quad \text{for all } n = 0, 1, \dots, N-1, \tag{11.96}$$

where P_N^n denotes the orthogonal projection onto the closed subspace $\overline{Q_N^n}^H$ of the Hilbert space H.

The sequence $\boldsymbol{u}_N$ constructed in this chapter, however, *does not satisfy this property*. Nevertheless, $\boldsymbol{u}_N$ satisfy the following *generalized Property B*, which implies the desired equicontinuity under which the generalized Aubin-Lions theorem from [270] still holds:

Generalized Property B There exist a constant C independent of n and N, an exponent p, $1 \le p < 2$, and a sequence of nonnegative numbers $\{a_N^n\}_{n=0}^{N-1}$ for each N, satisfying $(\Delta t)\sum_{n=0}^{N-1} |a_N^n|^2 \le C$ uniformly in N, such that

$$\left\| P_N^n \frac{\boldsymbol{u}_N^{n+1} - \boldsymbol{u}_N^n}{\Delta t} \right\|_{(Q_N^n)'} \le C\left(a_N^n + \|\boldsymbol{u}_N^n\|_{V_N^n} + \|\boldsymbol{u}_N^{n+1}\|_{V_N^{n+1}}\right)^p, \quad \text{for all } n = 0, 1, \dots, N-1. \tag{11.97}$$

□

Theorem 11.7.3 (Generalized Aubin-Lions Compactness Result II) *Assume that properties A1–3 and C1–3 of Theorem 2.10.1 are satisfied. Furthermore, assume that Generalized Property B above is satisfied. Then conclusion of Theorem 2.10.1 holds, namely,* $\{\boldsymbol{u}_N\}$ *is relatively compact in* $L^2(0, T; H)$.

Proof We just need to prove that the essential equicontinuity estimate in the proof of [270, Theorem 3.1.] still holds under the modified assumption. In particular, for the original form of Property B in (11.96), one has from Lemma 3.1 in [270] the following equicontinuity estimate for a constant $C > 0$ that is independent of N:

$$\|P_{\Delta t}^{n,l}(\boldsymbol{u}_N^{n+l} - \boldsymbol{u}_N^n)\|_{(Q_N^{n,l})'} \le C\sqrt{l\Delta t}.$$

With the generalized form of Property B that we use above in (11.97), the same arguments as in the proof of Lemma 3.1 in [270] will still give rise to the following equicontinuity estimate for a constant $C > 0$ that is independent of N:

$$\|P_{\Delta t}^{n,l}(\boldsymbol{u}_N^{n+l} - \boldsymbol{u}_N^n)\|_{(Q_N^{n,l})'} \le C(l\Delta t)^{1-\frac{p}{2}},$$

where the generalized Aubin-Lions compactness theorem on moving domains still holds with this new equicontinuity estimate. This is because $1 \le p < 2$, and hence, $C(l\Delta t)^{1-\frac{p}{2}}$ still converges to zero as $\Delta t \to 0$. □

We can now complete the proof of Proposition 11.7.3 by verifying that our sequence $\boldsymbol{u}_N^n$ indeed satisfies the Generalized Property B.

Verification that $\boldsymbol{u}_N^n$ *satisfies the Generalized Property B.* First, recall that by definition,

$$\left\| P_N^n \frac{\boldsymbol{u}_N^{n+1} - \boldsymbol{u}_N^n}{\Delta t} \right\|_{(Q_N^n)'} = \max_{\|\boldsymbol{v}\|_{Q_N^n} \le 1} \left| \int_{\Omega_{f,N}^n} \frac{\boldsymbol{u}_N^{n+1} - \boldsymbol{u}_N^n}{\Delta t} \cdot \boldsymbol{v} d\boldsymbol{x} \right|. \tag{11.98}$$

To estimate the right-hand side, we use

$$\left|\int_{\Omega^n_{f,N}} \frac{\boldsymbol{u}_N^{n+1} - \boldsymbol{u}^n}{\Delta t} \cdot \boldsymbol{v} d\boldsymbol{x}\right| \leq \left|\int_{\Omega^n_{f,N}} \frac{\boldsymbol{u}_N^{n+1} - \tilde{\boldsymbol{u}}_N^n}{\Delta t} \cdot \boldsymbol{v} d\boldsymbol{x}\right| + \left|\int_{\Omega^n_{f,N}} \frac{\tilde{\boldsymbol{u}}_N^n - \boldsymbol{u}_N^n}{\Delta t} \cdot \boldsymbol{v} d\boldsymbol{x}\right|. \tag{11.99}$$

To estimate the first term on the right-hand side, we use the semidiscrete formulation for the fluid velocity on the physical domain given by (11.94) to obtain

$$\begin{aligned}
&\left|\int_{\Omega^n_{f,N}} \frac{\boldsymbol{u}_N^{n+1} - \tilde{\boldsymbol{u}}_N^n}{\Delta t} \cdot \boldsymbol{v} d\boldsymbol{x}\right| \leq 2\nu \left|\int_{\Omega^n_{f,N}} \boldsymbol{D}(\boldsymbol{u}_N^{n+1}) : \boldsymbol{D}(\boldsymbol{v})\right| \\
&+ \frac{1}{2}\left|\int_{\Omega^n_{f,N}} \left[\left(\left(\tilde{\boldsymbol{u}}_N^n - \zeta_N^{n+\frac{1}{2}} \frac{R+y}{R+\omega_N^n}\boldsymbol{e}_y\right) \cdot \nabla \boldsymbol{u}_N^{n+1}\right) \cdot \boldsymbol{v}\right.\right. \\
&\left.\left. - \left(\left(\tilde{\boldsymbol{u}}_N^n - \zeta_N^{n+\frac{1}{2}} \frac{R+y}{R+\omega_N^n}\boldsymbol{e}_y\right) \cdot \nabla \boldsymbol{v}\right) \cdot \boldsymbol{u}_N^{n+1}\right]\right| \\
&+ \frac{1}{2R}\left|\int_{\Omega^n_{f,N}} \frac{R}{R+\omega_N^n} \zeta_N^{n+\frac{1}{2}} \boldsymbol{u}_N^{n+1} \cdot \boldsymbol{v}\right| + \frac{1}{2}\left|\int_{\Gamma_N^n} (\boldsymbol{u}_N^{n+1} - \dot{\boldsymbol{\eta}}_N^{n+1}) \cdot \boldsymbol{n}(\tilde{\boldsymbol{u}}_N^n \cdot \boldsymbol{v})\right| \\
&+ \left|\int_{\Gamma_N^n} \left(\frac{1}{2}\boldsymbol{u}_N^{n+1} \cdot \tilde{\boldsymbol{u}}_N^n - p_N^{n+1}\right)(\boldsymbol{v} \cdot \boldsymbol{n})\right| + \beta \left|\int_{\Gamma_N^n} (\dot{\boldsymbol{\eta}}_N^{n+1} - \boldsymbol{u}_N^{n+1}) \cdot \boldsymbol{\tau}(\boldsymbol{v} \cdot \boldsymbol{\tau})\right|.
\end{aligned} \tag{11.100}$$

We can bound the terms on the right-hand side uniformly in n, N, and $\|\boldsymbol{v}\|_{Q_N^n} \leq 1$ as follows. By the boundedness of $\boldsymbol{u}_N^{n+1}$ in the uniform energy estimates, we immediately have

$$2\nu \left|\int_{\Omega^n_{f,N}} \boldsymbol{D}(\boldsymbol{u}_N^{n+1}) : \boldsymbol{D}(\boldsymbol{v})\right| \leq C\|\boldsymbol{u}_N^{n+1}\|_{H^1(\Omega^n_{f,N})}.$$

The second term on the right-hand side of the above inequality is bounded as follows. First notice that because $\|\boldsymbol{v}\|_{Q_N^n} \leq 1$, and by the definition of Q_N^n in (11.93), we have that $\boldsymbol{v}$ is bounded in $H^3(\Omega^n_{f,N})$, and hence, $\boldsymbol{v}$ and $\nabla \boldsymbol{v}$ are bounded in $L^\infty(\Omega^n_{f,N})$. Furthermore, by the boundedness of the fluid velocity $\boldsymbol{u}_N^n$ on the reference domain due to the uniform energy estimate and by the uniform boundedness of the Jacobian of the ALE map $\Phi_f^{\omega_N^n}$, we obtain the following bound:

$$\begin{aligned}
&\frac{1}{2}\left|\int_{\Omega^n_{f,N}} \left[\left(\left(\tilde{\boldsymbol{u}}_N^n - \zeta_N^{n+\frac{1}{2}} \frac{R+y}{R+\omega_N^n}\boldsymbol{e}_y\right) \cdot \nabla \boldsymbol{u}_N^{n+1}\right) \cdot \boldsymbol{v}\right.\right. \\
&\left.\left. - \left(\left(\tilde{\boldsymbol{u}}_N^n - \zeta_N^{n+\frac{1}{2}} \frac{R+y}{R+\omega_N^n}\boldsymbol{e}_y\right) \cdot \nabla \boldsymbol{v}\right) \cdot \boldsymbol{u}_N^{n+1}\right]\right|
\end{aligned}$$

$$\leq C\left(\|\tilde{\boldsymbol{u}}_N^n\|_{L^2(\Omega_{f,N}^n)} + \|\zeta^{n+\frac{1}{2}}\|_{L^2(\Gamma)}\right)\|\boldsymbol{u}_N^{n+1}\|_{H^1(\Omega_{f,N}^n)} \cdot \|\boldsymbol{v}\|_{H^3(\Omega_{f,N}^n)}$$
$$\leq C\|\boldsymbol{u}_N^{n+1}\|_{H^1(\Omega_{f,N}^n)}.$$

Similarly, the next term in (11.100) is bounded as follows:

$$\frac{1}{2R}\left|\int_{\Omega_{f,N}^n}\frac{R}{R+\omega_N^n}\zeta_N^{n+\frac{1}{2}}\boldsymbol{u}_N^{n+1}\cdot\boldsymbol{v}\right| \leq C\|\zeta_N^{n+\frac{1}{2}}\|_{L^2(\Gamma)}\|\boldsymbol{u}_N^{n+1}\|_{L^2(\Omega_{f,N}^n)}\cdot\|\boldsymbol{v}\|_{H^3(\Omega_{f,N}^n)}$$
$$\leq C\|\boldsymbol{u}_N^{n+1}\|_{L^2(\Omega_{f,N}^n)}.$$

To bound the next term, we observe that $\|\dot{\boldsymbol{\eta}}_N^{n+1}\|_{L^2(\Gamma)}$ is bounded uniformly, and furthermore, the arc length element on Γ_N^n is uniformly bounded pointwise since η_N^n is uniformly bounded in $H_0^2(\Gamma)$. Therefore, by using the trace inequality on Ω_f, we have the following estimate:

$$\frac{1}{2}\left|\int_{\Gamma_N^n}(\boldsymbol{u}_N^{n+1}-\dot{\boldsymbol{\eta}}_N^{n+1})\cdot\boldsymbol{n}(\tilde{\boldsymbol{u}}_N^n\cdot\boldsymbol{v})\right|$$
$$\leq C\left(\|\boldsymbol{u}_N^{n+1}\|_{L^4(\Gamma)}\cdot\|\boldsymbol{u}_N^n\|_{L^4(\Gamma)}\cdot\|\boldsymbol{v}\|_{L^2(\Gamma)}\right.$$
$$\left.+\|\dot{\boldsymbol{\eta}}_N^{n+1}\|_{L^2(\Gamma)}\cdot\|\boldsymbol{u}_N^n\|_{L^4(\Gamma)}\cdot\|\boldsymbol{v}\|_{L^4(\Gamma)}\right)$$
$$\leq C\left(\|\boldsymbol{u}_N^{n+1}\|_{H^{1/4}(\Gamma)}\cdot\|\boldsymbol{u}_N^n\|_{H^{1/4}(\Gamma)}\cdot\|\boldsymbol{v}\|_{H^1(\Omega_f)}\right.$$
$$\left.+\|\dot{\boldsymbol{\eta}}_N^{n+1}\|_{L^2(\Gamma)}\cdot\|\boldsymbol{u}_N^n\|_{H^{1/4}(\Gamma)}\cdot\|\boldsymbol{v}\|_{H^{1/4}(\Gamma)}\right)$$
$$\leq C\left(\|\boldsymbol{u}_N^{n+1}\|_{H^{3/4}(\Omega_f)}\cdot\|\boldsymbol{u}_N^n\|_{H^{3/4}(\Omega_f)}\right.$$
$$\left.+\|\dot{\boldsymbol{\eta}}_N^{n+1}\|_{L^2(\Gamma)}\cdot\|\boldsymbol{u}_N^n\|_{H^{3/4}(\Omega_f)}\right)\cdot\|\boldsymbol{v}\|_{H^1(\Omega_f)}$$
$$\leq C\left(\|\boldsymbol{u}_N^{n+1}\|_{L^2(\Omega_f)}^{1/4}\|\boldsymbol{u}_N^{n+1}\|_{H^1(\Omega_f)}^{3/4}\|\boldsymbol{u}_N^n\|_{L^2(\Omega_f)}^{1/4}\|\boldsymbol{u}_N^n\|_{H^1(\Omega_f)}^{3/4}\right.$$
$$\left.+\|\dot{\boldsymbol{\eta}}_N^{n+1}\|_{L^2(\Gamma)}\|\boldsymbol{u}_N^n\|_{L^2(\Omega_f)}^{1/4}\|\boldsymbol{u}_N^n\|_{H^1(\Omega_f)}^{3/4}\right)$$
$$\leq C\left(\|\boldsymbol{u}_N^{n+1}\|_{H^1(\Omega_f)}^{3/4}\cdot\|\boldsymbol{u}_N^n\|_{H^1(\Omega_f)}^{3/4}+\|\boldsymbol{u}_N^n\|_{H^1(\Omega_f)}^{3/4}\right)$$
$$\leq C\left[1+\left(\|\boldsymbol{u}_N^n\|_{V_N^n}+\|\boldsymbol{u}_N^{n+1}\|_{V_N^n}\right)^{3/2}\right].$$

More precisely, in the above chain of inequalities, we used the following results to justify the steps:

- In the first inequality, we used Hölder's inequality with the exponents 2, 4, and 4, the fact that $\tilde{\boldsymbol{u}}_N^n$ and $\boldsymbol{u}_N^n$ have the same trace along the interface Γ by (11.95), and the uniform boundedness of the Jacobian $\mathcal{J}_\Gamma^{\omega_N^n}$ defined in (11.21).
- In the second inequality, we use Sobolev embedding on the one-dimensional domain Γ.
- In the third inequality, we use the trace theorem on Ω_f to bound the trace along Γ.
- In the fourth inequality, we use interpolation, and the fact that since the test function satisfies $\|\boldsymbol{v}\|_{Q_N^n} \leq 1$ where the test space Q_N^n is defined in (11.93), we have that $\|\boldsymbol{v}\|_{H^1(\Omega_f)}$ is uniformly bounded on the reference domain independently of n and N by the uniform bounds on ω_N. The uniform bounds on ω_N ensure uniform boundedness of $\mathcal{J}_f^\omega$ defined in (11.21), which is the term that appears when transforming integrals back to the reference domain.
- In the fifth inequality, we use uniform boundedness of $\|\boldsymbol{u}_N^n\|_{L^2(\Omega_f)}$ and $\|\dot{\boldsymbol{\eta}}_N^n\|_{L^2(\Gamma)}$, which is implied by the previous uniform energy estimates (see Lemma 11.6.1).
- We use the inequality $ab \leq \frac{1}{2}(a^2 + b^2)$.

Next, we estimate the second-to-last term in (11.100) in a way similar to the immediately preceding term, using similar justifications as given directly above:

$$
\begin{aligned}
&\left|\int_{\Gamma_N^n} \left(\frac{1}{2}\boldsymbol{u}_N^{n+1} \cdot \tilde{\boldsymbol{u}}_N^n - p_N^{n+1}\right)(\boldsymbol{v} \cdot \boldsymbol{n})\right| \\
&\quad \leq C\left(\|\boldsymbol{u}_N^{n+1}\|_{L^4(\Gamma)} \cdot \|\boldsymbol{u}_N^n\|_{L^4(\Gamma)} \cdot \|\boldsymbol{v}\|_{L^2(\Gamma)} + \|p_N^{n+1}\|_{L^2(\Gamma)} \cdot \|\boldsymbol{v}\|_{L^2(\Gamma)}\right) \\
&\quad \leq C\left(\|\boldsymbol{u}_N^{n+1}\|_{H^{1/4}(\Gamma)} \cdot \|\boldsymbol{u}_N^n\|_{H^{1/4}(\Gamma)} \cdot \|\boldsymbol{v}\|_{H^1(\Omega_f)} + \|p_N^{n+1}\|_{H^1(\Omega_b)} \cdot \|\boldsymbol{v}\|_{H^1(\Omega_f)}\right) \\
&\quad \leq C\left(\|\boldsymbol{u}_N^{n+1}\|_{H^{3/4}(\Omega_f)} \cdot \|\boldsymbol{u}_N^n\|_{H^{3/4}(\Omega_f)} + \|p_N^{n+1}\|_{H^1(\Omega_b)}\right) \\
&\quad \leq C\left(\|\boldsymbol{u}_N^{n+1}\|_{L^2(\Omega_f)}^{1/4} \|\boldsymbol{u}_N^{n+1}\|_{H^1(\Omega_f)}^{3/4} \cdot \|\boldsymbol{u}_N^n\|_{L^2(\Omega_f)}^{1/4} \|\boldsymbol{u}_N^n\|_{H^1(\Omega_f)}^{3/4} + \|p_N^{n+1}\|_{H^1(\Omega_b)}\right) \\
&\quad \leq C\left[1 + \left(\|p_N^{n+1}\|_{H^1(\Omega_b)} + \|\boldsymbol{u}_N^n\|_{V_N^n} + \|\boldsymbol{u}_N^{n+1}\|_{V_N^{n+1}}\right)^{3/2}\right].
\end{aligned}
$$

Finally, we estimate the last term

$$
\begin{aligned}
&\beta\left|\int_{\Gamma_N^n} (\dot{\boldsymbol{\eta}}_N^{n+1} - \boldsymbol{u}_N^{n+1}) \cdot \boldsymbol{\tau}(\boldsymbol{v} \cdot \boldsymbol{\tau})\right| \\
&\quad \leq C\left(\|\dot{\boldsymbol{\eta}}_N^{n+1}\|_{L^2(\Gamma)} \cdot \|\boldsymbol{v}\|_{L^2(\Gamma)} + \|\boldsymbol{u}_N^{n+1}\|_{L^2(\Gamma)} \cdot \|\boldsymbol{v}\|_{L^2(\Gamma)}\right) \\
&\quad \leq C\left(1 + \|\boldsymbol{u}_N^{n+1}\|_{H^1(\Omega_f)}\right).
\end{aligned}
$$

Therefore, we obtain the final estimate of the first term in (11.99), which implies the existence of a constant C independent of n and N, such that

$$\max_{\|\boldsymbol{v}\|_{Q_N^n}\leq 1}\left|\int_{\Omega_{f,N}^n}\frac{\boldsymbol{u}_N^{n+1}-\tilde{\boldsymbol{u}}_N^n}{\Delta t}\cdot\boldsymbol{v}d\boldsymbol{x}\right|\leq C\left(a_N^n+\|\boldsymbol{u}_N^n\|_{V_N^n}+\|\boldsymbol{u}_N^{n+1}\|_{V_N^{n+1}}\right)^{3/2},$$

$$\text{for } a_N^n := 1+\|p_N^{n+1}\|_{H^1(\Omega_b)}, \text{ where}$$

$$(\Delta t)\sum_{n=0}^{N-1}|a_N^n|^2\leq 2\left[(\Delta t)N+\|p_N\|^2_{L^2(0,T;H^1(\Omega_b))}\right]\leq C. \tag{11.101}$$

To complete the estimate (11.99), it remains to show that the second term $\left|\int_{\Omega_{f,N}^n}\frac{\tilde{\boldsymbol{u}}_N^n-\boldsymbol{u}_N^n}{\Delta t}\cdot\boldsymbol{v}d\boldsymbol{x}\right|$ is uniformly bounded. This follows from the same estimates as those presented in [270], which show that there exists a constant C independent of n and N, such that

$$\max_{\|\boldsymbol{v}\|_{Q_N^n}\leq 1}\left|\int_{\Omega_{f,N}^n}\frac{\tilde{\boldsymbol{u}}_N^n-\boldsymbol{u}_N^n}{\Delta t}\cdot\boldsymbol{v}d\boldsymbol{x}\right|\leq C. \tag{11.102}$$

Combining (11.101) and (11.102) with (11.98) and (11.99) establishes Generalized Property B and completes the proof of Proposition 11.7.3.

11.8 Passing to the Limit in the Regularized Weak Formulation

We have so far established the following strong convergence results:

$$\begin{aligned}
&\boldsymbol{\eta}_N\to\boldsymbol{\eta}, && \text{in } C(0,T;L^2(\Omega_b)),\\
&\omega_N\to\omega, && \text{in } L^\infty(0,T;H^s(\Gamma)) \text{ for } 0<s<2,\\
&\zeta_N^*\to\zeta, && \text{in } L^2(0,T;H^{-s}(\Gamma)), \text{ for } -1/2<s<0,\\
&\zeta_N\to\zeta, && \text{in } L^2(0,T;H^{-s}(\Gamma)), \text{ for } -1/2<s<0,\\
&\boldsymbol{\xi}_N\to\boldsymbol{\xi}, && \text{in } L^2(0,T;H^{-s}(\Omega_b)), \text{ for } -1/2<s<0,\\
&\boldsymbol{u}_N\to\boldsymbol{u}, && \text{in } L^2(0,T;L^2(\Omega_f^M)), \qquad p_N\to p, \quad \text{in } L^2(0,T;L^2(\Omega_b)),
\end{aligned}$$

where ζ_N^* and ζ_N converge to the same limit in $L^2(0, T; H^{-s}(\Gamma))$ for $-1/2 < s < 0$ due to the numerical dissipation estimates $\sum_{n=1}^{N} \|\zeta_N^n - \zeta_N^{n-\frac{1}{2}}\|^2_{L^2(\Gamma)} \leq C$, which imply that $\|\zeta_N - \zeta_N^*\|_{L^2(0,T;L^2(\Gamma))} \to 0$.

These strong convergence results will be used to pass to the limit in the semidiscrete formulation of the coupled problem (11.64) and show that the limit satisfies the weak formulation of the regularized problem. Before we can do this, there are two more convergence results that need to be established. One is a strong convergence result for the traces for the fluid velocity on the boundary of the fluid domain, and the other is a convergence result for the test functions, which are defined on approximate moving domains.

We start with the convergence result for the trace of the fluid velocity $\hat{\boldsymbol{u}}_N|_\Gamma$ along Γ.

11.8.1 Strong Convergence of the Fluid Velocity Traces on Γ

Proposition 11.8.1 *The traces $\hat{\boldsymbol{u}}_N|_\Gamma$ of the approximate fluid velocities on Γ converge to the trace of the limiting fluid velocity on Γ as $N \to \infty$:*

$$\hat{\boldsymbol{u}}_N|_\Gamma \to \hat{\boldsymbol{u}}|_\Gamma, \qquad \text{in } L^2(0, T; H^{s-\frac{1}{2}}(\Gamma)), \qquad \text{for } s \in (0, 1),$$

where $\hat{\boldsymbol{u}}_N = \boldsymbol{u}_N \circ \Phi_f^{\tau_{\Delta t}\omega_N}$ and $\hat{\boldsymbol{u}} = \boldsymbol{u} \circ \Phi_f^{\omega}$.

To prove Proposition 11.8.1, we will use the following elementary lemma.

Lemma 11.8.1 *Suppose that the functions $\{f_n\}_{n=1}^{\infty}$ and f are all uniformly bounded in $L^2(0, T; H^1(\Omega_f))$ and $f_n \to f$ in $L^2(0, T; L^2(\Omega_f))$. Then, $f_n \to f$ in $L^2(0, T; H^s(\Omega_f))$ for $s \in (0, 1)$, and hence $f_n|_\Gamma \to f|_\Gamma$ in $L^2(0, T; H^{s-\frac{1}{2}}(\Gamma))$ for $s \in (1/2, 1)$.*

Proof (Proof of Lemma 11.8.1) For $s \in (0, 1)$, we compute using Sobolev interpolation that

$$\|f_n - f\|^2_{L^2(0,T;H^s(\Omega_f))} = \int_0^T \|(f_n - f)(t)\|^2_{H^s(\Omega_f)} dt$$

$$\leq \int_0^T \|(f_n - f)(t)\|^{2(1-s)}_{L^2(\Omega_f)} \cdot \|(f_n - f)(t)\|^{2s}_{H^1(\Omega_f)} dt \leq \|f_n - f\|^{2(1-s)}_{L^2(0,T;L^2(\Omega_f))}$$

$$\cdot \|f_n - f\|^{2s}_{L^2(0,T;H^1(\Omega_f))}.$$

The result then follows from the fact that $\|f_n - f\|_{L^2(0,T;H^1(\Omega_f))} \leq C$ for a constant C that does not depend on N, the assumption that $\|f_n - f\|_{L^2(0,T;L^2(\Omega_f))} \to 0$ as

$N \to \infty$, and the trace embedding that gives that $\|f_n|_\Gamma - f|_\Gamma\|^2_{L^2(0,T;H^{s-\frac{1}{2}}(\Gamma))} \leq \|f_n - f\|^2_{L^2(0,T;H^s(\Omega_f))}$ for $s \in (1/2, 1)$. □

We can use the elementary lemma above to show the desired strong convergence of the fluid velocity traces.

Proof (Proof of Proposition 11.8.1) We would like to combine the fact that $\boldsymbol{u}_N \to \boldsymbol{u}$ in $L^2(0, T; L^2(\Omega_f^M))$ with the fact that $\boldsymbol{u}_N$ and $\boldsymbol{u}$ are all uniformly bounded in $L^2(0, T; H^1(\Omega_f(t)))$ for all N, to deduce strong convergence of the traces of the fluid velocities using Lemma 11.8.1. We do this in the following steps.

Step 1. We show that $\hat{\boldsymbol{u}}_N \to \hat{\boldsymbol{u}}$ on $L^2(0, T; L^2(\Omega_f))$, for $\hat{\boldsymbol{u}}_N$ and $\hat{\boldsymbol{u}}$ defined on the reference fluid domain.

To prove this, we compute $\|\hat{\boldsymbol{u}}_N - \hat{\boldsymbol{u}}\|^2_{L^2(0,T;L^2(\Omega_f))}$ using the functions $\boldsymbol{u}_N$ and $\boldsymbol{u}$, which are defined on the maximal domain Ω_f^M:

$$\|\hat{\boldsymbol{u}}_N - \hat{\boldsymbol{u}}\|^2_{L^2(0,T;L^2(\Omega_f))} = \int_0^T \int_{\Omega_f} \Big|\boldsymbol{u}_N \left(t, x, y + \left(1 + \frac{y}{R}\right) \tau_{\Delta t}\omega_N\right)$$

$$-\boldsymbol{u}\left(t, x, y + \left(1 + \frac{y}{R}\right)\omega\right)\Big|^2 \leq 2(I_1 + I_2),$$

where

$$I_1 = \int_0^T \int_{\Omega_f} \Big|\boldsymbol{u}_N \left(t, x, y + \left(1 + \frac{y}{R}\right) \tau_{\Delta t}\omega_N\right)$$

$$-\boldsymbol{u}\left(t, x, y + \left(1 + \frac{y}{R}\right) \tau_{\Delta t}\omega_N\right)\Big|^2,$$

$$I_2 = \int_0^T \int_{\Omega_f} \Big|\boldsymbol{u}\left(t, x, y + \left(1 + \frac{y}{R}\right) \tau_{\Delta t}\omega_N\right) - \boldsymbol{u}\left(t, x, y + \left(1 + \frac{y}{R}\right)\omega\right)\Big|^2.$$

We show that $I_1 \to 0$ as $N \to \infty$ by using the fact that $1 + \frac{\omega_N^n}{R}$ is uniformly bounded from above by a positive constant and the fact that Ω_f^M contains all of the domains $\Omega_{f,N}^n$, so that we can estimate:

$$\begin{aligned} I_1 &= \sum_{n=0}^{N-1} \int_{n\Delta t}^{(n+1)\Delta t} \int_{\Omega_{f,N}^n} \left(1 + \frac{\omega_N^n}{R}\right) |\boldsymbol{u}_N^{n+1} - \boldsymbol{u}|^2 \\ &\leq C \sum_{n=0}^{N-1} \int_{n\Delta t}^{(n+1)\Delta t} \int_{\Omega_{f,N}^n} |\boldsymbol{u}_N^{n+1} - \boldsymbol{u}|^2 \\ &\leq C\|\boldsymbol{u}_N - \boldsymbol{u}\|^2_{L^2(0,T;L^2(\Omega_f^M))} \to 0. \end{aligned}$$

For I_2, we break up the integral into two parts:

$$I_2 = I_{2,1} + I_{2,2},$$

where

$$I_{2,1} = \int_0^T \int_0^L \int_{-R}^{\min(0,y^*(t,x))} \left| \boldsymbol{u}\left(t, x, y + \left(1 + \frac{y}{R}\right) \tau_{\Delta t}\omega_N\right) \right.$$
$$\left. -\boldsymbol{u}\left(t, x, y + \left(1 + \frac{y}{R}\right)\omega\right)\right|^2,$$
$$I_{2,2} = \int_0^T \int_0^L \int_{\min(0,y^*(t,x))}^{0} \left|\boldsymbol{u}\left(t, x, y + \left(1 + \frac{y}{R}\right) \tau_{\Delta t}\omega_N\right)\right|^2,$$

for $y^*(t,x) = \frac{\omega - \tau_{\Delta t}\omega_N}{R + \tau_{\Delta t}\omega_N}$. We can interpret $y^*(t,x)$ as the y value for which $y + \left(1 + \frac{y}{R}\right)\tau_{\Delta t}\omega_N = \omega$. Now, note that

$$I_{2,1} \le \int_0^T \int_0^L \int_{-R}^{\min(0,y^*(t,x))} \left(\int_{y+(1+\frac{y}{R})\tau_{\Delta t}\omega_N}^{y+(1+\frac{y}{R})\omega} |\partial_y \boldsymbol{u}(t,x,y')| dy'\right)^2$$
$$\le \int_0^T \int_0^L \int_{-R}^{\min(0,y^*(t,x))} \left(\int_{y+(1+\frac{y}{R})\tau_{\Delta t}\omega_N}^{y+(1+\frac{y}{R})\omega} |\partial_y \boldsymbol{u}(t,x,y')|^2 dy'\right)$$
$$\cdot \left(1 + \frac{y}{R}\right) \cdot |\omega - \tau_{\Delta t}\omega_N|,$$

where we applied Cauchy-Schwarz to the inner dy' integral. We note that $\tau_{\Delta t}\omega_N \to \omega$ pointwise uniformly on $[0,T] \times \Gamma$ as $N \to \infty$ by the convergence $\overline{\omega}_N \to \omega$ in $C(0,T; H^s(\Gamma))$ for $0 < s < 2$ (from Proposition 11.7.1) and the estimate (11.84). Combining this with the fact that $\|\nabla \boldsymbol{u}\|_{L^2(0,T;L^2(\Omega_f^\omega(t)))}$ is bounded, we have that $I_{2,1} \to 0$ as $N \to \infty$.
Next, by the fundamental theorem of calculus and Jensen's inequality,

$$I_{2,2} \le \int_0^T \int_0^L |\min(0, y^*(t,x))| \cdot \max_{w \in [-R, \omega(t,x)]} |\boldsymbol{u}(t,x,w)|^2$$
$$\le C \int_0^T \int_0^L |\min(0, y^*(t,x))| \cdot \int_{-R}^{\omega(t,x)} |\partial_r \boldsymbol{u}(t,x,y')|^2 dy',$$

for a constant C that is independent of N, so we conclude that $I_{2,2} \to 0$ as $N \to \infty$ by the fact that $|\min(0, y^*(t,x))| \to 0$ uniformly on $[0,T] \times \Gamma$ and by the boundedness of $\|\nabla \boldsymbol{u}\|_{L^2(0,T;L^2(\Omega_f^\omega(t))}$. Thus, we have that $\|\hat{\boldsymbol{u}}_N - \hat{\boldsymbol{u}}\|_{L^2(0,T;L^2(\Omega_f)))} \to 0$. Note that this maximum is well defined for almost every

(t, x), since Sobolev functions are absolutely continuous for almost every line $\{(t, x, y) : -R \leq y \leq \omega(t, x)\}$ with fixed (t, x) (see [133, Theorem 2, Section 4.9]).

Step 2. We claim that the functions $\hat{\boldsymbol{u}}_N$ for positive integers N and $\hat{\boldsymbol{u}}$ are all uniformly bounded in $L^2(0, T; H^1(\Omega_f))$. Recall from Lemma 11.6.1 that the approximate solutions $\hat{\boldsymbol{u}}_N$ are uniformly bounded in $L^2(0, T; H^1(\Omega_f))$. Since $\hat{\boldsymbol{u}}$ is the strong limit of $\hat{\boldsymbol{u}}_N$ in $L^2(0, T; L^2(\Omega_f))$ and $\hat{\boldsymbol{u}}_N$ converge weakly in $L^2(0, T; H^1(\Omega_f))$ along a subsequence to a weak limit that hence must also be $\hat{\boldsymbol{u}}$, we conclude that $\hat{\boldsymbol{u}}$ is also in $L^2(0, T; H^1(\Omega_f))$, which establishes the desired result of this step.

Step 3. From Step 1, we have that $\hat{\boldsymbol{u}}_N \to \hat{\boldsymbol{u}}$ in $L^2(0, T; L^2(\Omega_f))$, and from Step 2, the functions $\hat{\boldsymbol{u}}_N$ and $\hat{\boldsymbol{u}}$ are bounded in $L^2(0, T; H^1(\Omega_f))$ independently of N, so we can conclude the proof of Proposition 11.8.1 by using Lemma 11.8.1.

□

11.8.2 Convergence of the Test Functions on Approximate Fluid Domains

The main difficulty in passing to the limit will be the test functions for the fluid velocity. In particular, on the fixed reference domain Ω_f for the fluid, we note that the test functions for the fluid velocity in (11.42) satisfy $\nabla_f^{\omega} \cdot \boldsymbol{v} = 0$ on Ω_f, where ω is the solution for the plate displacement. However, the test functions for the fluid velocity in the semidiscrete formulation in the semidiscrete test space $\mathcal{Q}_N^{n+1}$, defined by (11.56), satisfy $\nabla^{\omega_N^n} \cdot \boldsymbol{v} = 0$ on Ω_f. Hence, we need a way of comparing test functions in $\mathcal{Q}_N^{n+1}$ to test functions in the actual test space $\mathcal{V}_{\text{test}}^{\omega}$.

To do this, recall that we have defined the maximal domain Ω_f^M that contains all of the numerical fluid domains $\Omega_{f,N}^n$. We then propose to work with the test functions that are defined on Ω_f^M and are constructed in such a way that the restrictions of those test functions to the domain defined by the plate displacement ω, and composed with the ALE mapping Φ_f^{ω} defined in (11.19), give a space of test functions $\mathcal{X}_f^{\omega}$ that is dense in the fluid velocity test space $\mathcal{V}_f^{\omega}$. The space of all such test functions defined on Ω_f^M is denoted by $\mathcal{X}$, and it is defined as follows.

The Test Space $\mathcal{X}$ The test space $\mathcal{X}$ consists of functions $\boldsymbol{v} \in C_c^1([0, T); H^1(\Omega_f^M))$ satisfying the following properties for each $t \in [0, T)$:

1. For each $t \in [0, T)$, $\boldsymbol{v}(t)$ is a smooth vector-valued function on Ω_f^M.
2. $\nabla \cdot \boldsymbol{v}(t) = 0$ on Ω_f^M for all $t \in [0, T)$.
3. $\boldsymbol{v}(t) = 0$ on $\partial\Omega_f^M \setminus \Gamma_M$ for all $t \in [0, T)$, where $\Gamma_M = \{(x, M(x)) : 0 \leq x \leq L\}$ is the top boundary of the maximal fluid domain Ω_f^M.

Given $\boldsymbol{v} \in \mathcal{X}$, define

$$\tilde{\boldsymbol{v}} = \boldsymbol{v}|_{\Omega_f^\omega} \circ \boldsymbol{\Phi}_f^\omega \quad \text{and} \quad \tilde{\boldsymbol{v}}_N = \boldsymbol{v}|_{\Omega_f^{\omega_N}} \circ \boldsymbol{\Phi}_f^{\omega_N}. \tag{11.103}$$

The test functions $\tilde{\boldsymbol{v}}$ are dense in the fluid velocity test space $\mathcal{V}_f^\omega$ associated with the fixed domain formulation, and the test functions $\tilde{\boldsymbol{v}}_N$ restricted to $[n\Delta t, (n+1)\Delta t)$ are dense in $V_f^{\omega_N^n}$, where $V_f^{\omega_N^n}$ is the velocity test space for the semidiscretized problem(s) given in (11.56). Therefore, for each fixed N, we can consider the semidiscrete formulation with the test function $\tilde{\boldsymbol{v}}_N$, which we emphasize is discontinuous in time, due to the jumps in ω_N at each $n\Delta t$. To pass to the limit as $N \to \infty$, we can use the same approach as in Lemma 7.1 in [265] and Lemma 2.8 in [333], to obtain the following strong convergence results of the velocity test functions $\tilde{\boldsymbol{v}}_N$ and their gradients, which will allow us to pass to the limit in the semidiscrete weak formulations:

Proposition 11.8.2 *Consider* $\boldsymbol{v} \in \mathcal{X}$, $\tilde{\boldsymbol{v}}$, *and* $\tilde{\boldsymbol{v}}_N$ *defined in* (11.103). *Then*

$$\tilde{\boldsymbol{v}}_N \to \boldsymbol{v}, \qquad \nabla \tilde{\boldsymbol{v}}_N \to \nabla \boldsymbol{v},$$

pointwise, uniformly on $[0, T] \times \Omega_f$, *as* $N \to \infty$.

Remark 11.8.1 We emphasize that we were able to construct such a test space $\mathcal{X}$ because in the definition of the full test space $\mathcal{V}_{\text{test}}^\omega$ in (11.42), the only component of the test space whose definition depends on the plate displacement is the fluid velocity, and fortunately, this fluid velocity component of the test space is *decoupled from the other components*. This is a feature of fluid-poroelastic structure interaction problems. In the purely elastic case of FSI, the fluid velocity test space is coupled to that of the structure, and the construction of the test functions that converge on the approximate fluid domains is more involving (see, e.g., [81, 265, 267] and Chap. 5).

11.8.3 Passing to the Limit

We are now in a position to pass to the limit in the semidiscrete formulation. To do this, we consider any test function $(\tilde{\boldsymbol{v}}_N, \varphi, \boldsymbol{\psi}, r)$ for a given $\boldsymbol{v} \in \mathcal{X}$, and for each $n = 0, 1, \ldots, N-1$, we test the semidiscrete formulation (11.64) with $(\tilde{\boldsymbol{v}}_N(t), \varphi(t), \boldsymbol{\psi}(t), r(t))$ for each $t \in [n\Delta t, (n+1)\Delta t)$, integrate the resulting expressions in time from $t = n\Delta t$ to $t = (n+1)\Delta t$, and then finally sum over $n = 0, 1, \ldots, N-1$ to get an integral over the entire time interval $[0, T]$. Then, using the definition of the approximate solutions from Sect. 11.6, we thus obtain that for all $(\tilde{\boldsymbol{v}}_N, \varphi, \boldsymbol{\psi}, r)$ in the test space with $\boldsymbol{v} \in \mathcal{X}$, the following holds:

$$\int_0^T \int_{\Omega_f} \left(1 + \frac{\tau_{\Delta t}\omega_N}{R}\right) \partial_t \overline{\boldsymbol{u}}_N \cdot \tilde{\boldsymbol{v}}_N + \frac{1}{2} \int_0^T \int_{\Omega_f} \left(1 + \frac{\tau_{\Delta t}\omega_N}{R}\right)$$

$$\Big[\Big(\Big(\tau_{\Delta t}\boldsymbol{u}_N - \zeta_N \frac{R+y}{R}\boldsymbol{e}_y\Big)\cdot \nabla_f^{\tau_{\Delta t}\omega_N}\boldsymbol{u}_N\Big)\cdot \tilde{\boldsymbol{v}}_N$$

$$-\Big(\Big(\tau_{\Delta t}\boldsymbol{u}_N - \zeta_N \frac{R+y}{R}\boldsymbol{e}_y\Big)\cdot \nabla_f^{\tau_{\Delta t}\omega_N}\tilde{\boldsymbol{v}}_N\Big)\cdot \boldsymbol{u}_N\Big] + \frac{1}{2R}\int_0^T\int_{\Omega_f}\zeta_N \boldsymbol{u}_N\cdot\tilde{\boldsymbol{v}}_N$$

$$+\frac{1}{2}\int_0^T\int_\Gamma (\boldsymbol{u}_N - \zeta_N^*\boldsymbol{e}_y)\cdot \boldsymbol{n}^{\tau_{\Delta t}\omega_N}(\tau_{\Delta t}\boldsymbol{u}_N\cdot\tilde{\boldsymbol{v}}_N)$$

$$+2\nu\int_0^T\int_{\Omega_f}\Big(1+\frac{\tau_{\Delta t}\omega_N}{R}\Big)\boldsymbol{D}_f^{\tau_{\Delta t}\omega_N}(\boldsymbol{u}_N):\boldsymbol{D}_f^{\tau_{\Delta t}\omega_N}(\tilde{\boldsymbol{v}}_N)$$

$$+\int_0^T\int_\Gamma\Big(\frac{1}{2}\boldsymbol{u}_N\cdot\tau_{\Delta t}\boldsymbol{u}_N - p_N\Big)(\boldsymbol{\psi}-\tilde{\boldsymbol{v}}_N)\cdot\boldsymbol{n}^{\tau_{\Delta t}\omega_N}$$

$$+\beta\int_0^T\int_\Gamma(\mathcal{J}_\Gamma^{\tau_{\Delta t}\omega_N})^{-1}(\zeta_N^*\boldsymbol{e}_y-\boldsymbol{u}_N)\cdot\boldsymbol{\tau}^{\tau_{\Delta t}\omega_N}(\boldsymbol{\psi}-\tilde{\boldsymbol{v}}_N)\cdot\boldsymbol{\tau}^{\tau_{\Delta t}\omega_N}$$

$$+\rho_b\int_0^T\int_{\Omega_b}\Big(\frac{\boldsymbol{\xi}_N-\tau_{\Delta t}\boldsymbol{\xi}_N}{\Delta t}\Big)\cdot\boldsymbol{\psi}+\rho_p\int_0^T\int_\Gamma\partial_t\overline{\zeta}_N\cdot\varphi$$

$$+2\mu_e\int_0^T\int_{\Omega_b}\boldsymbol{D}(\boldsymbol{\eta}_N):\boldsymbol{D}(\boldsymbol{\psi})+\lambda_e\int_0^T\int_{\Omega_b}(\nabla\cdot\boldsymbol{\eta}_N)(\nabla\cdot\boldsymbol{\psi})$$

$$+2\mu_v\int_0^T\int_{\Omega_b}\boldsymbol{D}(\boldsymbol{\xi}_N):\boldsymbol{D}(\boldsymbol{\psi})+\lambda_v\int_0^T\int_{\Omega_b}(\nabla\cdot\boldsymbol{\xi}_N)(\nabla\cdot\boldsymbol{\psi})$$

$$-\alpha\int_0^T\int_{\Omega_b}\mathcal{J}_b^{(\tau_{\Delta t}\eta_N)^\delta}p_N\nabla_b^{(\tau_{\Delta t}\eta_N)^\delta}\cdot\boldsymbol{\psi}+c_0\int_0^T\int_{\Omega_b}\partial_t\overline{p}_N\cdot r$$

$$-\alpha\int_0^T\int_{\Omega_b}\mathcal{J}_b^{(\tau_{\Delta t}\eta_N)^\delta}\boldsymbol{\xi}_N\cdot\nabla_b^{(\tau_{\Delta t}\eta_N)^\delta}r$$

$$-\alpha\int_0^T\int_\Gamma(\zeta_N^*\boldsymbol{e}_y\cdot\boldsymbol{n}^{(\tau_{\Delta t}\omega_N)^\delta})r+\kappa\int_0^T\int_{\Omega_b}\mathcal{J}_b^{(\tau_{\Delta t}\eta_N)^\delta}\nabla_b^{(\tau_{\Delta t}\eta_N)^\delta}p_N\cdot\nabla_b^{(\tau_{\Delta t}\eta_N)^\delta}r$$

$$-\int_0^T\int_\Gamma[(\boldsymbol{u}_N-\zeta_N^*\boldsymbol{e}_y)\cdot\boldsymbol{n}^{\tau_{\Delta t}\omega_N}]r+\int_0^T\int_\Gamma\Delta\omega_N\cdot\Delta\varphi=0.$$

For additional details about the limit passage in the semidiscrete formulation (11.64), we refer the reader to the discussion in Sect. 5.8 of Chap. 5.

Using the strong convergence results established above, combined with the previously established weak convergence results in Proposition 11.6.1, we can pass to the limit in all of the terms in the semidiscrete weak formulation except those involving time derivatives. However, we can handle these by a discrete integration by parts. For example, for the first integral, we can use a discrete integration by parts to obtain:

$$\int_0^T \int_{\Omega_f} \left(1 + \frac{\tau_{\Delta t}\omega_N}{R}\right) \partial_t \overline{\boldsymbol{u}}_N \cdot \tilde{\boldsymbol{v}}_N \to - \int_0^T \int_{\Omega_f} \left(1 + \frac{\omega}{R}\right) \boldsymbol{u} \cdot \partial_t \tilde{\boldsymbol{v}}$$

$$- \frac{1}{R} \int_0^T \int_{\Omega_f} (\partial_t \omega) \boldsymbol{u} \cdot \tilde{\boldsymbol{v}} - \int_{\Omega_f} \left(1 + \frac{\omega_0}{R}\right) \boldsymbol{u}(0) \cdot \tilde{\boldsymbol{v}}(0),$$

where $\tilde{\boldsymbol{v}}_N = \boldsymbol{v} \circ \Phi_f^{\tau_{\Delta t}\omega_N}$ and $\tilde{\boldsymbol{v}} = \boldsymbol{v} \circ \Phi_f^{\omega}$ for $\boldsymbol{v} \in \mathcal{X}$. See, for example, pg. 79–81 in [333].

The limiting weak formulation holds for all velocity test functions in the smooth test space, which can be extended to the general test space $\mathcal{V}_{\text{test}}^{\omega}$ defined in (11.42) by using a density argument. Therefore, we have shown that the approximate weak solutions converge, up to a subsequence, to a weak solution to the regularized problem, as stated in Theorem 11.4.1.

This completes the main result of this chapter, stated in Theorem 11.4.1, providing existence of a weak solution to the nonlinearly coupled, regularized fluid-poroviscoelastic structure interaction problem, given in Definition 11.4.5.

We conclude this section by making the important observation that the weak solution that we have constructed to the regularized FPSI problem satisfies the desired energy estimate. This will be important for showing weak-classical consistency in the next section and can be shown easily by using the discrete energy estimate for the approximate solutions.

Proposition 11.8.3 (Energy Estimate for the Limiting Solution to the Regularized Problem) *The weak solution* $(\boldsymbol{u}, \boldsymbol{\eta}, p, \omega)$ *constructed from the splitting scheme as the limit of approximate solutions satisfies the following energy estimate for almost every* $t \in [0, T]$*:*

$$\frac{1}{2} \int_{\Omega_f(t)} |\boldsymbol{u}|^2 + \frac{1}{2} \rho_b \int_{\Omega_b} |\boldsymbol{\xi}|^2 + \frac{1}{2} c_0 \int_{\Omega_b} |p|^2 + \mu_e \int_{\Omega_b} |\boldsymbol{D}(\boldsymbol{\eta})|^2$$

$$+ \frac{1}{2} \lambda_e \int_{\Omega_b} |\nabla \cdot \boldsymbol{\eta}|^2 + \frac{1}{2} \rho_p \int_{\Gamma} |\zeta|^2 + \frac{1}{2} \int_{\Gamma} |\Delta \omega|^2 + 2\nu \int_0^t \int_{\Omega_f(s)} |\boldsymbol{D}(\boldsymbol{u})|^2$$

$$+ 2\mu_v \int_0^t \int_{\Omega_b} |\boldsymbol{D}(\boldsymbol{\xi})|^2 + \lambda_v \int_0^t \int_{\Omega_b} |\nabla \cdot \boldsymbol{\xi}|^2 + \kappa \int_0^t \int_{\Omega_b^{\delta}(s)} |\nabla p|^2$$

$$+ \beta \int_0^t \int_{\Gamma(s)} |(\zeta \boldsymbol{e}_y - \boldsymbol{u}) \cdot \boldsymbol{\tau})|^2 \leq E_0, \tag{11.104}$$

where E_0 *is the initial energy of the problem.*

Proof The approximate solutions $(\boldsymbol{u}_N, \boldsymbol{\eta}_N, p_N, \omega_N)$ satisfy the following energy inequality:

$$\frac{1}{2}\int_{\Omega_{f,N}(t)}|\boldsymbol{u}_N|^2+\frac{1}{2}\rho_b\int_{\Omega_b}|\boldsymbol{\xi}_N|^2+\frac{1}{2}c_0\int_{\Omega_b}|p_N|^2+\mu_e\int_{\Omega_b}|\boldsymbol{D}(\boldsymbol{\eta}_N)|^2$$

$$+\frac{1}{2}\lambda_e\int_{\Omega_b}|\nabla\cdot\boldsymbol{\eta}_N|^2+\frac{1}{2}\rho_p\int_{\Gamma}|\zeta_N|^2+\frac{1}{2}\int_{\Gamma}|\Delta\omega_N|^2+2\nu\int_0^t\int_{\Omega_{f,N}(s)}|\boldsymbol{D}(\boldsymbol{u}_N)|^2$$

$$+2\mu_v\int_0^t\int_{\Omega_b}|\boldsymbol{D}(\boldsymbol{\xi}_N)|^2+\lambda_v\int_0^t\int_{\Omega_b}|\nabla\cdot\boldsymbol{\xi}_N|^2+\kappa\int_0^t\int_{\Omega_{b,N}^{\delta}(s)}|\nabla p_N|^2$$

$$+\beta\int_0^t\int_{\Gamma(s)}|(\zeta_N^*\boldsymbol{e}_y-\boldsymbol{u}_N)\cdot\boldsymbol{\tau}|^2\leq E_0.$$

By using the weak and weak-star convergences of the approximate solutions, stated in Proposition 11.6.1, and the weak lower semicontinuity property of the norms, we can pass to the limit in the energy inequality, and one recovers the energy inequality (11.104). □

11.9 Weak-Classical Consistency

We have now shown the existence of weak solutions to the regularized FPSI problem (11.43). However, it is not clear that the solutions to this regularized problem are physically relevant, since the regularized weak formulation is not equivalent to the original weak formulation without the regularization. However, we will demonstrate the following weak-classical consistency result: given a spatially and temporally smooth solution $(\boldsymbol{u},\boldsymbol{\eta},p,\omega)$ to the FPSI problem, then the weak solutions to the regularized problem with regularization parameter δ, which we will denote by $(\boldsymbol{u}_\delta,\boldsymbol{\eta}_\delta,p_\delta,\omega_\delta)$, converge to the smooth solution as $\delta\to 0$.

11.9.1 Notation

Since we will have to use spatial convolution of the solution to the regularized problem $(\boldsymbol{u}_\delta,\boldsymbol{\eta}_\delta,p_\delta,\omega_\delta)$, and spatial convolution of the smooth solution $(\boldsymbol{u},\boldsymbol{\eta},p,\omega)$, we introduce the following notation to avoid additional superscripts involving δ.

1. Recall that $(\boldsymbol{u}_\delta,\boldsymbol{\eta}_\delta,p_\delta,\omega_\delta)$ denotes the weak solutions to the regularized problem (11.43).
2. We will use $(\boldsymbol{u},\boldsymbol{\eta},p,\omega)$ to denote a spatially and temporally smooth solution to the original FPSI problem (11.5), (11.6), (11.9), (11.10), and (11.14)–(11.18).
3. We will use the superscript δ notation

$$\boldsymbol{\eta}_\delta^\delta=(\boldsymbol{\eta}_\delta)^\delta:=\delta^{-2}\boldsymbol{\eta}_\delta*\sigma(x/\delta)$$

to denote the spatial convolution defined by (11.27) of the weak solution to the regularized problem with the smooth convolution δ kernel.

4. Similarly, in the same spirit as in 3, we will use

$$\boldsymbol{\eta}^{\delta} = (\boldsymbol{\eta})^{\delta} = \delta^{-2}\boldsymbol{\eta} * \sigma(x/\delta) \tag{11.105}$$

to denote the spatial convolution of the classical solution $\boldsymbol{\eta}$ with the convolution kernel.

5. We will use superscript δ to denote the physical Biot domain under the regularized displacement:

$$\Omega_{b,\delta}^{\delta}(t) = (\boldsymbol{I} + \boldsymbol{\eta}_{\delta}^{\delta}(t))(\Omega_b). \tag{11.106}$$

Weak Formulations Reformulated We note that even though the weak formulation (11.26) and the regularized weak formulation (11.43) are stated up until a fixed final time T, we can reformulate the weak formulation for almost every time $t \in [0, T]$ by using a cutoff function (see, e.g., the proof of Lemma 11.10.2 in the Appendix where this is done explicitly).

Thus, the **classical (temporally and spatially smooth)** solution $(\boldsymbol{u}, \boldsymbol{\eta}, p, \omega)$ satisfies the following *non-regularized weak formulation* for almost all $t \in [0, T]$, for all test functions $(\boldsymbol{v}, \varphi, \boldsymbol{\psi}, r) \in \mathcal{V}_{\text{test}}$ with the (moving domain) test space $\mathcal{V}_{\text{test}}$ defined in (11.41):

$$\begin{aligned}
&-\int_0^t \int_{\Omega_f(s)} \boldsymbol{u} \cdot \partial_t \boldsymbol{v} + \frac{1}{2}\int_0^t \int_{\Omega_f(s)} [((\boldsymbol{u} \cdot \nabla)\boldsymbol{u}) \cdot \boldsymbol{v} - ((\boldsymbol{u} \cdot \nabla)\boldsymbol{v}) \cdot \boldsymbol{u}] \\
&+ \frac{1}{2}\int_0^t \int_{\Gamma_1(s)} (\boldsymbol{u} \cdot \boldsymbol{n} - 2\boldsymbol{\xi}_1 \cdot \boldsymbol{n})\boldsymbol{u} \cdot \boldsymbol{v} \\
&+ 2\nu \int_0^t \int_{\Omega_f(s)} \boldsymbol{D}(\boldsymbol{u}) : \boldsymbol{D}(\boldsymbol{v}) + \int_0^t \int_{\Gamma_1(s)} \left(\frac{1}{2}|\boldsymbol{u}|^2 - p\right)(\psi_n - v_n) \\
&+ \beta \int_0^t \int_{\Gamma_1(s)} (\boldsymbol{\xi} - \boldsymbol{u}) \cdot \boldsymbol{t}(\psi_t - v_t) \\
&- \rho_p \int_0^t \int_{\Gamma} \partial_t \omega \cdot \partial_t \varphi + \int_0^t \int_{\Gamma} \Delta\omega \cdot \Delta\varphi - \rho_b \int_0^t \int_{\Omega_b} \partial_t \boldsymbol{\eta}_1 \cdot \partial_t \boldsymbol{\psi} \\
&+ 2\mu_e \int_0^t \int_{\Omega_b} \boldsymbol{D}(\boldsymbol{\eta}_1) : \boldsymbol{D}(\boldsymbol{\psi}) + \lambda_e \int_0^t \int_{\Omega_b} (\nabla \cdot \boldsymbol{\eta}_1)(\nabla \cdot \boldsymbol{\psi}) \\
&+ 2\mu_v \int_0^t \int_{\Omega_b} \boldsymbol{D}(\partial_t \boldsymbol{\eta}) : \boldsymbol{D}(\boldsymbol{\psi}) + \lambda_v \int_0^t \int_{\Omega_b} (\nabla \cdot \partial_t \boldsymbol{\eta})(\nabla \cdot \boldsymbol{\psi}) \\
&- \alpha \int_0^t \int_{\Omega_b(s)} p\nabla \cdot \boldsymbol{\psi} - c_0 \int_0^t \int_{\Omega_b} p\partial_t r - \alpha \int_0^t \int_{\Omega_b(s)} \frac{D}{Dt}\boldsymbol{\eta} \cdot \nabla r
\end{aligned}$$

$$-\alpha\int_0^t\int_{\Gamma(s)}(\boldsymbol{\xi}\cdot\boldsymbol{n})r+\kappa\int_0^t\int_{\Omega_b(s)}\nabla p\cdot\nabla r-\int_0^t\int_{\Gamma_1(s)}((\boldsymbol{u}-\boldsymbol{\xi})\cdot\boldsymbol{n})r$$
$$=-\int_{\Omega_f(t)}\boldsymbol{u}(t)\cdot\boldsymbol{v}(t)-\rho_p\int_\Gamma\zeta(t)\cdot\boldsymbol{\psi}(t)-\rho_b\int_{\Omega_b}\boldsymbol{\xi}(t)\cdot\boldsymbol{\psi}(t)-c_0\int_{\Omega_b}p(t)\cdot r(t)$$
$$+\int_{\Omega_f(0)}\boldsymbol{u}_0\cdot\boldsymbol{v}(0)+\rho_p\int_\Gamma\beta_0\cdot\boldsymbol{\psi}(0)+\rho_b\int_{\Omega_b}\boldsymbol{\xi}_0\cdot\boldsymbol{\psi}(0)+c_0\int_{\Omega_b}p_0\cdot r(0). \tag{11.107}$$

Similarly, the **solution to the regularized FPSI problem** $(\boldsymbol{u}_\delta, \boldsymbol{\eta}_\delta, p_\delta, \omega_\delta)$ satisfies the following *regularized weak formulation* for every test function $(\boldsymbol{v}, \varphi, \boldsymbol{\psi}, r) \in \mathcal{V}_{\text{test}}$, and for almost every $t \in [0, T_\delta]$ where the final time T_δ potentially depends on δ:

$$-\int_0^t\int_{\Omega_{f,\delta}(s)}\boldsymbol{u}_\delta\cdot\partial_t\boldsymbol{v}+\frac{1}{2}\int_0^t\int_{\Omega_{f,\delta}(s)}[((\boldsymbol{u}_\delta\cdot\nabla)\boldsymbol{u}_\delta)\cdot\boldsymbol{v}-((\boldsymbol{u}_\delta\cdot\nabla)\boldsymbol{v})\cdot\boldsymbol{u}_\delta]$$
$$+\frac{1}{2}\int_0^t\int_{\Gamma_\delta(s)}(\boldsymbol{u}_\delta\cdot\boldsymbol{n}-2\boldsymbol{\xi}_\delta\cdot\boldsymbol{n})\boldsymbol{u}_\delta\cdot\boldsymbol{v}+2\nu\int_0^t\int_{\Omega_{f,\delta}(s)}\boldsymbol{D}(\boldsymbol{u}_\delta):\boldsymbol{D}(\boldsymbol{v})$$
$$+\int_0^t\int_{\Gamma_\delta(s)}\left(\frac{1}{2}|\boldsymbol{u}_\delta|^2-p_\delta\right)(\psi_n-v_n)+\beta\int_0^t\int_{\Gamma_\delta(s)}(\boldsymbol{\xi}_\delta-\boldsymbol{u}_\delta)\cdot\boldsymbol{t}(\psi_t-v_t)$$
$$-\rho_p\int_0^t\int_\Gamma\partial_t\omega_\delta\cdot\partial_t\varphi+\int_0^t\int_\Gamma\Delta\omega_\delta\cdot\Delta\varphi-\rho_b\int_0^t\int_{\Omega_b}\partial_t\boldsymbol{\eta}_\delta\cdot\partial_t\boldsymbol{\psi}$$
$$+2\mu_e\int_0^t\int_{\Omega_b}\boldsymbol{D}(\boldsymbol{\eta}_\delta):\boldsymbol{D}(\boldsymbol{\psi})+\lambda_e\int_0^t\int_{\Omega_b}(\nabla\cdot\boldsymbol{\eta}_\delta)(\nabla\cdot\boldsymbol{\psi})$$
$$+2\mu_v\int_0^t\int_{\Omega_b}\boldsymbol{D}(\partial_t\boldsymbol{\eta}_\delta):\boldsymbol{D}(\boldsymbol{\psi})+\lambda_v\int_0^t\int_{\Omega_b}(\nabla\cdot\partial_t\boldsymbol{\eta}_\delta)(\nabla\cdot\boldsymbol{\psi})$$
$$-\alpha\int_0^t\int_{\Omega^\delta_{b,\delta}(s)}p_\delta\nabla\cdot\boldsymbol{\psi}-c_0\int_0^t\int_{\Omega_b}p_\delta\partial_t r-\alpha\int_0^t\int_{\Omega^\delta_{b,\delta}(s)}\frac{D^\delta}{Dt}\boldsymbol{\eta}_\delta\cdot\nabla r$$
$$-\alpha\int_0^t\int_{\Gamma_\delta(s)}(\boldsymbol{\xi}_\delta\cdot\boldsymbol{n})r+\kappa\int_0^t\int_{\Omega^\delta_{b,\delta}(s)}\nabla p_\delta\cdot\nabla r-\int_0^t\int_{\Gamma_\delta(s)}((\boldsymbol{u}_\delta-\boldsymbol{\xi}_\delta)\cdot\boldsymbol{n})r$$
$$=-\int_{\Omega_{f,\delta}(t)}\boldsymbol{u}_\delta(t)\cdot\boldsymbol{v}(t)-\rho_p\int_\Gamma\zeta_\delta(t)\cdot\varphi(t)-\rho_b\int_{\Omega_b}\boldsymbol{\xi}_\delta(t)\cdot\boldsymbol{\psi}(t)$$
$$-c_0\int_{\Omega_b}p_\delta(t)\cdot r(t)$$
$$+\int_{\Omega_f(0)}\boldsymbol{u}_0\cdot\boldsymbol{v}(0)+\rho_p\int_\Gamma\beta_0\cdot\varphi(0)+\rho_b\int_{\Omega_b}\boldsymbol{\xi}_0\cdot\boldsymbol{\psi}(0)+c_0\int_{\Omega_b}p_0\cdot r(0), \tag{11.108}$$

where $\frac{D^\delta}{Dt}$ is the material derivative with respect to the regularized Biot displacement. We remark that while our existence proof in the previous sections holds for both a purely elastic and viscoelastic Biot medium, our weak-classical consistency result will hold in the specific case of a Biot *poroviscoelastic* medium so that the viscoelasticity parameters μ_v and λ_v are strictly positive, and hence, the plate velocity $\zeta_\delta \boldsymbol{e}_y$ in the weak formulation is equivalently the trace of the Biot medium velocity $\boldsymbol{\xi}_\delta \in L^2(0, T; H^1(\Omega_b))$ along Γ.

11.9.2 Statement of the Result

In the remainder of the chapter, we will prove the weak-classical consistency result. Before stating the result, we need to introduce some additional notation. Namely, to prove the weak-classical consistency, we will subtract the weak formulations for the two solutions $\boldsymbol{u}$ and $\boldsymbol{u}_\delta$ and test formally with the difference of the two solutions $\boldsymbol{v} = \boldsymbol{u} - \boldsymbol{u}_\delta$. However, the functions $\boldsymbol{u}$ and $\boldsymbol{u}_\delta$ are defined on different domains, and the test functions on the physical domains are required to be divergence-free, so we need to be able to transfer $\boldsymbol{u}_\delta$ to the physical domain for $\boldsymbol{u}$ in a way that preserves the divergence-free condition. To do this, we emphasize that we cannot use the usual ALE mapping $\Phi_f^\omega : \Omega_f \to \Omega_f(t)$ defined in (11.19). This is because given two structure displacements ω_1 and ω_2, which define two respective fluid domains $\Omega_f^{\omega_1}$ and $\Omega_f^{\omega_2}$ and a divergence-free function $\boldsymbol{u}_1$ on $\Omega_f^{\omega_1}$, the function $\boldsymbol{u} \circ \hat{\boldsymbol{\Phi}}_f^{\omega_1} \circ (\hat{\boldsymbol{\Phi}}_f^{\omega_2})^{-1}$ transformed via the ALE mapping is not necessarily divergence-free on $\Omega_f^{\omega_2}$. Therefore, we will have to use a different transformation to bring a divergence-free function defined on one fluid domain to a divergence-free function defined on another fluid domain.

For this purpose, consider the two fluid domains

$$\Omega_f(t) = \{(x, y) \in \mathbb{R}^2 : 0 \le x \le L, -R \le y \le \omega(t, x)\},$$

$$\Omega_{f,\delta}(t) = \{(x, y) \in \mathbb{R}^2 : 0 \le x \le L, -R \le y \le \omega_\delta(t, x)\},$$

that are associated to the plate displacements ω and ω_δ.

We define a map between $\Omega_f(t)$ and $\Omega_{f,\delta}(t)$ and a transformation that sends functions on one domain to functions on the other domain as follows. Let $\psi_\delta(t) : \Omega_{f,\delta}(t) \to \Omega_f(t)$ be the mapping defined by

$$\psi_\delta(t, x, y) = (t, x, \gamma_\delta(t, x)(R + y) - R), \text{ where } \gamma_\delta(t, x) = \frac{R + \omega(t, x)}{R + \omega_\delta(t, x)}. \tag{11.109}$$

This mapping, unfortunately, does not preserve the divergence-free condition. However, if we calculate the gradient of the composite mapped function, we get

$$\nabla(\boldsymbol{u} \circ \psi_\delta) = [(\nabla \boldsymbol{u}) \circ \psi_\delta] J_\delta \tag{11.110}$$

where

$$J_\delta(t, x, y) = \begin{pmatrix} 1 & 0 \\ (R + y)\partial_x \gamma_\delta(t, x) & \gamma_\delta(t, x) \end{pmatrix}. \tag{11.111}$$

Similarly, for the regularized problem, we define

$$\tilde{J}_\delta = J_\delta \circ \psi_\delta^{-1} = \begin{pmatrix} 1 & 0 \\ (R + y)\gamma_\delta^{-1}\partial_x \gamma_\delta(t, x) & \gamma_\delta(t, x) \end{pmatrix}. \tag{11.112}$$

These Jacobian matrices will now be used to define the transformations that map divergence-free functions to divergence-free functions.

Definition 11.9.1

Part I: Given a divergence-free function $\boldsymbol{u}$ on $\Omega_f(t)$, the following transformation $\widehat{\ }: \boldsymbol{u} \mapsto \widehat{\boldsymbol{u}}$ maps $\boldsymbol{u}$ to a divergence-free function $\widehat{\boldsymbol{u}}$ on $\Omega_{f,\delta}(t)$:

$$\widehat{\boldsymbol{u}} = \gamma_\delta J_\delta^{-1} \cdot (\boldsymbol{u} \circ \psi_\delta). \tag{11.113}$$

Part II: Given a divergence-free function $\boldsymbol{u}_\delta$ on $\Omega_{f,\delta}(t)$ the following transformation $\check{\ }: \boldsymbol{u}_\delta \mapsto \check{\boldsymbol{u}}_\delta$ maps $\boldsymbol{u}_\delta$ to a divergence-free function $\check{\boldsymbol{u}}_\delta$ on $\Omega_f(t)$:

$$\check{\boldsymbol{u}}_\delta = \gamma_\delta^{-1} \tilde{J}_\delta \cdot (\boldsymbol{u}_\delta \circ \psi_\delta^{-1}). \tag{11.114}$$

Remark 11.9.1 Both transformations preserve the trace of functions along Γ.

Note that even though the definition of $\widehat{\boldsymbol{u}}$ depends on δ, we will not explicitly notate this dependence, as δ will be clear from the context. We now state the weak-classical consistency result.

Theorem 11.9.1 (Weak-Classical Consistency) *Let* $(\boldsymbol{\eta}_0, \boldsymbol{\xi}_0, \omega_0, \zeta_0, p_0, \boldsymbol{u}_0)$ *be smooth initial data for the nonlinearly coupled FPSI problem* (11.5), (11.6), (11.9), (11.10), *and* (11.14)–(11.18). *Suppose* $(\boldsymbol{\eta}, \omega, p, \boldsymbol{u})$ *is a classical (temporally and spatially smooth) solution to this FPSI problem on the time interval* $[0, T]$. *Let* $(\boldsymbol{\eta}_\delta, \omega_\delta, p_\delta, \boldsymbol{u}_\delta)$ *denote the weak solution to the regularized FPSI problem* (11.43) *with regularity parameter* δ.

Then the following holds true:

1. $(\boldsymbol{\eta}_\delta, \omega_\delta, p_\delta, \boldsymbol{u}_\delta)$ *is defined on the time interval* $[0, T]$ *for all* $\delta > 0$, *where the final time* T *is independent of* δ.
2. *The energy norm of the difference between the two solutions* $E_\delta(t)$ *converges to zero as* $\delta \to 0$, *for all* $t \in [0, T]$, *where*

$$
\begin{aligned}
E_\delta(t) := &\|(\hat{\boldsymbol{u}} - \boldsymbol{u}_\delta)(t)\|^2_{L^2(\Omega_{f,\delta}(t))} + \int_0^t \|\boldsymbol{D}(\hat{\boldsymbol{u}} - \boldsymbol{u}_\delta)(s)\|^2_{L^2(\Omega_{f,\delta}(s))} ds \\
&+ \|(\boldsymbol{\xi} - \boldsymbol{\xi}_\delta)(t)\|^2_{L^2(\Gamma)} + \|(\omega - \omega_\delta)(t)\|^2_{H^2(\Gamma)} + \|(\boldsymbol{\xi} - \boldsymbol{\xi}_\delta)(t)\|^2_{L^2(\Omega_b)} \\
&+ \|\boldsymbol{D}(\boldsymbol{\eta} - \boldsymbol{\eta}_\delta)(t)\|^2_{L^2(\Omega_b)} + \|(\nabla \cdot (\boldsymbol{\eta} - \boldsymbol{\eta}_\delta))(t)\|^2_{L^2(\Omega_b)} \\
&+ \int_0^t \|\boldsymbol{D}(\boldsymbol{\xi} - \boldsymbol{\xi}_\delta)(s)\|^2_{L^2(\Omega_b)} ds \\
&+ \int_0^t \|\nabla \cdot (\boldsymbol{\xi} - \boldsymbol{\xi}_\delta)(s)\|^2_{L^2(\Omega_b)} ds + \|(p - p_\delta)(t)\|^2_{L^2(\Omega_b)} \\
&+ \int_0^t \|\nabla (p - p_\delta)(s)\|^2_{L^2(\Omega^\delta_{b,\delta}(s))} ds.
\end{aligned}
\tag{11.115}
$$

Preview of the Main Steps of the Proof of Weak-Classical Consistency The proof is based on Gronwall's inequality for $E_\delta(t)$. However, there are several obstacles to applying Gronwall's inequality due to the fact that we are working on a moving domain problem. We summarize those main obstacles and the main ideas behind their resolution here.

The main idea is to estimate the energy difference between $(\boldsymbol{u}, \boldsymbol{\eta}, p, \omega)$ and $(\boldsymbol{u}_\delta, \boldsymbol{\eta}_\delta, p_\delta, \omega_\delta)$, defined in (11.115), and obtain an estimate for $E_\delta(t)$ in terms of $E_\delta(0)$, the integral of $E_\delta(s)$ for times $s \in [0, t]$, and other terms that have sufficiently strong convergence in δ as $\delta \to 0$:

$$
E_\delta(t) \leq C \left(\int_0^t \|\nabla \boldsymbol{\eta} - \nabla \boldsymbol{\eta}^\delta\|^2_{L^2(\Omega_b)} ds + \int_0^t E_\delta(s) ds \right)
\tag{11.116}
$$

and then apply Gronwall's inequality to obtain

$$
E_\delta(t) \leq C\delta^3 e^{Ct},
\tag{11.117}
$$

where C is independent of δ, and conclude that $E_\delta(t) \to 0$ as $\delta \to 0$. We remark that the factor of δ^3 appearing in the Gronwall estimate comes from an estimate of the convergence rate of the spatial convolution $\boldsymbol{\eta}^\delta$ to $\boldsymbol{\eta}$ in $H^1(\Omega_b)$, which we establish in the upcoming Lemma 11.9.2.

To do this, we will test the weak formulations for $\boldsymbol{u}$ and $\boldsymbol{u}_\delta$ with appropriate test functions and use the energy inequality (11.104) from Proposition 11.8.3. More precisely, the main steps in the proof are:

1. Test the non-regularized weak formulation (11.107) for the classical solution $(\boldsymbol{u}, \boldsymbol{\eta}, p, \omega)$ with the "difference" of $(\boldsymbol{u}, \partial_t \boldsymbol{\eta}, p, \partial_t \omega)$ and $(\boldsymbol{u}_\delta, \partial_t \boldsymbol{\eta}_\delta, p_\delta, \partial_t \omega_\delta)$, where the notion of the difference between these two solutions will be made precise in Sect. 11.9.3

2. Test the regularized weak formulation (11.108) for $(\boldsymbol{u}_\delta, \boldsymbol{\eta}_\delta, p_\delta, \omega_\delta)$ with $(\boldsymbol{u}, \partial_t \boldsymbol{\eta}, p, \partial_t \omega)$
3. Rewrite the energy inequality (11.104) for $(\boldsymbol{u}_\delta, \boldsymbol{\eta}_\delta, p_\delta, \omega_\delta)$ so that it parallels the terms in the weak formulation (11.108)
4. Combine the equations from Step 1, Step 2, and Step 3. This will give us an expression that we can analyze term by term in order to obtain estimate (11.116) for the energy difference $E_\delta(t)$. Details will be presented in section "Proof of Gronwall's Estimate".
5. Now, we have that the inequality (11.116) and the resulting Gronwall estimate (11.117) are proven locally in time, namely, on the interval $[0, T_\delta]$, along which certain boundedness assumptions on the Lagrangian map hold for the solution of the regularized problem. However, we want the estimate (11.117) to hold along the *entire* time interval $[0, T]$, along which the classical solution is defined. Hence, we will use a bootstrap argument on boundedness assumptions of the Lagrangian map in order to propagate the estimate (11.116) to the entire time interval $[0, T]$ (see Sect. 11.9.4).
6. Apply Gronwall's inequality to (11.116) holding on $[0, T]$ to obtain the following bound for $E_\delta(t)$:

$$E_\delta(t) \leq C\delta^3 e^{Ct},$$

where C is independent of δ, and conclude that $E_\delta(t) \to 0$ as $\delta \to 0$.

Before we start with the proof of weak-classical consistency, we emphasize that there are two main **mathematical difficulties** that need to be addressed in the proof:

1. In step 1 above, we want to test (11.107) with the difference of $(\boldsymbol{u}, \partial_t \boldsymbol{\eta}, p, \partial_t \omega)$ and $(\boldsymbol{u}_\delta, \partial_t \boldsymbol{\eta}_\delta, p_\delta, \partial_t \omega_\delta)$. This is formal because the test functions in $\mathcal{V}_{\text{test}}$, defined in (11.41), must be continuously differentiable in time, and furthermore, for the fluid velocities, the difference between $\boldsymbol{u}$ and $\boldsymbol{u}_\delta$ does not make sense, since these functions are defined on different fluid domains. Thus, we must carefully define which test functions we will use. This is addressed at the beginning of Sect. 11.9.3.
2. As mentioned in step 5 above, the regularized weak formulation involves integrals on the physical time-dependent Biot domain $\Omega^\delta_{b,\delta}(t)$, which gives an extra factor of $\det(\boldsymbol{I} + \nabla \boldsymbol{\eta}^\delta_\delta)$ in the integrand from the Jacobian, when the integrals are transferred to the fixed reference Biot domain Ω_b. This factor cannot be estimated in the finite energy space where $\boldsymbol{\eta}_\delta$ is only bounded uniformly in δ in the function space $L^\infty(0, T; H^1(\Omega_b))$. To obtain pointwise estimates of this term that hold on the time interval $[0, T]$, where T is independent of δ, we need to use a *bootstrap argument* to get from the local pointwise estimates on $[0, T_\delta]$, where T_δ depends on δ, to the global, uniform estimates on $[0, T]$. This is addressed in Sect. 11.9.4.

11.9.3 Gronwall's Inequality

We show that the following Gronwall's inequality holds for almost all $t \in [0, T_\delta]$, where T_δ depends on δ. Later on, we will use a bootstrap argument to show that the weak-classical consistency holds uniformly, on the entire interval $[0, T]$ on which the classical solution exists.

Lemma 11.9.1 (Gronwall's Estimate) *Suppose that there exist constants $c > 0$ and $C > 0$, which are independent of δ such that the following estimates hold for almost all $t \in [0, T_\delta]$:*

$$\det(\boldsymbol{I} + \nabla \boldsymbol{\eta}_\delta^\delta) \geq c > 0, \tag{11.118}$$

$$0 < c \leq |\boldsymbol{I} + \nabla \boldsymbol{\eta}_\delta^\delta| \leq C, \qquad \text{pointwise in } \overline{\Omega_b}, \tag{11.119}$$

$$|\nabla \boldsymbol{\eta}_\delta^\delta| \leq C, \qquad \text{pointwise in } \overline{\Omega_b}, \tag{11.120}$$

where the final time T_δ potentially depends on δ. Furthermore, let $\boldsymbol{\eta}$ and $\boldsymbol{\eta}^\delta$ denote the smooth solution and its regularization, defined on $[0, T]$, and E_δ be the energy norm difference (11.115). *Then the following inequality hold:*

$$E_\delta(t) \leq C \left(\int_0^t \|\nabla \boldsymbol{\eta} - \nabla \boldsymbol{\eta}^\delta\|^2_{L^2(\Omega_b)} ds + \int_0^t E_\delta(s) ds \right), \tag{11.121}$$

where $E_\delta(t)$ is defined by (11.115). *Furthermore,*

$$E_\delta(t) \leq C\delta^3 e^{Ct}.$$

To prove Gronwall's inequality, we want to test the non-regularized weak formulation formally with the difference between $(\boldsymbol{u}, \partial_t \boldsymbol{\eta}, p, \partial_t \omega)$ and $(\boldsymbol{u}_\delta, \partial_t \boldsymbol{\eta}_\delta, p_\delta, \partial_t \omega_\delta)$. However, there are two reasons why this is not rigorously justified. First, $\partial_t \boldsymbol{\eta} - \partial_t \boldsymbol{\eta}_\delta$ is not a continuously differentiable function in time as is required for the test functions, and hence, we will use a *convolution in time* and pass to the limit as the convolution parameter goes to zero. Second, the fluid velocities give an additional difficulty, as the fluid velocities are defined on *time-dependent moving domains*. Thus, we must transfer the fluid velocities between different time-dependent domains in order to make sense of the "difference" between $\boldsymbol{u}$ and $\boldsymbol{u}_\delta$ as a test function. Furthermore, the way in which we do this transformation and the way in which we perform the convolution in time must both respect the divergence-free nature of the fluid velocity on the time-dependent domain. We will address both of these difficulties as follows.

Construction of Appropriate Test Functions $(\boldsymbol{u}, \partial_t \boldsymbol{\eta}, p, \partial_t \omega) - (\boldsymbol{u}_\delta, \partial_t \boldsymbol{\eta}_\delta, p_\delta, \partial_t \omega_\delta)$:

Difficulty 1: Lack of Regularity in Time We address the first difficulty by defining a convolution in time. This will allow us to regularize $\partial_t(\boldsymbol{\eta}-\boldsymbol{\eta}_\delta) = \boldsymbol{\xi}-\boldsymbol{\xi}_\delta$, $p-p_\delta$, and $\partial_t(\omega-\omega_\delta) = \zeta-\zeta_\delta$ so that these functions are continuously differentiable in time. Since the classical solution is already continuously differentiable in time, we only need to regularize the weak solutions to the regularized problem. Because these differences are all defined on fixed domains, we can use a standard convolution in time.

Convolution in Time Let $j(\cdot):\mathbb{R}\to\mathbb{R}$ be a compactly supported even function with $\text{supp}(j)\subset[-1,1]$ and $\int_{\mathbb{R}} j = 1$, and we define $j_\nu(t) = \nu^{-1}j(\nu^{-1}t)$, where $\nu>0$ is the convolution parameter in time.

Consider $\nu>0$. Extend $\boldsymbol{\xi}_\delta$, p_δ, and ζ_δ to the larger interval $[-\nu, T+\nu]$ by reflecting across $t=0$ and $t=T$. For example, define:

$$\boldsymbol{\xi}_\delta(t) = \boldsymbol{\xi}_\delta(-t), \quad \text{for } t\in[-\nu,0],$$
$$\boldsymbol{\xi}_\delta(t) = \boldsymbol{\xi}_\delta(2T-t), \quad \text{for } t\in[T,T+\nu].$$

Convolution in time is then defined by:

$$(\boldsymbol{\xi}_\delta)_\nu(t) = \boldsymbol{\xi}_\delta(t,\cdot) * j_\nu = \int_{\mathbb{R}} \boldsymbol{\xi}_\delta(s) j_\nu(t-s)ds, \quad \text{for } t\in[0,T].$$

The convolutions $(p_\delta)_\nu$ and $(\zeta_\delta)_\nu$ are defined similarly. With these definitions, we can now test with $\boldsymbol{\xi}-(\boldsymbol{\xi}_\delta)_\nu$, $p-(p_\delta)_\nu$, and $\zeta-(\zeta_\delta)_\nu$.

Difficulty 2: Velocities Are Defined on Moving Domains Because the fluid velocities are defined on moving time-dependent domains, we cannot directly apply a convolution in time. We must first be able to transform fluid velocities from one domain to another while preserving the divergence-free condition and then convolve in time. The transformation of fluid velocities from one domain to another while preserving the divergence-free condition will be performed using the following matrix, which was inspired by the transformation introduced in [194] (see also [298]):

$$K(s,t,x,y) = \begin{pmatrix} \frac{R+\omega(s,x)}{R+\omega(t,x)} & 0 \\ -(R+y)\partial_x\left(\frac{R+\omega(s,x)}{R+\omega(t,x)}\right) & 1 \end{pmatrix}. \tag{11.122}$$

This matrix has the following essential property: if $\boldsymbol{u}(x,y)$ is a divergence-free function on the domain $\Omega_f(s)$ defined by the structure displacement $\omega(s,x)$, then the function

$$K(s,t,x,y)\boldsymbol{u}\left(x, \frac{R+\omega(s,x)}{R+\omega(t,x)}(R+y)-R\right)$$

is a divergence-free vector field on the domain $\Omega_f(t)$ defined by the structure displacement $\omega(t,x)$.

Combined Transformation of Fluid Velocities and Convolution in Time We can now use this transformation to convolve in time, as follows. We extend $\boldsymbol{u}_\delta$ to $[-\nu, T+\nu]$ by reflection, as above, and define, for $t \in [0,T]$,

$$(\boldsymbol{u}_\delta)_\nu(t) = \int_{\mathbb{R}} K_{2,\delta}(s,t,x,y)\boldsymbol{u}_\delta\left(s, x, \frac{R+\omega_\delta(s,x)}{R+\omega_\delta(t,x)}(R+y) - R\right) j_\nu(t-s)ds. \tag{11.123}$$

For a divergence-free function $\boldsymbol{v}$, extended as above in time to $[-\nu, T+\nu]$, we can define $\boldsymbol{v}_\nu$ on $\Omega_f(t)$ analogously by

$$\boldsymbol{v}_\nu(t) = \int_{\mathbb{R}} K_1(s,t,x,y)\boldsymbol{v}\left(s, x, \frac{R+\omega(s,x)}{R+\omega(t,x)}(R+y) - R\right) j_\nu(t-s)ds.$$

Here, $K_1(s,t,x,y)$ and $K_{2,\delta}(s,t,x,y)$ are defined as $K(s,t,x,y)$ with the choices of $\omega = \omega$ and $\omega = \omega_\delta$, respectively. An example of such a function $\boldsymbol{v}$, which will be convenient to consider on $\Omega_f(t)$, is the function $\breve{\boldsymbol{u}}_\delta$ defined on $\Omega_f(t)$, which is the function $\boldsymbol{u}_\delta$ defined on $\Omega_{f,\delta}(t)$ transferred in a divergence-free manner, as described above, onto the domain $\Omega_f(t)$. Specifically,

$$\breve{\boldsymbol{u}}_\delta(t,x,y) = \begin{pmatrix} \frac{R+\omega_\delta(t,x)}{R+\omega(t,x)} & 0 \\ -(R+y)\partial_x\left(\frac{R+\omega_\delta(t,x)}{R+\omega(t,x)}\right) & 1 \end{pmatrix} \cdot \boldsymbol{u}_\delta\left(t, x, \frac{R+\omega_\delta(t,x)}{R+\omega(t,x)}(R+y) - R\right).$$

We present the main properties of $(\boldsymbol{u}_\delta)_\nu$ in the proposition below, which are a specific case of Lemma 2.6 in [298].

Proposition 11.9.1 *Fix an arbitrary $\delta > 0$. Given $\boldsymbol{u}_\delta \in L^2(0,T; H^1(\Omega_b(t))$ and $\omega, \omega_\delta \in H_0^2(\Gamma)$, the following properties hold:*

- *Divergence-free condition: div $[(\boldsymbol{u}_\delta)_\nu] = 0$ and div$[(\breve{\boldsymbol{u}}_\delta)_\nu] = 0$, $\forall \nu > 0$, and $\forall t \in [0,T]$.*
- *Convergence properties:*

$$\begin{aligned}
&(\boldsymbol{u}_\delta)_\nu \to \boldsymbol{u}_\delta \quad \textit{strongly in } L^p(0,T; L^q(\Omega_{f,\delta}(t))), \quad \textit{for all } p \in [1,\infty), q \in [1,2),\\
&(\breve{\boldsymbol{u}}_\delta)_\nu \to \breve{\boldsymbol{u}}_\delta \quad \textit{strongly in } L^p(0,T; L^q(\Omega_{f,1}(t))), \quad \textit{for all } p \in [1,\infty), q \in [1,2),\\
&(\boldsymbol{u}_\delta)_\nu \rightharpoonup \boldsymbol{u}_\delta \quad \textit{weakly in } L^2(0,T; W^{1,p}(\Omega_{f,\delta}(t))), \quad \textit{for all } p \in [1,2),\\
&(\breve{\boldsymbol{u}}_\delta)_\nu \rightharpoonup \breve{\boldsymbol{u}}_\delta \quad \textit{weakly in } L^2(0,T; W^{1,p}(\Omega_{f,1}(t))), \quad \textit{for all } p \in [1,2).
\end{aligned}$$

Proof (Proof of Gronwall's Estimate) We begin by testing the weak formulation (11.107) for the classical solution $(\boldsymbol{u}, \boldsymbol{\eta}, p, \omega)$ to the original non-regularized problem with

$$\boldsymbol{v} = \boldsymbol{u} - (\check{\boldsymbol{u}}_\delta)_\nu, \quad \varphi = \zeta - (\zeta_\delta)_\nu, \quad \boldsymbol{\psi} = \boldsymbol{\xi} - (\boldsymbol{\xi}_\delta)_\nu, \quad r = p - (p_\delta)_\nu, \tag{11.124}$$

and then test the regularized weak formulations (11.108) for the weak solutions $(\boldsymbol{u}_\delta, \boldsymbol{\eta}_\delta, p_\delta, \omega_\delta)$ with

$$\boldsymbol{v} = \widehat{\boldsymbol{u}}, \quad \varphi = \zeta, \quad \boldsymbol{\psi} = \boldsymbol{\xi}, \quad r = p. \tag{11.125}$$

Next, we rewrite the energy estimate in Proposition 11.8.3, which holds for the function $\boldsymbol{u}_\delta$, in a more convenient form by adding extra terms that will cancel out, in order to have the energy inequality parallel the weak formulation term by term. In particular, we have that for almost every $t \in [0, T_\delta]$,

$$\begin{aligned}
\frac{1}{2}\int_{\Omega_{f,\delta}(t)} |\boldsymbol{u}_\delta|^2 + \frac{1}{2}\int_0^t \int_{\Omega_f(s)} [((\boldsymbol{u}_\delta \cdot \nabla)\boldsymbol{u}_\delta) \cdot \boldsymbol{u}_\delta - ((\boldsymbol{u}_\delta \cdot \nabla)\boldsymbol{u}_\delta]& \\
+ \frac{1}{2}\int_0^t \int_{\Gamma_\delta(s)} (\boldsymbol{u}_\delta \cdot \boldsymbol{n} - 2\boldsymbol{\xi}_\delta \cdot \boldsymbol{n})\boldsymbol{u}_\delta \cdot \boldsymbol{u}_\delta& \\
+ 2\nu \int_0^t \int_{\Omega_{f,\delta}(s)} |\boldsymbol{D}(\boldsymbol{u}_\delta)|^2 + \int_0^t \int_{\Gamma_\delta(s)} \left(\frac{1}{2}|\boldsymbol{u}_\delta|^2 - p_\delta\right)(\boldsymbol{\xi}_\delta - \boldsymbol{u}_\delta) \cdot \boldsymbol{n}& \\
+ \beta \int_0^t \int_{\Gamma_\delta(s)} |(\boldsymbol{\xi}_\delta - \boldsymbol{u}_\delta) \cdot \boldsymbol{t})|^2& \\
+ \frac{1}{2}\rho_p \int_\Gamma |\boldsymbol{\xi}_\delta|^2 + \frac{1}{2}\int_\Gamma |\Delta\omega_\delta|^2 + \frac{1}{2}\rho_b \int_{\Omega_b} |\boldsymbol{\xi}_\delta|^2 + \mu_e \int_0^t \int_{\Omega_b} |\boldsymbol{D}(\boldsymbol{\eta}_\delta)(s)|^2& \\
+ \frac{1}{2}\lambda_e \int_{\Omega_b} |\nabla \cdot \boldsymbol{\eta}_\delta(s)|^2 + 2\mu_v \int_0^t \int_{\Omega_b} |\boldsymbol{D}(\boldsymbol{\xi}_\delta)|^2 + \lambda_v \int_0^t \int_{\Omega_b} |\nabla \cdot \boldsymbol{\xi}_\delta|^2& \\
- \alpha \int_0^t \int_{\Omega_{b,\delta}^\delta(s)} p_\delta \nabla \cdot \boldsymbol{\xi}_\delta + \frac{1}{2}c_0 \int_0^t \int_{\Omega_b} |p_\delta(s)|^2 - \alpha \int_0^t \int_{\Omega_{b,\delta}^\delta(s)} \frac{D^\delta}{Dt}\boldsymbol{\eta}_\delta \cdot \nabla p_\delta& \\
- \alpha \int_0^t \int_{\Gamma_\delta^\delta(s)} (\boldsymbol{\xi}_\delta \cdot \boldsymbol{n}) p_\delta + \kappa \int_0^t \int_{\Omega_{b,\delta}^\delta(s)} |\nabla p_\delta|^2 - \int_0^t \int_{\Gamma_\delta(s)} ((\boldsymbol{u}_\delta - \boldsymbol{\xi}_\delta) \cdot \boldsymbol{n}) p_\delta& \\
\leq \frac{1}{2}\int_{\Omega_f(0)} |\boldsymbol{u}_0|^2 + \frac{1}{2}\rho_p \int_\Gamma |\boldsymbol{\xi}_0|^2 + \frac{1}{2}\int_\Gamma |\Delta\omega_0|^2 + \frac{1}{2}\rho_b \int_{\Omega_b} |\boldsymbol{\xi}_0|^2& \\
+ \mu_e \int_{\Omega_b} |\boldsymbol{D}(\boldsymbol{\eta}_0)|^2 + \frac{1}{2}\lambda_e \int_{\Omega_b} |\nabla \cdot \boldsymbol{\eta}_0|^2 + \frac{1}{2}c_0 \int_{\Omega_b} |p_0|^2.&
\end{aligned} \tag{11.126}$$

Finally, we combine the weak formulation for $\boldsymbol{u}$ tested with (11.124), subtract the regularized weak formulation for $\boldsymbol{u}_\delta$ tested with (11.125), and add the energy estimate (11.126) for $\boldsymbol{u}_\delta$ to obtain an expression of the form

$$\sum_{i=1}^{18} T_i \leq 0, \tag{11.127}$$

where the terms T_i are given below. We have to estimate each term, and the combined estimate will give the Gronwall's inequality (11.121). To make this section more concise, we summarize the final estimates here and present details of the derivation of these terms and the estimates in Appendix 11.10.2.

As a notational note, in many of the estimates on the terms T_i that follow, we will use Cauchy's inequality with ϵ often: $|ab| \leq \epsilon|a|^2 + C(\epsilon)|b|^2$, where $\epsilon > 0$ is any parameter and $C(\epsilon)$ is a constant that depends on the final choice of $\epsilon > 0$. In the inequalities that appear, $\epsilon > 0$ will hence be a parameter appearing on dissipative terms that will, at the conclusion of the estimates, be chosen small enough so that the dissipative terms from the estimates on T_i with ϵ can be absorbed by the dissipative terms in (11.115) to give the final inequality (11.116).

Term T1. Term T_1 is defined as follows:

$$
\begin{aligned}
T_1 = & -\int_0^t \int_{\Omega_f(s)} \boldsymbol{u} \cdot \partial_t \left[\boldsymbol{u} - (\breve{\boldsymbol{u}}_\delta)_\nu\right] - \frac{1}{2}\int_0^t \int_{\Gamma(s)} (\boldsymbol{\xi} \cdot \boldsymbol{n})\boldsymbol{u} \cdot [\boldsymbol{u} - (\breve{\boldsymbol{u}}_\delta)_\nu] \\
& + \int_{\Omega_f(t)} \boldsymbol{u}(t) \cdot [\boldsymbol{u} - (\breve{\boldsymbol{u}}_\delta)_\nu](t) \qquad (11.128) \\
& - \int_{\Omega_f(0)} \boldsymbol{u}(0) \cdot [\boldsymbol{u} - (\breve{\boldsymbol{u}}_\delta)_\nu](0) - \int_0^t \int_{\Omega_{f,\delta}(s)} \boldsymbol{u}_\delta \cdot \partial_t \widehat{\boldsymbol{u}} \\
& - \frac{1}{2}\int_0^t \int_{\Gamma_\delta(s)} (\boldsymbol{\xi}_\delta \cdot \boldsymbol{n}_\delta)\boldsymbol{u}_\delta \cdot \widehat{\boldsymbol{u}} \\
& + \int_{\Omega_{f,\delta}(t)} \boldsymbol{u}_\delta(t) \cdot \widehat{\boldsymbol{u}}(t) - \int_{\Omega_f(0)} \boldsymbol{u}_\delta(0) \cdot \widehat{\boldsymbol{u}}(0) + \frac{1}{2}\int_{\Omega_{f,\delta}(t)} |\boldsymbol{u}_\delta(t)|^2 \\
& - \frac{1}{2}\int_{\Omega_{f,\delta}(0)} |\boldsymbol{u}_0|^2. \qquad (11.129)
\end{aligned}
$$

This term is estimated so that after taking the limit as $\nu \to 0$, the contribution of this term becomes

$$
T_1 = \frac{1}{2}\int_{\Omega_{f,\delta}(t)} |(\widehat{\boldsymbol{u}} - \boldsymbol{u}_\delta)(t)|^2 + R_1,
$$

where

$$
\begin{aligned}
|R_1| \leq & \epsilon \int_0^t \|\widehat{\boldsymbol{u}} - \boldsymbol{u}_\delta\|^2_{H^1(\Omega_{f,\delta}(s))} \\
& + C(\epsilon)\left(\int_0^t \|\omega - \omega_\delta\|^2_{H^2(\Gamma)} + \int_0^t \|\partial_t\omega - \partial_t\omega_\delta\|^2_{L^2(\Gamma)}\right. \\
& \left. + \int_0^t \|\widehat{\boldsymbol{u}} - \boldsymbol{u}_\delta\|^2_{L^2(\Omega_{f,\delta}(s))}\right).
\end{aligned}
$$

Term T2. Term T_2 is defined as follows:

$$\begin{aligned}T_2 =&\frac{1}{2}\int_0^t\int_{\Omega_f(s)}((\boldsymbol{u}\cdot\nabla)\boldsymbol{u})\cdot[\boldsymbol{u}-(\breve{\boldsymbol{u}}_\delta)_\nu]-\frac{1}{2}\int_0^t\int_{\Omega_f(s)}(\boldsymbol{u}\cdot\nabla)[\boldsymbol{u}-(\breve{\boldsymbol{u}}_\delta)_\nu]\cdot\boldsymbol{u}\\&-\frac{1}{2}\int_0^t\int_{\Omega_{f,\delta}(s)}((\boldsymbol{u}_\delta\cdot\nabla)\boldsymbol{u}_\delta)\cdot(\widehat{\boldsymbol{u}}-\boldsymbol{u}_\delta)\\&+\frac{1}{2}\int_0^t\int_{\Omega_{f,\delta}(s)}((\boldsymbol{u}_\delta\cdot\nabla)(\widehat{\boldsymbol{u}}-\boldsymbol{u}_\delta))\cdot\boldsymbol{u}_\delta.\end{aligned}\tag{11.130}$$

After taking the limit $\nu\to 0$, term T_2 can be estimated as follows:

$$\begin{aligned}|T_2|\le\epsilon\int_0^t\|\nabla(\widehat{\boldsymbol{u}}-\boldsymbol{u}_\delta)\|^2_{L^2(\Omega_{f,\delta}(s))}+C(\epsilon)\left(\int_0^t\|\omega-\omega_\delta\|^2_{H^2(\Gamma)}\right.\\\left.+\int_0^t\|\widehat{\boldsymbol{u}}-\boldsymbol{u}_\delta\|^2_{L^2(\Omega_{f,\delta}(s))}\right).\end{aligned}$$

Term T3. Term T_3 is defined as follows:

$$\begin{aligned}T_3 =&\frac{1}{2}\int_0^t\int_{\Gamma(s)}(\boldsymbol{u}\cdot\boldsymbol{n}-\boldsymbol{\xi}\cdot\boldsymbol{n})\boldsymbol{u}\cdot[\boldsymbol{u}-(\breve{\boldsymbol{u}}_\delta)_\nu]\\&-\frac{1}{2}\int_0^t\int_{\Gamma_\delta(s)}(\boldsymbol{u}_\delta\cdot\boldsymbol{n}_\delta-\boldsymbol{\xi}_\delta\cdot\boldsymbol{n}_\delta)\boldsymbol{u}_\delta\cdot\widehat{\boldsymbol{u}}\\&+\frac{1}{2}\int_0^t\int_{\Gamma(s)}|\boldsymbol{u}|^2(\boldsymbol{\xi}\cdot\boldsymbol{n}-\boldsymbol{u}\cdot\boldsymbol{n})-\frac{1}{2}\int_0^t\int_{\Gamma(s)}|\boldsymbol{u}|^2[(\boldsymbol{\xi}_\delta)_\nu\cdot\boldsymbol{n}-(\breve{\boldsymbol{u}}_\delta)_\nu\cdot\boldsymbol{n}]\\&-\frac{1}{2}\int_0^t\int_{\Gamma_\delta(s)}|\boldsymbol{u}_\delta|^2(\boldsymbol{\xi}\cdot\boldsymbol{n}_\delta-\widehat{\boldsymbol{u}}\cdot\boldsymbol{n}_\delta).\end{aligned}\tag{11.131}$$

After taking the limit $\nu\to 0$, term $T3$ can be estimated as follows:

$$\begin{aligned}|T_3|\le&\epsilon\int_0^t\|\widehat{\boldsymbol{u}}-\boldsymbol{u}_\delta\|^2_{H^1(\Omega_{f,\delta}(s))}\\&+C(\epsilon)\left(\int_0^t\|\omega-\omega_\delta\|^2_{H^2(\Gamma)}+\int_0^t\|\boldsymbol{\xi}-\boldsymbol{\xi}_\delta\|^2_{L^2(\Gamma)}\right.\\&\left.+\int_0^t\|\widehat{\boldsymbol{u}}-\boldsymbol{u}_\delta\|^2_{L^2(\Omega_{f,\delta}(s))}\right).\end{aligned}$$

Term T4. Term T_4 is defined as follows:

$$T_4 = 2\nu \int_0^t \int_{\Omega_f(s)} \boldsymbol{D}(\boldsymbol{u}) : \boldsymbol{D}(\boldsymbol{u} - (\check{\boldsymbol{u}}_\delta)_\nu) - 2\nu \int_0^t \int_{\Omega_{f,\delta}(s)} \boldsymbol{D}(\boldsymbol{u}_\delta) : \boldsymbol{D}(\widehat{\boldsymbol{u}} - \boldsymbol{u}_\delta). \tag{11.132}$$

After taking the limit $\nu \to 0$, term $T4$ can be estimated as follows:

$$T_4 = 2\nu \int_0^t \int_{\Omega_{f,\delta}(s)} |\boldsymbol{D}(\widehat{\boldsymbol{u}} - \boldsymbol{u}_\delta)|^2 + R_4,$$

where

$$|R_4| \leq \epsilon \int_0^t \|\boldsymbol{D}(\widehat{\boldsymbol{u}} - \boldsymbol{u}_\delta)\|^2_{L^2(\Omega_{f,\delta}(s))} + C(\epsilon) \left(\int_0^t \|\omega - \omega_\delta\|^2_{H^2(\Gamma)} + \int_0^t \|\widehat{\boldsymbol{u}} - \boldsymbol{u}_\delta\|^2_{L^2(\Omega_{f,\delta}(s))} \right).$$

Term T5. Term T_5 is defined as follows:

$$T_5 = \beta \int_0^t \int_{\Gamma(s)} (\boldsymbol{\xi} - \boldsymbol{u}) \cdot \boldsymbol{\tau}(s)[(\boldsymbol{\xi} - (\boldsymbol{\xi}_\delta)_\nu) \cdot \boldsymbol{\tau}(s) - (\boldsymbol{u} - (\check{\boldsymbol{u}}_\delta)_\nu) \cdot \boldsymbol{\tau}(s)] - \beta \int_0^t \int_{\Gamma_\delta(s)} (\boldsymbol{\xi}_\delta - \boldsymbol{u}_\delta) \cdot \boldsymbol{\tau}_\delta(s)[(\boldsymbol{\xi} - \boldsymbol{\xi}_\delta) \cdot \boldsymbol{\tau}_\delta(s) - (\widehat{\boldsymbol{u}} - \boldsymbol{u}_\delta) \cdot \boldsymbol{\tau}_\delta(s)],$$

where $\boldsymbol{\tau}(s)$ is the unit tangent vector to $\Gamma(s)$ and $\boldsymbol{\tau}_\delta(s)$ is the unit tangent vector to $\Gamma_\delta(s)$.
After taking the limit $\nu \to 0$, term T_5 can be estimated as follows:

$$T_5 = \beta \int_0^t \int_{\Gamma_\delta(s)} |(\boldsymbol{\xi} - \boldsymbol{\xi}_\delta) \cdot \boldsymbol{\tau}_\delta(s) - (\widehat{\boldsymbol{u}} - \boldsymbol{u}_\delta) \cdot \boldsymbol{\tau}_\delta(s)|^2 + R_5,$$

where

$$|R_5| \leq \epsilon \int_0^t \|\boldsymbol{D}(\widehat{\boldsymbol{u}} - \boldsymbol{u}_\delta)\|_{L^2(\Omega_{f,\delta}(s))} + C(\epsilon) \left(\int_0^t \|\omega - \omega_\delta\|^2_{H^2(\Gamma)} + \int_0^t \|\boldsymbol{\xi} - \boldsymbol{\xi}_\delta\|^2_{L^2(\Gamma)} \right).$$

Terms T6–T8. Terms T_6–T_8 are defined as follows:

$$
\begin{aligned}
T_6 =& -\rho_p \int_0^t \int_\Gamma \zeta \cdot \partial_t [\zeta - (\zeta_\delta)_\nu] + \rho_p \int_\Gamma \zeta(s) \cdot [\zeta(t) - (\zeta_\delta)_\nu(t)] \\
& - \rho_p \int_\Gamma \zeta(0) \cdot [\zeta(0) - (\zeta_\delta)_\nu(0)] \\
& + \rho_p \int_0^t \int_\Gamma \zeta_\delta \cdot \partial_t \zeta - \rho_p \int_\Gamma \zeta_\delta(t) \cdot \zeta(t) + \rho_p \int_\Gamma \zeta_\delta(0) \cdot \zeta(0) \\
& + \frac{1}{2}\rho_p \int_\Gamma |\zeta_\delta(t)|^2 - \frac{1}{2}\rho_p \int_\Gamma |\zeta_0|^2. \qquad (11.133)
\end{aligned}
$$

$$
\begin{aligned}
T_7 =& \int_0^t \int_\Gamma \Delta\omega \cdot \Delta [\zeta - (\zeta_\delta)_\nu] - \int_0^t \int_\Gamma \Delta\omega_\delta \cdot \Delta\zeta + \frac{1}{2}\int_\Gamma |\Delta\omega_\delta(t)|^2 \\
& - \frac{1}{2}\int_\Gamma |\Delta\omega_0|^2. \qquad (11.134)
\end{aligned}
$$

$$
\begin{aligned}
T_8 =& -\rho_b \int_0^t \int_{\Omega_b} \partial_t \boldsymbol{\eta} \cdot \partial_t \left[\boldsymbol{\xi} - (\boldsymbol{\xi}_\delta)_\nu\right] + \rho_b \int_{\Omega_b} \boldsymbol{\xi}(t) \cdot \left[\boldsymbol{\xi}(s) - (\boldsymbol{\xi}_\delta)_\nu(s)\right] \\
& - \rho_b \int_{\Omega_b} \boldsymbol{\xi}(0) \cdot \left[\boldsymbol{\xi}(0) - (\boldsymbol{\xi}_\delta)_\nu(0)\right] \\
& + \rho_b \int_0^t \int_{\Omega_b} \partial_t \boldsymbol{\eta}_\delta \cdot \partial_t \boldsymbol{\xi} - \rho_b \int_{\Omega_b} \boldsymbol{\xi}_\delta(t) \cdot \boldsymbol{\xi}(t) + \rho_b \int_{\Omega_b} \boldsymbol{\xi}_\delta(0) \cdot \boldsymbol{\xi}(0) \\
& + \frac{1}{2}\rho_b \int_{\Omega_b} |\boldsymbol{\xi}_\delta(t)|^2 - \frac{1}{2}\rho_b \int_{\Omega_b} |\boldsymbol{\xi}_0|^2. \qquad (11.135)
\end{aligned}
$$

After taking the limit $\nu \to 0$, the terms T_6–T_8 become:

$$
\begin{aligned}
T_6 &= \frac{1}{2}\rho_p \int_\Gamma |(\zeta - \zeta_\delta)(t)|^2, \quad T_7 = \frac{1}{2}\int_\Gamma |\Delta(\omega - \omega_\delta)(t)|^2, \\
T_8 &= \frac{1}{2}\rho_b \int_{\Omega_b} |(\boldsymbol{\xi} - \boldsymbol{\xi}_\delta)(t)|^2.
\end{aligned}
$$

Terms T9–T12. Terms T_9–T_{12} are defined as follows:

$$
\begin{aligned}
T_9 =& 2\mu_e \int_0^t \int_{\Omega_b} \boldsymbol{D}(\boldsymbol{\eta}) : \boldsymbol{D}\left[\boldsymbol{\xi} - (\boldsymbol{\xi}_\delta)_\nu\right] - 2\mu_e \int_0^t \int_{\Omega_b} \boldsymbol{D}(\boldsymbol{\eta}_\delta) : \boldsymbol{D}(\boldsymbol{\xi}) \\
& + \mu_e \int_{\Omega_b} |\boldsymbol{D}(\boldsymbol{\eta}_\delta)(t)|^2 - \mu_e \int_{\Omega_b} |\boldsymbol{D}(\boldsymbol{\eta}_0)|^2.
\end{aligned}
$$

$$
T_{10} = \lambda_e \int_0^t \int_{\Omega_b} (\nabla \cdot \boldsymbol{\eta}) \left(\nabla \cdot \left[\boldsymbol{\xi} - (\boldsymbol{\xi}_\delta)_\nu\right]\right) - \lambda_e \int_0^t \int_{\Omega_b} (\nabla \cdot \boldsymbol{\eta}_\delta)(\nabla \cdot \boldsymbol{\xi})
$$

$$
\begin{aligned}
&+\frac{1}{2}\lambda_e \int_{\Omega_b} |\nabla\cdot\boldsymbol{\eta}_\delta(t)|^2 - \frac{1}{2}\lambda_e \int_{\Omega_b} |\nabla\cdot\boldsymbol{\eta}_0|^2.\\
T_{11} =&2\mu_v \int_0^t \int_{\Omega_b} \boldsymbol{D}(\boldsymbol{\xi}) : \boldsymbol{D}\left[\boldsymbol{\xi} - (\boldsymbol{\xi}_\delta)_\nu\right] - 2\mu_v \int_0^t \int_{\Omega_b} \boldsymbol{D}(\boldsymbol{\xi}_\delta) : \boldsymbol{D}(\boldsymbol{\xi})\\
&+ 2\mu_v \int_0^t \int_{\Omega_b} |\boldsymbol{D}(\boldsymbol{\xi}_\delta)|^2.\\
T_{12} =&\lambda_v \int_0^t \int_{\Omega_b} (\nabla\cdot\boldsymbol{\xi})\left(\nabla\cdot\left[\boldsymbol{\xi} - (\boldsymbol{\xi}_\delta)_\nu\right]\right) - \lambda_v \int_0^t \int_{\Omega_b} (\nabla\cdot\boldsymbol{\xi}_\delta)(\nabla\cdot\boldsymbol{\xi})\\
&+ \lambda_v \int_0^t \int_{\Omega_b} |\nabla\cdot\boldsymbol{\xi}_\delta|^2.
\end{aligned}
\tag{11.136}
$$

Because $\boldsymbol{\xi}_\delta \in L^2(0,T;H^1(\Omega_b))$ where Ω_b is a fixed domain, we have that $(\boldsymbol{\xi}_\delta)_\nu \to \boldsymbol{\xi}_\delta$ strongly in $L^2(0,T;H^1(\Omega_b))$. Hence, as $\nu \to 0$, we have that Terms 9–12 converge to the following:

$$
T_9 = \mu_e \int_{\Omega_b} |\boldsymbol{D}(\boldsymbol{\eta} - \boldsymbol{\eta}_\delta)(t)|^2, \qquad T_{10} = \frac{1}{2}\lambda_e \int_{\Omega_b} |\nabla\cdot(\boldsymbol{\eta} - \boldsymbol{\eta}_\delta)(t)|^2,
$$

$$
T_{11} = 2\mu_v \int_0^t \int_{\Omega_b} |\boldsymbol{D}(\boldsymbol{\xi} - \boldsymbol{\xi}_\delta)|^2, \qquad T_{12} = \lambda_v \int_0^t \int_{\Omega_b} |\nabla\cdot(\boldsymbol{\xi} - \boldsymbol{\xi}_\delta)|^2.
$$

Term T13. Term T_{13} is defined as follows:

$$
T_{13} = -\alpha \int_0^t \int_{\Omega_b(s)} p\left(\nabla\cdot\left[\boldsymbol{\xi} - (\boldsymbol{\xi}_\delta)_\nu\right]\right) + \alpha \int_0^t \int_{\Omega^\delta_{b,\delta}(s)} p_\delta\left(\nabla\cdot(\boldsymbol{\xi} - \boldsymbol{\xi}_\delta)\right).
$$

After taking the limit $\nu \to 0$, term T_{13} can be estimated as follows:

$$
\begin{aligned}
|T_{13}| \le& C(\epsilon) \int_0^t \|\nabla\boldsymbol{\eta} - \nabla\boldsymbol{\eta}^\delta\|^2_{L^2(\Omega_b)} + \epsilon \int_0^t \|\nabla(\boldsymbol{\xi} - \boldsymbol{\xi}_\delta)\|^2_{L^2(\Omega_b)}\\
&+ C(\epsilon)\left(\int_0^t \|\nabla\boldsymbol{\eta} - \nabla\boldsymbol{\eta}_\delta\|^2_{L^2(\Omega_b)}\right.\\
&\left.+ \int_0^t \|\omega - \omega_\delta\|^2_{H^2(\Gamma)} + \int_0^t \|p - p_\delta\|^2_{L^2(\Omega_b)}\right).
\end{aligned}
$$

Term T14. Term T_{14} is defined as follows:

$$
T_{14} = -c_0 \int_0^t \int_{\Omega_b} p\cdot\partial_t\left[p - (p_\delta)_\nu\right] + c_0 \int_{\Omega_b} p(t)\cdot\left[p(t) - (p_\delta)_\nu(t)\right]
$$

$$
\begin{aligned}
&- c_0 \int_{\Omega_b} p_0 \cdot [p(0) - (p_\delta)_\nu(0)] \\
&+ c_0 \int_0^t \int_{\Omega_b} p_\delta \cdot \partial_t p - c_0 \int_{\Omega_b} p_\delta(t) \cdot p(t) + c_0 \int_{\Omega_b} |p_0|^2 \\
&+ \frac{1}{2} c_0 \int_{\Omega_b} |p_\delta(t)|^2 - \frac{1}{2} c_0 \int_{\Omega_b} |p_0|^2. \qquad (11.137)
\end{aligned}
$$

This term can be handled in the same way as Terms 6–8. In the limit as $\nu \to 0$, the contribution from this term is

$$
T_{14} = \frac{1}{2} c_0 \int_{\Omega_b} |(p - p_\delta)(t)|^2.
$$

Term T15. Term T_{15} is defined as follows:

$$
T_{15} = -\alpha \int_0^t \int_{\Omega_b(s)} \boldsymbol{\xi} \cdot \nabla [p - (p_\delta)_\nu] + \alpha \int_0^t \int_{\Omega^\delta_{b,\delta}(s)} \boldsymbol{\xi}_\delta \cdot \nabla (p - p_\delta). \qquad (11.138)
$$

After taking the limit $\nu \to 0$, term T_{15} can be estimated as follows:

$$
\begin{aligned}
|T_{15}| \leq & \epsilon \int_0^t \|\nabla(p - p_\delta)\|^2_{L^2(\Omega^\delta_{b,\delta}(s))} + C(\epsilon) \int_0^t \|\nabla \boldsymbol{\eta} - \nabla \boldsymbol{\eta}^\delta\|^2_{L^2(\Omega_b)} \\
& + C(\epsilon) \left(\int_0^t \|\nabla \boldsymbol{\eta} - \nabla \boldsymbol{\eta}_\delta\|^2_{L^2(\Omega_b)} + \int_0^t \|\omega - \omega_\delta\|^2_{H^2(\Gamma)} \right. \\
& \left. + \int_0^t \|\partial_t \boldsymbol{\eta} - \partial_t \boldsymbol{\eta}_\delta\|^2_{L^2(\Omega_b)} \right).
\end{aligned}
$$

Term T16. Term T_{16} is defined as follows:

$$
T_{16} = -\alpha \int_0^t \int_{\Gamma(s)} (\boldsymbol{\xi} \cdot \boldsymbol{n}) [p - (p_\delta)_\nu] + \alpha \int_0^t \int_{\Gamma^\delta_\delta(s)} (\boldsymbol{\xi}_\delta \cdot \boldsymbol{n})(p - p_\delta). \qquad (11.139)
$$

After passing to the limit as $\nu \to 0$, this term can be estimated as follows:

$$
\begin{aligned}
|T_{16}| \leq \epsilon & \left(\int_0^t \|\nabla \boldsymbol{\xi} - \nabla \boldsymbol{\xi}_\delta\|^2_{L^2(\Omega_b)} + \int_0^t \|\nabla p - \nabla p_\delta\|^2_{L^2(\Omega^\delta_{b,\delta}(s))} \right) \\
& + C(\epsilon) \left(\int_0^t \|p - p_\delta\|^2_{L^2(\Omega_b)} \right.
\end{aligned}
$$

$$+ \int_0^t \|\nabla \boldsymbol{\eta} - \nabla \boldsymbol{\eta}^\delta\|^2_{L^2(\Omega_b)} + \int_0^t \|\nabla \boldsymbol{\eta} - \nabla \boldsymbol{\eta}_\delta\|^2_{L^2(\Omega_b)} + \int_0^t \|\omega - \omega_\delta\|^2_{H^2(\Gamma)} + \int_0^t \|\boldsymbol{\xi} - \boldsymbol{\xi}_\delta\|^2_{L^2(\Omega_b)} \Bigg).$$

Term T17. Term T_{17} is defined as follows:

$$T_{17} = \kappa \int_0^t \int_{\Omega_b(s)} \nabla p \cdot \nabla [p - (p_\delta)_\nu] - \kappa \int_0^t \int_{\Omega^\delta_{b,\delta}(s)} \nabla p_\delta \cdot \nabla (p - p_\delta). \tag{11.140}$$

This term can be estimated as follows:

$$T_{17} \leq \kappa \int_0^t \int_{\Omega^\delta_{b,\delta}(s)} |\nabla (p - p_\delta)|^2 + R_{17},$$

where the remainder is bounded by

$$\begin{aligned} |R_{17}| \leq & \epsilon \int_0^t \|\nabla (p - p_\delta)\|^2_{L^2(\Omega^\delta_{b,\delta}(s))} \\ & + C(\epsilon) \Bigg(\int_0^t \|\nabla \boldsymbol{\eta} - \nabla \boldsymbol{\eta}^\delta\|^2_{L^2(\Omega_b)} + \int_0^t \|\nabla \boldsymbol{\eta} - \nabla \boldsymbol{\eta}_\delta\|^2_{L^2(\Omega_b)} \\ & + \int_0^t \|\omega - \omega_\delta\|^2_{H^2(\Gamma)} \Bigg). \end{aligned}$$

Term T18. Term T_{18} is defined as follows:

$$\begin{aligned} T_{18} = & \int_0^t \int_{\Gamma(s)} p(\boldsymbol{u} - \boldsymbol{\xi}) \cdot \boldsymbol{n} - \int_0^t \int_{\Gamma(s)} p[(\boldsymbol{u}_\delta)_\nu - (\boldsymbol{\xi}_\delta)_\nu] \cdot \boldsymbol{n} \\ & - \int_0^t \int_{\Gamma_\delta(s)} p_\delta (\boldsymbol{u} - \boldsymbol{\xi}) \cdot \boldsymbol{n} \\ & + \int_0^t \int_{\Gamma_\delta(s)} p_\delta (\boldsymbol{u}_\delta - \boldsymbol{\xi}_\delta) \cdot \boldsymbol{n} - \int_0^t \int_{\Gamma(s)} ((\boldsymbol{u} - \boldsymbol{\xi}) \cdot \boldsymbol{n})[p - (p_\delta)_\nu] \\ & + \int_0^t \int_{\Gamma_\delta(s)} ((\boldsymbol{u}_\delta - \boldsymbol{\xi}_\delta) \cdot \boldsymbol{n})(p - p_\delta). \end{aligned}$$

This term can be estimated as follows:

$$
\begin{aligned}
|T_{18}| \leq & \epsilon \left(\int_0^t \|\boldsymbol{D}(\widehat{\boldsymbol{u}} - \boldsymbol{u}_\delta)\|^2_{L^2(\Omega_{f,\delta}(s))} + \int_0^t \|\nabla(\boldsymbol{\xi} - \boldsymbol{\xi}_\delta)\|_{L^2(\Omega_b)} \right. \\
& \left. + \int_0^t \|\nabla(p - p_\delta)\|^2_{L^2(\Omega^\delta_{b,\delta}(s))} \right) \\
& + C(\epsilon) \int_0^t \|\omega - \omega_\delta\|^2_{H^2(\Gamma)}.
\end{aligned}
$$

The combined estimates for the terms T_1–T_{18} give the estimate:

$$
\begin{aligned}
E_\delta(t) \leq & \epsilon \left(\int_0^t \|\widehat{\boldsymbol{u}} - \boldsymbol{u}_\delta\|^2_{H^1(\Omega_{f,\delta}(s))} + \int_0^t \|\boldsymbol{D}(\widehat{\boldsymbol{u}} - \boldsymbol{u}_\delta)\|^2_{L^2(\Omega_{f,\delta}(s))} \right. \\
& \left. + \int_0^t \|\nabla(\boldsymbol{\xi} - \boldsymbol{\xi}_\delta)\|^2_{L^2(\Omega_b)} + \int_0^t \|\nabla(p - p_\delta)\|^2_{L^2(\Omega^\delta_{b,\delta}(s))} \right) \\
& + C(\epsilon) \left(\int_0^t \|\omega - \omega_\delta\|^2_{H^2(\Gamma)} + \int_0^t \|\boldsymbol{\xi} - \boldsymbol{\xi}_\delta\|^2_{L^2(\Gamma)} + \int_0^t \|\widehat{\boldsymbol{u}} - \boldsymbol{u}_\delta\|^2_{L^2(\Omega_{f,\delta}(s))} \right. \\
& + \int_0^t \|\nabla\boldsymbol{\eta} - \nabla\boldsymbol{\eta}^\delta\|^2_{L^2(\Omega_b)} \\
& \left. + \int_0^t \|\nabla(\boldsymbol{\eta} - \boldsymbol{\eta}_\delta)\|^2_{L^2(\Omega_b)} + \int_0^t \|p - p_\delta\|^2_{L^2(\Omega_b)} + \int_0^t \|\boldsymbol{\xi} - \boldsymbol{\xi}_\delta\|^2_{L^2(\Omega_b)} \right),
\end{aligned}
$$

where $\epsilon > 0$ remains to be chosen and $C(\epsilon)$ is a constant depending only on ϵ, that is, independent of t and δ. By using Korn's inequality and Poincaré's inequality and by choosing ϵ sufficiently small to absorb the terms on the right-hand side into the dissipation terms in $E_\delta(t)$ defined by (11.115), we then obtain the final inequality (11.121). This finishes the proof of the Gronwall estimate presented in Lemma 11.9.1. □

All that is left to show to complete the proof of weak-classical consistency stated in Theorem 11.9.1 is to argue that Gronwall's inequality (11.121) holds for all $t \in [0, T]$ where T is independent of δ. This will also imply the first statement in the theorem, which states that $(\boldsymbol{\eta}_\delta, \omega_\delta, p_\delta, \boldsymbol{u}_\delta)$ is uniformly defined on the time interval $[0, T]$ for all $\delta > 0$. In order to do this, we use a bootstrap argument presented in the next subsection.

11.9.4 Bootstrap Argument

To obtain the desired Gronwall estimate as stated in Lemma 11.9.1, we need the following uniform bounds (11.118), (11.119), and (11.120) on the factor $\det(\boldsymbol{I} +$

$\nabla\boldsymbol{\eta}_\delta^\delta$), which appears in the regularized weak formulation (11.43) defined on the fixed reference domain Ω_b:

$$\begin{aligned} \det(\boldsymbol{I} + \nabla\boldsymbol{\eta}_\delta^\delta) &\geq c > 0, \\ 0 < c \leq |\boldsymbol{I} + \nabla\boldsymbol{\eta}_\delta^\delta| &\leq C, \qquad \text{pointwise in } \overline{\Omega_b}, \\ |\nabla\boldsymbol{\eta}_\delta^\delta| &\leq C, \qquad \text{pointwise in } \overline{\Omega_b}, \end{aligned}$$

which need to hold *for all* $t \in [0, T]$ *where* $T > 0$ *is independent of* δ. Notice that we only have uniform boundedness of $\boldsymbol{\eta}_\delta$ with respect to δ in $L^\infty(0, T; H^1(\Omega_b))$, which implies that $\det(\boldsymbol{I} + \nabla\boldsymbol{\eta}_\delta^\delta)$ is uniformly bounded with respect to δ only in $L^\infty(0, T; L^1(\Omega_b))$, which is insufficient for estimating any integrands with this factor.

To get around this difficulty, we use the following strategy. Recall that by the way the weak solution to the regularized problem was constructed using the splitting scheme, we have that there exists a sufficiently small constant c (uniform in δ) such that

$$\det(\boldsymbol{I} + \nabla\boldsymbol{\eta}_\delta^\delta) \geq c > 0, \tag{11.141}$$

for all $t \in [0, T_\delta]$ where $T_\delta > 0$ *may depend on* δ. This estimate holds at least locally, although not locally uniformly, for each $\delta > 0$. In fact, similarly, the following three estimates, (11.118), (11.119), and (11.120) from Lemma 11.9.1, hold locally, for $t \in [0, T_\delta]$, where T_δ *may depend on* δ, with positive constants c and C that are independent of δ:

$$\begin{aligned} \det(\boldsymbol{I} + \nabla\boldsymbol{\eta}_\delta^\delta) &\geq c > 0, \\ 0 < c \leq |\boldsymbol{I} + \nabla\boldsymbol{\eta}_\delta^\delta| &\leq C, \qquad \text{pointwise in } \overline{\Omega_b}, \\ |\nabla\boldsymbol{\eta}_\delta^\delta| &\leq C, \qquad \text{pointwise in } \overline{\Omega_b}. \end{aligned}$$

These estimates imply that for sufficiently small $c > 0$, the following inequality also holds locally, for all $t \in [0, T_\delta]$:

$$0 < C^{-1} \leq |(\boldsymbol{I} + \nabla\boldsymbol{\eta}_\delta^\delta)^{-1}| \leq c^{-1}. \tag{11.142}$$

Let $[0, T]$ denote the time interval on which the classical solution $\boldsymbol{\eta}$ exists. Then, we can choose $c > 0$ and $C > 0$ so that the inequalities (11.141)–(11.142) also hold for the classical solution *for all* $t \in [0, T]$.

We will now show that the time interval on which estimates (11.141)–(11.142) hold for the weak solution of the regularized problem $\boldsymbol{\eta}_\delta^\delta$ can, in fact, be extended to the entire interval $[0, T]$, namely, that estimates (11.141)–(11.142) hold globally, uniformly in δ, where T is independent of δ. We will do this by using a classical bootstrap argument, the steps of which we present below. This bootstrap argument will propagate the desired estimates, which hold a priori locally on $[0, T_\delta]$ (for a time T_δ possibly depending on δ), to an entire uniform time interval $[0, T]$.

The global uniform estimates will follow if we can show that $\nabla\boldsymbol{\eta}$ and $\nabla\boldsymbol{\eta}_\delta^\delta$ are pointwise uniformly "close," i.e.,

$$|(\nabla\boldsymbol{\eta}-\nabla\boldsymbol{\eta}_\delta^\delta)(t,x)| \to 0 \text{ pointwise uniformly in } [0,T]\times\Omega_b \text{ as } \delta\to 0. \quad (11.143)$$

To obtain this estimate, we start with the main proof of Gronwall's inequality *under the assumptions that* (11.118), (11.119), *and* (11.120) *are locally valid for* $t \in [0, T_\delta]$:

$$E_\delta(t) \le C_1 \int_0^t \|(\boldsymbol{\eta}^\delta - \boldsymbol{\eta})(s)\|^2_{H^1(\Omega_b)} ds + C_2 \int_0^t E_\delta(s)ds, \quad (11.144)$$

where the constants C_1 and C_2 are independent of δ. Then, by Lemma 11.9.2, we obtain that the first term on the right-hand side above can be estimated as follows:

$$\|\boldsymbol{\eta}^\delta - \boldsymbol{\eta}\|_{H^1(\Omega_b)} \le C\delta^{3/2}, \qquad \text{for all } t\in[0,T], \quad (11.145)$$

since the classical solution $\boldsymbol{\eta}$ is spatially smooth and $\boldsymbol{\eta}^\delta$ is the convolution of $\boldsymbol{\eta}$ with the smooth δ kernel, defined in (11.105). With this essential observation, the Gronwall estimate based on (11.144) gives

$$\begin{aligned} E_\delta(t) &\le C_1\left(\int_0^t \|(\boldsymbol{\eta}^\delta - \boldsymbol{\eta})(s)\|^2_{H^1(\Omega_b)} ds\right) e^{C_2 t} \\ &\le C_1\left(\int_0^T \|(\boldsymbol{\eta}^\delta - \boldsymbol{\eta})(s)\|^2_{H^1(\Omega_b)} ds\right) e^{C_2 t} \le C\delta^3 e^{C_2 t}. \end{aligned}$$

By the definition of $E_\delta(t)$ and an application of Poincare's and Korn's inequalities on Ω_b (see Proposition 11.5.1), this implies that the following terms in the definition of $E_\delta(t)$

$$\|(\boldsymbol{\eta}-\boldsymbol{\eta}_\delta)(t)\|_{H^1(\Omega_b)} \le C\delta^{3/2}, \text{ and } \|(\omega-\omega_\delta)(t)\|_{H^2(\Gamma)} \le C\delta^{3/2} \to 0 \text{ as } \delta\to 0$$

converge to zero as $\delta \to 0$ at a rate of $\delta^{3/2}$, as long as the assumptions (11.118), (11.119), and (11.120) hold. Therefore, by Hölder's inequality, for sufficiently small $\delta > 0$, we can prove that the following estimate holds:

$$|(\nabla\boldsymbol{\eta}^\delta - \nabla\boldsymbol{\eta}_\delta^\delta)(t,x)| = \left|\int_{\tilde{\Omega}_b} (\nabla\boldsymbol{\eta} - \nabla\boldsymbol{\eta}_\delta)(t,y)\sigma_\delta(x-y)dy\right| \le C\delta^{3/2}\cdot\delta^{-1} \to 0, \quad (11.146)$$

$$\text{pointwise uniformly in } [0,T]\times\Omega_b \text{ as } \delta\to 0,$$

where C is independent of δ. More precisely, notice that the convolution integral in (11.146) is defined on the domain $\tilde{\Omega}_b$, which is triple the size of the domain Ω_b. Furthermore, we recall that the convolution and the larger domain $\tilde{\Omega}_b$ are defined using odd extensions as in Definition 11.4.1. Thus, by the definition of the odd extensions of η and η_δ to the larger domain $\tilde{\Omega}_b$, we get

$$\|\eta - \eta_\delta\|_{H^1(\tilde{\Omega}_b)} \leq C\left(\|\eta - \eta_\delta\|_{H^1(\Omega_b)} + \|\omega - \omega_\delta\|_{H^1(\Gamma)}\right).$$

In addition, since we have extended the functions η and η_δ to the larger domain $\tilde{\Omega}_b$, the estimate (11.146) holds for all δ such that $\{(x, y) \in \mathbb{R}^2 : \text{dist}((x, y), \Omega_b) \leq \delta\} \subset \tilde{\Omega}_b$. Thus, $|(\nabla\eta^\delta - \nabla\eta^\delta_\delta)(t, x)| \to 0$ pointwise uniformly in $[0, T]$ as $\delta \to 0$.

To obtain (11.143), it suffices to show that $|(\nabla\eta - \nabla\eta^\delta)(t, x)| \to 0$ pointwise uniformly in $[0, T]$ as $\delta \to 0$. This follows from Lemma 11.9.2. Namely, Lemma 11.9.2 implies

$$|(\nabla\eta - \nabla\eta^\delta)(t, x)| \leq C\delta \to 0 \text{ pointwise uniformly in } [0, T] \times \Omega_b \text{ as } \delta \to 0. \tag{11.147}$$

So combining (11.147) with (11.146), we get (11.143).

To conclude the bootstrap argument, we combine the fact that $\det(\boldsymbol{I} + \nabla\eta) \geq c > 0$ on a time interval $[0, T]$ with the estimate (11.143) (which states that $\nabla\eta^\delta_\delta$ and $\nabla\eta$ are uniformly close on the full time interval $[0, T]$ as $\delta \to 0$), to deduce that $\det(\boldsymbol{I} + \nabla\eta^\delta_\delta) \geq c > 0$ also uniformly on $[0, T]$, for all sufficiently small δ. Similarly, for all sufficiently small δ, the assumptions (11.119) and (11.120) will also hold up to the final time $T > 0$, as we can also propagate the estimates (11.119) and (11.120) similarly by combining estimate (11.143) with the fact that these estimates hold for the classical solution η on some time interval $[0, T]$. This closes the bootstrap argument, and so we obtain that the estimate (11.118) and similarly the estimates (11.119) and (11.120) hold uniformly up to the final time $T > 0$ uniformly in δ.

We end this section by proving the following lemma, which establishes convergence of the spatial convolution of the classical solution η in $H^1(\Omega_b)$, which is needed for the argument described above, in the estimates (11.145) and (11.147).

Lemma 11.9.2 *Let $\eta \in L^\infty(0, T; V_d)$ be an arbitrary but fixed smooth function in time and space on $[0, T] \times \overline{\Omega_b}$, where V_d is defined in* (11.35). *Then, there exists a constant C independent of $\delta > 0$, depending only on η, such that*

$$\max_{t\in[0,T]} \|\eta^\delta - \eta\|_{H^1(\Omega_b)} \leq C\delta^{3/2}, \text{ and } |\nabla\eta^\delta - \nabla\eta| \leq C\delta \quad \forall x \in \overline{\Omega}_b \text{ and } \forall t \in [0, T].$$

Remark 11.9.2 More generally, if f is a smooth function on $\mathbb{R}^2$ with sufficient decay at infinity, such as a Schwartz function, then the argument below shows that the function $\tilde{f}$ defined by

$$\tilde{f} = f * \sigma_\delta \qquad \text{on } \mathbb{R}^2$$

would satisfy $\|\tilde{f} - f\|_{H^1(\Omega_b)} \leq C\delta^2$ for a constant C. However, because we are working on a bounded domain Ω_b, we must use an odd extension to define the spatial convolution of $\boldsymbol{\eta}$. Since the odd extension of $\boldsymbol{\eta}$ to the larger domain $\tilde{\Omega}_b$ is not necessarily smooth on $\tilde{\Omega}_b$ even if $\boldsymbol{\eta}$ is a smooth function on $\overline{\Omega_b}$, we incur a loss in our estimate due to potential irregularities of the odd extension due to the behavior of the initial function $\boldsymbol{\eta}$ near the boundary $\partial\Omega_b$, which gives rise to the convergence rate $\delta^{3/2}$ instead of the optimal rate of convergence δ^2.

Proof Separate the domain $\Omega_b = (0, L) \times (0, R)$ into two parts:

$$\Omega_{b,1} = (\delta, L - \delta) \times (\delta, R - \delta), \qquad \Omega_{b,2} = \Omega_b \setminus \Omega_{b,1}.$$

For $\boldsymbol{x} \in \Omega_{b,1}$, we note that because the convolution kernel σ_δ is radially symmetric,

$$(\boldsymbol{\eta}^\delta - \boldsymbol{\eta})(\boldsymbol{x}) = \int_{\Omega_b} \left(\frac{1}{2}\boldsymbol{\eta}(\boldsymbol{x} + \boldsymbol{x}') - \boldsymbol{\eta}(\boldsymbol{x}) + \frac{1}{2}\boldsymbol{\eta}(\boldsymbol{x} - \boldsymbol{x}')\right) \sigma_\delta(\boldsymbol{x}')d\boldsymbol{x}',$$

$$(\nabla\boldsymbol{\eta}^\delta - \nabla\boldsymbol{\eta})(\boldsymbol{x}) = \int_{\Omega_b} \left(\frac{1}{2}\nabla\boldsymbol{\eta}(\boldsymbol{x} + \boldsymbol{x}') - \nabla\boldsymbol{\eta}(\boldsymbol{x}) + \frac{1}{2}\nabla\boldsymbol{\eta}(\boldsymbol{x} - \boldsymbol{x}')\right) \sigma_\delta(\boldsymbol{x}')d\boldsymbol{x}'.$$

For $\boldsymbol{x} \in \Omega_{b,1}$, these points are at least δ away from the boundary. Therefore, we have the following estimate for the discretized second derivative:

$$\left|\frac{1}{2}\boldsymbol{\eta}(\boldsymbol{x} + \boldsymbol{x}') - \boldsymbol{\eta}(\boldsymbol{x}) + \frac{1}{2}\boldsymbol{\eta}(\boldsymbol{x} - \boldsymbol{x}')\right| \leq C\delta^2 \quad \text{for } |\boldsymbol{x}'| \leq \delta,$$

and similarly for $\nabla\boldsymbol{\eta}$, by using the fact that $\boldsymbol{\eta}$ is spatially smooth in $\overline{\Omega_b}$. Therefore,

$$|(\boldsymbol{\eta}^\delta - \boldsymbol{\eta})(\boldsymbol{x})| \leq C\delta^2, \quad |(\nabla\boldsymbol{\eta}^\delta - \nabla\boldsymbol{\eta})(\boldsymbol{x})| \leq C\delta^2, \qquad \text{for } \boldsymbol{x} \in \Omega_{b,1}, \tag{11.148}$$

for a constant C depending only on $\boldsymbol{\eta}$.

For $\boldsymbol{x} \in \Omega_{b,2}$, we cannot use the same estimate, since after extending $\boldsymbol{\eta}$ to the larger domain $\tilde{\Omega}_b$, the extended function on $\tilde{\Omega}_b$ does not necessarily have a continuous second derivative, as a result of the properties of odd extension, and in fact, there may be discontinuities of the second derivative along the boundary $\partial\Omega_b$. However, $\nabla\boldsymbol{\eta}$ on the larger domain $\tilde{\Omega}_b$ is still *Lipschitz continuous*. Thus, we instead use the equations:

$$(\boldsymbol{\eta}^\delta - \boldsymbol{\eta})(\boldsymbol{x}) = \int_{\Omega_b} (\boldsymbol{\eta}(\boldsymbol{x} + \boldsymbol{x}') - \boldsymbol{\eta}(\boldsymbol{x}))\sigma_\delta(\boldsymbol{x}')d\boldsymbol{x}',$$

$$(\nabla\boldsymbol{\eta}^\delta - \nabla\boldsymbol{\eta})(\boldsymbol{x}) = \int_{\Omega_b} (\nabla\boldsymbol{\eta}(\boldsymbol{x} + \boldsymbol{x}') - \nabla\boldsymbol{\eta}(\boldsymbol{x}))\sigma_\delta(\boldsymbol{x}')d\boldsymbol{x}'.$$

Since $\boldsymbol{x} \in \Omega_{b,2}$, even if $|\boldsymbol{x}'| \leq \delta$, we may have that $\boldsymbol{x} + \boldsymbol{x}'$ is outside of Ω_b. However, due to the Lipschitz continuity of $\nabla \boldsymbol{\eta}$ on the larger domain $\tilde{\Omega}_b$, we still have the estimates

$$|\boldsymbol{\eta}(\boldsymbol{x}+\boldsymbol{x}')-\boldsymbol{\eta}(\boldsymbol{x})| \leq C\delta, \quad |\nabla\boldsymbol{\eta}(\boldsymbol{x}+\boldsymbol{x}')-\nabla\boldsymbol{\eta}(\boldsymbol{x})| \leq C\delta, \qquad \text{for } \boldsymbol{x} \in \Omega_{b,2}, |\boldsymbol{x}'| \leq \delta,$$

which give

$$|(\boldsymbol{\eta}^\delta - \boldsymbol{\eta})(\boldsymbol{x})| \leq C\delta, \quad |(\nabla\boldsymbol{\eta}^\delta - \nabla\boldsymbol{\eta})(\boldsymbol{x})| \leq C\delta, \qquad \text{for } \boldsymbol{x} \in \Omega_{b,2}. \tag{11.149}$$

The area of $\Omega_{b,2}$ is bounded by $(2R + 2L)\delta$, so by (11.148) and (11.149), we have $\|\boldsymbol{\eta}^\delta - \boldsymbol{\eta}\|_{H^1(\Omega_b)} \leq C\delta^{3/2}$ for a spatially smooth function $\boldsymbol{\eta}$ on $\overline{\Omega_b}$, where C depends only on the norms of up to the second spatial derivative of $\boldsymbol{\eta}$ on $\overline{\Omega_b}$. The generalization of this result to a function $\boldsymbol{\eta}$ that also depends on time and is spatially smooth in both space and time follows analogously. □

This completes the proof of the weak-classical consistency results. This proof effectively shows that the weak solutions that we have constructed to the regularized FPSI problem converge (in the energy norm on a uniform time interval) as the regularization parameter goes to zero to a classical solution of the original (non-regularized) FPSI problem when such a classical solution to the original FPSI problem exists.

11.10 Appendix

11.10.1 Weak Continuity of Solutions to the Regularized FPSI Problem

In this appendix, we show a result related to weak continuity of solutions to the regularized FPSI problem, namely, we will show that as $\nu \to 0$:

$$\int_{\Omega_{f,\delta}(0)} \widehat{\boldsymbol{u}}(0) \cdot (\boldsymbol{u}_\delta)_\nu(0) \to \int_{\Omega_{f,\delta}(0)} |\boldsymbol{u}_0|^2, \quad \text{and}$$

$$\int_{\Omega_{f,\delta}(t)} \widehat{\boldsymbol{u}}(t) \cdot (\boldsymbol{u}_\delta)_\nu(t) \to \int_{\Omega_{f,\delta}(t)} \widehat{\boldsymbol{u}}(t) \cdot \boldsymbol{u}_\delta(t),$$

for almost all points $0 < t \leq T$.

This result will be used in Sect. 11.10.2 to estimate the first term T_1 in (11.127) in the Gronwall's estimate. We will show weak continuity through the following series of lemmas.

Lemma 11.10.1 *Let* $\omega \in L^\infty(0, T; H_0^2(\Gamma)) \cap W^{1,\infty}(0, T; L^2(\Gamma))$ *with*

$$\min_{t\in[0,T],x\in[0,L]} R+\omega(t,x)>0,$$

define the moving fluid domain $\Omega_f^\omega(t)$. *Then, given* $\boldsymbol{u} \in L^2(0,T;H^1(\Omega_f^\omega(t))) \cap L^\infty(0,T;L^2(\Omega_f^\omega(t)))$ *where* $\Omega_f^\omega(t) = \{(x,y)\in\mathbb{R}^2 : 0\le x\le L, -R\le y\le \omega(t,x)\}$, *we have that*

$$\|\boldsymbol{u}_\nu(t,x,y)-\boldsymbol{u}(t,x,y)\|_{L^2(\Omega_f^\omega(t))}\to 0 \qquad \text{as } \nu\to 0,$$

for almost all $t\in[0,T]$.

Proof Recall that in the case of real-valued functions, one shows convergence of the convolution to the function itself almost everywhere by using the Lebesgue differentiation theorem [132]. To apply the theorem in this context, we need to apply it to a function taking values in a *fixed* Banach space rather than a time-dependent Banach space.

As a result, we consider the following function,

$$\boldsymbol{v}(t,x,y)=K(t,0,x,y)\boldsymbol{u}\left(t,x,\frac{R+\omega(t,x)}{R+\omega(0,x)}(R+y)-R\right),$$

where we have pulled the fluid velocity back to the fixed initial domain $\Omega_f^\omega(0)$. We recall the definition of $K(s,t,z,r)$ from (11.122) and its inverse:

$$K(s,t,x,y)=\begin{pmatrix}\frac{R+\omega(s,x)}{R+\omega(t,x)} & 0\\ -(R+y)\partial_x\left(\frac{R+\omega(s,x)}{R+\omega(t,x)}\right) & 1\end{pmatrix},$$

$$K^{-1}(s,t,x,y)=\begin{pmatrix}\frac{R+\omega(t,x)}{R+\omega(s,x)} & 0\\ (R+y)\frac{R+\omega(t,x)}{R+\omega(s,x)}\partial_x\left(\frac{R+\omega(s,x)}{R+\omega(t,x)}\right) & 1\end{pmatrix}.$$

By the uniform boundedness of $R+\omega(t,x)$ and $|\partial_x\omega(t,x)|$, and $\min_{t\in[0,T],x\in[0,L]} R+\omega(t,x)>0$, it is immediate to see that $\boldsymbol{v}(t,z,r)$ is in $L^\infty(0,T;L^2(\Omega_f^\omega(0)))$, where we emphasize that $L^2(\Omega_f^\omega(0))$ is a fixed function space that no longer depends on time.

By Lebesgue's differentiation theorem, almost every $t\in[0,T]$ is a Lebesgue point satisfying

$$\lim_{\nu\to 0}\frac{1}{2\nu}\int_{t-\nu}^{t+\nu}\|\boldsymbol{v}(t,\cdot)-\boldsymbol{v}(s,\cdot)\|_{L^2(\Omega_f^\omega(0))}ds\to 0. \tag{11.150}$$

Recall that by Definition (11.123),

$$\boldsymbol{u}_\nu(t,x,y)=\int_{\mathbb{R}}K(s,t,x,y)\boldsymbol{u}\left(s,x,\frac{R+\omega(s,x)}{R+\omega(t,x)}(R+y)-R\right)j_\nu(t-s)ds.$$

Thus, we compute

$$\begin{aligned} \boldsymbol{u}_\nu(t,x,y)-\boldsymbol{u}(t,x,y) = \int_{\mathbb{R}} \Bigg(K(s,t,x,y)\boldsymbol{u}\left(s,x,\frac{R+\omega(s,x)}{R+\omega(t,x)}(R+y)-R\right) \\ -\boldsymbol{u}(t,x,y)\Bigg)\cdot j_\nu(t-s)ds \\ := I_1 + I_2, \end{aligned}$$

where

$$\begin{aligned} I_1 = \int_{\mathbb{R}} K^{-1}\left(t,0,x,\frac{R+\omega(0,x)}{R+\omega(t,x)}(R+y)-R\right)\cdot \\ \left(\boldsymbol{v}\left(s,x,\frac{R+\omega(0,x)}{R+\omega(t,x)}(R+y)-R\right)\right. \\ \left.-\boldsymbol{v}\left(t,x,\frac{R+\omega(0,x)}{R+\omega(t,x)}(R+y)-R\right)\right) j_\nu(t-s)ds, \end{aligned}$$

$$\begin{aligned} I_2 = \int_{\mathbb{R}} \left(K(s,t,x,y)K^{-1}\left(s,0,x,\frac{R+\omega(0,x)}{R+\omega(t,x)}(R+y)-R\right)\right. \\ \left.-K^{-1}\left(t,0,x,\frac{R+\omega(0,x)}{R+\omega(t,x)}(R+y)-R\right)\right)\cdot \\ \boldsymbol{v}\left(s,x,\frac{R+\omega(0,x)}{R+\omega(t,x)}(R+y)-R\right) j_\nu(t-s)ds. \end{aligned}$$

We estimate each of these terms as follows. For I_1, we compute that

$$\begin{aligned} K^{-1}\left(t,0,x,\frac{R+\omega(0,x)}{R+\omega(t,x)}(R+y)-R\right) \\ = \begin{pmatrix} \frac{R+\omega(0,x)}{R+\omega(t,x)} & 0 \\ (R+y)\left(\frac{R+\omega(0,x)}{R+\omega(t,x)}\right)^2 \partial_x\left(\frac{R+\omega(t,x)}{R+\omega(0,x)}\right) & 1 \end{pmatrix}, \end{aligned}$$

which we note is uniformly bounded on $[0,T]$. Hence, using the fact that $|j_\nu(t-s)| \le \frac{1}{\nu}$, we get

$$\begin{aligned} \|I_1\|_{L^2(\Omega_f^\omega(t))} \le C\cdot\frac{1}{\nu}\int_{t-\nu}^{t+\nu} \Bigg\| \boldsymbol{v}\left(s,x,\frac{R+\omega(0,x)}{R+\omega(t,x)}(R+y)-R\right) \\ -\boldsymbol{v}\left(t,x,\frac{R+\omega(0,x)}{R+\omega(t,x)}(R+y)-R\right)\Bigg\|_{L^2(\Omega_f^\omega(t))} ds \end{aligned}$$

$$\leq C \cdot \frac{1}{\nu} \int_{t-\nu}^{t+\nu} \left(\frac{R+\omega(t,x)}{R+\omega(0,x)} \right)^{1/2} \|\boldsymbol{v}(s,x,y) - \boldsymbol{v}(t,x,y)\|_{L^2(\Omega_f^\omega(0))} ds \to 0,$$

as $\nu \to 0$ if t is a Lebesgue point, by (11.150) and the uniform boundedness of $\frac{R+\omega(t,x)}{R+\omega(0,x)}$ on $[0,T]$.

To estimate I_2, we can use the continuity in time of ω and $\partial_x \omega$ to calculate that

$$\left| K(s,t,x,y) K^{-1}\left(s,0,x,\frac{R+\omega(0,x)}{R+\omega(t,x)}(R+y)-R\right) - K^{-1}\left(t,0,x,\frac{R+\omega(0,x)}{R+\omega(t,x)}(R+y)-R\right) \right| \to 0,$$

uniformly in (x,y) as $s \to t$. Now, we estimate

$$\begin{aligned}
\|I_2\|_{L^2(\Omega_f^\omega(t))} &\leq \int_{\mathbb{R}} \max_{x,y \in \Omega_f^\omega(t)} \left| K(s,t,x,y) K^{-1}\left(s,0,x,\frac{R+\omega(0,x)}{R+\omega(t,x)}(R+y)-R\right) \right. \\
&\quad \left. - K^{-1}\left(t,0,x,\frac{R+\omega(0,x)}{R+\omega(t,x)}(R+y)-R\right) \right| \\
&\quad \cdot \left\| \boldsymbol{v}\left(s,x,\frac{R+\omega(0,x)}{R+\omega(t,x)}(R+y)-R\right) \right\|_{L^2(\Omega_f^\omega(t))} \cdot j_\nu(t-s)ds \\
&\leq \int_{\mathbb{R}} \max_{x,y \in \Omega_f^\omega(t)} \left| K(s,t,x,y) K^{-1}\left(s,0,x,\frac{R+\omega(0,x)}{R+\omega(t,x)}(R+y)-R\right) \right. \\
&\quad \left. - K^{-1}\left(t,0,x,\frac{R+\omega(0,x)}{R+\omega(t,x)}(R+y)-R\right) \right| \\
&\quad \cdot \left(\frac{R+\omega(t,x)}{R+\omega(0,x)} \right)^{1/2} \cdot ||\boldsymbol{v}(s,x,y)||_{L^2(\Omega_f^\omega(0))} \cdot j_\nu(t-s)ds \\
&\leq C \int_{\mathbb{R}} \max_{x,y \in \Omega_f^\omega(t)} \left| K(s,t,x,y) K^{-1}\left(s,0,x,\frac{R+\omega(0,x)}{R+\omega(t,x)}(R+y)-R\right) \right. \\
&\qquad \left. - K^{-1}\left(t,0,x,\frac{R+\omega(0,x)}{R+\omega(t,x)}(R+y)-R\right) \right| \cdot j_\nu(t-s)ds,
\end{aligned}$$

where we used the fact that $\boldsymbol{v} \in L^\infty(0,T;L^2(\Omega_f^\omega(0)))$. Thus, we conclude that $\|I_2\|_{L^2(\Omega_f^\omega(t))} \to 0$ as $\nu \to 0$. This completes the proof. □

We also have a weak continuity lemma, which states that the value of $\boldsymbol{u}_\delta$ tested against any function in the fluid function space has a continuity property as $t \to 0$.

Lemma 11.10.2 *Consider an arbitrary $\boldsymbol{q} \in C^1(0, T; V_{f,\delta}(t))$ and the weak solution $\boldsymbol{u}_\delta$ to the regularized problem for arbitrary δ, where $V_{f,\delta}(t)$ is defined by the displacement ω_δ and* (11.30). *There exists a measure zero subset S of* $[0, T]$ *(depending on δ) such that*

$$\lim_{t \to 0, t \in [0,T] \cap S^c} \int_{\Omega_{f,\delta}(t)} \boldsymbol{u}_\delta(t) \cdot \boldsymbol{q}(t) = \int_{\Omega_{f,\delta}(0)} \boldsymbol{u}_0 \cdot \boldsymbol{q}(0).$$

Proof Consider the following function for each $\tau \in [0, T]$ and $\alpha > 0$, given by

$$J_{\tau,\nu}(t) = 1 - \int_0^t j_\nu(s - \tau) ds, \tag{11.151}$$

and note that $J'_{\tau,\nu}(t) = -j_\nu(t - \tau)$. We want to test the regularized weak formulation for $\boldsymbol{u}_\delta$ with the test function $J_{\tau,\nu}(t)\boldsymbol{q}$ for certain admissible choices of τ. To see which τ we want to choose, we define the function

$$\begin{aligned} \boldsymbol{w}(t, x, y) = & \frac{R + \omega_\delta(t)}{R + \omega_\delta(0)} \cdot \boldsymbol{u}_\delta \left(t, x, \frac{R + \omega_\delta(t)}{R + \omega_\delta(0)}(R + y) - R \right) \\ & \cdot \boldsymbol{q} \left(t, x, \frac{R + \omega_\delta(t)}{R + \omega_\delta(0)}(R + y) - R \right). \end{aligned}$$

We claim that $\boldsymbol{w} \in L^\infty(0, T; L^1(\Omega_{f,\delta}(0)))$. To see this, we compute by a change of variables that

$$\|\boldsymbol{w}(t, x, y)\|_{L^1(\Omega_{f,\delta}(0))} = \int_{\Omega_{f,\delta}(t)} |\boldsymbol{u}_\delta(t, x, y) \cdot \boldsymbol{q}(t, x, y)|,$$

and we then use the fact that $\boldsymbol{u}_\delta, \boldsymbol{q} \in L^\infty(0, T; L^2(\Omega_{f,\delta}(t)))$.

Hence, by the Lebesgue differentiation theorem (see Theorem 2.7.4), there exists a measurable subset $S \subset [0, T]$ of measure zero such that every point in $[0, T] \cap S^c$ is a Lebesgue point of $\boldsymbol{w}$, in the sense that

$$\lim_{\nu \to 0} \frac{1}{2\nu} \int_{\tau-\nu}^{\tau+\nu} \|\boldsymbol{w}(\tau, \cdot) - \boldsymbol{w}(s, \cdot)\|_{L^1(\Omega_{f,\delta}(0))} ds \to 0. \tag{11.152}$$

for every $\tau \in [0, T] \cap S^c$. These are the τ for which we will consider the test function $J_{\tau,\nu}(t)\boldsymbol{q}$. For the test functions for the Biot medium and the plate, we will take these test functions to be zero. Hence, in the regularized weak formulation (11.108), we will test with $(\boldsymbol{v}, \varphi, \boldsymbol{\psi}, r) = (J_{\tau,\nu}(t)\boldsymbol{q}, 0, 0, 0)$.

Hence, we obtain the following equality:

$$-\int_0^T\int_{\Omega_{f,\delta}(t)} \boldsymbol{u}_\delta \cdot \partial_t(J_{\tau,\nu}(t)\boldsymbol{q}) + \frac{1}{2}\int_0^T\int_{\Omega_{f,\delta}(t)} [((\boldsymbol{u}_\delta\cdot\nabla)\boldsymbol{u}_\delta)\cdot(J_{\tau,\nu}(t)\boldsymbol{q})$$
$$-((\boldsymbol{u}_\delta\cdot\nabla)(J_{\tau,\nu}(t)\boldsymbol{q}))\cdot\boldsymbol{u}_\delta]$$
$$+\frac{1}{2}\int_0^T\int_{\Gamma_\delta(t)} (\boldsymbol{u}_\delta\cdot\boldsymbol{n} - 2\boldsymbol{\xi}_\delta\cdot\boldsymbol{n})\boldsymbol{u}_\delta\cdot(J_{\tau,\nu}(t)\boldsymbol{q})$$
$$+2\nu\int_0^T\int_{\Omega_{f,\delta}(t)} \boldsymbol{D}(\boldsymbol{u}_\delta):\boldsymbol{D}(J_{\tau,\nu}(t)\boldsymbol{q})$$
$$-\int_0^T\int_{\Gamma_\delta(t)}\left(\frac{1}{2}|\boldsymbol{u}_\delta|^2 - p_\delta\right)J_{\tau,\nu}(t)q_n - \beta\int_0^T\int_{\Gamma_\delta(t)}[(\xi_\delta)_\tau - (u_\delta)_\tau]\cdot J_{\tau,\nu}(t)q_\tau$$
$$=\int_{\Omega_{f,\delta}(0)} \boldsymbol{u}_0\cdot J_{\tau,\nu}(0)\boldsymbol{q}(0).$$

Consider $\tau \in (0,T)\cap S^c$. We want to pass to the limit as $\nu \to 0$ and then pass to the limit as $\tau \to 0$, in order to obtain the desired result.

First, we pass to the limit as $\nu \to 0$. We handle the convergences as follows:

First term: We will show that because τ is a Lebesgue point of $\boldsymbol{w}$,

$$-\int_0^T\int_{\Omega_{f,\delta}(t)} \boldsymbol{u}_\delta\cdot\partial_t(J_{\tau,\nu}(t)\boldsymbol{q}) \to \int_{\Omega_{f,\delta}(\tau)} \boldsymbol{u}_\delta(\tau)\boldsymbol{q}(\tau)$$
$$-\int_0^t\int_{\Omega_{f,\delta}(t)} \boldsymbol{u}_\delta\cdot\partial_t\boldsymbol{q}, \qquad \text{as } \nu\to 0.$$

We compute that

$$-\int_0^T\int_{\Omega_{f,\delta}(t)} \boldsymbol{u}_\delta\cdot\partial_t(J_{\tau,\nu}(t)\boldsymbol{q}) = \int_0^T\int_{\Omega_{f,\delta}(t)} \boldsymbol{u}_\delta\cdot j_\nu(t-\tau)\boldsymbol{q}$$
$$-\int_0^T\int_{\Omega_{f,\delta}(t)} \boldsymbol{u}_\delta\cdot J_{\tau,\nu}(t)\partial_t\boldsymbol{q}.$$

It is easy to see that

$$\int_0^T\int_{\Omega_{f,\delta}(t)} \boldsymbol{u}_\delta J_{\tau,\nu}(t)\partial_t\boldsymbol{q} \to \int_0^t\int_{\Omega_{f,\delta}(t)} \boldsymbol{u}_\delta\partial_t\boldsymbol{q}.$$

So it remains to show that

$$\int_0^T\int_{\Omega_{f,\delta}(t)} \boldsymbol{u}_\delta\cdot j_\nu(t-\tau)\boldsymbol{q} \to \int_{\Omega_{f,\delta}(\tau)} \boldsymbol{u}_\delta(\tau)\boldsymbol{q}(\tau), \qquad \text{as } \nu\to 0.$$

By a change of variables, we compute that

$$\begin{aligned}
&\int_0^T \int_{\Omega_{f,\delta}(t)} \boldsymbol{u}_\delta \cdot j_\nu(t-\tau)\boldsymbol{q} \\
&\quad = \int_0^T \int_{\Omega_{f,\delta}(\tau)} \frac{R+\omega_\delta(t)}{R+\omega_\delta(\tau)} \cdot \boldsymbol{u}_\delta\left(t, x, \frac{R+\omega_\delta(t)}{R+\omega_\delta(\tau)}(R+y)-R\right) \\
&\qquad \cdot j_\nu(t-\tau)\boldsymbol{q}\left(t, x, \frac{R+\omega_\delta(t)}{R+\omega_\delta(\tau)}(R+y)-R\right) \\
&\quad = \int_0^T \int_{\Omega_{f,\delta}(\tau)} \frac{R+\omega_\delta(0)}{R+\omega_\delta(\tau)} \boldsymbol{w}\left(t, x, \frac{R+\omega_\delta(0)}{R+\omega_\delta(\tau)}(R+y)-R\right) \\
&\qquad\qquad \cdot j_\nu(t-\tau) = \int_0^T \int_{\Omega_{f,\delta}(0)} \boldsymbol{w}(t,x,y) \cdot j_\nu(t-\tau).
\end{aligned}$$

By (11.152), we have that

$$\int_0^T \int_{\Omega_{f,\delta}(0)} \boldsymbol{w}(t,x,y) \cdot j_\nu(t-\tau) \to \int_{\Omega_{f,\delta}(0)} \boldsymbol{w}(\tau,x,y) = \int_{\Omega_{f,\delta}(\tau)} \boldsymbol{u}_\delta(\tau) \cdot \boldsymbol{q}(\tau),$$

which establishes the desired convergence.

Final term: It is immediate to see that for all sufficiently small $\nu > 0$,

$$\int_{\Omega_{f,\delta}(0)} \boldsymbol{u}_0 \cdot J_{\tau,\nu}(0)\boldsymbol{q}(0) = \int_{\Omega_{f,\delta}(0)} \boldsymbol{u}_0 \cdot \boldsymbol{q}(0).$$

We can now easily take $\nu \to 0$ in the remaining terms to obtain that for any $\tau \in (0,T) \cap S^c$,

$$\begin{aligned}
&\int_{\Omega_{f,\delta}(\tau)} \boldsymbol{u}_\delta(\tau) \cdot \boldsymbol{q}(\tau) - \int_0^t \int_{\Omega_{f,\delta}(t)} \boldsymbol{u}_\delta \cdot \partial_t \boldsymbol{q} \\
&\qquad + \frac{1}{2}\int_0^t \int_{\Omega_{f,\delta}(t)} [((\boldsymbol{u}_\delta \cdot \nabla)\boldsymbol{u}_\delta) \cdot \boldsymbol{q} - ((\boldsymbol{u}_\delta \cdot \nabla)\boldsymbol{q}) \cdot \boldsymbol{u}_\delta] \\
&\quad + \frac{1}{2}\int_0^t \int_{\Gamma_\delta(t)} (\boldsymbol{u}_\delta \cdot \boldsymbol{n} - 2\boldsymbol{\xi}_\delta \cdot \boldsymbol{n})\boldsymbol{u}_\delta \cdot \boldsymbol{q} + 2\nu \int_0^t \int_{\Omega_{f,\delta}(t)} \boldsymbol{D}(\boldsymbol{u}_\delta) : \boldsymbol{D}(\boldsymbol{q}) \\
&- \int_0^t \int_{\Gamma_\delta(t)} \left(\frac{1}{2}|\boldsymbol{u}_\delta|^2 - p_\delta\right) q_n - \beta \int_0^t \int_{\Gamma_\delta(t)} [(\xi_\delta)_\tau - (u_\delta)_\tau] \cdot q_\tau = \int_{\Omega_{f,\delta}(0)} \boldsymbol{u}_0 \cdot \boldsymbol{q}(0).
\end{aligned}$$

Passing to the limit as $\tau \to 0$ with $\tau \in (0,T) \cap S^c$ gives the desired result.

□

Lemma 11.10.3 *Let $\boldsymbol{u}_0$ be divergence-free and smooth on $\overline{\Omega_f(0)}$. Define*

$$\tilde{\boldsymbol{q}}(t, x, y) = K_\delta(0, t, x, y)\boldsymbol{u}_0\left(x, \frac{R + \omega_\delta(0, x)}{R + \omega_\delta(t, x)}(R + y) - R\right), \tag{11.153}$$

where K_δ is given by (11.122). *Then, there exists a sequence of functions $\tilde{\boldsymbol{q}}_m \in C_c^1(0, T; V_{f,\delta}(t))$, with $V_{f,\delta}(t)$ determined by the plate displacement ω_δ via the Definition* (11.30), *such that*

$$\max_{0\le t\le T} \|\tilde{\boldsymbol{q}} - \tilde{\boldsymbol{q}}_m\|_{L^2(\Omega_{f,\delta}(t))} \to 0, \qquad \text{as } m \to \infty.$$

Proof There exists a rectangular two-dimensional maximal domain Ω_M of the form $[0, L]\times[-R, R_{max}]$ for some positive constant R_{max} that contains all of the domains $\Omega_{f,\delta}(t)$ for $t \in [0, T]$. We will extend $\tilde{\boldsymbol{q}}$ to the maximal spacetime domain $[0, T] \times \Omega_M$ by extending vertically in the radial direction by the trace of $\tilde{\boldsymbol{q}}$ along $\Gamma_\delta(t)$. In particular, we define

$$\begin{aligned} &\tilde{\boldsymbol{q}}(t, x, y) = K(0, t, x, \omega_\delta(t, x))\boldsymbol{u}_0\left(x, \omega_\delta(0, x)\right), \\ &\text{for } (t, x, y) \in ([0, T] \times \Omega_M) - ([0, T] \times \Omega_{f,\delta}(t)). \end{aligned} \tag{11.154}$$

Note that this extension preserves the *divergence-free* property.

We have the following two claims about the extended function, considered as a function on the *fixed* maximal domain Ω_M. First, we claim that $\tilde{\boldsymbol{q}} \in L^\infty(0, T; H^1(\Omega_M))$. Second, we claim that $\tilde{\boldsymbol{q}} \in C(0, T; L^2(\Omega_M))$. To see that $\tilde{\boldsymbol{q}} \in L^\infty(0, T; H^1(\Omega_M))$, we note that ω_δ and $\partial_x\omega_\delta$ are bounded uniformly pointwise, and furthermore, $\boldsymbol{u}_0$ and its first spatial derivatives are bounded by assumption. In addition, $\partial_x^2\omega_\delta \in L^\infty(0, T; L^2(\Gamma))$, which allows us to conclude that $\tilde{\boldsymbol{q}} \in L^\infty(0, T; H^1(\Omega_M))$.

Next, we want to verify that $\tilde{\boldsymbol{q}} \in C(0, T; L^2(\Omega_M))$. Consider any $t \in [0, T]$, and consider any $s \in [0, T]$ with $s \neq t$. We define the following regions:

$$\begin{aligned} A(s, t) &= \Omega_f^M \cap (\Omega_{f,\delta}(s) \cup \Omega_{f,\delta}(t))^c, \\ B(s, t) &= [\Omega_{f,\delta}(s) \cap (\Omega_{f,\delta}(t))^c] \cup [(\Omega_{f,\delta}(s))^c \cap \Omega_{f,\delta}(t)], \\ C(s, t) &= \Omega_{f,\delta}(s) \cap \Omega_{f,\delta}(t). \end{aligned}$$

Consider $\epsilon > 0$. We want to find $h > 0$ such that

$$\|\tilde{\boldsymbol{q}}(t, \cdot) - \tilde{\boldsymbol{q}}(s, \cdot)\|_{L^2(\Omega_M)}^2 \le \epsilon, \qquad \text{for all } s \in (t - h, t + h) \cap [0, T]. \tag{11.155}$$

We compute that

$$
\begin{aligned}
\|\tilde{\boldsymbol{q}}(t,\cdot) - \tilde{\boldsymbol{q}}(s,\cdot)\|^2_{L^2(\Omega_M)} &= \int_{A(s,t)} |\tilde{\boldsymbol{q}}(t,x,y) - \tilde{\boldsymbol{q}}(s,x,y)|^2 \\
&+ \int_{B(s,t)} |\tilde{\boldsymbol{q}}(t,x,y) - \tilde{\boldsymbol{q}}(s,x,y)|^2 \\
&+ \int_{C(s,t)} |\tilde{\boldsymbol{q}}(t,x,y) - \tilde{\boldsymbol{q}}(s,x,y)|^2 = I_A + I_B + I_C.
\end{aligned}
\tag{11.156}
$$

We estimate each of the terms I_A, I_B, and I_C separately.

For I_A, we recall that we are extending by the trace as in (11.154) on $A(s,t)$, so we have that

$$
I_A = \int_{A(s,t)} |K_\delta(0,t,x,\omega_\delta(t,x)) - K_\delta(0,s,x,\omega_\delta(s,x))|^2 \cdot |\boldsymbol{u}_0(x,\omega_\delta(0,x))|^2.
$$

We have that $|\boldsymbol{u}_0(x,\omega_\delta(0,x))| \leq M_1$ for some constant M_1 by the fact that $\boldsymbol{u}_0$ is continuous on $\overline{\Omega_f(0)}$. By continuity, we can choose $h > 0$ sufficiently small so that

$$
\begin{aligned}
&|K_\delta(0,t,x,\omega_\delta(t,x)) - K_\delta(0,s,x,\omega_\delta(s,x))|^2 \\
&< \frac{\epsilon}{3M_1^2(R+R_{max})L}, \qquad \text{for all } s \in (t-h,t+h) \cap [0,T].
\end{aligned}
$$

Thus, for all $s \in (t-h,t+h) \cap [0,T]$,

$$
I_A \leq |A(s,t)| \cdot \frac{\epsilon}{3(R+R_{max})L} \leq \frac{\epsilon}{3}.
$$

For I_B, we will use the fact that ω_δ does not change much in time over small time intervals, by continuity. We note that there exists a uniform constant M_2 such that $|\tilde{\boldsymbol{q}}| \leq M_2$ on $[0,T] \times \Omega_M$. Hence,

$$
\begin{aligned}
I_B &= \int_{B(s,t)} |\tilde{\boldsymbol{q}}(t,z,r) - \tilde{\boldsymbol{q}}(s,z,r)|^2 \leq |B(s,t)| \cdot 4M_2^2 \\
&= 4M_2^2 \int_0^L |\omega_\delta(t,x) - \omega_\delta(s,x)| dx.
\end{aligned}
$$

Because $\omega_\delta \in L^\infty(0,T;H_0^2(\Gamma)) \cap W^{1,\infty}(0,T;L^2(\Gamma))$, there exists $h > 0$ sufficiently small such that

$$
|\omega_\delta(t,x) - \omega_\delta(s,x)| \leq \frac{\epsilon}{12M_2^2 L},
$$

$$
\text{for all } x \in [0,L] \text{ and } s \in (t-h,t+h) \cap [0,T].
$$

This allows us to conclude that $I_B \leq \frac{\epsilon}{3}$, for all $s \in (t-h, t+h) \cap [0, T]$.

For I_C, we refer to the definition of $\tilde{\boldsymbol{q}}$ in (11.153), and note that $K_\delta(0, t, x, y)$ is continuous in time uniformly in $(x, y) \in [0, L] \times [-R, R_{max}]$, $\boldsymbol{u}_0$ is uniformly continuous as a function on $\overline{\Omega_f(0)}$, and $\omega_\delta(t, x)$ is continuous in time uniformly in $x \in [0, L]$. Hence, there exists $h > 0$ sufficiently small such that

$$|\tilde{\boldsymbol{q}}(t, x, y) - \tilde{\boldsymbol{q}}(s, x, y)|^2 \leq \frac{\epsilon}{3(R + R_{max})L},$$

$$\text{for all } (x, y) \in C(s, t) \text{ and } s \in (t-h, t+h) \cap [0, T],$$

which gives the desired result that $I_C \leq \frac{\epsilon}{3}$ for all $s \in (t-h, t+h) \cap [0, T]$. Thus, by using (11.156), we have established (11.155).

Since $\tilde{\boldsymbol{q}} \in L^\infty(0, T; H^1(\Omega_M)) \cap C(0, T; L^2(\Omega_M))$, we can extend $\tilde{\boldsymbol{q}}$ to a continuous function on all of $\mathbb{R}$ as follows. We can find an increasing sequence T_m with $T_m \to T$ as $m \to \infty$, such that $\tilde{\boldsymbol{q}}(T_m) \in H^1(\Omega_M)$ for all m. Define an extension $\hat{\boldsymbol{q}}_m$ for each m to all of $\mathbb{R}$ by $\hat{\boldsymbol{q}}_m = \tilde{\boldsymbol{q}}$ if $t \in [0, T_m]$,

$$\hat{\boldsymbol{q}}_m = \tilde{\boldsymbol{q}}(0), \qquad \text{if } t < 0, \qquad \hat{\boldsymbol{q}}_m = \tilde{\boldsymbol{q}}(T_m), \qquad \text{if } t > T_m.$$

Define

$$\tilde{\boldsymbol{q}}_m = \hat{\boldsymbol{q}}_m * j_{1/m},$$

where the convolution is a convolution in time with j_ν for $\alpha = 1/m$. Because $\hat{\boldsymbol{q}}_m \in L^\infty(0, T; H^1(\Omega_M)) \cap C(0, T; L^2(\Omega_M))$ with $\hat{\boldsymbol{q}}_m$ being divergence-free for every $t \in [0, T]$, we have that $\tilde{\boldsymbol{q}}_m$ restricted to $\bigcup_{t\in[0,T]}\{t\} \times \Omega_{f,\delta}(t)$ gives a function in $C^1([0, T); V_{f,\delta}(t))$, where $V_{f,\delta}(t)$ is the space defined in (11.30) with the plate displacement ω_δ. The fact that

$$\max_{0\leq t\leq T} \|\tilde{\boldsymbol{q}} - \tilde{\boldsymbol{q}}_m\|_{L^2(\Omega_{f,\delta}(t))} \to 0, \qquad \text{as } m \to \infty,$$

follows from the uniform continuity of $\tilde{\boldsymbol{q}}$ on $[0, T]$ as a function taking values in $L^2(\Omega_M)$, convergence properties of convolutions, and the fact that $\tilde{\boldsymbol{q}} \in C(0, T; L^2(\Omega_M))$, which gives the convergence

$$\max_{t\in[T_m,T]} \|\tilde{\boldsymbol{q}}(T) - \tilde{\boldsymbol{q}}(t)\|_{L^2(\Omega_M)} \to 0, \qquad \text{as } m \to \infty.$$

□

Lemma 11.10.4 *For the function $\tilde{\boldsymbol{q}}$ defined in* (11.153), *there exists a measure zero subset S of* $[0, T]$ *such that*

$$\lim_{t\to 0, t\in[0,T]\cap S^c} \int_{\Omega_{f,\delta}(t)} \boldsymbol{u}_\delta(t) \cdot \tilde{\boldsymbol{q}}(t) = \int_{\Omega_{f,\delta}(0)} \boldsymbol{u}_0 \cdot \tilde{\boldsymbol{q}}(0).$$

Proof Note that because $\partial_t \tilde{\boldsymbol{q}}$ is not necessarily in $H^1(\Omega_{f,\delta}(t))$, $\tilde{\boldsymbol{q}}$ is not a valid test function. Thus, we use the sequence $\tilde{\boldsymbol{q}}_m \in C^1(0, T; \mathcal{V}_{f,\delta}(t))$ from Lemma 11.10.3, which satisfies

$$\max_{0\leq t\leq T} \|\tilde{\boldsymbol{q}} - \tilde{\boldsymbol{q}}_m\|_{L^2(\Omega_{f,\delta}(t))} \to 0, \qquad \text{as } m \to \infty.$$

We can then apply Lemma 11.10.2 to each of the test functions $\tilde{\boldsymbol{q}}_m$, to deduce that there exists a measure zero subset S_m of $[0, T]$ such that

$$\lim_{t\to 0, t\in[0,T]\cap S_m^c} \int_{\Omega_{f,\delta}(t)} \boldsymbol{u}_\delta(t) \cdot \tilde{\boldsymbol{q}}_m(t) = \int_{\Omega_{f,\delta}(0)} \boldsymbol{u}_0 \cdot \tilde{\boldsymbol{q}}_m(0).$$

In addition, by uniform boundedness, $\boldsymbol{u}_\delta \in L^\infty(0, T; L^2(\Omega_{f,\delta}(t)))$, and hence, there exists a measure zero subset S_0 of $[0, T]$, and a positive constant C such that $\|\boldsymbol{u}_0\|_{L^2(\Omega_{f,\delta}(0))} \leq C$, and

$$\|\boldsymbol{u}_\delta(t)\|_{L^2(\Omega_{f,\delta}(t))} \leq C, \qquad \text{for all } t \in S_0^c. \tag{11.157}$$

Define $S = S_0 \cup \bigcup_{m\geq 1} S_m$, which is also a measure zero subset of $[0, T]$. Then, for each m,

$$\lim_{t\to 0, t\in[0,T]\cap S^c} \int_{\Omega_{f,\delta}(t)} \boldsymbol{u}_\delta(t) \cdot \tilde{\boldsymbol{q}}_m(t) = \int_{\Omega_{f,\delta}(0)} \boldsymbol{u}_0 \cdot \tilde{\boldsymbol{q}}_m(0). \tag{11.158}$$

By passing to the limit in m, we claim that in addition,

$$\lim_{t\to 0, t\in[0,T]\cap S^c} \int_{\Omega_{f,\delta}(t)} \boldsymbol{u}_\delta(t) \cdot \tilde{\boldsymbol{q}}(t) = \int_{\Omega_{f,\delta}(0)} \boldsymbol{u}_0 \cdot \tilde{\boldsymbol{q}}(0).$$

To see this, consider $\epsilon > 0$. We claim that there exists $h > 0$ sufficiently small such that for all $t \in (0, h) \cap S^c$,

$$\left| \int_{\Omega_{f,\delta}(t)} \boldsymbol{u}_\delta(t) \cdot \tilde{\boldsymbol{q}}(t) - \int_{\Omega_{f,\delta}(0)} \boldsymbol{u}_0 \cdot \tilde{\boldsymbol{q}}(0) \right| < \epsilon.$$

We can choose M sufficiently large such that $\max_{0\leq t\leq T} \|\tilde{\boldsymbol{q}} - \tilde{\boldsymbol{q}}_M\|_{L^2(\Omega_{f,\delta}(t))} < \dfrac{\epsilon}{3C}$, where C is defined by (11.157). Therefore, for all $t \in [0, T] \cap S^c$,

$$\left| \int_{\Omega_{f,\delta}(t)} \boldsymbol{u}_\delta(t) \cdot \tilde{\boldsymbol{q}}(t) - \int_{\Omega_{f,\delta}(t)} \boldsymbol{u}_\delta(t) \cdot \tilde{\boldsymbol{q}}_M(t) \right| < \frac{\epsilon}{3}.$$

In addition,

$$\left| \int_{\Omega_{f,\delta}(0)} \boldsymbol{u}_0 \cdot \tilde{\boldsymbol{q}}(0) - \int_{\Omega_{f,\delta}(0)} \boldsymbol{u}_0 \cdot \tilde{\boldsymbol{q}}_M(0) \right| < \frac{\epsilon}{3}.$$

By applying (11.158) with $m = M$, we can choose $h > 0$ sufficiently small such that for all $t \in (0, h) \cap S^c$,

$$\left| \int_{\Omega_{f,\delta}(t)} \boldsymbol{u}_\delta(t) \cdot \tilde{\boldsymbol{q}}_M(t) - \int_{\Omega_{f,\delta}(0)} \boldsymbol{u}_0 \cdot \tilde{\boldsymbol{q}}_M(0) \right| < \frac{\epsilon}{3}.$$

Thus, by applying the triangle inequality, we have that for all $t \in (0, h) \cap S^c$,

$$\left| \int_{\Omega_{f,\delta}(t)} \boldsymbol{u}_\delta(t) \cdot \tilde{\boldsymbol{q}}(t) - \int_{\Omega_{f,\delta}(0)} \boldsymbol{u}_0 \cdot \tilde{\boldsymbol{q}}(0) \right| < \epsilon,$$

which establishes the desired result. □

We can now prove the final result of this Appendix. We recall the definition of $\widehat{\boldsymbol{u}}$ from (11.113).

Lemma 11.10.5 *In the limit as* $\nu \to 0$, *we have the following convergence results:*

$$\int_{\Omega_{f,\delta}(0)} \widehat{\boldsymbol{u}}(0) \cdot (\boldsymbol{u}_\delta)_\nu(0) \to \int_{\Omega_{f,\delta}(0)} |\boldsymbol{u}_0|^2, \quad \textit{and}$$

$$\int_{\Omega_{f,\delta}(t)} \widehat{\boldsymbol{u}}(t) \cdot (\boldsymbol{u}_\delta)_\nu(t) \to \int_{\Omega_{f,\delta}(t)} \widehat{\boldsymbol{u}}(t) \cdot \boldsymbol{u}_\delta(t),$$

for almost all points $t \in (0, T]$.

Proof The second convergence for almost all points $t \in (0, T]$ follows directly from Lemma 11.10.1 and the fact that $\widehat{\boldsymbol{u}} \in L^\infty(0, T; L^2(\Omega_{f,\delta}(t)))$.

So we just need to verify the convergence at $t = 0$. To do this, we note that $\widehat{\boldsymbol{u}}(0) = \boldsymbol{u}_0$. Hence,

$$\begin{aligned}
&\int_{\Omega_{f,\delta}(0)} \widehat{\boldsymbol{u}}(0) \cdot (\boldsymbol{u}_\delta)_\nu(0) \\
&= \int_{\Omega^{\omega_0}} \left(\int_{\mathbb{R}} K_\delta(s, 0, x, y) \boldsymbol{u}_\delta \left(s, x, \frac{R + \omega_\delta(s, x)}{R + \omega_\delta(0, x)}(R + y) - R \right) \right. \\
&\qquad \left. j_\delta(t - s) ds \right) \boldsymbol{u}_0(x, y) dx dy \\
&= \int_{\mathbb{R}} \left(\int_{\Omega^{\omega_0}} K_\delta(s, 0, x, y) \boldsymbol{u}_\delta \left(s, x, \frac{R + \omega_\delta(s, x)}{R + \omega_\delta(0, x)}(R + y) - R \right) \right.
\end{aligned}$$

$$
\left. \cdot \boldsymbol{u}_0(x, y)dxdy \right) j_\nu(t-s)ds
$$

$$
= \int_{\mathbb{R}} \left(\int_{\Omega_{f,\delta}(s)} \boldsymbol{u}_\delta(s, x, y) \cdot \frac{R + \omega_\delta(0, x)}{R + \omega_\delta(s, x)} \right.
$$

$$
K_\delta^t \left(s, 0, x, \frac{R + \omega_\delta(0, x)}{R + \omega_\delta(s, x)}(R+y) - R \right)
$$

$$
\left. \cdot \boldsymbol{u}_0 \left(x, \frac{R + \omega_\delta(0, x)}{R + \omega_\delta(s, x)}(R + y) - R \right) dxdy \right) j_\nu(t-s)ds.
$$

We compute

$$
\begin{aligned}
&\frac{R + \omega_\delta(0, x)}{R + \omega_\delta(s, x)} \cdot K_\delta^t \left(s, 0, x, \frac{R + \omega_\delta(0, x)}{R + \omega_\delta(s, x)}(R + y) - R \right) \\
&= \begin{pmatrix} 1 & (R + y)\nabla \left(\frac{R+\omega_\delta(0,x)}{R+\omega_\delta(s,x)} \right) \\ 0 & \frac{R+\omega_\delta(0,x)}{R+\omega_\delta(s,x)} \end{pmatrix} \\
&= \begin{pmatrix} \frac{R+\omega_\delta(0,x)}{R+\omega_\delta(s,x)} & 0 \\ -(R + y)\nabla \left(\frac{R+\omega_\delta(0,x)}{R+\omega_\delta(s,x)} \right) & 1 \end{pmatrix} \\
&+ \begin{pmatrix} 1 - \frac{R+\omega_\delta(0,x)}{R+\omega_\delta(s,x)} & (R + y)\nabla \left(\frac{R+\omega_\delta(0,x)}{R+\omega_\delta(s,x)} \right) \\ (R + y)\nabla \left(\frac{R+\omega_\delta(0,x)}{R+\omega_\delta(s,x)} \right) & \frac{R+\omega_\delta(0,x)}{R+\omega_\delta(s,x)} - 1 \end{pmatrix} \\
&:= K_\delta(0, s, x, y) + R_\delta(0, s, x, y).
\end{aligned}
$$

Hence,

$$
\int_{\Omega_{f,\delta}(0)} \widehat{\boldsymbol{u}}(0) \cdot (\boldsymbol{u}_\delta)_\nu(0)
$$

$$
= \int_{\mathbb{R}} \left(\int_{\Omega_{f,\delta}(s)} \boldsymbol{u}_\delta(s, x, y) \cdot K_\delta(0, s, x, y) \right.
$$

$$
\left. \boldsymbol{u}_0 \left(x, \frac{R + \omega_\delta(0, x)}{R + \omega_\delta(s, x)}(R + y) - R \right) dxdy \right) j_\nu(t-s)ds
$$

$$
+ \int_{\mathbb{R}} \left(\int_{\Omega_{f,\delta}(s)} \boldsymbol{u}_\delta(s, x, y) \cdot R_\delta(0, s, x, y) \right.
$$

$$
\left. \boldsymbol{u}_0 \left(x, \frac{R + \omega_\delta(0, x)}{R + \omega_\delta(s, x)}(R + y) - R \right) dxdy \right) j_\nu(t-s)ds = I_{K,\delta} + I_{R,\delta}.
$$

Note that

$$I_{K,\delta} = \int_{\mathbb{R}} \left(\int_{\Omega_{f,\delta}(s)} \boldsymbol{u}_\delta(s,x,y) \cdot \tilde{\boldsymbol{q}}(s,x,y)dxdy \right) j_\nu(t-s)ds$$

where $\tilde{\boldsymbol{q}}$ is defined by (11.153). Since $\boldsymbol{u}_\delta(s) = \boldsymbol{u}_\delta(-s)$ so that $\omega_\delta(s) = \omega_\delta(-s)$ for $s \leq 0$ (see the extension procedure), we conclude by Lemma 11.10.4 that

$$I_{K,\delta} \to \int_{\Omega_{f,\delta}(0)} \boldsymbol{u}_0 \cdot \tilde{\boldsymbol{q}}(0) = \int_{\Omega_{f,\delta}(0)} |\boldsymbol{u}_0|^2, \qquad \text{as } \nu \to 0.$$

So it suffices to show that $I_{R,\delta} \to 0$ as $\nu \to 0$. This follows from the fact that $|R_\delta| \to 0$ uniformly as $s \to 0$. In particular,

$$\int_{\Omega_{f,\delta}(s)} \left| \boldsymbol{u}_\delta(s,x,y) \cdot \boldsymbol{u}_0 \left(x, \frac{R+\omega_\delta(0,x)}{R+\omega_\delta(s,x)}(R+y) - R \right) \right| dxdy \leq C,$$
$$\text{for almost all } s \in [0,T],$$

by the boundedness of $\boldsymbol{u}_\delta \in L^\infty(0,T; L^2(\Omega_{f,\delta}(t))$ and the fact that $\boldsymbol{u}_0$ is uniformly bounded. In addition, by the continuity properties of ω_δ in time, we have that

$$\max_{(x,y)\in\overline{\Omega_{f,\delta}(s)}} |R_\delta(0,s,x,y)| \to 0, \qquad \text{as } s \to 0,$$

which implies that $I_{R,\delta} \to 0$ as $\nu \to 0$. This completes the proof. □

We will use this result in the next section to estimate the first term T_1 (see (11.127) in the Gronwall's estimate).

11.10.2 Gronwall's Terms Estimates

In this Appendix, we provide details of the derivation of the terms appearing in (11.127) and the calculations providing the desired estimates of the terms in (11.127) used to prove Gronwall's estimate in Sect. 11.9.3.

Term T1. To derive term T_1, defined in (11.129), we first multiply the weak formulation (11.107) for $\boldsymbol{u}$ with the test function $\boldsymbol{v} = \boldsymbol{u} - (\breve{\boldsymbol{u}}_\delta)_\nu$ to obtain the terms:

$$T_{1,1} = -\int_0^t \int_{\Omega_f(s)} \boldsymbol{u} \cdot \partial_t \left[\boldsymbol{u} - (\breve{\boldsymbol{u}}_\delta)_\nu \right] - \frac{1}{2} \int_0^t \int_{\Gamma(s)} (\boldsymbol{\xi} \cdot \boldsymbol{n}) \boldsymbol{u} \cdot [\boldsymbol{u} - (\breve{\boldsymbol{u}}_\delta)_\nu]$$
$$+ \int_{\Omega_f(t)} \boldsymbol{u}(t) \cdot [\boldsymbol{u} - (\breve{\boldsymbol{u}}_\delta)_\nu](t) - \int_{\Omega_f(0)} \boldsymbol{u}(0) \cdot [\boldsymbol{u} - (\breve{\boldsymbol{u}}_\delta)_\nu](0),$$

where $\Omega_f(0)$ is the fluid domain corresponding to the initial structure displacement ω_0. We note that $\boldsymbol{u}$ is smooth in time and $(\breve{\boldsymbol{u}}_\delta)_\nu$ is differentiable in time as a result of the time convolution. Thus, by the Reynold's transport theorem,

$$T_{1,1} = \int_0^t \int_{\Omega_f(s)} \partial_t \boldsymbol{u} \cdot [\boldsymbol{u} - (\breve{\boldsymbol{u}}_\delta)_\nu] + \frac{1}{2} \int_0^t \int_{\Gamma(s)} (\boldsymbol{\xi} \cdot \boldsymbol{n})\boldsymbol{u} \cdot [\boldsymbol{u} - (\breve{\boldsymbol{u}}_\delta)_\nu].$$

Because $\boldsymbol{u}$ is smooth and by the weak convergence properties of $(\breve{\boldsymbol{u}}_\delta)_\nu$ in Proposition 11.9.1,

$$T_{1,1} = \int_0^t \int_{\Omega_f(s)} \partial_t \boldsymbol{u} \cdot [\boldsymbol{u} - \breve{\boldsymbol{u}}_\delta] + \frac{1}{2} \int_0^t \int_{\Gamma(s)} (\boldsymbol{\xi} \cdot \boldsymbol{n})\boldsymbol{u} \cdot [\boldsymbol{u} - \breve{\boldsymbol{u}}_\delta] + K_{1,1,\nu},$$

where $K_{1,1,\nu} \to 0$ as $\nu \to 0$. Using estimates as found in [298], we can transfer the first integral from $\Omega_1(s)$ to $\Omega_{f,\delta}(s)$ at the cost of an additional term, so that

$$T_{1,1} = \int_0^t \int_{\Omega_{f,\delta}(s)} \partial_t \widehat{\boldsymbol{u}} \cdot (\widehat{\boldsymbol{u}} - \boldsymbol{u}_\delta) + \frac{1}{2} \int_0^t \int_{\Gamma(s)} (\boldsymbol{\xi} \cdot \boldsymbol{n})\boldsymbol{u} \cdot (\boldsymbol{u} - \breve{\boldsymbol{u}}_\delta) + \tilde{R}_1 + K_{1,1,\nu},$$

where

$$|\tilde{R}_1| \le \epsilon \int_0^t \|\widehat{\boldsymbol{u}} - \boldsymbol{u}_\delta\|^2_{H^1(\Omega_{f,\delta}(s))}$$
$$+C(\epsilon)\left(\int_0^t \|\omega - \omega_\delta\|^2_{H^2(\Gamma)} + \int_0^t \|\partial_t\omega - \partial_t\omega_\delta\|^2_{L^2(\Gamma)} + \int_0^t \|\widehat{\boldsymbol{u}} - \boldsymbol{u}_\delta\|^2_{L^2(\Omega_{f,\delta}(s))}\right).$$

Thus, by using Proposition 11.9.1 again,

$$T_{1,1} = \int_0^t \int_{\Omega_{f,\delta}(s)} \partial_t \widehat{\boldsymbol{u}} \cdot (\widehat{\boldsymbol{u}} - (\boldsymbol{u}_\delta)_\nu) + \frac{1}{2} \int_0^t \int_{\Gamma(s)} (\boldsymbol{\xi} \cdot \boldsymbol{n})\boldsymbol{u} \cdot (\boldsymbol{u} - (\breve{\boldsymbol{u}}_\delta)_\nu) + \tilde{R}_1 + K_{1,1,\nu}, \tag{11.159}$$

where $K_{1,1,\nu} \to 0$ as $\nu \to 0$.
Next, we test the regularized weak formulation for $\boldsymbol{u}_\delta$ with $\widehat{\boldsymbol{u}}$ and obtain the following terms:

$$T_{1,2} = -\int_0^t \int_{\Omega_{f,\delta}(s)} \boldsymbol{u}_\delta \cdot \partial_t \widehat{\boldsymbol{u}} - \frac{1}{2} \int_0^t \int_{\Gamma_\delta(s)} (\boldsymbol{\xi}_\delta \cdot \boldsymbol{n}_\delta)\boldsymbol{u}_\delta \cdot \widehat{\boldsymbol{u}} + \int_{\Omega_{f,\delta}(t)} \boldsymbol{u}_\delta(t) \cdot \widehat{\boldsymbol{u}}(t)$$
$$- \int_{\Omega_f(0)} \boldsymbol{u}_\delta(0) \cdot \widehat{\boldsymbol{u}}(0).$$

We want to integrate by parts in time, but $\boldsymbol{u}_\delta$ is not necessarily smooth in time. Thus, we replace $\boldsymbol{u}_\delta$ by its time regularization $(\boldsymbol{u}_\delta)_\nu$ at the cost of a term $K_{1,2,\nu}$, which goes to zero as $\nu \to 0$ by Proposition 11.9.1. Combining this with the Reynold's transport theorem, we get:

$$T_{1,2} = \int_0^t \int_{\Omega_{f,\delta}(s)} \partial_t \left[(\boldsymbol{u}_\delta)_\nu \right] \cdot \widehat{\boldsymbol{u}} + \frac{1}{2} \int_0^t \int_{\Gamma_\delta(s)} (\boldsymbol{\xi}_\delta \cdot \boldsymbol{n}_\delta)(\boldsymbol{u}_\delta)_\nu \cdot \widehat{\boldsymbol{u}} + K_{1,2,\nu}, \tag{11.160}$$

where $K_{1,2,\nu} \to 0$ as $\nu \to 0$.
Now, from the energy inequality, we obtain the terms

$$T_{1,3} = \frac{1}{2} \int_{\Omega_{f,\delta}(t)} |\boldsymbol{u}_\delta(t)|^2 - \frac{1}{2} \int_{\Omega_{f,\delta}(0)} |\boldsymbol{u}_\delta(0)|^2. \tag{11.161}$$

Using the Reynold's transport theorem, the total contribution $T_1 = T_{1,1} - T_{1,2} + T_{1,3}$ is

$$\begin{aligned} T_1 &= \frac{1}{2} \int_{\Omega_{f,\delta}(t)} |\widehat{\boldsymbol{u}}(t)|^2 - \frac{1}{2} \int_{\Omega_{f,\delta}(0)} |\widehat{\boldsymbol{u}}(0)|^2 - \int_{\Omega_{f,\delta}(t)} (\widehat{\boldsymbol{u}} \cdot (\boldsymbol{u}_\delta)_\nu)(t) \\ &+ \int_{\Omega_{f,\delta}(0)} (\widehat{\boldsymbol{u}} \cdot (\boldsymbol{u}_\delta)_\nu)(0) \\ &+ \frac{1}{2} \int_{\Omega_{f,\delta}(t)} |\boldsymbol{u}_\delta(t)|^2 - \frac{1}{2} \int_{\Omega_{f,\delta}(0)} |\boldsymbol{u}_\delta(0)|^2 - \frac{1}{2} \int_0^t \int_{\Gamma_\delta(s)} (\boldsymbol{\xi}_\delta \cdot \boldsymbol{n}_\delta) \widehat{\boldsymbol{u}} \cdot (\widehat{\boldsymbol{u}} - (\boldsymbol{u}_\delta)_\nu) \\ &+ \frac{1}{2} \int_0^t \int_{\Gamma(s)} (\boldsymbol{\xi} \cdot \boldsymbol{n}) \boldsymbol{u} \cdot (\boldsymbol{u} - (\breve{\boldsymbol{u}}_\delta)_\nu) + \tilde{R}_1 + K_{1,1,\nu} + K_{1,2,\nu}. \end{aligned}$$

By Proposition 11.9.1, $(\boldsymbol{u}_\delta)_\nu$ and $(\breve{\boldsymbol{u}}_\delta)_\nu$ converge weakly to $\boldsymbol{u}_\delta$ and $\breve{\boldsymbol{u}}_\delta$, respectively, weakly in $L^2(0, T, W^{1,p}(\Omega_{f,\delta}(t)))$ and $L^2(0, T, W^{1,p}(\Omega_{f,1}(t)))$ for all $p \in [1, 2)$. Furthermore, by Lemma 11.10.5 proved in the Appendix above, we have that

$$\begin{aligned} \int_{\Omega_{f,\delta}(0)} (\widehat{\boldsymbol{u}} \cdot (\boldsymbol{u}_\delta)_\nu)(0) \to \int_{\Omega_{f,\delta}(0)} (\widehat{\boldsymbol{u}} \cdot \boldsymbol{u}_\delta)(0), \quad & \int_{\Omega_{f,\delta}(t)} (\widehat{\boldsymbol{u}} \cdot (\boldsymbol{u}_\delta)_\nu)(t) \\ \to \int_{\Omega_{f,\delta}(t)} (\widehat{\boldsymbol{u}} \cdot \boldsymbol{u}_\delta)(t). & \end{aligned} \tag{11.162}$$

Thus, taking the limit as $\nu \to 0$, the contribution of this term is now

$$\begin{aligned} T_1 &= \frac{1}{2} \int_{\Omega_{f,\delta}(s)} |(\widehat{\boldsymbol{u}} - \boldsymbol{u}_\delta)(t)|^2 - \frac{1}{2} \int_{\Omega_{f,\delta}(0)} |(\widehat{\boldsymbol{u}} - \boldsymbol{u}_\delta)(0)|^2 \\ &- \frac{1}{2} \int_0^t \int_{\Gamma_\delta(s)} (\boldsymbol{\xi}_\delta \cdot \boldsymbol{n}_\delta) \widehat{\boldsymbol{u}} \cdot (\widehat{\boldsymbol{u}} - \boldsymbol{u}_\delta) + \frac{1}{2} \int_0^t \int_{\Gamma(s)} (\boldsymbol{\xi} \cdot \boldsymbol{n}) \boldsymbol{u} \cdot (\boldsymbol{u} - \breve{\boldsymbol{u}}_\delta) + \tilde{R}_1. \end{aligned}$$

Since $\widehat{\boldsymbol{u}}(0) = \boldsymbol{u}_\delta(0) = \boldsymbol{u}_0$, we obtain after some standard estimates that

$$T_1 = \frac{1}{2}\int_{\Omega_{f,\delta}(t)} |(\widehat{\boldsymbol{u}} - \boldsymbol{u}_\delta)(t)|^2 + R_1,$$

where

$$\begin{aligned}|R_1| \le \epsilon \int_0^t \|\widehat{\boldsymbol{u}} - \boldsymbol{u}_\delta\|^2_{H^1(\Omega_{f,\delta}(s))} \\ + C(\epsilon)\left(\int_0^t \|\omega - \omega_\delta\|^2_{H^2(\Gamma)} + \int_0^t \|\partial_t\omega - \partial_t\omega_\delta\|^2_{L^2(\Gamma)} \right.\\ \left. + \int_0^t \|\widehat{\boldsymbol{u}} - \boldsymbol{u}_\delta\|^2_{L^2(\Omega_{f,\delta}(s))}\right).\end{aligned}$$

This completes the calculations associated with term T_1.

Term T2. To estimate term T_2, defined in (11.130), we notice that since $(\breve{\boldsymbol{u}}_\delta)_\nu$ converges weakly to $\breve{\boldsymbol{u}}_\delta$ in $L^2(0, T; W^{1,p}(\Omega_{f,\delta}(t)))$ for $p \in [1, 2)$ by Proposition 11.9.1, and because $\boldsymbol{u}$ is smooth, as $\nu \to 0$, we have that T_2 converges to

$$\begin{aligned}T_2 := \frac{1}{2}\int_0^t\int_{\Omega_f(s)} ((\boldsymbol{u}\cdot\nabla)\boldsymbol{u})\cdot(\boldsymbol{u} - \breve{\boldsymbol{u}}_\delta) - \frac{1}{2}\int_0^t\int_{\Omega_f(s)} ((\boldsymbol{u}\cdot\nabla)(\boldsymbol{u} - \breve{\boldsymbol{u}}_\delta))\cdot\boldsymbol{u} \\ - \frac{1}{2}\int_0^t\int_{\Omega_{f,\delta}(s)} ((\boldsymbol{u}_\delta\cdot\nabla)\boldsymbol{u}_\delta)\cdot(\widehat{\boldsymbol{u}} - \boldsymbol{u}_\delta) + \frac{1}{2}\int_0^t\int_{\Omega_{f,\delta}(s)} ((\boldsymbol{u}_\delta\cdot\nabla)(\widehat{\boldsymbol{u}} - \boldsymbol{u}_\delta))\cdot\boldsymbol{u}_\delta.\end{aligned}$$

We note that the quantity $\frac{1}{2}\int_0^t\int_{\Omega_{f,\delta}(s)} ((\boldsymbol{u}_\delta\cdot\nabla)\boldsymbol{u}_\delta)\cdot\boldsymbol{u}_\delta$, is well defined because $\boldsymbol{u}_\delta \in L^\infty(0, T; L^2(\Omega_{f,\delta}(t))) \cap L^2(0, T; H^1(\Omega_{f,\delta}(t)))$, which by interpolation is in $L^4(0, T; H^{1/2}(\Omega_{f,\delta}(t)))$ and hence by Sobolev inequalities embeds into $L^4(0, T; L^4(\Omega_{f,\delta}(t)))$.

We want to transfer the integrals

$$\int_0^t\int_{\Omega_f(s)} ((\boldsymbol{u}\cdot\nabla)\boldsymbol{u})\cdot(\boldsymbol{u} - \breve{\boldsymbol{u}}_\delta), \qquad \int_0^t\int_{\Omega_f(s)} ((\boldsymbol{u}\cdot\nabla)(\boldsymbol{u} - \breve{\boldsymbol{u}}_\delta))\cdot\boldsymbol{u}, \tag{11.163}$$

to integrals on $\Omega_{f,\delta}(s)$ by using the map $\psi_\delta(s,\cdot) : \Omega_{f,\delta}(s) \to \Omega_f(s)$ defined by (11.109). We use

$$\widehat{\boldsymbol{u}} = \gamma_\delta J_\delta^{-1}\cdot(\boldsymbol{u}\circ\psi_\delta), \qquad \widehat{\boldsymbol{u}} - \boldsymbol{u}_\delta = \gamma_\delta J_\delta^{-1}\cdot((\boldsymbol{u} - \breve{\boldsymbol{u}}_\delta)\circ\psi_\delta),$$

where we recall the definitions of the appropriate terms from (11.109), (11.111), (11.113), and (11.114).

Following arguments found in [298], we obtain the following estimates. We have, using (11.110), that

$$\int_0^t \int_{\Omega_f(s)} ((\boldsymbol{u} \cdot \nabla)\boldsymbol{u}) \cdot (\boldsymbol{u} - \breve{\boldsymbol{u}}_\delta)$$

$$= \int_0^t \int_{\Omega_{f,\delta}(s)} \gamma_\delta[(\nabla(\boldsymbol{u} \circ \psi_\delta))J_\delta^{-1}(\boldsymbol{u} \circ \psi_\delta)] \cdot (\boldsymbol{u} - \breve{\boldsymbol{u}}_\delta) \circ \psi_\delta$$

$$= \int_0^t \int_{\Omega_{f,\delta}(s)} [(\nabla(\boldsymbol{u} \circ \psi_\delta))\widehat{\boldsymbol{u}}] \cdot [\gamma_\delta^{-1} J_\delta(\widehat{\boldsymbol{u}} - \boldsymbol{u}_\delta)]$$

$$= \int_0^t \int_{\Omega_{f,\delta}(s)} [(\nabla(\boldsymbol{u} \circ \psi_\delta))\widehat{\boldsymbol{u}}] \cdot (\widehat{\boldsymbol{u}} - \boldsymbol{u}_\delta) - \int_0^t \int_{\Omega_{f,\delta}(s)} [(\nabla(\boldsymbol{u} \circ \psi_\delta))\widehat{\boldsymbol{u}}]$$
$$\cdot [(I - \gamma_\delta^{-1} J_\delta)(\widehat{\boldsymbol{u}} - \boldsymbol{u}_\delta)]$$

$$= \int_0^t \int_{\Omega_{f,\delta}(s)} ((\nabla\widehat{\boldsymbol{u}})\widehat{\boldsymbol{u}}) \cdot (\widehat{\boldsymbol{u}} - \boldsymbol{u}_\delta)$$

$$+ \int_0^t \int_{\Omega_{f,\delta}(s)} (\nabla((I - \gamma_\delta J_\delta^{-1})(\boldsymbol{u} \circ \psi_\delta))\widehat{\boldsymbol{u}}) \cdot (\widehat{\boldsymbol{u}} - \boldsymbol{u}_\delta)$$

$$- \int_0^t \int_{\Omega_{f,\delta}(s)} [(\nabla(\boldsymbol{u} \circ \psi_\delta))\widehat{\boldsymbol{u}}] \cdot [(I - \gamma_\delta^{-1} J_\delta)(\widehat{\boldsymbol{u}} - \boldsymbol{u}_\delta)]$$

$$= \int_0^t \int_{\Omega_{f,\delta}(s)} [(\widehat{\boldsymbol{u}} \cdot \nabla)\widehat{\boldsymbol{u}}] \cdot (\widehat{\boldsymbol{u}} - \boldsymbol{u}_\delta) + R_{2,1}, \tag{11.164}$$

where

$$R_{2,1} = \int_0^t \int_{\Omega_{f,\delta}(s)} (\nabla((I - \gamma_\delta J_\delta^{-1})(\boldsymbol{u} \circ \psi_\delta))\widehat{\boldsymbol{u}}) \cdot (\widehat{\boldsymbol{u}} - \boldsymbol{u}_\delta)$$
$$- \int_0^t \int_{\Omega_{f,\delta}(s)} [(\nabla(\boldsymbol{u} \circ \psi_\delta))\widehat{\boldsymbol{u}}] \cdot [(I - \gamma_\delta^{-1} J_\delta)(\widehat{\boldsymbol{u}} - \boldsymbol{u}_\delta)].$$

In the following estimates, we will repeatedly use the following inequalities, which hold for a constant C that is independent of δ:

$$|\gamma_\delta^{-1} J_\delta - I| \leq C(|\gamma_\delta^{-1} - 1| + |\nabla\gamma_\delta|) \leq C\|\omega - \omega_\delta\|_{H^2(\Gamma)},$$

$$|\gamma_\delta J_\delta^{-1} - I| \leq C(|\gamma_\delta - 1| + |\nabla\gamma_\delta|) \leq C\|\omega - \omega_\delta\|_{H^2(\Gamma)},$$

$$|\nabla(\gamma_\delta J_\delta^{-1})| \leq C(|\partial_x \gamma_\delta| + |\partial_{xx}\gamma_\delta|) \leq C(\|\omega - \omega_\delta\|_{H^2(\Gamma)} + |\partial_{xx}(\omega - \omega_\delta)|), \tag{11.165}$$

so that

$$\|\nabla(\gamma_\delta J_\delta^{-1})\|_{L^2(\Omega_{f,\delta}(t))} \leq C\|\omega - \omega_\delta\|_{H^2(\Gamma)}. \tag{11.166}$$

To obtain (11.165), we estimate $|\partial_{xx}\gamma_\delta|$ by using the fact that ω is smooth so that $|\partial_{xx}\omega| \leq C$ and a direct computation of $\partial_{xx}\gamma_\delta$. Using these estimates, the Leibniz rule, and the smoothness of $\boldsymbol{u}$, we get

$$\left|\int_0^t \int_{\Omega_{f,\delta}(s)} (\nabla((I - \gamma_\delta J_\delta^{-1})(\boldsymbol{u} \circ \psi_\delta))\hat{\boldsymbol{u}}) \cdot (\hat{\boldsymbol{u}} - \boldsymbol{u}_\delta)\right|$$
$$\leq C \int_0^t \|\omega - \omega_\delta\|_{H^2(\Gamma)} \|\hat{\boldsymbol{u}} - \boldsymbol{u}_\delta\|_{L^2(\Omega_{f,\delta}(s))}$$
$$\leq C\left(\int_0^t \|\omega - \omega_\delta\|^2_{H^2(\Gamma)} + \int_0^t \|\hat{\boldsymbol{u}} - \boldsymbol{u}_\delta\|^2_{L^2(\Omega_{f,\delta}(s))}\right).$$

By using (11.110), and the fact that $|J_\delta| \leq C$ is uniformly bounded, due to the fact that $|J_\delta| \leq C(1 + \|\omega - \omega_\delta\|_{H^2(\Gamma)}) \leq C$ is uniformly bounded, we obtain a similar estimate:

$$\int_0^t \int_{\Omega_{f,\delta}(s)} [(\nabla(\boldsymbol{u} \circ \psi_\delta))\hat{\boldsymbol{u}}] \cdot [(I - \gamma_\delta^{-1} J_\delta)(\hat{\boldsymbol{u}} - \boldsymbol{u}_\delta)]$$
$$\leq C\left(\int_0^t \|\omega - \omega_\delta\|^2_{H^2(\Gamma)} + \int_0^t \|\hat{\boldsymbol{u}} - \boldsymbol{u}_\delta\|^2_{L^2(\Omega_{f,\delta}(s))}\right).$$

Thus, we obtain

$$|R_1| \leq C\left(\int_0^t \|\omega - \omega_\delta\|^2_{H^2(\Gamma)} + \int_0^t \|\hat{\boldsymbol{u}} - \boldsymbol{u}_\delta\|^2_{L^2(\Omega_{f,\delta}(s))}\right). \tag{11.167}$$

We now focus on the second integral in (11.163). By using (11.110), we obtain

$$\int_0^t \int_{\Omega_f(s)} ((\boldsymbol{u} \cdot \nabla)(\boldsymbol{u} - \check{\boldsymbol{u}}_\delta)) \cdot \boldsymbol{u}$$
$$= \int_0^t \int_{\Omega_{f,\delta}(s)} \gamma_\delta[(\nabla((\boldsymbol{u} - \check{\boldsymbol{u}}_\delta) \circ \psi_\delta))J_\delta^{-1}(\boldsymbol{u} \circ \psi_\delta)] \cdot (\boldsymbol{u} \circ \psi_\delta)$$
$$= \int_0^t \int_{\Omega_{f,\delta}(s)} [(\nabla((\boldsymbol{u} - \check{\boldsymbol{u}}_\delta) \circ \psi_\delta))\hat{\boldsymbol{u}}] \cdot (\gamma_\delta^{-1} J_\delta \hat{\boldsymbol{u}})$$
$$= \int_0^t \int_{\Omega_{f,\delta}(s)} [(\nabla((\boldsymbol{u} - \check{\boldsymbol{u}}_\delta) \circ \psi_\delta))\hat{\boldsymbol{u}}] \cdot \hat{\boldsymbol{u}}$$

$$- \int_0^t \int_{\Omega_{f,\delta}(s)} [(\nabla((\boldsymbol{u} - \breve{\boldsymbol{u}}_\delta) \circ \psi_\delta))\widehat{\boldsymbol{u}}] \cdot [(I - \gamma_\delta^{-1} J_\delta)\widehat{\boldsymbol{u}}]$$

$$= \int_0^t \int_{\Omega_{f,\delta}(s)} (\nabla(\widehat{\boldsymbol{u}} - \boldsymbol{u}_\delta)\widehat{\boldsymbol{u}}) \cdot \widehat{\boldsymbol{u}}$$

$$+ \int_0^t \int_{\Omega_{f,\delta}(s)} (\nabla[(I - \gamma_\delta J_\delta^{-1})((\boldsymbol{u} - \breve{\boldsymbol{u}}_\delta) \circ \psi_\delta)]\widehat{\boldsymbol{u}}) \cdot \widehat{\boldsymbol{u}}$$

$$- \int_0^t \int_{\Omega_{f,\delta}(s)} [(\nabla((\boldsymbol{u} - \breve{\boldsymbol{u}}_\delta) \circ \psi_\delta))\widehat{\boldsymbol{u}}] \cdot [(I - \gamma_\delta^{-1} J_\delta)\widehat{\boldsymbol{u}}]$$

$$= \int_0^t \int_{\Omega_{f,\delta}(s)} ((\widehat{\boldsymbol{u}} \cdot \nabla)(\widehat{\boldsymbol{u}} - \boldsymbol{u}_\delta)) \cdot \widehat{\boldsymbol{u}} + R_{2,2}, \tag{11.168}$$

where

$$R_{2,2} := \int_0^t \int_{\Omega_{f,\delta}(s)} (\nabla[(I - \gamma_\delta J_\delta^{-1})((\boldsymbol{u} - \breve{\boldsymbol{u}}_\delta) \circ \psi_\delta)]\widehat{\boldsymbol{u}}) \cdot \widehat{\boldsymbol{u}}$$
$$- \int_0^t \int_{\Omega_{f,\delta}(s)} [(\nabla((\boldsymbol{u} - \breve{\boldsymbol{u}}_\delta) \circ \psi_\delta))\widehat{\boldsymbol{u}}] \cdot [(I - \gamma_\delta^{-1} J_\delta)\widehat{\boldsymbol{u}}].$$

To estimate $R_{2,2}$, we will use the following inequalities:

$$|(\boldsymbol{u} - \breve{\boldsymbol{u}}_\delta) \circ \psi| = |\gamma_\delta^{-1} J_\delta \cdot (\widehat{\boldsymbol{u}} - \boldsymbol{u}_\delta)| \le C|\widehat{\boldsymbol{u}} - \boldsymbol{u}_\delta|,$$

$$\begin{aligned} |\nabla((\boldsymbol{u} - \breve{\boldsymbol{u}}_\delta) \circ \psi_\delta)| = |\nabla(\gamma_\delta^{-1} J_\delta \cdot (\widehat{\boldsymbol{u}} - \boldsymbol{u}_\delta))| &\le |\nabla(\gamma_\delta^{-1} J_\delta)| \cdot |\widehat{\boldsymbol{u}} - \boldsymbol{u}_\delta| \\ &\quad + |\gamma_\delta^{-1} J_\delta| \cdot |\nabla(\widehat{\boldsymbol{u}} - \boldsymbol{u}_\delta)| \\ &\le C(|\nabla(\gamma_\delta^{-1} J_\delta)| \cdot |\widehat{\boldsymbol{u}} - \boldsymbol{u}_\delta| + |\nabla(\widehat{\boldsymbol{u}} - \boldsymbol{u}_\delta)|). \end{aligned}$$

From the fact that $\max\left(|I - \gamma_\delta^{-1} J_\delta|, |I - \gamma_\delta J_\delta^{-1}|\right) \le C \min\left(1, \|\omega - \omega_\delta\|_{H^2(\Gamma)}\right)$, we obtain:

$$\begin{aligned} |R_{2,2}| \le C\Bigg(&\int_0^t \int_{\Omega_{f,\delta}(s)} |\nabla(\gamma_\delta J_\delta^{-1})| \cdot |(\boldsymbol{u} - \breve{\boldsymbol{u}}_\delta) \circ \psi_\delta| \\ &+ \int_0^t \int_{\Omega_{f,\delta}(s)} |I - \gamma_\delta J_\delta^{-1}| \cdot |\nabla((\boldsymbol{u} - \breve{\boldsymbol{u}}_\delta) \circ \psi_\delta)| \\ &+ \int_0^t \int_{\Omega_{f,\delta}(s)} |I - \gamma_\delta^{-1} J_\delta| \cdot |\nabla((\boldsymbol{u} - \breve{\boldsymbol{u}}_\delta) \circ \psi_\delta)| \Bigg) \end{aligned}$$

$$
\begin{aligned}
&\le C\Bigg(\int_0^t\int_{\Omega_{f,\delta}(s)}\Big(|\nabla(\gamma_\delta J_\delta^{-1})|+|\nabla(\gamma_\delta^{-1}J_\delta)|\Big)\cdot|\widehat{\boldsymbol{u}}-\boldsymbol{u}_\delta| \\
&+\int_0^t\int_{\Omega_{f,\delta}(s)}\|\omega-\omega_\delta\|_{H^2(\Gamma)}\cdot|\nabla(\widehat{\boldsymbol{u}}-\boldsymbol{u}_\delta)|\Bigg) \\
&\le \epsilon\int_0^t\|\nabla(\widehat{\boldsymbol{u}}-\boldsymbol{u}_\delta)\|^2_{L^2(\Omega_{f,\delta}(s))}+C(\epsilon)\left(\int_0^t\|\omega-\omega_\delta\|^2_{H^2(\Gamma)}\right. \\
&\left.+\int_0^t\|\widehat{\boldsymbol{u}}-\boldsymbol{u}_\delta\|^2_{L^2(\Omega_{f,\delta}(s))}\right). \qquad (11.169)
\end{aligned}
$$

In the last line, we use the following estimates, derived similarly as for (11.166),

$$
\begin{aligned}
|\nabla(\gamma_\delta^{-1}J_\delta)|\le C(|\partial_x(\gamma_\delta^{-1})|+|\partial_x\gamma_\delta|+|\partial_{xx}\gamma_\delta|)&\le C(\|\omega-\omega_\delta\|_{H^2(\Gamma)} \\
&\quad+|\partial_{xx}(\omega-\omega_\delta)|), \\
\|\nabla(\gamma_\delta^{-1}J_\delta)\|_{L^2(\Omega_{f,\delta}(t))}&\le C\|\omega-\omega_\delta\|_{H^2(\Gamma)}.
\end{aligned}
$$

Therefore, for the expression in (11.130), after transferring the integrals (11.164) and (11.168) and estimating $R_{2,1}$ (11.167) and $R_{2,2}$ (11.169), the remaining terms are:

$$
\begin{aligned}
&\frac{1}{2}\int_0^t\int_{\Omega_{f,\delta}(s)}[(\widehat{\boldsymbol{u}}\cdot\nabla)\widehat{\boldsymbol{u}}]\cdot(\widehat{\boldsymbol{u}}-\boldsymbol{u}_\delta)-[(\widehat{\boldsymbol{u}}\cdot\nabla)(\widehat{\boldsymbol{u}}-\boldsymbol{u}_\delta)]\cdot\widehat{\boldsymbol{u}} \\
&-\frac{1}{2}\int_0^t\int_{\Omega_{f,\delta}(s))}[(\boldsymbol{u}_\delta\cdot\nabla)\boldsymbol{u}_\delta]\cdot(\widehat{\boldsymbol{u}}-\boldsymbol{u}_\delta)-[(\boldsymbol{u}_\delta\cdot\nabla)(\widehat{\boldsymbol{u}}-\boldsymbol{u}_\delta)]\cdot\boldsymbol{u}_\delta \\
&=\frac{1}{2}\int_0^t\int_{\Omega_{f,\delta}(s)}[((\widehat{\boldsymbol{u}}-\boldsymbol{u}_\delta)\cdot\nabla)\boldsymbol{u}_\delta]\cdot\widehat{\boldsymbol{u}}-\frac{1}{2}\int_0^t\int_{\Omega_{f,\delta}(s)}[((\widehat{\boldsymbol{u}}-\boldsymbol{u}_\delta)\cdot\nabla)\widehat{\boldsymbol{u}}]\cdot\boldsymbol{u}_\delta.
\end{aligned}
$$

In absolute values, the right-hand side can be bounded as follows:

$$
\le \epsilon\int_0^t\|\nabla(\widehat{\boldsymbol{u}}-\boldsymbol{u}_\delta)\|^2_{L^2(\Omega_{f,\delta}(s))}+C(\epsilon)\int_0^t\|\widehat{\boldsymbol{u}}-\boldsymbol{u}_\delta\|^2_{L^2(\Omega_{f,\delta}(s))}.
$$

Combining this estimate with (11.167) and (11.169), we obtain

$$
\begin{aligned}
|T_2|\le \epsilon\int_0^t\|\nabla(\widehat{\boldsymbol{u}}-\boldsymbol{u}_\delta)\|^2_{L^2(\Omega_{f,\delta}(s))}&+C(\epsilon)\left(\int_0^t\|\omega-\omega_\delta\|^2_{H^2(\Gamma)}\right. \\
&\left.+\int_0^t\|\widehat{\boldsymbol{u}}-\boldsymbol{u}_\delta\|^2_{L^2(\Omega_{f,\delta}(s))}\right).
\end{aligned}
$$

Term T3. To estimate term T_3 defined in (11.131), we start by noting that because $\boldsymbol{u}$ and $\boldsymbol{\xi}$ are smooth, we can pass to the limit as $\nu \to 0$ using Proposition 11.9.1 and the fact that $(\boldsymbol{\xi}_\delta)_\nu \to \boldsymbol{\xi}_\delta$ strongly in $L^2(0, T; H^1(\Omega_b))$, so that we can ultimately just test with $\boldsymbol{v} = \boldsymbol{u} - \breve{\boldsymbol{u}}_\delta$ and $\boldsymbol{\psi} = \boldsymbol{\xi} - \boldsymbol{\xi}_\delta$. In the regularized weak formulation for $\boldsymbol{u}_\delta$, we test with $\boldsymbol{u}$ and $\boldsymbol{\xi}$. Note that both test functions $\boldsymbol{u} - \breve{\boldsymbol{u}}_\delta$ and $\widehat{\boldsymbol{u}} - \boldsymbol{u}_\delta$ have the same trace along $\Gamma(t)$ and $\Gamma_\delta(t)$, respectively, which we will formally denote by $\boldsymbol{u} - \boldsymbol{u}_\delta$ along the reference configuration of the interface Γ. Combining the resulting expressions, we have the following contribution of T_3 in the limit as $\nu \to 0$:

$$
\begin{aligned}
T_3 =& \frac{1}{2}\int_0^t \int_{\Gamma(s)} (\boldsymbol{u} \cdot \boldsymbol{n} - \boldsymbol{\xi} \cdot \boldsymbol{n})\boldsymbol{u} \cdot (\boldsymbol{u} - \breve{\boldsymbol{u}}_\delta) - \frac{1}{2}\int_0^t \int_{\Gamma_\delta(s)} (\boldsymbol{u}_\delta \cdot \boldsymbol{n}_\delta - \boldsymbol{\xi}_\delta \cdot \boldsymbol{n}_\delta)\boldsymbol{u}_\delta \cdot \widehat{\boldsymbol{u}} \\
&+ \frac{1}{2}\int_0^t \int_{\Gamma(s)} |\boldsymbol{u}|^2(\boldsymbol{\xi} \cdot \boldsymbol{n} - \boldsymbol{u} \cdot \boldsymbol{n}) - \frac{1}{2}\int_0^t \int_{\Gamma(s)} |\boldsymbol{u}|^2(\boldsymbol{\xi}_\delta \cdot \boldsymbol{n} - \breve{\boldsymbol{u}}_\delta \cdot \boldsymbol{n}) \\
&- \frac{1}{2}\int_0^t \int_{\Gamma_\delta(s)} |\boldsymbol{u}_\delta|^2(\boldsymbol{\xi} \cdot \boldsymbol{n}_\delta - \widehat{\boldsymbol{u}} \cdot \boldsymbol{n}_\delta) = \frac{1}{2}\int_0^t \int_{\Gamma(s)} (\boldsymbol{\xi} \cdot \boldsymbol{n} - \boldsymbol{u} \cdot \boldsymbol{n})\boldsymbol{u} \cdot \breve{\boldsymbol{u}}_\delta \\
&- \frac{1}{2}\int_0^t \int_{\Gamma(s)} (\boldsymbol{\xi}_\delta \cdot \boldsymbol{n} - \breve{\boldsymbol{u}}_\delta \cdot \boldsymbol{n})|\boldsymbol{u}|^2 - \frac{1}{2}\int_0^t \int_{\Gamma_\delta(s)} (\boldsymbol{\xi} \cdot \boldsymbol{n}_\delta - \widehat{\boldsymbol{u}} \cdot \boldsymbol{n}_\delta)|\boldsymbol{u}_\delta|^2 \\
&+ \frac{1}{2}\int_0^t \int_{\Gamma_\delta(s)} (\boldsymbol{\xi}_\delta \cdot \boldsymbol{n}_\delta - \boldsymbol{u}_\delta \cdot \boldsymbol{n}_\delta)\boldsymbol{u}_\delta \cdot \widehat{\boldsymbol{u}} = R_{3,1} + R_{3,2},
\end{aligned}
$$

where

$$
R_{3,1} = \frac{1}{2}\int_0^t \int_\Gamma [(\boldsymbol{\xi} - \boldsymbol{u}) \cdot \boldsymbol{e}_y]\boldsymbol{u}_\delta \cdot (\boldsymbol{u} - \boldsymbol{u}_\delta) - \frac{1}{2}\int_0^t \int_\Gamma [(\boldsymbol{\xi}_\delta - \boldsymbol{u}_\delta) \cdot \boldsymbol{e}_y]\boldsymbol{u} \cdot (\boldsymbol{u} - \boldsymbol{u}_\delta),
$$

$$
\begin{aligned}
R_{3,2} = & \frac{1}{2}\int_0^t \int_\Gamma \partial_x \omega (\boldsymbol{u} \cdot \boldsymbol{e}_x)\boldsymbol{u} \cdot \boldsymbol{u}_\delta - \frac{1}{2}\int_0^t \int_\Gamma \partial_x \omega (\boldsymbol{u}_\delta \cdot \boldsymbol{e}_x)|\boldsymbol{u}|^2 \\
&- \frac{1}{2}\int_0^t \int_\Gamma \partial_x \omega_\delta (\boldsymbol{u} \cdot \boldsymbol{e}_x)|\boldsymbol{u}_\delta|^2 \\
&+ \frac{1}{2}\int_0^t \int_\Gamma \partial_x \omega_\delta (\boldsymbol{u}_\delta \cdot \boldsymbol{e}_x)\boldsymbol{u} \cdot \boldsymbol{u}_\delta.
\end{aligned}
$$

We estimate $R_{3,1}$ as follows: decompose $R_{3,1}$ as $R_{3,1} = R_{3,1,1} + R_{3,1,2}$, where

$$
R_{3,1,1} = -\frac{1}{2}\int_0^t \int_\Gamma (\boldsymbol{\xi} \cdot \boldsymbol{e}_y)(\boldsymbol{u} - \boldsymbol{u}_\delta) \cdot (\boldsymbol{u} - \boldsymbol{u}_\delta) + \frac{1}{2}\int_0^t \int_\Gamma [(\boldsymbol{\xi} - \boldsymbol{\xi}_\delta) \cdot \boldsymbol{e}_y]\boldsymbol{u} \cdot (\boldsymbol{u} - \boldsymbol{u}_\delta),
$$

$$
R_{3,1,2} = \frac{1}{2}\int_0^t \int_\Gamma (\boldsymbol{u} \cdot \boldsymbol{e}_y)(\boldsymbol{u} - \boldsymbol{u}_\delta) \cdot (\boldsymbol{u} - \boldsymbol{u}_\delta) - \frac{1}{2}\int_0^t \int_\Gamma [(\boldsymbol{u} - \boldsymbol{u}_\delta) \cdot \boldsymbol{e}_y]\boldsymbol{u} \cdot (\boldsymbol{u} - \boldsymbol{u}_\delta).
$$

By interpolation,

$$|R_{3,1,1}| \le C\left(\int_0^t \|\widehat{\boldsymbol{u}}-\boldsymbol{u}_\delta\|_{L^2(\Omega_{f,\delta}(s))}^{1/2}\|\widehat{\boldsymbol{u}}-\boldsymbol{u}_\delta\|_{H^1(\Omega_{f,\delta}(s))}^{3/2} + \int_0^t \|\boldsymbol{\xi}-\boldsymbol{\xi}_\delta\|_{L^2(\Gamma)}\|\widehat{\boldsymbol{u}}-\boldsymbol{u}_\delta\|_{H^1(\Omega_{f,\delta}(s))}\right)$$

$$\le \epsilon \int_0^t \|\widehat{\boldsymbol{u}}-\boldsymbol{u}_\delta\|_{H^1(\Omega_{f,\delta}(s))}^2 + C(\epsilon) \left(\int_0^t \|\widehat{\boldsymbol{u}}-\boldsymbol{u}_\delta\|_{L^2(\Omega_{f,\delta}(s))}^2 + \int_0^t \|\boldsymbol{\xi}-\boldsymbol{\xi}_\delta\|_{L^2(\Gamma)}^2\right).$$

By using the same interpolation inequality, we obtain the following estimate for $R_{3,1,2}$.

$$|R_{3,1,2}| \le \epsilon \int_0^t \|\widehat{\boldsymbol{u}}-\boldsymbol{u}_\delta\|_{H^1(\Omega_{f,\delta}(s))}^2 + C(\epsilon)\int_0^t \|\widehat{\boldsymbol{u}}-\boldsymbol{u}_\delta\|_{L^2(\Omega_{f,\delta}(s))}^2.$$

We estimate $R_{3,2}$ by first rewriting $R_{3,2}$ as follows:

$$R_{3,2} = -\frac{1}{2}\int_0^t\int_\Gamma (\partial_x\omega - \partial_x\omega_\delta)(\boldsymbol{u})_x\boldsymbol{u}\cdot(\boldsymbol{u}-\boldsymbol{u}_\delta) - \frac{1}{2}\int_0^t\int_\Gamma \partial_x\omega_\delta(\boldsymbol{u})_x(\boldsymbol{u}-\boldsymbol{u}_\delta)\cdot(\boldsymbol{u}-\boldsymbol{u}_\delta) + \frac{1}{2}\int_0^t\int_\Gamma (\partial_x\omega - \partial_x\omega_\delta)(\boldsymbol{u}-\boldsymbol{u}_\delta)_x|\boldsymbol{u}|^2 + \frac{1}{2}\int_0^t\int_\Gamma \partial_x\omega_\delta(\boldsymbol{u}-\boldsymbol{u}_\delta)_x\boldsymbol{u}\cdot(\boldsymbol{u}-\boldsymbol{u}_\delta).$$

By interpolation, by the boundedness of $|\partial_x\omega|$ and $|\partial_x\omega_\delta|$, and by the smoothness of $\boldsymbol{u}$, we get:

$$|R_{3,2}| \le \epsilon \int_0^t \|\widehat{\boldsymbol{u}}-\boldsymbol{u}_\delta\|_{H^1(\Omega_{f,\delta}(s))}^2 + C(\epsilon)\left(\int_0^t \|\omega-\omega_\delta\|_{H^2(\Gamma)}^2 + \int_0^t \|\widehat{\boldsymbol{u}}-\boldsymbol{u}_\delta\|_{L^2(\Omega_{f,\delta}(s))}^2\right).$$

Hence, by combining the two estimates, we get the final estimate for T_3:

$$|T_3| \leq \epsilon \int_0^t \|\hat{\boldsymbol{u}} - \boldsymbol{u}_\delta\|^2_{H^1(\Omega_{f,\delta}(s))}$$
$$+ C(\epsilon)\left(\int_0^t \|\omega - \omega_\delta\|^2_{H^2(\Gamma)} + \int_0^t \|\boldsymbol{\xi} - \boldsymbol{\xi}_\delta\|^2_{L^2(\Gamma)} + \int_0^t \|\hat{\boldsymbol{u}} - \boldsymbol{u}_\delta\|^2_{L^2(\Omega_{f,\delta}(s))}\right).$$

Term T4. To estimate term T_4, defined in (11.132), we again use Proposition 11.9.1 to pass to the limit as $\nu \to 0$ so that the contribution from T_4 is

$$T_4 := 2\nu \int_0^t \int_{\Omega_f(s)} \boldsymbol{D}(\boldsymbol{u}) : \boldsymbol{D}(\boldsymbol{u} - \breve{\boldsymbol{u}}_\delta) - 2\nu \int_0^t \int_{\Omega_{f,\delta}(s)} \boldsymbol{D}(\boldsymbol{u}_\delta) : \boldsymbol{D}(\hat{\boldsymbol{u}} - \boldsymbol{u}_\delta). \tag{11.170}$$

We want to transfer the integral on $\Omega_1(t)$ to $\Omega_{f,\delta}(t)$. Recalling (11.110), we have that

$$\int_0^t \int_{\Omega_f(s)} \boldsymbol{D}(\boldsymbol{u}) : \boldsymbol{D}(\boldsymbol{u} - \breve{\boldsymbol{u}}_\delta)$$
$$= \int_0^t \int_{\Omega_{f,\delta}(s)} \gamma_\delta[\nabla(\boldsymbol{u} \circ \psi_\delta)\boldsymbol{J}_\delta^{-1}]^{sym} : [\nabla((\boldsymbol{u} - \breve{\boldsymbol{u}}_\delta) \circ \psi_\delta)\boldsymbol{J}_\delta^{-1}]^{sym},$$

where the superscript 'sym' notation denotes a symmetrization. Following the procedure in [298], we break up the integral as

$$\int_0^t \int_{\Omega_f(s)} \boldsymbol{D}(\boldsymbol{u}) : \boldsymbol{D}(\boldsymbol{u} - \breve{\boldsymbol{u}}_\delta) = \int_0^t \int_{\Omega_{f,\delta}(s)} \boldsymbol{D}(\hat{\boldsymbol{u}}) : \boldsymbol{D}(\hat{\boldsymbol{u}} - \boldsymbol{u}_\delta) + R_{4,1}$$
$$+ R_{4,2} + R_{4,3} + R_{4,4}, \tag{11.171}$$

where

$$R_{4,1} = \int_0^t \int_{\Omega_{f,\delta}(s)} (\nabla(\boldsymbol{u} \circ \psi_\delta)\boldsymbol{J}_\delta^{-1})^{sym} : [\nabla(\hat{\boldsymbol{u}} - \boldsymbol{u}_\delta)(\boldsymbol{J}_\delta^{-1} - \boldsymbol{I})$$
$$+ (\boldsymbol{J}_\delta - \boldsymbol{I})\nabla(\hat{\boldsymbol{u}} - \boldsymbol{u}_\delta)\boldsymbol{J}_\delta^{-1}]^{sym},$$

$$R_{4,2} = \int_0^t \int_{\Omega_{f,\delta}(s)} [(\boldsymbol{I} - \gamma_\delta \boldsymbol{J}_\delta^{-1})\nabla(\boldsymbol{u} \circ \psi_\delta)$$
$$+ \nabla(\boldsymbol{u} \circ \psi_\delta)(\boldsymbol{J}_\delta^{-1} - \boldsymbol{I})]^{sym} : \boldsymbol{D}(\hat{\boldsymbol{u}} - \boldsymbol{u}_\delta),$$

$$R_{4,3} = \int_0^t \int_{\Omega_{f,\delta}(s)} (\nabla(\boldsymbol{u} \circ \psi_\delta)\boldsymbol{J}_\delta^{-1})^{sym} : (\gamma_\delta \nabla(\gamma_\delta^{-1}\boldsymbol{J}_\delta)(\hat{\boldsymbol{u}} - \boldsymbol{u}_\delta)\boldsymbol{J}_\delta^{-1})^{sym},$$

$$R_{4,4} = -\int_0^t \int_{\Omega_{f,\delta}(s)} [(\nabla(\gamma_\delta \boldsymbol{J}_\delta^{-1}))\boldsymbol{u} \circ \psi_\delta]^{sym} : \boldsymbol{D}(\hat{\boldsymbol{u}} - \boldsymbol{u}_\delta).$$

To verify this equality, one can use the Leibniz rule, the definition $\widehat{\boldsymbol{u}} = \gamma_\delta J_\delta^{-1} \cdot (\boldsymbol{u} \circ \psi_\delta)$, and the identity $\widehat{\boldsymbol{u}} - \boldsymbol{u}_\delta = \gamma_\delta J_\delta^{-1} \cdot ((\boldsymbol{u} - \breve{\boldsymbol{u}}_\delta) \circ \psi_\delta)$.
We now estimate the terms $R_{4,1}$-$R_{4,4}$. For this purpose, we will use the following inequalities:

$$|J_\delta^{-1}| \leq C(1 + |\partial_x \gamma_\delta|), \quad |J_\delta^{-1} - I| \leq C(|\gamma_\delta^{-1} - 1| + |\partial_x \gamma_\delta|),$$
$$|J_\delta - I| \leq C(|\gamma_\delta - 1| + |\partial_x \gamma_\delta|), \quad |\gamma_\delta J_\delta^{-1} - I| \leq C(|\gamma_\delta - 1| + |\partial_x \gamma_\delta|).$$

Recalling the definition of γ_δ in (11.109), we have the following inequalities:

$$|\gamma_\delta - 1| \leq C\|\omega - \omega_\delta\|_{H^2(\Gamma)}, \qquad |\gamma_\delta^{-1} - 1| \leq C\|\omega - \omega_\delta\|_{H^2(\Gamma)},$$

$$|\partial_x \gamma_\delta| \leq C\|\omega - \omega_\delta\|_{H^2(\Gamma)}, \qquad |\partial_x(\gamma_\delta^{-1})| \leq C\|\omega - \omega_\delta\|_{H^2(\Gamma)}.$$

Because $|J_\delta^{-1}| \leq C(1 + |\partial_x \gamma_\delta|) \leq C$ since $|\partial_x \gamma_\delta|$ is bounded and because $\boldsymbol{u}$ is smooth,

$$\begin{aligned}|R_{4,1}| \leq C &\int_0^t \|\nabla(\widehat{\boldsymbol{u}} - \boldsymbol{u}_\delta)\|_{L^2(\Omega_{f,\delta}(s))}(\|\gamma_\delta^{-1} - 1\|_{L^2(\Omega_{f,\delta}(s))} \\ &+ \|\gamma_\delta - 1\|_{L^2(\Omega_{f,\delta}(s))} + \|\partial_x \gamma_\delta\|_{L^2(\Omega_{f,\delta}(s))}) \\ \leq \epsilon &\int_0^t \|\nabla(\widehat{\boldsymbol{u}} - \boldsymbol{u}_\delta)\|^2_{L^2(\Omega_{f,\delta}(s))} + C(\epsilon) \int_0^t \|\omega - \omega_\delta\|^2_{H^2(\Gamma)}.\end{aligned}$$

We also have that

$$|R_{4,2}| \leq \epsilon \int_0^t \|\boldsymbol{D}(\widehat{\boldsymbol{u}} - \boldsymbol{u}_\delta)\|^2_{L^2(\Omega_{f,\delta}(s))} + C(\epsilon) \int_0^t \|\omega - \omega_\delta\|^2_{H^2(\Gamma)}.$$

For $R_{4,3}$ and $R_{4,4}$, we compute that

$$\nabla(\gamma_\delta^{-1} J_\delta) = \nabla \begin{pmatrix} \gamma_\delta^{-1} & 0 \\ (R + y)\gamma_\delta^{-1}\partial_x \gamma_\delta & 1 \end{pmatrix}, \qquad \nabla(\gamma_\delta J_\delta^{-1}) = \nabla \begin{pmatrix} \gamma_\delta & 0 \\ -(R + y)\partial_x \gamma_\delta & 1 \end{pmatrix}.$$

Therefore,

$$|\nabla(\gamma_\delta^{-1} J_\delta)| \leq C(|\partial_x(\gamma_\delta^{-1})| + |\partial_x \gamma_\delta| + |\partial_{xx} \gamma_\delta|),$$
$$|\nabla(\gamma_\delta J_\delta^{-1})| \leq C(|\partial_x \gamma_\delta| + |\partial_{xx} \gamma_\delta|),$$

where we can estimate

$$|\partial_{xx} \gamma_\delta| \leq C(\|\omega - \omega_\delta\|_{H^2(\Gamma)}|\partial_{xx}\omega| + |\partial_{xx}(\omega - \omega_\delta)| + \|\omega - \omega_\delta\|_{H^2(\Gamma)}).$$

So since $\|\partial_{xx}\omega\|_{L^2(\Omega_{f,\delta}(t))} \leq C$ since ω is uniformly bounded in $H^2(\Gamma)$, we have that

$$|R_{4,3}| \leq C \int_0^t \|\nabla(\gamma_\delta^{-1} J_\delta)\|_{L^2(\Omega_{f,\delta}(s))} \|\widehat{\boldsymbol{u}} - \boldsymbol{u}_\delta\|_{L^2(\Omega_{f,\delta}(s))}$$
$$\leq C \int_0^t \|\omega - \omega_\delta\|_{H^2(\Gamma)} \|\widehat{\boldsymbol{u}} - \boldsymbol{u}_\delta\|_{L^2(\Omega_{f,\delta}(s))}$$
$$\leq C \left(\int_0^t \|\omega - \omega_\delta\|^2_{H^2(\Gamma)} + \int_0^t \|\widehat{\boldsymbol{u}} - \boldsymbol{u}_\delta\|^2_{L^2(\Omega_{f,\delta}(s))} \right).$$

Similarly, using $\|\nabla(\gamma_\delta J_\delta^{-1})\|_{L^2(\Omega_{f,\delta}(t))} \leq C\|\omega - \omega_\delta\|_{H^2(\Gamma)}$, we have the following estimate for R_4:

$$|R_{4,4}| \leq C \int_0^t \|\nabla(\gamma_\delta J_\delta^{-1})\|_{L^2(\Omega_{f,\delta}(s))} \|\boldsymbol{D}(\widehat{\boldsymbol{u}} - \boldsymbol{u}_\delta)\|_{L^2(\Omega_{f,\delta}(s))}$$
$$\leq \epsilon \int_0^t \|\boldsymbol{D}(\widehat{\boldsymbol{u}} - \boldsymbol{u}_\delta)\|^2_{L^2(\Omega_{f,\delta}(s))} + C(\epsilon) \int_0^t \|\omega - \omega_\delta\|^2_{H^2(\Gamma)}.$$

We now have the final estimate of T_4, obtained after using (11.170) and (11.171) as follows:

$$T_4 = 2\nu \int_0^t \int_{\Omega_{f,\delta}(s)} |\boldsymbol{D}(\widehat{\boldsymbol{u}} - \boldsymbol{u}_\delta)|^2 + R_4,$$

where

$$|R_4| \leq \epsilon \int_0^t \|\boldsymbol{D}(\widehat{\boldsymbol{u}} - \boldsymbol{u}_\delta)\|^2_{L^2(\Omega_{f,\delta}(s))}$$
$$+ C(\epsilon) \left(\int_0^t \|\omega - \omega_\delta\|^2_{H^2(\Gamma)} + \int_0^t \|\widehat{\boldsymbol{u}} - \boldsymbol{u}_\delta\|^2_{L^2(\Omega_{f,\delta}(s))} \right).$$

Term T5. Similarly as before, after passing to the limit as $\nu \to 0$ in term T_5, defined by (11.172), the contribution of this term is

$$T_5 = \beta \int_0^t \int_{\Gamma(s)} (\boldsymbol{\xi} - \boldsymbol{u}) \cdot \boldsymbol{\tau}(s)[(\boldsymbol{\xi} - \boldsymbol{\xi}_\delta) \cdot \boldsymbol{\tau}(s) - (\boldsymbol{u} - \breve{\boldsymbol{u}}_\delta) \cdot \boldsymbol{\tau}(s)]$$
$$- \beta \int_0^t \int_{\Gamma_\delta(s)} (\boldsymbol{\xi}_\delta - \boldsymbol{u}_\delta) \cdot \boldsymbol{\tau}_\delta(s)[(\boldsymbol{\xi} - \boldsymbol{\xi}_\delta) \cdot \boldsymbol{\tau}_\delta(s) - (\widehat{\boldsymbol{u}} - \boldsymbol{u}_\delta) \cdot \boldsymbol{\tau}_\delta(s)]. \tag{11.172}$$

We note that when we test the weak formulation for $\boldsymbol{u}$ with $\boldsymbol{v} = \boldsymbol{u} - (\breve{\boldsymbol{u}}_\delta)_\nu$ and $\boldsymbol{\psi} = \boldsymbol{\xi} - (\boldsymbol{\xi}_\delta)_\nu$, we can pass to the limit as $\nu \to 0$ to obtain the first term in T_5 above, by using similar arguments involving Proposition 11.9.1, as for the previously considered terms. This term can now be rewritten as follows:

$$T_5 = \beta \int_0^t \int_{\Gamma_\delta(s)} |(\boldsymbol{\xi} - \boldsymbol{\xi}_\delta) \cdot \boldsymbol{\tau}_\delta(s) - (\widehat{\boldsymbol{u}} - \boldsymbol{u}_\delta) \cdot \boldsymbol{\tau}_\delta(s)|^2 + R_5,$$

where

$$R_5 = \beta \int_0^t \int_{\Gamma(s)} (\boldsymbol{\xi} - \boldsymbol{u}) \cdot \boldsymbol{\tau}(s)[(\boldsymbol{\xi} - \boldsymbol{\xi}_\delta) \cdot \boldsymbol{\tau}(s) - (\boldsymbol{u} - \breve{\boldsymbol{u}}_\delta) \cdot \boldsymbol{\tau}(s)]$$
$$- \beta \int_0^t \int_{\Gamma_\delta(s)} (\boldsymbol{\xi} - \widehat{\boldsymbol{u}}) \cdot \boldsymbol{\tau}_\delta(s)[(\boldsymbol{\xi} - \boldsymbol{\xi}_\delta) \cdot \boldsymbol{\tau}_\delta(s) - (\widehat{\boldsymbol{u}} - \boldsymbol{u}_\delta) \cdot \boldsymbol{\tau}_\delta(s)].$$

Denote the arc length elements of $\Gamma(t)$ and $\Gamma_\delta(t)$, respectively, by $\mathcal{J}_\Gamma^\omega = \sqrt{1 + |\partial_x \omega|^2}$ and $\mathcal{J}_\Gamma^{\omega_\delta} = \sqrt{1 + |\partial_x \omega_\delta|^2}$, and we recall that we denote the tangent vectors to $\Gamma(t)$ and $\Gamma_\delta(t)$, respectively, by $\boldsymbol{\tau}(t) = \frac{1}{\mathcal{J}_\Gamma^\omega}(1, \partial_x \omega)$ and $\boldsymbol{\tau}_\delta(t) = \frac{1}{\mathcal{J}_\Gamma^{\omega_\delta}}(1, \partial_x \omega_\delta)$. We can now rewrite R_5 by writing everything in terms of the x and y components. For this purpose, recall that $\boldsymbol{\xi}$ and $\boldsymbol{\xi}_\delta$ along the interface displace in only the y direction. We formally express the common trace of $\boldsymbol{u} - \breve{\boldsymbol{u}}_\delta$ and $\widehat{\boldsymbol{u}} - \boldsymbol{u}_\delta$ along the reference configuration of the interface Γ by $\boldsymbol{u} - \boldsymbol{u}_\delta$. Thus,

$$R_5 = \beta \int_0^t \int_\Gamma (\boldsymbol{\xi} - \boldsymbol{u}) \cdot (1, \partial_x \omega)[(\boldsymbol{\xi} - \boldsymbol{\xi}_\delta) - (\boldsymbol{u} - \boldsymbol{u}_\delta)] \cdot \boldsymbol{\tau}(s)$$
$$- \beta \int_0^t \int_\Gamma (\boldsymbol{\xi} - \boldsymbol{u}) \cdot (1, \partial_x \omega_\delta)[(\boldsymbol{\xi} - \boldsymbol{\xi}_\delta) - (\boldsymbol{u} - \boldsymbol{u}_\delta)] \cdot \boldsymbol{\tau}_\delta(s).$$

In the previous step, we used the fact that when transferred back to the reference configuration Ω_f, $\widehat{\boldsymbol{u}} - \boldsymbol{u}_\delta$ and $\boldsymbol{u} - \breve{\boldsymbol{u}}_\delta$ have the same trace along Γ. Thus, $R_5 = R_{5,1} + R_{5,2}$, where

$$R_{5,1} = \beta \int_0^t \int_\Gamma (\boldsymbol{\xi} - \boldsymbol{u}) \cdot \boldsymbol{e}_y(\partial_x \omega - \partial_x \omega_\delta)[(\boldsymbol{\xi} - \boldsymbol{\xi}_\delta) - (\boldsymbol{u} - \boldsymbol{u}_\delta)] \cdot \boldsymbol{\tau}(s),$$
$$R_{5,2} = \beta \int_0^t \int_\Gamma (\boldsymbol{\xi} - \boldsymbol{u}) \cdot (1, \partial_x \omega_\delta)[(\boldsymbol{\xi} - \boldsymbol{\xi}_\delta) - (\boldsymbol{u} - \boldsymbol{u}_\delta)] \cdot (\boldsymbol{\tau}(s) - \boldsymbol{\tau}_\delta(s)).$$

We can use the fact that $|\partial_x \omega|$ and $|\partial_x \omega_\delta|$ are uniformly bounded to obtain the following estimates:

$$|R_{5,1}| \leq \epsilon \int_0^t \|\boldsymbol{D}(\widehat{\boldsymbol{u}} - \boldsymbol{u}_\delta)\|_{L^2(\Omega_{f,\delta}(s))}$$
$$+ C(\epsilon) \left(\int_0^t \|\omega - \omega_\delta\|^2_{H^2(\Gamma)} + \int_0^t \|\boldsymbol{\xi} - \boldsymbol{\xi}_\delta\|^2_{L^2(\Gamma)} \right),$$

where we used the trace inequality, Poincare's inequality, and Korn's inequality for the fluid. For the second term $R_{5,2}$, we use the estimate $|\boldsymbol{\tau}(s) - \boldsymbol{\tau}_\delta(s)| \leq C|\partial_x \omega - \partial_x \omega_\delta|$ to obtain

$$
\begin{aligned}
|R_{5,2}| \leq \epsilon \int_0^t \|\boldsymbol{D}(\widehat{\boldsymbol{u}} - \boldsymbol{u}_\delta)\|_{L^2(\Omega_{f,\delta}(s))} \\
+ C(\epsilon) \left(\int_0^t \|\omega - \omega_\delta\|^2_{H^2(\Gamma)} + \int_0^t \|\boldsymbol{\xi} - \boldsymbol{\xi}_\delta\|^2_{L^2(\Gamma)} \right).
\end{aligned}
$$

Hence,

$$
T_5 = \beta \int_0^t \int_{\Gamma_\delta(s)} |(\boldsymbol{\xi} - \boldsymbol{\xi}_\delta) \cdot \boldsymbol{\tau}_\delta(s) - (\widehat{\boldsymbol{u}} - \boldsymbol{u}_\delta) \cdot \boldsymbol{\tau}_\delta(s)|^2 + R_5,
$$

where

$$
\begin{aligned}
|R_5| \leq \epsilon \int_0^t \|\boldsymbol{D}(\widehat{\boldsymbol{u}} - \boldsymbol{u}_\delta)\|_{L^2(\Omega_{f,\delta}(s))} \\
+ C(\epsilon) \left(\int_0^t \|\omega - \omega_\delta\|^2_{H^2(\Gamma)} + \int_0^t \|\boldsymbol{\xi} - \boldsymbol{\xi}_\delta\|^2_{L^2(\Gamma)} \right).
\end{aligned}
$$

Terms T6–T8. We will present estimates only for term T_6, defined in (11.133), as the same procedure will hold for T_7 and T_8. Since ζ and ζ_δ are weakly continuous in $L^2(\Gamma)$, by the weak formulation, we get:

$$
\begin{aligned}
&\int_0^t \int_\Gamma \zeta \cdot \partial_t \left[(\zeta_\delta)_\nu\right] + \int_0^t \int_\Gamma \zeta_\delta \cdot \partial_t \zeta \\
&= \int_0^t \int_\Gamma \zeta \cdot \partial_t \left[(\zeta_\delta)_\nu\right] + \int_0^t \int_\Gamma \zeta_\delta \cdot \partial_t \left[(\zeta)_\nu\right] - \int_0^t \int_\Gamma \zeta_\delta \cdot \partial_t \left[(\zeta)_\nu - \zeta\right] \\
&\to \int_\Gamma \zeta(t) \cdot \zeta_\delta(t) - \int_\Gamma |\zeta_0|^2.
\end{aligned}
$$

This follows from Lemma 2.5 in [298], which implies:

$$
\int_0^t \int_\Gamma \zeta \cdot \partial_t \left[(\zeta_\delta)_\nu\right] + \int_0^t \int_\Gamma \zeta_\delta \cdot \partial_t \left[(\zeta)_\nu\right] \to \int_\Gamma \zeta(t) \cdot \zeta_\delta(t) - \int_\Gamma |\zeta_0|^2, \qquad \text{as } \nu \to 0,
$$

and from the fact that ζ is smooth in space and time, which implies

$$
\int_0^t \int_\Gamma \zeta_\delta \cdot \partial_t \left[(\zeta)_\nu - \zeta\right] \to 0, \qquad \text{as } \nu \to 0.
$$

Furthermore, because $\zeta(0) = \zeta_\delta(0) = \zeta_0$ weak continuity of ζ_δ at $t = 0$ implies that $\int_\Gamma \zeta(0) \cdot [\zeta(0) - (\zeta_\delta)_\nu(0)] \to 0$ as $\nu \to 0$. Similarly, $\int_\Gamma \zeta(s) \cdot [\zeta(s) - (\zeta_\delta)_\nu(s)] \to 0$ as $\nu \to 0$ for almost every $s \in [0, T]$. Hence, as $\nu \to 0$, the contribution from T_6 is

$$T_6 = \frac{1}{2}\rho_p \int_\Gamma |(\zeta - \zeta_\delta)(t)|^2.$$

Similarly, the contributions from T_7 and T_8 as $\nu \to 0$ are

$$T_7 = \frac{1}{2}\int_\Gamma |\Delta(\omega - \omega_\delta)(t)|^2, \qquad T_8 = \frac{1}{2}\rho_b \int_{\Omega_b} |(\boldsymbol{\xi} - \boldsymbol{\xi}_\delta)(t)|^2.$$

Terms T9–T12. Since these calculations are straightforward, a discussion about the limiting expressions as $\nu \to 0$ for terms T_9–T_{12} was presented earlier, just under (11.136).

Term T13. Similarly as before, by taking the limit as $\nu \to 0$, we have that

$$T_{13} = -\alpha \int_0^t \int_{\Omega_b(s)} p\nabla \cdot (\boldsymbol{\xi} - \boldsymbol{\xi}_\delta) + \alpha \int_0^t \int_{\Omega^\delta_{b,\delta}(s)} p_\delta \nabla \cdot (\boldsymbol{\xi} - \boldsymbol{\xi}_\delta).$$

To estimate this term, we use (11.24) and the matrix identity $\boldsymbol{B}^{-1} = \frac{1}{\det(\boldsymbol{B})}\boldsymbol{B}^C$ to obtain

$$\begin{aligned}
|T_{13}| &= \alpha \left| \int_0^t \int_{\Omega_b} \mathcal{J}_b^\eta p \nabla_b^\eta \cdot (\boldsymbol{\xi} - \boldsymbol{\xi}_\delta) - \int_0^t \int_{\Omega_b} \mathcal{J}_b^{\eta_\delta^\delta} p_\delta \nabla_b^{\eta_\delta^\delta} \cdot (\boldsymbol{\xi} - \boldsymbol{\xi}_\delta) \right| \\
&= \alpha \left| \int_0^t \int_{\Omega_b} p \cdot \mathrm{tr}\left(\nabla(\boldsymbol{\xi} - \boldsymbol{\xi}_\delta) \cdot (\boldsymbol{I} + \nabla\boldsymbol{\eta})^C\right) \right. \\
&\quad \left. - \int_0^t \int_{\Omega_b} p_\delta \cdot \mathrm{tr}\left(\nabla(\boldsymbol{\xi} - \boldsymbol{\xi}_\delta) \cdot (\boldsymbol{I} + \nabla\boldsymbol{\eta}_\delta^\delta)^C\right) \right| \leq R_{13,1} + R_{13,2},
\end{aligned}$$

where the superscript "C" denotes the cofactor matrix. The integrals $R_{13,1}$ and $R_{13,2}$ are defined as follows:

$$R_{13,1} = \alpha \left| \int_0^t \int_{\Omega_b} p \cdot \mathrm{tr}\left(\nabla(\boldsymbol{\xi} - \boldsymbol{\xi}_\delta) \cdot (\nabla(\boldsymbol{\eta} - \boldsymbol{\eta}_\delta^\delta))^C\right) \right|,$$

$$R_{13,2} = \alpha \left| \int_0^t \int_{\Omega_b} (p - p_\delta) \cdot \mathrm{tr}\left(\nabla(\boldsymbol{\xi} - \boldsymbol{\xi}_\delta) \cdot (\boldsymbol{I} + \nabla\boldsymbol{\eta}_\delta^\delta)^C\right) \right|.$$

In the previous calculations, we observe that the cofactor matrix operation is linear when the matrices are two by two. Using the fact that p is smooth, the assumption (11.120), and the fact that

$$\|\nabla \boldsymbol{\eta}^{\delta} - \nabla \boldsymbol{\eta}_{\delta}^{\delta}\|_{L^2(\Omega_b)} \leq C\|\nabla \boldsymbol{\eta} - \nabla \boldsymbol{\eta}_{\delta}\|_{L^2(\tilde{\Omega}_b)}$$
$$\leq C\left(\|\nabla \boldsymbol{\eta} - \nabla \boldsymbol{\eta}_{\delta}\|_{L^2(\Omega_b)} + \|\omega - \omega_{\delta}\|_{H^2(\Gamma)}\right) \tag{11.173}$$

for a constant C independent of δ, by Young's convolution inequality and the definition of odd extension to the larger domain $\tilde{\Omega}_b$ in Definition 11.4.1, we obtain the estimates on $R_{13,1}$ and $R_{13,2}$:

$$R_{13,1} \leq \epsilon \int_0^t \|\nabla(\boldsymbol{\xi} - \boldsymbol{\xi}_{\delta})\|_{L^2(\Omega_b)}^2 + C(\epsilon)\left(\int_0^t \|\nabla \boldsymbol{\eta} - \nabla \boldsymbol{\eta}_{\delta}^{\delta}\|_{L^2(\Omega_b)}^2\right)$$
$$\leq C(\epsilon)\int_0^t \|\nabla \boldsymbol{\eta} - \nabla \boldsymbol{\eta}^{\delta}\|_{L^2(\Omega_b)}^2 + \epsilon \int_0^t \|\nabla(\boldsymbol{\xi} - \boldsymbol{\xi}_{\delta})\|_{L^2(\Omega_b)}^2$$
$$+ C(\epsilon)\left(\int_0^t \|\nabla \boldsymbol{\eta} - \nabla \boldsymbol{\eta}_{\delta}\|_{L^2(\Omega_b)}^2 + \int_0^t \|\omega - \omega_{\delta}\|_{H^2(\Gamma)}^2\right),$$
$$R_{13,2} \leq \epsilon \int_0^t \|\nabla(\boldsymbol{\xi} - \boldsymbol{\xi}_{\delta})\|_{L^2(\Omega_b)}^2 + C(\epsilon)\left(\int_0^t \|p - p_{\delta}\|_{L^2(\Omega_b)}^2\right).$$

Therefore, the final estimate for T_{13} is as follows:

$$|T_{13}| \leq C(\epsilon)\int_0^t \|\nabla \boldsymbol{\eta} - \nabla \boldsymbol{\eta}^{\delta}\|_{L^2(\Omega_b)}^2 + \epsilon \int_0^t \|\nabla(\boldsymbol{\xi} - \boldsymbol{\xi}_{\delta})\|_{L^2(\Omega_b)}^2$$
$$+ C(\epsilon)\left(\int_0^t \|\nabla \boldsymbol{\eta} - \nabla \boldsymbol{\eta}_{\delta}\|_{L^2(\Omega_b)}^2\right.$$
$$\left.+ \int_0^t \|\omega - \omega_{\delta}\|_{H^2(\Gamma)}^2 + \int_0^t \|p - p_{\delta}\|_{L^2(\Omega_b)}^2\right).$$

Term T14. This term can be handled in the same way as terms T_6–T_8.
Term T15. We pass to the limit as $\nu \to 0$ in (11.138) to obtain:

$$T_{15} = -\alpha \int_0^t \int_{\Omega_b(s)} \frac{D}{Dt}\boldsymbol{\eta} \cdot \nabla(p - p_{\delta}) + \alpha \int_0^t \int_{\Omega_{b,\delta}^{\delta}(s)} \frac{D^{\delta}}{Dt}\boldsymbol{\eta}_{\delta} \cdot \nabla(p - p_{\delta}).$$

To estimate this term, we pull back to the reference domain and use (11.24) and the cofactor formula for the matrix inverse to obtain:

$$|T_{15}| = \alpha \left| \int_0^t \int_{\Omega_b} \mathcal{J}_b^{\eta} \partial_t \boldsymbol{\eta} \cdot \nabla_b^{\eta}(p - p_{\delta}) - \int_0^t \int_{\Omega_b} \mathcal{J}_b^{\eta_{\delta}^{\delta}} \partial_t \boldsymbol{\eta}_{\delta} \cdot \nabla_b^{\eta_{\delta}^{\delta}}(p - p_{\delta}) \right|$$
$$= \alpha \left| \int_0^t \int_{\Omega_b} \partial_t \boldsymbol{\eta} \cdot \left[\nabla(p - p_{\delta}) \cdot (\boldsymbol{I} + \nabla \boldsymbol{\eta})^C\right] \right.$$
$$\left. - \int_0^t \int_{\Omega_b} \partial_t \boldsymbol{\eta}_{\delta} \cdot \left[\nabla(p - p_{\delta}) \cdot (\boldsymbol{I} + \nabla \boldsymbol{\eta}_{\delta}^{\delta})^C\right] \right| \leq R_{15,1} + R_{15,2},$$

where

$$R_{15,1} = \alpha \left| \int_0^t \int_{\Omega_b} \partial_t \boldsymbol{\eta} \cdot \left[\nabla(p - p_\delta) \cdot (\nabla \boldsymbol{\eta} - \nabla \boldsymbol{\eta}_\delta^\delta)^C \right] \right|,$$

$$R_{15,2} = \alpha \left| \int_0^t \int_{\Omega_b} (\partial_t \boldsymbol{\eta} - \partial_t \boldsymbol{\eta}_\delta) \cdot \left[\nabla(p - p_\delta) \cdot (\boldsymbol{I} + \nabla \boldsymbol{\eta}_\delta^\delta)^C \right] \right|.$$

To estimate $R_{15,1}$, we use (11.118), (11.119), and the convolution inequality (11.173) to obtain:

$$R_{15,1} \leq \epsilon \int_0^t \|\nabla(p - p_\delta)\|^2_{L^2(\Omega_{b,\delta}^\delta(s))} + C(\epsilon) \int_0^t \|\nabla \boldsymbol{\eta} - \nabla \boldsymbol{\eta}^\delta\|^2_{L^2(\Omega_b)}$$
$$+ C(\epsilon) \left(\int_0^t \|\nabla \boldsymbol{\eta} - \nabla \boldsymbol{\eta}_\delta\|^2_{L^2(\Omega_b)} + \int_0^t \|\omega - \omega_\delta\|^2_{H^2(\Gamma)} \right).$$

Here, we also used the following estimate on the norm of the gradient of the pressure on the reference domain and on the moving domain, which is obtained by using (11.118), (11.119), and (11.24):

$$\|\nabla(p - p_\delta)(t)\|^2_{L^2(\Omega_b)}$$
$$= \int_{\Omega_b} |\nabla(p - p_\delta)|^2 = \int_{\Omega_b} \mathcal{J}_b^{\eta_\delta^\delta} |\nabla_b^{\eta_\delta^\delta}(p - p_\delta) \cdot (\boldsymbol{I} + \nabla \boldsymbol{\eta}_\delta^\delta)|^2 \cdot (\mathcal{J}_b^{\eta_\delta^\delta})^{-1}$$
$$\leq C \int_{\Omega_b} \mathcal{J}_b^{\eta_\delta^\delta} |\nabla_b^{\eta_\delta^\delta}(p - p_\delta)|^2 = C \|\nabla(p - p_\delta)(t)\|^2_{L^2(\Omega_{b,\delta}^\delta(t))}, \tag{11.174}$$

where constant C is independent of δ and $t \in [0, T_\delta]$.
The estimate of $R_{15,2}$ is straightforward:

$$R_{15,2} \leq \epsilon \int_0^t \|\nabla(p - p_\delta)\|^2_{L^2(\Omega_{b,\delta}^\delta(s))} + C(\epsilon) \int_0^t \|\partial_t \boldsymbol{\eta} - \partial_t \boldsymbol{\eta}_\delta\|^2_{L^2(\Omega_b)}.$$

From here we get the final estimate of T_{15}:

$$|T_{15}| \leq \epsilon \int_0^t \|\nabla(p - p_\delta)\|^2_{L^2(\Omega_{b,\delta}^\delta(s))} + C(\epsilon) \int_0^t \|\nabla \boldsymbol{\eta} - \nabla \boldsymbol{\eta}^\delta\|^2_{L^2(\Omega_b)}$$
$$+ C(\epsilon) \left(\int_0^t \|\nabla \boldsymbol{\eta} - \nabla \boldsymbol{\eta}_\delta\|^2_{L^2(\Omega_b)} \right.$$
$$\left. + \int_0^t \|\omega - \omega_\delta\|^2_{H^2(\Gamma)} + \int_0^t \|\partial_t \boldsymbol{\eta} - \partial_t \boldsymbol{\eta}_\delta\|^2_{L^2(\Omega_b)} \right).$$

Term T16. To estimate T_{16} defined in (11.139), we start by passing to the limit as $\nu \to 0$ to obtain

$$T_{16} = -\alpha \int_0^t \int_{\Gamma(s)} (\boldsymbol{\xi} \cdot \boldsymbol{n})(p - p_\delta) + \alpha \int_0^t \int_{\Gamma_\delta^\delta(s)} (\boldsymbol{\xi}_\delta \cdot \boldsymbol{n}_\delta^\delta)(p - p_\delta),$$

where $\boldsymbol{n}_\delta^\delta$ is the upward pointing normal vector to $\Gamma_\delta^\delta(t)$. We integrate by parts to obtain that $|T_{16}| \le R_{16,1} + R_{16,2}$, where

$$R_{16,1} := \alpha \left| \int_0^t \int_{\Omega_b(s)} (\nabla \cdot \boldsymbol{\xi})(p - p_\delta) - \int_0^t \int_{\Omega_{b,\delta}^\delta(s)} (\nabla \cdot \boldsymbol{\xi}_\delta)(p - p_\delta) \right|,$$

$$R_{16,2} := \alpha \left| \int_0^t \int_{\Omega_b(s)} \boldsymbol{\xi} \cdot \nabla(p - p_\delta) - \int_0^t \int_{\Omega_{b,\delta}^\delta(s)} \boldsymbol{\xi}_\delta \cdot \nabla(p - p_\delta) \right|.$$

By using (11.24) and the bootstrap assumption (11.120), we have that

$$\begin{aligned}
R_{16,1} &= \alpha \left| \int_0^t \int_{\Omega_b} \mathcal{J}_b^\eta (\mathrm{tr}(\nabla_b^\eta \boldsymbol{\xi}))(p - p_\delta) - \int_0^t \int_{\Omega_b} \mathcal{J}_b^{\eta_\delta^\delta} (\mathrm{tr}(\nabla_b^{\eta_\delta^\delta} \boldsymbol{\xi}_\delta))(p - p_\delta) \right| \\
&= \alpha \left| \int_0^t \int_{\Omega_b} \mathrm{tr}(\nabla \boldsymbol{\xi} \cdot (\boldsymbol{I} + \nabla \boldsymbol{\eta})^C)(p - p_\delta) \right. \\
&\quad \left. - \int_0^t \int_{\Omega_b} \mathrm{tr}(\nabla \boldsymbol{\xi}_\delta \cdot (\boldsymbol{I} + \nabla \boldsymbol{\eta}_\delta^\delta)^C)(p - p_\delta) \right| \\
&\le \alpha \left| \int_0^t \int_{\Omega_b} \mathrm{tr}(\nabla \boldsymbol{\xi} \cdot (\nabla \boldsymbol{\eta} - \nabla \boldsymbol{\eta}_\delta^\delta)^C)(p - p_\delta) \right| \\
&\quad + \alpha \left| \int_0^t \int_{\Omega_b} \mathrm{tr}(\nabla(\boldsymbol{\xi} - \boldsymbol{\xi}_\delta) \cdot (\boldsymbol{I} + \nabla \boldsymbol{\eta}_\delta^\delta)^C)(p - p_\delta) \right| \\
&\le C \int_0^t \|\nabla \boldsymbol{\eta} - \nabla \boldsymbol{\eta}_\delta^\delta\|_{L^2(\Omega_b)} \cdot \|p - p_\delta\|_{L^2(\Omega_b)} \\
&\quad + C \int_0^t \|\nabla \boldsymbol{\xi} - \nabla \boldsymbol{\xi}_\delta\|_{L^2(\Omega_b)} \cdot \|p - p_\delta\|_{L^2(\Omega_b)}.
\end{aligned}$$

For $R_{16,2}$, we compute

$$\begin{aligned}
R_{16,2} = \alpha \Bigg| &\int_0^t \int_{\Omega_b} \boldsymbol{\xi} \cdot \left[\nabla(p - p_\delta) \cdot (\boldsymbol{I} + \nabla \boldsymbol{\eta})^C \right] \\
&- \int_0^t \int_{\Omega_b} \boldsymbol{\xi}_\delta \cdot \left[\nabla(p - p_\delta) \cdot (\boldsymbol{I} + \nabla \boldsymbol{\eta}_\delta^\delta)^C \right] \Bigg|
\end{aligned}$$

$$
\begin{aligned}
&\leq \alpha \left| \int_0^t \int_{\Omega_b} \boldsymbol{\xi} \cdot \left[\nabla(p - p_\delta) \cdot (\nabla \boldsymbol{\eta} - \nabla \boldsymbol{\eta}_\delta^\delta)^C \right] \right| \\
&\quad + \alpha \left| \int_0^t \int_{\Omega_b} (\boldsymbol{\xi} - \boldsymbol{\xi}_\delta) \cdot \left[\nabla(p - p_\delta) \cdot (\boldsymbol{I} + \nabla \boldsymbol{\eta}_\delta^\delta)^C \right] \right| \\
&\leq C \int_0^t \|\nabla p - \nabla p_\delta\|_{L^2(\Omega_b)} \cdot \|\nabla \boldsymbol{\eta} - \nabla \boldsymbol{\eta}_\delta^\delta\|_{L^2(\Omega_b)} \\
&\quad + C \int_0^t \|\boldsymbol{\xi} - \boldsymbol{\xi}_\delta\|_{L^2(\Omega_b)} \cdot \|\nabla p - \nabla p_\delta\|_{L^2(\Omega_b)}.
\end{aligned}
$$

By the convolution inequality (11.173) and the previous estimate on the gradient of the pressure (11.174), we conclude that

$$
\begin{aligned}
|T_{16}| \leq \epsilon &\left(\int_0^t \|\nabla \boldsymbol{\xi} - \nabla \boldsymbol{\xi}_\delta\|_{L^2(\Omega_b)}^2 + \int_0^t \|\nabla p - \nabla p_\delta\|_{L^2(\Omega_{b,\delta}^\delta(s))}^2 \right) \\
&\quad + C(\epsilon) \left(\int_0^t \|p - p_\delta\|_{L^2(\Omega_b)}^2 \right. \\
&+ \int_0^t \|\nabla \boldsymbol{\eta} - \nabla \boldsymbol{\eta}^\delta\|_{L^2(\Omega_b)}^2 + \int_0^t \|\nabla \boldsymbol{\eta} - \nabla \boldsymbol{\eta}_\delta\|_{L^2(\Omega_b)}^2 \\
&\quad \left. + \int_0^t \|\omega - \omega_\delta\|_{H^2(\Gamma)}^2 + \int_0^t \|\boldsymbol{\xi} - \boldsymbol{\xi}_\delta\|_{L^2(\Omega_b)}^2 \right).
\end{aligned}
$$

Term T17. To estimate term T_{17} defined in (11.140), we use (11.24) to compute

$$
\begin{aligned}
T_{17} &= \kappa \int_0^t \int_{\Omega_b} \mathcal{J}_b^\eta \nabla_b^\eta p \cdot \nabla_b^\eta (p - p_\delta) - \kappa \int_0^t \int_{\Omega_b} \mathcal{J}_b^{\eta_\delta^\delta} \nabla_b^{\eta_\delta^\delta} p_\delta \cdot \nabla_b^{\eta_\delta^\delta} (p - p_\delta) \\
&= \kappa \int_0^t \int_{\Omega_b} \mathcal{J}_b^{\eta_\delta^\delta} \nabla_b^{\eta_\delta^\delta} (p - p_\delta) \cdot \nabla_b^{\eta_\delta^\delta} (p - p_\delta) + I_1 + I_2 \\
&= \kappa \int_0^t \int_{\Omega_{b,\delta}^\delta(t)} |\nabla(p - p_\delta)|^2 + R_{17,1} + R_{17,2},
\end{aligned}
$$

where

$$
\begin{aligned}
R_{17,1} &= \kappa \int_0^t \int_{\Omega_b} \mathcal{J}_b^\eta \nabla_b^\eta p \cdot \nabla_b^\eta (p - p_\delta) - \kappa \int_0^t \int_{\Omega_b} \mathcal{J}_b^{\eta_\delta^\delta} \nabla_b^\eta p \cdot \nabla_b^{\eta_\delta^\delta} (p - p_\delta), \\
R_{17,2} &= \kappa \int_0^t \int_{\Omega_b} \mathcal{J}_b^{\eta_\delta^\delta} (\nabla_b^\eta p - \nabla_b^{\eta_\delta^\delta} p) \cdot \nabla_b^{\eta_\delta^\delta} (p - p_\delta).
\end{aligned}
$$

To estimate $R_{17,1}$, we use (11.24) to obtain

$$R_{17,1} = \kappa \int_0^t \int_{\Omega_b} \nabla_b^\eta p \cdot \left(\nabla(p - p_\delta) \cdot \left[(\boldsymbol{I} + \nabla\boldsymbol{\eta})^C - (\boldsymbol{I} + \nabla\boldsymbol{\eta}_\delta^\delta)^C\right]\right).$$

Because $\boldsymbol{\eta}$ is smooth, $|\nabla_b^\eta p| \le C$ uniformly in space and time. Therefore,

$$|R_{17,1}| \le C \int_0^t \|\nabla(p - p_\delta)\|_{L^2(\Omega_b)} \cdot \|(\nabla\boldsymbol{\eta} - \nabla\boldsymbol{\eta}_\delta^\delta)^C\|_{L^2(\Omega_b)}.$$

Using the estimate in (11.174), we obtain the desired estimate that

$$|R_{17,1}| \le \epsilon \int_0^t \|\nabla(p - p_\delta)\|^2_{L^2(\Omega^\delta_{b,\delta}(s))} + C(\epsilon) \int_0^t \|\nabla\boldsymbol{\eta} - \nabla\boldsymbol{\eta}_\delta^\delta\|^2_{L^2(\Omega_b)}.$$

To estimate $R_{17,2}$, we use the bootstrap assumption (11.120) that there exists a constant C (independent of δ) such that $|\nabla\boldsymbol{\eta}_\delta^\delta| \le C$ pointwise for $t \in [0, T_\delta]$. Therefore, $|(\boldsymbol{I} + \nabla\boldsymbol{\eta}_\delta)^C|$ is pointwise uniformly bounded in space and time on the time interval $[0, T_\delta]$. Thus, by (11.24),

$$R_{17,2} = \kappa \int_0^t \int_{\Omega_b} (\nabla_b^\eta p - \nabla_b^{\eta_\delta^\delta} p) \cdot \left[\nabla(p - p_\delta) \cdot (\boldsymbol{I} + \nabla\boldsymbol{\eta}_\delta^\delta)^C\right]$$

and hence

$$|R_{17,2}| \le C \int_0^t \|\nabla_b^\eta p - \nabla_b^{\eta_\delta^\delta} p\|_{L^2(\Omega_b)} \cdot \|\nabla(p - p_\delta)\|_{L^2(\Omega_b)}.$$

We estimate the first pressure term by using (11.24) to obtain

$$\begin{aligned}
\|\nabla_b^\eta p - \nabla_b^{\eta_\delta^\delta} p\|^2_{L^2(\Omega_b)} &= \int_{\Omega_b} \left|\nabla p \cdot \left[(\boldsymbol{I} + \nabla\boldsymbol{\eta})^{-1} - (\boldsymbol{I} + \nabla\boldsymbol{\eta}_\delta^\delta)^{-1}\right]\right|^2 \\
&= \int_{\Omega_b} \left|\nabla p \cdot (\boldsymbol{I} + \nabla\boldsymbol{\eta}_\delta^\delta)^{-1}[(\boldsymbol{I} + \nabla\boldsymbol{\eta}_\delta^\delta)(\boldsymbol{I} + \nabla\boldsymbol{\eta})^{-1} - \boldsymbol{I}]\right|^2 \\
&= \int_{\Omega_b} \left|\nabla p \cdot (\boldsymbol{I} + \nabla\boldsymbol{\eta}_\delta^\delta)^{-1}[(\boldsymbol{I} + \nabla\boldsymbol{\eta}_\delta^\delta)\right. \\
&\quad \left. -(\boldsymbol{I} + \nabla\boldsymbol{\eta})](\boldsymbol{I} + \nabla\boldsymbol{\eta})^{-1}\right|^2 \\
&= \int_{\Omega_b} \left|\nabla p \cdot (\boldsymbol{I} + \nabla\boldsymbol{\eta}_\delta^\delta)^{-1}(\nabla\boldsymbol{\eta}_\delta^\delta - \nabla\boldsymbol{\eta})(\boldsymbol{I} + \nabla\boldsymbol{\eta})^{-1}\right|^2.
\end{aligned}$$

Using the fact that p is smooth and the bootstrap assumption (11.119), we have that

$$\|\nabla_b^\eta p - \nabla_b^{\eta_\delta^\delta} p\|_{L^2(\Omega_b)}^2 \leq C\|\nabla \boldsymbol{\eta}_\delta^\delta - \nabla \boldsymbol{\eta}\|_{L^2(\Omega_b)}^2.$$

Therefore, combining this with (11.174), we obtain

$$R_{17,2} \leq \epsilon \int_0^t \|\nabla(p - p_\delta)\|_{L^2(\Omega_{b,\delta}^\delta(s))}^2 + C(\epsilon) \int_0^t \|\nabla \boldsymbol{\eta} - \nabla \boldsymbol{\eta}_\delta^\delta\|_{L^2(\Omega_b)}^2.$$

The final estimate of T_{17} now follows after the application of the convolution inequality (11.173):

$$T_{17} \leq \kappa \int_0^t \int_{\Omega_{b,\delta}^\delta(s)} |\nabla(p - p_\delta)|^2 + R_{17},$$

where the remainder is bounded by

$$\begin{aligned}
|R_{17}| &\leq \epsilon \int_0^t \|\nabla(p - p_\delta)\|_{L^2(\Omega_{b,\delta}^\delta(s))}^2 \\
&\quad + C(\epsilon) \left(\int_0^t \|\nabla \boldsymbol{\eta} - \nabla \boldsymbol{\eta}^\delta\|_{L^2(\Omega_b)}^2 \right. \\
&\quad \left. + \int_0^t \|\nabla \boldsymbol{\eta} - \nabla \boldsymbol{\eta}_\delta\|_{L^2(\Omega_b)}^2 + \int_0^t \|\omega - \omega_\delta\|_{H^2(\Gamma)}^2 \right).
\end{aligned}$$

Term 18. Here we want to estimate

$$\begin{aligned}
T_{18} &= \int_0^t \int_{\Gamma(s)} p(\boldsymbol{u} - \boldsymbol{\xi}) \cdot \boldsymbol{n} - \int_0^t \int_{\Gamma(s)} p(\boldsymbol{u}_\delta - \boldsymbol{\xi}_\delta) \cdot \boldsymbol{n} \\
&\quad - \int_0^t \int_{\Gamma_\delta(s)} p_\delta(\boldsymbol{u} - \boldsymbol{\xi}) \cdot \boldsymbol{n}_\delta + \int_0^t \int_{\Gamma_\delta(s)} p_\delta(\boldsymbol{u}_\delta - \boldsymbol{\xi}_\delta) \cdot \boldsymbol{n}_\delta \\
&\quad - \int_0^t \int_{\Gamma(s)} ((\boldsymbol{u} - \boldsymbol{\xi}) \cdot \boldsymbol{n})(p - p_\delta) + \int_0^t \int_{\Gamma_\delta(s)} ((\boldsymbol{u}_\delta - \boldsymbol{\xi}_\delta) \cdot \boldsymbol{n}_\delta)(p - p_\delta) \\
&= - \int_0^t \int_{\Gamma(s)} p(\boldsymbol{u}_\delta - \boldsymbol{\xi}_\delta) \cdot \boldsymbol{n} - \int_0^t \int_{\Gamma_\delta(s)} p_\delta(\boldsymbol{u} - \boldsymbol{\xi}) \cdot \boldsymbol{n}_\delta \\
&\quad + \int_0^t \int_{\Gamma(s)} ((\boldsymbol{u} - \boldsymbol{\xi}) \cdot \boldsymbol{n}) p_\delta + \int_0^t \int_{\Gamma_\delta(s)} ((\boldsymbol{u}_\delta - \boldsymbol{\xi}_\delta) \cdot \boldsymbol{n}_\delta) p.
\end{aligned}$$

By mapping all of the integrals back to the reference domain Γ, we obtain

$$T_{18} = - \int_0^t \int_\Gamma p(\boldsymbol{u}_\delta - \boldsymbol{\xi}_\delta) \cdot (-\partial_x \omega, 1) - \int_0^t \int_\Gamma p_\delta(\boldsymbol{u} - \boldsymbol{\xi}) \cdot (-\partial_x \omega_\delta, 1)$$

$$
\begin{aligned}
&+\int_0^t\int_\Gamma p_\delta(\boldsymbol{u}-\boldsymbol{\xi})\cdot(-\partial_x\omega,1)+\int_0^t\int_\Gamma p(\boldsymbol{u}_\delta-\boldsymbol{\xi}_\delta)\cdot(-\partial_x\omega_\delta,1)\\
&=\int_0^t\int_\Gamma p(\boldsymbol{u}_\delta-\boldsymbol{\xi}_\delta)\cdot\boldsymbol{e}_x(\partial_x\omega-\partial_x\omega_\delta)-\int_0^t\int_\Gamma p_\delta(\boldsymbol{u}-\boldsymbol{\xi})\cdot\boldsymbol{e}_x(\partial_x\omega-\partial_x\omega_\delta)\\
&=-\int_0^t\int_\Gamma p[(\boldsymbol{u}-\boldsymbol{\xi})\cdot\boldsymbol{e}_x-(\boldsymbol{u}_\delta-\boldsymbol{\xi}_\delta)\cdot\boldsymbol{e}_x](\partial_x\omega-\partial_x\omega_\delta)\\
&+\int_0^t\int_\Gamma(p-p_\delta)(\boldsymbol{u}-\boldsymbol{\xi})\cdot\boldsymbol{e}_x(\partial_x\omega-\partial_x\omega_\delta).
\end{aligned}
$$

The absolute value is bounded as follows:

$$
\begin{aligned}
&\left|\int_0^t\int_\Gamma p[(\boldsymbol{u}-\boldsymbol{\xi})\cdot\boldsymbol{e}_x-(\boldsymbol{u}_\delta-\boldsymbol{\xi}_\delta)\cdot\boldsymbol{e}_x](\partial_x\omega-\partial_x\omega_\delta)\right|\\
&\qquad+\left|\int_0^t\int_\Gamma(p-p_\delta)(\boldsymbol{u}-\boldsymbol{\xi})\cdot\boldsymbol{e}_x(\partial_x\omega-\partial_x\omega_\delta)\right|\\
&\le C\Bigg(\int_0^t\|(\boldsymbol{u}-\boldsymbol{\xi})\cdot\boldsymbol{e}_x-(\boldsymbol{u}_\delta-\boldsymbol{\xi}_\delta)\cdot\boldsymbol{e}_x\|_{L^2(\Gamma)}\|\partial_x\omega-\partial_x\omega_\delta\|_{L^2(\Gamma)}\\
&\qquad+\int_0^t\|p-p_\delta\|_{L^2(\Gamma)}\|\partial_x\omega-\partial_x\omega_\delta\|_{L^2(\Gamma)}\Bigg).
\end{aligned}
$$

After the application of the trace theorem, Poincare's inequality, and Korn's inequality, we obtain the final estimate:

$$
\begin{aligned}
|T_{18}|&\le\epsilon\Bigg(\int_0^t\|\boldsymbol{D}(\widehat{\boldsymbol{u}}-\boldsymbol{u}_\delta)\|^2_{L^2(\Omega_{f,\delta}(s))}\\
&+\int_0^t\|\nabla(\boldsymbol{\xi}-\boldsymbol{\xi}_\delta)\|^2_{L^2(\Omega_b)}+\int_0^t\|\nabla(p-p_\delta)\|^2_{L^2(\Omega^\delta_{b,\delta}(s))}\Bigg)\\
&+C(\epsilon)\int_0^t\|\omega-\omega_\delta\|^2_{H^2(\Gamma)}.
\end{aligned}
$$

11.11 Bibliographical Notes

Here, we give a brief literature review of the well-posedness analysis for the Biot equations and on the mathematical well-posedness for fluid-poroelastic structure interaction. We start with the Biot equations. One of the first well-posedness analyses of the Biot equations was carried out in [345], where the quasi-static Biot equations with $\rho_b = 0$ in addition to $c_0 = 0$ were analyzed from the perspective of solutions satisfying a variational formulation. The existence of weak (variational) solutions that are in addition unique was established in this work by discretizing the problem in time using a backward Euler scheme and by discretizing the problem in space using finite elements and then passing to the limit in the time and space approximation parameters. The result in [345] was later extended in [278], where the same quasistatic Biot system with $\rho_b = c_0 = 0$ was considered, but more regularity was established for the resulting weak solutions. The Biot displacement and pore pressure were shown to be continuous as functions taking values in $H^1(\Omega_b)$, with the Biot displacement additionally having a time derivative in $L^2(0, T; H^1(\Omega_b))$ in [278], via a Galerkin method combined with energy estimates.

Variational solutions to the Biot equations were also considered in the later work [27], where the authors consider the Biot equations with so-called secondary consolidation. In this case, the second equation, (11.2), remained the same, and the first equation had a secondary consolidation term involving a parameter $\lambda^* \geq 0$ that read

$$\rho_b \boldsymbol{\eta}_{tt} - \nabla(\lambda^*(\nabla \cdot \boldsymbol{\eta})_t) - \nabla \cdot \boldsymbol{\sigma}(\nabla \boldsymbol{\eta}, p) = F \quad \text{on } \Omega_b,$$

where F was an external forcing term. This work [27] considered several regimes and showed the existence of a weak solution via a Galerkin approach if $\rho > 0$ and $\lambda^* > 0$. By taking the limit as $\lambda^* = 0$, this work also obtained results for the so-called thermoelastic case where $\rho > 0$ and $\lambda^* = 0$, which has also been considered in the different context of linear thermoelasticity in [105] (where in thermoelasticity, the elastodynamics equation takes a similar form to the first equation in the Biot equations, and the temperature equation takes a similar form to the pressure equations in the Biot equations). Finally, the work in [27] also considered the limiting case where $\rho = 0$, and $\lambda^* > 0$ by taking the limit as ρ goes to zero to obtain results for the quasi-static Biot equations with secondary consolidation. Work on variational solutions to Biot equations was extended to nonlinear Biot equations on fixed domains in [28], where a semilinear Biot equation in 1D is considered. This equation involves the same second equation, (11.2), but adds an additional semilinear nonlinearity in addition to secondary consolidation to the first equation, (11.1), which now reads

$$\rho_b \partial_{tt}\eta - \lambda^* \partial_t \partial_x^2 \eta - \partial_x \sigma(\nabla \eta, p) - \mu^* \partial_x(|\partial_x \eta|^{q-2} \partial_x u) = f \quad \text{on } \Omega_b,$$

for some $q > 1$.

Another approach to studying the Biot equations is to reframe the Biot equations as an abstract evolution equation, for which the tools of implicit/degenerate evolution equation theory can be applied in order to obtain well-posedness results. This was the approach taken in [303], where a quasistatic Biot model with $c_0 \geq 0$, and potentially with secondary consolidation, was considered. This framework of considering abstract evolution equations was extended in the work [305] to a quasistatic nonlinear Biot model on a fixed domain, where there were nonlinear relationships between the pore pressure and physical quantities in the problem. The work in [305] studied a problem with the same relation for the first equation (11.1) with $\rho_b = 0$, but the second equation was replaced by the nonlinear equation

$$(b(p) + \alpha\nabla \cdot \boldsymbol{\eta})_t - \nabla \cdot (\kappa \nabla p + \boldsymbol{g}(p)) = F \quad \text{on } \Omega_b,$$

where b and $\boldsymbol{g}$ were nonlinear functions describing the relationship between pore pressure and density and gravitational effects, respectively.

Finally, there has been recent work on challenging Biot models involving nonlinear permeability in Darcy's law (see [42]), where quasistatic Biot models with $c_0 = 0$ were considered with poroelasticity and poroviscoelasticity. The model here was

$$\nabla \cdot \boldsymbol{\sigma}(\nabla \boldsymbol{\eta}, p) = -F,$$

$$\alpha(\nabla \cdot \boldsymbol{\eta})_t - \nabla \cdot (\kappa(\nabla \cdot \boldsymbol{\eta})\nabla p) = S,$$

on a fixed domain Ω_b, where the permeability κ depends on $\nabla \cdot \boldsymbol{\eta}$. The stress tensor was given by

$$\boldsymbol{\sigma}(\boldsymbol{\eta}, p) = 2\mu_e \boldsymbol{D}(\boldsymbol{\eta}) + \lambda_e(\nabla \cdot \boldsymbol{\eta})\boldsymbol{I} + 2\mu_v \boldsymbol{D}(\boldsymbol{\eta}_t) + \lambda_v(\nabla \cdot \boldsymbol{\eta}_t) - \alpha p \boldsymbol{I}.$$

This problem is purely poroelastic when $\mu_v = 0, \lambda_v = 0$ and poroviscoelastic when μ_v and λ_v are positive. The approach in [42] was to discretize in space and time, as in [345], and then pass to the limit in the discretization parameters, in order to establish existence of a solution under appropriate assumptions on the nonlinear permeability. This work was extended in [44], where a more general quasistatic case where $c_0 \geq 0$ was considered, rather than the case of $c_0 = 0$ in [42]. In this case, the permeability κ was not just a function of $\nabla \cdot \boldsymbol{\eta}$, but it was more generally a function of $c_0 p + \alpha\nabla \cdot \boldsymbol{\eta}$. The well-posedness analysis in this work [44] was based on only spatial discretization rather than discretization in both time and space, and it relied on analyzing a linear problem and using a fixed point argument for multi-valued maps, since it was not known if weak solutions were unique for the specific linear system that was needed for this fixed point argument.

In terms of literature for fluid-poroelastic structure interaction (FPSI), despite the importance of FPSI in engineering applications [79, 162, 237, 350], there are few works analyzing the well-posedness of fluids interacting with poroelastic structures.

While the study of fluids interacting with poroelastic materials is an emerging, yet important, field of mathematical research, we note that the study of FPSI has arisen naturally from past studies of incompressible Newtonian fluids interacting with porous materials that are not elastic [121, 232, 292] and more generally non-Newtonian fluids interacting with porous materials [78, 124, 130, 131], where the porous material dynamics is modeled using the Darcy equation for porous media flow.

The abstract evolution equation approach from the Biot equations on a fixed domain [303, 305] has been considered for linearly coupled FPSI in [304], where an FPSI system involving the Biot equations interacting with slightly compressible (rather than incompressible) Stokes equations is considered from an abstract evolution equation framework. Next, a linearly coupled FPSI system involving a poroelastic material modeled by the Biot equations interacting with an incompressible fluid modeled by the Navier-Stokes equations driven by inlet flow was considered in [80], and it was shown that there exists a unique weak solution to this problem, given sufficiently small external forcing and inlet pressure. To show this result, a Galerkin method was used to obtain existence of solutions to a weak formulation, where the test functions for the fluid velocity were divergence-free test functions in $H^1(\Omega_f)$. The fluid pressure was recovered using an inf-sup argument, where the fluid pressure along with the other quantities satisfied a weak formulation where the fluid velocity test functions were not necessarily divergence-free functions in $H^1(\Omega_f)$ and the divergence-free condition on the fluid velocity was enforced by testing this divergence-free condition with test functions in $L^2(\Omega_f)$. The analysis of linearly coupled FPSI has been extended to the interaction between Biot poroelastic materials and non-Newtonian fluids in [7], where the generalized Stokes equations model the fluid dynamics. In this case, the fluid viscosity was a nonlinear function of the strain $\boldsymbol{D}(\boldsymbol{u})$, which is a behavior that can be observed in non-Newtonian fluids such as shear-thinning fluids. In addition, the effective viscosity in the Biot poroelastic material was also a nonlinear function of the filtration velocity $\boldsymbol{q}$, so that Darcy's law (11.3) now reads

$$\nu_{eff}(\boldsymbol{q})\kappa^{-1}\boldsymbol{q} + \nabla p = 0.$$

Finally, we mention the recent work in [43], which considers a novel FPSI model involving the coupled interaction between an incompressible fluid modeled by the Stokes equations and a *multilayered poroelastic structure* consisting of a thin poroelastic plate and a thick Biot poroelastic material. This model was motivated by applications to biomedical engineering in the context of developing bioartificial organs. A well-posedness result in the context of weak solutions was obtained for this problem by discretizing the problem in time in order to construct approximate solutions and then passing to the limit in the time discretization. Chap. 10 of this book summarizes the results from this work. Extension of this work to a nonlinearly coupled problem, discussed in Chap. 11 of this book, was presented in [218] and [219]. In this chapter, we showed how to prove the existence of a weak solution to a fluid-structure interaction problem between the flow of an incompressible, viscous fluid and a multi-layered poroelastic/poroviscoelastic structure consisting of the Biot equations of poro(visco)elasticity and a thin, reticular interface with mass and elastic energy, which is transparent to fluid flow. The fluid and multilayered structure was nonlinearly coupled, giving rise to significant difficulties in the existence proof, associated with the geometric nonlinearity of the coupled problem. The existence proof was constructive, and it consisted of two major steps. In the first step, we proved the existence of a weak solution to a regularized problem in the class of finite energy solutions. In the second step, we showed that the solution of this regularized problem converges to a classical solution to the original, nonregularized problem as the regularization parameter tends to zero, as long as the original problem possesses a classical solution. While the proof of the existence of a weak solution to the regularized problem only requires that the Biot structure is poroelastic, additional regularity of the Biot poroelastic medium was required to prove the weak-classical consistency—the Biot structure is assumed to be poroviscoelastic. This weak-classical consistency result also shows that the solution we constructed is unique in the sense of weak-classical uniqueness.

An interesting extension of this work is to consider the singular limit as the thin interface thickness converges to zero and investigate the existence of a weak solution to the FSI problem between the Navier-Stokes equations for an incompressible, viscous fluid and the Biot equations of poroviscoelasticity, nonlinearly coupled over the moving interface. Preliminary results indicate that this will be possible under certain assumptions.

Bibliography

1. R. A. Adams. *Sobolev spaces*, volume 65 of *Pure and Applied Mathematics*. Academic Press, New York-London, 1975.
2. R. A. Adams and J. J. F. Fournier. *Sobolev spaces*, volume 140 of *Pure and Applied Mathematics (Amsterdam)*. Elsevier/Academic Press, Amsterdam, second edition, 2003.
3. M. Akköse, S. Adanur, A. Bayraktar, and A. A. Dumanoğlu. Elasto-plastic earthquake response of arch dams including fluid-structure interaction by the Lagrangian approach. *Appl. Math. Model.*, 32(11):2396–2412, 2008.
4. H. Al Baba, N. V. Chemetov, Š. Nečasová, and B. Muha. Strong solutions in L^2 framework for fluid-rigid body interaction problem. Mixed case. *Topol. Methods Nonlinear Anal.*, 52(1):337–350, 2018.
5. A. Alphonse, D. Caetano, A. Djurdjevac, and C. M. Elliott. Function spaces, time derivatives and compactness for evolving families of Banach spaces with applications to PDEs. *J. Differential Equations*, 353:268–338, 2023.
6. H. Amann. Compact embeddings of vector-valued Sobolev and Besov spaces. *Glas. Mat. Ser. III*, 35(55)(1):161–177, 2000. Dedicated to the memory of Branko Najman.
7. I. Ambartsumyan, V. J. Ervin, T. Nguyen, and I. Yotov. A nonlinear Stokes-Biot model for the interaction of a non-Newtonian fluid with poroelastic media. *ESAIM Math. Model. Numer. Anal.*, 53(6):1915–1955, 2019.
8. T. B. Amina, B. Mohamed, L. André, and B. Abdelmalek. Fluid-structure interaction of Brezina arch dam: 3D modal analysis. *Eng. Struct.*, 84:19–28, 2015.
9. D. Applebaum. *Lévy processes and stochastic calculus*, volume 116 of *Cambridge Studies in Advanced Mathematics*. Cambridge University Press, Cambridge, second edition, 2009.
10. N. Ataei and J. E. Padgett. Influential fluid-structure interaction modelling parameters on the response of bridges vulnerable to coastal storms. *Struct. Infrastruct. Eng.*, 11(3):321–333, 2015.
11. J.-P. Aubin. Un théorème de compacité. *C. R. Acad. Sci. Paris*, 256(24):5042–5044, 1963.
12. J.-P. Aubin and H. Frankowska. *Set-valued analysis*. Modern Birkhäuser Classics. Birkhäuser Boston, Inc., Boston, MA, 2009. Reprint of the 1990 edition [MR1048347].
13. G. Avalos and F. Bucci. Rational rates of uniform decay for strong solutions to a fluid-structure PDE system. *J. Differential Equations*, 258(12):4398–4423, 2015.
14. G. Avalos, P. G. Geredeli, and B. Muha. Wellposedness, spectral analysis and asymptotic stability of a multilayered heat-wave-wave system. *J. Differential Equations*, 269(9):7129–7156, 2020.
15. G. Avalos, P. G. Geredeli, and B. Muha. Rational decay of a multilayered structure-fluid PDE system. *J. Math. Anal. Appl.*, 514(2):Paper No. 126284, 16, 2022.

S. Čanić et al., *Deterministic and Stochastic Fluid-Structure Interaction*, Advances in Mathematical Fluid Mechanics, https://doi.org/10.1007/978-3-032-00898-5

16. G. Avalos, I. Lasiecka, and R. Triggiani. Higher regularity of a coupled parabolic-hyperbolic fluid-structure interactive system. *Georgian Math. J.*, 15(3):403–437, 2008.
17. G. Avalos, I. Lasiecka, and R. Triggiani. Heat-wave interaction in 2–3 dimensions: optimal rational decay rate. *J. Math. Anal. Appl.*, 437(2):782–815, 2016.
18. G. Avalos and R. Triggiani. The coupled PDE system arising in fluid/structure interaction. Part I: Explicit semigroup generator and its spectral properties. In F. Botelho, T. Hagen, and J. Jamison, editors, *Fluids and waves*, volume 440 of *Contemporary Mathematics*, pages 15–54. American Mathematical Society, Providence, Rhode Island, 2007.
19. G. Avalos and R. Triggiani. Uniform stabilization of a coupled PDE system arising in fluid-structure interaction with boundary dissipation at the interface. *Discrete Contin. Dyn. Syst.*, 22(4):817–833, 2008.
20. G. Avalos and R. Triggiani. Boundary feedback stabilization of a coupled parabolic-hyperbolic Stokes-Lamé PDE system. *J. Evol. Equ.*, 9(2):341–370, 2009.
21. G. Avalos and R. Triggiani. Semigroup well-posedness in the energy space of a parabolic-hyperbolic coupled Stokes-Lamé PDE system of fluid-structure interaction. *Discrete Contin. Dyn. Syst. Ser. S*, 2(3):417–447, 2009.
22. S. Badia, A. Quaini, and A. Quarteroni. Coupling Biot and Navier-Stokes equations for modelling fluid-poroelastic media interaction. *J. Comput. Phys.*, 228(21):7986–8014, 2009.
23. A. Balakrishna, I. Kukavica, B. Muha, and A. Tuffaha. Inviscid fluid interacting with a nonlinear two-dimensional plate. *Interfaces Free Bound.*, 27(1):141–175, 2025.
24. A. Balakrishna, I. Lasiecka, and J. T. Webster. Elastic stabilization of an intrinsically unstable hyperbolic flow-structure interaction on the 3D half-space. *Math. Models Methods Appl. Sci.*, 33(3):505–545, 2023.
25. V. Barbu, Z. Grujić, I. Lasiecka, and A. Tuffaha. Existence of the energy-level weak solutions for a nonlinear fluid-structure interaction model. In F. Botelho, T. Hagen, and J. Jamison, editors, *Fluids and waves*, volume 440 of *Contemporary Mathematics*, pages 55–82. American Mathematical Society, Providence, Rhode Island, 2007.
26. V. Barbu, Z. Grujić, I. Lasiecka, and A. Tuffaha. Smoothness of weak solutions to a nonlinear fluid-structure interaction model. *Indiana Univ. Math. J.*, 57(3):1173–1207, 2008.
27. H. Barucq, M. Madaune-Tort, and P. Saint-Macary. Theoretical aspects of wave propagation for Biot's consolidation problem. *Monografías del Seminario Matemático García de Galdeano*, 31:449–458, 2004.
28. H. Barucq, M. Madaune-Tort, and P. Saint-Macary. On nonlinear Biot's consolidation models. *Nonlinear Anal.*, 63(5):e985–e995, 2005.
29. A. M. Bavo, G. Rocatello, F. Iannaccone, J. Degroote, J. Vierendeels, and P. Segers. Fluid-structure interaction simulation of prosthetic aortic valves: comparison between immersed boundary and Arbitrary Lagrangian-Eulerian techniques for the mesh representation. *PLoS ONE*, 11(4):e0154517, 2016.
30. Y. Bazilevs, M.-C. Hsu, Y. Zhang, W. Wang, X. Liang, T. Kvamsdal, R. Brekken, and J. G. Isaksen. A fully-coupled fluid-structure interaction simulation of cerebral aneurysms. *Comput. Mech.*, 46(1):3–16, 2010.
31. H. Beirão da Veiga. On the existence of strong solutions to a coupled fluid-structure evolution problem. *J. Math. Fluid Mech.*, 6(1):21–52, 2004.
32. B. Benešová, M. Kampschulte, and S. Schwarzacher. A variational approach to hyperbolic evolutions and fluid-structure interactions. *J. Eur. Math. Soc. (JEMS)*, 26(12):4615–4697, 2024.
33. B. Benešová, Š. Nečasová, J. Scherz, and A. Schlömerkemper. Fluid-rigid body interaction in an incompressible electrically conducting fluid. *SIAM J. Math. Anal.*, 55(2):929–965, 2023.
34. A. Bensoussan. Stochastic Navier-Stokes equations. *Acta Appl. Math.*, 38(3):267–304, 1995.
35. A. Bensoussan, R. Glowinski, and A. Răşcanu. Approximation of some stochastic differential equations by the splitting up method. *Appl. Math. Optim.*, 25(1):81–106, 1992.
36. A. Bensoussan and R. Temam. Équations stochastiques du type Navier-Stokes. *J. Functional Analysis*, 13:195–222, 1973.

37. F. Berthelin and J. Vovelle. Stochastic isentropic Euler equations. *Ann. Sci. Éc. Norm. Supér. (4)*, 52(1):181–254, 2019.
38. P. Billingsley. *Probability and measure*. Wiley Series in Probability and Mathematical Statistics. John Wiley & Sons, Inc., New York, third edition, 1995. A Wiley-Interscience Publication.
39. P. Billingsley. *Convergence of probability measures*. Wiley Series in Probability and Statistics: Probability and Statistics. John Wiley & Sons, Inc., New York, second edition, 1999. A Wiley-Interscience Publication.
40. M. A. Biot. General theory of three-dimensional consolidation. *J. Appl. Phys.*, 12(2):155–164, 1941.
41. M. A. Biot. Theory of elasticity and consolidation for a porous anisotropic solid. *J. Appl. Phys.*, 26(2):182–185, 1955.
42. L. Bociu, G. Guidoboni, R. Sacco, and J. T. Webster. Analysis of nonlinear poro-elastic and poro-visco-elastic models. *Arch. Ration. Mech. Anal.*, 222(3):1445–1519, 2016.
43. L. Bociu, S. Čanić, B. Muha, and J. T. Webster. Multilayered poroelasticity interacting with Stokes flow. *SIAM J. Math. Anal.*, 53(6):6243–6279, 2021.
44. L. Bociu and J. T. Webster. Nonlinear quasi-static poroelasticity. *J. Differential Equations*, 296:242–278, 2021.
45. D. Bonheure and G. P. Galdi. Global weak solutions to a time-periodic body-liquid interaction problem. *Ann. Inst. Henri Poincaré C*, 2024.
46. D. Bonheure, G. P. Galdi, and F. Gazzola. Stability of equilibria and bifurcations for a fluid-solid interaction problem. *arXiv preprint arXiv:2406.04162*, 2024.
47. M. Boulakia. Existence of weak solutions for the three-dimensional motion of an elastic structure in an incompressible fluid. *J. Math. Fluid Mech.*, 9(2):262–294, 2007.
48. M. Boulakia and S. Guerrero. Regular solutions of a problem coupling a compressible fluid and an elastic structure. *J. Math. Pures Appl.*, 94(4):341–365, 2010.
49. M. Boulakia and S. Guerrero. On the interaction problem between a compressible fluid and a Saint-Venant Kirchhoff elastic structure. *Adv. Differential Equations*, 22(1–2):1–48, 2017.
50. M. Bravin. Energy equality and uniqueness of weak solutions of a "viscous incompressible fluid + rigid body" system with Navier slip-with-friction conditions in a 2D bounded domain. *J. Math. Fluid Mech.*, 21(2):Paper No. 23, 31, 2019.
51. H. Breckner. Galerkin approximation and the strong solution of the Navier-Stokes equation. *J. Appl. Math. Stochastic Anal.*, 13(3):239–259, 2000.
52. D. Breit, E. Feireisl, and M. Hofmanová. *Stochastically forced compressible fluid flows*, volume 3 of *De Gruyter Series in Applied and Numerical Mathematics*. De Gruyter, Berlin, 2018.
53. D. Breit and M. Hofmanova. Stochastic Navier-Stokes equations for compressible fluids. *Indiana Univ. Math. J.*, 65(4):1183–1250, 2016.
54. D. Breit and P. R. Mensah. An incompressible polymer fluid interacting with a Koiter shell. *J. Nonlinear Sci.*, 31(1):Paper No. 25, 56, 2021.
55. D. Breit, P. R. Mensah, and T. C. Moyo. Martingale solutions in stochastic fluid-structure interaction. *J. Nonlinear Sci.*, 34:34, 2024.
56. D. Breit, P. R. Mensah, S. Schwarzacher, and P. Su. Ladyzhenskaya-Prodi-Serrin condition for fluid-structure interaction systems. *arXiv preprint arXiv:2307.12273*, 2023.
57. D. Breit and S. Schwarzacher. Compressible fluids interacting with a linear-elastic shell. *Arch. Ration. Mech. Anal.*, 228(2):495–562, 2018.
58. D. Breit and S. Schwarzacher. Navier-Stokes-Fourier fluids interacting with elastic shells. *Ann. Sc. Norm. Super. Pisa Cl. Sci. (5)*, 24(2):619–690, 2023.
59. S. C. Brenner and L. R. Scott. *The Mathematical Theory of Finite Element Methods*, volume 15 of *Texts in Applied Mathematics*. Springer Science+Business Media, LLC, New York, third edition, 2008.
60. H. Brezis. *Functional analysis, Sobolev spaces and partial differential equations*. Universitext. Springer, New York, 2011.

61. J. Broniszewski and J.R. Piechna. Fluid-structure interaction analysis of a competitive car during brake-in-turn manoeuvre. *Energies*, 15(8):2917, 2022.
62. Z. Brzeźniak, E. Hausenblas, and P. Razafimandimby. Stochastic reaction-diffusion equations driven by jump processes. *Potential Anal.*, 49(1):131–201, 2018.
63. Z. Brzeźniak and M. Ondreját. Stochastic geometric wave equations with values in compact Riemannian homogeneous spaces. *Ann. Probab.*, 41(3B):1938–1977, 2013.
64. M. Bukal and B. Muha. Rigorous derivation of a linear sixth-order thin-film equation as a reduced model for thin fluid-thin structure interaction problems. *Appl. Math. Optim.*, 84(2):2245–2288, 2021.
65. M. Bukal and B. Muha. Justification of a nonlinear sixth-order thin-film equation as the reduced model for a fluid-structure interaction problem. *Nonlinearity*, 35(8):4695–4726, 2022.
66. M. Bukač and S. Čanić. Longitudinal displacement in viscoelastic arteries: a novel fluid-structure interaction computational model, and experimental validation. *Journal Mathematical Biosciences and Engineering*, 10(2):258–388, 2013.
67. M. Bukač, S. Čanić, and B. Muha. A partitioned scheme for fluid-composite structure interaction problems. *J. Comput. Phys.*, 281:493–517, 2015.
68. M. Bukač, S. Čanić, and B. Muha. A nonlinear fliud-structure interaction problem in compliant arteries treated with vascular stents. *Appl. Math. Optim.*, 73(3):433–473, 2016.
69. M. Bukač, S. Čanić, B. Muha, and Y. Wang. A computational algorithm for optimal design of a bioartificial organ scaffold architecture. *PLOS Computational Biology, to appear*, 2024.
70. M. Bukač, S. Čanić, J. Tambača, and Y. Wang. Fluid-structure interaction between pulsatile blood flow and a curved stented coronary artery on a beating heart: a four stent computational study. *Comput. Methods Appl. Mech. Engrg.*, 350:679–700, 2019.
71. M. Bukač, P. Zunino, and I. Yotov. Explicit partitioning strategies for the interaction between a fluid and a multilayered poroelastic structure: an operator-splitting approach. *Journal of Computational Physics*, 228(21):7986–8014, 2013.
72. L. Caffarelli and L. Silvestre. An extension problem related to the fractional Laplacian. *Comm. Partial Differential Equations*, 32(8):1245–1260, 2007.
73. C. Canuto and D. Fransos. Numerical solution of partial differential equations in random domains: An application to wind engineering. *Comm. Comp. Phys.*, 5(2–4):515–531, 2009.
74. Y. Cao, S. Chen, and A. J. Meir. Analysis and numerical approximations of equations of nonlinear poroelasticity. *Discrete Contin. Dyn. Syst. Ser. B*, 18(5):1253–1273, 2013.
75. M. Capiński and N. Cutland. Stochastic Navier-Stokes equations. *Acta Appl. Math.*, 25(1):59–85, 1991.
76. J.-J. Casanova. Existence of time-periodic strong solutions to a fluid-structure system. *Discrete Contin. Dyn. Syst.*, 39(6):3291–3313, 2019.
77. J.-J. Casanova, C. Grandmont, and M. Hillairet. On an existence theory for a fluid-beam problem encompassing possible contacts. *J. Éc. Polytech. Math.*, 8:933–971, 2021.
78. S. Caucao, G. N. Gatica, R. Oyarzua, and I. Sebestova. A fully-mixed finite element method for the Navier-Stokes/Darcy coupled problem with nonlinear viscosity. *J. Numer. Math.*, 25(2):55–88, 2017.
79. P. Causin, G. Guidoboni, A. Harris, D. Prada, R. Sacco, and S. Terragni. A poroelastic model for the perfusion of the lamina cribrosa in the optic nerve head. *Math. Biosci.*, 257:33–41, 2014.
80. A. Cesmelioglu. Analysis of the coupled Navier-Stokes/Biot problem. *J. Math. Anal. Appl.*, 456(2):970–991, 2017.
81. A. Chambolle, B. Desjardins, M. J. Esteban, and C. Grandmont. Existence of weak solutions for the unsteady interaction of a viscous fluid with an elastic plate. *J. Math. Fluid Mech.*, 7(3):368–404, 2005.
82. N. V. Chemetov and Š. Nečasová. The motion of the rigid body in the viscous fluid including collisions. Global solvability result. *Nonlinear Anal. Real World Appl.*, 34:416–445, 2017.
83. N. V. Chemetov, Š. Nečasová, and B. Muha. Weak-strong uniqueness for fluid-rigid body interaction problem with slip boundary condition. *J. Math. Phys.*, 60(1):011505, 2019.

84. C. H. A. Cheng, D. Coutand, and S. Shkoller. Navier-Stokes equations interacting with a nonlinear elastic biofluid shell. *SIAM J. Math. Anal.*, 39(3):742–800, 2007.
85. C. H. A. Cheng and S. Shkoller. The interaction of the 3D Navier-Stokes equations with a moving nonlinear Koiter elastic shell. *SIAM J. Math. Anal.*, 42(3):1094–1155, 2010.
86. N. Cheraghi-Shirazi, A. R. Kabiri-Samani, and B. Boroomand. Numerical analysis of rubber dams using fluid-structure interactions. *Flow Meas. Instrum.*, 40:91–98, 2014.
87. I. Chueshov. Dynamics of a nonlinear elastic plate interacting with a linearized compressible viscous fluid. *Nonlinear Anal.*, 95:650–665, 2014.
88. I. Chueshov and I. Lasiecka. *Von Karman evolution equations*. Springer Monographs in Mathematics. Springer, New York, 2010. Well-posedness and long-time dynamics.
89. I. Chueshov, I. Lasiecka, and J. Webster. Flow-plate interactions: well-posedness and long-time behavior. *Discrete Contin. Dyn. Syst. Ser. S*, 7(5):925–965, 2014.
90. I. Chueshov and I. Ryzhkova. Unsteady interaction of a viscous fluid with an elastic shell modeled by full von Karman equations. *J. Differential Equations*, 254(4):1833–1862, 2013.
91. P. G. Ciarlet. *Mathematical Elasticity. Volume I: Three-Dimensional Elasticity*, volume 20 of *Studies in Mathematics and Its Applications*. Elsevier Science Publishers B.V., Amsterdam, 1988.
92. P. G. Ciarlet. *Mathematical elasticity. Vol. II*, volume 27 of *Studies in Mathematics and its Applications*. North-Holland Publishing Co., Amsterdam, 1997. Theory of plates.
93. P. G. Ciarlet. *Mathematical elasticity. Vol. III*, volume 29 of *Studies in Mathematics and its Applications*. North-Holland Publishing Co., Amsterdam, 2000. Theory of shells.
94. D. Cioranescu and J. Saint Jean Paulin. *Homogenization of reticulated structures*, volume 136 of *Applied Mathematical Sciences*. Springer-Verlag, New York, 1999.
95. E. Clark, N. Katzourakis, and B. Muha. Vectorial variational problems in L^∞ constrained by the Navier-Stokes equations. *Nonlinearity*, 35(1):470–491, 2022.
96. C. Conca, J. San Martín H., and M. Tucsnak. Existence of solutions for the equations modelling the motion of a rigid body in a viscous fluid. *Comm. Partial Differential Equations*, 25(5–6):1019–1042, 2000.
97. C. Conca, F. Murat, and O. Pironneau. The Stokes and Navier-Stokes equations with boundary conditions involving the pressure. *Japan. J. Math.*, 20(2):279–318, 1994.
98. D. Conus and R. C. Dalang. The non-linear stochastic wave equation in high dimensions. *Electron. J. Probab.*, 13:629–670, 2008.
99. D. Coutand and S. Shkoller. Motion of an elastic solid inside an incompressible viscous fluid. *Arch. Ration. Mech. Anal.*, 176(1):25–102, 2005.
100. D. Coutand and S. Shkoller. The interaction between quasilinear elastodynamics and the Navier-Stokes equations. *Arch. Ration. Mech. Anal.*, 179(3):303–352, 2006.
101. S. C Cowin. Bone poroelasticity. *J. Biomech.*, 32(3):217–238, 1999.
102. P. Cumsille and T. Takahashi. Wellposedness for the system modelling the motion of a rigid body of arbitrary form in an incompressible viscous fluid. *Czechoslovak Math. J.*, 58(133)(4):961–992, 2008.
103. G. Da Prato and A. Debussche. Two-dimensional Navier-Stokes equations driven by a space-time white noise. *J. Funct. Anal.*, 196(1):180–210, 2002.
104. G. Da Prato and J. Zabczyk. *Stochastic equations in infinite dimensions*, volume 152 of *Encyclopedia of Mathematics and its Applications*. Cambridge University Press, Cambridge, second edition, 2014.
105. C. M. Dafermos. On the existence and the asymptotic stability of solutions to the equations of linear thermoelasticity. *Arch. Ration. Mech. Anal.*, 29:241–271, 1968.
106. R. C. Dalang. Extending martingale measure stochastic integral with applications to spatially homogeneous S.P.D.E.'s. *Electron. J. Probab.*, 4:1–29, 1999.
107. R. C. Dalang. The stochastic wave equation. In D. Khoshnevisan and F. Rassoul-Agha, editors, *A minicourse on stochastic partial differential equations*, volume 1962 of *Lecture Notes in Mathematics*. Springer-Verlag, Berlin Heidelberg, 2009.
108. R. C. Dalang and N. E. Frangos. The stochastic wave equation in two spatial dimensions. *Ann. Probab.*, 26(1):187–212, 1998.

109. R. C. Dalang, D. Khoshnevisan, C. Mueller, D. Nualart, and Y. Xiao. *A minicourse on stochastic partial differential equations*, volume 1962 of *Lecture Notes in Mathematics*. Springer-Verlag, Berlin Heidelberg, 2009.
110. R. C. Dalang and M. Sanz-Solé. Hölder-Sobolev regularity of the solution to the stochastic wave equation in dimension three. *Mem. Amer. Math. Soc.*, 199(931):vi–70, 2009.
111. D. Daub, S. Willems, and A. Gülhan. Experiments on aerothermoelastic fluid-structure interaction in hypersonic flow. *J Sound Vibration*, 531:116714, 2022.
112. R. Dautray and J-L. Lions. *Mathematical analysis and numerical methods for science and technology. Vol. 5*. Springer-Verlag, Berlin, 1992. Evolution problems. I, With the collaboration of Michel Artola, Michel Cessenat and Hélène Lanchon, Translated from the French by Alan Craig.
113. A. Debussche, N. Glatt-Holtz, and R. Temam. Local martingale and pathwise solutions for an abstract fluids model. *Phys. D*, 240(14–15):1123–1144, 2011.
114. F. Demengel and G. Demengel. *Functional spaces for the theory of elliptic partial differential equations*. Universitext. Springer, London; EDP Sciences, Les Ulis, 2012. Translated from the 2007 French original by Reinie Erné.
115. A. Demir, A. E. Dincer, Z. Bozkus, and A. S. Tijsseling. Numerical and experimental investigation of damping in a dam-break problem with fluid-structure interaction. *J. Zhejiang Univ.-Sci. A (Appl. Phys. & Eng.)*, 20(4):258–271, 2019.
116. B. Desjardins and M. J. Esteban. Existence of weak solutions for the motion of rigid bodies in a viscous fluid. *Arch. Ration. Mech. Anal.*, 146(1):59–71, 1999.
117. B. Desjardins and M. J. Esteban. On weak solutions for fluid-rigid structure interaction: compressible and incompressible models. *Comm. Partial Differential Equations*, 25(7–8):263–285, 2000.
118. B. Desjardins, M. J. Esteban, C. Grandmont, and P. Le Tallec. Weak solutions for a fluid-elastic structure interaction model. *Rev. Mat. Complut.*, 14(2):523–538, 2001.
119. E. Di Nezza, G. Palatucci, and E. Valdinoci. Hitchhiker's guide to the fractional Sobolev spaces. *Bull. Sci. Math.*, 136(5):521–573, 2012.
120. J. Diestel and J. J. Uhl, Jr. *Vector measures*, volume No. 15 of *Mathematical Surveys*. American Mathematical Society, Providence, RI, 1977. With a foreword by B. J. Pettis.
121. M. Discacciati, E. Miglio, and A. Quarteroni. Mathematical and numerical models for coupling surface and groundwater flows. *Appl. Numer. Math.*, 43(1–2):57–74, 2002.
122. M. Discacciati and A. Quarteroni. Navier-Stokes/Darcy coupling: modeling, analysis, and numerical approximation. *Rev. Mat. Complut.*, 22(2):315–426, 2009.
123. K. Disser, G. P. Galdi, G. Mazzone, and P. Zunino. Inertial motions of a rigid body with a cavity filled with a viscous liquid. *Arch. Ration. Mech. Anal.*, 221(1):487–526, Jul 2016.
124. S. Domínguez, G. N. Gatica, A. Márquez, and S. Meddahi. A primal-mixed formulation for the strong coupling of quasi-Newtonian fluids with porous media. *Adv. Comput. Math.*, 42(3):675–720, 2016.
125. S. S. Dragomir. *Some Gronwall type inequalities and applications*. Nova Science Publishers, Inc., Hauppauge, NY, 2003.
126. M. Dreher and A. Jüngel. Compact families of piecewise constant functions in $L^p(0, T; B)$. *Nonlinear Anal.*, 75(6):3072–3077, 2012.
127. Q. Du, M. D. Gunzburger, L. S. Hou, and J. Lee. Analysis of a linear fluid-structure interaction problem. *Discrete Contin. Dyn. Syst.*, 9(3):633–650, 2003.
128. R. Durrett. *Probability: Theory and examples*, volume 49 of *Cambridge Series in Statistical and Probabilistic Mathematics*. Cambridge University Press, Cambridge, fifth edition edition, 2019.
129. T. Duyckaerts. Optimal decay rates of the energy of a hyperbolic-parabolic system coupled by an interface. *Asymptot. Anal.*, 51(1):17–45, 2007.
130. V. J. Ervin, E. W. Jenkins, and S. Sun. Coupled generalized nonlinear Stokes flow with flow through a porous medium. *SIAM J. Numer. Anal.*, 47(2):929–952, 2009.
131. V. J. Ervin, E. W. Jenkins, and S. Sun. Coupling nonlinear Stokes and Darcy flow using mortar finite elements. *Appl. Numer. Math.*, 61(11):1198–1222, 2011.

132. L. C. Evans. *Partial differential equations*, volume 19 of *Graduate Studies in Mathematics*. American Mathematical Society, Providence, Rhode Island, second edition edition, 2010.
133. L. C. Evans and R. F. Gariepy. *Measure theory and fine properties of functions*. Studies in Advanced Mathematics. CRC Press, Boca Raton, FL, 1992.
134. C. Farhat, M. Lesoinne, and P. Le Tallec. Load and motion transfer algorithms for fluid/structure interaction problems with non-matching discrete interfaces: momentum and energy conservation, optimal discretization and application to aeroelasticity. *Comput. Methods Appl. Mech. Engrg.*, 157(1–2):95–114, 1998.
135. I. K. Fathallah. Logarithmic decay of the energy for an hyperbolic-parabolic coupled system. *ESAIM Control Optim. Calc. Var.*, 17(3):801–835, 2011.
136. R.A. Febrianto, S. Haidi, R.L.G. Hidayat, D.M. Bae, B. Cao, and A.R. Prabowo. Implementation of fluid-structure interaction (fsi) in marine design: Calculation review on hull structures. *AIP Conf. Proc.*, 2674(1), 2023.
137. E. Feireisl. On the motion of rigid bodies in a viscous fluid. *Appl. Math.*, 47(6):463–484, 2002. Mathematical theory in fluid mechanics (Paseky, 2001).
138. E. Feireisl. On the motion of rigid bodies in a viscous compressible fluid. *Arch. Ration. Mech. Anal.*, 167(4):281–308, 2003.
139. F. Flandoli and D. Gatarek. Martingale and stationary solutions for stochastic Navier-Stokes equations. *Probab. Theory Related Fields*, 102(3):367–391, 1995.
140. G. B. Folland. *Real analysis: modern techniques and their applications*. Pure and Applied Mathematics (New York). John Wiley & Sons, Inc., New York, second edition, 1999.
141. L. Formaggia, J.-F. Gerbeau, F. Nobile, and A. Quarteroni. On the coupling of 3D and 1D Navier-Stokes equations for flow problems in compliant vessels. *Comput. Methods Appl. Mech. Engrg.*, 191(6):561–582, 2001.
142. Y. C. Fung. *Foundation of Solid Mechanics*. Prentice-Hall, 1965.
143. G. P. Galdi. On the steady self-propelled motion of a body in a viscous incompressible fluid. *Arch. Ration. Mech. Anal.*, 148(1):53–88, 1999.
144. G. P Galdi. On the motion of a rigid body in a viscous liquid: a mathematical analysis with applications. In *Handbook of mathematical fluid dynamics*, volume 1, pages 653–791. Elsevier, 2002.
145. G. P. Galdi. *An introduction to the mathematical theory of the Navier-Stokes equations*. Springer Monographs in Mathematics. Springer, New York, second edition, 2011. Steady-state problems.
146. G. P. Galdi. On the self-propulsion of a rigid body in a viscous liquid by time-periodic boundary data. *J. Math. Fluid Mech.*, 22(4):61, 2020.
147. G. P. Galdi. Navier–Stokes flow past a rigid body that moves by time-periodic motion. *J. Math. Fluid Mech.*, 24(2):30, 2022.
148. G. P. Galdi. Large-time behavior of a rigid body of arbitrary shape in a viscous fluid under the action of prescribed forces and torques. *J. Math. Fluid. Mech.*, 25(3):43, 2023.
149. G. P. Galdi and M. Kyed. Steady flow of a Navier-Stokes liquid past an elastic body. *Arch. Ration. Mech. Anal.*, 194(3):849–875, 2009.
150. G. P. Galdi, V. Mácha, and Š. Nečasová. On weak solutions to the problem of a rigid body with a cavity filled with a compressible fluid, and their asymptotic behavior. *Int. J. Nonlinear Mech.*, 121:103431, 2020.
151. G. P. Galdi, G. Mazzone, and M. Mohebbi. On the motion of a liquid-filled heavy body around a fixed point. *Quart. Appl. Math.*, 76(1):113–145, 2018.
152. G. P. Galdi and A. L. Silvestre. Existence of time-periodic solutions to the Navier-Stokes equations around a moving body. *Pacific J. Math.*, 223(2):251–267, 2006.
153. M. Galić. Regularity of a weak solution to a linear fluid-composite structure interaction problem. *Glas. Mat.*, 56(2):407–440, 2021.
154. X. Gao, Q. Zhang, and Q. Tang. Fluid-structure interaction analysis of parachute finite mass inflation. *Int. J. Aerosp. Eng.*, 2016:1–8, 2016.
155. L. Garelli, R. R. Paz, and M. A. Storti. Fluid-structure interaction study of the start-up of a rocket engine nozzle. *Comput. & Fluids*, 39(7):1208–1218, 2010.

156. F. Gazzola, V. Pata, and C. Patriarca. Attractors for a fluid-structure interaction problem in a time-dependent phase space, Preprint, 2022.
157. M. Geissert, K. Götze, and M. Hieber. L^p-theory for strong solutions to fluid-rigid body interaction in Newtonian and generalized Newtonian fluids. *Trans. Amer. Math. Soc.*, 365(3):1393–1439, 2013.
158. D. Gérard-Varet and M. Hillairet. Regularity issues in the problem of fluid structure interaction. *Arch. Ration. Mech. Anal.*, 195(2):375–407, 2010.
159. D. Gérard-Varet and M. Hillairet. Existence of weak solutions up to collision for viscous fluid-solid systems with slip. *Comm. Pure Appl. Math.*, 67(12):2022–2075, 2014.
160. D. Gérard-Varet, M. Hillairet, and C. Wang. The influence of boundary conditions on the contact problem in a 3D Navier-Stokes flow. *J. Math. Pures Appl. (9)*, 103(1):1–38, 2015.
161. D. Gilbarg and N. Trudinger. *Elliptic partial differential equations of second order*, volume 224 of *Grundlehren der mathematischen Wissenschaften [Fundamental Principles of Mathematical Sciences]*. Springer-Verlag, Berlin, second edition, 1983.
162. V. Girault, M. F. Wheeler, B. Ganis, and M. E. Mear. A lubrication fracture model in a poro-elastic medium. *Math. Models Methods Appl. Sci.*, 25(4):587–645, 2015.
163. O. Glass and F. Sueur. Uniqueness results for weak solutions of two-dimensional fluid-solid systems. *Arch. Ration. Mech. Anal.*, 218(2):907–944, 2015.
164. N. Glatt-Holtz and M. Ziane. Strong pathwise solutions of the stochastic Navier-Stokes system. *Adv. Differential Equations*, 14(5–6):567–600, 2009.
165. R. Glowinski. Finite element methods for incompressible viscous flow. In *Handbook of numerical analysis*, volume IX, pages 3–1176. North-Holland, Amsterdam, 2003.
166. C. Grandmont. Existence for a three-dimensional steady state fluid-structure interaction problem. *J. Math. Fluid Mech.*, 4(1):76–94, 2002.
167. C. Grandmont. Existence of weak solutions for the unsteady interaction of a viscous fluid with an elastic plate. *SIAM J. Math. Anal.*, 40(2):716–737, 2008.
168. C. Grandmont and M. Hillairet. Existence of global strong solutions to a beam-fluid interaction system. *Arch. Ration. Mech. Anal.*, 220(3):1283–1333, 2016.
169. C. Grandmont, M. Hillairet, and J. Lequeurre. Existence of local strong solutions to fluid-beam and fluid-rod interaction systems. *Ann. Inst. H. Poincaré C Anal. Non Linéaire*, 36(4):1105–1149, 2019.
170. C. Grandmont, M. Lukáčová-Medvid'ová, and Š. Nečasová. Mathematical and numerical analysis of some FSI problems. In T. Bodnár, G. P. Galdi, and Š. Nečasová, editors, *Fluid-structure interaction and biomedical applications*, Advances in Mathematical Fluid Mechanics, pages 1–77. Springer Basel AG, Switzerland, 2014.
171. C. Grandmont and Y. Maday. Existence for an unsteady fluid-structure interaction problem. *ESAIM Math. Model. Numer. Anal.*, 34(3):609–636, 2000.
172. G. Gravina, S. Schwarzacher, O. Souček, and K. Tuma. Contactless rebound of elastic bodies in a viscous incompressible fluid. *J. Fluid Mech.*, 942:A34, 2022.
173. G. Griso. Asymptotic behavior of structures made of curved rods. *Anal. Appl.*, 6(1):11–22, 2008.
174. P. Grisvard. *Elliptic problems in nonsmooth domains*, volume 69 of *Classics in Applied Mathematics*. Society for Industrial and Applied Mathematics (SIAM), Philadelphia, PA, 2011. Reprint of the 1985 original [MR0775683], With a foreword by Susanne C. Brenner.
175. G. Guidoboni, R. Glowinski, N. Cavallini., and S. Čanić. Stable loosely-coupled-type algorithm for fluid-structure interaction in blood flow. *J. Comput. Phys.*, 228(18):6916–6937, 2009.
176. M. D. Gunzburger, H.-C. Lee, and G. A. Seregin. Global existence of weak solutions for viscous incompressible flows around a moving rigid body in three dimensions. *J. Math. Fluid Mech.*, 2(3):219–266, 2000.
177. I. Gyöngy and N. Krylov. Existence of strong solutions for Itô's stochastic equations via approximations. *Probab. Theory Related Fields*, 105(2):143–158, 1996.
178. M. Hairer. An introduction to stochastic PDEs. *Electronic Lecture Notes:* https://arxiv.org/abs/0907.4178, 2009.

179. J. Heinonen, P. Koskela, N. Shanmugalingam, and J. T. Tyson. *Sobolev spaces on metric measure spaces: an approach based on upper gradients*, volume 27 of *New Mathematical Monographs*. Cambridge University Press, Cambridge, 2015.
180. J. G. Heywood and R. Rannacher. Finite-element approximation of the nonstationary Navier-Stokes problem. Part IV: error analysis for second-order time discretization. *SIAM J. Numer. Anal.*, 27(2):353–384, 1990.
181. M. Hillairet. Lack of collision between solid bodies in a 2D incompressible viscous flow. *Comm. Partial Differential Equations*, 32(7–9):1345–1371, 2007.
182. M. Hillairet and T. Takahashi. Collisions in three-dimensional fluid structure interaction problems. *SIAM J. Math. Anal.*, 40(6):2451–2477, 2009.
183. T. Hishida. Stability of time-dependent motions for fluid–rigid ball interaction. *J. Math. Fluid Mech.*, 26(2):17, 2024.
184. K.-H. Hoffmann and V. N. Starovoitov. On a motion of a solid body in a viscous fluid. Two-dimensional case. *Adv. Math. Sci. Appl.*, 9:633–648, 1999.
185. E. Hopf. Über die anfangswertaufgabe für die hydrodynamischen grundgleichungen. *Math. Nachr.*, 4(1–6):213–231, 1951.
186. C. T. Hsu and P. Cheng. Thermal dispersion in a porous media. *Int. J. Heat Mass. Transf.*, 33(8):1587–1597, 1990.
187. M.-C. Hsu, D. Kamensky, Y. Bazilevs, M. S. Sacks, and T. J. R. Hughes. Fluid-structure interaction analysis of bioprosthetic heart valves: significance of arterial wall deformation. *Comput. Mech.*, 54(4):1055–1071, 2014.
188. B. Huang, V. Mácha, and Š. Nečasová. On the motion of a body with a cavity filled with magnetohydrodynamic fluid. *J. Differential Equations*, 398:218–270, 2024.
189. G. Huang, Y. Xia, Y. Dai, C. Yang, and Y. Wu. Fluid-structure interaction in piezoelectric energy harvesting of a membrane wing. *Phys Fluid*, 33(6):063610, 2021.
190. A. Hundertmark-Zaušková, M. Lukáčová-Medviďová, and Š. Nečasová. On the existence of weak solution to the coupled fluid-structure interaction problem for non-Newtonian shear-dependent fluid. *J. Math. Soc. Japan*, 68(1):193–243, 2016.
191. T. Hytönen, J. van Neervan, M. Veraar, and L. Weis. *Analysis in Banach spaces. Volume I: Martingales and Littlewood-Paley theory*, volume 63 of *A Series of Modern Surveys in Mathematics*. Springer International Publishing, Cham, Switzerland, 2016.
192. M. Ignatova, I. Kukavica, I. Lasiecka, and A. Tuffaha. On well-posedness for a free boundary fluid-structure model. *J. Math. Phys.*, 53(11):115624, 2012.
193. M. Ignatova, I. Kukavica, I. Lasiecka, and A. Tuffaha. On well-posedness and small data global existence for an interface damped free boundary fluid-structure model. *Nonlinearity*, 27(3):467–499, 2014.
194. A. Inoue and M. Wakimoto. On existence of solutions of the Navier-Stokes equation in a time dependent domain. *J. Fac. Sci. Univ. Tokyo Sect. IA Math.*, 24(2):303–319, 1977.
195. W. Jäger and A. Mikelić. On the boundary conditions at the contact interface between a porous medium and a free fluid. *Ann. Scuola Norm. Sup. Pisa Cl. Sci. (4)*, 23(3):403–465, 1996.
196. W. Jäger and A. Mikelić. On the interface boundary condition of Beavers, Joseph, and Saffman. *SIAM J. Appl. Math.*, 60(4):1111–1127, 2000.
197. A. Jakubowski. The almost sure skorokhod representation for subsequences in nonmetric spaces. *Teor. Veroyatnost. i Primenen.*, 42(1):209–216, 1997.
198. R. Jamal and É. Sanchez-Palencia. Théorie asymptotique des tiges courbes anisotropes. *C. R. Acad. Sci. Paris, Série I*, 322(11):1099–1106, 1996.
199. M. Jurak and J. Tambača. Derivation and justification of a curved rod model. *Math. Models Methods Appl. Sci.*, 9(7):991–1014, 1999.
200. M. Jurak and J. Tambača. Linear curved rod model: General curve. *Math. Models Methods Appl. Sci.*, 11(7):1237–1252, 2001.
201. O. Kallenberg. *Foundations of modern probability*, volume 99 of *Probability theory and stochastic modelling*. Springer, Switzerland, third edition edition, 2021.

202. M. Kalousek, S. Mitra, and Š. Nečasová. The existence of a weak solution for a compressible multicomponent fluid structure interaction problem. *J. Math. Pures Appl. (9)*, 184:118–189, 2024.
203. B. Kaltenbacher, I. Kukavica, I. Lasiecka, R. Triggiani, A. Tuffaha, and J. T. Webster. *Mathematical theory of evolutionary fluid-flow structure interactions*, volume 48 of *Oberwolfach Seminars*. Birkhäuser/Springer, Cham, 2018. Lecture notes from Oberwolfach seminars, November 20–26, 2016.
204. R. Kamakoti and W. Shyy. Fluid-structure interaction for aeroelastic applications. *Prog. Aerosp. Sci.*, 40(8):535–558, 2004.
205. M. Kampschulte, B. Muha, and S. Trifunović. Global weak solutions to a 3D/3D fluid-structure interaction problem including possible contacts. *J. Differential Equations*, 385:280–324, 2024.
206. M. Kampschulte, S. Schwarzacher, and G. Sperone. Unrestricted deformations of thin elastic structures interacting with fluids. *J. Math. Pures Appl. (9)*, 173:96–148, 2023.
207. A. Karczewska and J. Zabczyk. Stochastic PDEs with function-valued solutions. In Ph. Clément, F. den Hollander, J. van Neerven, and B. de Pagter, editors, *Infinite dimensional stochastic analysis*, Proceedings of the Colloquium of the Royal Netherlands Academy of Arts and Sciences, pages 197–216, Amsterdam, 1999.
208. A. Karimi, M. Navidbakhsh, R. Razaghi, and M. Haghpanahi. A computational fluid-structure interaction model for plaque vulnerability assessment in atherosclerotic human coronary arteries. *J. Appl. Phys.*, 115(14):144702, 2014.
209. D. Khoshnevisan. *Multiparameter processes: an introduction to random fields*. Springer, New York, 2002.
210. D. Khoshnevisan. A primer on stochastic partial differential equations. In D. Khoshnevisan and F. Rassoul-Agha, editors, *A minicourse on stochastic partial differential equations*, volume 1962 of *Lecture Notes in Mathematics*. Springer-Verlag, Berlin Heidelberg, 2009.
211. D. Kowollik, V. Tini, S. Reese, and M. Haupt. 3D fluid-structure interaction analysis of a typical liquid rocket engine cycle based on a novel viscoplastic damage model. *Int. J. Numer. Methods Eng.*, 94(13):1165–1190, 2013.
212. O. Kreml, V. Mácha, Š. Nečasová, and S. Trifunović. On time-periodic solutions to an interaction problem between compressible viscous fluids and viscoelastic beams. *arXiv preprint arXiv:2307.02687*, 2023.
213. J. Kuan and S. Čanić. Deterministic ill-posedness and probabilistic well-posedness of the viscous nonlinear wave equation describing fluid-structure interaction. *Trans. Amer. Math. Soc.*, 374(8):5925–5994, 2021.
214. J. Kuan, T. Oh, and S. Čanić. Probabilistic global well-posedness for a viscous nonlinear wave equation modeling fluid-structure interaction. *Appl. Anal.*, 101(12):4349–4373, 2022.
215. J. Kuan and K. Tawri. Existence of solutions to a stochastic fluid-structure interaction problem of compressible fluid flow interacting with an elastic shell. *J. Differential Equations*, 449:113–669, 2025.
216. J. Kuan and S. Čanić. A stochastically perturbed fluid-structure interaction problem modeled by a stochastic viscous wave equation. *J. Differential Equations*, 310:45–98, 2022.
217. J. Kuan and S. Čanić. Well-posedness of solutions to stochastic fluid-structure interaction. *J. Math. Fluid Mech.*, 26:4, 2024.
218. J. Kuan, S. Čanić, and B. Muha. Existence of a weak solution to a regularized moving boundary fluid-structure interaction problem with poroelastic media. *C. R. Mec.*, 351(S1):505–534, 2023.
219. J. Kuan, S. Čanić, and B. Muha. Fluid-poroviscoelastic structure interaction problem with nonlinear geometric coupling. *J. Math. Pures Appl.*, 188:345–445, 2024.
220. I. Kukavica and W. S. Ożański. On a model of an elastic body fully immersed in a viscous incompressible fluid with small data. *SIAM J. Math. Anal.*, 56(1):746–761, 2024.
221. I. Kukavica and A. Tuffaha. Solutions to a fluid-structure interaction free boundary problem. *Discrete Contin. Dyn. Syst.*, 32(4):1355–1389, 2012.

222. I. Kukavica and A. Tuffaha. An inviscid free boundary fluid-wave model. *J. Evol. Equ.*, 23(2):Paper No. 41, 18, 2023.
223. I. Kukavica, A. Tuffaha, and M. Ziane. Strong solutions to a nonlinear fluid structure interaction system. *J. Differential Equations*, 247(5):1452–1478, 2009.
224. I. Kukavica, A. Tuffaha, and M. Ziane. Strong solutions for a fluid structure interaction system. *Adv. Differential Equations*, 15(3–4):231–254, 2010.
225. I. Kukavica, A. Tuffaha, and M. Ziane. Strong solutions to a Navier-Stokes-Lamé system on a domain with a non-flat boundary. *Nonlinearity*, 24(1):159–176, 2011.
226. I. Kukavica and F. Xu. Local existence of strong solutions to the stochastic Navier-Stokes equations with L^p data. *J. Differential Equations*, 359:183–210, 2023.
227. I. Kukavica, F. Xu, and M Ziane. Global existence for the stochastic Navier-Stokes equations with small L^p data. *Stoch. Partial Differ. Equ. Anal. Comput.*, 10(1):160–189, 2022.
228. S.S. Kulkarni, L. Wang, N. Golsby, and M. Lande. Fluid-structure interaction based optimisation in tidal turbines: A perspective review. *Journal of Ocean Engineering and Science*, 7(5):449–461, 2022.
229. J. S.-H. Kuo. Fluid-structure interactions: added mass computations for incompressible fluid. Technical Report UCB/EERC-82/09, University of California, Berkeley, California, August 1982.
230. O. A. Ladyzhenskaia. Solutions "in the large" of the nonstationary boundary value problem for the Navier-Stokes system with two space variables. *Comm. Pure Appl. Math.*, 12(3):427–433, 1959.
231. O. A. Ladyzhenskaya. Initial-boundary problem for Navier-Stokes equations in domains with time-varying boundaries. In *In Boundary Value Problems of Mathematical Physics and Related Aspects of Function Theory*, pages 35–46. Springer, 1970.
232. W. J. Layton, F. Schieweck, and I. Yotov. Coupling fluid flow with porous media flow. *SIAM J. Numer. Anal.*, 40(6):2195–2218, 2003.
233. D. Lengeler and M. Růžička. Weak solutions for an incompressible Newtonian fluid interacting with a Koiter type shell. *Arch. Ration. Mech. Anal.*, 211(1):205–255, 2014.
234. J. Lequeurre. Existence of strong solutions to a fluid-structure system. *SIAM J. Math. Anal.*, 43(1):389–410, 2011.
235. J. Lequeurre. Existence of strong solutions for a system coupling the Navier-Stokes equations and a damped wave equation. *J. Math. Fluid Mech.*, 15(2):249–271, 2013.
236. J. Leray. Sur le mouvement d'un liquide visqueux emplissant l'espace. *Acta Math.*, 63(1):193–248, 1934.
237. M. Lesinigo, C. D'Angelo, and A. Quarteroni. A multiscale Darcy-Brinkman model for fluid flow in fractured porous media. *Numer. Math.*, 117(4):717–752, 2011.
238. Y. Li, C. Han, Z. Ya'nan, and L. Shaoteng. Study of parachute inflation process using fluid-structure interaction method. *Chin. J. Aeronaut.*, 27(2):272–279, 2014.
239. J. Link, P. Nguyen, and R. Temam. Local martingale solutions to the stochastic one layer shallow water equations. *J. Math. Anal. Appl.*, 448(1):93–139, 2017.
240. J.-L. Lions. *Quelques méthodes de résolution des problèmes aux limites non linéaires*, volume 31. Dunod, Paris, 1969.
241. Q. Ma, V. Sriram, and S. Yan. Fluid-structure interactions in marine engineering. *Comp. Num. Anal.*, 2017.
242. V. Mácha, B. Muha, Š. Nećasová, A. Roy, and S. Trifunović. Existence of a weak solution to a nonlinear fluid-structure interaction problem with heat exchange. *Comm. Partial Differential Equations*, 47(8):1591–1635, 2022.
243. D. Maity, J.-P. Raymond, and A. Roy. Maximal-in-time existence and uniqueness of strong solution of a 3D fluid-structure interaction model. *SIAM J. Math. Anal.*, 52(6):6338–6378, 2020.
244. D. Maity and T. Takahashi. Existence and uniqueness of strong solutions for the system of interaction between a compressible Navier-Stokes-Fourier fluid and a damped plate equation. *Nonlinear Anal. Real World Appl.*, 59:Paper No. 103267, 34, 2021.

245. D. Maity and T. Takahashi. L^p theory for the interaction between the incompressible Navier-Stokes system and a damped plate. *J. Math. Fluid Mech.*, 23(4):Paper No. 103, 23, 2021.
246. D. Maity and T. Takahashi. Regularity of weak solutions of a fluid-rigid body interaction system in a bounded domain under the prodi-serrin condition. *Preprint*, 2024.
247. D. Maity and T. Takahashi. Uniqueness and regularity of weak solutions of a fluid-rigid body interaction system under the Prodi-Serrin condition. *Discrete Contin. Dyn. Syst.*, 44(3):718–742, 2024.
248. D. Maity and M. Tucsnak. L^p-L^q maximal regularity for some operators associated with linearized incompressible fluid-rigid body problems. In *Mathematical analysis in fluid mechanics—selected recent results*, volume 710 of *Contemp. Math.*, pages 175–201. Amer. Math. Soc., Providence, RI, 2018.
249. A. J. Majda and A. L. Bertozzi. *Vorticity and incompressible flow*, volume 27 of *Cambridge Texts in Applied Mathematics*. Cambridge University Press, Cambridge, 2002.
250. A. Marciniak-Czochra and A. Mikelić. A rigorous derivation of the equations for the clamped Biot-Kirchhoff-Love poroelastic plate. *Arch. Ration. Mech. Anal.*, 215(3):1035–1062, 2015.
251. J. S. Martín, J.-F. Scheid, T. Takahashi, and M. Tucsnak. An initial and boundary value problem modeling of fish-like swimming. *Arch. Ration. Mech. Anal.*, 188:429–455, 2008.
252. E. S. Di Martino, G. Guadagni, A. Fumero, G. Ballerini, R. Spirito, P. Biglioli, and A. Redaelli. Fluid-structure interaction within realistic three-dimensional models of the aneurysmatic aorta as a guidance to assess the risk of rupture of the aneurysm. *Med. Eng. Phys.*, 23(9):647–655, 2001.
253. M. Maywerk, F. Kecker, and J. Cordes. Fluid-structure interaction in crash simulation. *Proc Inst Mech Eng, Part:D, J Auto Eng*, 214(7), 2000.
254. W. McLean. *Strongly elliptic systems and boundary integral equations*. Cambridge University Press, Cambridge, 2000.
255. J.-L. Menaldi and S. S. Sritharan. Stochastic 2-D Navier-Stokes equation. *Appl. Math. Optim.*, 46(1):31–53, 2002.
256. M. Métivier. *Stochastic partial differential equations in infinite-dimensional spaces*. Scuola Normale Superiore di Pisa. Quaderni. [Publications of the Scuola Normale Superiore of Pisa]. Scuola Normale Superiore, Pisa, 1988. With a preface by G. Da Prato.
257. S. E. Mikhailov. Traces, extensions and co-normal derivatives for elliptic systems on Lipschitz domains. *J. Math. Anal. Appl.*, 378(1):324–342, 2011.
258. R. Mikulevicius and B. L. Rozovskii. Global L_2-solutions of stochastic Navier-Stokes equations. *Ann. Probab.*, 33(1):137–176, 2005.
259. C. Mindrila and S. Schwarzacher. Time-periodic weak solutions for an incompressible newtonian fluid interacting with an elastic plate. *SIAM J. Math. Anal.*, 54(4):4139–4162, 2022.
260. P. Moireau, N. Xiao, M. Astorino, C. A. Figueroa, D. Chapelle, C. A. Taylor, and J.-F. Gerbeau. External tissue support and fluid-structure simulation in blood flows. *Biomech. Model. Mechanobiol.*, 11(1–2):1–18, 2012.
261. G. Morgenthal and A. McRobie. A comparative study of numerical methods for fluid structure interaction analysis in long-span bridge design. *Wind Struct.*, 5(2–4):101–114, 2002.
262. S. Mosný, B. Muha, S. Schwarzacher, and J. Webster. Time-periodic solutions for hyperbolic-parabolic systems, 2024.
263. B. Muha. A note on the Trace Theorem for domains which are locally subgraph of a hölder continuous function. *Netw. Heterog. Media*, 9(1):191–196, 2014.
264. B. Muha. A note on optimal regularity and regularizing effects of point mass coupling for a heat-wave system. *J. Math. Anal. Appl.*, 425(2):1134–1147, 2015.
265. B. Muha and S. Čanić. Existence of a weak solution to a nonlinear fluid-structure interaction problem modeling the flow of an incompressible, viscous fluid in a cylinder with deformable walls. *Arch. Ration. Mech. Anal.*, 207(3):919–968, 2013.
266. B. Muha and S. Čanić. A nonlinear, 3D fluid-structure interaction problem driven by the time-dependent dynamic pressure data: a constructive existence proof. *Commun. Inf. Syst.*, 13(3):357–397, 2013.

267. B. Muha and S. Čanić. Existence of a solution to a fluid-multi-layered-structure interaction problem. *J. Differential Equations*, 256(2):658–706, 2014.
268. B. Muha and S. Čanić. Fluid-structure interaction between an incompressible, viscous 3D fluid and an elastic shell with nonlinear Koiter membrane energy. *Interfaces Free Bound.*, 17(4):465–495, 2015.
269. B. Muha and S. Čanić. Existence of a weak solution to a fluid-elastic structure interaction problem with the Navier slip boundary condition. *J. Differential Equations*, 260(12):8550–8589, 2016.
270. B. Muha and S. Čanić. A generalization of the Aubin-Lions-Simon compactness lemma for problems on moving domains. *J. Differential Equations*, 266(12):8370–8418, 2019.
271. B. Muha, Š. Nečasová, and A. Radošević. A uniqueness result for 3D incompressible fluid-rigid body interaction problem. *J. Math. Fluid Mech.*, 23(1):Paper No. 1, 39, 2021.
272. B. Muha, Š. Nečasová, and A. Radošević. On the regularity of weak solutions to the fluid-rigid body interaction problem. *Math. Ann.*, 389(2):1007–1052, 2024.
273. B. Muha and S. Schwarzacher. Existence and regularity of weak solutions for a fluid interacting with a non-linear shell in three dimensions. *Ann. Inst. H. Poincaré Anal. Non Linéaire*, 39(6):1369–1412, 2022.
274. B. Muha and S. Čanić. Existence of a weak solution to a nonlinear fluid-structure interaction problem modeling the flow of an incompressible, viscous fluid in a cylinder with deformable walls. *Arch. Ration. Mech. Anal.*, 207(3):919–968, 2013.
275. P Nägele. *Monotone operator theory for unsteady problems on non-cylindrical domains [dissertation]*. Freiburg: Albert-Ludwigs-Universität Freiburg, 2015.
276. P. Nägele, M. Růžička, and D. Lengeler. Functional setting for unsteady problems in moving domains and applications. *Complex Var. Elliptic Equ.*, 62(1):66–97, 2017.
277. O. Omidi and V. Lotfi. A symmetric implementation of pressure-based fluid-structure interaction for nonlinear dynamic analysis of arch dams. *J. Fluids Struct.*, 69:34–55, 2017.
278. S. Owczarek. A Galerkin method for Biot consolidation model. *Math. Mech. Solids*, 15(1):42–56, 2010.
279. O. Oyekole, C. Trenchea, and M. Bukač. A second-order in time approximation of fluid-structure interaction problem. *SIAM J. Numer. Anal.*, 56(1):590–613, 2018.
280. C.A.J. Pantula, J. K. Calautit, and Y. Wu. A fluid-structure interaction (FSI) and energy generation modelling for roof mounted renewable energy installations in buildings for extreme weather and typhoon resilience. *Renewable Ener*, 160:770–787, 2020.
281. G. Pisier. Martingales with values in uniformly convex spaces. *Israel J. Math.*, 20(3–4):326–350, 1975.
282. G. Pontrelli. A mathematical model of flow in a liquid-filled viscoelastic tube. *Med. Biol. Eng. Comput.*, 40(5):550–556, 2002.
283. M. Potomkin, V. Gyrya, I. Aranson, and L. Berlyand. Collision of microswimmers in a viscous fluid. *Phys. Rev. E*, 87(5):053005, 2013.
284. Z. Qu, G. Hu, A. Garfinkel, and J.N. Weiss. Nonlinear and stochastic dynamics in the heart. *Phys. Rep.*, 543(2):61–162, 2014.
285. A. Quarteroni and L. Formaggia. Mathematical modelling and numerical simulation of the cardiovascular system. In *Handbook of numerical analysis. Vol. XII*, volume XII of *Handb. Numer. Anal.*, pages 3–127. North-Holland, Amsterdam, 2004.
286. A. Quarteroni, M. Tuveri, and A. Veneziani. Computational vascular fluid dynamics: problems, models and methods. *Comput. Vis. Sci.*, 2(4):163–197, 2000.
287. J.-P. Raymond and M. Vanninathan. A fluid-structure model coupling the Navier-Stokes equations and the Lamé system. *J. Math. Pures Appl.*, 102(3):546–596, 2014.
288. A. Remond-Tiedrez and I. Tice. The viscous surface wave problem with generalized surface energies. *SIAM J. Math. Anal.*, 51(6):4894–4952, 2019.
289. M. Renardy and R. C. Rogers. *An introduction to partial differential equations*, volume 13 of *Texts in Applied Mathematics*. Springer, New York, second edition edition, 2004.

290. D. Revuz and M. Yor. *Continuous martingales and Brownian motion*, volume 293 of *A Series of Comprehensive Studies in Mathematics*. Springer-Verlag, Berlin Heidelberg, third edition edition, 1999.
291. T. Richter. *Fluid-structure interactions*, volume 118 of *Lecture Notes in Computational Science and Engineering*. Springer, Cham, 2017. Models, analysis and finite elements.
292. B. Rivière and I. Yotov. Locally conservative coupling of Stokes and Darcy flows. *SIAM J. Numer. Anal.*, 42(5):1959–1977, 2005.
293. L. Sabbagh. On the motion of rigid bodies in viscoelastic fluids. *Acta Appl. Math.*, 178:Paper No. 8, 43, 2022.
294. J. Sanchez-Hubert and É. Sanchez-Palencia. Statics of curved rods on account of torsion and flexion. *Eur. J. Mech. A Solids*, 18(3):365–390, 1999.
295. M. Sanz-Solé and M. Sarrà. Hölder continuity for the stochastic heat equation with spatially correlated noise. In R. C. Dalang, M. Dozzi, and F. Russo, editors, *Seminar on Stochastic Analysis, Random Fields and Applications III*, volume 52 of *Progress in Probability*, pages 259–268. Birkhäuser, Basel, 2002.
296. A. Scharf, S. Čanić, and Y. Wang. A partitioned scheme for fluid-structure interaction with multilayered poroelastic media. *In draft form.*, 2024.
297. J. Scherz. Fluid-rigid body interaction in a compressible electrically conducting fluid. *Math. Nachr.*, 296(12):5513–5550, 2023.
298. S. Schwarzacher and M. Sroczinski. Weak-strong uniqueness for an elastic plate interacting with the Navier-Stokes equation. *SIAM J. Math. Anal.*, 54(4):4104–4138, 2022.
299. S. Schwarzacher and P. Su. Regularity for an elastic beam interacting with 2D Navier-Stokes equations. *arXiv preprint arXiv:2308.04253*, 2023.
300. A. Seboldt, O. Oyekole, J. Tambača, and M. Bukač. Numerical modeling of the fluid-porohyperelastic structure interaction. *SIAM J. Sci. Comput.*, 43(4):A2923–A2948, 2021.
301. D. Serre. Chute libre d'un solide dans un fluide visqueux incompressible. Existence. *Japan J. Appl. Math.*, 4(1):99–110, 1987.
302. R. E. Showalter. *Monotone operators in Banach space and nonlinear partial differential equations*, volume 49 of *Mathematical Surveys and Monographs*. American Mathematical Society, Providence, RI, 1997.
303. R. E. Showalter. Diffusion in poro-elastic media. *J. Math. Anal. Appl.*, 251(1):310–340, 2000.
304. R. E. Showalter. Poroelastic filtration coupled to Stokes flow. In G. Leugering, O. Imanuvilov, B.-Y. Zhang, and R. Triggiani, editors, *Control theory of partial differential equations*, volume 242 of *A Series of Lecture Notes in Pure and Applied Mathematics*, pages 229–242. Chapman and Hall/CRC, New York, first edition edition, 2005.
305. R. E. Showalter and N. Su. Partially saturated flow in a poroelastic medium. *Discrete Contin. Dyn. Syst. Ser. B*, 1(4):403–420, 2001.
306. A. L. Silvestre. On the self-propelled motion of a rigid body in a viscous liquid and on the attainability of steady symmetric self-propelled motions. *J. Math. Fluid Mech.*, 4:285–326, 2002.
307. J. Simon. Compact sets in the space $L^p(0, T; B)$. *Ann. Mat. Pura Appl. (4)*, 146:65–96, 1987.
308. A. V. Skorohod. Limit theorems for stochastic processes. *Teor. Veroyatnost. i Primenen.*, 1:289–319, 1956.
309. S. A. Smith. Random perturbations of viscous, compressible fluids: global existence of weak solutions. *SIAM J. Math. Anal.*, 49(6):4521–4578, 2017.
310. S. A. Smith and K. Trivisa. The stochastic Navier-Stokes equations for heat-conducting, compressible fluids: global existence of weak solutions. *J. Evol. Equ.*, 18(2):411–465, 2018.
311. V. N. Starovoitov. Nonuniqueness of a solution to the problem on motion of a rigid body in a viscous incompressible fluid. *J. Math. Sci. (N.Y.)*, 130:4893–4898, 2005.
312. E. M. Stein and R. Shakarchi. *Fourier analysis: An introduction*, volume 1 of *Princeton Lectures in Analysis*. Princeton University Press, Princeton, New Jersey, 2003.
313. E. M. Stein and R. Shakarchi. *Real analysis: Measure theory, integration, and Hilbert spaces*, volume 3 of *Princeton Lectures in Analysis*. Princeton University Press, Princeton, New Jersey, 2005.

314. E. M. Stein and R. Shakarchi. *Functional analysis: An introduction to further topics in analysis*, volume 4 of *Princeton Lectures in Analysis*. Princeton University Press, Princeton, New Jersey, 2011.
315. K. Stein, R. Benney, V. Kalro, T. E. Tezduyar, J. Leonard, and M. Accorsi. Parachute fluid-structure interactions: 3-D computation. *Comput. Methods Appl. Mech. Engrg.*, 190(3):373–386, 2000.
316. K. R. Stein, R. J. Benney, T. E. Tezduyar, J. W. Leonard, and M. L. Accorsi. Fluid-structure interactions of a round parachute: modeling and simulation techniques. *J. Aircr.*, 38(5):800–808, 2001.
317. T. Takahashi. Analysis of strong solutions for the equations modeling the motion of a rigid-fluid system in a bounded domain. *Adv. Diffential Equations*, 8(12):1499–1532, 2003.
318. T. Takahashi. Existence of strong solutions for the problem of a rigid-fluid system. *C. R. Math.*, 336(5):453–458, 2003.
319. T. Takahashi and M. Tucsnak. Global strong solutions for the two-dimensional motion of an infinite cylinder in a viscous fluid. *J. Math. Fluid Mech.*, 6(1):53–77, 2004.
320. J. Tambača, M. Kosor, S. Čanić, and D. Paniagua. Mathematical modeling of vascular stents. *SIAM J. Appl. Math.*, 70(6):1922–1952, 2010.
321. K. Tawri. A stochastic fluid-structure interaction problem with the Navier-slip boundary condition. *SIAM J. Math. Anal.*, 56(6):7508–7544, 2024.
322. K. Tawri. A 2D stochastic nonlinearly coupled fluid-structure interaction problem in compliant arteries with unrestricted structural displacement. *J. Differential Equations*, 431:113243, 2025.
323. K. Tawri and S. Čanić. Existence of martingale solutions to a nonlinearly coupled stochastic fluid-structure interaction problem. *Comm. Partial Differential Equations*, 50(3):353–406, 2025.
324. M. E. Taylor. *Partial differential equations I: basic theory*, volume 115 of *Applied Mathematical Sciences*. Springer Science+Business Media, New York, second edition, 2011.
325. R. Temam. *Navier-Stokes equations and nonlinear functional analysis*, volume 66 of *CBMS-NSF Regional Conference Series in Applied Mathematics*. Society for Industrial and Applied Mathematics (SIAM), Philadelphia, PA, second edition, 1995.
326. R. Temam. *Navier-Stokes equations: theory and numerical analysis*. AMS Chelsea Publishing, Providence, RI, 2001. Reprint of the 1984 edition.
327. T. Tezduyar and Y. Osawa. Fluid-structure interactions of a parachute crossing the far wake of an aircraft. *Comput. Methods Appl. Mech. Engrg.*, 191(6):717–726, 2001.
328. Z. Tian, F. Liu, L. Zhou, and C. Yian. Fluid-structure interaction analysis of offshore structures based on separation of transferred responses. *Ocen Eng.*, 195:106598, 2020.
329. S. Trifunović. Compressible fluids interacting with plates: regularity and weak-strong uniqueness. *J. Math. Fluid Mech.*, 25(1):Paper No. 13, 28, 2023.
330. S. Trifunović and Y.-G. Wang. Existence of a weak solution to the fluid-structure interaction problem in 3D. *J. Differential Equations*, 268(4):1495–1531, 2020.
331. F. Tröltzsch. *Optimal control of partial differential equations*, volume 112 of *Graduate Studies in Mathematics*. American Mathematical Society, Providence, RI, 2010. Theory, methods and applications, Translated from the 2005 German original by Jürgen Sprekels.
332. A. van der Vaart and J. Wellner. *Weak convergence and empirical processes*. Springer Series in Statistics. Springer-Verlag, New York, 1996. With applications to statistics.
333. S. Čanić. Fluid-structure interaction with incompressible fluids. In L. C. Berselli and M. Růžička, editors, *Progress in mathematical fluid dynamics*, volume 2272 of *Lecture Notes in Mathematics*, pages 15–87. Springer, 2020.
334. S. Čanić, M. Galić, M. Ljulj, B. Muha, J. Tambača, and Y. Wang. Analysis of a linear 3D fluid-mesh-shell interaction problem. *Z. Angew. Math. Phys.*, 70(2):1–38, 2019.
335. S. Čanić, M. Galić, and B. Muha. Analysis of a 3D nonlinear, moving boundary problem describing fluid-mesh-shell interaction. *Trans. Amer. Math. Soc.*, 373(9):6621–6681, 2020.
336. S. Čanić, M. Galović, M. Ljulj, and J. Tambača. A dimension-reduction based coupled model of mesh-reinforced shells. *SIAM J. Appl. Math.*, 77(2):744–769, 2017.

337. S. Čanić, L. Grubišić, D. Lacmanović, M. Ljulj, and J. Tambača. Optimal design of vascular stents using a network of 1D slender curved rods. *Comput. Methods Appl. Mech. Engrg.*, 394:114853, 2022.
338. S. Čanić, B. Muha, and K. Tawri. Existence and regularity results for a nonlinear fluid-structure interaction problem with three-dimensional structural displacement. *preprint at arXiv:2409.06939*, 2024.
339. S. Čanić and J. Tambača. Cardiovascular stents as PDE nets: 1D vs. 3D. *IMA J. Appl. Math.*, 77(6):748–770, 2012.
340. S. Čanić, J. Tambača, G. Guidoboni, A. Mikelić, C. J. Hartley, and D. Rosenstrauch. Modeling viscoelastic behavior of arterial walls and their interaction with pulsatile blood flow. *SIAM J. Appl. Math.*, 67(1):164–193, 2006.
341. S. Čanić, Y. Wang, and M. Bukač. A next-generation mathematical model for drug-eluting stents. *SIAM J. Appl. Math.*, 81(4):1503–1529, 2021.
342. I. Velčić. Nonlinear weakly curved rod by Γ-convergence. *J. Elasticity*, 108(2):125–150, 2012.
343. H. von Wahl, T. Richter, S. Frei, and T. Hagemeier. Falling balls in a viscous fluid with contact: Comparing numerical simulations with experimental data. *Phys. Fluids*, 33(3), 2021.
344. Š. Nečasová and J. Wolf. On the existence of global strong solutions to the equations modeling a motion of a rigid body around a viscous fluid. *Discrete Contin. Dyn. Syst.*, 36(3):1539–1562, 2016.
345. A. Ženíšek. The existence and uniqueness theorem in Biot's consolidation theory. *Aplikace Matematiky*, 29(3):194–211, 1984.
346. J. B. Walsh. An introduction to stochastic partial differential equations. In P. L. Hennequin, editor, *École d'Été de Probabilités de Saint Flour XIV - 1984*, volume 1180 of *Lecture Notes in Mathematics*, pages 265–439. Springer, Berlin, 1986.
347. K. Wang, Q. Zhang, X. Xia, L. Wang, and X. Liu. Analysis of hydraulic fracturing in concrete dam considering fluid-structure interaction using XFEM-FVM model. *Eng. Fail. Anal.*, 57:399–412, 2015.
348. Y. Wang, S. Čanić, M. Bukač, C. Blaha, and S. Roy. Mathematical and computational modeling of a poroelastic cell scaffold in a bioartificial pancreas. *Fluids*, 7(7):222, 2022.
349. H. Q. Yang, J. West, and R. E. Harris. Coupled fluid-structure interaction analysis of solid rocket motor with flexible inhibitors. *J. Spacecr. Rockets*, 55(2):303–314, 2018.
350. J. Young, B. Rivière, Jr. C. S. Cox, and K. Uray. A mathematical model of intestinal oedema formation. *Math. Med. Biol.*, 31(1):1–15, 2014.
351. R. Zakerzadeh and P. Zunino. A computational framework for fluid-porous structure interaction with large structural deformation. *Meccanica*, 54(1–2):101–121, 2019.
352. X. Zhang and E. Zuazua. Long-time behavior of a coupled heat-wave system arising in fluid-structure interaction. *Arch. Ration. Mech. Anal.*, 184(1):49–120, 2007.

Index

S. Čanić et al., *Deterministic and Stochastic Fluid-Structure Interaction*, Advances in Mathematical Fluid Mechanics, https://doi.org/10.1007/978-3-032-00898-5

GPSR Compliance

The European Union's (EU) General Product Safety Regulation (GPSR) is a set of rules that requires consumer products to be safe and our obligations to ensure this.

If you have any concerns about our products, you can contact us on ProductSafety@springernature.com

In case Publisher is established outside the EU, the EU authorized representative is:

Springer Nature Customer Service Center GmbH
Europaplatz 3
69115 Heidelberg, Germany

Batch number: 10755324

Printed by Printforce, the Netherlands